Introductory Circuit Analysis

5th Edition

Robert L. Boylestad

Merrill Publishing Company
A Bell & Howell Company
Columbus Toronto London Melbourne

To Else Marie,
Eric, Alison, and Stacey

Cover Photo: Larry Hamill

Published by Merrill Publishing Company
A Bell & Howell Company
Columbus, Ohio 43216

This book was set in Times Roman.

Administrative Editor: Tim McEwen
Developmental Editor: Don Thompson
Production Coordinator: Rex Davidson
Copy Editor: Linda M. Johnstone
Art Coordinator: Pete Robison
Cover Designer: Cathy Watterson
Text Designer: Cynthia Brunk

Library of Congress Catalog Card Number: 86–60483
International Standard Book Number: 0–675–20631–6
Printed in the United States of America
  2 3 4 5 6 7 8 9—91 90 89 88 87

Preface

Both the publisher and I were quite sure with the completion of the fourth edition that the text was finally in a form that would require only minor modification in the future and our primary effort would be to ensure the highest degree of accuracy. In fact, Merrill Publishing Co. felt quite secure in placing the entire manuscript in memory so that accuracy would be maintained for future editions and the next edition would be an easier, less expensive proposition. Low and behold, we find some five years later that the text is undergoing a major revision that includes redrawing all artwork (over 1400 figures), introducing a second color as a teaching aid, removing three chapters and adding two new chapters, expanding coverage in a number of vital areas, and adding an entire range of new examples and exercises. The input for these changes stems from three major sources: the first, and most important, is classroom experience, where improved teaching methods still surface after all these years. Student response continues to point out areas of the text that need modification and improvement. Second, and deeply appreciated by the author, are suggestions and criticisms from teaching associates and users of the text. The third is truly a function of the ability of the publisher to identify those individuals who can provide detailed reviews to identify problem areas and recommend changes that will benefit the majority of users. To this end, those individuals listed at the end of the preface have continued to provide input that has included last-minute changes to the text. The publisher and I would both appreciate your continued input on this edition for a file for the sixth edition, since a number of excellent suggestions could not

be incorporated in this edition without totally revising the text. For those current users of the text who are generally pleased with the content, please be aware that the core material has been maintained. The primary objective of this edition was to provide those additional comments, examples, and material that would clarify critical concepts and introduce subject areas that currently require coverage at an earlier stage in the student's development.

As indicated above, accuracy remained a top priority in the fifth edition. In addition to the author's review, a technical review was performed that examined all the examples and problems of the text to ensure accuracy and a proper transfer in the publishing process. A number of the examples and exercises now include standard resistor values to establish familiarity with commercially available units and reveal the effect on the required calculations.

One major change was to split the series and parallel circuits chapter into two chapters to provide additional support to these critical areas in the student's early development. The resulting chapters (Chapters 5 and 6) include expanded coverage of single- and double-subscript voltage notation and the impact of short and open circuits. Chapter 21 on pulse waveforms was added in response to a growing need for familiarity with the terminology and the response of R-C networks. The introduction of Chapter 1 now includes a brief history of the field to provide some background on the exponential growth of the field and to avoid a head-on collision with mathematical development in the first section of the text. Instrumentation now appears throughout the text rather than as two separate chapters. As a measurable quantity is introduced, the instrumentation for its display and measurement is also presented. Too often the chapters on instrumentation have been left as reading material because of their location or not covered until the student had already been exposed to the equipment in the laboratory session.

Throughout the ac section, an effort was made to expand on the frequency response of standard configurations rather than dwell on the response at a single frequency. In addition, the filter section was significantly expanded and the parallel resonance section was completely rewritten to remove any confusion stemming from different levels of quality factor. Logarithms, log plots, and the development and use of Bode plots were also added to this edition to improve on the frequency response coverage. Topics such as power factor correction (Chapter 19) were also added to contribute to the text's practical orientation.

A number of other sections of the text, such as the transient behavior of *R-C* and *R-L* networks, were completely rewritten to improve the presentation and upgrade material. Computer programs now appear throughout the text, not with the thought of introducing programming techniques, but simply to reveal how the program may appear, the format of the equations in computer language, the results that can be obtained, and how the computer can be an effective tool for analysis and design. For those students with computer backgrounds, problems are provided at the end of each chapter.

The laboratory manual, *Experiments in Circuit Analysis,* 5th Edition, which accompanies this text, has also been heavily revised. The first section of the dc and ac sequence now contains a math review that can be used to establish fundamental mathematical operations or simply to provide a spacer between the beginning of a semester and that point when the student will have sufficient lecture background to perform a meaningful laboratory exercise. The first few labs of both the dc and ac sections have been completely revised to provide an improved introduction to a number of fundamental measurement techniques.

Transparency masters that reflect coverage of all the major areas of the text are now available for classroom use. The study guide has also been totally revamped to reflect changes in the text presentation and the needs of the students. For students interested in improving their computer skills as applied to circuit analysis, the manual *BASIC Applied To Circuit Analysis* is available and covers the text material in the same order with additional examples and exercises.

There is no question that this edition has received more attention from the author and publisher than any previous edition. The new art, additions and deletions, modified sections, and the concern for accuracy have raised my awareness that this edition has been treated as a high-priority effort. To this end the author was very fortunate to have the support of a number of key individuals at Merrill Publishing Co: Don Thompson, for developing the manuscript and coordinating reviews; Rex Davidson, for his production editorial work; and Pete Robison for his art coordination and creative page layout. Additionally at Merrill Publishing I thank Tim McEwen, administrative editor; Bruce Johnson, art supervisor; Cynthia Brunk, text designer; Ann Castel, marketing coordinator; and Cathy White and Corinne Folino, buyers. A special thanks goes to Linda Johnstone, whose diligent copyediting greatly contributes to the accuracy of this text.

I would also like to acknowledge the following reviewers for the quality, depth, and care of their comments and suggestions: Richard Adler, DeVry Institute of Technology, Toronto; Alexander W. Avtgis, Wentworth Institute of Technology; James Brice, DeVry Institute of Technology, Phoenix; C. W. Cowan, Southern Technical Institute; James Evans, Paisley College of Technology; Steve Kalina, DeVry, Inc.; Russell E. Puckett, Texas A & M; Robert Reid, Broome Community College; Brian Sibbald, Paisley College of Technology; and Donald Szymanski, Owens Technical College.

Finally, I must extend my deepest appreciation to my good friends Professors Aidala and Katz for the continual flow of comments and suggestions throughout the editions that have had a marked impact on the content of the text.

Robert L. Boylestad

Contents

1

Introduction

$$\sum \begin{smallmatrix} S \\ I \end{smallmatrix}$$

1.1 THE ELECTRICAL/ELECTRONICS INDUSTRY

The foundation of modern-day society is particularly sensitive to a few areas of development, research, and interest. In recent years, it has become obvious that the electrical/electronics industry is one area that will have a broad impact on future development in a host of activities that affect our life style, general health, and capabilities. Can you think of a field today, even those headstrong to minimize technical ties, that does not, at the very least, seek to broaden its horizons through the use of some technical innovation such as recording, duplication, or data-handling instrumentation?

Every facet of our lives seems touched by developments that appear to surface at an ever increasing rate. For the layperson, the most obvious improvement of recent years has been the reduced size of electrical/electronics systems. TVs are now small enough to be hand held and have a battery capability which allows them to be more portable. Computers with significant memory capacity are now as small as portable typewriters. The size of radios is limited simply by the ability to read the numbers on the face of the dial. Hearing aids are no longer visible, and pacemakers are significantly smaller and more reliable. All the reduction in size is due primarily to a marvelous development of the last few decades—the integrated circuit (IC). First developed in the late 1950s, the IC industry has now reached a point where it can cut 1-micrometer lines. Consider that some 25,000 of these lines would fit

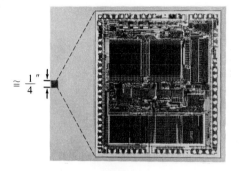

$\cong \frac{1}{4}''$

FIG. 1.1

Integrated circuit. (Courtesy of Motorola Semiconductor Products)

within one inch. Try to visualize breaking down an inch into 100 divisions and then consider 1000 or 25,000 divisions—an incredible achievement.

The integrated circuit of Fig. 1.1 has over 68,000 transistors in addition to thousands of other elements, yet is only about 1/4 in. on each side.

It is natural to wonder what the limits to growth may be when we consider the changes over the last few decades. Rather than following a steady growth curve that would be somewhat predictable, the industry is subject to surges that revolve around significant developments in the field. Present indications are that the level of miniaturization will continue but at a more moderate pace. Interest has turned toward increasing the quality and yield levels (percent of good integrated circuits in the production process).

History reveals that there have been peaks and valleys in industry growth but that revenues continue to rise at a steady rate and funds set aside for research and development continue to command an increasing share of the budget. The field changes at a rate that requires constant retraining of employees from the entry to the director level. Many companies have instituted their own training programs and have encouraged local universities to develop programs which will insure that the latest concepts and procedures are brought to the attention of their employees. A period of relaxation could be disastrous to a company dealing in competitive products.

No matter what the pressures on an individual in this field may be to keep up with the latest technology, there is one saving grace that becomes immediately obvious: Once a concept or procedure is clearly and correctly understood, it will bear fruit throughout the career of the individual at any level of the industry. For example, once a fundamental equation such as Ohm's law (Chapter 4) is understood, it will not be *replaced* by another equation as more advanced theory is considered. It is a relationship of fundamental quantities that can have application in the most advanced setting. In addition, once a procedure or method of analysis is understood, it usually can be applied to a wide (if not infinite) variety of problems, making it unnecessary to learn a different technique for each slight variation in the system. The content of this text is such that every morsel of information will have application in more advanced courses. It will not be replaced by a different set of equations and procedures unless required by the specific area of application. Even then, the new procedures will usually be an expanded application of concepts already presented in the text.

It is therefore paramount that the material presented in this introductory course be clearly and precisely understood. It is the foundation for the material to follow and will be applied throughout your working days in this growing and exciting field.

1.2 A BRIEF HISTORY

In the sciences, once a hypothesis is proven and accepted, it becomes one of the building blocks of that area of study, permitting further investigation and development. Naturally, the more pieces of a puzzle available, the more obvious the avenue toward a possible solution. In fact, history demonstrates that a single development may provide the key that will result in a mushroom effect that brings the science to a new plateau of understanding and impact.

If the opportunity presents itself, it would be time well spent to read one of the many publications reviewing the history of this field. Space requirements are such that only a brief review can be provided here. There are many more contributors than could be listed, and their efforts have often provided important keys to the solution of some very important concepts.

As noted earlier, there were periods characterized by what appeared to be an explosion of interest and development in particular areas. As you will see from the discussion of the late 1700s and the early 1800s, inventions, discoveries, and theories came fast and furiously. Each new concept broadened the possible areas of application until it becomes almost impossible to trace developments without picking a particular area of interest and following it through. In the review, as you read about the development of the radio, TV, and computer, keep in mind that similar progressive steps were occurring in the areas of the telegraph, the telephone, power generation, the phonograph, appliances, and so on.

There is a tendency when reading about the great scientists, inventors, and innovators to believe their contribution was a totally individual effort. In many instances, this was not the case. In fact, many of the great contributors were friends or associates and provided support and encouragement in their efforts to investigate various theories. At the very least, they were aware of one another's efforts to the degree possible in the days when a letter was often the best form of communication. In particular, note the closeness of the dates during periods of rapid development. One contributor seemed to spur on the efforts of the others or possibly provided the key needed to continue with the area of interest.

In the early stages, the contributors were not electrical, electronic, or computer engineers as we know them today. In most cases, they were physicists, chemists, mathematicians, or even philosophers. In addition, they were not from one or two communities of the Old World. The home country of many of the major contributors listed below is provided to show that almost every established community had some impact on the development of the fundamental laws of electrical circuits.

As you proceed through the remaining chapters of the text, you will find that a number of the units of measurement bear the name of major contributors in those areas—*volt* after Count Alessandro Volta, *ampere* after André Ampère, *ohm* after Georg Ohm, and so forth—fitting recognition for their important contribution to the birth of a major field of study.

Time charts indicating a limited number of major developments are provided in Fig. 1.2, primarily to identify specific periods of rapid development and to reveal how far we have come in the last few decades. In essence, the current state of the art is a result of efforts that

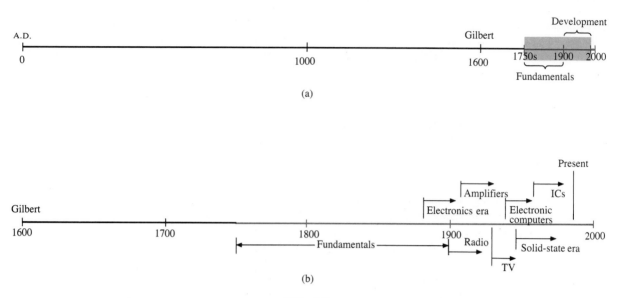

FIG. 1.2

Time charts. (a) Long-range; (b) expanded.

began in earnest some 250 years ago, with progress in the last 100 years almost exponential.

As you read through the following brief review, try to sense the growing interest in the field and the enthusiasm and excitement that must have accompanied each new revelation. Although you may find some of the terms used in the review new and essentially meaningless, the remaining chapters will explain them thoroughly.

The Beginning

The phenomenon of static electricity has been toyed with since antiquity. The Greeks called the fossil resin substance so often used to demonstrate the effects of static electricity *elektron,* but no extensive study was made of the subject until William Gilbert researched the event in 1600. In the years to follow, there was a continuing investigation of electrostatic charge by a number of individuals such as Otto von Guericke, who developed the first machine to generate large amounts of charge, and Stephen Gray, who was able to transmit electrical charge over long distances on silk threads. Charles Du Fay demonstrated that charges either attract or repel each other, leading him to believe that there were two types of charge—a theory we subscribe to today with our defined positive and negative charges.

There are many who believe the true beginnings of the electrical era lie with the efforts of Pieter van Musschenbroek and Benjamin Franklin. In 1745, van Musschenbroek introduced the *Leyden jar* for the storage of electrical charge (the first capacitor) and demonstrated electrical shock (and therefore the power of this new form of energy).

Franklin used the Leyden jar some seven years later to establish that lightning is simply an electrical discharge, and expanded on a number of other important theories including the definition of the two types of charge as *positive* and *negative*. From this point on, new discoveries and theories seemed to occur at an increasing rate as the number of individuals performing research in the area grew.

In 1784, Charles Coulomb demonstrated in Paris that the force between charges is inversely related to the square of the distance between the charges. In 1791, Luigi Galvani, Professor of Anatomy at the University of Bologna, in Italy, performed experiments revealing that electricity is present in every animal. The first *voltaic cell* with its ability to produce electricity through the chemical action of a metal dissolving in an acid was developed by another Italian, Alessandro Volta, in 1799.

The fever pitch continued into the early 1800s with Hans Christian Oersted, a Swedish professor of physics, announcing in 1820 a relationship between magnetism and electricity that serves as the foundation for the theory of *electromagnetism* as we know it today. In the same year, a French physicist, André Ampère, demonstrated that there are magnetic effects around every current-carrying conductor and that current-carrying conductors can attract and repel each other just like magnets. In the period 1826 to 1827, a German physicist, Georg Ohm, introduced an important relationship between potential, current, and resistance which we now refer to as Ohm's law. In 1831, an English physicist, Michael Faraday, demonstrated his theory of *electromagnetic induction,* whereby a changing current in one coil can induce a changing current in another coil even though the two coils are not directly connected. Professor Faraday also did extensive work on a storage device he called the condenser, which we refer to as a capacitor today. He introduced the idea of adding a dielectric between the plates of a capacitor to increase the storage capacity (Chapter 10). James Clerk Maxwell, a Scottish professor of natural philosophy, performed extensive mathematical analyses to develop what is currently called *Maxwell's equations,* which support the efforts of Faraday linking electric and magnetic effects. Maxwell also developed the *electromagnetic theory of light* in 1862, which among other things revealed that electromagnetic waves travel through air at the velocity of light (186,000 miles per second or 3×10^8 meters per second). In 1888, a German physicist, Heinrich Rudolph Hertz, through experimentation with lower-frequency electromagnetic waves (microwaves), substantiated Maxwell's predictions and equations. In the mid-1800s, Professor Gustav Robert Kirchhoff introduced a series of laws of voltages and currents that find application at every level and area of this field (Chapters 5 and 6). In 1895, another German physicist, Wilhelm Röntgen, discovered electromagnetic waves of high frequency commonly called X-rays today.

By the end of the 1800s, a significant number of the fundamental equations, laws, and relationships had been established and various fields of study including electronics, power generation, and calculating equipment started to develop in earnest.

The Age of Electronics

Radio The true beginning of the electronics era is open to debate and is sometimes attributed to efforts by early scientists in applying potentials across evacuated glass envelopes. However, many trace the beginning to Thomas Edison, who added a metallic electrode to the vacuum of the tube and discovered that a current was established between the metal electrode and the filament when a positive voltage was applied to the metal electrode. The phenomenon, demonstrated in 1883, was referred to as the *Edison effect*. In the period to follow, the transmission of radio waves and the development of the radio received widespread attention. In 1887, Heinrich Hertz in his efforts to verify Maxwell's equations transmitted radio waves for the first time in his laboratory. In 1896, an Italian scientist, Guglielmo Marconi (often called the father of the radio), demonstrated that telegraph signals could be sent through the air over long distances (2.5 kilometers) using a grounded antenna. In the same year, Aleksandr Popov sent what might have been the first radio message some 300 yards. The message was the name "*Heinrich Hertz*" in respect for Hertz's earlier contributions. In 1901, Marconi established radio communication across the Atlantic.

In 1904, John Ambrose Fleming expanded on the efforts of Edison to develop the first diode, commonly called *Fleming's valve*—actually the first of the *electronic devices*. The device had a profound impact on the design of detectors in the receiving section of radios. In 1906, Lee De Forest added a third element to the vacuum structure and created the first amplifier, the triode. Shortly thereafter, in 1912, Edwin Armstrong built the first regenerative circuit to improve receiver capabilities and then used the same contribution to develop the first nonmechanical oscillator. By 1915 radio signals were being transmitted across the United States, and in 1918 Armstrong applied for a patent for the superheterodyne circuit employed in virtually every TV and radio to permit amplification at one frequency rather than at the full range of incoming signals. The major components of the modern-day radio were now in place and sales in radios grew from a few million dollars in the early 1920s to over one billion by the 1930s. The 1930s were truly the golden years of radio, with a wide range of productions for the listening audience.

Television The 1930s were also the true beginnings of the television era, although development on the picture tube began in earlier years with Paul Nipkow and his *electrical telescope* in 1884 and John Baird and his long list of successes including the transmission of TV pictures over telephone lines in 1927 and over radio waves in 1928, and simultaneous transmission of pictures and sound in 1930. In 1932, NBC installed the first commercial TV antenna on top of the Empire State Building, and RCA began regular broadcasting in 1939. The war slowed development and sales, but in the mid 1940s the number of sets grew from a few thousand to a few million. Color TV became popular in the early 1960s.

Computers The earliest computer system can be traced back to Blaise Pascal in 1642 with his mechanical machine for adding and subtracting numbers. In 1673 Gottfried Wilhelm von Leibniz used the *Leibniz wheel* to add multiplication and division to the range of operations, and in 1823 Charles Babbage developed the *Difference Engine* to add the mathematical operations of sine, cosine, logs, and a number of others. In the years to follow, improvements were made, but the system remained primarily mechanical until the 1930s when electromechanical systems using components such as relays were introduced. It was not until the 1940s that totally electronic systems became the new wave. It is interesting to note that even though IBM was formed in 1924, it did not enter the computer industry until 1937. An entirely electronic system known as *Eniac* was dedicated at the University of Pennsylvania in 1946. It contained 18,000 tubes and weighed 30 tons but was several times faster than most electromechanical systems. Although other vacuum tube systems were built, it was not until the birth of the solid-state era that computer systems experienced a major change in size, speed, and capability.

The Solid-State Era

In 1947, physicists William Shockley, John Bardeen, and Walter H. Brattain of Bell Telephone Laboratories demonstrated the point-contact *transistor* (Fig. 1.3), an amplifier constructed entirely of solid-state materials with no requirement for a vacuum, glass envelope, or heater voltage for the filament. Although reluctant at first due to the vast amount of material available on the design, analysis, and synthesis of tube networks, the industry eventually accepted this new technology as the wave of the future. In 1958 the first *integrated circuit* (IC) was developed at Texas Instruments, and in 1961 the first commercial integrated circuit was manufactured by the Fairchild Corporation.

It is impossible to properly review the entire history of the electrical/electronics field in a few pages. The effort here, both through the discussion and the time graphs of Fig. 1.2, was to reveal the amazing progress of this field in the last 50 years. The growth appears to be truly exponential since the early 1900s, raising the interesting question as to where we go from here. The time chart suggests that the next few decades will probably contain a number of important innovative contributions that may cause an even faster growth curve than we are now experiencing.

FIG. 1.3
(*Courtesy of AT&T, Bell Laboratories*)

1.3 UNITS OF MEASUREMENT

It is vital that the importance of units of measurement be understood and appreciated early in the development of a technically oriented background. Too frequently their effect on the most basic substitution is ignored. Consider, for example, the following very fundamental physics equation:

$$\boxed{v = \frac{d}{t}} \qquad \begin{array}{l} v = \text{velocity} \\ d = \text{distance} \\ t = \text{time} \end{array} \qquad \textbf{(1.1)}$$

Assume, for the moment, that the following data are obtained:

$$d = 4000 \, \text{ft}$$

$$t = 1 \, \text{min}$$

and v is desired in miles per hour. Often, without a second thought or consideration, the numerical values are simply substituted into the equation, with the result here that

$$v = \frac{d}{t} = \frac{4000}{1} = \cancel{4000 \, \text{mi/h}}$$

As indicated above, the solution is incorrect. If the result is desired in *miles per hour,* the unit of measurement for distance must be *miles,* and that for time, *hours.* In a moment, when the problem is analyzed properly, the extent of the error will demonstrate the importance of *giving due consideration to the unit of measurement before substituting numerical values.*

The next question is normally how to convert the distance and time to the proper unit of measurement. A method will be presented in a later section of this chapter, but for now it is given that

$$1 \, \text{mi} = 5280 \, \text{ft}$$

$$4000 \, \text{ft} = 0.7576 \, \text{mi}$$

$$1 \, \text{min} = \tfrac{1}{60} \, \text{h} = 0.0167 \, \text{h}$$

Substituting into Eq. (1.1), we have

$$v = \frac{d}{t} = \frac{0.7576}{0.0167} = 45.37 \, \text{mi/h}$$

which is significantly different from the result obtained before.

To complicate the matter further, suppose the distance is given in kilometers, as is now the case on many road signs. First, we must realize that the prefix *kilo* stands for a multiplier of 1000 (to be introduced in Section 1.5), and then we must find the conversion factor between kilometers and miles. If this conversion factor is not readily available, we must be able to make the conversion between units using the conversion factors between meters and feet or inches as described in Section 1.6.

Before substituting numerical values into an equation, try to mentally establish a reasonable range of solutions for comparison purposes. For instance, if a car travels 4000 ft in 1 min, does it seem reasonable that the speed would be 4000 mi/h? Obviously not! This self-checking procedure is particularly important in this day of the hand-held calculator when ridiculous results may be accepted simply because they appear on the digital display of the instrument.

Finally, *if a unit of measurement is applicable to a result or piece of data, then it must be applied to the numerical value*. To state that $v = 45.37$ without including the unit of measurement *mi/h* is meaningless.

Equation (1.1) is not a difficult one. A simple algebraic manipulation will result in the solution for any one of the three variables. However, in light of the number of questions arising from this equation, the reader may wonder if the difficulty associated with an equation will increase at the same rate as the number of terms in the equation. In the broad sense, this will not be the case. There is, of course, more room for a mathematical error with a more complex equation, but once the proper system of units is chosen and each term properly found in that system, there should be very little added difficulty associated with an equation requiring an increased number of mathematical calculations.

In review, before substituting numerical values into an equation, be absolutely sure of the following:

1. Each quantity has the proper unit of measurement as defined by the equation.
2. The proper magnitude of each quantity as determined by the defining equation is substituted.
3. Each quantity is in the same system of units (or as defined by the equation).
4. The magnitude of the result is of a reasonable nature when compared to the level of the substituted quantities.
5. The proper unit of measurement is applied to the result.

1.4 SYSTEMS OF UNITS

In the past, the *systems of units* most commonly used were the English and metric, as outlined by Table 1.1. Note that while the English system is based on a single standard, the metric is subdivided into two interrelated standards: the MKS and CGS. Fundamental quantities of these systems are compared in Table 1.1 along with their abbreviations. The MKS and CGS systems draw their names from the units of measurement used with each system; the MKS system uses *M*eters, *K*ilograms, and *S*econds, while the CGS system uses *C*entimeters, *G*rams, and *S*econds.

Understandably, the use of more than one system of units in a world that finds itself continually shrinking in size, due to advanced technical developments in communications and transportation, would introduce unnecessary complications to the basic understanding of any technical data. The need for a standard set of units to be adopted by all nations has become increasingly obvious. The International Bureau of Weights and Measures located at Sèvres, France, has been the host for the General Conference of Weights and Measures, attended by representatives from all nations of the world. In 1960, the General Conference adopted a system called Le Système International d'Unités (International System of Units), which has the international abbreviation SI. Since then, it has

TABLE 1.1

Comparison of the English and metric systems of units.

English	Metric		SI
	MKS	**CGS**	**SI**
Length:			
Yard (yd)	Meter (m)	Centimeter (cm)	Meter (m)
(0.914 m)	(39.37 in.)	(2.54 cm = 1 in.)	
	(100 cm)		
Mass:			
Slug	Kilogram (kg)	Gram (g)	Kilogram (kg)
(14.6 kg)	(1000 g)		
Force:			
Pound (lb)	Newton (N)	Dyne	Newton (N)
(4.45 N)	(100,000 dynes)		
Temperature:			
Fahrenheit (°F)	Celsius or	Centigrade (°C)	Kelvin (K)
	Centigrade (°C)		K = 273.15 + °C
$\left(=\dfrac{9}{5}°C + 32\right)$	$\left(=\dfrac{5}{9}(°F - 32)\right)$		
Energy:			
Foot-pound	Newton-meter	Dyne-centimeter	Joule (J)
(ft-lb)	(N-m)	or Erg	
(1.356 joules)	or Joule (J)	(1 joule = 10^7 ergs)	
	(0.7378 ft-lb)		
Time:			
Second (s)	Second (s)	Second (s)	Second (s)

been adopted by the Institute of Electrical and Electronic Engineers, Inc. (IEEE) in 1965 and by the United States of America Standards Institute in 1967 as a standard for all scientific and engineering literature.

The inevitable changeover to the metric system has already resulted in the use of *both* miles per hour (mi/h) and kilometers per hour (km/h) on some road signs and the distribution of English-to-metric conversion charts as advertising literature by some firms. In fact, calculators are now available that are designed specifically to convert from one system to the other.

For comparison, the SI units of measurement and their abbreviations appear in Table 1.1. These abbreviations are those usually applied to each unit of measurement, and they were carefully chosen to be the most effective. Therefore, it is important that they be used whenever applicable to insure universal understanding. Note the similarities of the SI system to the MKS system. This text will employ, whenever possible and practical, all of the major units and abbreviations of the SI system in an effort to support the need for a universal system. For those readers requiring further information on the SI system, a complete kit has been

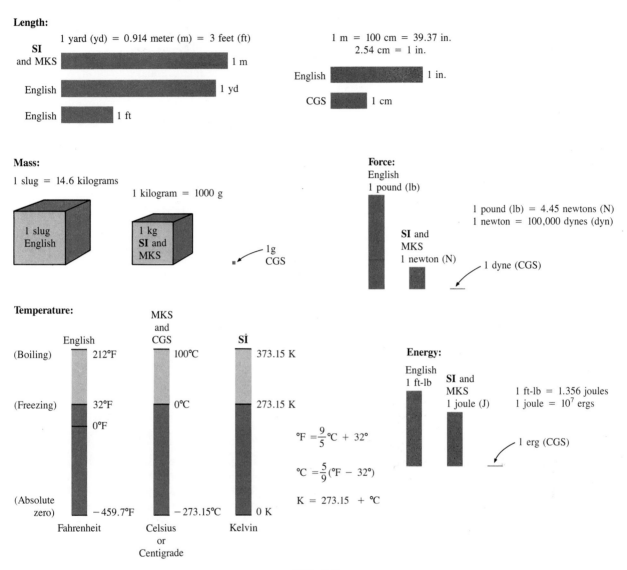

FIG. 1.4

Comparison of units of the various systems of units.

assembled for general distribution by the American Society for Engineering Education (ASEE).*

Figure 1.4 should help the reader develop some feeling for the relative magnitudes of the units of measurement of each system of units. Note in the figure the relatively small magnitude of the units of measurement for the CGS system.

A standard exists for each unit of measurement of each system. The standards of some units are quite interesting. The *meter* was defined in

---

*American Society for Engineering Education (ASEE), One Dupont Circle, Suite 400, Washington, D.C. 20036.

1960 as 1,650,763.73 wavelengths of the orange-red light of krypton 86. It was originally defined in 1790 to be 1/10,000,000 the distance between the equator and either pole at sea level. This length is preserved on a platinum-iridium bar at the International Bureau of Weights and Measures at Sèvres, France. The *kilogram* is defined as a mass equal to 1000 times the mass of one cubic centimeter of pure water at 4°C. This standard is preserved in the form of a platinum-iridium cylinder in Sèvres. The *second* was originally defined as 1/86,400 of the mean solar day. It was redefined in 1960 as 1/31,556,925.9747 of the tropical year 1900.

1.5 SCIENTIFIC NOTATION

It should be apparent from the relative magnitude of the various units of measurement that very large and very small numbers will frequently be encountered in the study of the sciences. To ease the difficulty of mathematical operations with numbers of extreme size, *scientific notation* is usually employed. This notation takes full advantage of the mathematical properties of powers of 10. The notation used to represent numbers that are integer powers of 10 is as follows:

$$1 = 10^0 \qquad 1/10 = \quad 0.1 = 10^{-1}$$
$$10 = 10^1 \qquad 1/100 = \quad 0.01 = 10^{-2}$$
$$100 = 10^2 \qquad 1/1000 = \quad 0.001 = 10^{-3}$$
$$1000 = 10^3 \qquad 1/10,000 = 0.0001 = 10^{-4}$$

A quick method of determining the proper power of 10 is to place a caret mark to the right of the numeral 1 wherever it may occur; then count from this point to the number of places to the right or left before arriving at the decimal point. Moving to the right indicates a positive power of 10, while moving to the left indicates a negative power. For example,

$$10,000.0 = 1\underbrace{0\,0\,0\,0}_{1\;2\;3\;4}. = 10^{+4}$$

$$0.00001 = 0.\underbrace{0\,0\,0\,0\,1}_{5\;4\;3\;2\;1} = 10^{-5}$$

Since some of these powers of 10 appear frequently, a written form of abbreviation has been adopted (as indicated in Table 1.2), which when written in conjunction with the unit of measurement eliminates the need to include the power of 10 in numerical form.

TABLE 1.2

Power of 10	Prefix	Abbreviation
10^{12}	Tera	T
10^9	Giga	G
10^6	Mega	M
10^3	Kilo	k
10^{-3}	Milli	m
10^{-6}	Micro	μ
10^{-9}	Nano	n
10^{-12}	Pico	p

EXAMPLES

$$1,000,000 \text{ ohms} = 1 \times 10^6 \text{ ohms}$$
$$= 1 \text{ megohm } (M\Omega)$$

$$100,000 \text{ meters} = 100 \times 10^3 \text{ meters}$$
$$= 100 \text{ kilometers (km)}$$

$$0.0001 \text{ second} = 0.1 \times 10^{-3} \text{ second}$$
$$= 0.1 \text{ millisecond (ms)}$$

$$0.000001 \text{ farad} = 1 \times 10^{-6} \text{ farad}$$
$$= 1 \text{ microfarad } (\mu F)$$

Some important mathematical equations and relationships applying to powers of 10 are listed below along with a few examples. In each case, n and m can be any positive or negative real number.

$$\frac{1}{10^n} = 10^{-n} \qquad \frac{1}{10^{-n}} = 10^n \qquad \textbf{(1.2)}$$

Equation (1.2) clearly reveals that shifting a power of 10 from the denominator to the numerator, or the reverse, requires simply changing the sign of the power.

EXAMPLES

$$\frac{1}{1000} = \frac{1}{10^{+3}} = 10^{-3}$$

$$\frac{1}{0.00001} = \frac{1}{10^{-5}} = 10^{+5}$$

The product of powers of 10:

$$(10^n)(10^m) = 10^{(n+m)} \qquad \textbf{(1.3)}$$

EXAMPLES

$$(1000)(10{,}000) = (10^3)(10^4) = 10^{(3+4)} = 10^7$$
$$(0.00001)(100) = (10^{-5})(10^2) = 10^{(-5+2)} = 10^{-3}$$

The division of powers of 10:

$$\frac{10^n}{10^m} = 10^{(n-m)} \qquad \textbf{(1.4)}$$

EXAMPLES

$$\frac{100{,}000}{100} = \frac{10^5}{10^2} = 10^{(5-2)} = 10^3$$

$$\frac{1000}{0.0001} = \frac{10^3}{10^{-4}} = 10^{(3-(-4))} = 10^{(3+4)} = 10^7$$

Note in the last example the use of parentheses to insure the proper sign is established between operators.

$$\sum_{}^{S}{}_{I}$$

The power of powers of 10:

$$\boxed{(10^n)^m = 10^{(nm)}}$$

(1.5)

EXAMPLES

$$(100)^4 = (10^2)^4 = 10^{(2)(4)} = 10^8$$
$$(1000)^{-2} = (10^3)^{-2} = 10^{(3)(-2)} = 10^{-6}$$
$$(0.01)^{-3} = (10^{-2})^{-3} = 10^{(-2)(-3)} = 10^6$$

Let us now consider a few examples demonstrating the use of powers of 10 with arbitrary numbers. When working with arbitrary numbers, we can separate the operations involving powers of 10 from those of the whole numbers.

Addition: In order to perform addition using powers of 10, the power of 10 of each term must be the same:

$$6300 + 75,000 = (6.3)(1000) + (75)(1000)$$
$$= 6.3 \times 10^3 + 75 \times 10^3$$
$$= (6.3 + 75) \times 10^3$$
$$= \mathbf{81.3 \times 10^3}$$

Subtraction: As with addition, the powers of 10 must be the same before the operation can be performed:

$$960,000 - 40,000 = (96)(10,000) - (4)(10,000)$$
$$= 96 \times 10^4 - 4 \times 10^4$$
$$= (96 - 4) \times 10^4$$
$$= \mathbf{92 \times 10^4}$$

Multiplication:

$$(0.0002)(0.000007) = [(2)(0.0001)](7)(0.000001)]$$
$$= (2 \times 10^{-4})(7 \times 10^{-6})$$
$$= (2)(7) \times (10^{-4})(10^{-6})$$
$$= \mathbf{14 \times 10^{-10}}$$
$$(340,000)(0.00061) = (3.4 \times 10^5)(61 \times 10^{-5})$$
$$= (3.4)(61) \times (10^5)(10^{-5})$$
$$= 207.4 \times 10^0$$
$$= \mathbf{207.4}$$

Division:

$$\frac{0.00047}{0.002} = \frac{47 \times 10^{-5}}{2 \times 10^{-3}} = \left(\frac{47}{2}\right) \times \left(\frac{10^{-5}}{10^{-3}}\right)$$
$$= \mathbf{23.5 \times 10^{-2}}$$
$$\frac{690,000}{0.00000013} = \frac{69 \times 10^4}{13 \times 10^{-8}} = \left(\frac{69}{13}\right) \times \left(\frac{10^4}{10^{-8}}\right)$$
$$= \mathbf{5.31 \times 10^{12}}$$

Powers:

$$(0.00003)^3 = (3 \times 10^{-5})^3 = (3)^3 \times (10^{-5})^3$$
$$= \mathbf{27 \times 10^{-15}}$$
$$(90{,}800{,}000)^2 = (9.08 \times 10^7)^2 = (9.08)^2 \times (10^7)^2$$
$$= \mathbf{82.4464 \times 10^{14}}$$

The following examples include units of measurement.

EXAMPLES

a. 41,200 m is equivalent to 41.2×10^3 m = **41.2 km.**

b. 0.00956 J is equivalent to 9.56×10^{-3} J = **9.56 mJ.**

c. 0.000768 s is equivalent to 768×10^{-6} s = **768 μs.**

d. $\dfrac{8400\,\text{m}}{0.06} = \dfrac{8.4 \times 10^3\,\text{m}}{6 \times 10^{-2}} = \left(\dfrac{8.4}{6}\right) \times \left(\dfrac{10^3}{10^{-2}}\right)\text{m}$
$$= 1.4 \times 10^5\,\text{m} = 140 \times 10^3\,\text{m} = \mathbf{140\,km}$$

e. $(0.0003)^4$ s $= (3 \times 10^{-4})^4$ s $= 81 \times 10^{-16}$ s
$$= 0.0081 \times 10^{-12}\,\text{s} = \mathbf{0.0081\,ps}$$

To demonstrate the amount of work saved and the reduced possibility of error that result by using powers of 10, consider finding the solution to the last example in the following manner:

$$
\begin{array}{r}
0.0003 \\
\times\ 0.0003 \\
\hline
0.00000009 \\
\times\ 0.0003 \\
\hline
0.000000000027 \\
\times\ 0.0003 \\
\hline
0.0000000000000081 = 81 \times 10^{-16}\,\text{s} = \mathbf{0.0081\ ps}
\end{array}
$$

1.6 CONVERSION WITHIN AND BETWEEN SYSTEMS OF UNITS

The conversion within and between systems of units is a process that cannot be avoided in the study of any technical field. It is an operation, however, that is performed incorrectly so often that this section was included to provide one approach which, if applied properly, will lead to the correct result.

There is more than one method to perform the conversion process. In fact, some people prefer to determine mentally whether the conversion factor is multiplied or divided. This approach is acceptable for some elementary conversions but is risky with more complex operations.

The procedure to be described here is best introduced by examining a relatively simple problem such as converting inches to meters. Specifically, let us convert 48 in. (4 ft) to meters.

If we multiply the 48 in. by a factor of 1, the magnitude of the quantity remains the same:

$$48 \text{ in.} = 48 \text{ in.}(1) \qquad \textbf{(1.6)}$$

The conversion factor is the following:

$$1 \text{ m} = 39.37 \text{ in.}$$

Dividing both sides of the conversion factor by 39.37 in. will result in

$$\frac{1 \text{ m}}{39.37 \text{ in.}} = \frac{39.37 \text{ in.}}{39.37 \text{ in.}} = (1)$$

Note that the end result is that the ratio 1 m/39.37 in. equals 1, as it should since they are equal quantities. If we now substitute this factor (1) into Eq. (1.6), we obtain

$$48 \text{ in.}(1) = 48 \text{ in.}\left(\frac{1 \text{ m}}{39.37 \text{ in.}}\right)$$

which results in the cancellation of inches as a unit of measurement and leaves meters as the unit of measure. In addition, since the 39.37 is in the denominator, it must be divided into the 48 to complete the operation:

$$\frac{48}{39.37} \text{ m} = \textbf{1.219 m}$$

Let us now review the method, which has the following sequence of steps:

1. Multiply the quantity to be converted by the factor (1).
2. Set up the conversion factor to form a numerical value of (1) with the unit of measurement to be removed in the denominator.
3. Perform the required mathematics to obtain the proper magnitude for the remaining unit of measurement.

EXAMPLES   Convert 6.8 min to seconds.

Step 1: $\qquad\qquad\qquad 6.8 \text{ min}(1)$

Step 2: $\qquad\qquad\qquad \left(\dfrac{60 \text{ s}}{1 \text{ min}}\right) = (1)$

Step 3: $\qquad 6.8 \text{ min}\left(\dfrac{60 \text{ s}}{1 \text{ min}}\right) = (6.8)(60) \text{ s}$

$$= \textbf{408 s}$$

Convert 0.24 m to centimeters.

Step 1: $\qquad\qquad\qquad 0.24 \text{ m}(1)$

Step 2: $\qquad\qquad\qquad \left(\dfrac{100 \text{ cm}}{1 \text{ m}}\right) = (1)$

Step 3:
$$0.24 \, \cancel{m}\left(\frac{100 \, cm}{1 \, \cancel{m}}\right) = (0.24)(100) \, cm$$
$$= 24 \, cm$$

The product (1)(1) or (1)(1)(1) is still 1. Using this fact, we can perform a series of conversions in the same operation.

EXAMPLES Determine the number of minutes in half a day.

$$0.5 \, \cancel{day}\left(\frac{24 \, \cancel{h}}{1 \, \cancel{day}}\right)\left(\frac{60 \, min}{1 \, \cancel{h}}\right) = (0.5)(24)(60) \, min$$
$$= 720 \, min$$

Convert 1/4 in. to millimeters.

$$\tfrac{1}{4} \, \cancel{in.}\left(\frac{1 \, \cancel{m}}{39.37 \, \cancel{in.}}\right)\left(\frac{10^3 \, mm}{1 \, \cancel{m}}\right) = \frac{0.25}{39.37}(10^3) \, mm$$
$$= 6.35 \, mm$$

The following examples are variations of the above to practical situations.

EXAMPLES In Europe, the speed limit is posted in kilometers per hour. How fast in miles per hour is 100 km/h?

$$\left(\frac{100 \, km}{h}\right)(1)(1)(1)(1)$$
$$= \left(\frac{100 \, \cancel{km}}{h}\right)\left(\frac{1000 \, \cancel{m}}{1 \, \cancel{km}}\right)\left(\frac{39.37 \, \cancel{in.}}{1 \, \cancel{m}}\right)\left(\frac{1 \, \cancel{ft}}{12 \, \cancel{in.}}\right)\left(\frac{1 \, mi}{5280 \, \cancel{ft}}\right)$$
$$= \frac{(100)(1000)(39.37)}{(12)(5280)} \, \frac{mi}{h}$$
$$= 62.14 \, mi/h$$

Determine the speed in miles per hour of a competitor who can run a 4-min mile. Inverting the factor 4 mi/1 mi to 1 mi/4 min, we can proceed as follows:

$$\left(\frac{1 \, mi}{4 \, \cancel{min}}\right)\left(\frac{60 \, \cancel{min}}{h}\right) = \frac{60}{4} \, mi/h = 15 \, mi/h$$

1.7 SYMBOLS

Throughout the text, various symbols will be employed that the reader may not have had occasion to use. Some are defined in Table 1.3, and others will be defined in the text as the need arises.

TABLE 1.3

Symbol	Meaning
$\neq$	Not equal to $6.12 \neq 6.13$
$>$	Greater than $4.78 > 4.20$
$\gg$	Much greater than $840 \gg 16$
$<$	Less than $430 < 540$
$\ll$	Much less than $0.002 \ll 46$
$\geq$	Greater than or equal to $x \geq y$ is satisfied for $y = 3$ and $x > 3$ or $x = 3$
$\leq$	Less than or equal to $x \leq y$ is satisfied for $y = 3$ and $x < 3$ or $x = 3$
$\cong$	Approximately equal to $3.14159 \cong 3.14$
Σ	Sum of $\Sigma \, (4 + 6 + 8) = 18$
$\| \, \|$	Absolute magnitude of $\|a\| = 4$, where $a = -4$ or $+4$
$\therefore$	Therefore $x = \sqrt{4} \quad \therefore x = \pm 2$

1.8 CONVERSION TABLES

Conversion tables such as those appearing in Appendix A can be very useful when time does not permit the application of methods described in this chapter. However, even though such tables appear easy to use, frequent errors occur because the operations appearing at the head of the table are not properly performed. In any case, when using such tables, try to establish mentally some order of magnitude for the quantity to be determined as compared to the magnitude of the quantity in its original set of units. This simple operation should prevent a number of the impossible results that may occur if the conversion operation is improperly applied.

For example, consider the following from such a conversion table:

To convert from	To	Multiply by
Miles	Meters	1.609×10^3

A conversion of 2.5 mi to meters would require that we multiply 2.5 by the conversion factor. That is,

$$2.5\,\text{mi}(1.609 \times 10^3) = 4.0225 \times 10^3\,\text{m}$$

A conversion from 4000 m to miles would require a division process:

$$\frac{4000\,\text{m}}{1.609 \times 10^3} = 2486.02 \times 10^{-3} = 2.48602\,\text{mi}$$

In each of the above, there should have been little difficulty realizing that 2.5 mi would convert to a few thousand meters, and 4000 m would be only a few miles. As indicated above, this kind of prior thinking will eliminate the possibility of ridiculous conversion results.

PROBLEMS

Note: More difficult problems are denoted by an asterisk (*) throughout the text.

Section 1.3

1. Determine the distance in feet traveled by a car moving at 50 mi/h for 1 min.

2. How many hours would it take a person to walk 12 mi if the average pace is 15 min/mile?

Section 1.4

3. How many foot-pounds of energy are associated with 1000 J?

4. A temperature of 212°F (boiling point of water) is what temperature in the CGS and SI systems?

5. How many centimeters are there in 1/2 yd?

Section 1.5

6. Express the following numbers as powers of 10:
  a. 10,000   **b.** 0.0001
  c. 1000   **d.** 1,000,000
  e. 0.0000001   **f.** 0.00001

7. Using only those powers of 10 listed in Table 1.2, express the following numbers in what seems to you the most logical form for future calculations:
  a. 15,000   **b.** 0.03000
  c. 7,400,000   **d.** 0.0000068
  e. 0.00040200   **f.** 0.0000000002

Perform each of the following operations and express the result as a power of 10:

8. a. $(100)(100)$ **b.** $(0.01)(1000)$
    c. $(10^3)(10^6)$ **d.** $(1000)(0.00001)$
    e. $(10^{-6})(10,000,000)$ **f.** $(10,000)(10^{-8})(10^{35})$ $=10^{31}$

9. a. $\dfrac{100}{1000}$ **b.** $\dfrac{0.01}{100}$

    c. $\dfrac{10,000}{0.00001}$ **d.** $\dfrac{0.0000001}{100}$

    e. $\dfrac{10^{38}}{0.000100}$ **f.** $\dfrac{(100)^{1/2}}{0.01}$

10. a. $(100)^3$ **b.** $(0.0001)^{1/2}$
    c. $(10,000)^8$ **d.** $(0.00000010)^9$

11. a. $(-0.001)^2$ **b.** $\dfrac{(100)(10^{-4})}{10}$

    c. $\dfrac{(0.01)^2(100)}{10,000}$ **d.** $\dfrac{(10^2)(10,000)}{0.001}$

    e. $\dfrac{(0.0001)^3(100)}{1,000,000}$ ***f.** $\dfrac{[(100)(0.01)]^{-3}}{[(100)^2][0.001]}$

Perform the following operations:

*12. a. $\dfrac{(300)^2(100)}{10^4}$ **b.** $[(40,000)^2][(20)^{-3}]$

      $(10^2)(10^{-6})$

    c. $\dfrac{(60,000)^2}{(0.02)^2}$ **d.** $\dfrac{(0.000027)^{1/3}}{210,000}$

    e. $\dfrac{[(4000)^2][300]}{0.02}$

    f. $[(0.000016)^{1/2}][(100,000)^5][0.02]$

    g. $\dfrac{[(0.003)^3][(0.00007)^2][(800)^2]}{[(100)(0.0009)]^{1/2}}$ (a challenge)

$$\sum {}^{S}_{I}$$

Section 1.6

Convert the following:

13. a. 1.5 min to seconds
   b. 0.04 h to seconds
   c. 0.05 s to microseconds
   d. 0.16 m to mm
   e. 0.00000012 s to nanoseconds
   f. 3,620,000 s to days
   g. 1020 mm to meters

14. a. 0.1 μF (microfarad) to picofarads
   b. 0.467 km to meters
   c. 63.9 mm to cm
   d. 69 cm to kilometers
   e. 3.2 h to milliseconds
   f. 0.016 mm to μm
   g. 60 sq cm (cm^2) to square meters (m^2)

*15. a. 100 in. to meters
   b. 4 ft to meters
   c. 6 lb to newtons
   d. 60,000 dyn to pounds
   e. 150,000 cm to feet
   f. 0.002 mi to meters (5280 ft = 1 mi)
   g. 7800 m to yards

16. Find the velocity in miles per hour of a mass that travels 50 ft in 20 s.

17. How long in seconds will it take a car traveling at 100 mi/h to travel the length of a football field (100 yd)?

*18. Find the distance in meters that a mass traveling at 600 cm/s will cover in 0.016 h.

19. Convert 6 mi/h to meters per second.

20. If an athlete can row at a rate of 50 m/min, how many days would it take to cross the Atlantic (3000 mi)?

21. How long would it take a runner to complete a 10-km race if a pace of 6.5 min/mi were maintained?

22. Quarters are about 1 inch in diameter. How many would be required to stretch from one end of a football field to the other (100 yd)?

Section 1.8

23. Using Appendix A, determine the number of
   a. Btu in 5 joules of energy.
   b. cubic meters in 24 ounces of a liquid.
   c. seconds in 1.4 days.
   d. pints in 1 cubic meter of a liquid.

GLOSSARY

CGS system The system of units employing the *C*entimeter, *G*ram, and *S*econd as its fundamental units of measure.

Difference Engine One of the first mechanical calculators.

Edison effect Establishing a flow of charge between two elements in an evacuated tube.

Electromagnetism The relationship between magnetic and electrical effects.

Eniac The first totally electronic computer.

Fleming's valve The first of the electronic devices, the diode.

Integrated circuit (IC) A subminiature structure containing a vast number of electronic devices designed to perform a particular set of functions.

Joule (J) A unit of measurement for energy in the SI or MKS system. Equal to 0.7378 foot-pound in the English system and 10^7 ergs in the CGS system.

Kelvin (K) A unit of measurement for temperature in the SI system. Equal to $273.15 + {}^{\circ}C$ in the MKS and CGS systems.

Kilogram (kg) A unit of measure for mass in the SI and MKS systems. Equal to 1000 grams in the CGS system.

Leyden jar One of the first charge storage devices.

Meter (m) A unit of measure for length in the SI and MKS systems. Equal to 1.094 yards in the English system and 100 centimeters in the CGS system.

MKS system The system of units employing the *M*eter, *K*ilogram, and *S*econd as its fundamental units of measure.

Newton (N) A unit of measurement for force in the SI and MKS systems. Equal to 100,000 dynes in the CGS system.

Pound (lb) A unit of measurement for force in the English system. Equal to 4.45 newtons in the SI or MKS system.

Scientific notation A method for describing very large and very small numbers through the use of powers of 10.

Second (s) A unit of measurement for time in the SI, MKS, English, and CGS systems.

SI system The system of units adopted by the IEEE in 1965 and the USASI in 1967 as the International System of Units (*Système I*nternational d'Unités).

Slug A unit of measure for mass in the English system. Equal to 14.6 kilograms in the SI or MKS system.

Static electricity Stationary charge in a state of equilibrium.

Transistor The first semiconductor amplifier.

Voltaic cell A storage device that converts chemical to electrical energy.

2
Current and Voltage

2.1 ATOMS AND THEIR STRUCTURE

A basic understanding of the fundamental concepts of current and voltage requires a degree of familiarity with the atom and its structure. The simplest of all atoms is the hydrogen atom, made up of two basic particles, the *proton* and the *electron*, in the relative positions shown in Fig. 2.1(a). The *nucleus* of the hydrogen atom is the proton, a positively charged particle. *The orbiting electron carries a negative charge that is equal in magnitude to the positive charge of the proton.* In all other elements, the nucleus also contains *neutrons*, which are slightly heavier than protons and have no electrical charge. The helium atom, for example, has two neutrons in addition to two electrons and two protons as shown in Fig. 2.1(b). *In all neutral atoms the number of electrons is equal to the number of protons.* The mass of the electron is 9.11×10^{-28} g, and that of the proton and neutron is 1.672×10^{-24} g. The mass of the proton (or neutron) is therefore approximately 1836 times that of the electron. The radii of the proton, neutron, and electron are all of the order of magnitude of 2×10^{-15} m.

For the hydrogen atom, the radius of the smallest orbit followed by the electron is about 5×10^{-11} m. The radius of this orbit is approximately 25,000 times that of the basic constituents of the atom. This is equivalent to a sphere the size of a dime rotating about another sphere of the same size more than a quarter of a mile away.

Different atoms will have various numbers of electrons in the concentric shells about the nucleus. The first shell, which is closest to the nucleus, can contain only two electrons. If an atom should have three

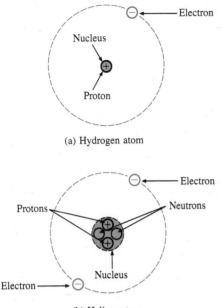

(a) Hydrogen atom

(b) Helium atom

FIG. 2.1

The hydrogen and helium atoms.

electrons, the third must go to the next shell. The second shell can contain a maximum of eight electrons, the third 18, and the fourth 32, as determined by the equation $2n^2$, where n is the shell number. These shells are usually denoted by a number ($n = 1, 2, 3, \ldots$) or letter ($n = k, l, m, \ldots$).

Each shell is then broken down into subshells, where the first subshell can contain a maximum of two electrons, the second subshell six electrons, the third 10 electrons, and the fourth 14, as shown in Fig. 2.2. The subshells are usually denoted by the letters s, p, d, and f, in that order, outward from the nucleus.

It has been determined by experimentation that *unlike charges attract, and like charges repel.* The force of attraction or repulsion between two charged bodies Q_1 and Q_2 can be determined by Coulomb's law:

$$F \text{ (attraction or repulsion)} = \frac{kQ_1Q_2}{r^2} \qquad (2.1)$$

where F is in newtons, k = constant = 9.0×10^9, Q_1 and Q_2 are the charges in coulombs (to be introduced in Section 2.2), and r is the distance in meters between the two charges. In particular, note the squared r term in the denominator, resulting in rapidly decreasing levels of F for increasing values of r.

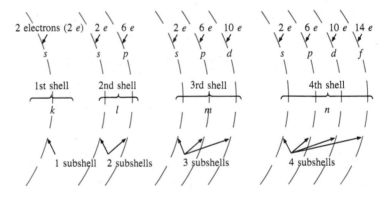

Nucleus

FIG. 2.2

Shells and subshells of the atomic structure.

In the atom, therefore, electrons will repel each other, and protons and electrons will attract each other. Since the nucleus consists of many positive charges (protons), a strong attractive force exists for the electrons in orbits close to the nucleus [note the effects of a large charge Q and a small distance r in Eq. (2.1)]. As the distance between the nucleus and the orbital electrons increases, the binding force diminishes until it reaches its lowest level at the outermost subshell (largest r). Due to the weaker binding forces, less energy must be expended to remove an electron from an outer subshell than from an inner subshell. Also, it is generally true that electrons are more readily removed from atoms having outer subshells that are incomplete *and,* in addition, possess few electrons. These properties of the atom that permit the removal of electrons under certain conditions are essential if motion of charge is to be created. Without this motion, this text could venture no further—our basic quantities rely on it.

Copper is the most commonly used metal in the electrical/electronics industry. An examination of its atomic structure will help identify why it has such widespread applications. The copper atom (Fig. 2.3) has one

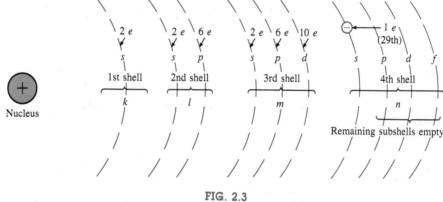

FIG. 2.3

The copper atom.

more electron than needed to complete the first three shells. This incomplete outermost subshell, possessing only one electron, and the distance between this electron and the nucleus, reveal that the twenty-ninth electron is loosely bound to the copper atom. If this twenty-ninth electron gains sufficient energy from the surrounding medium to leave its parent atom, it is called a *free electron*. In one cubic inch of copper at room temperature there are approximately $1.4 \times 10^{+24}$ free electrons. Copper also has the advantage of being able to be drawn into long thin wires (ductility) or worked into many different shapes (malleability). Other metals that exhibit the same properties as copper, but to a different degree, are silver, gold, platinum, and aluminum. Gold is used extensively in integrated circuits where the performance level and amount of material required balance the cost factor. Aluminum has found some commercial use but suffers from being more temperature sensitive (expansion and contraction) than copper.

2.2 CURRENT

Consider a short length of copper wire cut with an imaginary perpendicular plane, producing the circular cross section shown in Fig. 2.4. At room temperature with no external forces applied, there exists within the copper wire the random motion of free electrons created by the thermal energy that the electrons gain from the surrounding medium. When an atom loses its free electron, it acquires a net positive charge

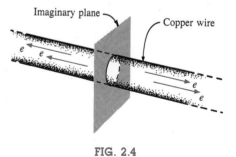

FIG. 2.4

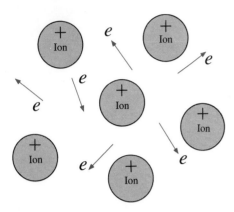

FIG. 2.5

Random motion of free electrons in an atomic structure.

and is referred to as a *positive ion*. The free electron is able to move within these positive ions and leave the general area of the parent atom, while the positive ions only oscillate in a mean fixed position. For this reason, *the free electron is the charge carrier in a copper wire or in any other solid conductor of electricity.*

An array of positive ions and free electrons is depicted in Fig. 2.5. Within this array, the free electrons find themselves continually gaining or losing energy by virtue of their changing direction and velocity. Some of the factors responsible for this random motion include (1) the collisions with positive ions and other electrons, (2) the attractive forces for the positive ions, and (3) the force of repulsion that exists between electrons. This random motion of free electrons is such that over a period of time, the number of electrons moving to the right across the circular cross section of Fig. 2.4 is exactly equal to the number passing over to the left. *With no external forces applied, the net flow of charge in any one direction is zero.*

Let us now connect this copper wire between two battery terminals as shown in Fig. 2.6. The battery, at the expense of chemical energy,

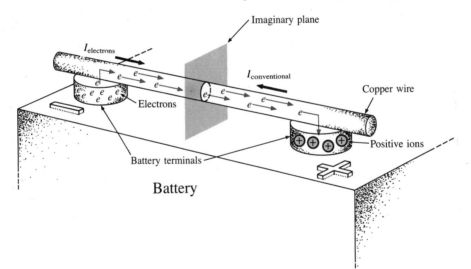

FIG. 2.6

places a net positive charge on one terminal and a net negative charge on the other. The instant the wire is connected between these two terminals, the free electrons of the copper wire will drift toward the positive terminal, while the positive ions will simply oscillate in a mean fixed position. The negative terminal is a supply of electrons to be drawn from when the electrons of the copper wire drift toward the positive terminal. The chemical activity of the battery will absorb the electrons at the positive terminal and maintain a steady supply of electrons at the negative terminal.

If 6.242×10^{18} *electrons* drift at uniform velocity through the imaginary circular cross section of Fig. 2.6 in 1 *second*, the flow of charge, or *current*, is said to be 1 *ampere* (A). The discussion of Chapter 1 revealed that this is an enormous number of electrons passing through the surface in 1 second. The current associated with only a few electrons per second would be inconsequential and of little practical value. To establish numerical values that permit immediate comparisons between levels, a coulomb (C) of charge was defined as the total charge associated with 6.242×10^{18} electrons. The charge associated with one electron can then be determined from

$$\text{Charge/electron} = Q_e = \frac{1 \text{ C}}{6.242 \times 10^{18}} = 1.6 \times 10^{-19} \text{ C}$$

The current in amperes can now be calculated using the following equation:

$$\boxed{I = \frac{Q}{t}} \qquad \begin{array}{l} I = \text{amperes (A)} \\ Q = \text{coulombs (C)} \\ t = \text{seconds (s)} \end{array} \qquad \textbf{(2.2)}$$

The capital letter I was chosen from the French word for current: *intensité*. The SI abbreviation for each quantity in Eq. (2.2) is provided to the right of the equation. The equation clearly reveals that for equal time intervals, the more charge that flows through the wire, the heavier the current.

Through algebraic manipulations, the other two quantities can be determined as follows:

$$\boxed{Q = It} \qquad \text{(coulombs, C)} \qquad \textbf{(2.3)}$$

and

$$\boxed{t = \frac{Q}{I}} \qquad \text{(seconds, s)} \qquad \textbf{(2.4)}$$

EXAMPLE 2.1. The charge flowing through the imaginary surface of Fig. 2.6 is 0.16 C every 64 ms. Determine the current in amperes.
Solution: Eq. (2.2):

$$I = \frac{Q}{t} = \frac{0.16}{64 \times 10^{-3}} = \frac{160 \times 10^{-3}}{64 \times 10^{-3}} = \textbf{2.50 A}$$

EXAMPLE 2.2. Determine the time required for 4×10^{16} electrons to pass through the imaginary surface of Fig. 2.6 if the current is 5 mA.
Solution: Determine Q:

$$4 \times 10^{16} \text{ electrons} \left(\frac{1 \text{ coulomb}}{6.242 \times 10^{18} \text{ electrons}} \right) = 0.641 \times 10^{-2} \text{ C}$$
$$= 0.00641 \text{ C} = 6.41 \text{ mC}$$

Calculate t [Eq. (2.4)]:

$$t = \frac{Q}{I} = \frac{6.41 \times 10^{-3}}{5 \times 10^{-3}} = \textbf{1.282 s}$$

A second glance at Fig. 2.6 will reveal that two directions of charge flow have been indicated. One is called *conventional flow* while the other is called *electron flow*. This text will deal only with conventional flow for a variety of reasons, including the fact that it is the most widely used at educational institutions and in industry, is employed in the design of all electronic device symbols, and is the popular choice for all major computer software packages. The flow controversy is a result of an assumption made at the time electricity was discovered that the positive charge was the moving particle in metallic conductors. Be assured that the choice of conventional flow will not create great difficulty and confusion in the chapters to follow. Once the direction of I is established, the issue is dropped and the analysis can continue.

2.3 VOLTAGE

The flow of charge described in the previous section is established by an external "pressure" derived from the energy that a mass has by virtue of its position: *potential energy*.

Energy, by definition, is the *capacity to do work*. If a mass (m) is raised to some height (h) above a reference plane, it has a measure of potential energy expressed in *joules* (J) that is determined by

$$\boxed{\text{Potential energy (PE)} = mgh} \qquad \text{(joules, J)} \qquad \textbf{(2.5)}$$

where g is the gravitational acceleration (9.754 m/s²). This mass now has the ability to do work such as crush an object placed on the reference plane. If the weight is raised further, it has an increased measure of potential energy and can do additional work. There is an obvious *difference in potential* between the two heights above the reference plane.

In the battery of Fig. 2.6, the internal chemical action will establish (through an expenditure of energy) an accumulation of negative charges (electrons) on one terminal (the negative terminal) and positive charges (positive ions) on the other (the positive terminal). A "positioning" of the charges has been established that will result in a *potential difference* between the terminals. If a conductor is connected between the terminals of the battery, the electrons at the negative terminal have sufficient potential energy to overcome collisions with other particles in the conductor and the repulsion from similar charges to reach the positive terminal to which they are attracted.

Charge can be raised to a higher potential level through the expenditure of energy from an external source, or it can lose potential energy as it travels through an electrical system. In any case, by definition:

A potential difference of 1 volt (V) exists between two points if 1 joule (J) of energy is exchanged in moving 1 coulomb (C) of charge between the two points.

In general, the potential difference between two points is determined by

$$V = \frac{W}{Q} \qquad \text{(volts)} \qquad\qquad \textbf{(2.6)}$$

Through algebraic manipulations, we have

$$W = QV \qquad \text{(joules)} \qquad\qquad \textbf{(2.7)}$$

and

$$Q = \frac{W}{V} \qquad \text{(coulombs)} \qquad\qquad \textbf{(2.8)}$$

EXAMPLE 2.3. Find the potential difference between two points in an electrical system if 60 J of energy are expended by a charge of 20 C between these two points.

Solution: Eq. (2.6):

$$V = \frac{W}{Q} = \frac{60}{20} = \textbf{3 V}$$

EXAMPLE 2.4. Determine the energy expended moving a charge of 50 μC through a potential difference of 6 V.

Solution: Eq. (2.7):

$$W = QV = (50 \times 10^{-6})(6) = 300 \times 10^{-6} \text{ J} = \textbf{300 } \boldsymbol{\mu}\textbf{J}$$

Notation plays a very important role in the analysis of electrical and electronic systems. To distinguish between sources of voltage (batteries and the like) and losses in potential across dissipative elements, the following notation will be used:

E for voltage sources (volts)
V for voltage drops (volts)

In summary, the applied potential difference (in volts) of a voltage source in an electric circuit is the "pressure" to set the system in motion and "cause" the flow of charge or current through the electrical system. A mechanical analogy of the applied voltage is the pressure applied to the water in a main. The resulting flow of water through the system is likened to the flow of charge through an electric circuit.

2.4 FIXED (dc) SUPPLIES

The terminology *dc* employed in the heading of this section is an abbreviation for *direct current,* which encompasses the various electrical systems in which there is a *unidirectional* ("one direction") flow of charge. A great deal more will be said about this terminology in the chapters to follow. For now, we will consider only those supplies that provide a fixed voltage or current.

dc Voltage Sources

Since the dc voltage source is the more familiar of the two types of supplies, it will be examined first. The symbol used for all dc voltage supplies in this text appears in Fig. 2.7. The relative lengths of the bars indicate the terminals they represent.

DC voltage sources can be divided into three broad categories: (1) batteries (chemical action), (2) generators (electromechanical), and (3) power supplies (rectification).

Batteries For the layperson, the battery is the most common of the dc sources. By definition, a battery (derived from the expression "battery of cells") consists of a combination of two or more similar *cells,* a cell being the fundamental source of electrical energy developed through the conversion of chemical or solar energy. All cells can be divided into the *primary* or *secondary* types. The secondary is rechargeable, whereas the primary is not. That is, the chemical reaction of the secondary cell can be reversed to restore its capacity. The two most common rechargeable batteries include the lead-acid unit (used primarily in automobiles) and the nickel-cadmium battery (used in calculators, tools, photoflash units, shavers, and so on). The obvious advantage of the rechargeable unit is the reduced costs associated with not having to continually replace discharged primary cells.

All of the cells appearing in this chapter except the *solar cell,* which absorbs energy from incident light in the form of photons, establish a potential difference at the expense of chemical energy. In addition, each has a positive and a negative *electrode* and an *electrolyte* to complete the circuit between electrodes within the battery. The electrolyte is the contact element and the source of ions for conduction between the terminals.

The popular carbon-zinc primary battery uses a zinc can as its negative electrode, a manganese dioxide mix and carbon rod as its positive electrode, and an electrolyte that is a mix of ammonium and zinc chlo-

FIG. 2.7

rides, flour, and starch, as shown in Fig. 2.8. Figure 2.9 shows a number of other types of primary units with an area of application and a rating to be considered later in this section.

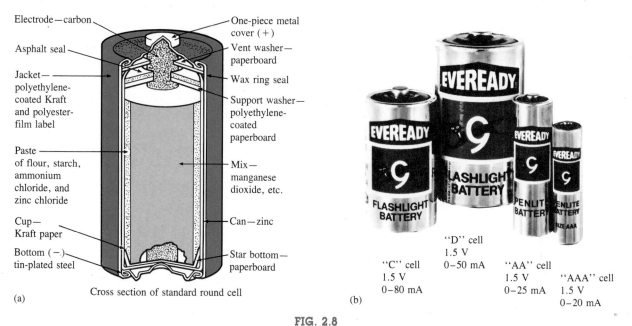

(a)

Cross section of standard round cell

(b)

"D" cell
1.5 V
0-50 mA

"C" cell
1.5 V
0-80 mA

"AA" cell
1.5 V
0-25 mA

"AAA" cell
1.5 V
0-20 mA

FIG. 2.8

Carbon-zinc primary battery. (*a*) *Construction;* (*b*) *appearance and ratings.*
(*Courtesy of Eveready Batteries*)

(a) Lithiode™ lithium-iodine cell
2.8 V, 870 mAh
Long-life power sources with printed circuit
board mounting capability

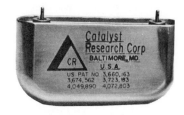

(b) Lithium-iodine pacemaker cell
2.8 V, 2.0 Ah

(c) Eveready transistor battery
9 V, 450 mAh

FIG. 2.9

Primary cells. (*Parts* (*a*) *and* (*b*) *courtesy of Catalyst Research Corp.; part*
(*c*) *courtesy of Eveready Batteries*)

For the secondary lead-acid unit appearing in Fig. 2.10, the electrolyte is sulfuric acid and the electrodes are spongy lead (Pb) and lead peroxide (PbO_2). When a load is applied to the battery terminals, there is a transfer of electrons from the spongy lead electrode to the lead peroxide electrode through the load. This transfer of electrons will

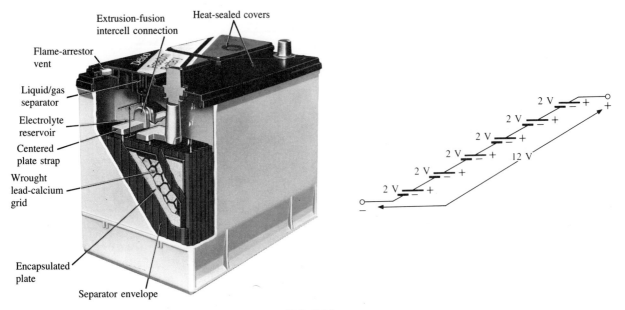

Extrusion-fusion intercell connection

Heat-sealed covers

Flame-arrestor vent

Liquid/gas separator

Electrolyte reservoir

Centered plate strap

Wrought lead-calcium grid

Encapsulated plate

Separator envelope

FIG. 2.10

Maintenance-free 12-V lead-acid battery. (Courtesy of Delco-Remy, a division of General Motors Corp.)

(a)

Eveready® BH 500 cell
1.2 V, 500 mAh
App: Where vertical height is severe limitation

(b)

Printed circuit board mountable battery
2.4 V, 70 mAh

(c)

FIG. 2.11

Rechargeable nickel-cadmium batteries. (Parts (a) and (b) courtesy of Eveready Batteries; part (c) courtesy of General Electric Co.)

continue until the battery is completely discharged. The discharge time is determined by how diluted the acid has become and how heavy the coating of lead sulfate is on each plate. The state of discharge of a lead storage cell can be determined by measuring the specific gravity of the electrolyte with a hydrometer. The specific gravity of a substance is defined to be the ratio of the weight of a given volume of the substance to the weight of an equal volume of water at 4°C. For fully charged batteries, the specific gravity should be somewhere between 1.28 and 1.30. When the specific gravity drops to about 1.1, the battery should be recharged.

Since the lead storage cell is a secondary cell, it can be recharged at any point during the discharge phase simply by applying an external dc source across the cell that will pass current through the cell in a direction opposite to that in which the cell supplied current to the load. This will remove the lead sulfate from the plates and restore the concentration of sulfuric acid.

The output of a lead storage cell over most of the discharge phase is about 2 V. In the commercial lead storage batteries used in the automobile, the 12-V can be produced by six cells in series, as shown in Fig. 2.10. the use of a grid made from a wrought lead-calcium alloy strip rather than the lead-antimony cast grid commonly used has resulted in maintenance-free batteries such as that appearing in the same figure. The lead-antimony structure was susceptible to corrosion, overcharge, gassing, water usage, and self-discharge. Improved design with the lead-calcium grid has either eliminated or substantially reduced most of these problems.

The nickel-cadmium battery is a rechargeable battery that has been receiving enormous interest and development in recent years. A number of such batteries manufactured by the Union Carbide Corporation and the General Electric Company appear in Fig. 2.11. The internal construction of the cylindrical-type cell appears in Fig. 2.12. In the fully charged condition the positive electrode is nickel hydroxide [$Ni(OH)_2$]; the negative electrode, metallic cadmium (Cd); and the electrolyte, potassium hydroxide (KOH). The oxidation (increased oxygen content) of the negative electrode occurring simultaneously with the reduction of the positive electrode provides the required electrical energy. The separator is required to isolate the two electrodes and maintain the location of the electrolyte. The advantage of such cells is that the active materials go through a change in oxidation state necessary to establish the required ion level without a change in the physical state. This establishes an excellent recovery mechanism for the recharging phase.

A high-density, 40-W solar cell appears in Fig. 2.13 with some of its associated data and areas of application. Since the maximum available wattage in an average bright sunlit day is 100 mW/cm² and conversion efficiencies are currently between 10% and 14%, the maximum available power per square centimeter from most commercial units is between 10 mW and 14 mW. For a square meter, however, the return would be 100 W to 140 W. A more detailed description of the solar cell

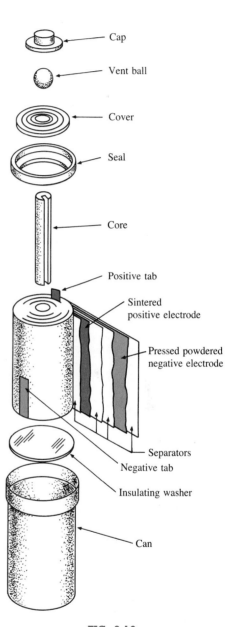

FIG. 2.12
Internal structure of the cylindrical-type nickel-cadmium rechargeable cell. (Courtesy of Eveready Batteries)

40-W, high-density solar module
100-mm × 100-mm (4″ × 4″) square cells are
used to provide maximum power in a minimum of
space. The 33 series cell module provides a strong
12-V battery charging current for a wide range of
temperatures (−40°C to 60°C)

FIG. 2.13
*Solar module. (Courtesy of Motorola
Semiconductor Products)*

will appear in your electronics courses. For now it is important to realize that a fixed illumination of the solar cell will provide a fairly steady
dc voltage for driving various loads, from watches to automobiles.

Batteries have a capacity rating given in ampere-hours (Ah) or milliampere-hours (mAh). Some of these ratings are included in the above
figures. A battery with an ampere-hour rating of 100 will theoretically
provide a steady current of 1 A for 100 h, 2 A for 50 h, 10 A for 10 h,
and so on, as determined by the following equation:

$$\text{Life (hours)} = \frac{\text{ampere-hour rating (Ah)}}{\text{amperes drawn (A)}} \qquad \textbf{(2.9)}$$

Two factors that affect this rating, however, are the temperature and
the rate of discharge. The disc-type EVEREADY® BH 500 cell appearing in Fig. 2.11 has the terminal characteristics appearing in Fig. 2.14.

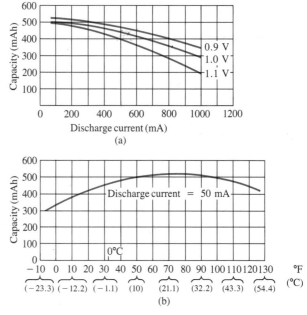

FIG. 2.14
*EVEREADY® BH 500 cell characteristics. (a) Capacity vs. discharge
current; (b) capacity vs. temperature. (Courtesy of Eveready Batteries)*

Note that for the 1-V unit the rating is above 500 mAh at a discharge
current of 100 mA [Fig. 2.14(a)] but drops to 300 mAh at about 1 A.
For a unit that is less than 1½ inches in diameter and less than 1/2 inch in
thickness, however, these are excellent terminal characteristics. Figure
2.14(b) reveals that the maximum mAh rating (at a current drain of
50 mA) occurs at about 75°F (≅ 24°C), or just above average room
temperature. Note how the curve drops to the right and left of this
maximum value. We are all aware of the reduced ''strength'' of a bat-

tery at low temperatures. Note that it has dropped to almost 300 mAh at −20°C.

Another curve of interest appears in Fig. 2.15. It provides the expected cell voltage at a particular drain over a period of hours of use. It is noteworthy that the loss in hours between 50 mA and 100 mA is much greater than between 100 mA and 150 mA, even though the increase in current is the same between levels.

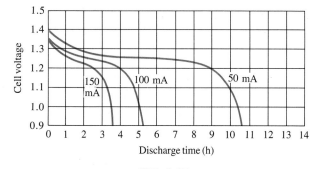
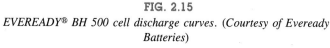

FIG. 2.15

EVEREADY® BH 500 cell discharge curves. (*Courtesy of Eveready Batteries*)

EXAMPLE 2.5.

a. Determine the capacity in milliampere-hours for the 0.9-V BH 500 cell of Fig. 2.14(a) if the discharge current is 600 mAh.
b. At what temperature will the mAh rating of the cell of Fig. 2.14(b) be 90% of its maximum value if the discharge current is 50 mA?

Solutions:

a. From Fig. 2.14(a), the capacity at 600 mA is about 450 mAh. Thus, from Eq. (2.9),

$$\text{Life} = \frac{450 \text{ mAh}}{600 \text{ mA}} = 0.75 \text{ h} = \textbf{45 min}$$

b. From Fig. 2.14(b), the maximum is approximately 520 mAh. The 90% level is therefore 468 mAh, which occurs just above freezing, or **1°C,** and at the higher temperature of **45°C.**

Generators The dc generator is quite different, both in construction (Fig. 2.16) and in mode of operation, from the battery. When the shaft of the generator is rotating at the nameplate speed due to the applied torque of some external source of mechanical power, a voltage of rated value will appear across the external terminals. The terminal voltage and power-handling capabilities of the dc generator are typically higher than those of most batteries, and its lifetime is determined only by its

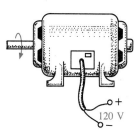

FIG. 2.16

dc generator.

FIG. 2.17

dc laboratory supply. (Courtesy of Lambda Electronics Corp.)

construction. Commercially used dc generators are typically of the 120-V or 240-V variety. As pointed out earlier in this section, for the purposes of this text no distinction will be made between the symbol for a battery and a generator.

Power supplies The dc supply encountered most frequently in the laboratory employs the rectification and filtering processes as its means toward obtaining a steady dc voltage. By this process, a time-varying voltage (such as ac voltage available from a home outlet) is converted to one of a fixed magnitude. This process will be covered in detail in the basic electronics courses. A dc laboratory supply of this type appears in Fig. 2.17.

Most dc laboratory supplies have a regulated, adjustable voltage output with three available terminals, as indicated in Figs. 2.17 and 2.18(a). The symbol for ground or zero potential (the reference) is also shown in Fig. 2.18(a). If 10 volts above ground potential are required, then the connections are made as shown in Fig. 2.18(b). If 15 volts

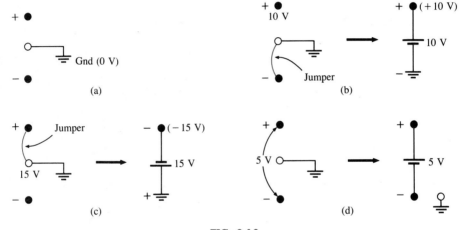

FIG. 2.18

Possible output connections of a dc laboratory supply.

below ground potential are required, then the connections are made as shown in Fig. 2.18(c). If connections are as shown in Fig. 2.18(d), we say we have a "floating" voltage of 5 volts since the reference level is not included. Seldom is the configuration of Fig. 2.18(d) employed since it fails to protect the operator by providing a direct low resistance path to ground and to establish a common ground for the system. In any case, the positive and negative terminals must be part of any circuit configuration.

dc Current Sources

The wide variety of types of and applications for the dc voltage source has resulted in its becoming a rather familiar device, the characteristics of which are understood, at least basically, by the layperson. For example, it is common knowledge that a 12-V car battery has a terminal voltage (at least approximately) of 12 V even though the current drain by the automobile may vary under different operating conditions. In other words, *a dc voltage source ideally will provide a fixed terminal voltage even though the current drain may vary,* as depicted in Fig. 2.19(a). A dc current source is the dual of the voltage source. That is,

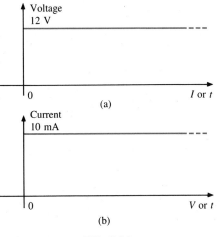

(a)

(b)

FIG. 2.19

Terminal characteristics. (a) Ideal voltage source; (b) ideal current source.

the current source will, ideally, supply a fixed current to a load even though there will be variations in the terminal voltage as determined by the load, as depicted in Fig. 2.19(b). (Do not become alarmed if the concept of a current source is strange and somewhat confusing at this point. It will be covered in great detail in later chapters.)

The introduction of semiconductor devices such as the transistor has accounted in large measure for the increasing interest in current sources. A representative commercially available dc current source appears in Fig. 2.20.

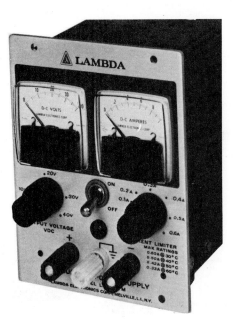

FIG. 2.20

dc current source. (Courtesy of Lambda Electronics Corp.)

2.5 CONDUCTORS AND INSULATORS

Different wires placed across the same two battery terminals will allow different amounts of charge to flow between the terminals. Many factors, such as the stability, density, and mobility of the material, account for these variations in charge flow. *Conductors are those materials that permit a generous flow of electrons with very little electromotive force applied.* Since copper is used most frequently, it serves as the standard of comparison for the relative conductivity in Table 2.1. Note that aluminum, which has lately seen some commercial use, has only 61% of the conductivity level of copper, but keep in mind that this must be weighed against the cost and weight factors.

Materials that have very few free electrons, high stability and density, and low mobility are called *insulators, since a large potential*

TABLE 2.1

Relative conductivity of various materials.

Metal	Relative Conductivity (%)
Silver	105
Copper	**100**
Gold	70.5
Aluminum	61
Tungsten	31.2
Nickel	22.1
Iron	14
Constantan	3.52
Nichrome	1.73
Calorite	1.44

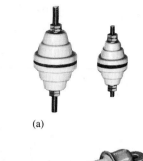

(a)

(b)

(c)

FIG. 2.21
Insulators. (a) Insulated thru-panel bushings; (b) antenna strain insulators; (c) porcelain stand-off insulators. (Courtesy of Herman H. Smith, Inc.)

TABLE 2.2
Breakdown strength of some common insulators.

Material	Average Breakdown Strength (kV/cm)
Air	30
Porcelain	70
Oils	140
Bakelite	150
Rubber	270
Paper (Paraffin-coated)	500
Teflon	600
Glass	900
Mica	2000

difference is required to produce any sizable current through such materials. A common use of insulating material is for covering current-carrying wire which, if uninsulated, could cause dangerous side effects. For example, workers on high-voltage power lines wear rubber gloves and stand on rubber mats as additional safety measures. A number of different types of insulators and their applications appear in Fig. 2.21.

It must be pointed out, however, that even the best insulator will break down (permit charge to flow around and through it) if a sufficiently large potential is applied across it. The breakdown strengths of some common insulators are listed in Table 2.2.

According to this table, for insulators with the same geometric shape, it would require 270/30 = 9 times as much potential to pass current through rubber as through air and approximately 67 times as much voltage to pass current through mica as through air.

2.6 SEMICONDUCTORS

Between the class of elements called *insulators* and those exhibiting conductor properties, there exists a group of elements of significant importance called *semiconductors.* The entire electronic industry is dependent on these materials since the diodes, transistors, and integrated circuits (ICs) that we hear so much about are constructed of semiconductor materials. Although *silicon* is the most extensively employed, *germanium* is also used in a number of devices. Both of these materials will be examined in some detail in your electronics courses. It will then be demonstrated why they are so appropriate for the applications noted above.

2.7 AMMETERS AND VOLTMETERS

It is important to be able to measure the current and voltage levels in the network in order to check its operation, isolate malfunctions, and investigate effects impossible to predict on paper. As the names imply, *ammeters* are used to measure current levels, and *voltmeters,* the potential difference between two points. If the current levels are usually of the order of milliamperes, the instrument will be referred to as a milliammeter, and if in the microampere range, as a microammeter. Similar statements can be made for voltage levels. Throughout the industry, voltage levels are measured more frequently than current levels primarily because the former does not require that the network connections be disturbed.

The potential difference between two points can be measured by simply connecting the leads of the meter *across the two points* as indicated in Fig. 2.22. An up-scale reading is obtained by placing the positive lead of the meter to the point of higher potential of the network and the common or negative lead to the point of lower potential. The reverse connection will result in a negative reading or a below-zero indication.

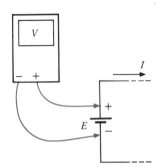

FIG. 2.22
Voltmeter connection for an up-scale reading.

Ammeters are connected in the *same branch* in which the current is to be measured, as shown in Fig. 2.23. Since ammeters measure the rate

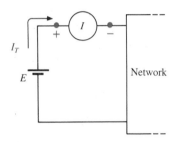

FIG. 2.23
Ammeter connection for an up-scale reading.

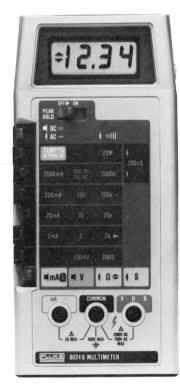

FIG. 2.24
Volt-Ohm-Milliammeter (VOM). (Courtesy of Simpson Electric Co.)

of flow of charge, the meter must be placed in the network such that the charge will flow through the meter. The only way this can be accomplished is to open the branch in which the current is to be measured and place the meter between the two resulting terminals. For the network of Fig. 2.23, the source lead must be disconnected from the network and the ammeter inserted as shown. An up-scale reading will be obtained if the polarities on the terminals of the ammeter are such that the current of the network enters the positive terminal or terminal at the higher potential.

The introduction of any meter into the network raises a concern about whether the meter will affect the behavior of the network. This question and others will be examined in Chapters 5 and 6 after additional terms and concepts have been introduced.

Instruments exist that are designed to measure just current or just voltage levels. However, the most common laboratory meters include the *Volt-Ohm-Milliammeter* (VOM) and the *Digital Multimeter* (DMM) of Figs. 2.24 and 2.25, respectively. Both instruments will measure voltage and current and a third quantity, resistance, to be introduced in the next chapter. The VOM uses an analog scale, which requires interpreting the position of a pointer on a continuous scale, while the DMM provides a display of numbers with decimal point accuracy determined

FIG. 2.25
Digital Multimeter (DMM). (Courtesy of John Fluke Mfg. Co. Inc.)

by the chosen scale. Comments on the characteristics and use of a variety of meters will be made throughout the text. However, the major study of meters will be left for the laboratory sessions.

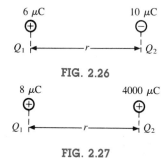

6 μC   10 μC

Q_1 |←———r———→| Q_2

FIG. 2.26

8 μC   4000 μC

Q_1 |←———r———→| Q_2

FIG. 2.27

PROBLEMS

Section 2.1

1. The number of orbiting electrons in aluminum and silver is 13 and 47, respectively. Draw the electronic configuration, including all the shells and subshells, and discuss briefly why each is a good conductor.

2. Find the force of attraction in newtons between the charges Q_1 and Q_2 in Fig. 2.26 when
   a. $r = 1$ m   **b.** $r = 3$ m
   c. $r = 10$ m
   (Note how quickly the force drops with increase in r.)

*3. Find the force of repulsion in newtons between Q_1 and Q_2 in Fig. 2.27 when
   a. $r = 1$ mi   **b.** $r = 0.01$ m
   c. $r = 1/16$ in.

4. Determine the distance between two charges of 20 μC if the force between the two charges is 3.6×10^4 N.

5. Two charged bodies, Q_1 and Q_2, when separated by a distance of 2 m, experience a force of repulsion equal to 1.8 N.
   a. What will the force of repulsion be when they are 10 m apart?
   b. If the ratio $Q_1/Q_2 = 1/2$, find Q_1 and Q_2 ($r = 10$ m).

Section 2.2

6. Find the current in amperes if 650 C of charge pass through a wire in 50 s.

7. If 465 C of charge pass through a wire in 2.5 min, find the current in amperes.

8. If a current of 40 A exists for 1 min, how many coulombs of charge have passed through the wire?

9. How many coulombs of charge pass through a lamp in 2 min if the current is constant at 750 mA?

10. If the current in a conductor is constant at 2 mA, how much time is required for 4600×10^{-6} C to pass through the conductor?

11. If $21.847 \times 10^{+18}$ electrons pass through a wire in 7 s, find the current.

12. How many electrons pass through a conductor in 1 min if the current is 1 A?

*13. If $0.784 \times 10^{+18}$ electrons pass through a wire in 643 ms, find the current.

14. Will a fuse rated at 1 A "blow" if 86 C pass through it in 1.2 min?

Section 2.3

15. If the potential difference between two points is 42 V, how much work is required to bring 6 C from one point to the other?

16. Find the charge Q that requires 96 J of energy to be moved through a potential difference of 16 V.

17. How much charge passes through a battery of 22.5 V if the energy expended is 90 J?

18. If a conductor with a current of 200 mA passing through it converts 40 J of electrical energy into heat in 30 s, what is the potential drop across the conductor?

19. Charge is flowing through a conductor at the rate of 420 C/min. If 742 J of electrical energy are converted to heat in 30 s, what is the potential drop across the conductor?

Section 2.4

20. What current will a battery with an Ah rating of 200 theoretically provide for 40 h?

21. What is the Ah rating of a battery that can provide 0.8 A for 76 h?

22. For how many hours will a battery with an Ah rating of 32 theoretically provide a current of 1.28 A?

23. Find the mAh rating of the EVEREADY® BH 500 battery at 100°F and 0°C at a discharge current of 50 mA using Fig. 2.14(b).

24. Find the mAh rating of the 1.0 V EVEREADY® BH 500 battery if the current drain is 550 mA using Fig. 2.14(a). How long will it supply this current?

25. For how long can 50 mA be drawn from the battery of Fig. 2.15 before its terminal voltage drops below 1 V? Determine the number of hours at a drain current of 150 mA, and compare the ratio of drain current to the resulting ratio of hours of availability.

26. Discuss briefly the difference between the three types of dc voltage supplies (batteries, rectification, and generators).

27. Indicate in a few sentences your concept of a current source. Employ its characteristics in your description.

Section 2.5

28. Discuss two properties of the atomic structure of copper that make it a good conductor.

29. Name two materials not listed in Table 2.1 that are good conductors of electricity.

30. Explain the terms *insulator* and *breakdown strength*.

31. List three uses of insulators not mentioned in Section 2.5.

Section 2.6

32. What is a semiconductor? How does it compare with a conductor and insulator?

33. Consult a semiconductor electronics text and note the extensive use of germanium and silicon semiconductor materials. Review the characteristics of each material.

Section 2.7

34. What are the significant differences in the way ammeters and voltmeters are connected?

35. If an ammeter reads 2.5 A for a period of 4 min, determine the charge that has passed through the meter.

36. Between two points in an electric circuit, a voltmeter reads 12.5 V for a period of 20 s. If the current measured by an ammeter is 10 mA, determine the energy expended and the charge that flowed between the two points.

GLOSSARY

Ammeter An instrument designed to read the current through elements in series with the meter.

Ampere (A) The SI unit of measurement applied to the flow of charge through a conductor.

Ampere-hour rating The rating applied to a source of energy that will reveal how long a particular level of current can be drawn from that source.

Cell A fundamental source of electrical energy developed through the conversion of chemical or solar energy.

Conductors Materials that permit a generous flow of electrons with very little voltage applied.

Copper A material possessing physical properties that make it particularly useful as a conductor of electricity.

Coulomb (C) The fundamental SI unit of measure for charge. It is equal to the charge carried by 6.242×10^{18} electrons.

Coulomb's law An equation defining the force of attraction or repulsion between two charges.

dc current source A source that will provide a fixed current level even though the load to which it is applied may cause its terminal voltage to change.

dc generator A source of dc voltage available through the turning of the shaft of the device by some external means.

Direct current Current in which the magnitude does not change over a period of time.

Ductility The property of a material that allows it to be drawn into long thin wires.

Electrolytes The contact element and the source of ions between the electrodes of the battery.

Electron The particle with negative polarity that orbits the nucleus of an atom.

Free electron An electron unassociated with any particular atom, relatively free to move through a crystal lattice structure under the influence of external forces.

Insulators Materials in which a very high voltage must be applied to produce any measurable current flow.

Malleability The property of a material that allows it to be worked into many different shapes.

Neutron The particle having no electrical charge, found in the nucleus of the atom.

Nucleus The structural center of an atom which contains both protons and neutrons.

Positive ion An atom having a net positive charge due to the loss of one of its negatively charged electrons.

Potential difference The difference in potential between two points in an electrical system.

Potential energy The energy that a mass possesses by virtue of its position.

Primary cell Sources of voltage that cannot be recharged.

Proton The particle of positive polarity found in the nucleus of the atom.

Rectification The process by which an ac signal is converted to one which has an average dc level.

Secondary cell Sources of voltage that can be recharged.

Semiconductor A material having a conductance value between that of an insulator and that of a conductor. Of significant importance in the manufacture of semiconductor electronic devices.

Solar cell Sources of voltage available through the conversion of light energy (photons) into electrical energy.

Specific gravity The ratio of the weight of a given volume of a substance to the weight of an equal volume of water at 4°C.

Volt (V) The unit of measurement applied to the difference in potential between two points. If one joule of energy is required to move one coulomb of charge between two points, the difference in potential is said to be one volt.

Voltmeter An instrument designed to read the voltage across an element or between any two points in a network.

3
Resistance

3.1 INTRODUCTION

The flow of charge through any material encounters an opposing force
similar in many respects to mechanical friction. This opposition, due to
the collisions between electrons and between electrons and other atoms
in the material, *which converts electrical energy into heat,* is called the
resistance of the material. The unit of measurement of resistance is the
ohm, for which the symbol is Ω, the capital Greek letter omega. The
circuit symbol for resistance appears in Fig. 3.1 with the graphic abbreviation for resistance (R).

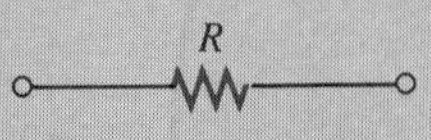

FIG. 3.1
Resistance symbol and notation.

The resistance of any material with a uniform cross-sectional area is
determined by the following four factors:

1. Material
2. Length
3. Cross-sectional area
4. Temperature

The chosen material, with its unique molecular structure, will react
differentially to pressures to establish current through its core. Conductors that permit a generous flow of charge with little external pressure

will have low resistance levels, while insulators will have high resistance characteristics.

As one might expect, the longer the path the charge must pass through, the higher the resistance level, whereas the larger the area (and therefore available room), the lower the resistance. Resistance is thus directly proportional to length and inversely proportional to area.

As the temperature of most conductors and resistive elements increases, the increased motion of the particles within the molecular structure makes it increasingly difficult for the "free" carriers to pass through, and the resistance level increases.

At a fixed temperature of 20°C (room temperature), the resistance is related to the other three factors by

$$ R = \rho \frac{l}{A} \qquad \text{(ohms, } \Omega \text{)} \qquad \textbf{(3.1)} $$

where ρ (Greek letter rho) is a characteristic of the material called the *resistivity*, l is the length of the sample, and A is the cross-sectional area of the sample.

The units of measurement substituted into Eq. (3.1) are related to the application. For circular wires, units of measurement are usually defined as in Section 3.2. For most other applications involving important areas such as integrated circuits, the units are as defined in Section 3.4.

3.2 RESISTANCE: CIRCULAR WIRES

For a circular wire, the quantities appearing in Eq. (3.1) are defined by Fig. 3.2. For two wires of the same physical size at the same temperature, as shown in Fig. 3.3(a), the relative resistances will be determined solely by the material. As indicated in Fig. 3.3(b), an increase in length will result in an increased resistance for similar areas, material, and temperature. Increased area [Fig. 3.3(c)] for remaining similar determining variables will result in a decrease in resistance. Finally, in-

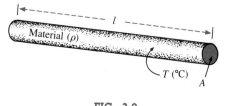

FIG. 3.2

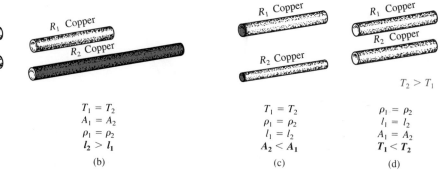

(a)	(b)	(c)	(d)
$T_1 = T_2$	$T_1 = T_2$	$T_1 = T_2$	$\rho_1 = \rho_2$
$A_1 = A_2$	$A_1 = A_2$	$\rho_1 = \rho_2$	$l_1 = l_2$
$l_1 = l_2$	$\rho_1 = \rho_2$	$l_1 = l_2$	$A_1 = A_2$
$\rho_2 > \rho_1$	$l_2 > l_1$	$A_2 < A_1$	$T_1 < T_2$

FIG. 3.3
Cases in which $R_2 > R_1$.

creased temperature [Fig. 3.3(d)] for metallic wires of identical construction and material will result in an increased resistance.

For circular wires, the quantities of Eq. (3.1) have the following units:

ρ—CM-ohms/ft at $T = 20°C$
l—feet
A—circular mils (CM)

Note that the area of the conductor is measured in *circular mils* and *not* in square meters, inches, and so on, as determined by the equation

$$\text{Area (circle)} = \pi r^2 = \frac{\pi d^2}{4}$$

$r = $ radius
$d = $ diameter

(3.2)

Recall from Chapter 1 that

$$1 \text{ mil} = \frac{1}{1000} \text{ in.} = 0.001 \text{ in.} = 10^{-3} \text{ in.}$$

or

$$1000 \text{ mils} = 1 \text{ in.}$$

A square mil will appear as shown in Fig. 3.4(a). By definition, *a wire that has a diameter of 1 mil, as shown in Fig. 3.4(b), has an area of 1 circular mil (CM)*. One square mil was superimposed on the 1-CM area of Fig. 3.4(b) to show clearly that the square mil has a larger surface area than the circular mil.

Applying the above definition to a wire having a diameter of 1 mil, we have

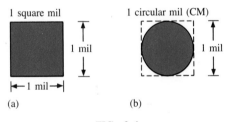

1 square mil
1 circular mil (CM)
1 mil
1 mil
1 mil
(a)
(b)

FIG. 3.4

$$A = \frac{\pi d^2}{4} = \frac{\pi}{4}(1)^2 = \frac{\pi}{4} \text{ sq mils} \overset{\text{by definition}}{\equiv} 1 \text{ CM}$$

Therefore,

$$1 \text{ CM} = \frac{\pi}{4} \text{ sq mils}$$

or

$$1 \text{ sq mil} = \frac{4}{\pi} \text{ CM}$$

For conversion purposes,

$$\text{CM} = \frac{4}{\pi} \times \text{no. of sq mils}$$

$$\text{sq mils} = \frac{\pi}{4} \times \text{no. of CM}$$

(3.3)

For a wire with a diameter of N mils (where N can be any positive number),

$$A = \frac{\pi d^2}{4} = \frac{\pi N^2}{4} \text{ sq mils}$$

Substituting the fact that $4/\pi$ CM = 1 sq mil, we have

$$A = \frac{\pi N^2}{4} \text{ (sq mils)} = \left(\frac{\pi N^2}{4}\right)\left(\frac{4}{\pi}\text{CM}\right) = N^2 \text{ CM}$$

Since $d = N$, the area in circular mils is simply equal to the diameter in mils square; that is,

$$\boxed{A_{CM} = (d_{mils})^2} \tag{3.4}$$

Therefore, in order to find the area in circular mils, the diameter must first be converted to mils. Since 1 mil = 0.001 in., if the diameter is given in inches, simply move the decimal point three places to the right. For example,

$$0.123 = 123.0 \text{ mils}$$

If in fractional form, first convert to decimal form and then proceed as above. For example,

$$\frac{1}{8} \text{ in.} = 0.125 \text{ in.} = 125 \text{ mils}$$

FIG. 3.5

The constant ρ (resistivity) is different for every material. Its value is the resistance of a length of wire 1 ft by 1 mil in diameter, measured at 20°C (Fig. 3.5). The unit of measurement for ρ can be determined from Eq. (3.1) as follows:

$$R = \rho \frac{l}{A}$$

$$\text{Ohms} = \rho \frac{\text{ft}}{\text{CM}}$$

$$\text{Units of } \rho = \frac{\text{CM-ohms}}{\text{ft}}$$

The resistivity ρ is also measured in ohms per mil-foot as determined by Fig. 3.5, or *ohm-meters* in the SI system of units.

Some typical values of ρ are listed in Table 3.1.

EXAMPLE 3.1. What is the resistance of a 100-ft length of copper wire with a diameter of 0.020 in. at 20°C?

Solution:

$$\rho = 10.37 \qquad 0.020 \text{ in.} = 20 \text{ mils}$$

$$A_{CM} = (d_{mils})^2 = (20)^2 = 400$$

TABLE 3.1

The resistivity of various materials.

Material	$\rho\left(\dfrac{\text{CM-ohms}}{\text{ft}}\right)$ @ 20°C
Silver	9.9
Copper	10.37
Gold	14.7
Aluminum	17.0
Tungsten	33.0
Nickel	47.0
Iron	74.0
Constantan	295.0
Nichrome	600.0
Calorite	720.0
Carbon	21,000.0

$$R = \rho\frac{l}{A} = \frac{(10.37)(100)}{400}$$

$$R = \mathbf{2.59\,\Omega}$$

EXAMPLE 3.2. An undetermined number of feet of wire have been used from the carton of Fig. 3.6. Find the length of the remaining copper wire if it has a diameter of 1/16 in. and a resistance of 0.5 Ω.

Solution:

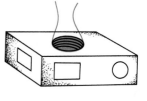

FIG. 3.6

$$\rho = 10.37 \qquad \frac{1}{16}\,\text{in.} = 0.0625\,\text{in.} = 62.5\,\text{mils}$$

$$A_{\text{CM}} = (d_{\text{mils}})^2 = (62.5)^2 = 3906.25\,\text{CM}$$

$$R = \rho\frac{l}{A} \Rightarrow l = \frac{RA}{\rho} = \frac{(0.5)(3906.25)}{10.37} = \frac{1953.125}{10.37}$$

$$l = \mathbf{188.34\,ft}$$

EXAMPLE 3.3. What is the resistance of a copper bus-bar as used in the power distribution panel of a high-rise office building with the dimensions indicated in Fig. 3.7?

Solution:

$$A_{\text{CM}} \begin{cases} 5.0\,\text{in.} = 5000\,\text{mils} \\[4pt] \dfrac{1}{2}\,\text{in.} = 500\,\text{mils} \\[4pt] A = (5000)(500) = 2.5 \times 10^6\,\text{sq mils} \\[4pt] \quad = 2.5 \times 10^6\,\text{sq mils}\left(\dfrac{4/\pi\,\text{CM}}{1\,\text{sq mil}}\right) \\[4pt] A = 3.185 \times 10^6\,\text{CM} \end{cases}$$

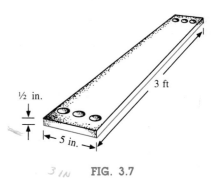

½ in.

5 in.

3 ft

FIG. 3.7

$$R = \rho\frac{l}{A} = \frac{(10.37)(3)}{3.185 \times 10^6} = \frac{31.110}{3.185 \times 10^6}$$

$$R = \mathbf{9.768 \times 10^{-6}\,\Omega}$$
$$\text{(quite small, } 0.000009768\,\Omega)$$

3.3 WIRE TABLES

The wire table was designed primarily to standardize the size of wire produced by manufacturers throughout the United States. As a result, the manufacturer has a larger market and the consumer knows that standard wire sizes will always be available. The table was designed to assist the user in every way possible; it usually includes such data as the cross-sectional area in circular mils, diameter in mils, ohms per 1000 feet at 20°C, and weight per 1000 feet.

The American Wire Gage (AWG) sizes are given in Table 3.2 for solid round copper wire. A column indicating the maximum allowable current in amperes, as determined by the National Fire Protection Association, has also been included.

The chosen sizes have an interesting relationship: For every drop in three gage numbers the area is doubled, and for every drop in 10 gage numbers the area increases by a factor of 10.

Examining Eq. (3.1), we note also that *doubling the area cuts the resistance in half, and increasing the area by a factor of 10 decreases the resistance to 1/10 the original,* everything else kept constant.

The actual sizes of some of the gage wires listed in Table 3.2 are shown in Fig. 3.8 with a few of their areas of application. A few examples using Table 3.2 follow.

EXAMPLE 3.4. Find the resistance of 650 ft of #8 copper wire ($T = 20°C$).

Solution: For #8 copper wire (solid), $\Omega/1000$ ft at $20°C = 0.6282\,\Omega$, and

$$650\,\cancel{ft}\left(\frac{0.6282\,\Omega}{1000\,\cancel{ft}}\right) = \mathbf{0.408\,\Omega}$$

EXAMPLE 3.5. What is the diameter, in inches, of a #12 copper wire?

Solution: For #12 copper wire (solid), $A = 6529.9$ CM, and

$$d_{\text{mils}} = \sqrt{A_{\text{CM}}} = \sqrt{6529.9} \cong 80.81\text{ mils}$$
$$d = \mathbf{0.0808\text{ in.}}\text{ (or close to 1/12 in.)}$$

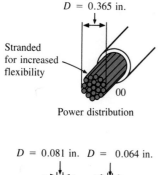

$D = 0.365$ in.

Stranded for increased flexibility

00

Power distribution

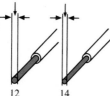

$D = 0.081$ in. $D = 0.064$ in.

12   14

Lighting, outlets, general home use

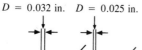

$D = 0.032$ in. $D = 0.025$ in.

20   22

Radio, television

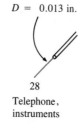

$D = 0.013$ in.

28

Telephone, instruments

FIG. 3.8

TABLE 3.2

American Wire Gage (AWG) sizes.

	AWG #	Area (CM)	$\Omega/1000$ ft at 20°C	Maximum Allowable Current for RHW Insulation (A)*
(4/0)	0000	211,600	0.0490	230
(3/0)	000	167,810	0.0618	200
(2/0)	00	133,080	0.0780	175
(1/0)	0	105,530	0.0983	150
	1	83,694	0.1240	130
	2	66,373	0.1563	115

AWG #	Area (CM)	Ω/1000 ft at 20°C	Maximum Allowable Current for RHW Insulation (A)*
3	52,634	0.1970	100
4	41,742	0.2485	85
5	33,102	0.3133	—
6	26,250	0.3951	65
7	20,816	0.4982	—
8	16,509	0.6282	45
9	13,094	0.7921	—
10	10,381	0.9989	30
11	8,234.0	1.260	—
12	6,529.0	1.588	20
13	5,178.4	2.003	—
14	4,106.8	2.525	15
15	3,256.7	3.184	
16	2,582.9	4.016	
17	2,048.2	5.064	
18	1,624.3	6.385	
19	1,288.1	8.051	
20	1,021.5	10.15	
21	810.10	12.80	
22	642.40	16.14	
23	509.45	20.36	
24	404.01	25.67	
25	320.40	32.37	
26	254.10	40.81	
27	201.50	51.47	
28	159.79	64.90	
29	126.72	81.83	
30	100.50	103.2	
31	79.70	130.1	
32	63.21	164.1	
33	50.13	206.9	
34	39.75	260.9	
35	31.52	329.0	
36	25.00	414.8	
37	19.83	523.1	
38	15.72	659.6	
39	12.47	831.8	
40	9.89	1049.0	

(INCREASING AWG #) (DECREASING DIAMETER (AND AREA))

*Not more than three conductors in raceway, cable, or direct burial.

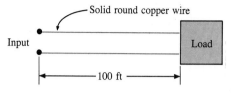

Solid round copper wire

Input

Load

100 ft

FIG. 3.9

EXAMPLE 3.6. For the system of Fig. 3.9, the total resistance of *each* power line cannot exceed $0.025\,\Omega$, and the maximum current to be drawn by the load is 95 A. What gage wire should be used?

Solution:

$$R = \rho\frac{l}{A} \Rightarrow A = \rho\frac{l}{R} = \frac{(10.37)(100)}{0.025} = 41{,}480\,\text{CM}$$

Using the wire table, we choose the wire with the next largest area, which is #4, to satisfy the resistance requirement. We note, however, that 95 A must flow through the line. This specification requires that #3 wire be used, since the #4 wire can carry a maximum current of only 85 A.

3.4 RESISTANCE: METRIC UNITS

The design of resistive elements for a variety of areas of application including thin-film resistors and integrated circuits uses metric units for the quantities of Eq. (3.1). In SI units, the resistivity would be measured in ohm-meters, the area in square meters, and the length in meters. However, the meter is generally too large a unit of measure for most applications, and so the centimeter is usually employed. The resulting dimensions for Eq. (3.1) are therefore

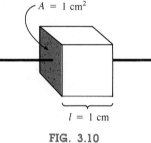

$A = 1\ \text{cm}^2$

$l = 1\ \text{cm}$

FIG. 3.10
Defining ρ.

| ρ—ohm-centimeters |
| l—centimeters |
| A—square centimeters |

The units for ρ can be derived from

$$R = \rho\frac{l}{A} \Rightarrow \rho = \frac{RA}{l} = \frac{\Omega\text{-cm}^2}{\text{cm}} = \Omega\text{-cm}$$

The resistivity of a material is actually the resistance of a sample such as appearing in Fig. 3.10. Table 3.3 provides a list of values of ρ in ohm-centimeters.

Note that the area now is expressed in square centimeters, which can be determined using the basic equation $A = \pi d^2/4$, eliminating the need to work with circular mils, the special unit of measure associated with circular wires.

TABLE 3.3
Resistivity (ρ) of various materials in ohm-centimeters.

Silver	1.629×10^{-6}
Copper	1.724×10^{-6}
Gold	2.44×10^{-6}
Aluminum	2.688×10^{-6}
Tungsten	5.5×10^{-6}
Nickel	7.8×10^{-6}
Iron	9.8×10^{-6}
Tantalum	15.5×10^{-6}
Nichrome	100×10^{-6}
Tin oxide	250×10^{-6}
Carbon	3500×10^{-6}

EXAMPLE 3.7. Determine the resistance of 100 ft of 28 copper telephone wire if the diameter is 0.0126 in.

Solution: Unit conversions:

$$l = 100\ \text{ft}\left(\frac{12\ \text{in.}}{1\ \text{ft}}\right)\left(\frac{2.54\ \text{cm}}{1\ \text{in.}}\right) = 3048\ \text{cm}$$

$$d = 0.0126 \, \text{in.} \left(\frac{2.54 \, \text{cm}}{1 \, \text{in.}} \right) = 0.032 \, \text{cm}$$

Therefore,

$$A = \frac{\pi d^2}{4} = \frac{(3.1416)(0.032)^2}{4} = 8.04 \times 10^{-4} \, \text{cm}^2$$

$$R = \rho \frac{l}{A} = \frac{(1.724 \times 10^{-6})(3048)}{8.04 \times 10^{-4}} \cong \mathbf{6.5 \, \Omega}$$

Using the units for circular wires and Table 3.2 for the area of a #28 wire, we find

$$R = \rho \frac{l}{A} = \frac{(10.37)(100)}{159.79} \cong \mathbf{6.5 \, \Omega}$$

EXAMPLE 3.8. Determine the resistance of the thin-film resistor of Fig. 3.11 if the sheet resistance R_S (defined by $R_S = \rho/d$) is $100 \, \Omega$.
Solution: For deposited materials of the same thickness, the sheet resistance factor is usually employed in the design of thin-film resistors.

Equation (3.1) can be written

$$R = \rho \frac{l}{A} = \rho \frac{l}{dw} = \left(\frac{\rho}{d} \right) \left(\frac{l}{w} \right) = R_S \frac{l}{w}$$

where l is the length of the sample and w is the width. Substituting into the above equation yields

$$R = R_S \frac{l}{w} = \frac{(100)(0.6)}{0.3} = \mathbf{200 \, \Omega}$$

as one might expect since $l = 2w$.

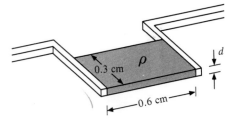

FIG. 3.11
Thin-film resistor (note Fig. 3.22).

3.5 TEMPERATURE EFFECTS

For most conductors, the resistance increases with increase in temperature, due to the increased molecular movement within the conductor which hinders the flow of charge. Figure 3.12 reveals that for copper (and most other metallic conductors), the resistance increases almost linearly (in a straight-line relationship) with increase in temperature. For the range of semiconductor materials such as employed in transistors, diodes, and so on, the resistance decreases with increase in temperature.

Since temperature can have such a pronounced effect on the resistance of a conductor, it is important that we have some method of determining the resistance at any temperature within operating limits. An equation for this purpose can be obtained by approximating the curve of

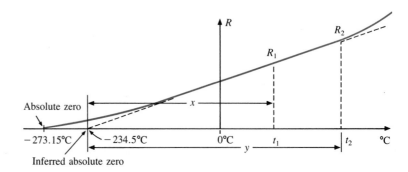

FIG. 3.12

Effect of temperature on the resistance of copper.

Fig. 3.12 by the straight dashed line that intersects the temperature scale at $-234.5°C$. Although the actual curve extends to *absolute zero* $(-273.15°C)$, the straight-line approximation is quite accurate for the normal operating temperature range. At two different temperatures, t_1 and t_2, the resistance of copper is R_1 and R_2, as indicated on the curve. Using a property of similar triangles, we may develop a mathematical relationship between these values of resistances at different temperatures. Let x equal the distance from $-234.5°C$ to t_1 and y the distance from $-234.5°C$ to t_2, as shown in Fig. 3.12. From similar triangles,

$$\frac{x}{R_1} = \frac{y}{R_2}$$

or

$$\boxed{\frac{234.5 + t_1}{R_1} = \frac{234.5 + t_2}{R_2}} \tag{3.5}$$

The temperature of $-234.5°C$ is called the *inferred absolute temperature* of copper. For different conducting materials, the intersection of the straight-line approximation will occur at different temperatures. A few typical values are listed in Table 3.4.

The minus sign does not appear with the inferred absolute temperature on either side of Eq. (3.5) because x and y are the *total distances* from $-234.5°C$ to t_1 and t_2, respectively, and therefore are always positive quantities. For t_1 and t_2 less than zero, x and y are less than $-234.5°C$ and the distances are the difference between the inferred absolute temperature and the temperature of interest.

Equation (3.5) can easily be adapted to any material by inserting the proper inferred absolute temperature. It may therefore be written as follows:

$$\boxed{\frac{|T| + t_1}{R_1} = \frac{|T| + t_2}{R_2}} \tag{3.6}$$

TABLE 3.4

Inferred absolute temperatures.

Material	Temperature (°C)
Silver	−243
Copper	−234.5
Gold	−274
Aluminum	−236
Tungsten	−204
Nickel	−147
Iron	−162
Constantan	−125,000
Nichrome	−2,250

where $|T|$ indicates that the inferred absolute temperature of the material involved is inserted as a positive value in the equation.

EXAMPLE 3.9. If the resistance of a copper wire is $50\,\Omega$ at 20°C, what is its resistance at 100°C (boiling point of water)?

Solution: Eq. (3.5):

$$\frac{234.5 + 20}{50} = \frac{234.5 + 100}{R_2}$$

$$R_2 = \frac{(50)(334.5)}{254.5} = \mathbf{65.72\,\Omega}$$

EXAMPLE 3.10. If the resistance of a copper wire at freezing (0°C) is $30\,\Omega$, what is its resistance at −40°C?

Solution: Eq. (3.5):

$$\frac{234.5 + 0}{30} = \frac{234.5 - 40}{R_2}$$

$$R_2 = \frac{(30)(194.5)}{234.5} = \mathbf{24.88\,\Omega}$$

There is a second popular equation for calculating the resistance of a conductor at different temperatures. Defining

$$\alpha_1 = \frac{1}{|T| + t_1}$$

as the *temperature coefficient of resistance* at a temperature t_1, we have

$$\boxed{R_2 = R_1[1 + \alpha_1(t_2 - t_1)]} \qquad \textbf{(3.7)}$$

The values of α_1 for different materials at a temperature of 20°C have been evaluated, and a few are listed in Table 3.5. As indicated in the table, carbon and the family of *semiconductor materials have negative temperature coefficients*. In other words, the resistance of the material will drop with increase in temperature and vice versa.

Equation (3.7) can be written in the following form:

$$m = \text{slope of the curve} = \frac{\Delta y}{\Delta x}$$

$$\alpha_1 = \frac{1}{R_1}\overbrace{\left(\frac{R_2 - R_1}{t_2 - t_1}\right)}$$

TABLE 3.5
Temperature coefficient of resistance for various materials at 20°C.

Material	Temperature Coefficient (α_1)
Silver	0.0038
Copper	0.00393
Gold	0.0034
Aluminum	0.00391
Tungsten	0.005
Nickel	0.006
Iron	0.0055
Constantan	0.000008
Nichrome	0.00044
Carbon	−0.0005

Referring to Fig. 3.12, we find that the temperature coefficient is directly proportional to the slope of the curve, so the greater the slope of the curve, the greater the value of α_1. We can then conclude that *the higher the value of α_1, the greater the rate of change of resistance with temperature*. Referring to Table 3.5, we find that copper is more sensitive to temperature variations than silver, gold, or aluminum, although the differences are quite small.

3.6 TYPES OF RESISTORS

Resistors are made in many forms, but all belong in either of two groups: fixed or variable. The most common of the low-wattage, fixed-type resistors is the molded carbon composition resistor. The basic construction is shown in Fig. 3.13.

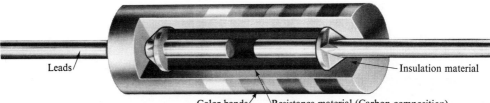

Leads — Color bands — Resistance material (Carbon composition) — Insulation material

FIG. 3.13
Fixed composition resistor. (Courtesy of Ohmite Manufacturing Co.)

The relative sizes of all fixed and variable resistors change with the wattage (power) rating, increasing in size for increased wattage ratings in order to withstand the higher currents and dissipation losses. The relative sizes of the molded composition resistors for different wattage ratings are shown in Fig. 3.14. Resistors of this type are readily available in values ranging from 2.7 Ω to 22 MΩ.

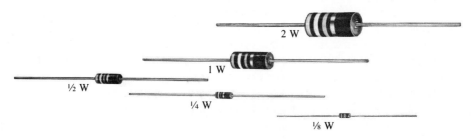

2 W

1 W

½ W

¼ W

⅛ W

FIG. 3.14
Fixed composition resistors of different wattage ratings. (Courtesy of Ohmite Manufacturing Co.)

The temperature-versus-resistance curve for a 10,000-Ω and 0.5-MΩ composition-type resistor is shown in Fig. 3.15. Note the small percent resistance change in the normal temperature operating range. Several other types of fixed resistors are shown in Fig. 3.16.

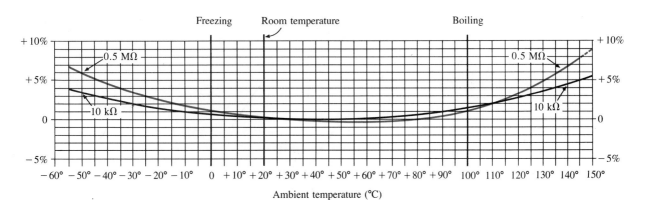

FIG. 3.15

Curves showing percent temporary resistance changes from +25°C values.
(Courtesy of Allen-Bradley Co.)

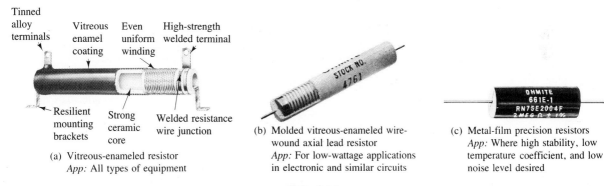

(a) Vitreous-enameled resistor
App: All types of equipment

(b) Molded vitreous-enameled wirewound axial lead resistor
App: For low-wattage applications
in electronic and similar circuits

(c) Metal-film precision resistors
App: Where high stability, low
temperature coefficient, and low
noise level desired

FIG. 3.16

Fixed resistors. (Courtesy of Ohmite Manufacturing Co.)

Variable resistors come in many forms, but basically they can be separated into the linear or nonlinear types. The symbol for a two-point linear or nonlinear *rheostat* is shown in Fig. 3.17. The three-point variable resistor may be called a *rheostat* or *potentiometer,* depending on how it is used. The symbol for a three-point variable resistor is shown in Fig. 3.18, along with the connections for its use as a rheostat or potentiometer. The arrow in the symbol of Fig. 3.18(a) is a contact that is movable on a continuous resistive element. As shown in Fig. 3.18(b), if the lug connected to the moving contact and a stationary lug are the only

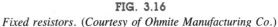

FIG. 3.17
Rheostat.

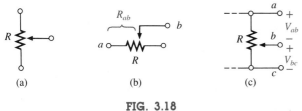

(a)   (b)   (c)

FIG. 3.18
Potentiometer.

(a) Linear winding

(b) Tapered winding

FIG. 3.19

Wirewound vitreous enameled potentiometers. (Courtesy of Ohmite Manufacturing Co.)

terminals used, the variable resistor is being used as a rheostat. The moving contact will determine whether R_{ab} is a minimum (zero ohms) or maximum value (R). If all three lugs are connected in the circuit as shown in Fig. 3.18(c), it is being employed as a potentiometer. The terminology *potentiometer* refers to the fact that the moving contact (wiper arm) will control by its position the *potential* differences V_{ab} and V_{bc} of Fig. 3.18(c).

Figure 3.19 shows both a linear and a tapered type of potentiometer. In the linear type of Fig. 3.19(a), the number of turns of the high-resistance wire per unit length of the core is uniform; therefore, the resistance will vary linearly with the position of the rotating contact. One-half of a turn will result in half the total resistance between either stationary lug and the moving contact. Three-quarters of a turn will establish three-quarters of the total across two terminals and one-quarter between the other stationary lug and the moving contact. If the number of turns is not uniform as in the tapered unit of Fig. 3.19(b), the resistance will vary nonlinearly with the position of the rotating contact. That is, a quarter turn may result in less or more than one-quarter the total resistance between a stationary lug and the moving contact. Potentiometers of both types in Fig. 3.19 are made in all sizes, with a range of maximum values from $200\,\Omega$ to $50,000\,\Omega$.

The molded composition linear potentiometer shown in Fig. 3.20 is the type used in circuits with smaller power demands than the one previously described. It is smaller in size but has maximum values ranging from $20\,\Omega$ to $22\,M\Omega$.

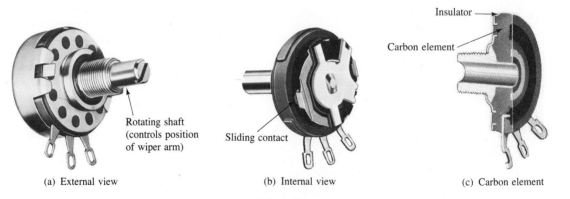

(a) External view — Rotating shaft (controls position of wiper arm)

(b) Internal view — Sliding contact

(c) Carbon element — Insulator, Carbon element

FIG. 3.20

Molded composition type potentiometer. (Courtesy of Allen-Bradley Co.)

The resistance of the screw-drive linear variable resistor of Fig. 3.21 is determined by the position of the contact arm, which can be moved by using the handwheel. The stationary terminal used with the movable contact determines whether the resistance increases or decreases with movement of the contact arm.

The miniaturization of parts—used quite extensively in computers—requires that resistances of different values be placed in very small packages. Two steps leading to the packaging of three resistors in a single module are shown in Fig. 3.22.

(a) Electrodes placed on module

(b) Resistance applied and adjusted to desired value by air-abrasion techniques

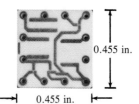

(c) Module completely encased

FIG. 3.22
Placement of resistors on a module. (Courtesy of International Business Machines Corp.)

FIG. 3.21
Screw-drive rheostat. (Courtesy of James G. Biddle Co.)

For use with printed circuit boards, resistor networks in a variety of configurations are available in miniature packages such as shown in Fig. 3.23 with a photograph of the casing and pins. The LDP is a coding for the production series, while the second number, 14, is the number of pins. The last two digits indicate the internal circuit configuration. The resistance range for the discrete elements in each chip is $10\,\Omega$ to $10\,M\Omega$.

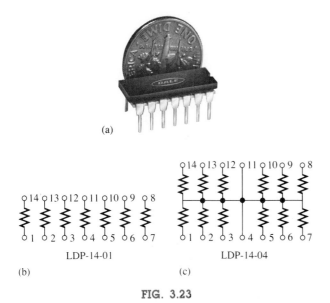

FIG. 3.23
Resistor configuration microcircuit. (Courtesy of Dale Electronics, Inc.)

3.7 COLOR CODING AND STANDARD RESISTOR VALUES

A wide variety of resistors, fixed or variable, are large enough to have their resistance in ohms printed on the casing. There are some,

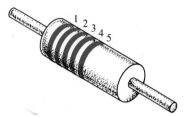

FIG. 3.24

Color coding—fixed molded composition resistor.

however, that are too small to have numbers printed on them, so a system of color coding is used. For the fixed molded composition resistor, four or five color bands are printed on one end of the outer casing as shown in Fig. 3.24. Each color has the numerical value indicated in Table 3.6. The color bands are always read from the end that has the band closest to it, as shown in Fig. 3.24. The first and second bands represent the first and second digits, respectively. The third band determines the power-of-10 multiplier for the first two digits (actually the number of zeros that follow the second digit), or a multiplying factor determined by the gold and silver bands. The fourth band is the manufacturer's tolerance, which is an indication of the precision by which the resistor was made. If the fourth band is omitted, the tolerance is assumed to be ±20%. The fifth band is a reliability factor which gives the percentage of failure per 1000 hours of use. For instance, a 1% failure rate would reveal that one out of every 100 (or 10 out of every 1000) will fail to fall within the tolerance range after 1000 hours of use.

TABLE 3.6

Resistor color coding.

Bands 1–3	Band 3		Band 4		Band 5	
0 Black	0.1 Gold	Multiplying	5%	Gold	1%	Brown
1 Brown	0.01 Silver	factors	10%	Silver	0.1%	Red
2 Red			20%	No band	0.01%	Orange
3 Orange					0.001%	Yellow
4 Yellow						
5 Green						
6 Blue						
7 Violet						
8 Gray						
9 White						

EXAMPLE 3.11. Find the range in which a resistor having the following color bands must exist to satisfy the manufacturer's tolerance:

a.

1st band	2nd band	3rd band	4th band	5th band
Gray	Red	Black	Gold	Brown
8	2	0	±5%	1%

$82 \, \Omega \pm 5\%$ (1% reliability)

Since 5% of 82 = 4.10, the resistor should be within the range $82 \, \Omega \pm 4.10 \, \Omega$, or *between 77.90 and 86.10 Ω*.

b.

1st band	2nd band	3rd band	4th band	5th band
Orange	White	Gold	Silver	No color
3	9	0.1	±10%	

$3.9 \, \Omega \pm 10\% = 3.9 \pm 0.39 \, \Omega$

The resistor should lie somewhere *between 3.51 and 4.29 Ω*.

Throughout the text material, resistor values in the network will be chosen to reduce the mathematical complexity of finding the solution. It was felt that the procedure or analysis technique was of primary importance and the mathematical exercise secondary. Many of the values appearing in the text are not *standard values*. That is, they are available only through special request. In the problem sections, however, standard values were frequently employed to make them more familiar and demonstrate their effect on the required calculations. A list of readily available standard values appears in Table 3.7. All the resistors appearing in Table 3.7 are available with 5% tolerance. Those in boldface are available with 5% and 10% tolerances while those in color are available with 5%, 10%, and 20% tolerances.

TABLE 3.7
Standard values of commercially available resistors.

Ohms (Ω)					Kilohms (kΩ)		Megohms (MΩ)	
0.10	1.0	10	100	1000	10	100	1.0	10.0
0.11	1.1	11	110	1100	11	110	1.1	11.0
0.12	**1.2**	**12**	**120**	**1200**	**12**	**120**	**1.2**	**12.0**
0.13	1.3	13	130	1300	13	130	1.3	13.0
0.15	1.5	15	150	1500	15	150	1.5	15.0
0.16	1.6	16	160	1600	16	160	1.6	16.0
0.18	**1.8**	**18**	**180**	**1800**	**18**	**180**	**1.8**	**18.0**
0.20	2.0	20	200	2000	20	200	2.0	20.0
0.22	2.2	22	220	2200	22	220	2.2	22.0
0.24	2.4	24	240	2400	24	240	2.4	
0.27	**2.7**	**27**	**270**	**2700**	**27**	**270**	**2.7**	
0.30	3.0	30	300	3000	30	300	3.0	
0.33	3.3	33	330	3300	33	330	3.3	
0.36	3.6	36	360	3600	36	360	3.6	
0.39	**3.9**	**39**	**390**	**3900**	**39**	**390**	**3.9**	
0.43	4.3	43	430	4300	43	430	4.3	
0.47	**4.7**	**47**	**470**	**4700**	**47**	**470**	**4.7**	
0.51	5.1	51	510	5100	51	510	5.1	
0.56	**5.6**	**56**	**560**	**5600**	**56**	**560**	**5.6**	
0.62	6.2	62	620	6200	62	620	6.2	
0.68	**6.8**	**68**	**680**	**6800**	**68**	**680**	**6.8**	
0.75	7.5	75	750	7500	75	750	7.5	
0.82	**8.2**	**82**	**820**	**8200**	**82**	**820**	**8.2**	
0.91	9.1	91	910	9100	91	910	9.1	

3.8 CONDUCTANCE

By finding the reciprocal of the resistance of a material, we have a measure of how well the material will conduct electricity. The quantity is called *conductance,* has the symbol G, and is measured in *siemens* (S).

In equation form, conductance is

$$G = \frac{1}{R} \qquad \text{(siemens, S)} \qquad (3.8)$$

A resistance of 1 MΩ is equivalent to a conductance of 10^{-6} S, and a resistance of 10 Ω is equivalent to a conductance of 10^{-1} S. The larger the conductance, therefore, the less the resistance and the greater the conductivity.

In equation form, the conductance is determined by

$$G = \frac{A}{\rho l} \qquad \text{(S)} \qquad (3.9)$$

indicating that increasing the area or decreasing either the length or the resistivity will increase the conductance.

EXAMPLE 3.12. What is the relative increase or decrease in conductivity of a conductor if the area is reduced by 80% and the length is increased by 40%? The resistivity is fixed.

Solution:

$$\frac{G_1 = \dfrac{A_1}{\rho_1 l_1}}{G_2 = \dfrac{A_2}{\rho_2 l_2}}$$

and for $\rho_1 = \rho_2$,

$$\frac{G_1}{G_2} = \frac{A_1 l_2}{A_2 l_1}$$

with $A_2 = (1/5)A_1$ and $l_2 = 1.4l_1$, resulting in

$$\frac{G_1}{G_2} = \frac{(A_1)(1.4l_1)}{(0.2A_1)(l_1)} = \frac{1.4}{0.2} = 7$$

and

$$G_2 = \frac{1}{7}G_1$$

3.9 OHMMETERS

The *ohmmeter* is an instrument used to perform the following tasks and a number of other useful functions:

1. Measure the resistance of individual or combined elements
2. Detect open-circuit (high-resistance) and short-circuit (low-resistance) situations
3. Check continuity of network connections and identify wires of a multilead cable
4. Test some semiconductor (electronic) devices

For most applications, the ohmmeters used most frequently appear as part of a VOM or DMM such as appearing in Figs. 2.24 and 2.25. The details of the internal circuitry and the method of using the meter will be left primarily for the laboratory exercise. In general, however, the resistance of a resistor can be measured by simply connecting the two leads of the meter across the resistor as shown in Fig. 3.25. There is no need to be concerned about which lead goes on which end; the result will be the same in either case since resistors offer the same resistance to the flow of charge (current) in either direction. If the VOM is employed, a switch must be set to the proper resistance range and a nonlinear scale (usually the top scale of the meter) must be properly read to obtain the resistance value. The DMM also requires choosing the best scale setting for the resistance to be measured but the result appears as a numerical display with the proper placement of the decimal point as determined by the chosen scale. When measuring the resistance of a single resistor, it is usually best to remove the resistor from the network before making the measurement. If this is difficult or impossible, at least one end of the resistor must not be connected to the network or the reading may include the effects of the other elements of the system.

If the two leads of the meter are touching in the ohmmeter mode, the resulting resistance is obviously zero. A connection can be checked as shown in Fig. 3.26 by simply hooking up the meter to either side of the connection. If the resistance is zero, the connection is secure. If other than zero, it could be a weak connection and, if infinite, there is no connection at all.

If one wire of a harness is known, a second can be found as shown in Fig. 3.27. Simply connect the end of the known lead to the end of any other lead. When the ohmmeter indicates zero ohms (or very low resistance), the second lead has been identified. The above procedure can

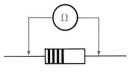

FIG. 3.25
Measuring the resistance of a single element.

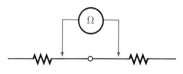

FIG. 3.26
Checking the continuity of a connection.

FIG. 3.27
Identifying the leads of a multilead cable.

also be used to determine the first known lead by simply connecting the meter to any wire at one end and then touching all the leads at the other end until a zero-ohm indication is obtained.

Preliminary measurements of the condition of some electronic devices such as the diode and transistor can be made using the ohmmeter. The meter can also be used to identify the terminals of such devices.

One important note about the use of any ohmmeter: *Never hook up an ohmmeter to a live circuit!* The reading will be meaningless and you may damage the instrument. The ohmmeter section of any meter is designed to pass a small sensing current through the resistance to be measured. Too large a current could damage the movement and would certainly throw off the calibration of the instrument. In addition, *never store an ohmmeter in the resistance mode.* The two leads of the meter could touch and the small sensing current could drain the internal battery. VOMs should be stored with the selector switch on the highest voltage range, and DMMs in the off position.

3.10 THERMISTORS

The *thermistor* is a two-terminal semiconductor device whose resistance, as the name suggests, is temperature sensitive. A representative characteristic appears in Fig. 3.28 with the graphic symbol for the device. Note the nonlinearity of the curve and the drop in resistance from about $5000\,\Omega$ to $100\,\Omega$ for an increase in temperature from 20°C to 100°C. The decrease in resistance with increase in temperature indicates a negative temperature coefficient.

The temperature of the device can be changed internally or externally. An increase in current through the device will raise its temperature, causing a drop in its terminal resistance. Any externally applied heat source will result in an increase in its body temperature and a drop in resistance. This type of action (internal or external) lends itself well to control mechanisms. A number of different types of thermistors are shown in Fig. 3.29. Materials employed in the manufacture of thermistors include oxides of cobalt, nickel, strontium, and manganese.

Note the use of a log scale in Fig. 3.28 for the vertical axis. The log scale permits the display of a wider range of specific resistance levels than a linear scale such as the horizontal axis. Note that it extends from $0.0001\,\Omega$-cm to $100,000,000\,\Omega$-cm over a very short interval. The log scale is used for both the vertical and the horizontal axis of Fig. 3.30, which appears in the next section.

3.11 PHOTOCONDUCTIVE CELL

The *photoconductive cell* is a two-terminal semiconductor device whose terminal resistance is determined by the intensity of the incident light on its exposed surface. As the applied illumination increases in intensity, the energy state of the surface electrons and atoms increases, with a resultant increase in the number of "free carriers" and a corresponding drop in resistance. A typical set of characteristics and its graphic symbol

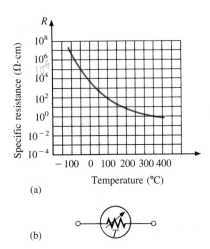

FIG. 3.28

Thermistor. (a) Characteristics; (b) symbol.

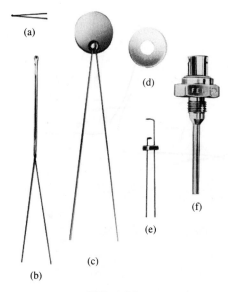

FIG. 3.29

Thermistors. (a) Beads; (b) glass probe; (c) disc; (d) washer; (e) specially mounted bead; (f) special probe assembly. (Courtesy of Fenwal Electronics, Inc.)

appear in Fig. 3.30. Note (as for the thermistor, which is also a semi-conductor device) the negative temperature coefficient. A number of cadmium sulfide photoconductive cells appear in Fig. 3.31.

3.12 VARISTORS

Varistors are voltage-dependent, nonlinear resistors used to suppress high-voltage transients. That is, their characteristics are such as to limit the voltage that can appear across the terminals of a sensitive device or system. A typical set of characteristics appears in Fig. 3.32(a) along with a linear resistance characteristic for comparison purposes. Note that at a particular "firing voltage," the current rises rapidly but the voltage is limited to a level just above this firing potential. In other

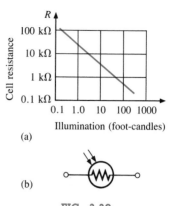

(a)

(b)

FIG. 3.30
Photoconductive cell. (a) Characteristics; (b) symbol.

FIG. 3.31
Photoconductive cells. (Courtesy of International Rectifier)

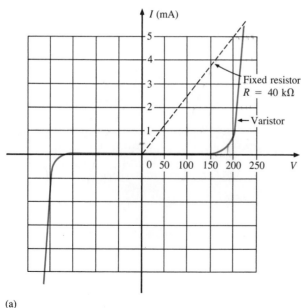

(a)

(b)

FIG. 3.32
Varistors. (a) Characteristics; (b) photograph. (Courtesy of General Electric Co.)

words, the magnitude of the voltage that can appear across this device cannot exceed that level defined by its characteristics. Through proper design techniques this device can therefore limit the voltage appearing across sensitive regions of a network. The current is simply limited by the network to which it is connected. A photograph of a number of commerical units appears in Fig. 3.32(b).

PROBLEMS

Section 3.2

1. Convert the following to mils:
   a. 0.5 in.   **b.** 0.01 in.
   c. 0.004 in.   **d.** 1 in.
   e. 0.02 ft   **f.** 0.01 cm

2. Calculate the area in circular mils (CM) of wires having the following diameters:
   a. 0.050 in.   **b.** 0.016 in.
   c. 0.30 in.   **d.** 0.1 cm
   e. 0.003 ft   **f.** 0.0042 m

3. The area in circular mils is
   a. 1600 CM   **b.** 900 CM
   c. 40,000 CM   **d.** 625 CM
   e. 7.75 CM   **f.** 81 CM
   What is the diameter of each wire in inches?

4. What is the resistance of a copper wire 200 ft long and 0.01 inch in diameter $(T = 20°C)$?

5. Find the resistance of a silver wire 50 yd long and 0.0045 inch in diameter $(T = 20°C)$.

6. a. What is the area in circular mils of an aluminum conductor that is 80 ft long with a resistance of 2.5 Ω?
   b. What is its diameter in inches?

7. a. What is the length of a copper wire with a diameter of 1/32 in. and a resistance of 0.004 kΩ $(T = 20°C)$?
   b. Repeat (a) for a silver wire and compare the results.

*8. a. What is the resistance of a copper bus-bar with the dimensions shown $(T = 20°C)$ in Fig. 3.33?
   b. Repeat (a) for aluminum and compare the results.
   c. Without working out the numerical solution, determine whether the resistance of the bar (aluminum or copper) will increase or decrease with increase in length. Explain your answer.
   d. Repeat (c) for increase in cross-sectional area.

9. a. What is the area in circular mils of a copper wire that has a resistance of 2.5 Ω and is 300 ft long $(T = 20°C)$?
   b. Without working out the numerical solution, determine whether the area of an aluminum wire will be smaller or larger than that of the copper wire. Explain.
   c. Repeat (b) for a silver wire.

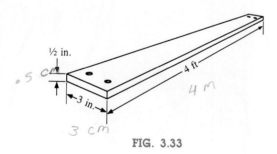

½ in.
.5 cm
4 ft
4 m
3 in.
3 cm

FIG. 3.33

10. In Fig. 3.34, three conductors of different materials are presented.
   a. Without working out the numerical solution, which section would appear to have the most resistance? Explain.
   b. Find the resistance of each section and compare with the result of (a) ($T = 20°C$).

11. A wire 1000 ft long has a resistance of 0.5 kΩ and an area of 94 CM. Of what material is the wire made ($T = 20°C$)?

*12. Determine the increase in resistance of a copper conductor if the area is reduced by a factor of 4 and the length doubled. The original resistance was 0.2 Ω. The temperature remains fixed. $1.6\,\Omega$

Section 3.3

13. a. Using Table 3.2, find the resistance of 450 ft of #11 and #14 AWG wires.
   b. Compare the resistances of the two wires.
   c. Compare the areas of the two wires.

14. a. Using Table 3.2, find the resistance of 1800 ft of #8 and #18 AWG wires.
   b. Compare the resistances of the two wires.
   c. Compare the areas of the two wires.

15. a. For the system of Fig. 3.35, the resistance of each line cannot exceed 0.006 Ω, and the maximum current drawn by the load is 110 A. What gage wire should be used?
   b. Repeat (a) for a maximum resistance of 0.003 Ω, $d = 30$ ft, and a maximum current of 110 A.

16. a. From Table 3.2, determine the maximum permissible current density (A/CM) for an AWG #0000 wire. $230\,A$ $211,600$
   b. Convert the result of (a) to A/in.^{2}.
   c. Using the result of (b), determine the cross-sectional area required to carry a current of 5000 A.

Section 3.4

17. Using metric units, determine the length of a copper wire that has a resistance of 0.2 Ω and a diameter of 1/10 in.

18. Repeat Problem 8 using metric units. That is, convert the given dimensions to metric units before determining the resistance.

19. If the sheet resistance of a tin oxide sample is 100 Ω, what is the thickness of the oxide layer?

20. Determine the width of a carbon resistor having a sheet resistance of 150 Ω if the length is 1/2 in. and the resistance is 500 Ω. P. 53

Section 3.5

21. The resistance of a copper wire is 2 Ω at 10°C. What is its resistance at 60°C?

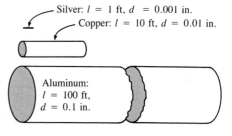

Silver: $l = 1$ ft, $d = 0.001$ in.
Copper: $l = 10$ ft, $d = 0.01$ in.
Aluminum: $l = 100$ ft, $d = 0.1$ in.

FIG. 3.34

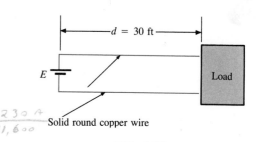

$d = 30$ ft

E Load

Solid round copper wire

FIG. 3.35

22. The resistance of an aluminum bus-bar is 0.02 Ω at 0°C. What is its resistance at 100°C?

23. The resistance of a copper wire is 4 Ω at 70°F. What is its resistance at 32°F?

24. The resistance of a copper wire is 0.76 Ω at 30°C. What is its resistance at −40°C?

25. a. The resistance of a copper wire is 0.002 Ω at room temperature (68°F). What is its resistance at 32°F (freezing) and 212°F (boiling)?
   b. For (a), determine the change in resistance for each 10° change in temperature between room temperature and 212°F.

26. If the resistance of a silver wire is 0.04 Ω at −30°C, what is its resistance at 0°C?

*27. a. The resistance of a copper wire is 0.92 Ω at 4°C. At what temperature (°C) will it be 1.06 Ω?
   b. At what temperature will it be 0.15 Ω?

28. Find the values of α_1 for copper and aluminum at 20°C, and compare them with those given in Table 3.5.

29. Using Eq. (3.7), find the resistance of a copper wire at 16°C if its resistance at 20°C is 40 Ω.

30. a. Find the value of α_1 at $t_1 = 40$°C for copper.
   b. Using the result of (a), find the resistance of a copper wire at 75°C if its resistance is 30 Ω at 40°C.

Section 3.6

31. a. What is the approximate increase in size from a 1-W to a 2-W carbon resistor?
   b. What is the approximate increase in size from a 1/2-W to a 2-W carbon resistor?

32. If the 10-kΩ resistor of Fig. 3.15 is exactly 10 kΩ at room temperature, what is its approximate resistance at −30°C?

33. Repeat Problem 32 at a temperature of 100°C.

34. If the resistance between the outside terminals of a linear potentiometer is 10 kΩ, what is its resistance between the wiper (movable) arm and an outside terminal if the resistance between the wiper arm and the other outside terminal is 3.5 kΩ?

35. If the wiper arm of a linear potentiometer is one-quarter the way around the contact surface, what is the resistance between the wiper arm and each terminal if the total resistance is 25 kΩ?

Section 3.7

36. Find the range in which a resistor having the following color bands must exist to satisfy the manufacturer's tolerance:

	1st band	2nd band	3rd band	4th band
a.	green	blue	orange	gold
b.	red	red	brown	silver
c.	brown	black	black	—

37. Is there an overlap in coverage between 20% resistors? That is, determine the tolerance range for a 10-Ω 20% resistor and a 15-Ω 20% resistor and note whether their tolerance ranges overlap.

38. Repeat Problem 37 for 10% resistors of the same value.

39. Repeat Problem 37 for a 47-Ω 20% resistor and a 68-Ω 20% resistor.

Section 3.8

40. Find the conductance of each of the following resistances:
a. 0.086 Ω **b.** 4000 Ω
c. 0.05 MΩ
Compare the three results.

*41. Find the conductance of 1000 ft of #18 AWG wire made of
a. copper
b. aluminum
c. iron

*42. The conductance of a wire is 100 S. If the area of the wire is increased by a factor of 2/3 and the length is reduced by the same factor, find the new conductance of the wire if the temperature remains fixed.

Section 3.9

43. How would you check the status of a fuse with an ohmmeter?

44. How would you determine the on and off states of a switch using an ohmmeter?

45. How would you use an ohmmeter to check the status of a light bulb?

Section 3.10

46. Find the resistance of the thermistor having the characteristics of Fig. 3.28 at $-50°C$, $50°C$, and $200°C$. Note that it is a log scale. If necessary, consult a reference with an expanded log scale.

Section 3.11

47. Using the characteristics of Fig. 3.30, determine the resistance of the photoconductive cell at 10 and 100 footcandle illumination. As in Problem 46, note that it is a log scale.

Section 3.12

48. a. Referring to Fig. 3.32(a), find the terminal voltage of the device at 0.5, 1, 3, and 5 mA.
   b. What is the total change in voltage for the indicated range of current levels?
   c. Compare the ratio of maximum to minimum current levels above to the corresponding ratio of voltage levels.

GLOSSARY

Absolute zero The temperature at which all molecular motion ceases; $-273.15°C$.

Circular mil (CM) The cross-sectional area of a wire having a diameter of one mil.

Color coding A technique employing bands of color to indicate the resistance levels and tolerance of resistors.

Conductance (G) An indication of the relative ease with which current can be established in a material. It is measured in siemens (S).

Inferred absolute temperature The temperature through which a straight-line approximation for the actual resistance-versus-temperature curve will intersect the temperature axis.

Negative temperature coefficient of resistance The value which reveals that the resistance of a material will decrease with increase in temperature.

Ohm (Ω) The unit of measurement applied to resistance.

Ohmmeter An instrument for measuring resistance levels.

Photoconductive cell A two-terminal semiconductor device whose terminal resistance is determined by intensity of the incident light on its exposed surface.

Positive temperature coefficient of resistance The value which reveals that the resistance of a material will increase with increase in temperature.

Potentiometer A three-terminal device through which potential levels can be varied in a linear or nonlinear manner.

Resistance A measure of the opposition to the flow of charge through a material.

Resistivity (ρ) A constant of proportionality between the resistance of a material and its physical dimensions.

Rheostat An element whose terminal resistance can be varied in a linear or nonlinear manner.

Sheet resistance Defined by ρ/d for thin-film and integrated circuit design.

Thermistor A two-terminal semiconductor device whose resistance is temperature sensitive.

Varistor A voltage-dependent, nonlinear resistor used to suppress high-voltage transients.

4

Ohm's Law, Power, and Energy

4.1 OHM'S LAW

Consider the following relationship:

$$\text{Effect} = \frac{\text{cause}}{\text{opposition}} \qquad \textbf{(4.1)}$$

Every conversion of energy from one form to another can be related to this equation. In electric circuits, the *effect* we are trying to establish is the flow of charge, or *current*. The *potential difference* between two points is the *cause* ("pressure"), and the opposition is the *resistance* encountered.

Substituting these terms into Eq. (4.1) results in

$$\text{Current} = \frac{\text{potential difference}}{\text{resistance}}$$

and

$$I = \frac{E}{R} \qquad \text{(amperes, A)} \qquad \textbf{(4.2)}$$

Equation (4.2), known as *Ohm's law*, clearly reveals that the greater the voltage across a resistor, the more the current, and the more the resistance for the same voltage, the less the current. In other words, the current is proportional to the applied voltage and inversely proportional to the resistance.

By simple mathematical manipulations, the voltage and resistance can be found in terms of the other two quantities:

$$E = IR \qquad \text{(volts, V)} \qquad \textbf{(4.3)}$$

$$R = \frac{E}{I} \qquad \text{(ohms, } \Omega) \qquad \textbf{(4.4)}$$

Recall from Chapter 2 that for voltage, the symbol E represents all sources of voltage such as the battery, and the symbol V represents the potential drop across a resistor or any other energy-converting device. In any case, E and V are interchangeable in Eqs. (4.2) through (4.4).

The three quantities of Eqs. (4.2) through (4.4) are defined by Fig. 4.1. The current I of Eq. (4.2) results from applying a dc supply of E volts across a network having a resistance R. Equation (4.3) determines the voltage E required to establish a current I through a network with a total resistance R, and Eq. (4.4) provides the resistance of a network that results in a current I due to an impressed voltage E.

EXAMPLE 4.1. Determine the current resulting from the application of a 9-V battery across a network with a resistance of $2.2\,\Omega$.

Solution: Eq. (4.2):

$$I = \frac{E}{R} = \frac{9}{2.2} = \textbf{4.09 A}$$

EXAMPLE 4.2. Calculate the resistance of a 60-W bulb if a current of 500 mA results from an applied voltage of 120 V.

Solution: Eq. (4.4):

$$R = \frac{E}{I} = \frac{120}{500 \times 10^{-3}} = \textbf{240 } \Omega$$

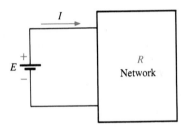

FIG. 4.1
Basic network.

For the resistive element, the polarity of the voltage drop is as shown in Fig. 4.2(a) for the indicated current direction. A reversal in current will reverse the polarity, as shown in Fig. 4.2(b). In general, the flow of charge is from a high (+) to a low (−) potential. Polarities as established by current direction will become increasingly important in the analysis to follow.

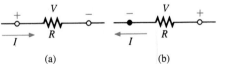

(a)   (b)

FIG. 4.2
Defining polarities.

EXAMPLE 4.3. Calculate the current through the 2-Ω resistor of Fig. 4.3 if the voltage drop across it is 16 V.
Solution:

$$I = \frac{V}{R} = \frac{16}{2} = \textbf{8 A}$$

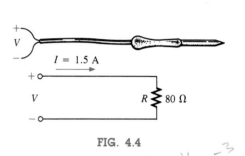

FIG. 4.3

EXAMPLE 4.4. Calculate the voltage that must be applied across the soldering iron of Fig. 4.4 to establish a current of 1.5 A through the iron if its internal resistance is 80 Ω.
Solution:

$$V = IR = 1.5 \cdot 80 = \textbf{120 V}$$

4.2 PLOTTING OHM'S LAW

Graphs, characteristics, plots, and the like, play an important role in every technical field as a mode through which the broad picture of the behavior or response of a system can be conveniently displayed. It is therefore critical to develop the skills necessary both to read data and to plot them in such a manner that they can be interpreted easily.

For most sets of characteristics of electronic devices, the current is represented by the vertical axis (ordinate), and the voltage by the horizontal axis (abscissa), as shown in Fig. 4.5. First note that the vertical axis is in amperes and the horizontal axis in volts. For some plots, I may be in milliamperes, microamperes, or whatever is appropriate for the range of interest. The same is true for the levels of voltage on the

FIG. 4.4

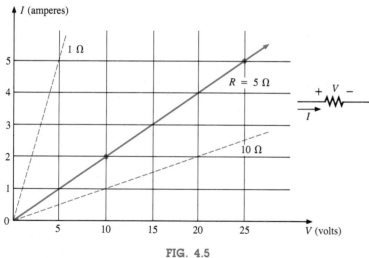

FIG. 4.5
Plotting Ohm's law.

horizontal axis. Note also that the chosen parameters require that the spacing between numerical values of the vertical axis be different from that of the horizontal axis. The linear (straight-line) graph reveals that the resistance is not changing with current or voltage level; rather, it is a fixed quantity throughout. At 25 V, $I = V/R = 25/5 = 5$ A; at 10 V, $I = V/R = 10/5 = 2$ A; and so on as shown on the graph. The current I is obviously equal to zero when the voltage is zero. For comparison purposes, a 1-Ω and 10-Ω resistor were plotted on the graph of Fig. 4.5. Note that the less the resistance, the steeper the slope (closer to the vertical axis) of the curve.

If we write Ohm's law in the following manner and relate it to the basic straight-line equation

$$I \; = \; \frac{1}{R} \cdot E \; + \; 0$$
$$\downarrow \qquad \downarrow \quad \downarrow \qquad \downarrow$$
$$y \; = \; m \cdot x \; + \; b$$

we find that the slope is equal to 1 divided by the resistance value, as indicated by the following:

$$m = \text{slope} = \frac{\Delta y}{\Delta x} = \frac{\Delta I}{\Delta V} = \frac{1}{R} \qquad \textbf{(4.5)}$$

where Δ signifies a small finite change in the variable.

Equation (4.5) clearly reveals that the greater the resistance, the less the slope. If written in the following form, Eq. (4.5) can be used to determine the resistance from the linear curve:

$$R = \frac{\Delta V}{\Delta I} \qquad \text{(ohms)} \qquad \textbf{(4.6)}$$

EXAMPLE 4.5. Determine the resistance associated with the curve of Fig. 4.6 using Eqs. (4.6) and (4.4), and compare results.

Solution: Using the interval between 6 V and 8 V [Eq. (4.6)],

$$R = \frac{\Delta V}{\Delta I} = \frac{8-6}{(16-12) \times 10^{-3}} = \frac{2}{4 \times 10^{-3}} = 0.5 \times 10^3$$
$$= \textbf{500 } \mathbf{\Omega}$$

For the interval between 0 V and 4 V [Eq. (4.6)],

$$R = \frac{\Delta V}{\Delta I} = \frac{4-0}{(8-0) \times 10^{-3}} = 0.5 \times 10^3$$
$$= \textbf{500 } \mathbf{\Omega}$$

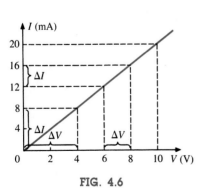

FIG. 4.6

Eq. (4.4): At 20 mA,

$$R = \frac{V}{I} = \frac{10}{20 \times 10^{-3}} = 0.5 \times 10^3$$

$$= \mathbf{500\ \Omega}$$

Before leaving the subject, let us first investigate the characteristics of a very important semiconductor device called the *diode,* which will be examined in detail in the basic electronics courses. This device will ideally act like a low resistance path to current in one direction and a high resistance path to current in the reverse direction, much like a switch that will pass current in only one direction. A typical set of characteristics appears in Fig. 4.7. Without any mathematical calcula-

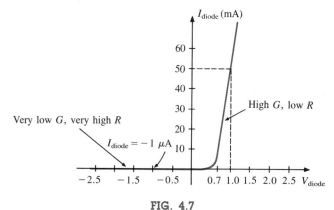

FIG. 4.7
Semiconductor diode characteristic.

tions, the closeness of the characteristic to the voltage axis for negative values of applied voltage indicates that this is the low conductance (high resistance, switch opened) region. Note that this region extends to approximately 0.7 V positive. However, for values of applied voltage greater than 0.7 V, the vertical rise in the characteristics indicates a high conductivity (low resistance, switch closed) region. Application of Ohm's law will now verify the above conclusions.

At $V = +1$ V,

$$R_{\text{diode}} = \frac{V}{I} = \frac{1\,\text{V}}{50\,\text{mA}} = \frac{1}{50 \times 10^{-3}}$$

$$= 20\,\Omega$$

(relatively low value for most applications)

At $V = -1$ V,

$$R_{\text{diode}} = \frac{V}{I} = \frac{1}{1\,\mu\text{A}}$$

$$= 1\,\text{M}\Omega$$

4.3 POWER

Power is an indication of how much work (the conversion of energy from one form to another) can be accomplished in a specified amount of time, that is, a *rate* of doing work. Since converted energy is measured in *joules* (J) and time in seconds (s), power is measured in joules/second (J/s). The electrical unit of measurement for power is the watt (W), defined by

$$1 \text{ watt (W)} = 1 \text{ joule/second (J/s)} \qquad (4.7)$$

In equation form, power is determined by

$$P = \frac{W}{t} \qquad \text{(watts, W, or joules/second, J/s)} \qquad (4.8)$$

with the energy W measured in joules and the time t in seconds.

Throughout the text, the abbreviation for energy (W) can be distinguished from that for the watt (W) by the fact that one is in italics while the other is in roman. In fact, all variables in the dc section appear in italics while the units appear in roman.

The unit of measurement, the watt, is derived from the surname of James Watt, who was instrumental in establishing the standards for power measurements. He introduced the *horsepower* (hp) as a measure of the average power of a strong dray horse over a full working day. It is approximately 50% more than can be expected from the average horse. The horsepower and watt are related in the following manner:

$$1 \text{ horsepower} \cong 746 \text{ watts}$$

The power delivered to, or absorbed by, an electrical device or system can be found in terms of the current and voltage by first substituting Eq. (2.3) into Eq. (4.5):

$$P = \frac{W}{t} = \frac{QV}{t} = V\frac{Q}{t}$$

But

$$I = \frac{Q}{t}$$

so that

$$P = VI \qquad \text{(watts)} \qquad (4.9)$$

By direct substitution of Ohm's law, the equation for power can be obtained in two other forms:

$$P = VI = V\left(\frac{V}{R}\right)$$

and

$$\boxed{P = \frac{V^2}{R}} \quad \text{(watts)} \qquad \textbf{(4.10)}$$

or

$$P = VI = (IR)I$$

and

$$\boxed{P = I^2R} \quad \text{(watts)} \qquad \textbf{(4.11)}$$

The result is that the power to the resistor of Fig. 4.8 can be found directly depending on the information available. In other words, if the current and resistance are known, it pays to use Eq. (4.11) directly, and if V and I are known, Eq. (4.9) is appropriate. It saves having to apply Ohm's law before determining the power.

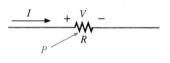

FIG. 4.8

EXAMPLE 4.6. Find the power delivered to the dc motor of Fig. 4.9.
Solution:

$$P = VI = (120)(5) = 600 \text{ W} = \textbf{0.6 kW}$$

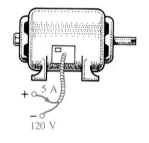

FIG. 4.9

EXAMPLE 4.7. What is the power dissipated by a 5-Ω resistor if the current is 4 A?
Solution:

$$P = I^2R = (4)^2 \cdot 5 = \textbf{80 W}$$

EXAMPLE 4.8. The I-V characteristics of a light bulb are provided in Fig. 4.10. Note the nonlinearity of the curve, indicating a wide range in resistance of the bulb with applied voltage. If the rated voltage is 120 V, find the wattage rating of the bulb. Also calculate the resistance of the bulb under rated conditions.
Solution: At 120 V,

$$I = 0.625 \text{ A}$$

and

$$P = VI = (120)(0.625) = \textbf{75 W}$$

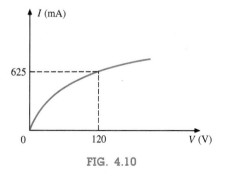

FIG. 4.10

At 120 V,

$$R = \frac{V}{I} = \frac{120}{0.625} = \mathbf{192\ \Omega}$$

As was stated in Section 4.2, the shape of the curve indicates that the resistance increases with increasing voltage levels.

The power delivered by an energy source is given by

$$\boxed{P = EI} \qquad \text{(watts)} \qquad \qquad \mathbf{(4.12)}$$

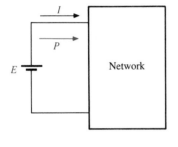

FIG. 4.11

where E is the source voltage and I is the current drain from the source, as shown in Fig. 4.11.

Sometimes the power is given and the current or voltage must be determined. Through algebraic manipulations, an equation for each variable is derived as follows:

$$P = I^2R \Rightarrow I^2 = \frac{P}{R}$$

and

$$\boxed{I = \sqrt{\frac{P}{R}}} \qquad \text{(amperes)} \qquad \qquad \mathbf{(4.13)}$$

$$P = \frac{V^2}{R} \Rightarrow V^2 = PR$$

and

$$\boxed{V = \sqrt{PR}} \qquad \text{(volts)} \qquad \qquad \mathbf{(4.14)}$$

EXAMPLE 4.9. Determine the current through a 5-kΩ resistor when the power dissipated by the element is 20 mW.

Solution: Eq. (4.13):

$$I = \sqrt{\frac{P}{R}} = \sqrt{\frac{20 \times 10^{-3}}{5 \times 10^3}} = \sqrt{4 \times 10^{-6}} = 2 \times 10^{-3}$$

$$= \mathbf{2\ mA}$$

Potential terminals   Current terminals

FIG. 4.12

Wattmeter. (Courtesy of Electrical Instrument Service, Inc.)

4.4 WATTMETERS

As one might expect, instruments exist that can measure the power delivered by a source and to a dissipative element. One such instrument appears in Fig. 4.12. Since power is a function of both the current and the voltage levels, four terminals must be connected as shown in Fig. 4.13 to measure the power to the resistor R.

If the current coils (CC) and potential coils (PC) of the wattmeter are connected as shown in Fig. 4.13, there will be an up-scale reading on the wattmeter. A reversal of either coil will result in a below-zero indication. Three voltage terminals may be available on the voltage side to permit a choice of voltage levels. On most wattmeters, the current terminals are physically larger than the voltage terminals for safety reasons and to insure a solid connection.

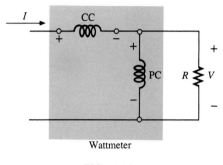

FIG. 4.13
Wattmeter connections.

4.5 EFFICIENCY

Any electrical system that converts energy from one form to another can be represented by the block diagram of Fig. 4.14 with an energy input and output terminal.

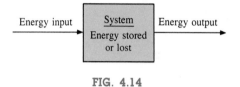

FIG. 4.14

Conservation of energy requires that

Energy input = Energy output + Energy lost or
stored in the system

Dividing both sides of the relationship by t gives

$$\frac{W_{in}}{t} = \frac{W_{out}}{t} + \frac{W_{\text{lost or stored by the system}}}{t}$$

Since $P = W/t$, we have the following:

$$\boxed{P_i = P_o + P_{\text{lost or stored}}} \qquad \textbf{(4.15)}$$

The efficiency (η) of the system is then determined by the following equation:

$$\text{Efficiency} = \frac{\text{power output}}{\text{power input}}$$

and

$$\boxed{\eta = \frac{P_o}{P_i}} \qquad \textbf{(4.16)}$$

where η (lowercase Greet letter eta) is a decimal number. Expressed as a percentage,

$$\boxed{\eta = \frac{P_o}{P_i} \times 100\%} \qquad \textbf{(4.17)}$$

In terms of the input and output energy, the efficiency in percent is given by

$$\boxed{\eta = \frac{W_o}{W_i} \times 100\%}$$ (4.18)

The maximum possible efficiency is 100%, which occurs when $P_o = P_i$, or when the power lost or stored in the system is zero. Obviously, the greater the internal losses of the system in generating the necessary output power or energy, the lower the net efficiency.

EXAMPLE 4.10. A 2-hp motor operates at an efficiency of 75%. What is the power input in watts? If the input current is 9.05 A, what is the input voltage?

Solution:

$$\eta = \frac{P_o}{P_i} \times 100\%$$

$$0.75 = \frac{(2)(746)}{P_i}$$

and

$$P_i = \frac{1492}{0.75} = \mathbf{1989.33 \ W}$$

$$P = EI \quad \text{or} \quad E = \frac{P}{I} = \frac{1990}{9.05} = 219.82 \ V \cong \mathbf{220 \ V}$$

EXAMPLE 4.11. What is the output in horsepower of a motor with an efficiency of 80% and an input current of 8 A at 120 V?

Solution:

$$\eta = \frac{P_o}{P_i} \times 100\%$$

$$0.80 = \frac{P_o}{(120)(8)}$$

and

$$P_o = (0.80)(120)(8) = 768 \ W$$

with

$$768 \ \cancel{W}\left(\frac{1 \ hp}{746 \ \cancel{W}}\right) = \mathbf{1.029 \ hp}$$

EXAMPLE 4.12. What is the efficiency in percent of a system in which the input energy is 50 J and the output energy is 42.5 J?

Solution:

$$\eta = \frac{W_o}{W_i} \times 100\% = \frac{42.5}{50} \times 100\% = \mathbf{85\%}$$

The very basic components of a generating (voltage) system are de-picted in Fig. 4.15. The source of mechanical power is a structure such

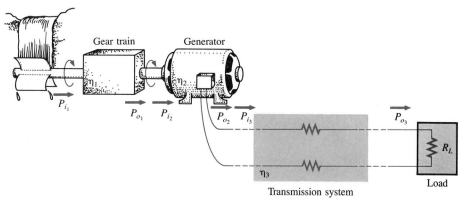

FIG. 4.15
Basic components of a generating system.

as a paddlewheel that is turned by the water rushing over the dam. The gear train will then insure that the rotating member of the generator is turning at rated speed. The output voltage must then be fed through a transmission system to the load. For each component of the system, an input and output power have been indicated. The efficiency of each system is given by

$$\eta_1 = \frac{P_{o_1}}{P_{i_1}} \qquad \eta_2 = \frac{P_{o_2}}{P_{i_2}} \qquad \eta_3 = \frac{P_{o_3}}{P_{i_3}}$$

If we form the product of these three efficiencies,

$$\eta_1 \cdot \eta_2 \cdot \eta_3 = \frac{P_{o_1}}{P_{i_1}} \cdot \frac{P_{o_2}}{P_{i_2}} \cdot \frac{P_{o_3}}{P_{i_3}}$$

and substitute the fact that $P_{i_2} = P_{o_1}$ and $P_{i_3} = P_{o_2}$, we find that the quantities indicated above will cancel, resulting in P_{o_3}/P_{i_1}, which is a measure of the efficiency of the entire system. In general, for the repre-sentative cascaded system of Fig. 4.16,

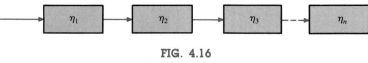

FIG. 4.16
Cascaded system.

$$\boxed{\eta_{\text{total}} = \eta_1 \cdot \eta_2 \cdot \eta_3 \cdot \cdots \cdot \eta_n} \qquad (4.19)$$

EXAMPLE 4.13. Find the overall efficiency of the system of Fig. 4.15 if $\eta_1 = 90\%$, $\eta_2 = 85\%$, and $\eta_3 = 95\%$.

Solution:

$$\eta_T = \eta_1 \cdot \eta_2 \cdot \eta_3 = (0.90)(0.85)(0.95) = 0.727 \text{ or } \textbf{72.7\%}$$

EXAMPLE 4.14. If the efficiency η_1 drops to 60%, find the new overall efficiency and compare the result with that obtained in Example 4.13.

Solution:

$$\eta_T = \eta_1 \cdot \eta_2 \cdot \eta_3 = (0.60)(0.85)(0.95) = 0.485 \text{ or } \textbf{48.5\%}$$

Certainly 48.5% is noticeably less than 75%. The total efficiency of a cascaded system is therefore determined primarily by the lowest efficiency (weakest link) and is less than (or equal to if the remaining efficiences are 100%) the least efficient link of the system.

4.6 ENERGY

In order for power, which is the rate of doing work, to produce an energy conversion of any form, it must be *used over a period of time*. For example, a motor may have the horsepower to run a heavy load, but unless the motor is *used* over a period of time, there will be no energy conversion. In addition, the longer the motor is used to drive the load, the greater will be the energy expended.

The energy lost or gained by any system is therefore determined by

$$\boxed{W = Pt} \qquad \text{(wattseconds, Ws, or joules)} \qquad (4.20)$$

Since power is measured in watts (or joules per second) and time in seconds, the unit of energy is the *wattsecond* or *joules* as indicated above. The wattsecond, however, is too small a quantity for most practical purposes, so the *watthour* (Wh) and *kilowatthour* (kWh) were defined, as follows:

$$\boxed{\text{Energy (Wh)} = \text{power (W)} \times \text{time (h)}} \qquad (4.21)$$

$$\boxed{\text{Energy (kWh)} = \frac{\text{power (W)} \times \text{time (h)}}{1000}} \qquad (4.22)$$

Note that the energy in kilowatthours is simply the energy in watthours divided by 1000.

To develop some sense for the kilowatthour energy level, consider that 1 kWh is the energy dissipated by a 100-W bulb in 10 h.

The *kilowatthour meter* is an instrument for measuring the energy supplied to the residential or commercial user of electricity. It is normally connected directly to the lines at a point just prior to entering the power distribution panel of the building. A typical set of dials is shown in Fig. 4.17 with a photograph of a kilowatthour meter. As indicated, each power of 10 below a dial is in kilowatthours. The more rapidly the aluminum disc rotates, the greater the energy demand. The dials are connected through a set of gears to the rotation of this disc.

EXAMPLE 4.15. For the dial positions of Fig. 4.17, calculate the electricity bill if the previous reading was 4650 kWh and the average cost is 7¢ per kilowatthour.
Solution:

$$5360 - 4650 = 710 \text{ kWh used}$$

$$710 \text{ kWh} \left(\frac{7¢}{\text{kWh}} \right) = \textbf{\$49.70}$$

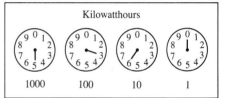

Kilowatthours

| 1000 | 100 | 10 | 1 |

FIG. 4.17
Kilowatthour meter. (Courtesy of Westinghouse Electric Corp.)

EXAMPLE 4.16. How much energy (in kilowatthours) is required to light a 60-W bulb continuously for 1 year (365 days)?
Solution:

$$W = \frac{P \cdot t}{1000} = \frac{(60)(24)(365)}{1000} = \frac{525,600}{1000} = \textbf{525.60 kWh}$$

EXAMPLE 4.17. How long can a 205-W television set be on before using more than 4 kWh of energy?
Solution:

$$W = \frac{P \cdot t}{1000} = \frac{(205)(t)}{1000} \Rightarrow t(\text{hours}) = \frac{(4)(1000)}{205} = \textbf{19.51 h}$$

EXAMPLE 4.18. What is the cost of using a 5-hp motor for 3 h if the cost is 7¢ per kilowatthour?
Solution:

$$W \text{ (kilowatthours)} = \frac{(5)(746 \times 3)}{1000} \times 11.2 \text{ kWh}$$

$$\text{Cost} = (11.2)(7) = \textbf{78.4¢}$$

EXAMPLE 4.19. What is the total cost of using the following at 7¢ per kilowatthour?
a. a 1200-W toaster for 30 min
b. six 50-W bulbs for 4 h
c. a 400-W washing machine for 45 min
d. a 4800-W electric clothes dryer for 20 min

Solution:

$$W = \frac{(1200)(\frac{1}{2}) + (6)(50)(4) + (400)(\frac{3}{4}) + (4800)(\frac{1}{3})}{1000}$$

$$= \frac{600 + 1200 + 300 + 1600}{1000} = \frac{3700}{1000}$$

$$W = 3.7 \text{ kWh}$$

$$\text{Cost} = (3.7)(7) = \mathbf{25.9¢}$$

The chart in Fig. 4.18 shows the average cost per kilowatthour as compared to the kilowatthours used per customer. Note that the cost today is about the same as in 1926 but that the average customer uses more than 20 times as much electrical energy in a year. Keep in mind that the chart of Fig. 4.18 is the average cost across the nation. Some states have average rates of 3¢ or 4¢ per kilowatthour while others pay over 11¢ per kilowatthour.

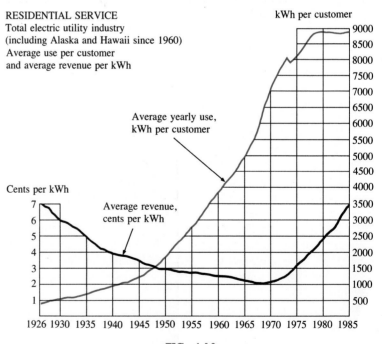

FIG. 4.18
(Courtesy of Edison Electric Institute)

Table 4.1 lists some common household items with their typical wattage ratings. It might prove interesting for the reader to calculate the cost of operating some of these appliances over a period of time using the preceding chart to find the cost per kilowatthour.

TABLE 4.1

Typical wattage ratings of some common household appliances.

Appliance	Wattage Rating	Appliance	Wattage Rating
Air conditioner	860	Microwave oven	800
Blow dryer	1,300	Phonograph	75
Cassette player/recorder	5	Projector	1,200
Clock	2	Radio	70
Clothes dryer (electric)	4,800	Range (self-cleaning)	12,200
Coffee maker	900	Refrigerator (automatic defrost)	1,800
Dishwasher	1,200	Shaver	15
Fan:		Stereo equipment	110
Portable	90	Sun lamp	280
Window	200	Toaster	1,200
Heater	1,322	Trash compactor	400
Heating equipment:		TV (color)	150
Furnace fan	320	Videocassette recorder	110
Oil-burner motor	230	Washing machine	400
Iron, dry or steam	1,100	Water heater	2,500

Courtesy of General Electric Co.

4.7 CIRCUIT BREAKERS AND FUSES

The incoming power to any large industrial plant, heavy equipment, simple circuit in the home, or meters used in the laboratory must be limited to insure that the current through the lines is not above the rated value. Otherwise, the electric or electronic equipment may be damaged or dangerous side effects such as fire or smoke may result. To limit the current level, fuses or circuit breakers are installed where the power enters the installation, such as in the panel in the basement of most homes at the point where the outside feeder lines enter the dwelling. The fuse (depicted in Fig. 4.19) has an internal bimetallic conductor through which the current will pass; it will begin to melt if the current through the system exceeds the rated value printed on the casing. Of course, if it melts through, the current path is broken and the load in its path protected.

In homes and industrial plants built in recent years, the fuse has been replaced by circuit breakers such as appearing in Fig. 4.20. When the current exceeds rated conditions, an electromagnet in the device will have sufficient strength to draw the connecting metallic link in the breaker out of the circuit and open the current path. When conditions

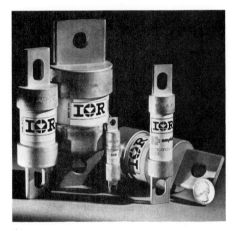

(a)

(b)

FIG. 4.19

Bimetallic fuses. (Part (a) courtesy of Bussman Manufacturing Co.; part (b) courtesy of International Rectifier Corp.)

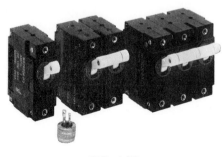

FIG. 4.20

Circuit breakers. (Courtesy of Potter and Brumfield Division, AMF, Inc.)

have been corrected, the breaker can be reset and used again, unlike the fuse, which has to be replaced.

4.8 COMPUTER ANALYSIS

In recent years the development of the personal computer (PC) has placed the system within financial reach of most individuals with a need for its capabilities and scope. Texts are beginning to surface dedicated solely to the software (written programs) that perform a wide variety of tasks with a high degree of accuracy in a very short period of time. Space constraints here do not permit an in-depth review of computers and the available software. However, a few typical programs have been interspersed throughout the text to demonstrate their range of application. In case you have a personal computer, or computer facilities are available for your use, exercises have been provided at the end of each chapter to test your programming skills.

The BASIC language is the communication link between the user (operator) and the computer in a wide range of personal computers manufactured today. You will note in Program 4.1 of Fig. 4.21 that the

```
10 REM ***** PROGRAM 4-1 *****
20 REM **********************************************
30 REM Program demonstrates selecting various forms
40 REM of equations
50 REM **********************************************
60 REM
100 PRINT:PRINT "Select which form of Ohm's law equation "
110 PRINT "you wish to use."
120 PRINT
130 PRINT TAB(10);"(1) V=I*R"
140 PRINT TAB(10);"(2) I=V/R"
150 PRINT TAB(10);"(3) R=V/I"
160 PRINT TAB(20);
170 INPUT "choice=";C
180 IF C<1 OR C>3 THEN GOTO 100
190 ON C GOSUB 400,600,800
200 PRINT:PRINT
210 INPUT "More (YES or NO)";A$
220 IF A$="YES" THEN 100
230 PRINT "Have a good day"
240 END
400 REM Accept input of I,R and output V
410 PRINT:PRINT "Enter the following data:"
420 INPUT "I=";I
430 INPUT "R=";R
440 V=I*R
450 PRINT "Voltage is ";V;"volts"
460 RETURN
600 REM Accept input of V,R and output I
610 PRINT "Enter the following data:"
620 INPUT "V=";V
630 INPUT "R=";R
640 I=V/R
650 PRINT "Current is ";I;"amperes"
660 RETURN
800 REM Accept input of V,I and output R
810 PRINT "Enter the following data:"
820 INPUT "V=";V
830 INPUT "I=";I
840 R=V/I
850 PRINT "Resistance is ";R;"ohms"
860 RETURN
```

Equation Selection (lines 100–200)

Continue? (lines 210–230)

$V = IR$ (lines 400–460)

$I = \dfrac{V}{R}$ (lines 600–660)

$R = \dfrac{V}{I}$ (lines 800–860)

```
READY

RUN


Select which form of Ohm's law equation
you wish to use.

              (1)  V=I*R
              (2)  I=V/R
              (3)  R=V/I
                        choice=? 2

Enter the following data:
V=? 12

R=? 4E3

Current is  3E-03 amperes


More (YES or NO)? YES


Select which form of Ohm's law equation
you wish to use.

              (1)  V=I*R
              (2)  I=V/R
              (3)  R=V/I
                        choice=? 1


Enter the following data:
I=? 2E-3

R=? 5.6E3

Voltage is  11.2 volts


More (YES or NO)? YES


Select which form of Ohm's law equation
you wish to use.

              (1)  V=I*R
              (2)  I=V/R
              (3)  R=V/I
                        choice=? 3

Enter the following data:
V=? 48

I=? 0.025

Resistance is  1920 ohms


More (YES or NO)? NO

Have a good day


READY
```

FIG. 4.21
Program 4.1.

commands, statements, and so on, found on each line use English words and phrases to indicate the operation to be performed. The REM statement (from the word *REMark*) simply indicates that a descriptive statement is being made about the program. The PRINT statement tells the system to print the characters between the quotes on the screen. The INPUT commands request the value of the variable appearing on the same line. The equations on lines 440, 640, and 840 carry out the operation to be performed in each module defined by the brackets at the left of the program. Lines 400 through 460 will request I and R and calculate the voltage. Line 450 will print the solution, while line 460 will "return" the program to line 210 to determine if a second calculation is to be performed. The module from line 600 to line 660 will calculate the current I, and the module from line 800 to line 860 will determine the resistance from the input voltage and current. Lines 100 through 200 permit a selection of the form of Ohm's law to be applied.

Three runs of the program are provided in the figure to reveal the format of the request for data and the output response.

The above description is by no means an attempt to make the reader an expert in the use of computer systems. Its intent is simply to introduce the format of a program, reveal the readability of a frequently applied language, and demonstrate the ability of a computer to perform calculations in an accurate, rapid, and efficient manner.

PROBLEMS

Section 4.1

1. What is the potential drop across a 6-Ω resistor if the current through it is 2.5 A?

2. What is the current through a 72-Ω resistor if the voltage drop across it is 12 V?

3. How much resistance is required to limit the current to 1.5 mA if the potential drop across the resistor is 6 V?

4. Find the current through a 3.4-MΩ resistance placed across a 125-V source.

5. If the current through a 0.02-MΩ resistor is 3.6 μA, what is the voltage drop across the resistor?

6. If a voltmeter has an internal resistance of 15 kΩ, find the current through the meter when it reads 62 V.

7. If a refrigerator draws 2.2 A at 120 V, what is its resistance?

8. If a clock has an internal resistance of 7.5 kΩ, find the current through the clock if it is plugged into a 120-V outlet.

9. What voltage is required to pass 42 mA through a resistance of 0.04 MΩ?

10. If a soldering iron draws 0.76 A at 120 V, what is its resistance?

11. If an electric heater draws 9.5 A when connected to a 120-V supply, what is the internal resistance of the heater?

12. The input current to a transistor is 20 μA. If the applied (input) voltage is 24 mV, determine the input resistance of the transistor.

13. The internal resistance of a dc generator is 0.5 Ω. Determine the loss in terminal voltage across this internal resistance if the current is 15 A.

Section 4.2

14. Plot the linear curves of a 100-Ω and 0.5-Ω resistor on the graph of Fig. 4.5.

15. Sketch the characteristics of a device that has an internal resistance of 20 Ω up to 10 V and an internal resistance of 2 Ω for higher voltages. Use the axes of Fig. 4.5.

16. Plot the linear curves of a 2-kΩ and 50-kΩ resistor on the graph of Fig. 4.5. Use a horizontal scale that extends from 0 V to 20 V and a vertical axis scaled off in milliamperes.

17. What is the change in voltage across a 2-kΩ resistor established by a change in current of 400 mA through the resistor?

Section 4.3

18. If 420 J of energy are absorbed by a resistor in 7 min, what is the power to the resistor?

19. The power to a device is 40 joules per second (J/s). How long will it take to deliver 640 J?

20. **a.** How many joules of energy does a 2-W nightlight dissipate in 8 h?
    b. How many kilowatthours does it dissipate?

21. A resistor of 10 Ω has charge flowing through it at the rate of 300 coulombs per minute (C/min). How much power is dissipated?

22. How long will a steady current of 2 A have to exist in a resistor that has 3 V across it to dissipate 12 J of energy?

23. What is the power delivered by a 6-V battery if the charge flows at the rate of 48 C/min?

24. The current through a 4-Ω resistor is 7 mA. What is the power delivered to the resistor?

25. The voltage drop across a 3-Ω resistor is 9 mV. What is the power input to the resistor?

26. If the power input to a 4-Ω resistor is 64 W, what is the current through the resistor?

27. A 1/2-W resistor has a resistance of 1000 Ω. What is the maximum current that it can safely handle?

28. A 2.2-kΩ resistor in a stereo system dissipates 42 mW of power. What is the voltage across the resistor?

29. A power supply can deliver 100 mA at 400 V. What is the power rating?

30. What are the resistance and current ratings of a 120-V, 100-W bulb?

31. What are the resistance and voltage ratings of a 450-W automatic washer that draws 3.75 A?

32. A 20-kΩ resistor has a rating of 100 W. What are the maximum current and the maximum voltage that can be applied to the resistor?

33. a. If a home is supplied with a 120-V, 100-A service, find the maximum power capability.
　　b. Can the homeowner safely operate the following loads at the same time?
　　　1. a 5-hp motor
　　　2. a 3000-W clothes dryer
　　　3. a 2400-W electric range
　　　4. a 1000-W steam iron

Section 4.5

34. What is the efficiency of a motor that has an output of 0.5 hp with an input of 450 W?

35. The motor of a power saw is rated 68.5% efficient. If 1.8 hp are required to cut a particular piece of lumber, what is the current drawn from a 120-V supply?

36. What is the efficiency of a dryer motor that delivers 1 hp when the input current and voltage are 4 A and 220 V, respectively?

37. If an electric motor having an efficiency of 87% and operating off a 220-V line delivers 3.6 hp, what input current does the motor draw?

38. A motor is rated to deliver 2 hp.
　　a. If it runs on 110 V and is 90% efficient, how many watts does it draw from the power line?
　　b. What is the input current?
　　c. What is the input current if the motor is only 70% efficient?

39. An electric motor used in an elevator system has an efficiency of 90%. If the input voltage is 220 V, what is the input current when the motor is delivering 15 hp?

40. A 2-hp motor drives a sanding belt. If the efficiency of the motor is 87% and that of the sanding belt 75% due to slippage, what is the overall efficiency of the system?

41. If two systems in cascade each have an efficiency of 80% and the input energy is 60 J, what is the output energy?

42. The overall efficiency of two systems in cascade is 72%. If the efficiency of one is 0.9, what is the efficiency in percent of the other?

43. If the total input and output power of two systems in cascade are 400 W and 128 W, respectively, what is the efficiency of each system if one has twice the efficiency of the other?

44. a. What is the total efficiency of three systems in cascade with efficiencies of 0.98, 0.87, and 0.21?

b. If the system with the least efficiency (0.21) were removed and replaced by one with an efficiency of 0.90, what would be the percent increase in total efficiency?

Section 4.6

45. A 10-Ω resistor is connected across a 15-V battery.

a. How many joules of energy will it dissipate in 1 min?

b. If the resistor is left connected for 2 min instead of 1 min, will the energy used increase? Will the power increase?

46. How much energy in kilowatthours is required to keep a 230-W oil-burner motor running 12 h a week for 5 months?

47. How long can a 1500-W heater be on before using more than 10 kWh of energy?

48. How much does it cost to use a 30-W radio for 3 h at 7¢ per kilowatthour?

49. What is the total cost of using the following at 7¢ per kilowatthour?

a. an 860-W air conditioner for 24 h

b. a 4800-W clothes dryer for 30 min

c. a 400-W washing machine for 1 h

d. a 1200-W dishwasher for 45 min

50. What is the total cost of using the following at 7¢ per kilowatthour?

a. a 110-W stereo set for 4 h

b. a 1200-W projector for 3 h

c. a 60-W tape recorder for 2 h

d. a 150-W color television set for 6 h

Section 4.8

51. Write a program to calculate the cost of using five different appliances for varying lengths of time if the cost is 7¢ per kilowatthour.

52. Request I, R, and t and determine V, P, and W. Print out the results with the proper units.

53. Given a resistance in kilohms, tabulate the voltage and power to the resistor for a range of current extending from 1 mA to 10 mA in increments of 1 mA.

54. Tabulate the time (in hours) and the cost (at 7¢ per kilowatthour) for the use of a particular system for $T = 1$ to N hours in increments of 1 h. N is an input quantity in the range 2–10.

GLOSSARY

Circuit breaker A two-terminal device designed to insure that current levels do not exceed safe levels. If "tripped," it can be reset with a switch or a reset button.

Diode A semiconductor device whose behavior is much like that of a simple switch; that is, it will pass current ideally in only one direction when operating within specified limits.

Efficiency (η) A ratio of output to input power that provides immediate information about the energy-converting characteristics of a system.

Energy (W) A quantity whose change in state is determined by the product of the rate of conversion (P) and the period involved (t). It is measured in joules (J) or wattseconds (Ws).

Fuse A two-terminal device whose sole purpose is to insure that current levels in a circuit do not exceed safe levels.

Horsepower (hp) Equivalent to 746 watts in the electrical system.

Kilowatthour meter An instrument for measuring kilowatt-hours of energy supplied to a residential or commercial user of electricity.

Ohm's law An equation that establishes a relationship among the current, voltage, and resistance of an electrical system.

Power An indication of how much work can be done in a specified amount of time; a *rate* of doing work. It is measured in joules/second (J/s) or watts (W).

Wattmeter An instrument capable of measuring the power delivered to an element by sensing both the voltage across the element and the current through the element.

5

Series Circuits

5.1 INTRODUCTION

Two types of current are readily available to the consumer today. One is
direct current (dc), in which ideally the flow of charge (current) does
not change in magnitude or direction. The other is *sinusoidal alternating current* (ac), in which the flow of charge is continually changing in
magnitude and direction. The next few chapters are an introduction to
circuit analysis purely from a dc approach. The methods and concepts
will be discussed in detail for direct current; and thus, when possible, a
short discussion will suffice to cover any variations we might encounter
when we consider ac in the later chapters.

The battery of Fig. 5.1, by virtue of the potential difference between
its terminals, has the ability to cause (or ''pressure'') charge to flow

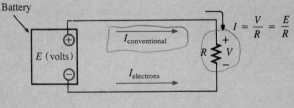

FIG. 5.1

through the simple circuit. The positive terminal attracts the electrons
through the wire at the same rate at which electrons are supplied by the
negative terminal. As long as the battery is connected in the circuit and

maintains its terminal characteristics, the current (dc) through the circuit will not change in magnitude or direction.

If we consider the wire to be an ideal conductor (that is, having no opposition to flow), the potential difference V across the resistor will equal the applied voltage of the battery: V (volts) = E (volts).

The current is limited only by the resistor R. The higher the resistance, the less the current, and conversely, as determined by Ohm's law.

By convention as discussed in Chapter 2, the direction of I as shown in Fig. 5.1 is opposite to that of electron flow. Also, the uniform flow of charge dictates that the direct current I be the same everywhere in the circuit. By following the direction of conventional flow, we notice that there is a rise in potential across the battery ($-$ to $+$), and a drop in potential across the resistor ($+$ to $-$). For single-voltage-source dc circuits, conventional flow always passes from a low potential to a high potential when passing through a voltage source, as shown in Fig. 5.2. However, conventional flow always passes from a high to a low potential when passing through a resistor for any number of voltage sources in the same circuit, as shown in Fig. 5.3.

The circuit of Fig. 5.1 is the simplest possible configuration. This chapter and the chapters to follow will add elements to the system in a very specific manner to introduce a range of concepts that will form a major part of the foundation required to analyze the most complex system. Be aware that the laws, rules, and so on, introduced in Chapters 5 and 6 will be used throughout your studies of electrical, electronic, or computer systems. They will not be dropped for a more advanced set as you progress to more sophisticated material. It is therefore critical that the concepts be understood thoroughly and that the various procedures and methods be applied with confidence.

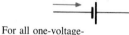

For all one-voltage-source dc circuits

FIG. 5.2

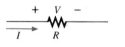

For any combination of voltage sources in the same dc circuit

FIG. 5.3

5.2 SERIES CIRCUITS

A *circuit* consists of any number of elements joined at terminal points, providing at least one closed path through which charge can flow. The circuit of Fig. 5.4(a) has three elements joined at three terminal points (a, b, and c) to provide a closed path for the current I.

Two elements are in series if they have only one point in common that is not connected to other current-carrying elements of the network.

In Fig. 5.4(a), the resistors R_1 and R_2 are in series because they have *only* point b in common. The other ends of the resistors are connected elsewhere in the circuit. For the same reason, the battery E and resistor R_1 are in series (terminal a in common) and the resistor R_2 and the battery E are in series (terminal c in common). Since all the elements are in series, the network is called a *series circuit*. Some common examples of series connections include the tying of small pieces of rope together to form a longer rope, the connecting of pipes to get water from one point to another, and the joining of hands of a group in a circle.

In a series circuit, the current is the same through each series element.

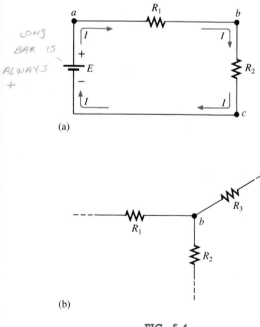

(a)

LONG BAR IS ALWAYS +

(b)

FIG. 5.4

For the circuit of Fig. 5.4(a), therefore, the current I through each resistor is the same as that through the battery.

If a third element, R_3, is added at terminal b as shown in Fig. 5.4(b), the resistors R_1 and R_2 are no longer in series because an additional element has been connected between the common terminal and some other point in the network.

A *branch* of a circuit is any portion of the circuit that has one or more elements in series. In Fig. 5.4(a), the resistor R_1 forms one branch of the circuit, the resistor R_2 another, and the battery E a third.

The total resistance of the circuit is determined by simply adding the values of the various resistors. In Fig. 5.4(a), for example, the total resistance (R_T) is equal to $R_1 + R_2$. Note that the total resistance is actually the resistance "seen" by the battery as it "looks" into the series combination of elements.

In general, to find the total resistance of N resistors in series, the following equation is applied:

$$R_T = R_1 + R_2 + R_3 + \cdots + R_N \qquad \textbf{(5.1)}$$

Once the total resistance is known, the current can be determined from

$$I = \frac{E}{R_T} \qquad \textbf{(5.2)}$$

and the voltage across each element can be determined from

$$V_1 = IR_1,\ V_2 = IR_2,\ V_3 = IR_3,\ \cdots,\ V_N = IR_N \qquad \textbf{(5.3)}$$

EXAMPLE 5.1.
a. Find the total resistance for the series circuit of Fig. 5.5.
b. Calculate the current I.
c. Determine the voltages V_1, V_2, and V_3.
Solutions:
a. $R_T = R_1 + R_2 + R_3 = 2 + 1 + 5 = \mathbf{8\,\Omega}$

b. $I = \dfrac{E}{R_T} = \dfrac{20}{8} = \mathbf{2.5\,A}$

c. $V_1 = IR_1 = (2.5)(2) = \mathbf{5\,V}$
$V_2 = IR_2 = (2.5)(1) = \mathbf{2.5\,V}$
$V_3 = IR_3 = (2.5)(5) = \mathbf{12.5\,V}$

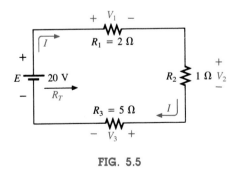

FIG. 5.5

To find the total resistance of N resistors of the same value in series, simply multiply the value of *one* of the resistors by the number in series. That is,

$$\boxed{R_T = NR} \tag{5.4}$$

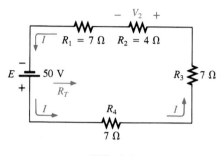

FIG. 5.6

EXAMPLE 5.2. Determine R_T, I, and V_2 for the circuit of Fig. 5.6.
Solution: Note the current direction as established by the battery and the polarity of the voltage drops across R_2 as determined by the current direction.
Since $R_1 = R_3 = R_4$,

$$R_T = NR_1 + R_2 = (3)(7) + 4 = 21 + 4 = \mathbf{25\,\Omega}$$

$$I = \frac{E}{R_T} = \frac{50}{25} = \mathbf{2\,A}$$

$$V_2 = IR_2 = (2)(4) = \mathbf{8\,V}$$

Examples 5.1 and 5.2 are straightforward substitution-type problems that are relatively easy to solve with some practice. Example 5.3, however, is evidence of another type of problem which requires a firm grasp of the fundamental equations and an ability to identify which equation to use first. The best preparation for this type of exercise is simply to work through as many problems of this kind as possible.

EXAMPLE 5.3. Given R_T and I, calculate R_1 and E for the circuit of Fig. 5.7.
Solution:

$$R_T = R_1 + R_2 + R_3$$
$$12\,\text{k}\Omega = R_1 + 4\,\text{k}\Omega + 6\,\text{k}\Omega$$
$$R_1 = 12\,\text{k}\Omega - 10\,\text{k}\Omega = \mathbf{2\,k\Omega}$$
$$E = IR_T = (6 \times 10^{-3})(12 \times 10^3) = \mathbf{72\,V}$$

FIG. 5.7

5.3 VOLTAGE SOURCES IN SERIES

Voltage sources can be connected in series as shown in Fig. 5.8 to increase or decrease the total voltage applied to a system. The net voltage is determined simply by summing the sources with the same polarity and subtracting the total of the sources with the opposite "pressure." The net polarity is the polarity of the larger sum.

In Fig. 5.8(a), for example, the sources are all "pressuring" current to the right, so the net voltage is

$$E_T = E_1 + E_2 + E_3 = 10 + 6 + 2 = 18\,V$$

as shown in the figure. In Fig. 5.8(b), however, the greater "pressure" is to the left, with a net voltage of

$$E_T = E_2 + E_3 - E_1 = 9 + 3 - 4 = 8\,V$$

and the polarity shown in the figure.

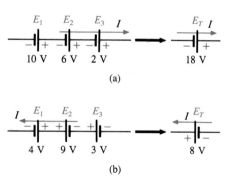

FIG. 5.8

5.4 KIRCHHOFF'S VOLTAGE LAW

Kirchhoff's voltage law (KVL) states that *the algebraic sum of the potential rises and drops around a closed loop (or path) is zero.*

A *closed loop* is any continuous connection of branches that allows us to trace a path that leaves a point in one direction and returns to that same point from another direction without leaving the circuit. In Fig. 5.9, by following the current, we can trace a continuous path that leaves point a through R_1 and returns through E without leaving the circuit. Therefore, *abca* is a closed loop. In order for us to be able to apply Kirchhoff's voltage law, the summation of potential rises and drops must be made in one direction around the closed loop.

For uniformity, the clockwise (CW) direction will be used throughout the text for all applications of Kirchhoff's voltage law. Be aware, however, that the same result will be obtained if the counterclockwise (CCW) direction is chosen and the law applied correctly.

A plus sign is assigned to a potential rise ($-$ to $+$) and a minus sign to a potential drop ($+$ to $-$). If we follow the current in Fig. 5.9 from point a, we first encounter a potential drop V_1 ($+$ to $-$) across R_1 and then another potential drop V_2 across R_2. Continuing through the voltage source, we have a potential rise E ($-$ to $+$) before returning to point a. In symbolic form, where Σ represents summation, $\circlearrowright$ the closed loop, and V the potential drops and rises, we have

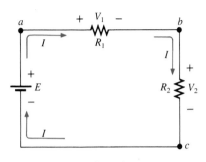

FIG. 5.9

$$\boxed{\Sigma_{\circlearrowright} V = 0}\qquad \text{(Kirchhoff's voltage law in symbolic form)}\qquad \textbf{(5.5)}$$

which for the circuit of Fig. 5.9 yields (clockwise direction, following the current I)

$$+E - V_1 - V_2 = 0$$

or

$$E = V_1 + V_2$$

revealing that the potential impressed on the circuit by the battery is equal to the potential drops within the circuit. To go a step further, the sum of the potential rises will equal the sum of the potential drops around a closed path. Therefore, another way of stating Kirchhoff's voltage law is the following:

$$\boxed{\Sigma_{\circlearrowright} V_{\text{rises}} = \Sigma_{\circlearrowright} V_{\text{drops}}}\qquad \textbf{(5.6)}$$

The text will emphasize the use of Eq. (5.5), however.

If the loop were taken in the counterclockwise direction, the following would result:

$$\Sigma_{\circlearrowright} V = 0$$
$$-E + V_2 + V_1 = 0$$

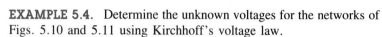

LOOK AT
THESE EXAMPLES

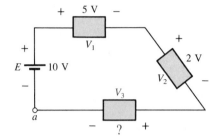

FIG. 5.10

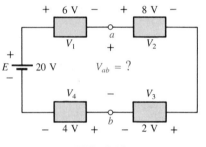

FIG. 5.11

or, as before,

$$E = V_1 + V_2$$

EXAMPLE 5.4. Determine the unknown voltages for the networks of Figs. 5.10 and 5.11 using Kirchhoff's voltage law.

Solution: In each case, note that we do not need to know whether there is a voltage source or dissipative element in each container. Once the magnitude and polarity of the voltage for each element are known, Kirchhoff's voltage law can be applied.

For Fig. 5.10, applying Kirchhoff's voltage law in the clockwise direction starting at point *a* will result in

$$E - V_1 - V_2 - V_3 = \,?$$

and

$$V_3 = E - V_1 - V_2 = 10 - 5 - 2 = 10 - 7 = \mathbf{3\,V}$$

In the case of Fig. 5.11, the voltage to be determined is not across a single element but between two points in the network. There are two routes to follow toward a solution.

First, Kirchhoff's voltage law can be applied in the clockwise direction around a closed loop including the voltage source E. That is,

$$+E - V_1 - V_{ab} - V_4 = 0$$

and

$$V_{ab} = E - V_1 - V_4 = 20 - 6 - 4 = 20 - 10 = \mathbf{10\,V}$$

The other alternative is to apply the law in a clockwise direction around a closed path that includes only V_2 and V_3. That is,

$$+V_{ab} - V_2 - V_3 = 0$$

and

$$V_{ab} = V_2 + V_3 = 8 + 2 = \mathbf{10\,V}$$

Note that both routes will result in the same solution.

EXAMPLE 5.5. Find V_1 and V_2 for the network of Fig. 5.12.

Solution: For path 1,

$$+25 - V_1 + 15 = 0$$

and

$$V_1 = \mathbf{40\,V}$$

For path 2,

$$-20 - V_2 = 0$$

and

$$V_2 = \mathbf{-20\,V}$$

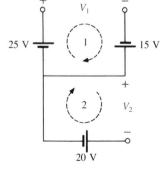

FIG. 5.12

The minus sign simply indicates that the actual polarities of the potential difference are opposite the assumed polarity indicated in Fig. 5.12.

Note in Fig. 5.12 that a potential difference can exist between two points not connected by an element. In other words, a potential difference can exist between two points even though there may not be a current or connecting element between the two points.

EXAMPLE 5.6. For the circuit of Fig. 5.13:
a. Find R_T.
b. Find I.
c. Find V_1 and V_2.
d. Find the power to the 4-Ω and 6-Ω resistors.
e. Find the power delivered by the battery, and compare it to that dissipated by the 4-Ω and 6-Ω resistors combined.
f. Verify Kirchhoff's voltage law (clockwise direction).

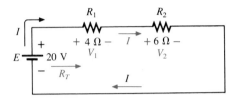

FIG. 5.13

Solutions:
a. $R_T = R_1 + R_2 = 4 + 6 = \mathbf{10\ \Omega}$

b. $I = \dfrac{E}{R_T} = \dfrac{20}{10} = \mathbf{2\ A}$

c. $V_1 = IR_1 = (2)(4) = \mathbf{8\ V}$
   $V_2 = IR_2 = (2)(6) = \mathbf{12\ V}$

d. $P_{4\Omega} = \dfrac{V_1^2}{R_1} = \dfrac{(8)^2}{4} = \dfrac{64}{4} = \mathbf{16\ W}$

   $P_{6\Omega} = I^2 R_2 = (2)^2(6) = (4)(6) = \mathbf{24\ W}$

e. $P_E = EI = (20)(2) = \mathbf{40\ W}$
   $P_E = P_{4\Omega} + P_{6\Omega}$
   $40 = 16 + 24$
   $40 = 40$ (checks)

f. $\Sigma_\circlearrowright V = +E - V_1 - V_2 = 0$
   $E = V_1 + V_2$
   $20 = 8 + 12$
   $20 = 20$ (checks)

[handwritten: $25 - V_1 + 15 - V_2 + 20 = 0$]

[handwritten: $V_1 = 40v$]

EXAMPLE 5.7. For the circuit of Fig. 5.14:
a. Determine V_2 using Kirchhoff's voltage law.
b. Determine I.
c. Find R_1 and R_3.

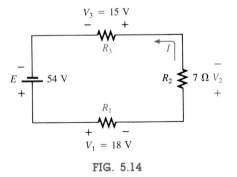

FIG. 5.14

Solutions:
a. Kirchhoff's voltage law (clockwise direction):

$$-E + V_3 + V_2 + V_1 = 0$$

or

$$E = V_1 + V_2 + V_3$$

and

$$V_2 = E - V_1 - V_3 = 54 - 18 - 15 = \mathbf{21\ V}$$

b. $I = \dfrac{V_2}{R_2} = \dfrac{21}{7} = \mathbf{3\ A}$

c. $R_1 = \dfrac{V_1}{I} = \dfrac{18}{3} = \mathbf{6\ \Omega}$

$R_3 = \dfrac{V_3}{I} = \dfrac{15}{3} = \mathbf{5\ \Omega}$

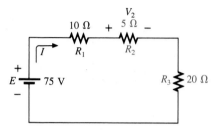

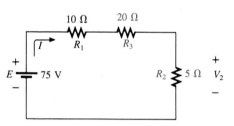

FIG. 5.15

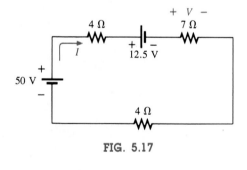

FIG. 5.16

5.5 INTERCHANGING SERIES ELEMENTS

The elements of a series circuit can be interchanged without affecting the total resistance, current, or power to each element. For instance, the network of Fig. 5.15 can be redrawn as shown in Fig. 5.16 without affecting I or V_2. Total resistance R_T is 25 Ω in both cases, and $I = 75/25 = 3$ A. The voltage $V_2 = IR_2 = (3)(5) = 15$ V for both configurations.

EXAMPLE 5.8. Determine I and the voltage across the 7-Ω resistor for the network of Fig. 5.17.

Solution: Network redrawn in Fig. 5.18:

$$R_T = (2)(4) + 7 = 15\ \Omega$$

$$I = \frac{E}{R_T} = \frac{37.5}{15} = \mathbf{2.5\ A}$$

$$V_{7\Omega} = IR = (2.5)(7) = \mathbf{17.5\ V}$$

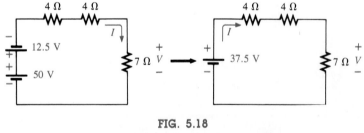

FIG. 5.17

FIG. 5.18

5.6 VOLTAGE DIVIDER RULE

The voltage across a single element or a combination of elements in a series circuit can be determined without first finding the current if the voltage divider rule is applied. The rule can be derived by analyzing the network of Fig. 5.19.

The total resistance is

$$R_T = R_1 + R_2$$

resulting in

$$I = \frac{E}{R_T}$$

and

$$V_1 = IR_1 = \left(\frac{E}{R_T}\right)R_1 = \frac{R_1 E}{R_T}$$

with

$$V_2 = IR_2 = \left(\frac{E}{R_T}\right)R_2 = \frac{R_2 E}{R_T}$$

Note that the format for V_1 and V_2 is

$$\boxed{V_x = \frac{R_x E}{R_T}} \qquad \text{(voltage divider rule)} \qquad \textbf{(5.7)}$$

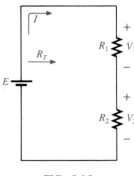

FIG. 5.19

where V_x is the voltage across R_x, E is the impressed voltage across the series elements, and R_T is the total resistance of the series circuit.

In words, the *voltage divider rule* states that *the voltage across a resistor in a series circuit is equal to the value of that resistor times the total impressed voltage across the series elements divided by the total resistance of the series elements.*

EXAMPLE 5.9. Determine the voltage V_1 for the network of Fig. 5.20.

Solution: Eq. (5.7):

$$V_1 = \frac{R_1 E}{R_T} = \frac{R_1 E}{R_1 + R_2} = \frac{(20)(64)}{20 + 60} = \frac{1280}{80} = \textbf{16 V}$$

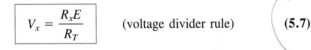

FIG. 5.20

EXAMPLE 5.10. Using the voltage divider rule, determine the voltages V_1 and V_3 for the series circuit of Fig. 5.21.

Solution:

$$V_1 = \frac{R_1 E}{R_T} = \frac{(2\text{ k}\Omega)(45)}{2\text{ k}\Omega + 5\text{ k}\Omega + 8\text{ k}\Omega} = \frac{(2\text{ k}\Omega)(45)}{15}$$

$$= \frac{(2 \times 10^3)(45)}{15 \times 10^3} = \frac{90}{15} = \textbf{6 V}$$

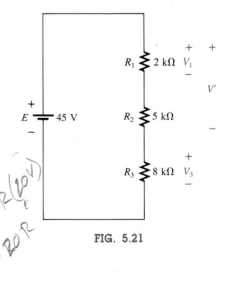

FIG. 5.21

$$V_3 = \frac{R_2 E}{R_T} = \frac{(8\,k\Omega)(45)}{15\,k\Omega} = \frac{(8 \times 10^3)(45)}{15 \times 10^3}$$

$$= \frac{360}{15} = 24\,V$$

The rule can be extended to the voltage across two or more series elements if the resistance in the numerator of Eq. (5.7) is expanded to include the total resistance of the series elements that the voltage is to be found across (R'). That is,

$$\boxed{V' = \frac{R'E}{R_T}} \qquad \text{(volts)} \qquad \textbf{(5.8)}$$

EXAMPLE 5.11.   Determine the voltage V' in Fig. 5.21 across resistors R_1 and R_2.

Solution:

$$V' = \frac{R'E}{R_T} = \frac{(2\,k\Omega + 5\,k\Omega)(45)}{15\,k\Omega} = \frac{(7)(45)}{15} = 21\,V$$

There is also no need for the voltage E in the equation to be the source voltage of the network. For example, if V is the total voltage across a number of series elements such as shown in Fig. 5.22, then

$$V_{2\Omega} = \frac{(2)(27)}{4 + 2 + 3} = \frac{54}{9} = 6\,V$$

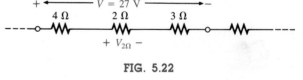

FIG. 5.22

An interesting conclusion can be derived by writing Eq. (5.7) in the following manner:

$$\boxed{\frac{V_x}{E} = \frac{R_x}{R_T}} \qquad \textbf{(5.9)}$$

In words, the ratio of the voltage across a particular element of a series circuit to the total voltage is the same as the ratio of the resistance across which the voltage is to be determined to the total resistance of the circuit. For example, if a resistor is 40% of the total resistance, then 40% of the total voltage will appear across it. In Example 5.9, R_1 is 1/3 of R_2, resulting in $V_1 = (1/3)(E) = (1/3)(64) = 16$ V, as obtained earlier.

EXAMPLE 5.12. Design the voltage divider of Fig. 5.23 such that $V_{R_1} = 4V_{R_2}$.
Solution: The total resistance is defined by

$$R_T = \frac{E}{I} = \frac{20 \text{ V}}{4 \text{ mA}} = 5 \text{ k}\Omega$$

Since $V_{R_1} = 4V_{R_2}$,

$$R_1 = 4R_2$$

Thus,

$$R_T = R_1 + R_2 = 4R_2 + R_2 = 5R_2$$

and

$$5R_2 = 5 \text{ k}\Omega$$
$$R_2 = \mathbf{1 \text{ k}\Omega}$$

and

$$R_1 = 4R_2 = \mathbf{4 \text{ k}\Omega}$$

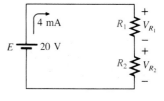

FIG. 5.23

5.7 NOTATION

Notation will play an increasingly important role in the analysis to follow. It is important, therefore, that we begin to consider some of the notation used throughout the industry.

Except for a few special cases, electrical and electronic systems are grounded for reference and safety purposes. The symbol for the ground connection appears in Fig. 5.24 with its defined potential level—zero volts. None of the circuits discussed thus far have contained the ground connection. If Fig. 5.4(a) were redrawn with a grounded supply, it might appear as shown in Fig. 5.25(a) or (b). In either case, it is understood that the negative terminal of the battery and the bottom of the resistor R_2 are at ground potential. Although Fig. 5.25(b) shows no connection between the two grounds, it is recognized that such a connection exists for the continuous flow of charge. If $E = 12$ V, then point a is 12 V positive with respect to ground potential and 12 V exist across the series combination of resistors R_1 and R_2. If a voltmeter placed from point b to ground reads 4 V, then the voltage across R_2 is 4 V, with the higher potential at point b.

FIG. 5.24
Ground potential.

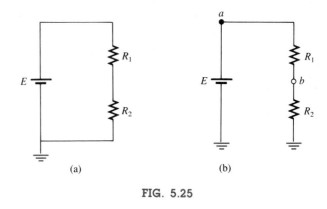

FIG. 5.25

On large schematics where space is at a premium and clarity is important, voltage sources may be indicated as shown in Figs. 5.26(a) and 5.27(a) rather than as illustrated in Figs. 5.26(b) and 5.27(b). In addi-

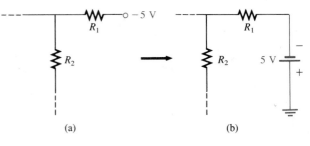

FIG. 5.26

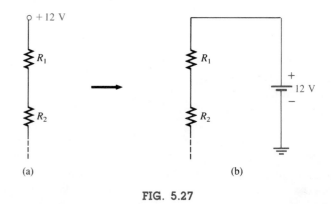

FIG. 5.27

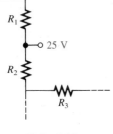

FIG. 5.28

tion, potential levels may be indicated as in Fig. 5.28, to permit a rapid check of the potential levels at various points in a network with respect to ground to insure the system is operating properly.

The fact that voltage is an *across* variable and exists between two points has resulted in a double-subscript notation that defines the first subscript as the higher potential. The notation V_{ab} for the potential difference appearing in Fig. 5.29 specifies that the potential at *a* is higher than the potential at *b* by 6 V. In fact, $V_{ab} = V_a - V_b = 6$ V. For a situation such as that shown in Fig. 5.30 where *a* is at a lesser potential than *b*, a negative sign is employed to reveal the polarity difference.

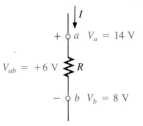

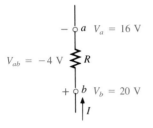

FIG. 5.30
Double-subscript notation ($-$ value, $V_b > V_a$).

FIG. 5.29
Double-subscript notation ($+$ value, $V_a > V_b$).

If a single-subscript notation such as V_a is used, it is understood that V_a is the potential from point *a* to ground, with *a* defined as the higher potential. Two applications of the single-subscript notation appear in Fig. 5.31 with the proper sign for the indicated voltage.

EXAMPLE 5.13. Using the voltage divider rule, determine the voltages V_1 and V_2 of Fig. 5.32.

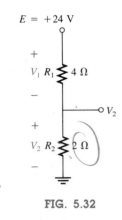

FIG. 5.32

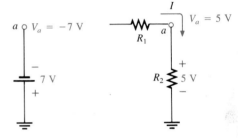

FIG. 5.31
Single-subscript notation.

Solution: Redrawing the network with the standard battery symbol will result in the network of Fig. 5.33. Applying the voltage divider rule,

$$V_1 = \frac{R_1 E}{R_1 + R_2} = \frac{(4)(24)}{4 + 2} = \mathbf{16\ V}$$

$$V_2 = \frac{R_2 E}{R_1 + R_2} = \frac{(2)(24)}{4 + 2} = \mathbf{8\ V}$$

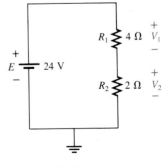

FIG. 5.33

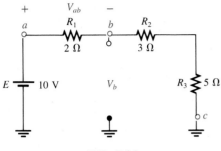

FIG. 5.34

EXAMPLE 5.14. For the network of Fig. 5.34:
a. Calculate V_{ab}.
b. Determine V_b.
c. Calculate V_c.

Solutions:
a. Voltage divider rule:

$$V_{ab} = \frac{R_1 E}{R_T} = \frac{(2)(10)}{2 + 3 + 5} = +2\,V$$

b. Voltage divider rule:

$$V_b = V_{R_2} + V_{R_3} = \frac{(R_2 + R_3)E}{R_T} = \frac{(3 + 5)(10)}{10} = 8\,V$$

or

$$V_b = E - V_{ab} = 10 - 2 = 8\,V$$

c. V_c = ground potential = **0 V**

EXAMPLE 5.15. Determine V_{ab}, V_{cb}, and V_b for the network of Fig. 5.35.

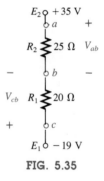

FIG. 5.35

Solution: Network redrawn in Fig. 5.36:

$$I = \frac{E_1 + E_2}{R_T} = \frac{E_1 + E_2}{R_1 + R_2} = \frac{19 + 35}{45} = \frac{54}{45} = 1.2\,A$$

$$V_{ab} = IR_2 = (1.2)(25) = 30\,V$$

$$V_{cb} = -IR_1 = -(1.2)(20) = -24\,V$$

$$V_c = -E_1 = -19\,V$$

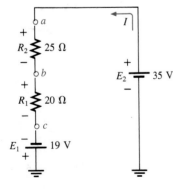

FIG. 5.36

5.8 INTERNAL RESISTANCE OF VOLTAGE SOURCES

Every source of voltage, whether it be a generator, battery, or laboratory supply as shown in Fig. 5.37(a), will have some internal resistance.

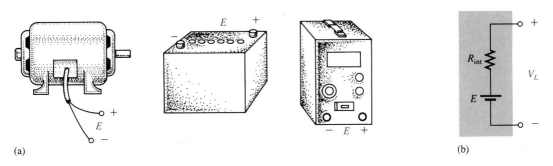

(a)

(b)

FIG. 5.37

The equivalent circuit of any source of voltage will therefore appear as shown in Fig. 5.37(b). In this section, we will examine the effect of the internal resistance on the output voltage so that any unexpected changes in terminal characteristics can be explained.

In all the circuit analyses to this point, the ideal voltage source (no internal resistance) was used [see Fig. 5.38(a)]. The ideal voltage source has no internal resistance and an output voltage of E volts with no load or full load. In the practical case [Fig. 5.38(b)], where we consider the effects of the internal resistance, the output voltage will be E volts only when no-load ($I_L = 0$) conditions exist. When a load is connected [Fig. 5.38(c)], the output voltage of the voltage source will decrease due to the voltage drop across the internal resistance.

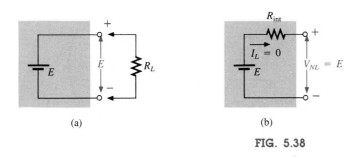

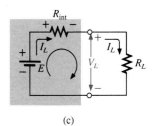

(a)

(b)

(c)

FIG. 5.38

By applying Kirchhoff's voltage law around the indicated loop of Fig. 5.38(c), we obtain

$$E - I_L R_{int} - V_L = 0$$

or, since

$$E = V_{NL}$$

we have

$$V_{NL} - I_L R_{int} - V_L = 0$$

and

$$\boxed{V_L = V_{NL} - I_L R_{int}}$$ **(5.10)**

If the value of R_{int} is not available, it can be found by first solving for R_{int} in the equation just derived for V_L. That is,

$$R_{int} = \frac{V_{NL} - V_L}{I_L} = \frac{V_{NL}}{I_L} - \frac{I_L R_L}{I}$$

and

$$\boxed{R_{int} = \frac{V_{NL}}{I_L} - R_L} \qquad\qquad (5.11)$$

A plot of the output voltage versus current appears in Fig. 5.39(b) for the circuit in Fig. 5.39(a). Note that an increase in load demand (I_L)

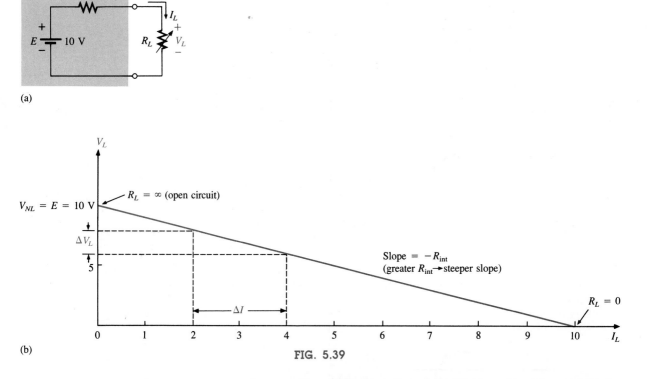

(a)

(b)

FIG. 5.39

increases the current through, and thereby the voltage drop across, the internal resistance of the source, resulting in a decrease in terminal voltage. Eventually, as the load resistance approaches zero ohms, all the generated voltage will appear across the internal resistance and none at the output terminals. The steeper the slope of the curve of Fig. 5.39(b), the greater the internal resistance. In fact, for any chosen interval of voltage or current, the magnitude of the internal resistance is given by

$$R_{\text{int}} = \frac{\Delta V_L}{\Delta I_L}$$ **(5.12)**

where Δ signifies finite change.

For the chosen interval of $2 \rightarrow 4$ A ($\Delta I_L = 2$ A) on Fig. 5.39(b), ΔV_L is 2 V, and so $R_{\text{int}} = 2/2 = 1\,\Omega$, as indicated in Fig. 5.39(a).

A direct consequence of the loss in output voltage is a loss in power delivered to the load. Multiplying both sides of Eq. (5.10) by the current I_L in the circuit, we obtain

$$\underset{\substack{\text{Power} \\ \text{to load}}}{I_L V_L} = \underset{\substack{\text{Power output} \\ \text{by battery}}}{I_L V_{NL}} - \underset{\substack{\text{Power loss in} \\ \text{the form of heat}}}{I_L^2 R_{\text{int}}}$$ **(5.13)**

EXAMPLE 5.16. Before a load is applied, the terminal voltage of the power supply of Fig. 5.40 is set to 40 V. When a load of 500 Ω is attached, as shown in Fig. 5.40(b), the terminal voltage drops to 36 V. What happened to the remainder of the no-load voltage, and what is the internal resistance of the source?

Solution: The difference of $40 - 36 = 4$ V now appears across the internal resistance of the source. The load current is $36/0.5\,\text{k}\Omega = 72\,\text{mA}$. Applying Eq. (5.11),

$$R_{\text{int}} = \frac{V_{NL}}{I_L} - R_L = \frac{40}{72\,\text{mA}} - 0.5\,\text{k}\Omega$$

$$= 555.55 - 500 = \mathbf{55.55\,\Omega}$$

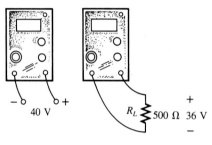

$-\circ$ $\circ+$
40 V

$R_L \lessgtr$ 500 Ω 36 V
$+$
$-$

FIG. 5.40

EXAMPLE 5.17. The battery of Fig. 5.41 has an internal resistance of 2 Ω. Find the voltage V_L and the power lost to the internal resistance if the applied load is a 13-Ω resistor.

Solution:

$$I_L = \frac{30}{2 + 13} = \frac{30}{15} = 2\,\text{A}$$

$$V_L = V_{NL} - I_L R_{\text{int}} = 30 - (2)(2) = \mathbf{26\,V}$$

$$P_{\text{lost}} = I_L^2 R_{\text{int}} = (2)^2(2) = (4)(2) = \mathbf{8\,W}$$

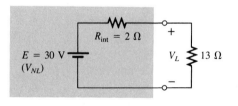

$R_{\text{int}} = 2\,\Omega$
$E = 30$ V
(V_{NL})
$+$
$V_L \lessgtr$ 13 Ω
$-$

FIG. 5.41

EXAMPLE 5.18. The terminal characteristics of a dc generator appear in Fig. 5.42. Rated (full-load) conditions are indicated at 120 V, 8 A.
a. Calculate the average internal resistance of the supply.

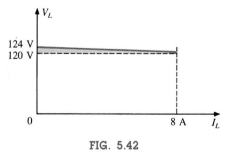

V_L

124 V
120 V

0   8 A I_L

FIG. 5.42

b. At what load current will the terminal voltage drop to 100 V?

Solutions:

a. $R_{\text{int}} = \dfrac{\Delta V_L}{\Delta I_L} = \dfrac{4}{8} = \mathbf{0.5\ \Omega}$

b. From Eq. (5.12),

$$\Delta I_L = \frac{\Delta V_L}{R_{\text{int}}}$$

Since

$$\Delta V_L = 124 - 100 = 24\ \text{V}$$

then

$$\Delta I_L = \frac{\Delta V_L}{R_{\text{int}}} = \frac{24}{0.5} = 48\ \text{A} \quad (\text{from } 0\ \text{A})$$

so that

$$I_L = \mathbf{48\ A}$$

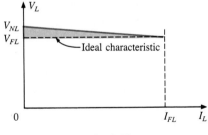

V_{NL}
V_{FL}
—Ideal characteristic

0 I_{FL} I_L

FIG. 5.43

5.9 VOLTAGE REGULATION

For any supply, ideal conditions dictate that for the range of load demand (I_L), the terminal voltage remain fixed in magnitude. In other words, if a supply is set for 12 V, it is desirable that it maintain this terminal voltage even though the current demand on the supply may vary. A measure of how close a supply will come to ideal conditions is given by the voltage regulation characteristic. By definition, the voltage regulation of a supply between the limits of full-load and no-load conditions (Fig. 5.43) is given by the following:

$$\boxed{\text{Voltage regulation } (VR)\% = \frac{V_{NL} - V_{FL}}{V_{FL}} \times 100\%} \quad \textbf{(5.14)}$$

For ideal conditions, $V_{FL} = V_{NL}$ and $VR\% = 0$. Therefore, *the smaller the voltage regulation, the less the variation in terminal voltage with change in load.*

It can be shown with a short derivation that the voltage regulation is also given by

$$\boxed{VR\% = \frac{R_{\text{int}}}{R_L} \times 100\%} \quad \textbf{(5.15)}$$

In other words, the smaller the internal resistance for the same load, the smaller the regulation and more ideal the output.

EXAMPLE 5.19. Calculate the voltage regulation of a supply having the characteristics of Fig. 5.42.

Solution:

$$VR\% = \frac{V_{NL} - V_{FL}}{V_{FL}} \times 100\% = \frac{124 - 120}{120} \times 100\%$$

$$= \frac{4}{120} \times 100\% = \mathbf{3.33\%}$$

EXAMPLE 5.20. Determine the voltage regulation of the supply of Fig. 5.41.

Solution:

$$VR\% = \frac{R_{\text{int}}}{R_L} \times 100\% = \frac{2}{13} \times 100\% \cong \mathbf{15.38\%}$$

5.10 AMMETERS: LOADING EFFECTS

In Chapter 2, it was noted that ammeters are inserted in the branch in which the current is to be measured. We now realize that such a condition specifies that *ammeters be placed in series with the branch in which the current is to be measured,* as shown in Fig. 5.44.

If the ammeter is to have minimal impact on the behavior of the network, its resistance should be very small (ideally zero ohms) compared to the other series elements of the branch such as the resistor R of Fig. 5.44. If the meter resistance approaches or exceeds 10% of R, it would naturally have a significant impact on the current level it is measuring. It is also noteworthy that the resistances of the separate current scales of the same meter are usually not the same. In fact, the meter resistance normally increases with decreasing current levels. However, for the majority of situations one can simply assume that the internal ammeter resistance is small enough compared to the other circuit elements that it can be ignored.

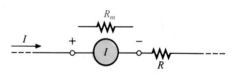

FIG. 5.44
Series connection of an ammeter.

5.11 COMPUTER SOLUTIONS

The true value of computer methods can be clearly demonstrated by Program 5.1 of Fig. 5.46, which calculates every quantity of significance for the network of Fig. 5.45 once the battery voltage and resistance values have been entered.

Note that R_T is calculated on line 150, I on line 180, the voltage across each resistor on line 230, and the power to each resistor on line 250. In addition, Kirchhoff's voltage law is applied on line 300 to show

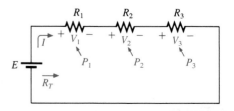

FIG. 5.45
Network for Program 5.1.

that the applied voltage equals the sum of the voltage drops. The total power supplied or dissipated is then calculated on line 310.

In the second run of the program, only two resistors were entered, with R_3 given a resistance of zero ohms (short circuit). Note that V_3 and P_3 are zero volts and zero watts, respectively.

```
        10 REM ***** PROGRAM 5-1 *****
        20 REM ****************************************************
        30 REM Analysis of a series resistor network
        40 REM ****************************************************
        50 REM
       100 PRINT:PRINT "Enter resistor values for up to 3 resistors"
       110 PRINT "in series (enter 0 if no resistor):"
       120 INPUT "R1=";R1
       130 INPUT "R2=";R2
       140 INPUT "R3=";R3
   R_T 150 RT=R1+R2+R3
       160 PRINT:PRINT "The total resistance is RT=";RT;"ohms"
       170 PRINT:INPUT "Enter value of supply voltage, E=";E
    I  180 I=E/RT
       190 PRINT
       200 PRINT "Supply current is, I=";I;"amperes"
       210 PRINT
       220 PRINT "The voltage drop across each resistor is:"
   V_x 230 V1=I*R1:V2=I*R2 :V3=I*R3
       240 PRINT "V1=";V1;"volts    V2=";V2;"volts    V3=";V3;"volts"
   P_x 250 P1=I^2*R1 :P2=I^2*R2 :P3=I^2*R3
       260 PRINT
       270 PRINT "The power dissipated by each resistor is:"
       280 PRINT "P1=";P1;"watts","P2=";P2;"watts","P3=";P3;"watts"
       290 PRINT
       300 PRINT "Total voltage around loop is, V1+V2+V3=";V1+V2+V3;"volts"
       310 PRINT "and total power dissipated, P1+P2+P3=";P1+P2+P3;"watts"
       320 END


READY

RUN


Enter resistor values for up to 3 resistors
in series (enter 0 if no resistor):
R1=? 6

R2=? 7

R3=? 5


The total resistance is RT= 18 ohms

Enter value of supply voltage, E=? 54


Supply current is, I= 3 amperes

The voltage drop across each resistor is:
V1= 18 volts    V2= 21 volts    V3= 15 volts

The power dissipated by each resistor is:
P1= 54 watts     P2= 63 watts     P3= 45 watts

Total voltage around loop is, V1+V2+V3= 54 volts
and total power dissipated, P1+P2+P3= 162 watts
```

```
READY

RUN


Enter resistor values for up to 3 resistors
in series (enter 0 if no resistor):
R1=? 1E3

R2=? 4E3

R3=? 0


The total resistance is RT= 5000 ohms

Enter value of supply voltage, E=? 50


Supply current is, I= .01 amperes

The voltage drop across each resistor is:
V1= 10 volts    V2= 40 volts    V3= 0 volts

The power dissipated by each resistor is:
P1= .1 watts     P2= .4 watts     P3= 0 watts

Total voltage around loop is, V1+V2+V3= 50 volts
and total power dissipated, P1+P2+P3= .5 watts


READY
```

FIG. 5.46

Program 5.1.

PROBLEMS

Section 5.2

1. Find the total resistance and current I for each circuit of
Fig. 5.47.

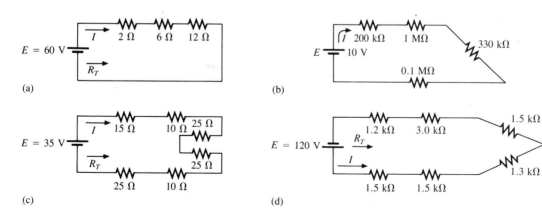

(a)

(b)

(c)

(d)

FIG. 5.47

2. For the circuits of Fig. 5.48, the total resistance is specified. Find the unknown resistances and the current I for
each circuit.

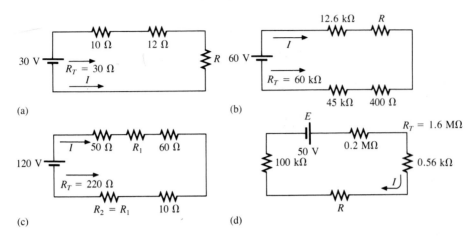

(a)

(b)

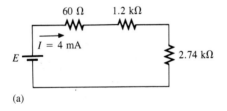

(c)

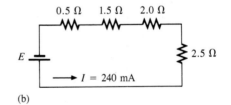

(d)

FIG. 5.48

3. Find the applied voltage E necessary to develop the current specified in each network of Fig. 5.49.

(a)

(b)

FIG. 5.49

4. For each network of Fig. 5.50, determine the current I,
the unknown resistance, and the voltage across each element.

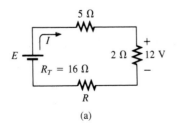

(a)

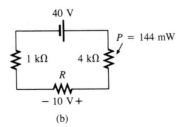

(b)

FIG. 5.50

Section 5.3

5. Determine the current I for each network of Fig. 5.51. Before solving for I, redraw each network with a single voltage source.

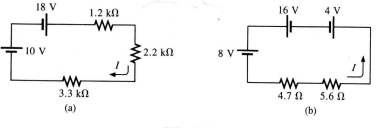

FIG. 5.51

Sections 5.4 and 5.5

6. Find V_{ab} with polarity for the circuits of Fig. 5.52. Each box can contain a load or a power supply, or a combination of both.

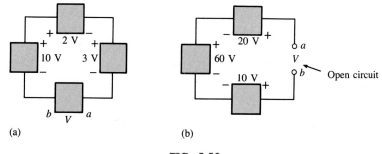

FIG. 5.52

7. Although the networks of Fig. 5.53 are not series circuits, determine the unknown voltages using Kirchhoff's voltage law.

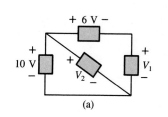
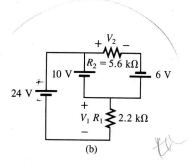

FIG. 5.53

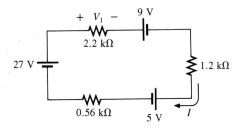

FIG. 5.54

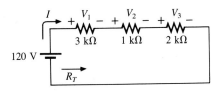

FIG. 5.55

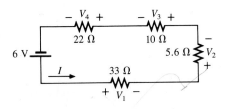

FIG. 5.56

8. Determine the current I and the voltage V_1 for the network of Fig. 5.54.

9. For the circuit of Fig. 5.55:
   a. Find the total resistance, current, and unknown voltage drops.
   b. Verify Kirchhoff's voltage law around the closed loop.
   c. Find the power dissipated by each resistor, and note whether the power delivered is equal to the power dissipated.
   d. If the resistors are available with wattage ratings of 1/2, 1, and 2 W, what minimum wattage rating can be used for each resistor in this circuit?

10. Repeat Problem 9 for the circuit of Fig. 5.56.

*11. Find the unknown quantities in the circuits of Fig. 5.57 using the information provided.

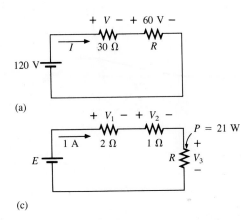

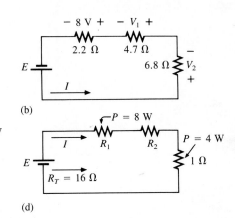

FIG. 5.57

12. There are eight Christmas tree lights connected in series as shown in Fig. 5.58.
  a. If the set is connected to a 120-V source, what is the current through the bulbs if each bulb has an internal resistance of $28\frac{1}{8}\,\Omega$?
  b. Determine the power delivered to each bulb.
  c. Calculate the voltage drop across each bulb.
  d. If one bulb burns out (that is, the filament opens), what is the effect on the remaining bulbs?

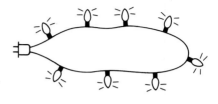

FIG. 5.58

Section 5.6

13. Using the voltage divider rule, find V_{ab} (with polarity) for the circuits of Fig. 5.59.

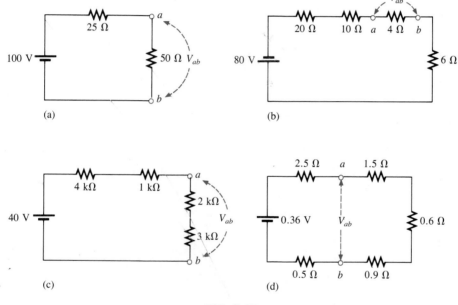

(a)

(b)

(c)

(d)

FIG. 5.59

14. Find the unknown quantities using the voltage divider rule and the information provided for the circuits of Fig. 5.60.

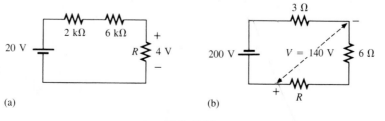

(a)

(b)

FIG. 5.60

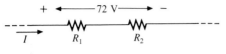

FIG. 5.61

15. Design the voltage divider of Fig. 5.61 such that $V_{R_1} = (1/5)V_{R_2}$ if $I = 4$ mA.

Section 5.7

16. a. Determine the voltages V_a and V_b for the networks of Fig. 5.62.
   b. Determine the voltage V_{ab} for each network of Fig. 5.62.

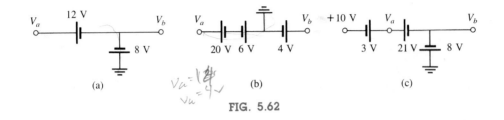

(a)   (b)   (c)

FIG. 5.62

17. Determine the voltages V_a and V_1 for the networks of Fig. 5.63.

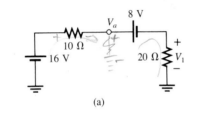

(a)

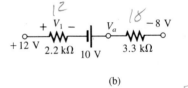

(b)

FIG. 5.63

$-10V + 16V + V_a = 0$

$+6V + 8V + V_1 = 0$

$V_1 =$

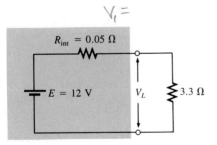

FIG. 5.64

Section 5.8

18. Find the internal resistance of a battery that has a no-load output voltage of 60 V and supplies a current of 2 A to a load of 28 Ω.

19. Find the voltage V_L and the power loss in the internal resistance for the configuration of Fig. 5.64.

20. Find the internal resistance of a battery that has a no-load output voltage of 6 V and supplies a current of 10 mA to a load of 1/2 kΩ.

Section 5.9

21. Determine the voltage regulation for the battery of Problem 18.

22. Calculate the voltage regulation for the supply of Fig. 5.64.

Section 5.11

23. Write a program to determine the total resistance of any number of resistors in series.

24. Write a program that will apply the voltage divider rule to either resistor of a series circuit with a single source and two series resistors.

25. Write a program to tabulate the current and power to the resistor R_L of the network of Fig. 5.65 for a range of values for R_L from 1 Ω to 20 Ω. Print out the value of R_L that results in maximum power to R_L.

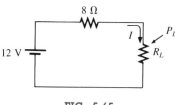

FIG. 5.65

GLOSSARY

Branch   The portion of a circuit consisting of one or more elements in series.

Circuit   A combination of a number of elements joined at terminal points providing at least one closed path through which charge can flow.

Closed loop   Any continuous connection of branches that allows tracing of a path that leaves a point in one direction and returns to that same point from another direction without leaving the circuit.

Conventional current flow   A defined direction for the flow of charge in an electrical system that is opposite to that of the motion of electrons.

Electron flow   The flow of charge in an electrical system having the same direction as the motion of electrons.

Internal resistance   The inherent resistance found internal to any source of energy.

Kirchhoff's voltage law   Law which states that the algebraic sum of the potential rises and drops around a closed loop (or path) is zero.

Series circuit   A circuit configuration in which the elements have only one point in common and each terminal is not connected to a third, current-carrying element.

Voltage divider rule   A method by which a voltage in a series circuit can be determined without first calculating the current in the circuit.

Voltage regulation (*VR*)   A value, given as a percent, that provides an indication of the change in terminal voltage of a supply with change in load demand.

6

Parallel Circuits

6.1 INTRODUCTION

There are two network configurations that form the framework for some
of the most complex network structures. A clear understanding of each
will pay enormous dividends as more complex methods and networks
are examined. The series connection was discussed in detail in the last
chapter. We will now examine the *parallel* connection and all the methods and laws associated with this important configuration.

6.2 PARALLEL ELEMENTS

*Two elements, branches, or networks are in parallel if they have two
points in common.*

In Fig. 6.1, for example, elements 1 and 2 have terminals *a* and *b* in
common; they are therefore in parallel.

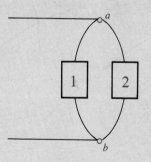

FIG. 6.1
Parallel elements.

In Fig. 6.2, all the elements are in parallel because they satisfy the above criterion. Three configurations are provided to demonstrate how the parallel networks can be drawn. Do not let the squaring of the connection at the top and bottom of Figs. 6.2(a) and (b) cloud the fact that all the elements are connected to one terminal point at top and bottom as shown in Fig. 6.2(c).

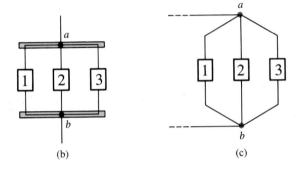

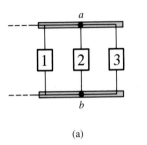

(a)　　　　(b)　　　　(c)

FIG. 6.2
Parallel configurations.

In Fig. 6.3, elements 1 and 2 are in parallel because they have terminals *a* and *b* in common. The parallel combination of 1 and 2 is then in series with element 3 due to the common terminal point *b*.

In Fig. 6.4, elements 1 and 2 are in series due to the common point *a*, but the series combination of 1 and 2 is in parallel with element 3 as defined by the common terminal connections at *b* and *c*.

In Figs. 6.1 through 6.4, the numbered boxes were used as a general symbol representing either single resistive elements, batteries, or complex network configurations.

Common examples of parallel elements include the rungs of a ladder, the tying of more than one rope between two points to increase the strength of the connection, and the use of pipes between two points to split the water between the two points at a ratio determined by the area of the pipes.

FIG. 6.3

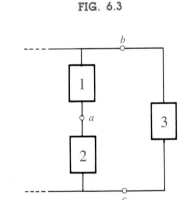

FIG. 6.4

6.3 TOTAL CONDUCTANCE AND RESISTANCE

For series resistors, the total resistance is the sum of the resistor values.

For parallel elements, the total conductance is the sum of the individual conductances.

That is, for the parallel network of Fig. 6.5, we write

$$G_T = G_1 + G_2 + G_3 + \cdots + G_N \tag{6.1}$$

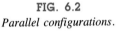

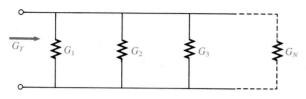

FIG. 6.5
Parallel conductances.

Since increasing levels of conductance will establish higher current levels, the more terms appearing in Eq. (6.1), the higher the input current level. In other words, as the number of resistors in parallel increases, the input current level will increase for the same applied voltage—the opposite effect of increasing the number of resistors in series.

Substituting resistor values for the network of Fig. 6.5 will result in the network of Fig. 6.6. Since $G = 1/R$, the total resistance for the network can be determined by direct substitution into Eq. (6.1):

$$\frac{1}{R_T} = \frac{1}{R_1} + \frac{1}{R_2} + \frac{1}{R_3} + \cdots + \frac{1}{R_N} \qquad \textbf{(6.2)}$$

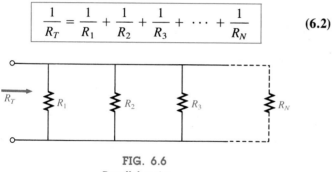

FIG. 6.6
Parallel resistors.

Note that the equation is for 1 divided by the total resistance rather than the total resistance. Once the sum of the terms to the right of the equal sign has been determined, it will then be necessary to divide the result into 1 to determine the total resistance. The following examples will demonstrate the additional calculations introduced by the inverse relationship.

EXAMPLE 6.1. Determine the total conductance and resistance for the parallel network of Fig. 6.7.
Solution:

$$G_T = G_1 + G_2 = \frac{1}{3} + \frac{1}{6} = 0.333 + 0.167 = \textbf{0.5 S}$$

and

$$R_T = \frac{1}{G_T} = \frac{1}{0.5} = \textbf{2 \Omega}$$

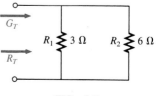

FIG. 6.7

EXAMPLE 6.2. Determine the effect on the total conductance and resistance of the network of Fig. 6.7 if another resistor of $10\,\Omega$ were added in parallel with the other elements.

Solution:

$$G_T = 0.5 + \frac{1}{10} = 0.5 + 0.1 = \mathbf{0.6\,S}$$

$$R_T = \frac{1}{G_T} = \frac{1}{0.6} \cong \mathbf{1.67\,\Omega}$$

Note, as mentioned above, that adding additional terms increases the conductance level and decreases the resistance level.

EXAMPLE 6.3. Determine the total resistance for the network of Fig. 6.8.

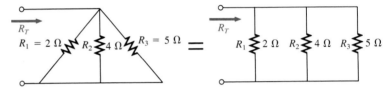

FIG. 6.8

Solution:

$$\frac{1}{R_T} = \frac{1}{R_1} + \frac{1}{R_2} + \frac{1}{R_3}$$

$$= \frac{1}{2} + \frac{1}{4} + \frac{1}{5} = 0.5 + 0.25 + 0.2$$

$$= 0.95$$

and

$$R_T = \frac{1}{0.95} = \mathbf{1.053\,\Omega}$$

The above examples demonstrate an interesting and useful (for checking purposes) characteristic of parallel resistors:

The total resistance of parallel resistors is always less than the value of the smallest resistor.

In addition, the wider the spread in numerical value between two parallel resistors, the closer the total resistance will be to the smaller resistor. For instance, the total resistance of $3\,\Omega$ in parallel with $6\,\Omega$ is $2\,\Omega$, as demonstrated in Example 6.1. However, the total resistance of $3\,\Omega$ in parallel with $60\,\Omega$ is $2.85\,\Omega$, which is much closer to the value of the smaller resistor.

For *equal* resistors in parallel, the equation becomes significantly easier to apply. For N equal resistors in parallel, Eq. (6.2) becomes

$$\frac{1}{R_T} = \underbrace{\frac{1}{R} + \frac{1}{R} + \frac{1}{R} + \cdots + \frac{1}{R}}_{N}$$

$$= N\left(\frac{1}{R}\right)$$

and

$$\boxed{R_T = \frac{R}{N}} \qquad \textbf{(6.3)}$$

In other words, the total resistance of N equal parallel resistors is the resistance of *one* resistor divided by the number (N) of parallel elements.

For conductance levels, we have

$$\boxed{G_T = NG} \qquad \textbf{(6.4)}$$

EXAMPLE 6.4.
a. Find the total resistance of the network of Fig. 6.9.
b. Calculate the total resistance for the network of Fig. 6.10.

Solutions:
a. Fig. 6.9 redrawn in Fig. 6.11:

$$R_T = \frac{R}{N} = \frac{12}{3} = \textbf{4}\,\boldsymbol{\Omega}$$

b. Fig. 6.10 redrawn in Fig. 6.12:

$$R_T = \frac{R}{N} = \frac{2}{4} = \textbf{0.5}\,\boldsymbol{\Omega}$$

In the vast majority of situations, only two or three parallel resistive elements need be combined. With this in mind, the following equations were developed to reduce the negative effects of the inverse relationship when determining R_T.

For two parallel resistors, we write

$$\frac{1}{R_T} = \frac{1}{R_1} + \frac{1}{R_2}$$

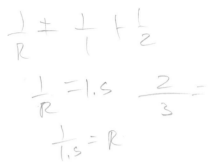

FIG. 6.9

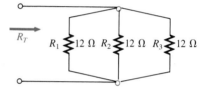

FIG. 6.10

FIG. 6.11

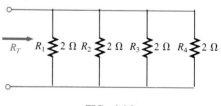

FIG. 6.12

Multiplying the top and bottom of each term of the right side of the equation by the other resistor will result in

$$\frac{1}{R_T} = \left(\frac{R_2}{R_2}\right)\frac{1}{R_1} + \left(\frac{R_1}{R_1}\right)\frac{1}{R_2} = \frac{R_2}{R_1R_2} + \frac{R_1}{R_1R_2}$$

$$= \frac{R_2 + R_1}{R_1R_2}$$

and

$$\boxed{R_T = \frac{R_1R_2}{R_1 + R_2}} \qquad \textbf{(6.5)}$$

In words, *the total resistance of two parallel resistors is the product of the two divided by their sum.*

For three parallel resistors, the equation becomes

$$\boxed{R_T = \frac{R_1R_2R_3}{R_1R_2 + R_1R_3 + R_2R_3}} \qquad \textbf{(6.6)}$$

with the denominator showing all the possible product combinations of the resistors taken two at a time.

EXAMPLE 6.5. Repeat Example 6.1 using Eq. (6.5).
Solution:

$$R_T = \frac{R_1R_2}{R_1 + R_2} = \frac{(3)(6)}{3 + 6} = \frac{18}{9} = \textbf{2 } \Omega$$

EXAMPLE 6.6. Repeat Example 6.3 using Eq. (6.6).
Solution:

$$R_T = \frac{R_1R_2R_3}{R_1R_2 + R_1R_3 + R_2R_3} = \frac{(2)(4)(5)}{(2)(4) + (2)(5) + (4)(5)}$$

$$= \frac{40}{8 + 10 + 20} = \frac{40}{38} = \textbf{1.053 } \Omega$$

Recall that series elements can be interchanged without affecting the magnitude of the total resistance or current.

In parallel networks, parallel elements can be interchanged without changing the total resistance or input current.

Note in the next example how redrawing the network can often clarify which operations and equations should be applied.

EXAMPLE 6.7. Calculate the total resistance of the parallel network of Fig. 6.13.

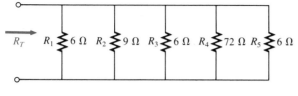

FIG. 6.13

Solution: Network redrawn in Fig. 6.14:

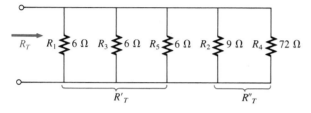

FIG. 6.14

$$R'_T = \frac{R}{N} = \frac{6}{3} = 2\ \Omega$$

$$R''_T = \frac{R_1 R_2}{R_1 + R_2} = \frac{(9)(72)}{9 + 72} = \frac{648}{81} = 8\ \Omega$$

and

$$R_T = R'_T \parallel R''_T$$
$$\qquad\qquad \underset{\text{in parallel with}}{\big\uparrow}$$

$$R_T = \frac{R'_T R''_T}{R'_T + R''_T} = \frac{(2)(8)}{2 + 8} = \frac{16}{10} = \mathbf{1.6\ \Omega}$$

6.4 PARALLEL NETWORKS

The network of Fig. 6.15 is the simplest parallel network. All the elements have terminals a and b in common. The total resistance is determined by $R_T = R_1 R_2/(R_1 + R_2)$, and the source or total current by $I_T = E/R_T$. Since the terminals of the battery are connected directly across the resistors R_1 and R_2, the following should be fairly obvious:

The voltage across parallel elements is the same.

Using this fact will result in

$$V_1 = V_2 = E \qquad\qquad P = \frac{V^2}{R^2}$$

and

$$I_1 = \frac{V_1}{R_1} = \frac{E}{R_1}$$

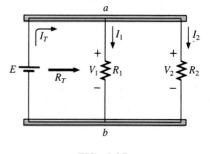

FIG. 6.15

with

$$I_2 = \frac{V_2}{R_2} = \frac{E}{R_2}$$

If we take the equation for the total resistance and multiply both sides by the applied voltage, we obtain

$$E\left(\frac{1}{R_T}\right) = E\left(\frac{1}{R_1} + \frac{1}{R_2}\right)$$

and

$$\frac{E}{R_T} = \frac{E}{R_1} + \frac{E}{R_2}$$

Substituting the Ohm's law relationships appearing above, we find that

$$I_T = I_1 + I_2$$

permitting the following conclusion:

For parallel networks, the source current is equal to the sum of the individual branch currents.

The power dissipated by the resistors and delivered by the source can be determined from

$$P_1 = V_1 I_1 = I_1^2 R_1 = \frac{V_1^2}{R_1}$$

$$P_2 = V_2 I_2 = I_2^2 R_2 = \frac{V_2^2}{R_2}$$

$$P_s = EI_T = I_T^2 R_T = \frac{E^2}{R_T}$$

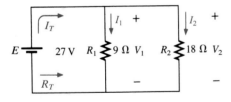

FIG. 6.16

EXAMPLE 6.8. For the parallel network of Fig. 6.16:
a. Calculate R_T.
b. Determine I_T.
c. Calculate I_1 and I_2 and demonstrate that $I_T = I_1 + I_2$.
d. Determine the power to each resistive load.
e. Determine the power delivered by the source and compare it to the total power dissipated by the resistive elements.

Solutions:

a. $R_T = \dfrac{R_1 R_2}{R_1 + R_2} = \dfrac{(9)(18)}{9 + 18} = \dfrac{162}{27} = \mathbf{6\,\Omega}$

b. $I_T = \dfrac{E}{R_T} = \dfrac{27}{6} = \mathbf{4.5\,A}$

c. $I_1 = \dfrac{V_1}{R_1} = \dfrac{E}{R_1} = \dfrac{27}{9} = \mathbf{3\,A}$

   $I_2 = \dfrac{V_2}{R_2} = \dfrac{E}{R_2} = \dfrac{27}{18} = \mathbf{1.5\,A}$

$$I_T = I_1 + I_2$$
$$4.5 = 3 + 1.5$$
$$\underline{4.5 = 4.5 \quad \text{(checks)}}$$

d. $P_1 = V_1 I_1 = EI_1 = (27)(3) = \textbf{81 W}$
     $P_2 = V_2 I_2 = EI_2 = (27)(1.5) = \textbf{40.5 W}$
e. $P_s = EI_T = (27)(4.5) = \textbf{121.5 W}$
     $P_s = P_1 + P_2 = 81 + 40.5 = \textbf{121.5 W}$

EXAMPLE 6.9.   Given the information provided in Fig. 6.17:

a. Determine R_3.
b. Calculate E.
c. Find I_T.
d. Find I_2.
e. Determine P_2.

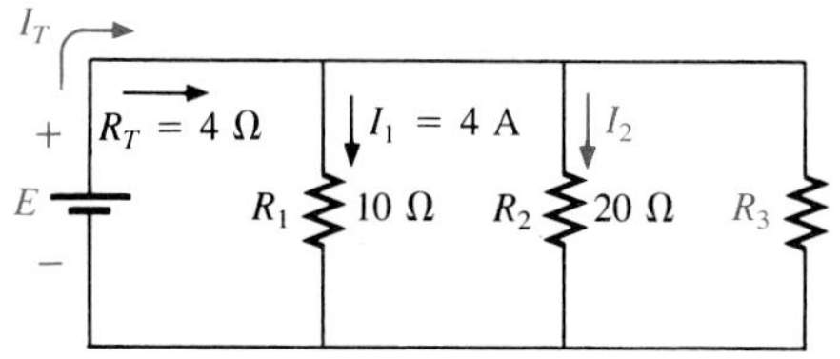

Solutions:

FIG. 6.17

a. $\dfrac{1}{R_T} = \dfrac{1}{R_1} + \dfrac{1}{R_2} + \dfrac{1}{R_3}$

     $\dfrac{1}{4} = \dfrac{1}{10} + \dfrac{1}{20} + \dfrac{1}{R_3}$

     $0.25 = 0.1 + 0.05 + \dfrac{1}{R_3}$

     $0.25 = 0.15 + \dfrac{1}{R_3}$

     $\dfrac{1}{R_3} = 0.1$

     $R_3 = \dfrac{1}{0.1} = \textbf{10 Ω}$

b. $E = V_1 = I_1 R_1 = (4)(10) = \textbf{40 V}$

c. $I_T = \dfrac{E}{R_T} = \dfrac{40}{4} = \textbf{10 A}$

d. $I_2 = \dfrac{V_2}{R_2} = \dfrac{E}{R_2} = \dfrac{40}{20} = \textbf{2 A}$

e. $P_2 = I_2^2 R_2 = (2)^2(20) = \textbf{80 W}$

6.5 KIRCHHOFF'S CURRENT LAW

Kirchhoff's current law (KCL) states that *the algebraic sum of the currents entering and leaving a node is zero.* (A *node* is a junction of two or more branches.) In other words:

The sum of the currents entering a node must equal the sum of the currents leaving a node.

In equation form, we have

$$\boxed{\Sigma\, I_{entering} = \Sigma\, I_{leaving}} \qquad\qquad (6.7)$$

In Fig. 6.18, the currents I_1, I_3, and I_4 enter the junction, while I_2 and I_5 leave. Applying Eq. (6.7), we obtain

$$I_1 + I_3 + I_4 = I_2 + I_5$$

Substituting values yields

$$2 + 4 + 3 = 8 + 1$$
$$9 = 9$$

The shaded region of Fig. 6.18 is not limited to a terminal point for the branches. As demonstrated by Fig. 6.19, it may include a very complex network. Applying Kirchhoff's current law to terminal a results in

$$3 = I_5 + I_{ab}$$
$$3 = 1 + I_{ab}$$

and

$$I_{ab} = 3 - 1 = 2\,\text{A}$$

At terminal b,

$$I_{ab} + I_4 = I_{bc}$$
$$2 + 3 = I_{bc}$$

and

$$I_{bc} = 5\,\text{A}$$

At terminal c,

$$I_{bc} + I_3 = I_{cd}$$
$$5 + 4 = I_{cd}$$

and

$$I_{cd} = 9\,\text{A}$$

At terminal d,

$$I_{cd} = I_2 + I_{de}$$
$$9 = 8 + I_{de}$$

and

$$I_{de} = 9 - 8 = 1\,\text{A}$$

At terminal e,

$$I_{de} + I_1 = I_{ea}$$
$$1 + 2 = I_{ea}$$

$I_5 = 1$ A
$I_1 = 2$ A
$I_4 = 3$ A
$I_2 = 8$ A
$I_3 = 4$ A

FIG. 6.18

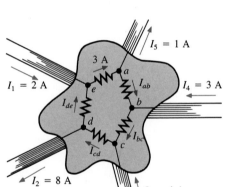

$I_5 = 1$ A
3 A
$I_1 = 2$ A
I_{ab}
$I_4 = 3$ A
I_{de}
I_{bc}
I_{cd}
$I_2 = 8$ A
$I_3 = 4$ A

FIG. 6.19

and

$$I_{ea} = 1 + 2 = 3 \, \text{A}$$

At terminal a,

$$I_{ea} = I_5 + I_{ab}$$
$$3 = 1 + 2$$
$$\underline{3 = 3 \quad \text{(checks)}}$$

EXAMPLE 6.10. Determine the currents I_3 and I_5 of Fig. 6.20 through applications of Kirchhoff's current law.

Solution: Note that since node b has two unknown quantities and node a only one, we must first apply Kirchhoff's current law to node a. The result can then be applied to node b. For node a,

$$I_1 + I_2 = I_3$$
$$4 + 3 = I_3$$

and

$$I_3 = \mathbf{7 \, A}$$

For node b,

$$I_3 = I_4 + I_5$$
$$7 = 1 + I_5$$

and

$$I_5 = 7 - 1 = \mathbf{6 \, A}$$

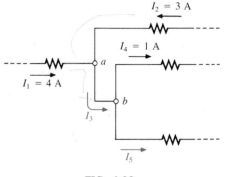

FIG. 6.20

EXAMPLE 6.11. Find the magnitude and direction of the currents I_3, I_4, I_6, and I_7 for the network of Fig. 6.21. Even though the elements are not in series or parallel, Kirchhoff's current law can be applied to determine all the unknown currents.

Solution: Considering the overall system, we know that the current entering must equal that leaving. Therefore,

$$I_7 = I_1 = \mathbf{10 \, A}$$

Since 10 A are entering node a and 12 A are leaving, I_3 must be supplying current to the node. Applying Kirchhoff's current law at node a,

$$I_1 + I_3 = I_2$$
$$10 + I_3 = 12$$

and

$$I_3 = 12 - 10 = \mathbf{2 \, A}$$

At node b, since 12 A are entering and 8 A are leaving, I_4 must be leaving. Therefore,

$$I_2 = I_4 + I_5$$
$$12 = I_4 + 8$$

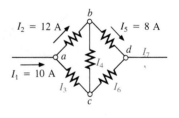

FIG. 6.21

and

$$I_4 = 12 - 8 = \mathbf{4\,A}$$

At node c, I_3 is leaving at 2 A and I_4 is entering at 4 A, requiring that I_6 be leaving. Applying Kirchhoff's current law at node c,

$$I_4 = I_3 + I_6$$
$$4 = 2 + I_6$$

and

$$I_6 = 4 - 2 = \mathbf{2\,A}$$

As a check at node d,

$$I_5 + I_6 = I_7$$
$$8 + 2 = 10$$
$$\underline{10 = 10 \qquad \text{(checks)}}$$

Looking back at Example 6.8, we find that the current entering the top node is 4.5 A and the current leaving the node is $I_1 + I_2 = 3 + 1.5 = 4.5$ A. For Example 6.9, we have

$$I_T = I_1 + I_2 + I_3$$
$$10 = 4 + 2 + I_3$$

and

$$I_3 = 10 - 6 = 4\,A$$

6.6 CURRENT DIVIDER RULE

As the name suggests, the *current divider rule* (CDR) will determine how the current entering a set of parallel branches will split between the elements. For two elements of equal value, the current will divide equally. For elements with different values, the smaller the resistance, the greater the share of input current.

The current divider rule will be derived through the use of the representative network of Fig. 6.22. The input current I equals V/R_T, where

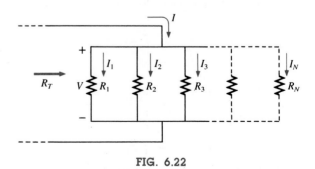

FIG. 6.22

R_T is the total resistance of the parallel branches. Substituting $V = I_x R_x$ into the above equation, where I_x refers to the current through a parallel branch of resistance R_x, we have

$$I = \frac{V}{R_T} = \frac{I_x R_x}{R_T}$$

and

$$\boxed{I_x = \frac{R_T}{R_x} I} \qquad\qquad \text{(6.8)}$$

which is the general form for the current divider rule. In words, the current through any parallel branch is equal to the product of the *total* resistance of the parallel branches and the input current divided by the resistance of the branch through which the current is to be determined.

For the current I_1,

$$I_1 = \frac{R_T}{R_1} I$$

and for I_2,

$$I_2 = \frac{R_T}{R_2} I$$

For the particular case of *two parallel resistors* as shown in Fig. 6.23,

$$R_T = \frac{R_1 R_2}{R_1 + R_2}$$

and

$$I_1 = \frac{R_T}{R_1} I = \frac{\dfrac{R_1 R_2}{R_1 + R_2}}{R_1} I$$

and

Note difference in subscripts.

$$\boxed{I_1 = \frac{R_2 I}{R_1 + R_2}} \qquad\qquad \text{(6.9)}$$

Similarly for I_2,

$$\boxed{I_2 = \frac{R_1 I}{R_1 + R_2}} \qquad\qquad \text{(6.10)}$$

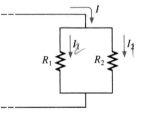

FIG. 6.23

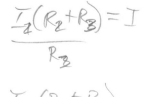

In words, for two parallel branches, the current through either branch is equal to the product of the *other* parallel resistor and the input current divided by the *sum* (not total parallel resistance) of the two parallel resistances.

EXAMPLE 6.12. Determine the current I_2 for the network of Fig. 6.24 using the current divider rule.
Solution:

$$I_2 = \frac{R_1 I_T}{R_1 + R_2} = \frac{(4\,\text{k}\Omega)(6\,\text{A})}{4\,\text{k}\Omega + 8\,\text{k}\Omega} = \frac{4}{12}(6) = \frac{1}{3}(6)$$

$$= \mathbf{2\,A}$$

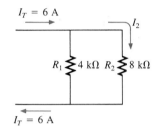

$I_T = 6\,\text{A}$

I_2

$R_1 \gtrless 4\,\text{k}\Omega$ $R_2 \gtrless 8\,\text{k}\Omega$

$I_T = 6\,\text{A}$

FIG. 6.24

EXAMPLE 6.13. Find the current I_1 for the network of Fig. 6.25.
Solution: By Eq. (6.8),

$$R_T = 6 \parallel 24 \parallel 24 = 6 \parallel 12 = 4\,\Omega$$

$$I_1 = \frac{R_T}{R_1}I = \frac{(4)(42 \times 10^{-3})}{6} = \mathbf{28\,mA}$$

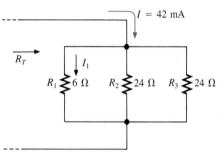

$I = 42\,\text{mA}$

R_T

I_1

$R_1 \gtrless 6\,\Omega$ $R_2 \gtrless 24\,\Omega$ $R_3 \gtrless 24\,\Omega$

FIG. 6.25

EXAMPLE 6.14. Determine the magnitude of the currents I_1 and I_2 for the network of Fig. 6.26.

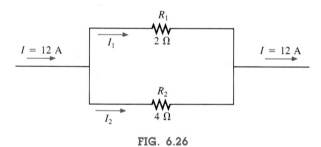

R_1

$I = 12\,\text{A}$ I_1 $2\,\Omega$ $I = 12\,\text{A}$

R_2

I_2 $4\,\Omega$

FIG. 6.26

Solution: By Eq. (6.9), the current divider rule,

$$I_1 = \frac{R_2 I}{R_1 + R_2} = \frac{(4)(12)}{2 + 4} = \mathbf{8\,A}$$

Applying Kirchhoff's current law,

$$I = I_1 + I_2$$

and

$$I_2 = I - I_1 = 12 - 8 = \mathbf{4\,A}$$

or, using the current divider rule again,

$$I_2 = \frac{R_1 I}{R_1 + R_2} = \frac{(2)(12)}{2 + 4} = \textbf{4 A}$$

EXAMPLE 6.15. Determine the resistance R_1 to effect the division of current in Fig. 6.27.

Solution: Applying the current divider rule,

$$I_1 = \frac{R_2 I}{R_1 + R_2}$$

and

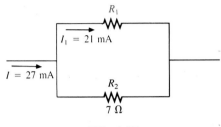

FIG. 6.27

$$21 \times 10^{-3} = \frac{(7)(27 \times 10^{-3})}{R_1 + 7}$$

$$(21 \times 10^{-3})R_1 + 147 \times 10^{-3} = 189 \times 10^{-3}$$

or

$$R_1 = \frac{(189 - 147) \times 10^{-3}}{21 \times 10^{-3}} = \frac{42}{21} = \textbf{2 } \Omega$$

An alternative approach is

$$I_2 = I - I_1 \quad \text{(Kirchhoff's current law)}$$
$$= 27\,\text{mA} - 21\,\text{mA} = 6\,\text{mA}$$
$$V_2 = I_2 R_2 = (6\,\text{mA})(7\,\Omega) = 42\,\text{mV}$$
$$V_1 = I_1 R_1 = V_2 = 42\,\text{mV}$$

and

$$R_1 = \frac{V_1}{I_1} = \frac{42\,\text{mV}}{21\,\text{mA}} = \textbf{2 } \Omega$$

From the examples just described, note the following:

1. More current passes through the smaller of two parallel resistors.
2. The current entering any number of parallel resistors divides into these resistors as the inverse ratio of their ohmic values. This relationship is depicted in Fig. 6.28.

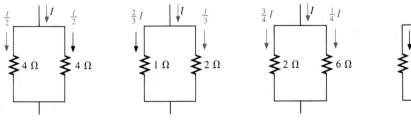
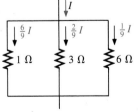

FIG. 6.28

A mechanical analogy often used to describe this division of current is the flow of water through pipes. The water represents the flow of charge, and the tubes or pipes represent conductors. In this analogy, the greater the resistance of the corresponding electrical element, the smaller the area of the tubing.

The total current I in Fig. 6.29(a) divides equally between the two equal resistors. The analogy just described is shown to the right. Obviously, for two pipes of equal diameter, the water will divide equally.

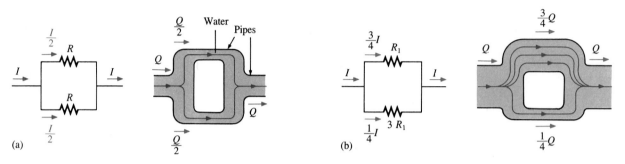

(a)   (b)

FIG. 6.29

In Fig. 6.29(b), one resistor is three times the other, resulting in the current dividing as shown. Its mechanical analogy is shown in the adjoining figure. Again, it should be obvious that three times as much water (current) will pass through one pipe as through the other. In both Figs. 6.29(a) and (b), the total water (current) entering the parallel systems from the left will equal that leaving to the right.

6.7 VOLTAGE SOURCES IN PARALLEL

Voltage sources are placed in parallel as shown in Fig. 6.30 only if they have the same voltage rating. Otherwise, Kirchhoff's voltage law

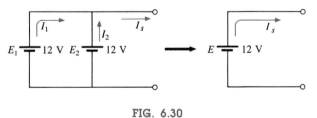

FIG. 6.30

Parallel voltage sources.

would be violated around the internal loop of the two batteries. The primary reason for placing two or more batteries in parallel of the same terminal voltage would be to increase the current rating of the source. As shown in Fig. 6.30, the current rating of the combination is determined by $I_s = I_1 + I_2$ at the same terminal voltage. The resulting power rating is twice that available with one supply.

6.8 OPEN AND SHORT CIRCUITS

Open circuits and short circuits can often cause more confusion and difficulty in the analysis of a system than standard series or parallel configurations. This will become more obvious in the chapters to follow when we apply some of the methods and theorems.

An *open circuit* is simply two isolated terminals not connected by an element of any kind. Consider the battery of Fig. 6.31. An open circuit exists between terminals *a* and *b*. There is a voltage of *E* volts between the two terminals, but the current between the two is zero due to the absence of a closed path for the flow of charge. In general:

An open circuit can have a potential difference (voltage) across its terminals but the current is always zero amperes.

A *short circuit* is a direct connection of zero ohms across an element or combination of elements. In Fig. 6.32(a), the current through the 2-Ω

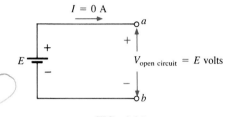

FIG. 6.31

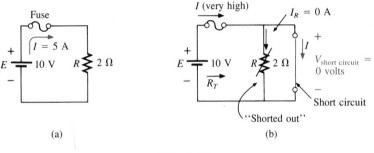

(a)

(b)

FIG. 6.32

resistor is 5 A. If a short circuit is established across the 2-Ω resistor as shown in Fig. 6.32(b) due to a faulty wire, connection, or other unexpected circumstance, the total resistance of the circuit R_T is now $R_T = (2)(0)/(2 + 0) = 0\,\Omega$ and the current will rise to very high levels. The 2-Ω resistor has effectively been "shorted out" by the low-resistance connection. The maximum current is limited only by the circuit breaker or fuse in series with the source. The resulting high current is often the cause of fire or smoke if the protective device fails to respond quickly enough. Since the resistance of a short circuit is zero ohms, there is no voltage drop across a short circuit, as determined by Ohm's law ($V = IR$). In general:

A short circuit can carry a current of any level but the potential difference (voltage) across its terminals is always zero volts.

EXAMPLE 6.16. Determine the voltage V_{ab} for the network of Fig. 6.33.

Solution: The open circuit requires that I be zero amperes. The voltage drop across both resistors is therefore zero volts since $V = IR = (0)R = 0\,V$. Applying Kirchhoff's voltage law around the closed loop,

$$V_{ab} = E = \textbf{20 V}$$

FIG. 6.33

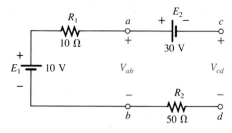

FIG. 6.34

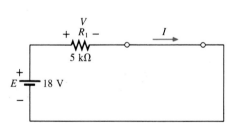

FIG. 6.35

FIG. 6.37

EXAMPLE 6.17. Determine the voltages V_{ab} and V_{cd} for the network of Fig. 6.34.

Solution: The current through the system is zero amperes due to the open circuit resulting in a 0-V drop across each resistor. Both resistors can therefore be replaced by short circuits as shown in Fig. 6.35. The voltage V_{ab} is then directly across the 10-V battery, and

$$V_{ab} = E_1 = \mathbf{10\,V}$$

The voltage V_{cd} requires an application of Kirchhoff's voltage law:

$$+E_1 - E_2 - V_{cd} = 0$$

or

$$V_{cd} = E_1 - E_2 = 10 - 30 = \mathbf{-20\,V}$$

EXAMPLE 6.18. Calculate the current I and the voltage V for the network of Fig. 6.36.

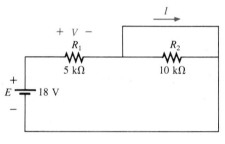

FIG. 6.36

Solution: The 10-Ω resistor has been effectively shorted out, resulting in the equivalent network of Fig. 6.37. Using Ohm's law,

$$I = \frac{E}{R_1} = \frac{18}{5\,\mathrm{k}\Omega} = \mathbf{3.6\,mA}$$

and

$$V = E = \mathbf{18\,V}$$

EXAMPLE 6.19. Determine V and I for the network of Fig. 6.38 if the resistor R_2 is shorted out.

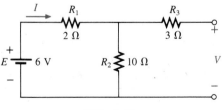

FIG. 6.38

Solution: The redrawn network appears in Fig. 6.39. The current through the 3-Ω resistor is zero due to the open circuit causing all of the current I to pass through the short circuit. Since $V_{3\Omega} = IR = (0)R = 0\,V$, the voltage V is directly across the short, and

$$V = 0\,V$$

with

$$I = \frac{E}{R_1} = \frac{6}{2} = 3\,A$$

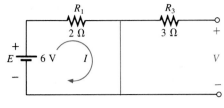

FIG. 6.39

6.9 VOLTMETERS: LOADING EFFECT

In Chapter 2, it was noted that voltmeters are always placed across an element to measure the potential difference. We now realize that this connection is synonymous with placing the voltmeter in parallel with the element. The insertion of a meter in parallel with a resistor results in a combination of parallel resistors as shown in Fig. 6.40. Since the resistance of two parallel branches is always less than the smaller parallel resistance, the resistance of the voltmeter should be as large as possible (ideally infinite). In Fig. 6.40, a DMM with an internal resistance of 11 MΩ is measuring the voltage across a 10-kΩ resistor. The total resistance of the combination is

$$R_T = 10\,k\Omega \parallel 11\,M\Omega = \frac{(10^4)(11 \times 10^6)}{10^4 + (11 \times 10^6)} = 9.99\,k\Omega$$

and we find that the network is essentially undisturbed. However, if we use a VOM with an internal resistance of 50 kΩ on the 2.5-V scale, the parallel resistance is

$$R_T = 10\,k\Omega \parallel 50\,k\Omega = \frac{(10^4)(50 \times 10^3)}{10^4 + (50 \times 10^3)} = 8.33\,k\Omega$$

and the behavior of the network will be altered somewhat since the 10-kΩ resistor will now appear to be 8.33 kΩ to the rest of the network.

The loading of a network by the insertion of meters is not to be taken lightly, especially in research efforts where accuracy is a primary consideration. It is good practice always to check the meter resistance level against the resistive elements of the network before making measurements. A factor of 10 between resistance levels will usually provide fairly accurate meter readings for a wide range of applications.

Most DMMs have internal resistance levels in excess of 10 MΩ on all voltage scales, while the internal resistance of VOMs is sensitive to the chosen scale. To determine the resistance of each scale setting of a VOM in the voltmeter mode, simply multiply the maximum voltage of the scale setting by the ohm/volt (Ω/V) rating of the meter, normally found at the bottom of the face of the meter.

For a typical ohm/volt rating of 20,000, the 2.5-V scale would have an internal resistance of

$$V_L = V_{OC} = E\,\frac{4.7}{4.7 + 2.9}$$

$$V_L = 6.13$$

$$9V\left(\frac{4.7}{4.7 + 2.9}\right)$$

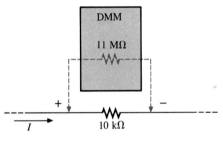

FIG. 6.40

Voltmeter loading.

$$(2.5)(20,000) = 50 \, k\Omega$$

while for the 100-V scale, it would be

$$(100)(20,000) = 2 \, M\Omega$$

and for the 250-V scale,

$$(250)(20,000) = 5 \, M\Omega$$

PROBLEMS

Section 6.2

1. For each configuration of Fig. 6.41, determine which elements are in series or parallel.

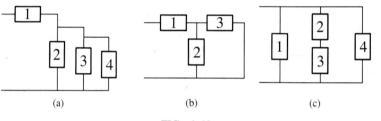

(a)   (b)   (c)

FIG. 6.41

Section 6.3

2. Find the total conductance and resistance for the networks of Fig. 6.42.

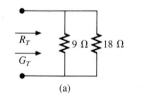

(a)

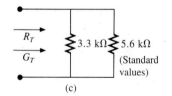

(c)

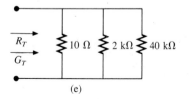

(e)

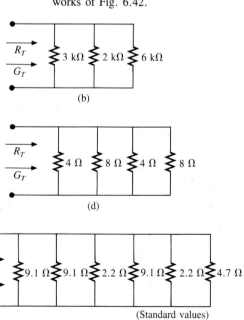

FIG. 6.42

3. The total conductance of the networks of Fig. 6.43 is specified. Find the value in ohms of the unknown resistances.

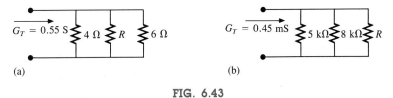

(a)   (b)

FIG. 6.43

4. The total resistance of the circuits of Fig. 6.44 is specified. Find the value in ohms of the unknown resistances.

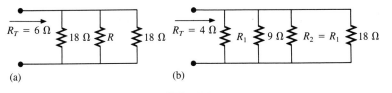

(a)   (b)

FIG. 6.44

Section 6.4

5. For the network of Fig. 6.45:
  a. Find the total conductance and resistance.
  b. Determine I_T and the current through each parallel branch.
  c. Verify that the source current equals the sum of the parallel branch currents.
  d. Find the power dissipated by each resistor, and note whether the power delivered is equal to the power dissipated.
  e. If the resistors are available with wattage ratings of 1/2, 1, 2, and 50 W, what is the minimum wattage rating for each resistor?

6. Repeat Problem 5 for the network of Fig. 6.46.

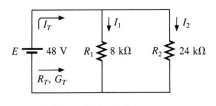

FIG. 6.45

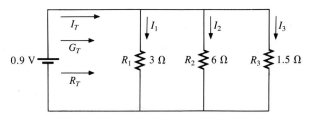

FIG. 6.46

7. Repeat Problem 5 for the network of Fig. 6.47 constructed of standard resistor values.

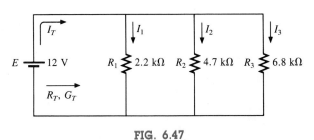

FIG. 6.47

FIG. 6.48

8. There are eight Christmas tree lights connected in parallel as shown in Fig. 6.48.
  a. If the set is connected to a 120-V source, what is the current through each bulb if each bulb has an internal resistance of $1.8\,k\Omega$?
  b. Determine the total resistance of the network.
  c. Find the power delivered to each bulb.
  d. If one bulb burns out (that is, the filament opens), what is the effect on the remaining bulbs?
  e. Compare the parallel arrangement of Fig. 6.48 to the series arrangement of Fig. 5.58. What are the relative advantages and disadvantages of the parallel system as compared to the series arrangement?

9. A portion of a residential service to a home is depicted in Fig. 6.49.
  a. Determine the current through each parallel branch of the network.
  b. Calculate the current drawn from the 120-V source. Will the 20-A circuit breaker trip?
  c. What is the total resistance of the network?
  d. Determine the power supplied by the 120-V source. How does it compare to the total power of the load?

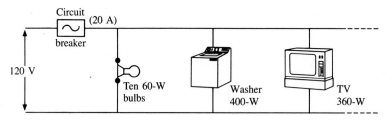

FIG. 6.49

10. Determine the currents I_1 and I_T for the networks of Fig. 6.50.

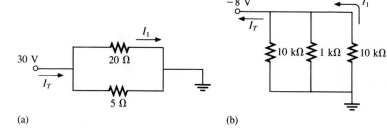

FIG. 6.50

Section 6.5

11. Find all unknown currents and their directions in the circuits of Fig. 6.51.

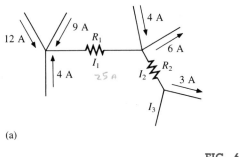

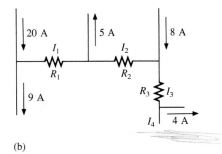

(a)   (b)

FIG. 6.51

12. Using Kirchhoff's current law, determine the unknown currents for the networks of Fig. 6.52.

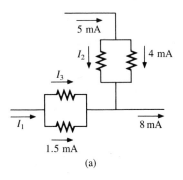

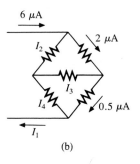

(a)   (b)

FIG. 6.52

*13. Find the unknown quantities for the circuits of Fig. 6.53 using the information provided.

$R = \dfrac{V}{R}$

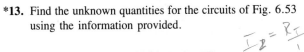

$I_2 = \dfrac{R_I}{I}$

(a)   (b)

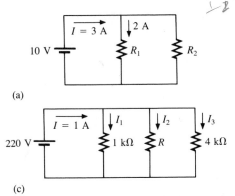

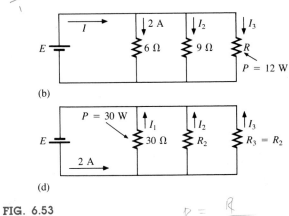

(c)   (d)

FIG. 6.53

$R = \dfrac{R}{\dfrac{1}{R}}$

Section 6.6

14. Using the current divider rule, find the unknown currents for the networks of Fig. 6.54.

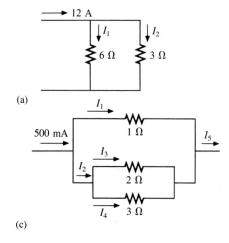

(a)

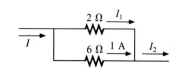

(c)

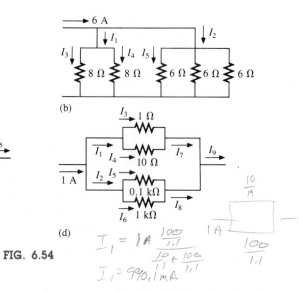

(b)

(d)

FIG. 6.54

$$I_1 = 1A \frac{100}{1.1}$$

$$\frac{10}{1.1} + \frac{100}{1.1}$$

$$I_1 = 990.1 \, mA$$

$$\frac{10}{11} \qquad 1A$$

$$\frac{100}{1.1}$$

15. Find the unknown quantities using the information provided for the networks of Fig. 6.55.

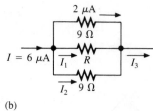

(a)

(b)

FIG. 6.55

16. Calculate the resistor R for the network of Fig. 6.56 that will insure the current $I_1 = 3I_2$.

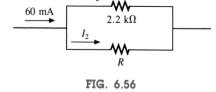

FIG. 6.56

Section 6.7

17. Determine the currents I_1 and I_2 for the network of Fig. 6.57.

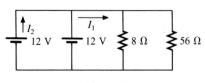

FIG. 6.57

Section 6.8

18. For the network of Fig. 6.58:
   a. Determine I_T and V_L.
   b. Determine I_T if R_L is shorted out.
   c. Determine V_L if R_L is replaced by an open circuit.

12 V

$\dfrac{10}{11}$

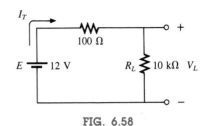

FIG. 6.58

19. For the network of Fig. 6.59:
   a. Determine the open-circuit voltage V_L.
   b. Place a short circuit across the output terminals and determine the current through the short circuit.
   c. If the 2.2-kΩ resistor is short circuited, what is the new value of V_L?
   d. Repeat part (b) with the 4.7-kΩ resistor replaced by an open circuit.

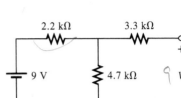

$I_{sc} = 9$

6.13

FIG. 6.59

Section 6.9

20. For the network of Fig. 6.60:
   a. Determine the voltage V_2.
   b. Determine the reading of a DMM having an internal resistance of 11 MΩ when used to measure V_2.
   c. Repeat part (b) with a VOM having an ohm/volt rating of 20,000 using the 10-V scale.
   d. Repeat part (c) with $R_1 = 100$ kΩ and $R_2 = 200$ kΩ.

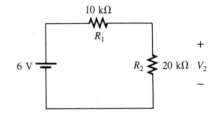

FIG. 6.60

$\dfrac{9}{5.5}$

Computer Problems

21. Write a program to determine the total resistance and conductance of any number of elements in parallel.

22. Write a program to provide a complete solution of a parallel network with a single source and two parallel resistors. That is, print out the total resistance, input current, current through each branch, and power to each element.

23. Write a program that will tabulate the voltage V_2 of Fig. 6.60 measured by a VOM with an internal resistance of 200 kΩ as R_2 varies from 10 kΩ to 200 kΩ in increments of 10 kΩ.

GLOSSARY

Current divider rule A method by which the current through parallel elements can be determined without first finding the voltage across those parallel elements.

Kirchhoff's current law The law which states that the algebraic sum of the currents entering and leaving a node is zero.

Node A junction of two or more branches.

Ohm/volt rating A rating used to determine both the current

sensitivity of the movement and the internal resistance of the meter.

Open circuit The absence of a direct connection between two points in a network.

Parallel circuit A circuit configuration in which the elements have two points in common.

Short circuit A direct connection of low resistive value that can significantly alter the behavior of an element or system.

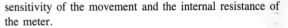

7

Series-Parallel Networks

7.1 SERIES-PARALLEL NETWORKS

A firm understanding of the basic principles associated with series and parallel circuits is a sufficient background to approach most complicated series-parallel networks (a network being a combination of any number of series and parallel elements) with *one* source of voltage. Multisource networks will be considered in Chapters 8 and 9. The variation introduced by series-parallel networks is that series and parallel circuit configurations appear within the same network.

In general, when working with series-parallel dc networks, the following is a natural sequence:

1. Study the problem and make a brief mental "sketch" of the overall approach you plan to use. The result may be time- and energy-saving shortcuts.
2. After you have determined the overall approach, examine each branch independently before tying them together in series-parallel combinations. This will eliminate many of the errors that might develop due to the lack of a systematic approach.
3. When you have a solution, check that it is reasonable by considering the magnitudes of the energy source and the elements in the network. If it does not seem reasonable, either solve the circuit using another approach or check over your work very carefully.

The block diagram approach will be employed to emphasize the fact that combinations of elements, not simply single resistive elements, can be in series or parallel. The approach will also reveal the number of seemingly different networks that have the same basic structure and therefore can involve similar analysis techniques.

The analysis of series-parallel dc networks with a single source usually requires that the resistance "seen" by the source (R_T) be determined by combining series and parallel elements until a single equivalent resistance remains. The source current can then be calculated and the remaining currents of the network determined by working back through the network.

Initially, there will be some concern about identifying series and parallel elements and branches and choosing the best procedure to follow toward a solution. However, as you progress through the examples and try a few problems, a common path toward most solutions will surface that can actually make the analysis of such systems an interesting, enjoyable experience.

In Fig. 7.1, blocks B and C are in parallel (points b and c in common), and the voltage source E is in series with block A (point a in common). The parallel combination of B and C is also in series with A and the voltage source E due to the common points b and c, respectively.

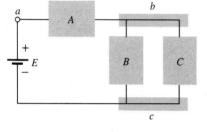

FIG. 7.1

To insure that the analysis to follow is as clear and uncluttered as possible, the following notation will be used for series and parallel combinations of elements. For series resistors R_1 and R_2, a comma will be inserted between their subscript notation as shown here:

$$R_{1,2} = R_1 + R_2$$

For parallel resistors R_1 and R_2, the parallel symbol will be inserted between their subscript notation as follows:

$$R_{1\|2} = R_1 \| R_2 = \frac{R_1 R_2}{R_1 + R_2}$$

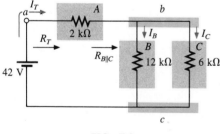

FIG. 7.2

EXAMPLE 7.1. If each block of Fig. 7.1 were a single resistive element, the network of Fig. 7.2 might result.

$$R_T = R_A + R_{B\|C} = 2\,\text{k}\Omega + \frac{(12\,\text{k}\Omega)(6\,\text{k}\Omega)}{12\,\text{k}\Omega + 6\,\text{k}\Omega}$$

$$= 2\,\text{k}\Omega + 4\,\text{k}\Omega = \mathbf{6\,k\Omega}$$

and the source current is

$$I_T = \frac{E}{R_T} = \frac{42}{6\,\text{k}\Omega} = \mathbf{7\,mA}$$

$$I_A = I_T = \mathbf{7\,mA}$$

The current divider rule must be employed to determine I_B and I_C:

$$I_B = \frac{6\,\text{k}\Omega(I_T)}{6\,\text{k}\Omega + 12\,\text{k}\Omega} = \frac{6}{18}I_T = \frac{1}{3}(7) = \mathbf{2\tfrac{1}{3}\,mA}$$

$$I_C = \frac{12\,\text{k}\Omega(I_T)}{12\,\text{k}\Omega + 6\,\text{k}\Omega} = \frac{12}{18}I_T = \frac{2}{3}(7) = \mathbf{4\tfrac{2}{3}\,mA}$$

or, applying Kirchhoff's current law,

$$I_C = I_T - I_B = 7 - 2\frac{1}{3} = \frac{21}{3} - \frac{7}{3} = \frac{14}{3} = \mathbf{4\tfrac{2}{3}\,mA}$$

EXAMPLE 7.2. It is also possible that the blocks A, B, and C of Fig. 7.1 contain the elements and configurations of Fig. 7.3.

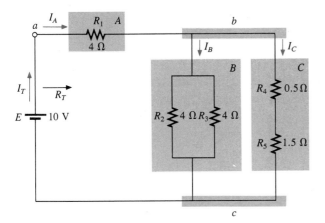

FIG. 7.3

$$R_A = 4\,\Omega$$

$$R_B = R_2 \parallel R_3 = R_{2\parallel3} = \frac{R}{N} = \frac{4}{2} = 2\,\Omega$$

$$R_C = R_4 + R_5 = R_{4,5} = 0.5 + 1.5 = 2\,\Omega$$

so that

$$R_{B\parallel C} = \frac{R}{N} = \frac{2}{2} = 1\,\Omega$$

with

$$R_T = R_A + R_{B\parallel C} \quad \text{\footnotesize (Note the similarity between this equation and that obtained for Example 7.1.)}$$
$$= 4 + 1 = \mathbf{5\,\Omega}$$

and

$$I_T = \frac{E}{R_T} = \frac{10}{5} = \mathbf{2\,A}$$

We can find the currents I_A, I_B, and I_C using the reduction of the network of Fig. 7.3 as found in Fig. 7.4:

$$I_A = I_T = \mathbf{2\,A}$$

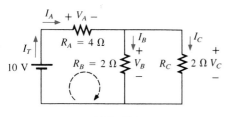

FIG. 7.4

and

$$I_B = I_C = \frac{I_A}{2} = \frac{I_T}{2} = \frac{2}{2} = \mathbf{1\,A}$$

Returning to the network of Fig. 7.3, we have

$$I_{R_2} = I_{R_3} = \frac{I_B}{2} = \mathbf{0.5\,A}$$

The voltages V_A, V_B, and V_C from either figure are

$$V_A = I_A R_A = (2)(4) = \mathbf{8\,V}$$
$$V_B = I_B R_B = (1)(2) = \mathbf{2\,V}$$
$$V_C = V_B = \mathbf{2\,V}$$

Applying Kirchhoff's voltage law for the loop indicated in Fig. 7.4, we obtain

$$\Sigma_\circlearrowright V = E - V_A - V_B = 0$$

or

$$E = V_A + V_B = 8 + 2$$
$$\underline{10 = 10 \qquad \text{(checks)}}$$

EXAMPLE 7.3. Another possible variation of Fig. 7.1 appears in Fig. 7.5.

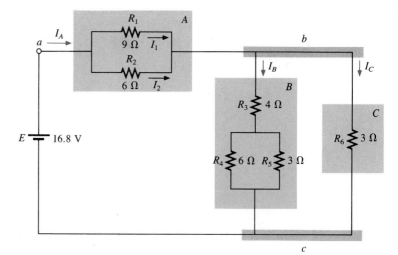

FIG. 7.5

$$R_A = R_{1\|2} = \frac{(9)(6)}{9 + 6} = \frac{54}{15} = 3.6\,\Omega$$

$$R_B = R_3 + R_{4\|5} = 4 + \frac{(6)(3)}{6+3} = 4 + 2 = 6\,\Omega$$

$$R_C = 3\,\Omega$$

The network of Fig. 7.5 can then be redrawn in reduced form as shown in Fig. 7.6. Note the similarities between this circuit and those of Figs. 7.2 and 7.4.

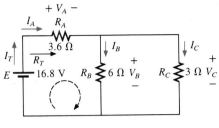

FIG. 7.6

$$R_T = R_A + R_{B\|C} = 3.6 + \frac{(6)(3)}{6+3}$$

$$= 3.6 + 2 = \textbf{5.6}\,\Omega$$

$$I_T = \frac{E}{R_T} = \frac{16.8}{5.6} = \textbf{3 A}$$

$$I_A = I_T = \textbf{3 A}$$

Applying the current divider rule yields

$$I_B = \frac{R_C I_A}{R_C + R_B} = \frac{(3)(3)}{3+6} = \frac{9}{9} = \textbf{1 A}$$

By Kirchhoff's current law,

$$I_C = I_A - I_B = 3 - 1 = \textbf{2 A}$$

By Ohm's law,

$$V_A = I_A R_A = (3)(3.6) = \textbf{10.8 V}$$

$$V_B = I_B R_B = V_C = I_C R_C = (2)(3) = \textbf{6 V}$$

Returning to the original network (Fig. 7.5) and applying the current divider rule,

$$I_1 = \frac{R_2 I_A}{R_2 + R_1} = \frac{(6)(3)}{6+9} = \frac{18}{15} = \textbf{1.2 A}$$

By Kirchhoff's current law,

$$I_2 = I_A - I_1 = 3 - 1.2 = \textbf{1.8 A}$$

Kirchhoff's voltage law for the loop indicated in the reduced network (Fig. 7.6) is

$$\Sigma_{\circlearrowleft} V = E - V_A - V_B = 0$$

$$E = V_A + V_B$$

$$16.8 = 10.8 + 6$$

$$\underline{16.8 = 16.8 \qquad \text{(checks)}}$$

Figures 7.2, 7.3, and 7.5 are only a few of the infinite variety of configurations that the network can assume starting with the basic arrangement of Fig. 7.1. They were included in our discussion to emphasize the importance of considering each branch independently before finding the solution for the network as a whole.

There are a variety of ways in which the blocks of Fig. 7.1 can be arranged. In fact, there is no limit on the number of series-parallel configurations that can appear within a given network. In reverse, the block diagram approach can be effectively used to reduce the apparent complexity of a system by identifying the major series and parallel components of the network. This approach will be demonstrated in the next few examples.

7.2 DESCRIPTIVE EXAMPLES

EXAMPLE 7.4. Find the current I_4 and the voltage V_2 for the network of Fig. 7.7.

Solution: With the block diagram approach, the network has the basic structure of Fig. 7.8, clearly indicating that the three branches are

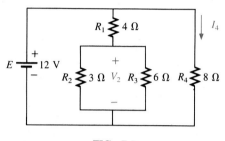

FIG. 7.7

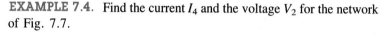
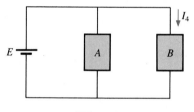

FIG. 7.8

in parallel and the voltage across A and B is the supply voltage. The current I_4 is now immediately obvious as simply the supply voltage divided by the resultant resistance for B. If desired, block A could be broken down further as shown in Fig. 7.9 to identify C and D as series elements with the voltage V_2 capable of being determined using the voltage divider rule once the resistance of C and D is reduced to a single value. This is an example of how a mental sketch of the approach might be made before applying laws, rules, and so on, to avoid dead ends and growing frustration.

Applying Ohm's law,

$$I_4 = \frac{E}{R_B} = \frac{E}{R_4} = \frac{12}{8} = \textbf{1.5 A}$$

Combining the resistors R_2 and R_3 of Fig. 7.7 will result in

$$R_D = R_2 \parallel R_3 = 3\,\Omega \parallel 6\,\Omega = \frac{(3)(6)}{3+6} = \frac{18}{9} = 2\,\Omega$$

and, applying the voltage divider rule,

$$V_2 = \frac{R_D E}{R_D + R_C} = \frac{(2)(12)}{2+4} = \frac{24}{6} = \textbf{4 V}$$

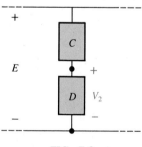

FIG. 7.9

EXAMPLE 7.5. Find the indicated currents and voltages for the network of Fig. 7.10.

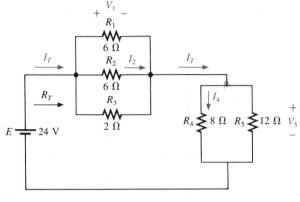

FIG. 7.10

Solution: The block diagram of the network may appear as shown in Fig. 7.11, clearly revealing that A and B are in series. Note in this form the number of unknowns that have been preserved. The voltage V_1 will be the same across the three parallel branches of Fig. 7.10, and V_5 will be the same across R_4 and R_5. Once V_1 and V_5 are known, the required currents can be found using Ohm's law.

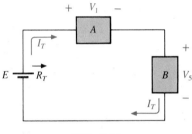

FIG. 7.11

$$R_{1\|2} = \frac{R}{N} = \frac{6}{2} = 3\,\Omega$$

$$R_A = R_{1\|2\|3} = \frac{(3)(2)}{3+2} = \frac{6}{5} = 1.2\,\Omega$$

$$R_B = R_{4\|5} = \frac{(8)(12)}{8+12} = \frac{96}{20} = 4.8\,\Omega$$

The reduced form of Fig. 7.10 will then appear as shown in Fig. 7.12, and

$$R_T = R_{1\|2\|3} + R_{4\|5} = 1.2 + 4.8 = \mathbf{6\,\Omega}$$

$$I_T = \frac{E}{R_T} = \frac{24}{6} = \mathbf{4\,A}$$

with

$$V_1 = I_T R_{1\|2\|3} = (4)(1.2) = \mathbf{4.8\,V}$$

$$V_5 = I_T R_{4\|5} = (4)(4.8) = \mathbf{19.2\,V}$$

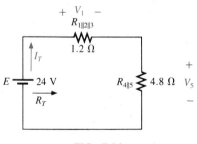

FIG. 7.12

Applying Ohm's law,

$$I_4 = \frac{V_5}{R_4} = \frac{19.2}{8} = \mathbf{2.4\,A}$$

$$I_2 = \frac{V_2}{R_2} = \frac{V_1}{R_2} = \frac{4.8}{6} = \mathbf{0.8\,A}$$

EXAMPLE 7.6.

a. Find the voltages V_1, V_3, and V_{ab} for the network of Fig. 7.13.

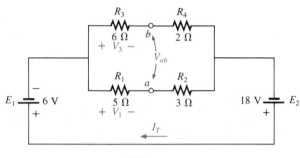

FIG. 7.13

b. Calculate the source current I_T.

Solutions: This is one of those situations where it might be best to redraw the network before beginning the analysis. Since combining both sources will not affect the unknowns, the network is redrawn as shown in Fig. 7.14. The net source voltage is the difference between the two with the polarity of the larger.

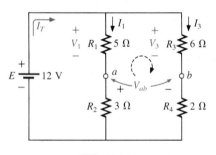

FIG. 7.14

a. Note the similarities with Fig. 7.9, permitting the use of the voltage divider rule to determine V_1 and V_3:

$$V_1 = \frac{R_1 E}{R_1 + R_2} = \frac{(5)(12)}{5 + 3} = \frac{60}{8} = \textbf{7.5 V}$$

$$V_3 = \frac{R_3 E}{R_3 + R_4} = \frac{(6)(12)}{6 + 2} = \frac{72}{8} = \textbf{9 V}$$

The open-circuit voltage V_{ab} is determined by applying Kirchhoff's voltage law around the indicated loop of Fig. 7.14 in the clockwise direction:

$$+V_1 - V_3 + V_{ab} = 0$$

and

$$V_{ab} = V_3 - V_1 = 9 - 7.5 = \textbf{1.5 V}$$

b. By Ohm's law,

$$I_1 = \frac{V_1}{R_1} = \frac{7.5}{5} = 1.5 \text{ A}$$

$$I_3 = \frac{V_3}{R_3} = \frac{9}{6} = 1.5 \text{ A}$$

By Kirchhoff's current law,

$$I_T = I_1 + I_3 = 1.5 + 1.5 = \textbf{3 A}$$

EXAMPLE 7.7. For the network of Fig. 7.15, determine the voltages V_1 and V_2 and the current I.

Solution: The network is redrawn in Fig. 7.16. Note the common connection of the grounds and the replacing of the terminal notation by actual supplies.

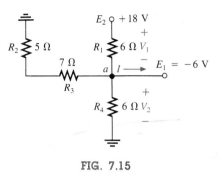

FIG. 7.15

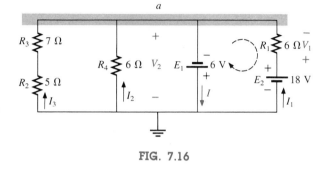

FIG. 7.16

It is now obvious that

$$V_2 = -E_1 = -6 \text{ V}$$

The minus sign simply indicates that the chosen polarity for V_2 in Fig. 7.15 is opposite to the actual voltage. Applying Kirchhoff's voltage law to the loop indicated, we obtain

$$-E_1 + V_1 - E_2 = 0$$

and

$$V_1 = E_2 + E_1 = 18 + 6 = \textbf{24 V}$$

Applying Kirchhoff's current law to node a yields

$$I = I_1 + I_2 + I_3$$

$$= \frac{V_1}{R_1} + \frac{E_1}{R_4} + \frac{E_1}{R_2 + R_3}$$

$$= \frac{24}{6} + \frac{6}{6} + \frac{6}{12}$$

$$= 4 + 1 + 0.5$$

$$I = \textbf{5.5 A}$$

EXAMPLE 7.8. For the transistor configuration of Fig. 7.17:
a. Determine the voltage V_E and the current I_E.
b. Calculate V_1.
c. Determine V_{BC} using the fact that the approximation $I_C = I_E$ is often applied to transistor networks.
d. Calculate V_{CE} using the information obtained in parts (a) through (c).

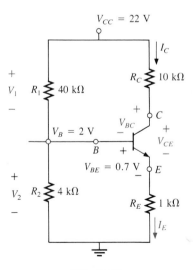

FIG. 7.17

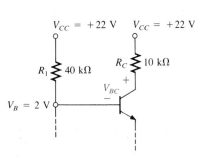

$V_{CC} = +22$ V $\qquad$ $V_{CC} = +22$ V

R_1 40 kΩ $\qquad$ R_C 10 kΩ

V_{BC}

$V_B = 2$ V

FIG. 7.18

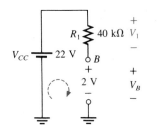

R_1 40 kΩ V_1

V_{CC} 22 V $\quad$ B

2 V $\quad V_B$

FIG. 7.19

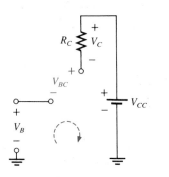

R_C V_C

V_{BC}

V_{CC}

V_B

FIG. 7.20

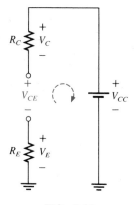

R_C V_C

V_{CE} V_{CC}

R_E V_E

FIG. 7.21

Solutions:

a. From Fig. 7.17, we find

$$V_2 = V_B = 2 \text{ V}$$

Writing Kirchhoff's voltage law around the lower loop yields

$$V_2 - V_{BE} - V_E = 0$$

or

$$V_E = V_2 - V_{BE} = 2 - 0.7 = \mathbf{1.3 \text{ V}}$$

and

$$I_E = \frac{V_E}{R_E} = \frac{1.3}{1000} = \mathbf{1.3 \text{ mA}}$$

b. The 22-V dc source can be split as shown in Fig. 7.18 as an aid in the required analysis. The section of interest can then be sketched as shown in Fig. 7.19, and Kirchhoff's law will result in

$$V_{CC} - V_1 - V_B = 0$$

or

$$V_1 = V_{CC} - V_B = 22 - 2 = \mathbf{20 \text{ V}}$$

c. Redrawing the section of the network of immediate interest will result in Fig. 7.20, where Kirchhoff's voltage law yields

$$V_B + V_{BC} + V_C - V_{CC} = 0$$

or

$$V_{BC} = V_{CC} - V_C - V_B$$

with

$$V_C = I_C R_C = (1.3 \text{ mA})(10 \text{ k}\Omega) = 13 \text{ V}$$

and

$$V_{BC} = 22 - 13 - 2 = \mathbf{7 \text{ V}}$$

d. The appropriate section appears in Fig. 7.21. Application of Kirchhoff's voltage law will result in

$$V_E + V_{CE} + V_C = V_{CC}$$

or

$$V_{CE} = V_{CC} - V_C - V_E = 22 - 13 - 1.3 = \mathbf{7.7 \text{ V}}$$

Note in the above analysis that there was no requirement to know any of the details regarding transistor behavior (except $I_C = I_E$ and $V_{BE} = 0.7$ V). Be assured that there will be a great deal of coverage of the type of analysis described above when electronic systems are examined.

EXAMPLE 7.9. Calculate the indicated currents and voltage of Fig. 7.22.

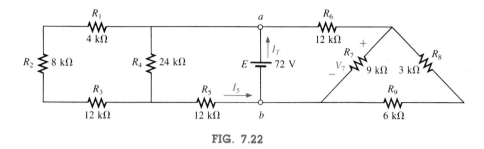

FIG. 7.22

Solution: Redrawing the network after combining series elements yields Fig. 7.23, and

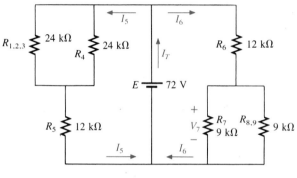

FIG. 7.23

$$I_5 = \frac{E}{R_{(1,2,3)\|4} + R_5} = \frac{72}{12\,\text{k}\Omega + 12\,\text{k}\Omega} = \frac{72}{24\,\text{k}\Omega} = \textbf{3 mA}$$

with

$$V_7 = \frac{R_{7\|(8,9)}E}{R_{7\|(8,9)} + R_6} = \frac{(4.5\,\text{k}\Omega)(72)}{4.5\,\text{k}\Omega + 12\,\text{k}\Omega} = \frac{324}{16.5} = \textbf{19.6 V}$$

$$I_6 = \frac{V_7}{R_{7\|(8,9)}} = \frac{19.6}{4.5\,\text{k}\Omega} = \textbf{4.35 mA}$$

and

$$I_T = I_5 + I_6 = 3 + 4.35 = \textbf{7.35 mA}$$

Since the potential difference between points a and b of Fig. 7.22 is fixed at E volts, the circuit to the right or left is unaffected if the network is reconstructed as shown in Fig. 7.24.

We can find each quantity required, except I_T, by analyzing each circuit independently. To find I_T, we must find the source current for each circuit and add it as in the above solution; that is, $I_T = I_5 + I_6$.

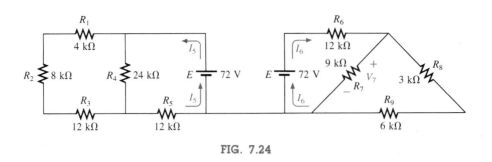

FIG. 7.24

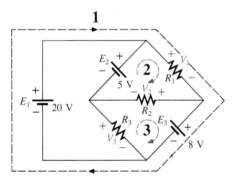

FIG. 7.25

FIG. 7.26

EXAMPLE 7.10. This example will demonstrate the power of Kirch-hoff's voltage law by determining the voltages V_1, V_2, and V_3 for the network of Fig. 7.25. For path 1 of Fig. 7.26,

$$E_1 - V_1 - E_3 = 0$$

and

$$V_1 = E_1 - E_3 = 20 - 8 = \mathbf{12\ V}$$

For path 2,

$$E_2 - V_1 - V_2 = 0$$

and

$$V_2 = E_2 - V_1 = 5 - 12 = \mathbf{-7\ V}$$

indicating that V_2 has a magnitude of 7 V but a polarity opposite to that appearing in Fig. 7.25. For path 3,

$$V_3 + V_2 - E_3 = 0$$

and

$$V_3 = E_3 - V_2 = 8 - (-7) = 8 + 7 = \mathbf{15\ V}$$

Note that the polarity of V_2 was maintained as originally assumed, requiring that -7 V be substituted for V_2.

7.3 LADDER NETWORKS

A three-section *ladder* network appears in Fig. 7.27. The reason for the terminology is quite obvious for the repetitive structure. There are basically two approaches used to solve networks of this type.

Method 1

Calculate the total resistance and resulting source current and then work back through the ladder until the desired current or voltage is obtained. This method is now employed to determine V_6 in Fig. 7.27.

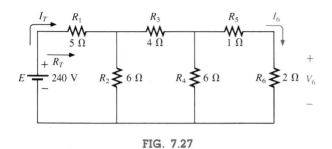

FIG. 7.27

Combining parallel and series elements as shown in Fig. 7.28 will result in the reduced network of Fig. 7.29, and

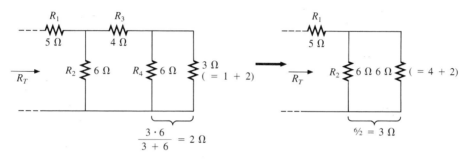

FIG. 7.28

$$R_T = 5 + 3 = 8 \, \Omega$$

$$I_T = \frac{E}{R_T} = \frac{240}{8} = 30 \, \text{A}$$

Working our way back to I_6 (Fig. 7.30), we find that

$$I_1 = I_T$$

and

$$I_3 = \frac{I_T}{2} = \frac{30}{2} = 15 \, \text{A}$$

and, finally (Fig. 7.31),

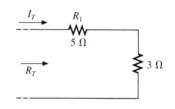

FIG. 7.29

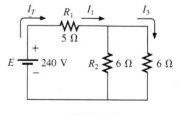

FIG. 7.30

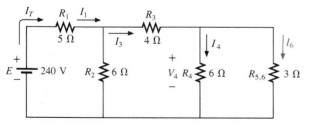

FIG. 7.31

$$I_6 = \frac{6I_3}{6+3} = \frac{6}{9}(15) = 10 \text{ A}$$

and

$$V_6 = I_6 R_6 = (10)(2) = \mathbf{20 \text{ V}}$$

Method 2

Assign a letter symbol to the last branch current and work back through the network to the source, maintaining this assigned current or other current of interest. The desired current can then be found directly. This method can best be described through the analysis of the same network considered above, redrawn in Fig. 7.32.

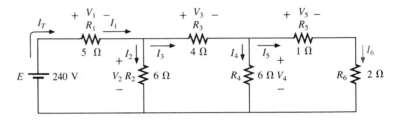

FIG. 7.32

The assigned notation for the current through the final branch is I_6:

$$I_6 = \frac{V_4}{R_5 + R_6} = \frac{V_4}{1+2} = \frac{V_4}{3}$$

or

$$V_4 = 3I_6$$

so that

$$I_4 = \frac{V_4}{R_4} = \frac{3I_6}{6} = 0.5I_6$$

and

$$I_3 = I_4 + I_6 = 0.5I_6 + I_6 = 1.5I_6$$
$$V_3 = I_3 R_3 = (1.5I_6)(4) = 6I_6$$

Also,

$$V_2 = V_3 + V_4 = 6I_6 + 3I_6 = 9I_6$$

so that

$$I_2 = \frac{V_2}{R_2} = \frac{9I_6}{6} = 1.5I_6$$

and

$$I_T = I_2 + I_3 = 1.5I_6 + 1.5I_6 = 3I_6$$

with

$$V_1 = I_1 R_1 = I_T R_1 = 5I_T$$

so that

$$E = V_1 + V_2 = 5I_T + 9I_6$$
$$= (5)(3I_6) + 9I_6 = 24I_6$$

and

$$I_6 = \frac{E}{24} = \frac{240}{24} = 10 \text{ A}$$

with

$$V_6 = I_6 R_6 = (10)(2) = \mathbf{20\ V}$$

as was obtained using method 1.

7.4 AMMETER, VOLTMETER, AND OHMMETER DESIGN

Now that the fundamentals of series, parallel, and series-parallel networks have been introduced, we are prepared to investigate the fundamental design of an ammeter, voltmeter, and ohmmeter. Our design of each will employ the d'Arsonval movement of Fig. 7.33. The movement consists basically of an iron-core coil mounted on bearings between a permanent magnet. The helical springs limit the turning motion of the coil and provide a path for the current to reach the coil. When a current is passed through the movable coil, the fluxes of the coil and permanent magnet will interact to develop a torque on the coil which will cause it to rotate on its bearings. The movement is adjusted to indicate zero deflection on a meter scale when the current through the coil is zero. The direction of current through the coil will then determine whether the pointer will display an up-scale or below-zero indication. For this reason, ammeters and voltmeters have an assigned polarity on their terminals to insure an up-scale reading.

D'Arsonval movements are usually rated by current and resistance. The specifications of a typical movement might be 1 mA, 50 Ω. The 1 mA is the *current sensitivity (CS)* of the movement, which is the current required for a full-scale deflection. It will be denoted by the symbol I_{CS}. The 50 Ω represents the internal resistance (R_m) of the movement. A common notation for the movement and its specifications is provided in Fig. 7.34.

The Ammeter

The maximum current that the d'Arsonval movement can read independently is equal to the current sensitivity of the movement. However, higher currents can be measured if additional circuitry is introduced.

FIG. 7.33
d'Arsonval movement. (Courtesy of Weston Instruments, Inc.)

1 mA, 50 Ω

FIG. 7.34
Movement notation.

som   as low as

10 μA

This additional circuitry, as shown in Fig. 7.35, results in the basic construction of an ammeter.

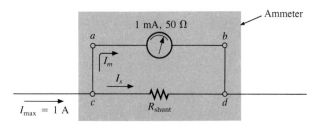

FIG. 7.35
Basic ammeter.

The resistance R_{shunt} is chosen for the ammeter of Fig. 7.35 to allow 1 mA to flow through the movement when a maximum current of 1 A enters the ammeter. If less than 1 A should flow through the ammeter, the movement will have less than 1 mA flowing through it and will indicate less than full-scale deflection.

Since the voltage across parallel elements must be the same, the potential drop across *a-b* in Fig. 7.35 must equal that across *c-d*; that is,

$$(1 \text{ mA})(50 \, \Omega) = R_{shunt}I_s$$

Also, I_s must equal 1 A $-$ 1 mA = 999 mA if the current is to be limited to 1 mA through the movement (Kirchhoff's current law). Therefore,

$$(1 \times 10^{-3})(50) = (999 \times 10^{-3})(R_{shunt})$$

$$R_{shunt} = \frac{(1 \times 10^{-3})(50)}{999 \times 10^{-3}}$$

$$\cong 0.05 \, \Omega$$

In general,

$$R_{shunt} = \frac{R_m I_{CS}}{I_{max} - I_{CS}} \qquad (7.1)$$

One method of constructing a multirange ammeter is shown in Fig. 7.36, where the rotary switch determines the R_{shunt} to be used for the

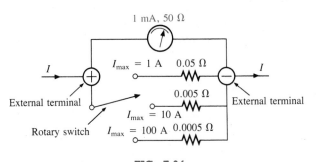

FIG. 7.36
Multirange ammeter.

maximum current indicated on the face of the meter. Most meters employ the same scale for various values of maximum current. If you read 375 on the 0–5 mA scale with the switch on the 5 setting, the current is 3.75 mA; on the 50 setting, the current is 37.5 mA; and so on.

The Voltmeter

A variation in the additional circuitry will permit the use of the d'Arsonval movement in the design of a voltmeter. The 1-mA, 50-Ω movement can also be rated as a 50-mV (1 mA $\times$ 50 Ω), 50-Ω movement, indicating that the maximum voltage that the movement can measure independently is 50 mV. The millivolt rating is sometimes referred to as the *voltage sensitivity (VS)*. The basic construction of the voltmeter is shown in Fig. 7.37.

The R_{series} is adjusted to limit the current through the movement to 1 mA when the maximum voltage is applied across the voltmeter. A lesser voltage would simply reduce the current in the circuit, and thereby the deflection of the movement.

Applying Kirchhoff's voltage law around the closed loop of Fig. 7.37, we obtain

$$[10 - (1 \times 10^{-3})(R_{\text{series}})] - 50 \times 10^{-3} = 0$$

or

$$R_{\text{series}} = \frac{10 - (50 \times 10^{-3})}{1 \times 10^{-3}} = 9950 \ \Omega$$

In general,

$$\boxed{R_{\text{series}} = \frac{V_{\text{max}} - V_{\text{VS}}}{I_{CS}}} \qquad (7.2)$$

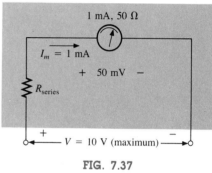

FIG. 7.37

Basic voltmeter.

One method of constructing a multirange voltmeter is shown in Fig. 7.38. If the rotary switch is at 10 V, $R_{\text{series}} = 9.950 \text{ k}\Omega$; at 50 V, $R_{\text{series}} = 40 \text{ k}\Omega + 9.950 \text{ k}\Omega = 49.950 \text{ k}\Omega$; and at 100 V, $R_{\text{series}} = 50 \text{ k}\Omega + 40 \text{ k}\Omega + 9.950 \text{ k}\Omega = 99.950 \text{ k}\Omega$.

The Ohmmeter

In general, ohmmeters are designed to measure resistance in the low, mid-, or high range. The most common is the *series ohmmeter*, designed to read resistance levels in the midrange. It employs the series configuration of Fig. 7.39. The design is quite different from that of the ammeter or voltmeter in that it will show a full-scale deflection for zero ohms and no deflection for infinite resistance.

To determine the series resistance R_s, the external terminals are shorted (a direct connection of zero ohms between the two) to simulate zero ohms, and the zero-adjust is set to half its maximum value. The resistance R_s is then adjusted to allow a current equal to the current sensitivity of the movement (1 mA) to flow in the circuit. The zero-

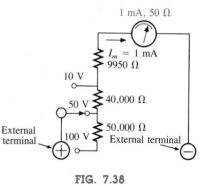

FIG. 7.38

Multirange voltmeter.

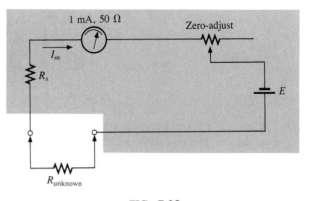

FIG. 7.39
Series ohmmeter.

adjust is set to half its value so that any variation in the components of the meter that may produce a current more or less than the current sensitivity can be compensated for. The current I_m is

$$I_m \text{ (full scale)} = I_{CS} = \frac{E}{R_s + R_m + \dfrac{\text{zero-adjust}}{2}} \qquad (7.3)$$

and

$$R_s = \frac{E}{I_{CS}} - R_m - \frac{\text{zero-adjust}}{2} \qquad (7.4)$$

If an unknown resistance is then placed between the external terminals, the current will be reduced, causing a deflection less than full scale. If the terminals are left open, simulating infinite resistance, the pointer will not deflect since the current through the circuit is zero.

An instrument designed to read very low values of resistance appears in Fig. 7.40. Because of its low range capability, the network design must be a great deal more sophisticated than described above. It employs electronic components that eliminate the inaccuracies introduced by lead and contact resistances. It is similar to the above system in the sense that it is completely portable and does require a dc battery to establish measurement conditions. Note the special leads designed to limit any introduced resistance levels. The maximum scale setting can be set as low as 0.00352 (3 mΩ).

The Megger tester is an instrument for measuring very high resistance values. The term *Megger* is derived from the fact that the device measures resistance values in the megohm range. Its primary function is to test the insulation found in power transmission systems, electrical

FIG. 7.40
Milliohmmeter. (Courtesy of Keithley Instruments, Inc.)

machinery, transformers, and so on. To measure the high-resistance values, a high dc voltage is established by a hand-driven generator. If the shaft is rotated above some set value, the output of the generator will be fixed at one selectable voltage, typically 250, 500, or 1000 V. A photograph of the commercially available Megger tester is shown in Fig. 7.41. The unknown resistance is connected between the terminals marked *Line* and *Earth*. For this instrument, the range is zero to 2000 MΩ.

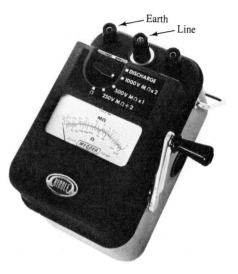

FIG. 7.41
The Megger® tester. (Courtesy of James G. Biddle Co.)

$$P = \frac{V^2}{R}$$

PROBLEMS

Section 7.2

1. For the network of Fig. 7.42:
   a. Does $I = I_3 = I_6$? Explain.
   b. If $I = 5$ A and $I_1 = 2$ A, find I_2.
   c. Does $I_1 + I_2 = I_4 + I_5$? Explain.
   d. If $V_1 = 6$ V and $E = 10$ V, find V_2.
   e. If $R_1 = 3\,\Omega$, $R_2 = 2\,\Omega$, $R_3 = 4\,\Omega$, and $R_4 = 1\,\Omega$, what is R_T?
   f. If the resistors have the values given in part (e) and $E = 10$ V, what is the value of I in amperes?
   g. Using values given in parts (e) and (f), find the power delivered by the battery E and dissipated by the resistors R_1 and R_2.

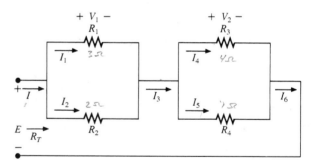

FIG. 7.42

2. For the network of Fig. 7.43:
   a. Calculate R_T.
   b. Determine I and I_1.
   c. Find V_3.

$$\frac{6}{5} + \frac{4}{5}$$

$$\underline{10}$$

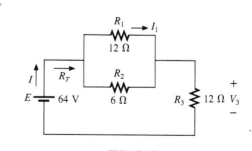

FIG. 7.43

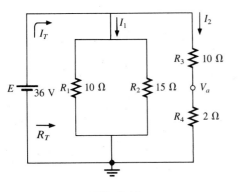

FIG. 7.44

3. For the network of Fig. 7.44:
  a. Determine R_T.
  b. Find I_T, I_1, and I_2.
  c. Calculate V_a.

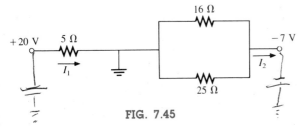

FIG. 7.45

4. Determine the currents I_1 and I_2 for the network of Fig. 7.45.

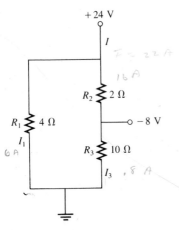

FIG. 7.46

5. a. Find the currents I, I_1, and I_3 for the network of Fig. 7.46.
  b. Indicate their direction on Fig. 7.46.

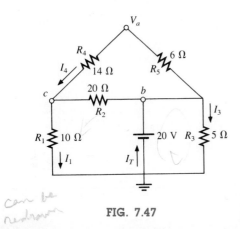

FIG. 7.47

*6. For the network of Fig. 7.47:
  a. Determine the currents I_T, I_1, I_3, and I_4.
  b. Calculate V_a and V_{bc}.

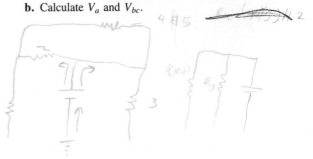

7. For the network of Fig. 7.48:
   a. Determine the current I_1.
   b. Calculate the currents I_2 and I_3.
   c. Determine the voltage levels V_a and V_b.

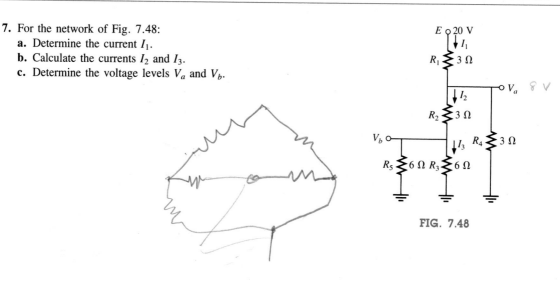

FIG. 7.48

8. For the network of Fig. 7.49:
   a. Find the currents I and I_6.
   b. Find the voltages V_1 and V_5.
   c. Find the power delivered to the 6-kΩ resistor.

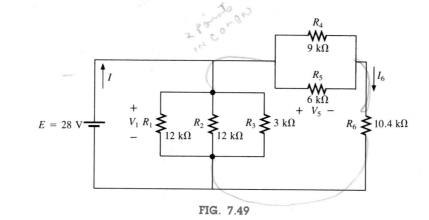

FIG. 7.49

*9. For the series-parallel network of Fig. 7.50:
   a. Find the current I.
   b. Find the currents I_3 and I_9.
   c. Find the current I_8.
   d. Find the voltage V_{ab}.

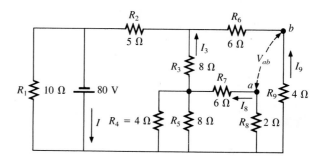

FIG. 7.50

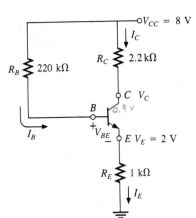

FIG. 7.51

*10. Determine the dc levels for the transistor network of Fig. 7.51 using the fact that $V_{BE} = 0.7\,$V, $V_E = 2\,$V, and $I_C = I_E$. That is,
   a. Determine I_E and I_C.
   b. Calculate I_B.
   c. Determine V_B and V_C.

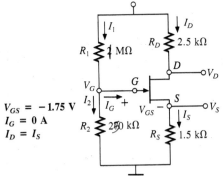

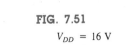

FIG. 7.52

*11. The network of Fig. 7.52 is the basic biasing arrangement for the *field-effect transistor* (FET), a device of increasing importance in electronic design. (*Biasing* simply means the application of dc levels to establish a particular set of operating conditions.) Even though you may be unfamiliar with the FET, you can perform the following analysis using only the basic laws introduced in this chapter and the information provided on the diagram.
   a. Determine the voltages V_G and V_S.
   b. Find the currents I_1, I_2, I_D, and I_S.
   c. Determine V_{DS}.
   d. Calculate V_{DG}.

EXTRA CREDIT

*12. The *difference amplifier* of Fig. 7.53 is a compound configuration that will establish an output (in the ac domain) that is the difference between the two input signals. Using the concepts introduced in this chapter and previous chapters, determine the following dc levels:
   a. V_E (given that $I_1 = 2\,$mA).
   b. I_2 (magnitude and direction).
   c. I_{E_1} and I_{E_2} using the fact that $I_C = I_E$ and $I_{E_1} = I_{E_2}$ (balanced system).
   d. V_C, V_{E_1}, and V_{E_2} if $V_{CE} = 10.7\,$V.
   e. V_{B_1} and V_{B_2} using the information obtained above.
   f. V_{CE_1} and V_{CE_2} if $I_{C_1} = I_{E_1}$ and $I_{C_2} = I_{E_2}$.

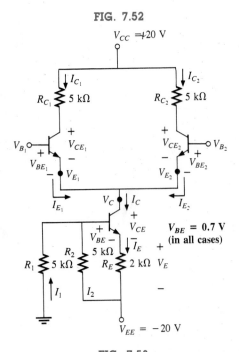

FIG. 7.53

*13. For the series-parallel configuration of Fig. 7.54:
  a. Find the current I.
  b. Find the currents I_1, I_3, and I_8.
  c. Find the power delivered to the 21-Ω resistor.

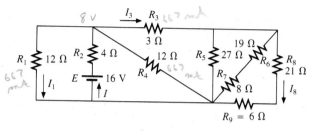

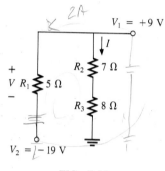

FIG. 7.54

14. For the network of Fig. 7.55:
  a. Determine the current I.
  b. Find V.

FIG. 7.55

15. For the configuration of Fig. 7.56:
  a. Find the currents I_2, I_6, and I_8.
  b. Find the voltages V_4 and V_8.

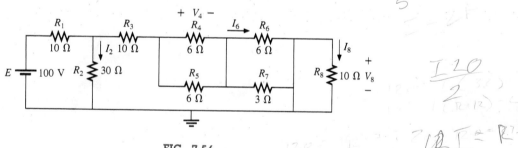

FIG. 7.56

*16. For the network of Fig. 7.57, find the resistance R_3 if the
current through it is 2 A.

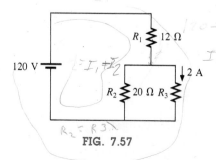

FIG. 7.57

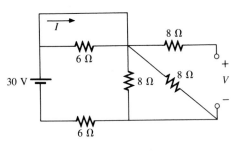

FIG. 7.58

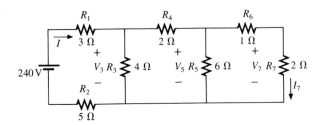

FIG. 7.59

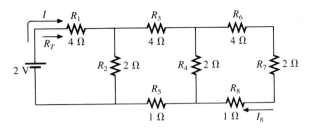

FIG. 7.60

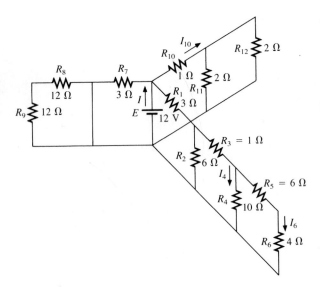

FIG. 7.61

17. Determine the voltage V and the current I for the network of Fig. 7.58.

Section 7.3

18. For the ladder network of Fig. 7.59:
   a. Find the current I.
   b. Find the current I_7.
   c. Determine the voltages V_3, V_5, and V_7.
   d. Calculate the power delivered to R_7 and compare it to the power delivered by the 240-V supply.

19. For the ladder network of Fig. 7.60:
   a. Determine R_T.
   b. Calculate I.
   c. Find I_8.

*20. For the multiple ladder configuration of Fig. 7.61:
   a. Determine I.
   b. Calculate I_4.
   c. Find I_6.
   d. Find I_{10}.

Section 7.4

21. A d'Arsonval movement is rated 1 mA, 100 Ω.
   a. What is the current sensitivity?
   b. Design a 20-A ammeter using the above movement. Show the circuit and component values.

22. Using a 50-μA, 1000-Ω d'Arsonval movement, design a multirange milliammeter having scales of 25, 50, and 100 mA. Show the circuit and component values.

23. A d'Arsonval movement is rated 50 μA, 1000 Ω.
   a. Design a 15-V dc voltmeter. Show the circuit and component values.
   b. What is the ohm/volt rating of the voltmeter?

24. Using a 1-mA, 100-Ω d'Arsonval movement, design a multirange voltmeter having scales of 5, 50, and 500 V. Show the circuit and component values.

25. A digital meter has an internal resistance of 10 MΩ on its 0.5-V range. If you had to build a voltmeter with a d'Arsonval movement, what current sensitivity would you need if the meter were to have the same internal resistance on the same voltage scale?

*26. **a.** Design a series ohmmeter using a 100-μA, 1000-Ω movement, a zero-adjust with a maximum value of 2 kΩ, a battery of 3 V, and a series resistor whose value is to be determined.

b. Find the resistance required for full-scale, $\frac{3}{4}$-scale, $\frac{1}{2}$-scale, and $\frac{1}{4}$-scale deflection.

c. Using the results of part (b), draw the scale to be used with the ohmmeter.

27. Describe the basic construction and operation of the Megger.

Computer Problems

28. Write a program that will find the complete solution for the network of Fig. 7.2. That is, given all the parameters of the network, calculate the current, voltage, and power to each element.

29. Write a program to find all the quantities of Example 7.8 given the network parameters.

30. Write a program to find R_T, I_T, and I_6 for the network of Fig. 7.27 given the network parameters.

GLOSSARY

d'Arsonval movement   An iron-core coil mounted on bearings between a permanent magnet. A pointer connected to the movable core indicates the strength of the current passing through the coil.

Field-effect transistor (FET)   A three-terminal electronic device capable of amplifying time-varying signals.

Ladder network   A network that consists of a cascaded set of series-parallel combinations and has the appearance of a ladder.

Megger® tester   An instrument for measuring very high resistance levels, such as in the megohm range.

Series ohmmeter   A resistance-measuring instrument in which the movement is placed in series with the unknown resistance.

Series-parallel network   A network consisting of a combination of both series and parallel branches.

Shunt ohmmeter   A resistance-measuring instrument in which the movement is placed in parallel with the unknown resistance.

Transistor   A three-terminal semiconductor electronic device that can be used for amplification and switching purposes.

8

Methods of Analysis and Selected Topics (dc)

8.1 INTRODUCTION

The circuits described in the previous chapters had only one source or two or more sources in series or parallel present. The step-by-step procedure outlined in those chapters cannot be applied if two or more sources in the same network are not in series or parallel. There will be an interaction of sources that will not permit the reduction technique used in Chapter 7 to find such quantities as the total resistance and source current.

Methods of analysis have been developed that allow us to approach, in a systematic manner, a network with any number of sources in any arrangement. Fortunately, these methods can also be applied to networks with only one source. The methods to be discussed in detail in this chapter include *branch-current analysis, mesh analysis,* and *nodal analysis.* Each can be applied to the same network. The "best" method cannot be defined by a set of rules but can be determined only by acquiring a firm understanding of the relative advantages of each. All of the methods will be described for *linear bilateral* networks only. The term *linear* indicates that the characteristics of the network elements (such as the resistors) are independent of the voltage across or current through them. The second term, *bilateral,* refers to the fact that there is no change in the behavior or characteristics of an element if the current through or voltage across the element is reversed. Before discussing these methods, we shall consider the current source and the use of determinants. At the end of the chapter we shall consider bridge networks and Δ-Y and Y-Δ conversions.

Chapter 9 will present the important theorems of network analysis that can also be employed to solve networks with more than one source.

8.2 CURRENT SOURCES

The concept of the current source was introduced in Section 2.4 with the photograph of a commercially available unit. We must now investigate its characteristics in greater detail so that we can properly determine its effect on the networks to be examined in this chapter.

The current source is often referred to as the *dual* of the voltage source. That is, while a battery supplies a *fixed* voltage and the source current can vary, the current source supplies a *fixed* current to the branch in which it is located, while its terminal voltage may vary as determined by the network to which it is applied. Note from the above that *duality* simply implies an interchange of current and voltage to distinguish the characteristics of one source from the other.

The increasing interest in the current source is due fundamentally to semiconductor devices such as the transistor. In the basic electronics courses, you will find that the transistor is a current-controlled device. In the physical model (equivalent circuit) of a transistor used in the analysis of transistor networks, there appears a current source as indicated in Fig. 8.1. The symbol for a current source appears in Fig. 8.1. The direction of the arrow within the circle indicates the direction in which current is being supplied.

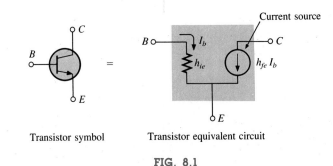

Transistor symbol   Transistor equivalent circuit

FIG. 8.1

For further comparison, the terminal characteristics of a *dc* voltage and current source are presented in Fig. 8.2. Note that for the voltage source the terminal voltage is fixed at E volts for the range of current values. For the region to the right of the voltage axis, the current will

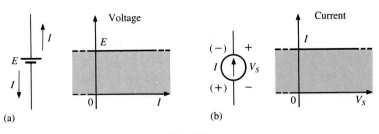

(a)   (b)

FIG. 8.2

have one direction through the source, while to the left of the same axis it is reversed. In other words, as indicated in the associated figure of Fig. 8.2(a), the terminal voltage of a source is unaffected by the direction of current through the source. The characteristics of the current source, shown in Fig. 8.2(b), indicate that the current source will supply a fixed current even though the voltage across the source may vary in magnitude or reverse its polarity. This is indicated in the associated figure of Fig. 8.2(b). For the voltage source, the current direction will be determined by the remaining elements of the network. For all one-voltage-source networks it will have the direction indicated to the right of the battery in Fig. 8.2(a). For the current source, the network to which it is connected will also determine the magnitude and polarity of the voltage across the source. For all single-current-source networks, it will have the polarity indicated to the right of the current source in Fig. 8.2(b).

EXAMPLE 8.1. Find the voltage V_S and the current I_1 for the circuit of Fig. 8.3.

Solution:

$$I_1 = I = \mathbf{10\,mA}$$
$$V_S = V_1 = I_1 R_1 = (10\,\text{mA})(20\,\text{k}\Omega) = \mathbf{200\,V}$$

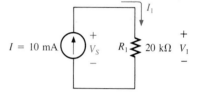

FIG. 8.3

EXAMPLE 8.2. Calculate the voltages V_1, V_2, and V_S for the circuit of Fig. 8.4.

Solution:

$$V_1 = I_1 R_1 = I R_1 = (5)(2) = \mathbf{10\,V}$$
$$V_2 = I_2 R_2 = I R_2 = (5)(3) = \mathbf{15\,V}$$

Applying Kirchhoff's voltage law in the clockwise direction, we obtain

$$\Sigma_\circlearrowright V = +V_S - V_1 - V_2 = 0$$

or

$$V_S = V_1 + V_2 = 10 + 15$$

and

$$V_S = \mathbf{25\,V}$$

Note the polarity of V_S for the single-source circuit.

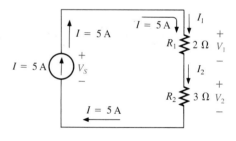

FIG. 8.4

EXAMPLE 8.3. Consider the series-parallel circuit of Fig. 8.5.
Solution: Using the current divider rule,

$$I_2 = \frac{R_3 I}{R_3 + R_2} = \frac{(1)I}{1 + 2} = \frac{1}{3}(6) = \mathbf{2\,A}$$

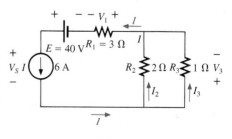

FIG. 8.5

By Ohm's law,

$$V_3 = I_3 R_3 = I_2 R_2 = (2)(2) = \textbf{4 V}$$
$$V_1 = I_1 R_1 = (6)(3) = \textbf{18 V}$$

By Kirchhoff's voltage law in the clockwise direction,

$$\Sigma_{\circlearrowright} V = +V_S - E + V_1 + V_3 = 0$$

or

$$V_S = -V_3 - V_1 + E = -4 - 18 + 40 = \textbf{18 V}$$

Note the polarity of V_S as determined by the multisource network.

8.3 SOURCE CONVERSIONS

It is often necessary or convenient to have a voltage source rather than a current source or a current source rather than a voltage source. If we consider the basic voltage source with its internal resistance as shown in Fig. 8.6, we find that

$$I_L = \frac{E}{R_S + R_L}$$

Or by multiplying the numerator of the equation by a factor of 1 which we choose to be R_S/R_S, we obtain

$$I_L = \frac{(1)(E)}{R_S + R_L} = \frac{(R_S/R_S)E}{R_S + R_L} = \frac{R_S(E/R_S)}{R_S + R_L} = \frac{R_S I}{R_S + R_L}$$

if we define $I = E/R_S$. The resulting equation is actually an application of the current divider rule to the network of Fig. 8.7.

For the load resistor R_L of Fig. 8.6 or 8.7, it is immaterial which source is applied as long as each element has the corresponding value. That is, the voltage across or current through R_L will be the same for each network. For clarity, the equivalent sources are repeated in Fig. 8.8 with the equations necessary for the conversion. Note that the resis-

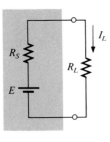

FIG. 8.6

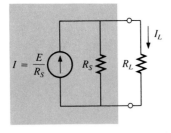

FIG. 8.7

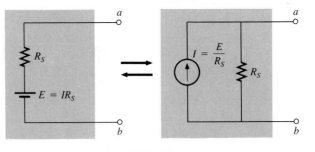

FIG. 8.8
Source conversion.

tor R_S is unchanged in magnitude and is simply brought from a series position for the voltage source to the parallel arrangement for the current source. It was pointed out in some detail in Chapter 6 that every source of voltage has some internal series resistance. *For the current source, some internal parallel resistance will always exist in the practical world.* However, in many cases, it is an excellent approximation to drop the internal resistance of a source due to the magnitude of the elements of the network to which it is applied. For this reason, in the analyses to follow, voltage sources may appear without a series resistor, and current sources may appear without a parallel resistance. Realize, however, that in order to perform a conversion from one type of source to another, a voltage source must have a resistor in series with it, and a current source must have a resistor in parallel.

EXAMPLE 8.4. Convert the voltage source of Fig. 8.9 to a current source and calculate the current through the load for each source.

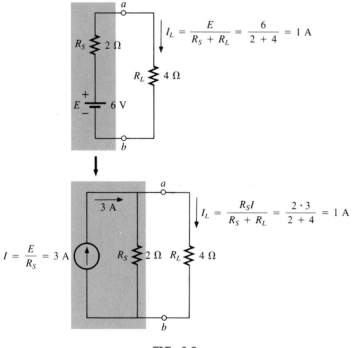

$$I_L = \frac{E}{R_S + R_L} = \frac{6}{2 + 4} = 1 \text{ A}$$

$$I_L = \frac{R_S I}{R_S + R_L} = \frac{2 \cdot 3}{2 + 4} = 1 \text{ A}$$

$$I = \frac{E}{R_S} = 3 \text{ A}$$

FIG. 8.9

Solution: See the right side of Fig. 8.9.

EXAMPLE 8.5. Convert the current source of Fig. 8.10 to a voltage source and find the current through the load for each source.

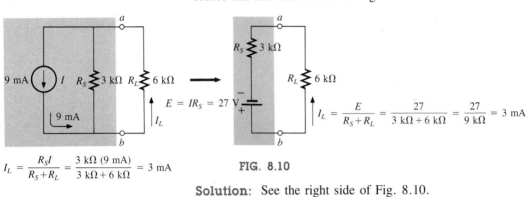

$$I_L = \frac{R_S I}{R_S + R_L} = \frac{3 \text{ k}\Omega \ (9 \text{ mA})}{3 \text{ k}\Omega + 6 \text{ k}\Omega} = 3 \text{ mA}$$

$$E = I R_S = 27 \text{ V}$$

$$I_L = \frac{E}{R_S + R_L} = \frac{27}{3 \text{ k}\Omega + 6 \text{ k}\Omega} = \frac{27}{9 \text{ k}\Omega} = 3 \text{ mA}$$

FIG. 8.10

Solution: See the right side of Fig. 8.10.

8.4 CURRENT SOURCES IN PARALLEL

If two or more current sources are in parallel, they may all be replaced by one current source having the magnitude and direction of the resultant, which can be found by summing the currents in one direction and subtracting the sum of the currents in the opposite direction. The new parallel resistance is determined by methods described in the discussion of parallel resistors in Chapter 5. Consider the following examples.

EXAMPLE 8.6. Reduce the left sides of Figs. 8.11 and 8.12 to a minimum number of elements.

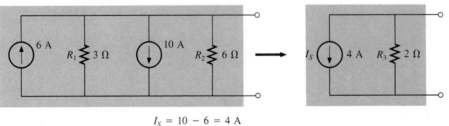

$$I_S = 10 - 6 = 4 \text{ A}$$
$$R_S = 3 \ \Omega \parallel 6 \ \Omega = 2 \ \Omega$$

FIG. 8.11

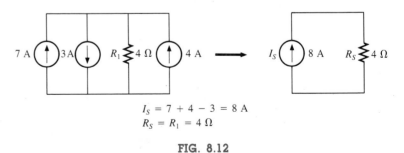

$$I_S = 7 + 4 - 3 = 8 \text{ A}$$
$$R_S = R_1 = 4 \ \Omega$$

FIG. 8.12

Solution: See the right side of the figures.

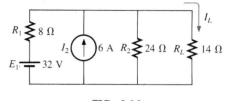

FIG. 8.13

EXAMPLE 8.7. Reduce the network of Fig. 8.13 to a single current source and calculate the current through R_L.

Solution: In this example, the voltage source will first be converted to a current source as shown in Fig. 8.14. Combining current sources,

$$I_S = I_1 + I_2 = 4 + 6 = \mathbf{10\,A}$$

and

$$R_S = R_1 \parallel R_2 = 8 \parallel 24 = \mathbf{6\,\Omega}$$

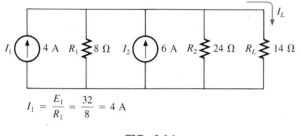

$$I_1 = \frac{E_1}{R_1} = \frac{32}{8} = 4\,\text{A}$$

FIG. 8.14

Applying the current divider rule to the network of Fig. 8.15,

$$I_L = \frac{R_S I_S}{R_S + R_L} = \frac{(6)(10)}{6 + 14} = \frac{60}{20} = \mathbf{3\,A}$$

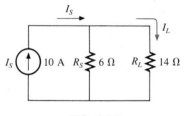

FIG. 8.15

EXAMPLE 8.8. Determine the current I_2 in the network of Fig. 8.16.
Solution: Although it might appear that the network cannot be solved using methods introduced thus far, one source conversion as shown in Fig. 8.17 will result in a simple series circuit:

$$E_S = I_1 R_1 = (4)(3) = 12\,\text{V}$$

and

$$R_S = R_1 = 3\,\Omega$$

and

$$I_2 = \frac{E_S + E_2}{R_S + R_2} = \frac{12 + 5}{3 + 2} = \frac{17}{5} = \mathbf{3.4\,A}$$

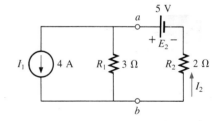

FIG. 8.16

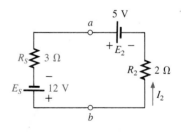

FIG. 8.17

8.5 CURRENT SOURCES IN SERIES

The current through any branch of a network can be only single-valued. For the situation indicated at point a in Fig. 8.18, we find by application

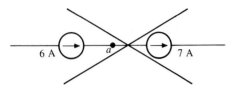

FIG. 8.18

of Kirchhoff's current law that the current leaving that point is greater than that entering—an impossible situation. Therefore, *current sources of different current ratings are not connected in series,* just as voltage sources of different voltage ratings are not connected in parallel.

8.6 DETERMINANTS

In the following analyses we will find it necessary to solve for the unknown variables of two or perhaps three simultaneous equations. Although a purely mathematical technique, the use of *determinants* is sufficiently important and helpful to warrant a complete section of the text. It is an approach that once understood will be appreciated as a time-saving method that usually results in fewer errors.

Consider the following equations, where x and y are the unknown variables and a_1, a_2, b_1, b_2, c_1, and c_2 are constants:

Col. 1		Col. 2		Col. 3	
$a_1 x$	$+$	$b_1 y$	$=$	c_1	**(8.1a)**
$a_2 x$	$+$	$b_2 y$	$=$	c_2	**(8.1b)**

It is certainly possible to solve for one variable in Eq. (8.1a) and substitute into Eq. (8.1b). That is, solving for x in Eq. (8.1a),

$$x = \frac{c_1 - b_1 y}{a_1}$$

and substituting the result in Eq. (8.1b),

$$a_2 \left(\frac{c_1 - b_1 y}{a_1} \right) + b_2 y = c_2$$

It is now possible to solve for y since it is the only variable remaining and then substitute into either equation for x. This is acceptable for two equations, but it becomes a very tedious and lengthy process for three or more simultaneous equations.

Using determinants to solve for x and y requires that the following formats be established for each variable:

$$x = \frac{\begin{vmatrix} c_1 & b_1 \\ c_2 & b_2 \end{vmatrix}}{\begin{vmatrix} a_1 & b_1 \\ a_2 & b_2 \end{vmatrix}} \qquad y = \frac{\begin{vmatrix} a_1 & c_1 \\ a_2 & c_2 \end{vmatrix}}{\begin{vmatrix} a_1 & b_1 \\ a_2 & b_2 \end{vmatrix}} \qquad \textbf{(8.2)}$$

with Col. 1 and Col. 2 headings over the determinants.

First note that only constants appear within the vertical brackets and that the denominator of each is the same. In fact, the denominator is simply the coefficients of x and y in the same arrangement as in Eqs. (8.1a) and (8.1b). When solving for x, the coefficients of x in the numerator are replaced by the constants to the right of the equal sign in Eqs. (8.1a) and (8.1b), while the coefficients of the y variable are simply repeated. When solving for y, the y coefficients in the numerator are replaced by the constants to the right of the equal sign and the coefficients of x are repeated.

Each configuration in the numerator and denominator of Eqs. (8.2) is referred to as a *determinant (D)*, which can be evaluated numerically in the following manner:

$$\text{Determinant} = D = \begin{matrix} \text{Col.} & \text{Col.} \\ 1 & 2 \end{matrix} \begin{vmatrix} a_1 & b_1 \\ a_2 & b_2 \end{vmatrix} = a_1 b_2 - a_2 b_1 \qquad (8.3)$$

The expanded value is obtained by first multiplying the top left element by the bottom right and then subtracting the product of the lower left and upper right elements. This particular determinant is referred to as a *second-order* determinant, since it contains two rows and two columns.

It is important to remember when using determinants that the columns of the equations, as indicated in Eqs. (8.1a) and (8.1b), be placed in the same order within the determinant configuration. That is, since a_1 and a_2 are in column 1 of Eqs. (8.1a) and (8.1b), they must be in column 1 of the determinant. (The same is true for b_1 and b_2.)

Expanding the entire expression for x and y, we have the following:

$$x = \frac{\begin{vmatrix} c_1 & b_1 \\ c_2 & b_2 \end{vmatrix}}{\begin{vmatrix} a_1 & b_1 \\ a_2 & b_2 \end{vmatrix}} = \frac{c_1 b_2 - c_2 b_1}{a_1 b_2 - a_2 b_1} \qquad (8.4a)$$

$$y = \frac{\begin{vmatrix} a_1 & c_1 \\ a_2 & c_2 \end{vmatrix}}{\begin{vmatrix} a_1 & b_1 \\ a_2 & b_2 \end{vmatrix}} = \frac{a_1 c_2 - a_2 c_1}{a_1 b_2 - a_2 b_1} \qquad (8.4b)$$

EXAMPLE 8.9. Evaluate the following determinants:

a. $\begin{vmatrix} 2 & 2 \\ 3 & 4 \end{vmatrix} = (2)(4) - (3)(2) = 8 - 6 = \mathbf{2}$

b. $\begin{vmatrix} 4 & -1 \\ 6 & 2 \end{vmatrix} = (4)(2) - (6)(-1) = 8 + 6 = \mathbf{14}$

c. $\begin{vmatrix} 0 & -2 \\ -2 & 4 \end{vmatrix} = (0)(4) - (-2)(-2) = 0 - 4 = \mathbf{-4}$

d. $\begin{vmatrix} 0 & 0 \\ 3 & 10 \end{vmatrix} = (0)(10) - (3)(0) = \mathbf{0}$

EXAMPLE 8.10. Solve for x and y:

$$2x + y = 3$$
$$3x + 4y = 2$$

Solution:

$$x = \frac{\begin{vmatrix} 3 & 1 \\ 2 & 4 \end{vmatrix}}{\begin{vmatrix} 2 & 1 \\ 3 & 4 \end{vmatrix}} = \frac{(3)(4) - (2)(1)}{(2)(4) - (3)(1)} = \frac{12 - 2}{8 - 3} = \frac{10}{5} = \mathbf{2}$$

$$y = \frac{\begin{vmatrix} 2 & 3 \\ 3 & 2 \end{vmatrix}}{5} = \frac{(2)(2) - (3)(3)}{5} = \frac{4 - 9}{5} = \frac{-5}{5} = \mathbf{-1}$$

Check:

$$2x + y = (2)(2) + (-1)$$
$$= 4 - 1 = 3 \quad \text{(checks)}$$
$$3x + 4y = (3)(2) + (4)(-1)$$
$$= 6 - 4 = 2 \quad \text{(checks)}$$

EXAMPLE 8.11. Solve for x and y:

$$-x + 2y = 3$$
$$3x - 2y = -2$$

Solution: In this example, note the effect of the minus sign and the use of parentheses to insure the proper sign is obtained for each product:

$$x = \frac{\begin{vmatrix} 3 & 2 \\ -2 & -2 \end{vmatrix}}{\begin{vmatrix} -1 & 2 \\ 3 & -2 \end{vmatrix}} = \frac{(3)(-2) - (-2)(2)}{(-1)(-2) - (3)(2)}$$

$$= \frac{-6 + 4}{2 - 6} = \frac{-2}{-4} = \mathbf{\frac{1}{2}}$$

$$y = \frac{\begin{vmatrix} -1 & 3 \\ 3 & -2 \end{vmatrix}}{-4} = \frac{(-1)(-2) - (3)(3)}{-4}$$

$$= \frac{2 - 9}{-4} = \frac{-7}{-4} = \mathbf{\frac{7}{4}}$$

EXAMPLE 8.12. Solve for x and y:

$$x = 3 - 4y$$
$$20y = -1 + 3x$$

Solution: In this case, the equations must first be placed in the format of Eqs. (8.1a) and (8.1b):

$$x + 4y = 3$$
$$-3x + 20y = -1$$

$$x = \frac{\begin{vmatrix} 3 & 4 \\ -1 & 20 \end{vmatrix}}{\begin{vmatrix} 1 & 4 \\ -3 & 20 \end{vmatrix}} = \frac{(3)(20) - (-1)(4)}{(1)(20) - (-3)(4)}$$

$$= \frac{60 + 4}{20 + 12} = \frac{64}{32} = \mathbf{2}$$

$$y = \frac{\begin{vmatrix} 1 & 3 \\ -3 & -1 \end{vmatrix}}{32} = \frac{(1)(-1) - (-3)(3)}{32}$$

$$= \frac{-1 + 9}{32} = \frac{8}{32} = \frac{\mathbf{1}}{\mathbf{4}}$$

The use of determinants is not limited to the solution of two simultaneous equations; determinants can be applied to any number of simultaneous linear equations. For our purposes, however, a shorthand method for solving the third-order (three-simultaneous-equation) determinant will be sufficient. Third- and higher-order determinants are considered in Appendix A.

Consider the three following simultaneous equations:

	Col. 1		Col. 2		Col. 3		Col. 4
	a_1x	+	b_1y	+	c_1z	=	d_1
	a_2x	+	b_2y	+	c_2z	=	d_2
	a_3x	+	b_3y	+	c_3z	=	d_3

in which x, y, and z are the variables, and $a_{1,2,3}$, $b_{1,2,3}$, $c_{1,2,3}$, and $d_{1,2,3}$ are constants.

The determinant configuration for x, y, and z can be found in a manner similar to that for two simultaneous equations. That is, to solve for x, find the determinant in the numerator by replacing column 1 with the elements to the right of the equal sign. The denominator is the determinant of the coefficients of the variables (the same applies to y and z). Again, the denominator is the same for each variable.

$$x = \frac{\begin{vmatrix} d_1 & b_1 & c_1 \\ d_2 & b_2 & c_2 \\ d_3 & b_3 & c_3 \end{vmatrix}}{\begin{vmatrix} a_1 & b_1 & c_1 \\ a_2 & b_2 & c_2 \\ a_3 & b_3 & c_3 \end{vmatrix}}, \quad y = \frac{\begin{vmatrix} a_1 & d_1 & c_1 \\ a_2 & d_2 & c_2 \\ a_3 & d_3 & c_3 \end{vmatrix}}{D}, \quad z = \frac{\begin{vmatrix} a_1 & b_1 & d_1 \\ a_2 & b_2 & d_2 \\ a_3 & b_3 & d_3 \end{vmatrix}}{D}$$

$$D = \begin{vmatrix} a_1 & b_1 & c_1 \\ a_2 & b_2 & c_2 \\ a_3 & b_3 & c_3 \end{vmatrix}$$

A shorthand method for evaluating the third-order determinant consists simply of repeating the first two columns of the determinant to the

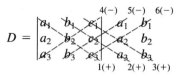

right of the determinant and then summing the products along specific diagonals as shown below:

$$D = \begin{vmatrix} a_1 & b_1 & c_1 \\ a_2 & b_2 & c_2 \\ a_3 & b_3 & c_3 \end{vmatrix} \begin{matrix} a_1 & b_1 \\ a_2 & b_2 \\ a_3 & b_3 \end{matrix}$$

The products of the diagonals 1, 2, and 3 are positive and have the following magnitudes:

$$+a_1 b_2 c_3 + b_1 c_2 a_3 + c_1 a_2 b_3$$

The products of the diagonals 4, 5, and 6 are negative and have the following magnitudes:

$$-a_3 b_2 c_1 - b_3 c_2 a_1 - c_3 a_2 b_1$$

The total solution is the sum of the diagonals 1, 2, and 3 minus the sum of the diagonals 4, 5, and 6:

$$+(a_1 b_2 c_3 + b_1 c_2 a_3 + c_1 a_2 b_3) - (a_3 b_2 c_1 + b_3 c_2 a_1 + c_3 a_2 b_1) \quad \textbf{(8.5)}$$

Warning: **This method of expansion is good only for third-order determinants!** It cannot be applied to fourth- and higher-order systems.

EXAMPLE 8.13. Evaluate the following determinant:

$$\begin{vmatrix} 1 & 2 & 3 \\ -2 & 1 & 0 \\ 0 & 4 & 2 \end{vmatrix} \Rightarrow \begin{vmatrix} 1 & 2 & 3 \\ -2 & 1 & 0 \\ 0 & 4 & 2 \end{vmatrix} \begin{matrix} 1 & 2 \\ -2 & 1 \\ 0 & 4 \end{matrix}$$

Solution:

$$[(1)(1)(2) + (2)(0)(0) + (3)(-2)(4)]$$
$$-[(0)(1)(3) + (4)(0)(1) + (2)(-2)(2)]$$
$$= (2 + 0 - 24) - (0 + 0 - 8) = (-22) - (-8)$$
$$= -22 + 8 = \mathbf{-14}$$

EXAMPLE 8.14. Solve for x, y, and z:

$$1x + 0y - 2z = -1$$
$$0x + 3y + 1z = +2$$
$$1x + 2y + 3z = 0$$

Solution:

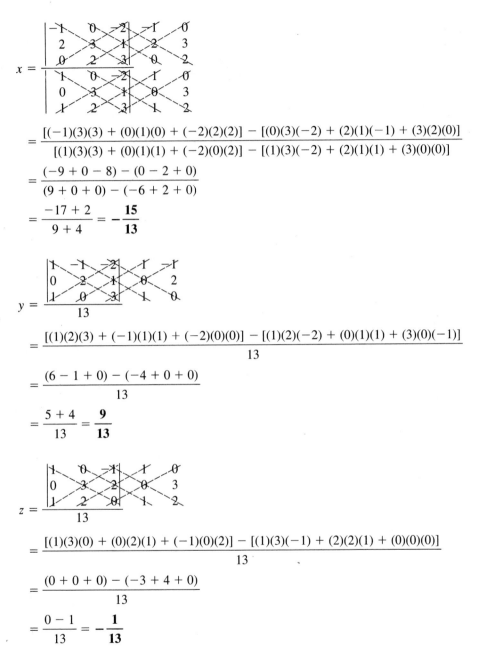

$$x = \frac{\begin{vmatrix} -1 & 0 & -2 \\ 2 & 3 & 1 \\ 0 & 2 & 3 \end{vmatrix}}{\begin{vmatrix} 1 & 0 & -2 \\ 0 & 3 & 1 \\ 1 & 2 & 3 \end{vmatrix}}$$

$$= \frac{[(-1)(3)(3) + (0)(1)(0) + (-2)(2)(2)] - [(0)(3)(-2) + (2)(1)(-1) + (3)(2)(0)]}{[(1)(3)(3) + (0)(1)(1) + (-2)(0)(2)] - [(1)(3)(-2) + (2)(1)(1) + (3)(0)(0)]}$$

$$= \frac{(-9 + 0 - 8) - (0 - 2 + 0)}{(9 + 0 + 0) - (-6 + 2 + 0)}$$

$$= \frac{-17 + 2}{9 + 4} = -\frac{15}{13}$$

$$y = \frac{\begin{vmatrix} 1 & -1 & -2 \\ 0 & 2 & 1 \\ 1 & 0 & 3 \end{vmatrix}}{13}$$

$$= \frac{[(1)(2)(3) + (-1)(1)(1) + (-2)(0)(0)] - [(1)(2)(-2) + (0)(1)(1) + (3)(0)(-1)]}{13}$$

$$= \frac{(6 - 1 + 0) - (-4 + 0 + 0)}{13}$$

$$= \frac{5 + 4}{13} = \frac{9}{13}$$

$$z = \frac{\begin{vmatrix} 1 & 0 & -1 \\ 0 & 3 & 2 \\ 1 & 2 & 0 \end{vmatrix}}{13}$$

$$= \frac{[(1)(3)(0) + (0)(2)(1) + (-1)(0)(2)] - [(1)(3)(-1) + (2)(2)(1) + (0)(0)(0)]}{13}$$

$$= \frac{(0 + 0 + 0) - (-3 + 4 + 0)}{13}$$

$$= \frac{0 - 1}{13} = -\frac{1}{13}$$

or from $0x + 3y + 1z = +2$,

$$z = 2 - 3y = 2 - 3\left(\frac{9}{13}\right) = \frac{26}{13} - \frac{27}{13} = -\frac{1}{13}$$

Check:

$$
\left. \begin{array}{l}
1x + 0y - 2z = -1 \\[6pt]
0x + 3y + 1z = +2 \\[6pt]
1x + 2y + 3z = 0
\end{array} \right|
\left. \begin{array}{l}
-\dfrac{15}{13} + 0 + \dfrac{2}{13} = -1 \\[6pt]
0 + \dfrac{27}{13} + \dfrac{-1}{13} = +2 \\[6pt]
-\dfrac{15}{13} + \dfrac{18}{13} + \dfrac{-3}{13} = 0
\end{array} \right|
\begin{array}{l}
-\dfrac{13}{13} = -1\checkmark \\[6pt]
\dfrac{26}{13} = +2\checkmark \\[6pt]
-\dfrac{18}{13} + \dfrac{18}{13} = 0\checkmark
\end{array}
$$

8.7 BRANCH-CURRENT METHOD

We will now consider the first in a series of methods for solving networks with two or more sources. Once this method is mastered, there is no linear bilateral dc network for which a solution cannot be found. Keep in mind that networks with two isolated voltage sources cannot be solved using the approach of Chapter 7. For further evidence of this fact, try solving for the unknown elements of Example 8.15 using the methods introduced in Chapter 7. The most direct introduction to a method of this type is to list the series of steps required for its application. There are four steps, as indicated below. Before continuing, understand that this method will produce the current through each branch of the network, the *branch current*. Once this is known, all other quantities, such as voltage or power, can be determined.

1. Assign a distinct current of *arbitrary* direction to each branch of the network.
2. Indicate the polarities for each resistor *as determined by the assumed current direction*.
3. Apply Kirchhoff's voltage law around *each* closed loop and Kirchhoff's current law *at* the minimum number of nodes that will include *all* of the branch currents of the network.
4. Solve the resulting simultaneous linear equations for assumed branch currents.

EXAMPLE 8.15. Apply the branch-current method to the network of Fig. 8.19.

Solution:

Step 1: Since there are three distinct branches (*cda, cba, ca*), three currents of arbitrary directions (I_1, I_2, I_3) are chosen as indicated in Fig. 8.19. The current directions for I_1 and I_2 were chosen to match the "pressure" applied by sources E_1 and E_2, respectively. Since both I_1 and I_2 enter node a, I_3 is leaving.

Step 2: Polarities for each resistor are drawn to agree with assumed current directions as indicated in Fig. 8.20.

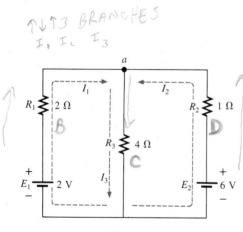

FIG. 8.19

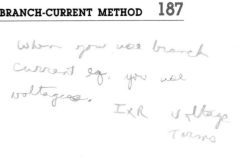

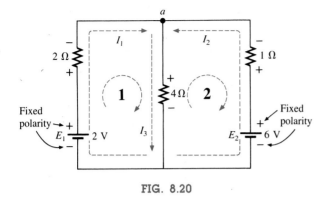

FIG. 8.20

Step 3: Kirchhoff's voltage law is applied around each closed loop (1 and 2) in the clockwise direction:

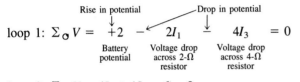

loop 1: $\Sigma_\circlearrowright V =$

Rise in potential $\downarrow$ +2 — Drop in potential $2I_1$ $\downarrow$ $4I_3$ $= 0$

Battery potential   Voltage drop across 2-Ω resistor   Voltage drop across 4-Ω resistor

loop 2: $\Sigma_\circlearrowright V = 4I_3 + 1I_2 - 6 = 0$

Applying Kirchhoff's current law at node a,

$$I_1 + I_2 = I_3$$

Step 4: There are three equations and three unknowns:

$$\begin{aligned} 2 - 2I_1 - 4I_3 &= 0 \\ 4I_3 + 1I_2 - 6 &= 0 \\ I_1 + I_2 &= I_3 \end{aligned} \qquad \text{Rewritten: } \begin{aligned} 2I_1 + 0 + 4I_3 &= +2 \\ 0 + I_2 + 4I_3 &= +6 \\ I_1 + I_2 - I_3 &= 0 \end{aligned}$$

Using third-order determinants, we have

$$I_1 = \frac{\begin{vmatrix} 2 & 0 & 4 \\ +6 & 1 & 4 \\ 0 & 1 & -1 \end{vmatrix}}{D} = +1\,\text{A}$$

A negative sign in front of a branch current indicates only that the actual current is in the direction opposite to that assumed.

$$D = \begin{vmatrix} 2 & 0 & 4 \\ 0 & 1 & 4 \\ 1 & 1 & -1 \end{vmatrix}$$

$$I_2 = \frac{\begin{vmatrix} 2 & 2 & 4 \\ 0 & 6 & 4 \\ 1 & 0 & -1 \end{vmatrix}}{D} = 2\,\text{A}$$

$$I_3 = \frac{\begin{vmatrix} 2 & 0 & 2 \\ 0 & 1 & 6 \\ 1 & 1 & 0 \end{vmatrix}}{D} = 1\,\text{A}$$

Instead of using third-order determinants, we could reduce the three equations to two by substituting the third equation in the first and second equations:

$$\left. \begin{array}{l} 2 - 2I_1 - 4(\overbrace{I_1 + I_2}^{I_3}) = 0 \\[4pt] 4(\overbrace{I_1 + I_2}^{I_3}) + I_2 - 6 = 0 \end{array} \right| \quad \begin{array}{l} 2 - 2I_1 - 4I_1 - 4I_2 = 0 \\[4pt] 4I_1 + 4I_2 + I_2 - 6 = 0 \end{array}$$

or

$$\begin{array}{l} -6I_1 - 4I_2 = -2 \\ +4I_1 + 5I_2 = +6 \end{array}$$

Multiplying through by -1 in the top equation yields

$$\begin{array}{l} 6I_1 + 4I_2 = +2 \\ 4I_1 + 5I_2 = +6 \end{array}$$

and

$$I_1 = \frac{\begin{vmatrix} 2 & 4 \\ 6 & 5 \end{vmatrix}}{\begin{vmatrix} 6 & 4 \\ 4 & 5 \end{vmatrix}} = \frac{10 - 24}{30 - 16} = \frac{-14}{14} = \mathbf{-1\,A}$$

$$I_2 = \frac{\begin{vmatrix} 6 & 2 \\ 4 & 6 \end{vmatrix}}{14} = \frac{36 - 8}{14} = \frac{28}{14} = \mathbf{2\,A}$$

$$I_3 = I_1 + I_2 = -1 + 2 = \mathbf{1\,A}$$

Note in Fig. 8.21 that the current through R_1 and E_1 is 1 A; through R_3, 1 A; and through R_2 and E_2, 2 A, in the directions shown. All the voltages and power levels can now be determined using Ohm's law and the appropriate power equation.

Applying Kirchhoff's voltage law around the loop indicated in Fig. 8.21,

$$\Sigma_{\circlearrowright} V = +4I_3 + 1I_2 - 6 = 0$$

or

$$4I_3 + I_2 = 6$$

and

$$(4)(1) + 2 = 6$$
$$\underline{6 = 6} \qquad \text{(checks)}$$

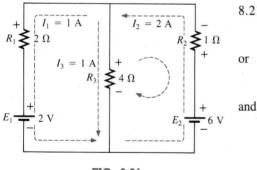

FIG. 8.21

EXAMPLE 8.16. Apply branch-current analysis to the network of Fig. 8.22.

Solution: Again, the current directions were chosen to match the "pressure" of each battery. The polarities are then added and Kirchhoff's voltage law is applied around each closed loop in the clockwise direction. The result is as follows:

loop 1: $+15 - 4I_1 + 10I_3 - 20 = 0$
loop 2: $+20 - 10I_3 - 5I_2 + 40 = 0$

Applying Kirchhoff's current law at node a,

$$I_1 + I_3 = I_2$$

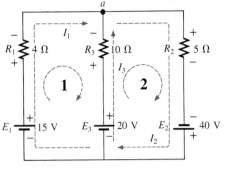

FIG. 8.22

Substituting the third equation into the other two yields

$$\left.\begin{array}{l} 15 - 4I_1 + 10I_3 - 20 = 0 \\ 20 - 10I_3 - 5(I_1 + I_3) + 40 = 0 \end{array}\right\}\ \text{Substituting for } I_2 \text{ (since it occurs only once in the two equations)}$$

or

$$-4I_1 + 10I_3 = 5$$
$$-5I_1 - 15I_3 = -60$$

Multiplying the lower equation by -1, we have

$$-4I_1 + 10I_3 = 5$$
$$5I_1 + 15I_3 = 60$$

$$I_1 = \frac{\begin{vmatrix} 5 & 10 \\ 60 & 15 \end{vmatrix}}{\begin{vmatrix} -4 & 10 \\ 5 & 15 \end{vmatrix}} = \frac{75 - 600}{-60 - 50} = \frac{-525}{-110} = \textbf{4.773 A}$$

$$I_3 = \frac{\begin{vmatrix} -4 & 5 \\ 5 & 60 \end{vmatrix}}{-110} = \frac{-240 - 25}{-110} = \frac{-265}{-110} = \textbf{2.409 A}$$

$$I_2 = I_1 + I_3 = 4.773 + 2.409 = \textbf{7.182 A}$$

8.8 MESH ANALYSIS (GENERAL APPROACH)

The second method of analysis to be described is called *mesh analysis*. The term *mesh* is derived from the similarities in appearance between the closed loops of a network and a wire mesh fence. Although this approach is on a more sophisticated plane than the branch-current method, it incorporates many of the ideas just developed. Of the two methods, mesh analysis is the one more frequently applied today. Branch-current analysis is introduced as a stepping stone to mesh analysis because branch currents are initially more "real" to the student than the loop currents employed in mesh analysis. Essentially, the mesh-analysis approach simply eliminates the need to substitute the results of Kirchhoff's current law into the equations derived from Kirchhoff's voltage law. It is now accomplished in the initial writing of the equations. The systematic approach outlined below should be followed when applying this method.

1. Assign a distinct current in the clockwise direction to each independent closed loop of the network. It is not absolutely necessary to choose the clockwise direction for each loop current. In fact, any direction can be chosen for each loop current with no loss in accuracy as long as the remaining steps are followed properly. However, by choosing the clockwise direction as a standard, we can develop a shorthand method (Section 8.9) for writing the required equations that will save time and possibly prevent some common errors.

This first step is most effectively accomplished by placing a loop current within each "window" of the network as demonstrated in the examples of this section. A window is simply any bounded area within the network. This insures that they are all independent. There are a variety of other loop currents that can be assigned. In each case, however, be sure that the information carried by any one loop equation is not included in a combination of the other network equations. This is the crux of the terminology: *independent*. No matter how you choose your loop currents, the number of loop currents required is always equal to the number of windows of a planar (no-crossovers) network. On occasion a network may appear to be nonplanar. However, a redrawing of the network may reveal that it is, in fact, planar. Such may be the case in one or two problems at the end of the chapter.

Before continuing to the next step, let us insure that the concept of a loop current is clear. For the network of Fig. 8.23, the loop current I_1 is the branch current of the branch containing the 2-Ω resistor and 2-V battery. The current through the 4-Ω resistor is not I_1, however, since there is also a loop current I_2 through it. Since they have opposite directions, $I_{4\Omega}$ equals $I_1 - I_2$, as pointed out in the example to follow. In other words, a loop current is a branch current only when it is the only loop current assigned to that branch.

2. Indicate the polarities *within* each loop for each resistor as determined by the assumed direction of loop current for that loop. Note the requirement that the polarities be placed within each loop. This requires, as shown in Fig. 8.23, that the 4-Ω resistor have two sets of polarities across it.

3. Apply Kirchhoff's voltage law around each closed loop in the clockwise direction. Again, the clockwise direction was chosen to establish uniformity and prepare us for the method to be introduced in the next section.

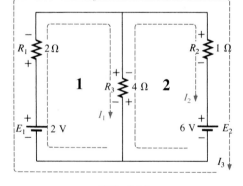

FIG. 8.23

a. If a resistor has two or more assumed currents through it, the total current through the resistor is the assumed current of the loop in which Kirchhoff's voltage law is being applied, plus the assumed currents of the other loops passing through in the *same* direction, minus the assumed currents through in the *opposite* direction.

b. The polarity of a voltage source is unaffected by the loop currents passing through it.

4. Solve the resulting simultaneous linear equations for the assumed loop currents.

EXAMPLE 8.17. Consider the same basic network as in Example 8.15 of the preceding section, now appearing in Fig. 8.23.

Solution:

Step 1: Two loop currents (I_1 and I_2) are assigned in the clockwise direction. A third loop (I_3) could have been included, but the information carried by this loop is already included in the other two.

Step 2: Polarities are drawn within each "window" to agree with assumed current directions. Note that for this case the polarities across the 4-Ω resistor are the opposite for each loop current.

Step 3: Kirchhoff's voltage law is applied around each loop in the clockwise direction. Keep in mind as this step is performed that the law is concerned only with the magnitude and polarity of the voltages around the closed loop and not with whether a voltage rise or drop is due to a battery or resistive element. The voltage across each resistor is determined by $V = IR$, and for a resistor with more than one current through it, the current is the loop current of the loop being examined plus or minus the other loop currents as determined by their directions. If clockwise applications of Kirchhoff's voltage law are always chosen, the other loop currents will always be subtracted from the loop current of the loop being analyzed.

loop 1: $E_1 - V_1 - V_3 = 0$

$$+2 - 2I_1 - 4(I_1 - I_2) = 0$$

Voltage drop across 4-Ω resistor. Total current through 4-Ω resistor. Subtracted since I_2 is opposite in direction to I_1.

loop 2: $-V_3 - V_2 - E_2 = 0$

$$-4(I_2 - I_1) - 1I_2 - 6 = 0$$

Step 4: The equations are then rewritten as follows:

loop 1: $+2 - 2I_1 - 4I_1 + 4I_2 = 0$
loop 2: $-4I_2 + 4I_1 - 1I_2 - 6 = 0$

and

loop 1: $+2 - 6I_1 + 4I_2 = 0$
loop 2: $-5I_2 + 4I_1 - 6 = 0$

or

loop 1: $-6I_1 + 4I_2 = -2$
loop 2: $+4I_1 - 5I_2 = +6$

Applying determinants will result in

$$I_1 = -1\,\text{A} \quad \text{and} \quad I_2 = -2\,\text{A}$$

The minus signs indicate that the currents have a direction opposite to that indicated by the assumed loop current.

The current through the 2-V source and 2-Ω resistor is therefore 1 A in the other direction, and the current through the 6-V source and 1-Ω resistor is 2 A in the opposite direction indicated on the circuit. The current through the 4-Ω resistor is determined by the following equation from the original network:

$$\text{loop 1: } I_{4\Omega} = I_1 - I_2 = -1 - (-2) = -1 + 2$$
$$= \mathbf{1\,A} \quad \text{(in the direction of } I_1)$$

The outer loop (I_3) and *one* inner loop (either I_1 or I_2) would also have produced the correct results. This approach, however, will often lead to errors since the loop equations may be more difficult to write. The best method of picking these loop currents is to use the window approach.

EXAMPLE 8.18. Find the current through each branch of the network of Fig. 8.24.

Solution:

Steps 1 and 2 are as indicated in the circuit. Note that the polarities of the 6-Ω resistor are different for each loop current.

Step 3: Kirchhoff's voltage law is applied around each closed loop in the clockwise direction:

loop 1: $E_1 - V_1 - V_2 - E_2 = 0$

Drop in potential

$$5 - 1I_1 - 6(I_1 - I_2) - 10 = 0$$

I_2 flows through the 6-Ω resistor in the direction opposite to I_1.

loop 2: $E_2 - V_2 - V_3 = 0$
$$10 - 6(I_2 - I_1) - 2I_2 = 0$$

The equations are rewritten as

$$\begin{aligned} 5 - I_1 - 6I_1 + 6I_2 - 10 = 0 \\ 10 - 6I_2 + 6I_1 - 2I_2 = 0 \end{aligned}\Bigg\} \quad \begin{aligned} -7I_1 + 6I_2 = 5 \\ +6I_1 - 8I_2 = -10 \end{aligned}$$

$$I_1 = \dfrac{\begin{vmatrix} 5 & 6 \\ -10 & -8 \end{vmatrix}}{\begin{vmatrix} -7 & 6 \\ 6 & -8 \end{vmatrix}} = \dfrac{-40 + 60}{56 - 36} = \dfrac{20}{20} = \mathbf{1\,A}$$

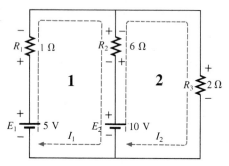

FIG. 8.24

$$I_2 = \frac{\begin{vmatrix} -7 & 5 \\ 6 & -10 \end{vmatrix}}{20} = \frac{70 - 30}{20} = \frac{40}{20} = \textbf{2 A}$$

Since I_1 and I_2 are positive and flow in opposite directions through the 6-Ω resistor and 10-V source, the total current in this branch is equal to the difference of the two currents in the direction of the larger one:

$$I_2 > I_1 \qquad (2 > 1)$$

Therefore, **1 A** $(2 - 1)$ flows in this branch in the direction of I_2.

It is sometimes impractical to draw all the branches of a circuit at right angles to one another. The next example demonstrates how a portion of a network may appear due to various constraints. The method of analysis does not change with this change in configuration.

EXAMPLE 8.19. Find the branch currents of the network of Fig. 8.25.

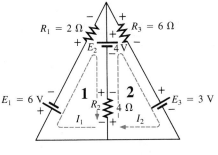

FIG. 8.25

Solution:

Steps 1 and 2 are as indicated in the circuit.

Step 3: Kirchhoff's voltage law is applied around each closed loop:

loop 1: $-E_1 - I_1 R_1 - E_2 - V_2 = 0$

$\qquad -6 - 2I_1 - 4 - 4(I_1 - I_2) = 0$

loop 2: $-V_2 + E_2 - V_3 - E_3 = 0$

$\qquad -4(I_2 - I_1) + 4 - 6I_2 - 3 = 0$

which are rewritten as

$$\left. \begin{array}{l} -10 - 4I_1 - 2I_1 + 4I_2 = 0 \\ +1 + 4I_1 - 4I_2 - 6I_2 = 0 \end{array} \right\} \quad \begin{array}{l} -6I_1 + 4I_2 = +10 \\ +4I_1 - 10I_2 = -1 \end{array}$$

or, by multiplying the top equation by -1, we obtain

$$6I_1 - 4I_2 = -10$$
$$4I_1 - 10I_2 = -1$$

and

$$I_1 = \frac{\begin{vmatrix} -10 & -4 \\ -1 & -10 \end{vmatrix}}{\begin{vmatrix} 6 & -4 \\ 4 & -10 \end{vmatrix}} = \frac{100 - 4}{-60 + 16} = \frac{96}{-44} = \textbf{-2.182 A}$$

$$I_2 = \frac{\begin{vmatrix} 6 & -10 \\ 3 & -1 \end{vmatrix}}{-44} = \frac{-6 + 40}{-44} = \frac{34}{-44} = \textbf{-0.773 A}$$

The current in the 4-Ω resistor and 4-V source for loop 1 is

$$I_1 - I_2 = -2.182 - (-0.773)$$
$$= -2.182 + 0.773$$
$$= \mathbf{-1.409 \ A}$$

indicating that it is 1.409 A in a direction opposite to I_1 in loop 1.

It may happen that current sources will be present in the network to which we wish to apply mesh analysis. The first step will then be to convert all current sources to voltage sources as in the next example. The example will also use standard 20% resistor values to demonstrate how the analysis can become clouded by a more extensive mathematical exercise.

EXAMPLE 8.20. Using mesh analysis, determine the current through the 9-V battery for the network of Fig. 8.26.

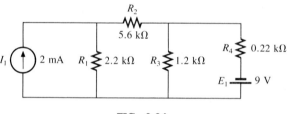

FIG. 8.26

Solution: First, the current source has to be converted to a voltage source as shown in Fig. 8.27 followed by a combination of series resistive elements as shown in Fig. 8.28.

FIG. 8.27

The loop currents were chosen as shown in Fig. 8.28 and Kirchhoff's voltage law was applied around each closed loop in the clockwise direction:

loop 1: $E_2 - I_1 R - V_3 = 0$

$$4.4 - 7.8 \, k\Omega(I_1) - 1.2 \, k\Omega(I_1 - I_2) = 0$$

loop 2: $-V_3 - I_2 R_4 + E_1 = 0$

$$-1.2 \, k\Omega(I_2 - I_1) - 0.22 \, k\Omega(I_2) + 9 = 0$$

which are rewritten as

$$9 \, k\Omega(I_1) - 1.2 \, k\Omega(I_2) = 4.4$$
$$-1.2 \, k\Omega(I_1) + 1.42 \, k\Omega(I_2) = 9$$

and

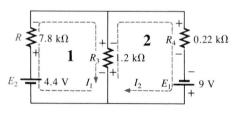

FIG. 8.28

$$I_{9V} = I_2 = \cfrac{\begin{vmatrix} 9 \, k\Omega & 4.4 \\ -1.2 \, k\Omega & 9 \end{vmatrix}}{\begin{vmatrix} 9 \, k\Omega & -1.2 \, k\Omega \\ -1.2 \, k\Omega & +1.42 \, k\Omega \end{vmatrix}}$$

$$= \frac{(9)(9\,\mathrm{k}\Omega) - (4.4)(-1.2\,\mathrm{k}\Omega)}{(9\,\mathrm{k}\Omega)(1.42\,\mathrm{k}\Omega) - (-1.2\,\mathrm{k}\Omega)(-1.2\,\mathrm{k}\Omega)}$$

$$= \frac{86.28}{11.34\,\mathrm{k}\Omega} = \mathbf{7.608\ mA}$$

If the current source has no resistance in parallel with it, as in the network of Fig. 8.29, there is no need to solve for I_1 since it is simply equal to that of the current source.

Applying Kirchhoff's voltage law around loop 2 yields

$$-2(I_2 - I_1) - 3I_2 - 4I_2 - 4 = 0$$

However, for loop 1, $I_1 = \mathbf{1.5\,A}$, and

$$
\begin{aligned}
-2(I_2 - 1.5) - 7I_2 &= 4 \\
-2I_2 + 3 - 7I_2 &= 4 \\
-9I_2 &= 1 \\
I_2 &= \mathbf{-0.111\,A}
\end{aligned}
$$

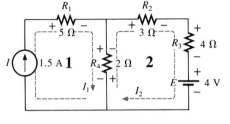

FIG. 8.29

THUR

8.9 MESH ANALYSIS (FORMAT APPROACH)

Now that the basis for the mesh-analysis approach has been established, in this section we will consider a technique for writing the mesh equations more rapidly and usually with fewer errors. As an aid in introducing the procedure, the network of Example 8.19 (Fig. 8.25) has been redrawn in Fig. 8.30 with the assigned loop currents. (Note that each loop current has a clockwise direction.)

The equations obtained are

$$
\begin{aligned}
-7I_1 + 6I_2 &= 5 \\
6I_1 - 8I_2 &= -10
\end{aligned}
$$

which can also be written as

$$
\begin{aligned}
7I_1 - 6I_2 &= -5 \\
8I_2 - 6I_1 &= 10
\end{aligned}
$$

and expanded as

Col. 1	Col. 2	Col. 3
$(1 + 6)I_1 -$	$6I_2$	$= (5 - 10)$
$(2 + 6)I_2 -$	$6I_1$	$= 10$

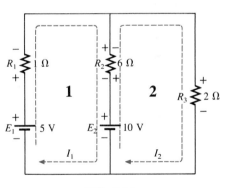

FIG. 8.30

Note in the above equations that column 1 is composed of a loop current times the sum of the resistors through which that loop current passes. Column 2 is the product of the resistors common to another loop current times that other loop current. Note that in each equation this column is subtracted from column 1. Column 3 is the *algebraic* sum of the voltage sources through which the loop current of interest passes. A source is assigned a positive sign if the loop current passes from the negative to the positive terminal, and a negative value if the polarities are reversed. The comments above are correct only for a standard direction of loop current in each window, the chosen being the clockwise direction.

The above statements can be extended to develop the following *format approach* to mesh analysis:

1. Assign a loop current to each independent closed loop (as in the previous section) in a *clockwise* direction.
2. The number of required equations is equal to the number of chosen independent closed loops. Column 1 of each equation is formed by summing the resistance values of those resistors through which the loop current of interest passes and multiplying the result by that loop current.
3. We must now consider the mutual terms which, as noted in the examples above, are always subtracted from the first column. It is possible to have more than one mutual term if the loop current of interest has an element in common with more than one other loop current. This will be demonstrated in an example to follow. Each term is the product of the mutual resistor and the other loop current passing through the same element.
4. The column to the right of the equality sign is the algebraic sum of the voltage sources through which the loop current of interest passes. Positive signs are assigned to those sources of voltage having a polarity such that the loop current passes from the negative to the positive terminal. A negative sign is assigned to those potentials for which the reverse is true.
5. Solve the resulting simultaneous equations for the desired loop currents.

Let us now consider a few examples.

EXAMPLE 8.21. Write the mesh equations for the network of Fig. 8.31 and find the current through the 7-Ω resistor.

Solution:

Step 1: As indicated in Fig. 8.31, each assigned loop current has a clockwise direction.

Steps 2 to 4:

$$I_1: (8 + 6 + 2)I_1 - (2)I_2 = 4$$
$$I_2: (7 + 2)I_2 - (2)I_1 = -9$$

or

$$16I_1 - 2I_2 = 4$$
$$9I_2 - 2I_1 = -9$$

which for determinants are

$$16I_1 - 2I_2 = 4$$
$$-2I_1 + 9I_2 = -9$$

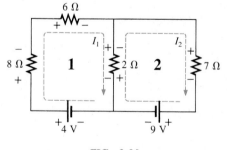

6 Ω

8 Ω **1**   2 Ω **2**   7 Ω

4 V   9 V

FIG. 8.31

and

$$I_2 = I_{7\Omega} = \frac{\begin{vmatrix} 16 & 4 \\ -2 & -9 \end{vmatrix}}{\begin{vmatrix} 16 & -2 \\ -2 & 9 \end{vmatrix}} = \frac{-144 + 8}{144 - 4} = \frac{-136}{140}$$

$$= -0.971\,A$$

EXAMPLE 8.22. Write the mesh equations for the network of Fig. 8.32.

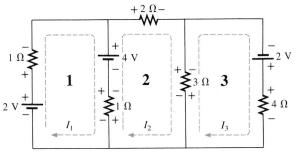

FIG. 8.32

Solution:

Step 1: Each window is assigned a loop current in the clockwise direction:

I_1 does not pass through an element mutual with I_3.
$\downarrow$

I_1: $(1 + 1)I_1 - (1)I_2 + 0 = 2 - 4$
I_2: $(1 + 2 + 3)I_2 - (1)I_1 - (3)I_3 = 4$
I_3: $(3 + 4)I_3 - (3)I_2 + 0 = 2$

$\uparrow$
I_3 does not pass through an element mutual with I_1.

Summing terms yields

$$2I_1 - I_2 + 0 = -2$$
$$6I_2 - I_1 - 3I_3 = 4$$
$$7I_3 - 3I_2 + 0 = 2$$

which are rewritten for determinants as

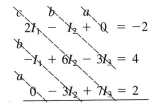

$$2I_1 - I_2 + 0 = -2$$
$$-I_1 + 6I_2 - 3I_3 = 4$$
$$0 - 3I_2 + 7I_3 = 2$$

which compares directly with the equations obtained in that example.

Note that the coefficients of the *a* and *b* diagonals are equal. This *symmetry* about the *c* axis will always be true for equations written using the format approach. It is a check on whether the equations were obtained correctly. Note the symmetry of the equations of Example 8.21.

We will now consider a network with only one source of voltage to point out that mesh analysis can be used to advantage in other than multisource networks.

EXAMPLE 8.23. Find the current through the 10-Ω resistor of the network of Fig. 8.33.

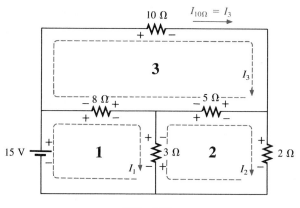

FIG. 8.33

Solution:

$$I_1: \quad (8 + 3)I_1 - (8)I_3 - (3)I_2 = 15$$
$$I_2: \quad (3 + 5 + 2)I_2 - (3)I_1 - (5)I_3 = 0$$
$$I_3: \quad (8 + 10 + 5)I_3 - (8)I_1 - (5)I_2 = 0$$

$$11I_1 - 8I_3 - 3I_2 = 15$$
$$10I_2 - 3I_1 - 5I_3 = 0$$
$$23I_3 - 8I_1 - 5I_2 = 0$$

or

$$11I_1 - 3I_2 - 8I_3 = 15$$
$$-3I_1 + 10I_2 - 5I_3 = 0$$
$$-8I_1 - 5I_2 + 23I_3 = 0$$

and

$$I_3 = I_{10\Omega} = \frac{\begin{vmatrix} 11 & -3 & 15 \\ -3 & 10 & 0 \\ -8 & -5 & 0 \end{vmatrix}}{\begin{vmatrix} 11 & -3 & -8 \\ -3 & 10 & -5 \\ -8 & -5 & 23 \end{vmatrix}} = \mathbf{1.220\ A}$$

The natural sequence of steps makes the mesh-analysis approach one that can easily be programmed on a computer, as demonstrated by Program 8.1 in Fig. 8.35. The network analyzed appears in Fig. 8.34, and the mesh equations obtained are the following:

$$I_1(R_1 + R_2) - I_2 R_3 = E_1$$
$$-I_1 R_3 + I_2(R_2 + R_3) = -E_2$$

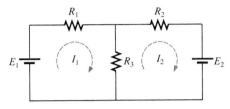

FIG. 8.34

```
10 REM ***** PROGRAM 8-1 *****
20 REM **************************************************
30 REM Program to evaluate the loop currents for a
40 REM 2-loop network.
50 REM **************************************************
60 REM
100 PRINT "For a 2-loop network"
110 PRINT "enter the following data:"
120 PRINT
130 INPUT "R1=";R1
140 INPUT "R2=";R2
150 INPUT "R3=";R3
160 INPUT "Voltage, E1=";E1
170 INPUT "Voltage, E2=";E2
180 PRINT
190 REM Calculate I1 and I2
200 D=R1*R2+R1*R3+R2*R3
210 I1=(E1*(R2+R3)-E2*R3)/D
220 I2=(-E2*(R1+R3)+E1*R3)/D
230 PRINT "The loop currents are:"
240 PRINT "I1=";I1;"amps"
250 PRINT "I2=";I2;"amps"
260 END


READY


For a 2-loop network
enter the following data:

R1=? 2

R2=? 1

R3=? 4

Voltage, E1=? 2

Voltage, E2=? 6


The loop currents are:
I1=-1 amps
I2=-2 amps
```

Input (lines 130–170)
Calc. (lines 200–220)
Output (lines 230–250)

```
READY

RUN


For a 2-loop network
enter the following data:

R1=? 1E3

R2=? 2.2E3

R3=? 3.3E3

Voltage, E1=? -5.4

Voltage, E2=? 8.6


The loop currents are:
I1=-4.5517E-03 amps
I2=-4.2947E-03 amps


READY
```

FIG. 8.35
Program 8.1.

The denominator determinant is calculated by line 200, and the currents I_1 and I_2 by lines 210 and 220, respectively. Note that the program requests parameter values on lines 130 through 170 to permit the analysis of the same network with a wide range of values. In fact, voltage sources can be reversed simply by entering a minus sign with the magnitude of the source.

The first run is for Example 8.17 while the second run includes resistors in the kilohm range and a reversed source.

8.10 NODAL ANALYSIS (GENERAL APPROACH)

Recall from the development of loop analysis that the general network equations were obtained by applying Kirchhoff's voltage law around each closed loop. We will now employ Kirchhoff's current law to develop a method referred to as *nodal analysis*.

A *node* is defined as a junction of two or more branches. If we now define one node of any network as a reference (that is, a point of zero potential or ground), the remaining nodes of the network will all have a fixed potential relative to this reference. For a network of N nodes, therefore, there will exist $(N - 1)$ nodes with a fixed potential relative to the assigned reference node. Equations relating these nodal voltages can be written by applying Kirchhoff's current law at each of the $(N - 1)$ nodes. To obtain the complete solution of a network, these nodal voltages are then evaluated in the same manner in which loop currents were found in loop analysis.

To facilitate the writing of the network equations, all voltage sources within the network will be converted to current sources before Kirchhoff's current law is applied. The nodal-analysis method is applied as follows:

1. Convert all voltage sources to current sources.
2. Determine the number of nodes within the network.
3. Pick a reference node and label each remaining node with a sub-scripted value of voltage: V_1, V_2, and so on.
4. Write Kirchhoff's current law at each node except the reference.
5. Solve the resulting equations for nodal voltages.

EXAMPLE 8.24. Apply nodal analysis to the network of Fig. 8.36.

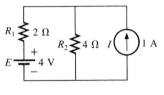

FIG. 8.36

Solution:

Step 1: Convert voltage sources to current sources (Fig. 8.37).

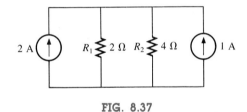

FIG. 8.37

Steps 2 and 3: See Fig. 8.38.

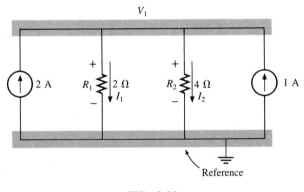

FIG. 8.38

Step 4 (One nodal equation at V_1): V_1 is positive with respect to the reference node. Therefore, current flows away from V_1 through the 2-Ω and 4-Ω resistors at a rate equal to

$$I_1 = \frac{V_1}{2} \quad \text{and} \quad I_2 = \frac{V_1}{4}$$

respectively. Applying Kirchhoff's current law yields

$$\underbrace{2+1}_{\text{Entering}} - \underbrace{\left(\frac{V_1}{2} + \frac{V_1}{4}\right)}_{\text{Leaving}} = 0$$

Step 5:

$$V_1\left(\frac{1}{2} + \frac{1}{4}\right) = 3 \quad \text{or} \quad V_1\left(\frac{3}{4}\right) = 3$$

$$V_1 = \frac{12}{3} = \mathbf{4\,V}$$

The potential across each current source and resistor is therefore 4 V, and

$$I_1 = \frac{V_1}{2} = \frac{4}{2} = \mathbf{2\,A}$$

$$I_2 = \frac{V_1}{4} = \frac{4}{4} = \mathbf{1\,A}$$

EXAMPLE 8.25. Determine the nodal voltages for the network of Fig. 8.39.

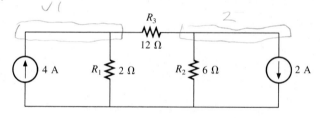

FIG. 8.39

Solution:
Steps 1 to 3: See Fig. 8.40.

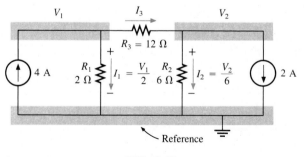

FIG. 8.40

Before attempting step 4, note that the current I_3 is determined by the difference in potential across the 12-Ω resistor. Either V_1 or V_2 can be assumed the larger so that some direction of flow is assigned to I_3, but this same assumption must be used when the equations are written at each node. Even though the assumption may be incorrect (the current I_3 may, in actuality, be flowing in the opposite direction), the values of V_1 and V_2 determined by the resulting equations will be correct.

Step 4: Assuming $V_1 > V_2$, I_3 will flow as indicated through the 12-Ω resistor. The resulting nodal equations are

node 1: $4 - I_1 - I_3 = 0$
node 2: $\underline{I_3 - I_2 - 2 = 0}$

Expanding in terms of V_1 and V_2, we have

node 1: $4 - \dfrac{V_1}{2} - \dfrac{(V_1 - V_2)}{12} = 0$

node 2: $\underline{\dfrac{(V_1 - V_2)}{12} - \dfrac{V_2}{6} - 2 = 0}$

or

$$V_1\left(\dfrac{1}{2} + \dfrac{1}{12}\right) - V_2\left(\dfrac{1}{12}\right) = +4$$

$$V_2\left(\dfrac{1}{12} + \dfrac{1}{6}\right) - V_1\left(\dfrac{1}{12}\right) = -2$$

$$(8.6)$$

producing

$$\left.\begin{array}{l} \dfrac{7}{12}V_1 - \dfrac{1}{12}V_2 = +4 \\[2mm] -\dfrac{1}{12}V_1 + \dfrac{3}{12}V_2 = -2 \end{array}\right| \quad \begin{array}{l} 7V_1 - V_2 = 48 \\[2mm] -1V_1 + 3V_2 = -24 \end{array}$$

and

$$V_1 = \dfrac{\begin{vmatrix} 48 & -1 \\ -24 & 3 \end{vmatrix}}{\begin{vmatrix} 7 & -1 \\ -1 & 3 \end{vmatrix}} = \dfrac{120}{20} = \textbf{+6 V}$$

$$V_2 = \dfrac{\begin{vmatrix} 7 & 48 \\ -1 & -24 \end{vmatrix}}{20} = \dfrac{-120}{20} = \textbf{-6 V}$$

Since V_1 is greater than V_2, the assumed direction of I_3 was correct. Its value is

$$I_3 = \dfrac{V_1 - V_2}{12} = \dfrac{6 - (-6)}{12} = \dfrac{12}{12} = \textbf{1 A}$$

$$I_1 = \dfrac{V_1}{2} = \dfrac{6}{2} = \textbf{3 A}$$

$$I_2 = \frac{V_2}{6} = \frac{-6}{6} = -1 \, A$$

A negative sign indicates that the current in the original network has the opposite direction.

8.11 NODAL ANALYSIS (FORMAT APPROACH)

A close examination of Eq. (8.6) appearing in Example 8.25 reveals that the subscripted voltage at the node in which Kirchhoff's current law is applied is multiplied by the sum of the conductances attached to that node. Note also that the other nodal voltages within the same equation are multiplied by the negative of the conductance between the two nodes. The current sources are represented to the right of the equal sign with a positive sign if they supply current to the node and with a negative sign if they draw current from the node.

These conclusions can be expanded to include networks with any number of nodes. This will allow us to write nodal equations rapidly and in a form that is convenient for the use of determinants. Note the parallelism between the following four steps of application and those required for mesh analysis in Section 8.9:

1. Choose a reference node and assign a subscripted voltage label to the $(N - 1)$ remaining nodes of the network.
2. The number of equations required for a complete solution is equal to the number of subscripted voltages $(N - 1)$. Column 1 of each equation is formed by summing the conductances tied to the node of interest and multiplying the result by that subscripted nodal voltage.
3. We must now consider the mutual terms which, as noted in the preceding example, are always subtracted from the first column. It is possible to have more than one mutual term if the nodal voltage of current interest has an element in common with more than one other nodal voltage. This will be demonstrated in an example to follow. Each mutual term is the product of the mutual conductance and the other nodal voltage tied to that conductance.
4. The column to the right of the equality sign is the algebraic sum of the current sources tied to the node of interest. A current source is assigned a positive sign if it supplies current to a node and a negative sign if it draws current from the node.
5. Solve the resulting simultaneous equations for the desired voltages.

Let us now consider a few examples.

EXAMPLE 8.26. Write the nodal equations for the network of Fig. 8.41.

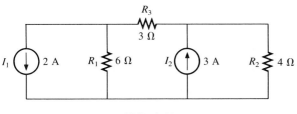

FIG. 8.41

Solution:

Step 1: The figure is redrawn with assigned subscripted voltages in Fig. 8,42.

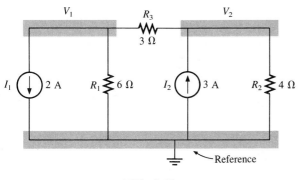

FIG. 8.42

Steps 2 to 4:

Drawing current from node 1

$$V_1: \left(\frac{1}{6} + \frac{1}{3}\right)V_1 - \left(\frac{1}{3}\right)V_2 = \overset{\downarrow}{-2}$$

Sum of conductances connected to node 1 — Mutual conductance

Supplying current to node 2

$$V_2: \left(\frac{1}{4} + \frac{1}{3}\right)V_2 - \left(\frac{1}{3}\right)V_1 = \overset{\downarrow}{+3}$$

Sum of conductances connected to node 2 — Mutual conductance

and

$$\frac{1}{2}V_1 - \frac{1}{3}V_2 = -2$$

$$-\frac{1}{3}V_1 + \frac{7}{12}V_2 = 3$$

EXAMPLE 8.27. Find the voltage across the 3-Ω resistor of Fig. 8.43 by nodal analysis.

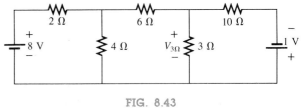

FIG. 8.43

Solution: Converting sources and choosing nodes (Fig. 8.44), we have

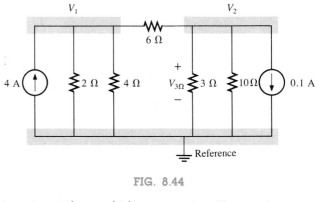

FIG. 8.44

$$\left(\frac{1}{2}+\frac{1}{4}+\frac{1}{6}\right)V_1 - \left(\frac{1}{6}\right)V_2 = +4 \qquad \frac{11}{12}V_1 - \frac{1}{6}V_2 = 4$$

$$\left(\frac{1}{10}+\frac{1}{3}+\frac{1}{6}\right)V_2 - \left(\frac{1}{6}\right)V_1 = -0.1 \qquad -\frac{1}{6}V_1 + \frac{3}{5}V_2 = -0.1$$

which is equivalent to

$$\frac{11}{12}V_1 - \frac{2}{12}V_2 = 4 \qquad 11V_1 - 2V_2 = +48$$

$$-\frac{5}{30}V_1 + \frac{18}{30}V_2 = -0.1 \qquad -5V_1 + 18V_2 = -3 \qquad \begin{array}{l} 11V_1 - 2V_2 = +48 \\ -5V_1 + 18V_2 = -3 \end{array}$$

$$V_2 = V_{3\Omega} = \frac{\begin{vmatrix} 11 & 48 \\ -5 & -3 \end{vmatrix}}{\begin{vmatrix} 11 & -2 \\ -5 & 18 \end{vmatrix}} = \frac{-33+240}{198-10} = \frac{207}{188} = \mathbf{1.101\ V}$$

As demonstrated for mesh analysis, nodal analysis can also be a very useful technique for solving networks with only one source.

EXAMPLE 8.28. Using nodal analysis, determine the potential across the 4-Ω resistor in Fig. 8.45.

Solution: The reference and four subscripted voltage levels were chosen as shown in Fig. 8.46. A moment of reflection should reveal that for any difference in potential between V_1 and V_3, the current through and the potential drop across each 5-Ω resistor will be the same. Therefore, V_4 is simply a midvoltage level between V_1 and V_3 and is known if V_1 and V_3 are available. We will therefore not include it in a nodal voltage and will redraw the network as shown in Fig. 8.47. Understand, however, that V_4 can be included if desired, although four nodal voltages will result rather than the three to be obtained in the solution of this problem.

$$V_1: \quad \left(\frac{1}{2}+\frac{1}{2}+\frac{1}{10}\right)V_1 - \left(\frac{1}{2}\right)V_2 - \left(\frac{1}{10}\right)V_3 = 0$$

$$V_2: \quad \left(\frac{1}{2}+\frac{1}{2}\right)V_2 - \left(\frac{1}{2}\right)V_1 - \left(\frac{1}{2}\right)V_3 = 3$$

$$V_3: \quad \left(\frac{1}{10}+\frac{1}{2}+\frac{1}{4}\right)V_3 - \left(\frac{1}{2}\right)V_2 - \left(\frac{1}{10}\right)V_1 = 0$$

which are rewritten as

$$1.1V_1 - 0.5V_2 - 0.1V_3 = 0$$
$$V_2 - 0.5V_1 - 0.5V_3 = 3$$
$$0.85V_3 - 0.5V_2 - 0.1V_1 = 0$$

For determinants,

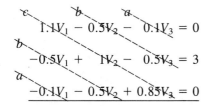

Before continuing, note the symmetry about the major diagonal in the equation above. Recall a similar result for mesh analysis. Examples 8.26 and 8.27 also exhibit this property in the resulting equations. Keep this thought in mind as a check on future applications of nodal analysis.

$$V_3 = V_{4\Omega} = \frac{\begin{vmatrix} 1.1 & -0.5 & 0 \\ -0.5 & +1 & 3 \\ -0.1 & -0.5 & 0 \end{vmatrix}}{\begin{vmatrix} 1.1 & -0.5 & -0.1 \\ -0.5 & +1 & -0.5 \\ -0.1 & -0.5 & +0.85 \end{vmatrix}} = \mathbf{4.645\ V}$$

Another example with only one source involves a ladder network.

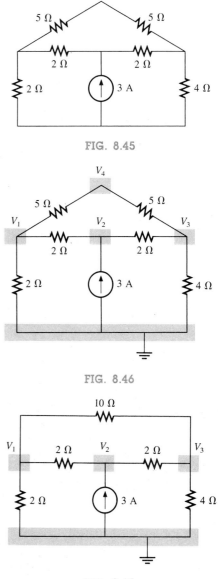

FIG. 8.45

FIG. 8.46

FIG. 8.47

EXAMPLE 8.29. Write the nodal equations and find the voltage across the 2-Ω resistor for the network of Fig. 8.48.

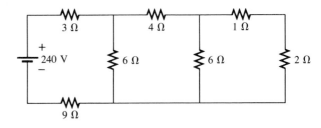

FIG. 8.48

Solution: The nodal voltages are chosen as shown in Fig. 8.49.

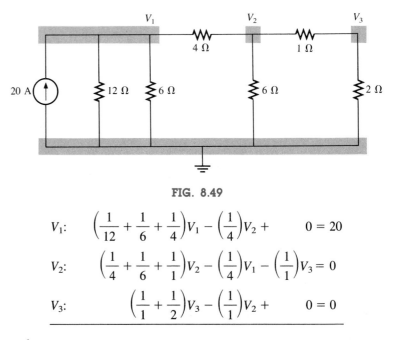

FIG. 8.49

$$V_1: \quad \left(\frac{1}{12} + \frac{1}{6} + \frac{1}{4}\right)V_1 - \left(\frac{1}{4}\right)V_2 + \qquad 0 = 20$$

$$V_2: \quad \left(\frac{1}{4} + \frac{1}{6} + \frac{1}{1}\right)V_2 - \left(\frac{1}{4}\right)V_1 - \left(\frac{1}{1}\right)V_3 = 0$$

$$V_3: \qquad \left(\frac{1}{1} + \frac{1}{2}\right)V_3 - \left(\frac{1}{1}\right)V_2 + \qquad 0 = 0$$

and

$$0.5V_1 - 0.25V_2 + \qquad 0 = 20$$

$$-0.25V_1 + \frac{17}{12}V_2 - \quad 1V_3 = 0$$

$$0 - \quad 1V_2 + 1.5V_3 = 0$$

Note the symmetry present about the major axis. Application of determinants reveals that

$$V_3 = V_{2\Omega} = \textbf{10.667 V}$$

Another example of general interest is the bridge network. This will be set aside for the section on bridge networks to follow in this chapter.

There are various situations in which it may appear impossible to apply nodal analysis. However, by eliminating components or introducing new components, we can often resolve this problem. For example, in the network of Fig. 8.50(a), the voltage source cannot be converted to a current source since it does not have a resistance in series with it.

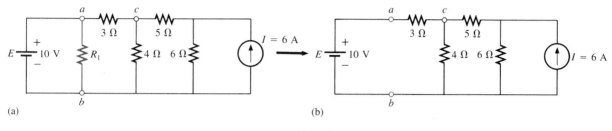

(a) (b)

FIG. 8.50

Note that the voltage V_{ab} across the resistor R_1 is always the source voltage $E = 10\,\text{V}$, no matter what value the resistance R_1 may be. The resistor R_1, therefore, does not affect the voltage V_{ab}, V_{ac}, or any other voltage within the network. Since we are solving only for voltage levels when we apply nodal analysis, the resistor R_1 can be removed without affecting our solution. Once the nodal analysis is complete (all nodal voltages known), the resistor R_1 must be considered if the current through it or through the battery is desired. After the resistor R_1 is removed, the circuit is as shown in the diagram of Fig. 8.50(b), and we can readily arrive at a solution by nodal analysis after we convert the source.

8.12 BRIDGE NETWORKS

This section will introduce the bridge network, a configuration that has a multitude of applications. In the chapters to follow, it will be employed in both dc and ac meters. In the electronics courses it will be encountered early in the discussion of rectifying circuits employed in converting a varying signal to one of a steady nature (such as dc). There are a number of other areas of application that require some knowledge of ac networks which will be discussed later.

The bridge network may appear in one of three forms as indicated in Fig. 8.51. The network of Fig. 8.51(c) is also called a symmetrical

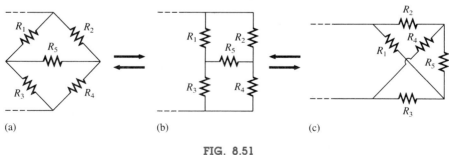

(a) (b) (c)

FIG. 8.51
Bridge network.

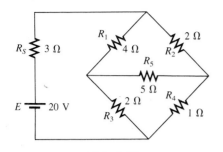

FIG. 8.52

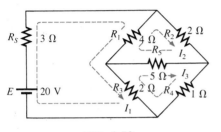

FIG. 8.53

FIG. 8.54

lattice network if $R_2 = R_3$ and $R_1 = R_4$. Figure 8.51(c) is an excellent example of how a planar network can be made to appear nonplanar. For the purposes of investigation, let us examine the network of Fig. 8.52 using mesh and nodal analysis.

Mesh analysis (Fig. 8.53) yields

$$(3 + 4 + 2)I_1 - (4)I_2 - (2)I_3 = 20$$
$$(4 + 5 + 2)I_2 - (4)I_1 - (5)I_3 = 0$$
$$(2 + 5 + 1)I_3 - (2)I_1 - (5)I_2 = 0$$

and

$$9I_1 - 4I_2 - 2I_3 = 20$$
$$-4I_1 + 11I_2 - 5I_3 = 0$$
$$-2I_1 - 5I_2 + 8I_3 = 0$$

with the result that

$$I_1 = \mathbf{4\,A}$$
$$I_2 = \mathbf{2\tfrac{2}{3}\,A}$$
$$I_3 = \mathbf{2\tfrac{2}{3}\,A}$$

The net current through the 5-Ω resistor is

$$I_{5\Omega} = I_2 - I_3 = 2\tfrac{2}{3} - 2\tfrac{2}{3} = \mathbf{0\,A}$$

Nodal analysis (Fig. 8.54) yields

$$\left(\frac{1}{3} + \frac{1}{4} + \frac{1}{2}\right)V_1 - \left(\frac{1}{4}\right)V_2 - \left(\frac{1}{2}\right)V_3 = \frac{20}{3}$$

$$\left(\frac{1}{4} + \frac{1}{2} + \frac{1}{5}\right)V_2 - \left(\frac{1}{4}\right)V_1 - \left(\frac{1}{5}\right)V_3 = 0$$

$$\left(\frac{1}{5} + \frac{1}{2} + \frac{1}{1}\right)V_3 - \left(\frac{1}{2}\right)V_1 - \left(\frac{1}{5}\right)V_2 = 0$$

and

$$\left(\frac{1}{3} + \frac{1}{4} + \frac{1}{2}\right)V_1 - \left(\frac{1}{4}\right)V_2 - \left(\frac{1}{2}\right)V_3 = \frac{20}{3}$$

$$-\left(\frac{1}{4}\right)V_1 + \left(\frac{1}{4} + \frac{1}{2} + \frac{1}{5}\right)V_2 - \left(\frac{1}{5}\right)V_3 = 0$$

$$-\left(\frac{1}{2}\right)V_1 - \left(\frac{1}{5}\right)V_2 + \left(\frac{1}{5} + \frac{1}{2} + \frac{1}{1}\right)V_3 = 0$$

Note the symmetry of the solution. The results are

$$V_1 = \mathbf{8\,V}$$
$$V_2 = \mathbf{2\tfrac{2}{3}\,V}$$
$$V_3 = \mathbf{2\tfrac{2}{3}\,V}$$

and the voltage across the 5-Ω resistor is

$$V_{5\Omega} = V_2 - V_3 = 2\tfrac{2}{3} - 2\tfrac{2}{3} = \mathbf{0\,V}$$

Since $V_{5\Omega} = 0\,\text{V}$, we can insert a short in place of the bridge arm without affecting the network behavior. (Certainly $V = IR = I \cdot 0 = 0\,\text{V}$.) In Fig. 8.55, a short circuit has replaced the resistor R_5, and the voltage across R_4 is to be determined. The network is redrawn in Fig. 8.56, and

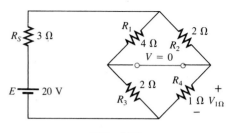

FIG. 8.55

$$V_{1\Omega} = \frac{(2 \parallel 1)20}{(2 \parallel 1) + (4 \parallel 2) + 3} \qquad \text{(voltage divider rule)}$$

$$= \frac{\dfrac{2}{3}(20)}{\dfrac{2}{3} + \dfrac{8}{6} + 3} = \frac{\dfrac{2}{3}(20)}{\dfrac{2}{3} + \dfrac{4}{3} + \dfrac{9}{3}}$$

$$= \frac{2(20)}{2 + 4 + 9} = \frac{40}{15} = 2\tfrac{2}{3}\,\text{V}$$

as obtained earlier.

We found through mesh analysis that $I_{5\Omega} = 0\,\text{A}$, which has as its equivalent an open circuit as shown in Fig. 8.57. (Certainly $I = V/R = 0/\infty = 0\,\text{A}$.) The voltage across the resistor R_4 will again be determined and compared with the result above.

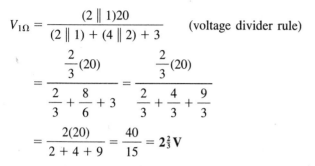

FIG. 8.56

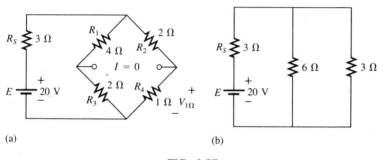

(a)   (b)

FIG. 8.57

The network is redrawn after combining series elements as shown in Fig. 8.57(b), and

$$V_{3\Omega} = \frac{(6 \parallel 3)(20)}{6 \parallel 3 + 3} = \frac{2(20)}{2 + 3} = 8\,\text{V}$$

and

$$V_{1\Omega} = \frac{1(8)}{1 + 2} = \frac{8}{3} = 2\tfrac{2}{3}\,\text{V}$$

as above.

The condition $V_{5\Omega} = 0\,\text{V}$ or $I_{5\Omega} = 0\,\text{A}$ exists only for a particular relationship between the resistors of the network. Let us now derive this relationship using the network of Fig. 8.58, in which it is indicated that $I = 0\,\text{A}$ and $V = 0\,\text{V}$. Note that resistor R_S of the network of Fig. 8.57 will not appear in the following analysis.

The bridge network is said to be *balanced* when the condition of $I = 0\,\text{A}$ or $V = 0\,\text{V}$ exists.

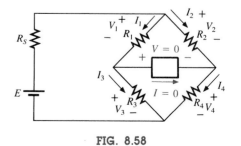

FIG. 8.58

If $V = 0\,V$ (short circuit between a and b), then

$$V_1 = V_2$$

and

$$I_1 R_1 = I_2 R_2$$

or

$$I_1 = \frac{I_2 R_2}{R_1}$$

In addition, when $V = 0\,V$,

$$V_3 = V_4$$

and

$$I_3 R_3 = I_4 R_4$$

If we set $I = 0\,A$, then $I_3 = I_1$ and $I_4 = I_2$ with the result that the above equation becomes

$$I_1 R_3 = I_2 R_4$$

Substituting for I_1 from above yields

$$\left(\frac{I_2 R_2}{R_1}\right) R_3 = I_2 R_4$$

or, rearranging, we have

$$\boxed{\frac{R_1}{R_3} = \frac{R_2}{R_4}} \tag{8.7}$$

This conclusion states that if the ratio of R_1 to R_3 is equal to that of R_2 to R_4, the bridge will be balanced, and $I = 0\,A$ or $V = 0\,V$. A method of memorizing this form is indicated in Fig. 8.59.

For the example above, $R_1 = 4\,\Omega$, $R_2 = 2\,\Omega$, $R_3 = 2\,\Omega$, $R_4 = 1\,\Omega$, and

$$\frac{R_1}{R_3} = \frac{R_2}{R_4} \Rightarrow \frac{4}{2} = \frac{2}{1}$$

The emphasis in this section has been on the balanced situation. Understand that if the ratio is not satisfied, there will be a potential drop across the balance arm and a current through it. The methods just described (mesh and nodal analysis) will yield any and all potentials or currents desired, just as they were applied for the balanced situation.

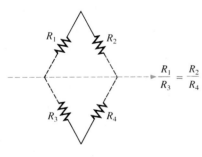

$$\frac{R_1}{R_3} = \frac{R_2}{R_4}$$

FIG. 8.59

8.13 Y-Δ (T-π) AND Δ-Y (π-T) CONVERSIONS

Circuit configurations are often encountered in which the resistors do not appear to be in series or parallel. Under these conditions, it may be necessary to convert the circuit from one form to another in order to solve for any unknown quantities if mesh or nodal analysis is not applied. Two circuit configurations that often account for these difficulties are the wye (Y) and delta (Δ), depicted in Fig. 8.60(a). They are also referred to as the tee (T) and pi (π), respectively, as indicated in Fig. 8.60(b). Note that the pi is actually an inverted delta.

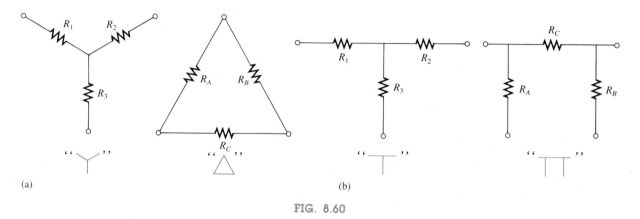

(a)   (b)

FIG. 8.60

The purpose of this section is to develop the equations for converting from Δ to Y or vice versa. This type of conversion will normally lead to a network that can be solved using techniques such as described in Chapter 7. In other words, in Fig. 8.60, with terminals a, b, and c held fast, if the wye (Y) configuration were desired *instead of* the inverted delta (Δ) configuration, all that would be necessary is a direct application of the equations to be derived. The phrase *instead of* is emphasized to insure that it is understood that only one of these configurations is to appear at one time between the indicated terminals.

It is our purpose (referring to Fig. 8.61) to find some expression for R_1, R_2, and R_3 in terms of R_A, R_B, and R_C and vice versa. If the two circuits are to be equivalent, the total resistance between any two terminals must be the same. Consider terminals a-c in the Δ-Y configurations of Fig. 8.62.

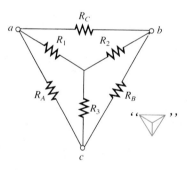

FIG. 8.61

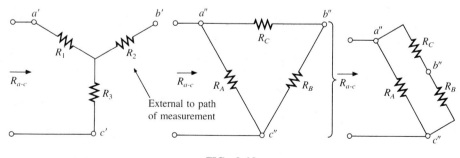

FIG. 8.62

Let us first assume that we want to convert the Δ (R_A, R_B, R_C) to the Y (R_1, R_2, R_3). This requires that we have a relationship for R_1, R_2, and R_3 in terms of R_A, R_B, and R_C. If the resistance is to be the same between terminals a-c for both the Δ and the Y, the following must be true:

$$R_{a\text{-}c} = R_{a'\text{-}c'} = R_{a''\text{-}c''}$$

so that

$$\boxed{R_{a\text{-}c} = R_1 + R_3 = \frac{R_A(R_B + R_C)}{R_A + (R_B + R_C)}} \qquad \textbf{(8.8a)}$$

Using the same approach for a-b and b-c, we obtain the following relationships:

$$\boxed{R_{a\text{-}b} = R_1 + R_2 = \frac{R_C(R_A + R_B)}{R_C + (R_A + R_B)}} \qquad \textbf{(8.8b)}$$

and

$$\boxed{R_{b\text{-}c} = R_2 + R_3 = \frac{R_B(R_A + R_C)}{R_B + (R_A + R_C)}} \qquad \textbf{(8.8c)}$$

Subtracting Eq. (8.8.a) from Eq. (8.8b), we have

$$(R_1 + R_2) - (R_1 + R_3) = \left(\frac{R_C R_A + R_C R_B}{R_A + R_B + R_C}\right) - \left(\frac{R_A R_B + R_A R_C}{R_A + R_B + R_C}\right)$$

so that

$$\boxed{R_2 - R_3 = \frac{R_B R_C - R_A R_B}{R_A + R_B + R_C}} \qquad \textbf{(8.8d)}$$

Subtracting Eq. (8.8d) from Eq. (8.8c) yields

$$(R_2 + R_3) - (R_2 - R_3) = \left(\frac{R_B R_A + R_B R_C}{R_A + R_B + R_C}\right) - \left(\frac{R_B R_C - R_A R_B}{R_A + R_B + R_C}\right)$$

so that

$$2R_3 = \frac{2R_A R_B}{R_A + R_B + R_C}$$

resulting in the following expression for R_3 in terms of R_A, R_B, and R_C:

$$\boxed{R_3 = \frac{R_A R_B}{R_A + R_B + R_C}} \qquad \textbf{(8.9a)}$$

Following the same procedure for R_1 and R_2, we have

$$\boxed{R_1 = \frac{R_A R_C}{R_A + R_B + R_C}} \qquad \textbf{(8.9b)}$$

and

$$R_2 = \frac{R_B R_C}{R_A + R_B + R_C} \qquad \textbf{(8.9c)}$$

Note that each resistor of the Y is equal to the product of the resistors in the two closest branches of the Δ divided by the sum of the resistors in the Δ.

To obtain the relationships necessary to convert from a Y to a Δ, first divide Eq. (8.9a) by Eq. (8.9b):

$$\frac{R_3}{R_1} = \frac{(R_A R_B)/(R_A + R_B + R_C)}{(R_A R_C)/(R_A + R_B + R_C)} = \frac{R_B}{R_C}$$

or

$$R_B = \frac{R_C R_3}{R_1}$$

Then divide Eq. (8.9a) by Eq. (8.9c):

$$\frac{R_3}{R_2} = \frac{(R_A R_B)/(R_A + R_B + R_C)}{(R_B R_C)/(R_A + R_B + R_C)} = \frac{R_A}{R_C}$$

or

$$R_A = \frac{R_3 R_C}{R_2}$$

Substituting for R_A and R_B in Eq. (8.9c) yields

$$R_2 = \frac{(R_C R_3/R_1)R_C}{(R_3 R_C/R_2) + (R_C R_3/R_1) + R_C}$$

$$= \frac{(R_3/R_1)R_C}{(R_3/R_2) + (R_3/R_1) + 1}$$

Placing these over a common denominator, we obtain

$$R_2 = \frac{(R_3 R_C/R_1)}{(R_1 R_2 + R_1 R_3 + R_2 R_3)/(R_1 R_2)}$$

$$= \frac{R_2 R_3 R_C}{R_1 R_2 + R_1 R_3 + R_2 R_3}$$

and

$$R_C = \frac{R_1 R_2 + R_1 R_3 + R_2 R_3}{R_3} \qquad \textbf{(8.10a)}$$

We follow the same procedure for R_B and R_A:

$$R_B = \frac{R_1 R_2 + R_1 R_3 + R_2 R_3}{R_1} \qquad \textbf{(8.10b)}$$

and

$$R_A = \frac{R_1R_2 + R_1R_3 + R_2R_3}{R_2}$$ (8.10c)

Note that the value of each resistor of the Δ is equal to the sum of the possible product combinations of the resistances of the Y divided by the resistance of the Y farthest from the resistor to be determined.

Let us consider what would occur if all the values of a Δ or Y were the same. If $R_A = R_B = R_C$, Eq. (8.9a) would become (using R_A only)

$$R_3 = \frac{R_A R_B}{R_A + R_B + R_C} = \frac{R_A R_A}{R_A + R_A + R_A} = \frac{R_A^2}{3R_A} = \frac{R_A}{3}$$

and, following the same procedure,

$$R_1 = \frac{R_A}{3} \qquad R_2 = \frac{R_A}{3}$$

In general, therefore,

$$R_Y = \frac{R_\Delta}{3}$$ (8.11a)

or

$$R_\Delta = 3R_Y$$ (8.11b)

which indicates that *for a Y of three equal resistors, the value of each resistor of the Δ is equal to three times the value of any resistor of the Y.* If only two elements of a Y or a Δ are the same, the corresponding Δ or Y of each will also have two equal elements. The converting of equations will be left as an exercise for the reader.

The Y and the Δ will often appear as shown in Fig. 8.63. They are then referred to as a *tee* (T) and *pi* (π) network. The equations used to convert from one form to the other are exactly the same as those developed for the Y and Δ transformation.

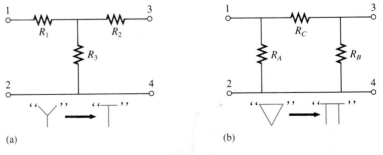

(a)   (b)

FIG. 8.63

EXAMPLE 8.30.   Convert the Δ of Fig. 8.64 to a Y.
Solution:

$$R_1 = \frac{R_A R_C}{R_A + R_B + R_C} = \frac{(20)(10)}{20 + 30 + 10} = \frac{200}{60} = 3\tfrac{1}{3}\,\Omega$$

$$R_2 = \frac{R_B R_C}{R_A + R_B + R_C} = \frac{(30)(10)}{60} = \frac{300}{60} = 5\,\Omega$$

$$R_3 = \frac{R_A R_B}{R_A + R_B + R_C} = \frac{(20)(30)}{60} = \frac{600}{60} = 10\,\Omega$$

The equivalent network is shown in Fig. 8.65.

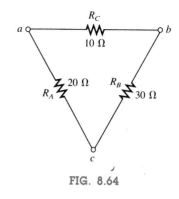

FIG. 8.64

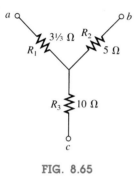

FIG. 8.65

EXAMPLE 8.31.   Convert the Y of Fig. 8.66 to a Δ.
Solution:

$$R_A = \frac{R_1 R_2 + R_1 R_3 + R_2 R_3}{R_2}$$

$$= \frac{(60)(60) + (60)(60) + (60)(60)}{60}$$

$$= \frac{3600 + 3600 + 3600}{60} = \frac{10,800}{60}$$

$$R_A = \mathbf{180\,\Omega}$$

However, the three resistors for the Y are equal, permitting the use of Eq. (8.11) and yielding

$$R_\Delta = 3R_Y = 3(60) = 180\,\Omega$$

and

$$R_B = R_C = \mathbf{180\,\Omega}$$

The equivalent network is shown in Fig. 8.67.

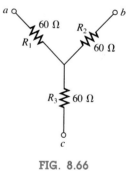

FIG. 8.66

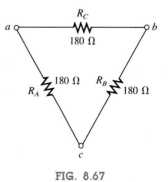

FIG. 8.67

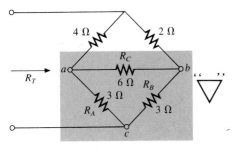

FIG. 8.68

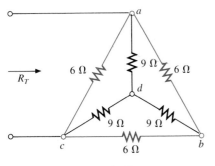

FIG. 8.69

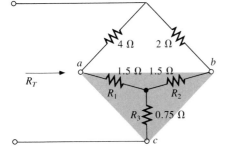

FIG. 8.70

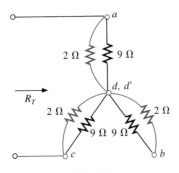

FIG. 8.72

EXAMPLE 8.32. Find the total resistance of the network of Fig. 8.68, where $R_A = 3\,\Omega$, $R_B = 3\,\Omega$, and $R_C = 6\,\Omega$.

Solution:

Two resistors of the Δ were equal; therefore, two resistors of the Y will be equal.

$$R_1 = \frac{R_A R_C}{R_A + R_B + R_C} = \frac{(3)(6)}{3+3+6} = \frac{18}{12} = \mathbf{1.5\,\Omega}$$

$$R_2 = \frac{R_B R_C}{R_A + R_B + R_C} = \frac{(3)(6)}{12} = \frac{18}{12} = \mathbf{1.5\,\Omega}$$

$$R_3 = \frac{R_A R_B}{R_A + R_B + R_C} = \frac{(3)(3)}{12} = \frac{9}{12} = \mathbf{0.75\,\Omega}$$

Replacing the Δ by the Y, as shown in Fig. 8.69, yields

$$R_T = 0.75 + \frac{(4+1.5)(2+1.5)}{(4+1.5)+(2+1.5)}$$

$$= 0.75 + \frac{(5.5)(3.5)}{5.5+3.5}$$

$$= 0.75 + 2.139$$

$$R_T = \mathbf{2.889\,\Omega}$$

EXAMPLE 8.33. Find the total resistance of the network of Fig. 8.70.
Solution: Since all the resistors of the Δ or Y are the same, Eqs. (8.11a) and (8.11b) can be used to convert either form to the other.
a. *Converting the Δ to a Y.* Note: When this is done, the resulting d' of the new Y will be the same as the point d shown in the original figure, only because both systems are "balanced." That is, the resistance in each branch of each system has the same value:

$$R_Y = \frac{R_\Delta}{3} = \frac{6}{3} = 2\,\Omega \qquad \text{(Fig. 8.71)}$$

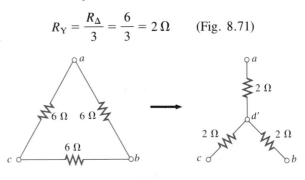

FIG. 8.71

The original circuit appears as shown in Fig. 8.72.

$$R_T = 2\left[\frac{(2)(9)}{2+9}\right] = \mathbf{3.2727\,\Omega}$$

b. *Converting the Y to a Δ.*

$$R_\Delta = 3R_Y = (3)(9) = 27\,\Omega \qquad \text{(Fig. 8.73)}$$

$$R'_T = \frac{(6)(27)}{6 + 27} = \frac{162}{33} = 4.9091\,\Omega$$

$$R_T = \frac{R'_T(R'_T + R'_T)}{R'_T + (R'_T + R'_T)} = \frac{R'_T 2R'_T}{3R'_T} = \frac{2R'_T}{3}$$

$$= \frac{2(4.9091)}{3} = \mathbf{3.2727\,\Omega}$$

which checks with the previous solution.

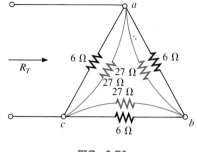

FIG. 8.73

PROBLEMS

Section 8.2

1. Find the voltage V_{ab} (with polarity) for the circuit of Fig. 8.74.

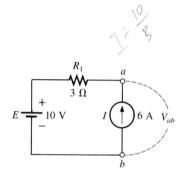

FIG. 8.74

2. For the network of Fig. 8.75:
  a. Find the voltages V_S and V_4.
  b. Find the current I_2.

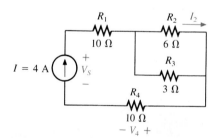

FIG. 8.75

3. For the network of Fig. 8.76:
  a. Find the currents I_1 and I_T.
  b. Find the voltages V_S and V_3.

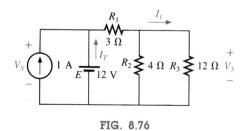

FIG. 8.76

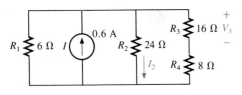

FIG. 8.77

4. Find the voltage V_3 and the current I_2 for the network of Fig. 8.77.

Section 8.3

5. Convert the voltage sources of Fig. 8.78 to current sources.

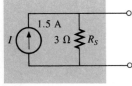

(a)   (b)

FIG. 8.78

6. Convert the current sources of Fig. 8.79 to voltage sources.

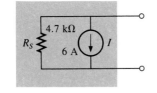

(a)   (b)

FIG. 8.79

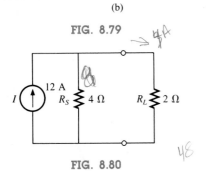

FIG. 8.80

7. For the network of Fig. 8.80:
  a. Find the current through the 2-Ω resistor.
  b. Convert the current source and 4-Ω resistor to a voltage source, and again solve for the current in the 2-Ω resistor. Compare the results.

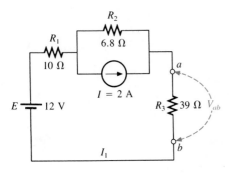

FIG. 8.81

8. For the configuration of Fig. 8.81:
  a. Convert the current source and 6.8-Ω resistor to a voltage source.
  b. Find the magnitude and direction of the current I_1.
  c. Find the voltage V_{ab} and the polarity of points a and b.

Section 8.4

9. Find the voltage V_2 and the current I_1 for the network of Fig. 8.82.

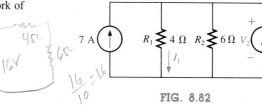

FIG. 8.82

10. a. Convert the voltage sources of Fig. 8.83 to current sources.
b. Find the voltage V_{ab} and the polarity of points a and b.
c. Find the magnitude and direction of the current I.

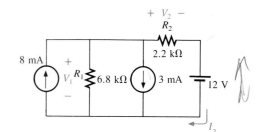

FIG. 8.83

11. For the network of Fig. 8.84:
a. Convert the voltage source to a current source.
b. Reduce the network to a single current source and determine the voltage V_1.
c. Using the results of part (b), determine V_2.
d. Calculate the current I_2.

FIG. 8.84

Section 8.6

12. Evaluate the following determinants:

a. $\begin{vmatrix} 2 & 4 \\ 1 & 2 \end{vmatrix}$ **b.** $\begin{vmatrix} 3 & 4 \\ 0 & -4 \end{vmatrix}$

c. $\begin{vmatrix} 4 & 0 \\ 0 & 7 \end{vmatrix}$ **d.** $\begin{vmatrix} 6 & -9 \\ 5 & 0 \end{vmatrix}$

13. Evaluate the following determinants:

a. $\begin{vmatrix} 3 & -4 & 0 \\ 2 & 7 & 0 \\ -1 & 0 & 1 \end{vmatrix}$ **b.** $\begin{vmatrix} 1 & 2 & 3 \\ 9 & 0 & 0 \\ 5 & 0 & -1 \end{vmatrix}$

14. Using determinants, solve for x and y.
a. $4x + 2y = 8$
  $1x + 3y = 6$
b. $3x - 6y = 1$
  $4x - 6y = 2$
c. $3x + 7y = 0$
  $5y - 4 = 15x$
d. $5x + 6y - 2x = -8$
  $5 + y - 6x = -4x$

*15. Solve for x, y, and z by determinants.
a. $2x + 1y + 0 = 6$
  $0 + 4y + 3z = 12$
  $8x + 0 + 8z = 18$
b. $1x + 0y + 2z = 1$
  $2x + 3y - 4z = 0$
  $-2x + 0y - 6z = 3$

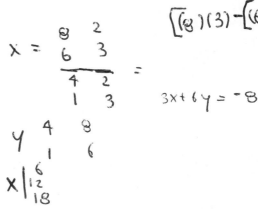

Section 8.7

16. Using branch-current analysis, find the current through each resistor of each network of Fig. 8.85.

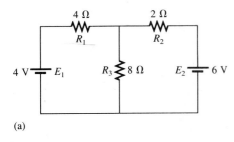

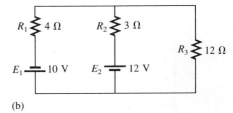

(a)

(b)

FIG. 8.85

*17. Repeat Problem 16 for the networks of Fig. 8.86.

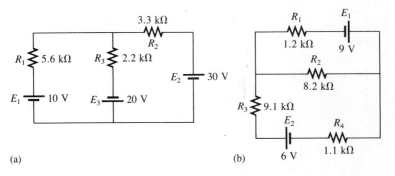

(a)

(b)

FIG. 8.86

*18. For the networks of Fig. 8.87:

a. Determine the current I_2 using branch-current analysis.

b. Find the voltage V_{ab}.

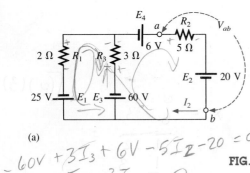

(a)

(b)

FIG. 8.87

Section 8.8

19. Repeat Problem 16 using mesh analysis.

20. Repeat Problem 17 using mesh analysis.

*21. Repeat Problem 18 using mesh analysis.

*22. Using mesh analysis, for each network of Fig. 8.88, determine the current through the 5-Ω resistor.

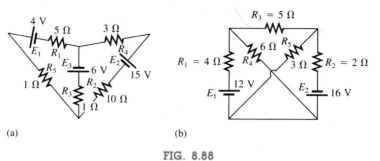

(a) (b)

FIG. 8.88

*23. Write the mesh equations for each of the networks of Fig. 8.89, and, using determinants, solve for the loop currents in each circuit.

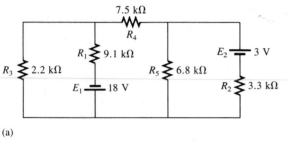

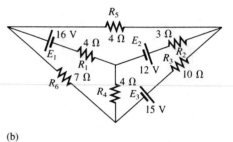

(a) (b)

FIG. 8.89

*24. Repeat Problem 23 for the networks of Fig. 8.90.

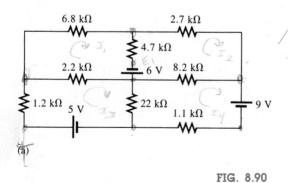

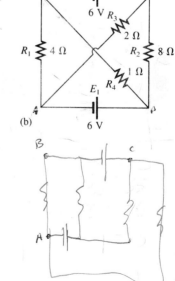

(a) (b)

FIG. 8.90

Section 8.9

25. Using the format approach, write the mesh equations for the networks of Fig. 8.85. Is symmetry present? Using determinants, solve for the mesh currents.

26. Repeat Problem 25 for the networks of Fig. 8.86.

*27. Repeat Problem 25 for the networks of Fig. 8.87.

*28. Repeat Problem 25 for the networks of Fig. 8.88.

*29. Repeat Problem 25 for the networks of Fig. 8.89.

*30. Repeat Problem 25 for the networks of Fig. 8.90 but do not solve for the mesh currents.

Section 8.11

31. Using the format approach, write the nodal equations for the networks of Fig. 8.91, and, using determinants, solve for the nodal voltages. Is symmetry present?

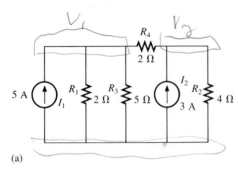

(a)

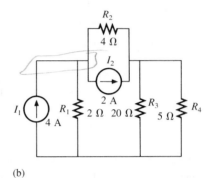

(b)

FIG. 8.91

*32. Repeat Problem 31 for the networks of Fig. 8.92.

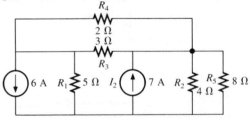

(a)

(b)

FIG. 8.92

33. Repeat Problem 31 for the networks of Fig. 8.89.

*34. For the networks of Fig. 8.93, using the format approach, write the nodal equations and solve for the nodal voltages.

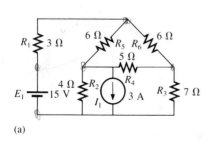

(a)

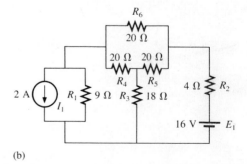

(b)

FIG. 8.93

*35. Repeat Problem 34 for the networks of Fig. 8.94.

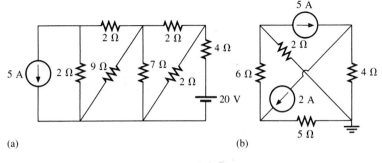

(a) (b)

FIG. 8.94

Section 8.12

*36. For the bridge network of Fig. 8.95:
  a. Using the format approach, write the mesh equations.
  b. Determine the current through R_5.
  c. Is the bridge balanced?
  d. Is Eq. (8.7) satisfied?

*37. For the network of Fig. 8.95:
  a. Using the format approach, write the nodal equations.
  b. Determine the voltage across R_5.
  c. Is the bridge balanced?
  d. Is Eq. (8.7) satisfied?

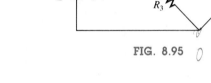

FIG. 8.95

38. Repeat Problem 36 for the network of Fig. 8.96.

39. Repeat Problem 37 for the network of Fig. 8.96.

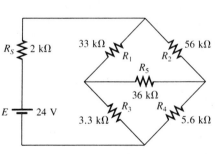

FIG. 8.96

40. Write the nodal equations for the bridge configuration of Fig. 8.97. Use the format approach.

$$V_1 \left(\tfrac{1}{2} + \tfrac{1}{4}\right)V_1 - \left(\tfrac{1}{4}\right)V_2 = 4\,A$$

$$V_2 \left(-\tfrac{1}{4}\right)V_1 + \left(\tfrac{1}{25} + \tfrac{1}{4} + \tfrac{1}{5}\right) = +2A$$

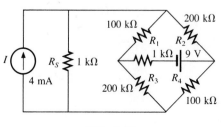

FIG. 8.97

*41. Determine the current through the resistor R_S of each network of Fig. 8.98 using either mesh or nodal analysis. Discuss why you chose one method over the other.

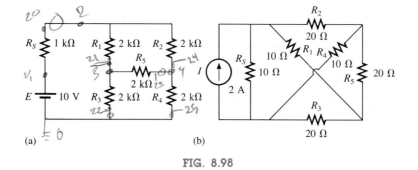

(a)

(b)

FIG. 8.98

Section 8.13

42. Using a Δ-Y or Y-Δ conversion, find the current I in each of the networks of Fig. 8.99.

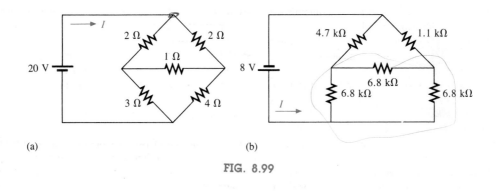

(a)

(b)

FIG. 8.99

*43. Repeat Problem 42 for the networks of Fig. 8.100.

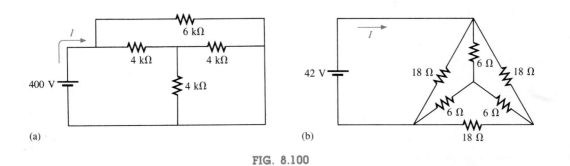

(a)

(b)

FIG. 8.100

*44. Determine the current I for the network of Fig. 8.101.

Computer Problems

45. Write a program to perform a source conversion of either type. That is, given a voltage source, convert to a current source, and given a current source, convert to a voltage source.

46. Given two simultaneous equations, write a program to solve for the unknown variables.

*47. Write a program to solve for both mesh currents of the network of Fig. 8.23 (for any component values) using mesh analysis and determinants.

*48. Write a program to solve for the nodal voltages of the network of Fig. 8.39 (for any component values) using nodal analysis and determinants.

*49. Write a program to determine the total resistance (resistance ''seen'' by the source) for a bridge configuration of any component values.

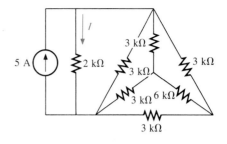

FIG. 8.101

GLOSSARY

Branch-current method   A technique for determining the branch currents of a multiloop network.

Bridge network   A network configuration typically having a diamond appearance in which no two elements are in series or parallel.

Current sources   Sources that supply a fixed current to a network and have a terminal voltage dependent on the network to which they are applied.

Delta (Δ), pi (π) configuration   A network structure that consists of three branches and has the appearance of the Greek letter delta (Δ) or pi (π).

Determinants method   A mathematical technique for finding the unknown variables of two or more simultaneous linear equations.

Mesh analysis   A technique for determining the mesh (loop) currents of a network that results in a reduced set of equations compared to the branch-current method.

Mesh (loop) current   A labeled current assigned to each distinct closed loop of a network that can individually or, in combination with other mesh currents, define all of the branch currents of a network.

Nodal analysis   A technique for determining the node voltages of a network.

Node   A junction of two or more branches in a network.

Wye (Y), tee (T) configuration   A network structure that consists of three branches and has the appearance of the capital letter Y or T.

9

Network Theorems

9.1 INTRODUCTION

This chapter will introduce the important fundamental theorems of network analysis. Included are the *superposition, Thevenin's, Norton's, maximum power transfer, substitution, Millman's,* and *reciprocity* theorems. We will consider a number of areas of application for each. A thorough understanding of each theorem is important because a number will be applied repeatedly in the material to follow.

9.2 SUPERPOSITION THEOREM

The superposition theorem, like the methods of the last chapter, can be used to find the solution to networks with two or more sources that are not in series or parallel. The most obvious advantage of this method is that it does not require the use of a mathematical technique such as determinants to find the required voltages or currents. Instead, each source is treated independently, and the algebraic sum is found to determine a particular unknown quantity of the network. In other words, for a network with *n* sources, *n* independent series-parallel networks would have to be considered before a solution could be obtained.

The superposition theorem states the following:

The current through, or voltage across, an element in a linear bilateral network is equal to the algebraic sum of the currents or voltages produced independently by each source.

To consider the effects of each source independently requires that sources be removed and replaced without affecting the final result. To remove a voltage source when applying this theorem, the difference in potential between the terminals of the voltage source must be set to zero (short circuited); removing a current source requires that its terminals be opened (open circuit). Any internal resistance or conductance associated with the displaced sources is not eliminated but must still be considered.

The total current through any portion of the network is equal to the algebraic sum of the currents produced independently by each source. That is, for a two-source network, if the current produced by one source is in one direction, while that produced by the other is in the opposite direction through the same resistor, *the resulting current is the difference of the two and has the direction of the larger*. If the individual currents are in the same direction, *the resulting current is the sum of two in the direction of either current*. This rule holds true for the voltage across a portion of a network as determined by polarities, and it can be extended to networks with any number of sources.

The superposition principle is not applicable to power effects since the power loss in a resistor varies as the square (nonlinear) of the current or voltage. For this reason, the power to an element cannot be calculated until the total current through (or voltage across) the element has been determined by superposition. This will be demonstrated in Example 9.4.

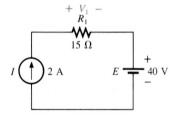

FIG. 9.1

EXAMPLE 9.1. Determine V_1 for the network of Fig. 9.1 using superposition.

Solution: For the voltage source set to zero volts (short circuit), the circuit of Fig. 9.2(a) will result, and

$$V'_1 = I_1 R_1 = I R_1 = (2)(15) = 30 \text{ V}$$

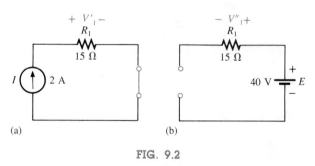

(a)   (b)

FIG. 9.2

The network of Fig. 9.2(b) will result for the current source set to zero (open circuit), and

$$V''_1 = I_1 R_1 = (0)R_1 = 0 \text{ V}$$

Note that V'_1 and V''_1 have opposite polarities. The voltage V_1, however, has the same polarity as V'_1, and therefore

$$V_1 = V'_1 - V''_1 = 30 - 0 = \textbf{30 V}$$

Note that the 40-V supply has no effect on V_1 since the current source determined the current through the 15-Ω resistor.

EXAMPLE 9.2. Determine I_1 for the network of Fig. 9.3.

Solution: Setting $E = 0$ for the network of Fig. 9.3 results in the network of Fig. 9.4(a), where a short-circuit equivalent has replaced the 30-V source.

As shown in Fig. 9.4(a), the source current will choose the short-circuit path, and $I'_1 = 0\,\text{A}$. If we applied the current divider rule,

$$I'_1 = \frac{R_{sc}I}{R_{sc} + R_1} = \frac{(0)I}{0 + 6} = 0\,\text{A}$$

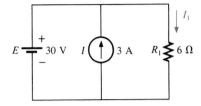

FIG. 9.3

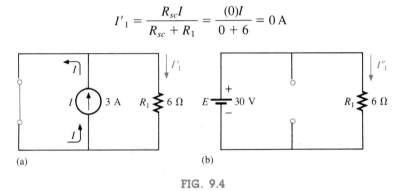

(a)   (b)

FIG. 9.4

Setting I to zero amperes will result in the network of Fig. 9.4(b) with the current source replaced by an open circuit. Applying Ohm's law,

$$I''_1 = \frac{E}{R_1} = \frac{30}{6} = 5\,\text{A}$$

Since I'_1 and I''_1 have the same defined direction in Figs. 9.4(a) and (b), the current I_1 is the sum of the two, and

$$I_1 = I'_1 + I''_1 = 0 + 5 = \mathbf{5\,A}$$

Note in this case that the current source has no effect on the current through the 6-Ω resistor since the voltage across the resistor must be fixed at 30 V because they are parallel elements—essentially the dual situation of Example 9.1.

EXAMPLE 9.3. Using superposition, determine the current through the 4-Ω resistor of Fig. 9.5. Note that this is a two-source network of the type considered in Chapter 8.

Solution: Considering the effects of a 54-V source (Fig. 9.6),

$$R_T = R_1 + R_2 \,\|\, R_3 = 24 + 12 \,\|\, 4 = 24 + 3 = 27\,\Omega$$

$$I = \frac{E_1}{R_T} = \frac{54}{27} = 2\,\text{A}$$

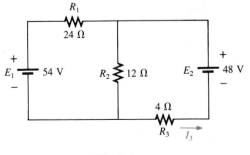

FIG. 9.5

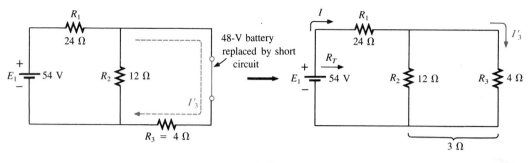

FIG. 9.6

By the current divider rule,

$$I'_3 = \frac{R_2 I}{R_2 + R_3} = \frac{(12)(2)}{12 + 4} = \frac{24}{16} = 1.5 \, A$$

Considering the effects of the 48-V source (Fig. 9.7),

$$R_T = R_3 + R_1 \parallel R_2 = 4 + 24 \parallel 12 = 4 + 8 = 12 \, \Omega$$

$$I''_3 = \frac{E_2}{R_T} = \frac{48}{12} = 4 \, A$$

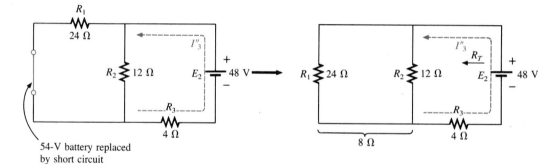

54-V battery replaced
by short circuit

FIG. 9.7

The total current through the 4-Ω resistor (Fig. 9.8) is

$$I_3 = I''_3 - I'_3 = 4 - 1.5 = \mathbf{2.5\,A} \qquad (\text{direction of } I''_3)$$

$I'_3 = 1.5 \, A$

$I''_3 = 4 \, A$

FIG. 9.8

EXAMPLE 9.4. Using superposition, find the current through the 6-Ω resistor of the network of Fig. 9.9.

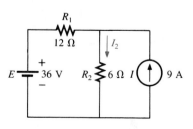

FIG. 9.9

Solution: Considering the effect of the 36-V source (Fig. 9.10),

$$I'_2 = \frac{E}{R_T} = \frac{E}{R_1 + R_2} = \frac{36}{12 + 6} = 2\,\text{A}$$

Considering the effect of the 9-A source (Fig. 9.11), by applying the current divider rule,

$$I''_2 = \frac{R_1 I}{R_1 + R_2} = \frac{(12)(9)}{12 + 6} = \frac{108}{18} = 6\,\text{A}$$

The total current through the 6-Ω resistor (Fig. 9.12) is

$$I_2 = I'_2 + I''_2 = 2 + 6 = \mathbf{8\,A}$$

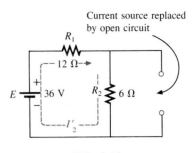

Current source replaced by open circuit

FIG. 9.10

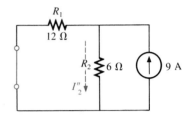

FIG. 9.11

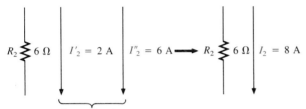

Same direction

FIG. 9.12

The power to the 6-Ω resistor is

$$\text{Power} = I^2 R = (8)^2(6) = \mathbf{384\,W}$$

The calculated power to the 6-Ω resistor due to each source, *misusing* the principle of superposition, is

$$P_1 = (I'_2)^2 R = (2)^2(6) = 24\,\text{W}$$
$$P_2 = (I''_2)^2 R = (6)^2(6) = 216\,\text{W}$$
$$P_1 + P_2 = 240\,\text{W} \neq 384\,\text{W}$$

This results because $2 + 6 = 8$, but

$$(2)^2 + (6)^2 \neq (8)^2$$

As mentioned previously, the superposition principle is not applicable to power effects, since power is proportional to the square of the current or voltage ($I^2 R$ or V^2/R).

Figure 9.13 is a plot of the power delivered to the 6-Ω resistor versus current.

Obviously, $x + y \neq z$, or $24 + 216 \neq 384$, and superposition does not hold. However, for a linear relationship, such as that between the voltage and current of the fixed-type 6-Ω resistor, superposition can be applied, as demonstrated by the graph of Fig. 9.14, where $a + b = c$, or $2 + 6 = 8$.

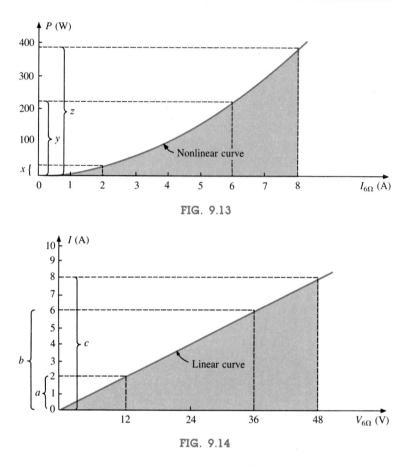

FIG. 9.13

FIG. 9.14

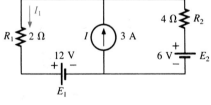

FIG. 9.15

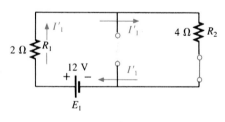

FIG. 9.16

EXAMPLE 9.5. Find the current through the 2-Ω resistor of the network of Fig. 9.15. The presence of three sources will result in three different networks to be analyzed.

Solution: Considering the effect of the 12-V source (Fig. 9.16),

$$I'_1 = \frac{E_1}{R_1 + R_2} = \frac{12}{2 + 4} = \frac{12}{6} = 2 \text{ A}$$

Considering the effect of the 6-V source (Fig. 9.17),

$$I''_1 = \frac{E_2}{R_1 + R_2} = \frac{6}{2 + 4} = \frac{6}{6} = 1 \text{ A}$$

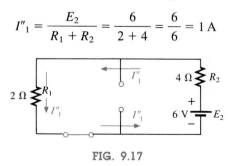

FIG. 9.17

Considering the effect of the 3-A source (Fig. 9.18), by applying the current divider rule,

$$I'''_1 = \frac{R_2 I}{R_1 + R_2} = \frac{(4)(3)}{2 + 4} = \frac{12}{6} = 2 \text{ A}$$

The total current through the 2-Ω resistor appears in Fig. 9.19, and

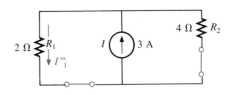

FIG. 9.18

$$I_1 = \overbrace{I''_1 + I'''_1}^{\text{Same direction as } I_1 \text{ in Fig. 9.15}} \overbrace{- I'_1}^{\text{Opposite direction to } I_1 \text{ in Fig. 9.15}}$$
$$= 1 + 2 - 2 = \mathbf{1} \text{ A}$$

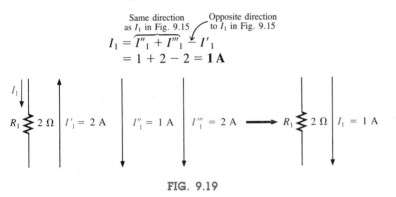

FIG. 9.19

9.3 THEVENIN'S THEOREM

Thevenin's theorem states the following:

Any two-terminal linear bilateral dc network can be replaced by an equivalent circuit consisting of a voltage source and a series resistor as shown in Fig. 9.20.

In Fig. 9.21, for example, the network between terminals *a* and *b* can be replaced by one resistor of 10 Ω and a single battery of 8 V, as

FIG. 9.20
Thevenin equivalent circuit.

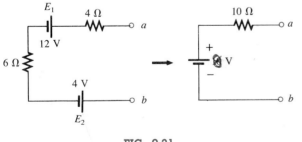

FIG. 9.21

shown in the adjoining figure. The method described below will allow us to extend the procedure just applied to more complex configurations and still end up with the relatively simple network of Fig. 9.20.

In most cases, there will be other elements connected to the right of terminals *a* and *b* in Fig. 9.21. To apply the theorem, however, the network to be reduced to the Thevenin equivalent form must be isolated as shown in Fig. 9.21, and the two "holding" terminals identified. Once the proper Thevenin equivalent circuit has been determined, the voltage, current, or resistance readings between the two "holding" terminals will be the same whether the original or Thevenin equivalent circuit is connected to the left of terminals *a* and *b* in Fig. 9.21. Any load connected to the right of terminals *a* and *b* of Fig. 9.21 will receive the same voltage or current with either network.

This theorem achieves two important objectives. First, as was true for all the methods previously described, it allows us to find any particular voltage or current in a linear network with one, two, or any other number of sources. Second, we can concentrate on a specific portion of a network by replacing the remaining network by an equivalent circuit. In Fig. 9.22, for example, by finding the Thevenin equivalent circuit for

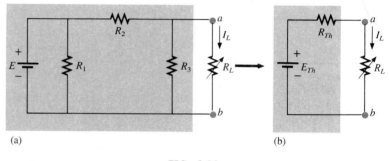

(a)   (b)

FIG. 9.22

the network in the shaded area, we can quickly calculate the change in current through or voltage across the variable resistor R_L for the various values that it may assume. This will be demonstrated in Example 9.6.

Before we examine the steps involved in applying this theorem, it is important that an additional word be included here to insure that the implications of the Thevenin equivalent circuit are clear. In Fig. 9.22, the entire network, except R_L, is to be replaced by a single series resistor and battery as shown in Fig. 9.20. The values of these two elements of the Thevenin equivalent circuit must be chosen to insure that the resistor R_L will react to the network of Fig. 9.22(a) in the same manner as to the network of Fig. 9.22(b). In other words, the current through or voltage across R_L must be the same for either network for any value of R_L.

The following sequence of steps will lead to the proper value of R_{Th} and E_{Th}.

Preliminary:

1. Remove that portion of the network across which the Thevenin equivalent circuit is to be found. In Fig. 9.22(a), this requires that the load resistor R_L be temporarily removed from the network.
2. Mark the terminals of the remaining two-terminal network. (The importance of this step will become obvious as we progress through some complex networks.)

R_{Th}:
3. Calculate R_{Th} by first setting all sources to zero (voltage sources are replaced by short circuits and current sources by open circuits) and then finding the resultant resistance between the two marked terminals. (If the internal resistance of the voltage and/or current sources is included in the original network, it must remain when the sources are set to zero.)

E_{Th}:
4. Calculate E_{Th} by first returning all sources to their original position and finding the *open-circuit* voltage between the marked terminals. (This step is invariably the one that will lead to the most confusion and errors. In *all* cases, keep in mind that it is the *open-circuit* potential between the two terminals marked in step 2 above.)

Conclusion:
5. Draw the Thevenin equivalent circuit with the portion of the circuit previously removed replaced between the terminals of the equivalent circuit. This step is indicated by the placement of the resistor R_L between the terminals of the Thevenin equivalent circuit as shown in Fig. 9.22(b).

EXAMPLE 9.6. Find the Thevenin equivalent circuit for the network in the shaded area of the network of Fig. 9.23. Then find the current through R_L for values of 2, 10, and 100 Ω.

Solution:
Steps 1 and 2 produce the network of Fig. 9.24. Note that the load resistor R_L has been removed and the two "holding" terminals defined as a and b.

Step 3: Replacing the voltage source E_1 by a short-circuit equivalent yields the network of Fig. 9.25, where

$$R_{Th} = R_1 \parallel R_2 = \frac{(3)(6)}{3 + 6} = \textbf{2 } \boldsymbol{\Omega}$$

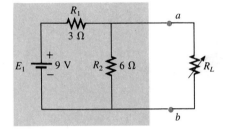

FIG. 9.23

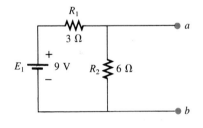

FIG. 9.24

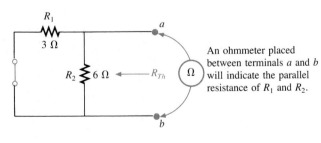

FIG. 9.25

An ohmmeter placed between terminals a and b will indicate the parallel resistance of R_1 and R_2.

Step 4: Replace the voltage source (Fig. 9.26). For this case, the open-circuit voltage E_{Th} is the same as the voltage drop across the 6-Ω resistor. Applying the voltage divider rule,

$$E_{Th} = \frac{R_2 E_1}{R_2 + R_1} = \frac{(6)(9)}{6 + 3} = \frac{54}{9} = \textbf{6 V}$$

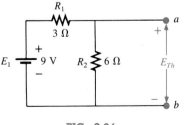

FIG. 9.26

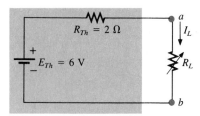

FIG. 9.27

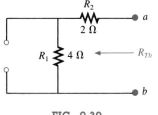

Step 5 (Fig. 9.27):

$$I_L = \frac{E_{Th}}{R_{Th} + R_L}$$

$$R_L = 2\,\Omega: \quad I_L = \frac{6}{2+2} = \mathbf{1.5\,A}$$

$$R_L = 10\,\Omega: \quad I_L = \frac{6}{2+10} = \mathbf{0.5\,A}$$

$$R_L = 100\,\Omega: \quad I_L = \frac{6}{2+100} = \mathbf{0.059\,A}$$

If Thevenin's theorem were unavailable, each change in R_L would require that the entire network of Fig. 9.23 be reexamined to find the new value of R_L.

EXAMPLE 9.7. Find the Thevenin equivalent circuit for the network in the shaded area of the network of Fig. 9.28.
Solution:
Steps 1 and 2 are shown in Fig. 9.29.

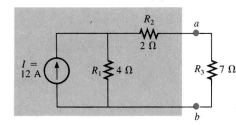

FIG. 9.28

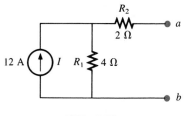

FIG. 9.29

Step 3 is shown in Fig. 9.30. The current source has been replaced by an open-circuit equivalent and the resistance determined between terminals a and b. In this case, an ohmmeter placed across terminals a-b would read the series resistance of R_1 and R_2.

$$R_{Th} = R_1 + R_2 = 4 + 2 = \mathbf{6\,\Omega}$$

FIG. 9.30

Step 4 (Fig. 9.31): In this case, since an open circuit exists between the two marked terminals, the current is zero between these terminals and

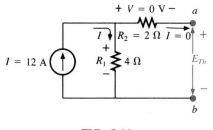

FIG. 9.31

through the 2-Ω resistor. The voltage drop across R_2 is therefore

$$V_2 = I_2 R_2 = (0)R_2 = 0\,V$$

and

$$E_{Th} = V_1 = I_1R_1 = IR_1 = (12)(4) = \mathbf{48\ V}$$

Step 5 is shown in Fig. 9.32.

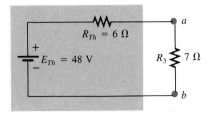

FIG. 9.32

EXAMPLE 9.8. Find the Thevenin equivalent circuit for the network in the shaded area of the network of Fig. 9.33. Note in this example that

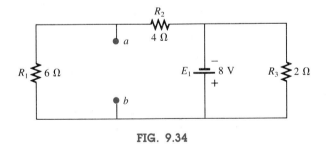

FIG. 9.33

there is no need for the section of the network to be preserved to be at the "end" of the configuration.

Solution:

Steps 1 and 2: See Fig. 9.34.

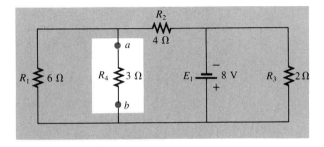

FIG. 9.34

Step 3: See Fig. 9.35. Steps 1 and 2 are relatively easy to apply, but now we must be careful to "hold" on to the terminals *a* and *b* as the

Circuit redrawn:

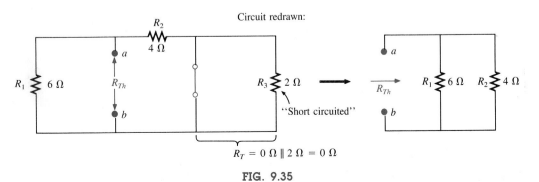

$$R_T = 0\ \Omega \parallel 2\ \Omega = 0\ \Omega$$

FIG. 9.35

Thevenin resistance and voltage are determined. In Fig. 9.35, all the remaining elements turn out to be in parallel, and the network can be redrawn as shown.

$$R_{Th} = R_1 \parallel R_2 = \frac{(6)(4)}{6+4} = \frac{24}{10} = \mathbf{2.4\,\Omega}$$

Step 4: See Fig. 9.36. In this case, the network can be redrawn as

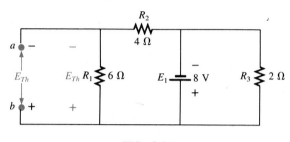

FIG. 9.36

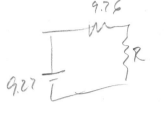

FIG. 9.37

shown in Fig. 9.37, and since the voltage is the same across parallel elements, the voltage across the series resistors R_1 and R_2 is E_1, or 8 V. Applying the voltage divider rule,

$$E_{Th} = \frac{R_1 E_1}{R_1 + R_2} = \frac{(6)(8)}{6+4} = \frac{48}{10} = \mathbf{4.8\,V}$$

Step 5: See Fig. 9.38.

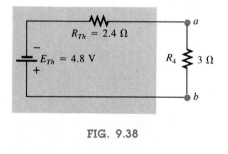

FIG. 9.38

The importance of marking the terminals should be obvious from Example 9.8. Note that there is no requirement that the Thevenin voltage have the same polarity as the equivalent circuit originally introduced.

EXAMPLE 9.9. Find the Thevenin equivalent circuit for the network in the shaded area of the bridge network of Fig. 9.39.

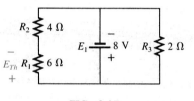

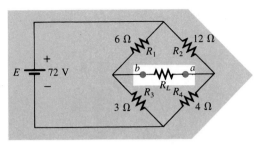

FIG. 9.39

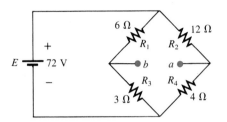

FIG. 9.40

Solution:

Steps 1 and 2 are shown in Fig. 9.40.

Step 3: See Fig. 9.41. In this case, the short-circuit replacement of the voltage source E provides a direct connection between c and c' of Fig.

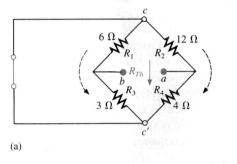

(a)

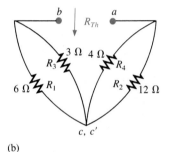

(b)

FIG. 9.41

9.41(a), permitting a "folding" of the network around the horizontal line of a-b to produce the configuration of Fig. 9.41(b).

$$R_{Th} = R_{a\text{-}b} = R_1 \parallel R_3 + R_2 \parallel R_4$$
$$= 6 \parallel 3 + 4 \parallel 12$$
$$= 2 + 3 = \mathbf{5\,\Omega}$$

Step 4: The circuit is redrawn in Fig. 9.42. The absence of a direct connection between a and b results in a network with three parallel

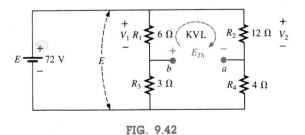

FIG. 9.42

branches. The voltages V_1 and V_2 can therefore be determined using the voltage divider rule:

$$V_1 = \frac{R_1 E}{R_1 + R_3} = \frac{(6)(72)}{6 + 3} = \frac{432}{9} = 48 \text{ V}$$

$$V_2 = \frac{R_2 E}{R_2 + R_4} = \frac{(12)(72)}{12 + 4} = \frac{864}{16} = 54 \text{ V}$$

Assuming the polarity shown for E_{Th} and applying Kirchhoff's voltage law to the top loop in the clockwise direction will result in

$$\Sigma_{\circlearrowright} V = +E_{Th} + V_1 - V_2 = 0$$

and

$$E_{Th} = V_2 - V_1 = 54 - 48 = \mathbf{6\ V}$$

Step 5 is shown in Fig. 9.43.

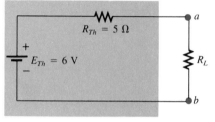

FIG. 9.43

Thevenin's theorem is not restricted to a single passive element, as shown in the preceding examples, but can be applied across sources, whole branches, portions of networks, or any circuit configuration, as shown in the following example. It is also possible that one of the methods previously described, such as mesh analysis or superposition, may have to be used to find the Thevenin equivalent circuit.

EXAMPLE 9.10. (Two sources) Find the Thevenin circuit for the network within the shaded area of Fig. 9.44.

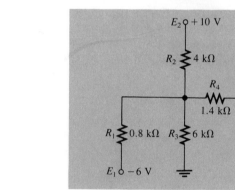

FIG. 9.44

Solution: The network is redrawn and steps 1 and 2 are applied as shown in Fig. 9.45.
Step 3: See Fig. 9.46.

$$\begin{aligned}
R_{Th} &= R_4 + R_1 \parallel R_2 \parallel R_3 \\
&= 1.4\,\text{k}\Omega + 0.8\,\text{k}\Omega \parallel 4\,\text{k}\Omega \parallel 6\,\text{k}\Omega \\
&= 1.4\,\text{k}\Omega + 0.8\,\text{k}\Omega \parallel 2.4\,\text{k}\Omega \\
&= 1.4\,\text{k}\Omega + 0.6\,\text{k}\Omega \\
&= \mathbf{2\,k\Omega}
\end{aligned}$$

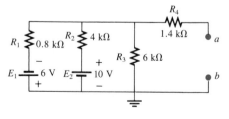

FIG. 9.45

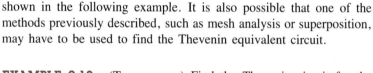

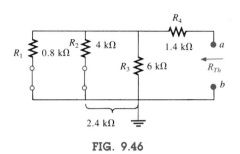

FIG. 9.46

Step 4: Applying superposition, we will consider the effects of the voltage source E_1 first. Note Fig. 9.47. The open circuit requires that $V_4 = I_4R_4 = (0)R_4 = 0\,\text{V}$, and

$$E'_{Th} = V_3$$
$$R'_T = R_2 \parallel R_3 = 4\,\text{k}\Omega \parallel 6\,\text{k}\Omega = 2.4\,\text{k}\Omega$$

Applying the voltage divider rule,

$$V_3 = \frac{R'_T E_1}{R'_T + R_1} = \frac{(2.4\,\text{k}\Omega)(6\,\text{V})}{2.4\,\text{k}\Omega + 0.8\,\text{k}\Omega} = \frac{14.4}{3.2}$$

and

$$E'_{Th} = V_3 = 4.5\,\text{V}$$

For the source E_2, the network of Fig. 9.48 will result. Again, $V_4 = I_4R_4 = (0)R_4 = 0\,\text{V}$, and

$$E''_{Th} = V_3$$
$$R'_T = R_1 \parallel R_3 = 0.8\,\text{k}\Omega \parallel 6\,\text{k}\Omega \cong 0.71\,\text{k}\Omega$$

and

$$V_3 = \frac{R'_T E_2}{R'_T + R_2} = \frac{(0.71\,\text{k}\Omega)(10\,\text{V})}{0.71\,\text{k}\Omega + 4\,\text{k}\Omega} = \frac{7.1}{4.71}$$

and

$$E''_{Th} = V_3 \cong 1.5\,\text{V}$$

Since E'_{Th} and E''_{Th} have opposite polarities,

$$\begin{aligned} E_{Th} &= E'_{Th} - E''_{Th} \\ &= 4.5 - 1.5 \\ &= \mathbf{3\,V} \qquad \text{(polarity of } E'_{Th}) \end{aligned}$$

Step 5: See Fig. 9.49.

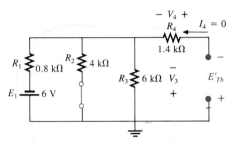

FIG. 9.47

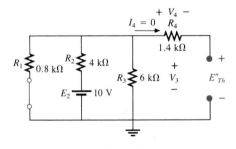

FIG. 9.48

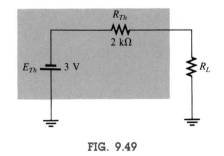

FIG. 9.49

9.4 NORTON'S THEOREM

It was demonstrated in Section 8.3 that every voltage source with a series internal resistance has a current source equivalent. The current source equivalent of the Thevenin network (which, you will note, satisfies the above conditions), as shown in Fig. 9.50, can be determined by Norton's theorem. It can also be found through the conversions of Section 8.3.

The theorem states the following:

Any two-terminal linear bilateral dc network can be replaced by an equivalent circuit consisting of a current source and a parallel resistor as shown in Fig. 9.50.

The discussion of Thevenin's theorem with respect to the equivalent circuit can also be applied to the Norton equivalent circuit.

The steps leading to the proper values of I_N and R_N are now listed.

Preliminary:
1. Remove that portion of the network across which the Norton equivalent circuit is found.
2. Mark the terminals of the remaining two-terminal network.

R_N:
3. Calculate R_N by first setting all sources to zero (voltage sources are replaced by short circuits and current sources by open circuits) and then finding the resultant resistance between the two marked terminals. (If the internal resistance of the voltage and/or current sources is included in the original network, it must remain when the sources are set to zero.) Since $R_N = R_{Th}$, the procedure and value obtained using the approach described for Thevenin's theorem will determine the proper value of R_N.

I_N:
4. Calculate I_N by first returning all sources to their original position and then finding the *short-circuit* current between the marked terminals. It is the same current that would be measured by an ammeter placed between the marked terminals.

Conclusion:
5. Draw the Norton equivalent circuit with the portion of the circuit previously removed replaced between the terminals of the equivalent circuit.

The Norton and Thevenin equivalent circuits can also be found from each other by using the source transformation discussed earlier in this chapter and reproduced in Fig. 9.51.

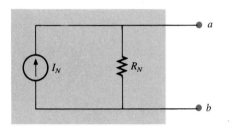

FIG. 9.50
Norton equivalent circuit.

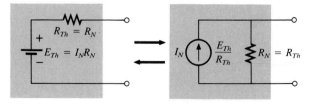

FIG. 9.51

EXAMPLE 9.11. Find the Norton equivalent circuit for the network in the shaded area of Fig. 9.52.

Solution:

Steps 1 and 2 are shown in Fig. 9.53.

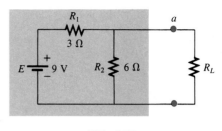

FIG. 9.52

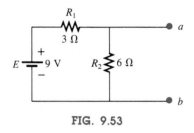

FIG. 9.53

Step 3 is shown in Fig. 9.54, and

$$R_N = R_1 \| R_2 = 3 \| 6 = \frac{(3)(6)}{3 + 6} = \frac{18}{9} = \mathbf{2\,\Omega}$$

Step 4 is shown in Fig. 9.55, clearly indicating that the short-circuit connection between terminals a and b is in parallel with R_2 and elimi-

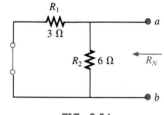

FIG. 9.54

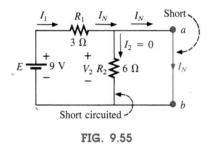

FIG. 9.55

nates its effect. I_N is therefore the same as through R_1, and the full battery voltage appears across R_1 since

$$V_2 = I_2 R_2 = (0)6 = 0\,V$$

Therefore,

$$I_N = \frac{E}{R_1} = \frac{9}{3} = \mathbf{3\,A}$$

Step 5: See Fig. 9.56.

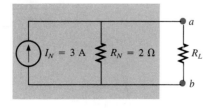

FIG. 9.56

This circuit is the same as the first one considered in the development of Thevenin's theorem. A simple conversion indicates that the Thevenin circuits are, in fact, the same (Fig. 9.57).

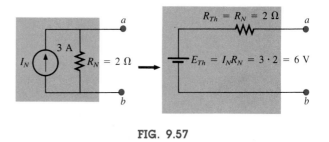

FIG. 9.57

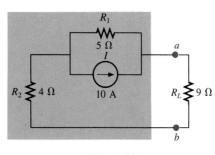

FIG. 9.58

EXAMPLE 9.12. Find the Norton equivalent circuit for the network external to the 9-Ω resistor in Fig. 9.58.

Solution:

Steps 1 and 2: See Fig. 9.59.

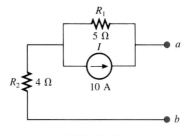

FIG. 9.59

Step 3: See Fig. 9.60, and

$$R_N = R_1 + R_2 = 5 + 4 = \mathbf{9\,\Omega}$$

Step 4: As shown in Fig. 9.61, the Norton current is the same as the current through the 4-Ω resistor. Applying the current divider rule,

$$I_N = \frac{R_1 I}{R_1 + R_2} = \frac{(5)(10)}{5 + 4} = \frac{50}{9} = \mathbf{5.556\,A}$$

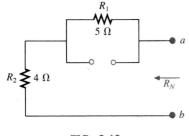

FIG. 9.60

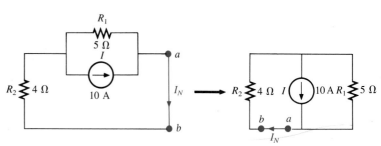

FIG. 9.61

Step 5: See Fig. 9.62.

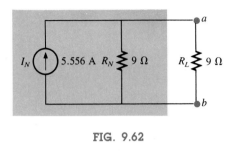

FIG. 9.62

EXAMPLE 9.13. (Two sources) Find the Norton equivalent circuit for the portion of the network to the left of *a-b* in Fig. 9.63.

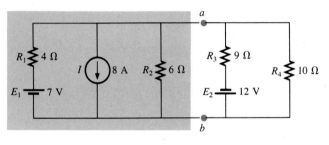

FIG. 9.63

Solution:
Steps 1 and 2: See Fig. 9.64.
Step 3 is shown in Fig. 9.65, and

$$R_N = R_1 \parallel R_2 = 4 \parallel 6 = \frac{(4)(6)}{4 + 6} = \frac{24}{10} = \mathbf{2.4\,\Omega}$$

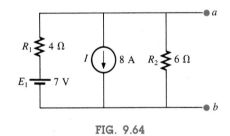

FIG. 9.64

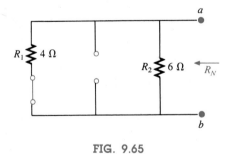

FIG. 9.65

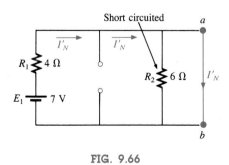

FIG. 9.66

Step 4: (Using superposition) For the 7-V battery (Fig. 9.66),

$$I'_N = \frac{E_1}{R_1} = \frac{7}{4} = 1.75\,\text{A}$$

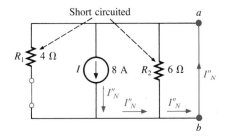

Short circuited

R_1 4 Ω I ⊗ 8 A R_2 6 Ω I''_N

I''_N I''_N I''_N

FIG. 9.67

For the 8-A source (Fig. 9.67), we find that both R_1 and R_2 have been "short circuited" by the direct connection between a and b, and

$$I''_N = I = 8 \text{ A}$$

The result is

$$I_N = I''_N - I'_N = 8 - 1.75 = \mathbf{6.25\ A}$$

Step 5: See Fig. 9.68.

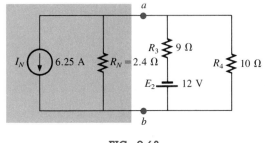

FIG. 9.68

9.5 MAXIMUM POWER TRANSFER THEOREM

The maximum power transfer theorem states the following:

A load will receive maximum power from a linear bilateral dc network when its total resistive value is exactly equal to the Thevenin resistance of the network as "seen" by the load.

For the network of Fig. 9.69, maximum power will be delivered to the load when

$$\boxed{R_L = R_{Th}} \tag{9.1}$$

From past discussions, we realize that a Thevenin equivalent circuit can be found across any element or group of elements in a linear bilateral dc network. Therefore, if we consider the case of the Thevenin equivalent circuit with respect to the maximum power transfer theorem, we are, in essence, considering the *total* effects of any network across a resistor R_L, such as in Fig. 9.69.

For the Norton equivalent circuit of Fig. 9.70, maximum power will be delivered to the load when

$$\boxed{R_L = R_N} \tag{9.2}$$

This result [Eq. (9.2)] will be used to its fullest advantage in the analysis of transistor networks where the most frequently applied transistor circuit model employs a current source rather than a voltage source.

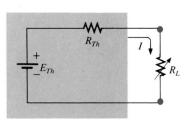

FIG. 9.69

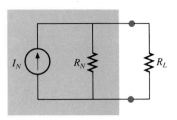

FIG. 9.70

For the network of Fig. 9.69,

$$I = \frac{E_{Th}}{R_{Th} + R_L}$$

and

$$P_L = I^2 R_L = \left(\frac{E_{Th}}{R_{Th} + R_L}\right)^2 R_L$$

so that

$$P_L = \frac{E_{Th}^2 R_L}{(R_{Th} + R_L)^2}$$

For $E_{Th} = 4$ V and $R_{Th} = 5\ \Omega$, the powers to R_L for different values of R_L are tabulated in Table 9.1. A plot of these data (Fig. 9.71) clearly reveals that maximum power is delivered to R_L when it is exactly equal to R_{Th}.

TABLE 9.1

R_L (Ohms)	$P_L = \dfrac{16 R_L}{(5 + R_L)^2}$ (Watts)	
1	0.444	
2	0.653	
3	0.750	Increase
4	0.790	
5	0.800 ←	Maximum
6	0.793	
7	0.778	
8	0.757	Decrease
9	0.735	
10	0.711	

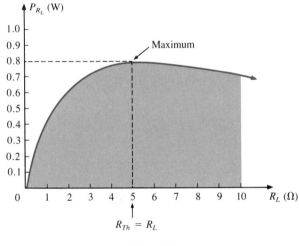

FIG. 9.71

Under maximum power transfer conditions, the operating efficiency is

$$\eta\% = \frac{P_o}{P_i} \times 100\% = \frac{V_L I_L}{E_{Th} I_L} \times 100\% = \frac{V_L}{E_{Th}} \times 100\%$$

with

$$R_L = R_{Th} \qquad V_L = \frac{E_{Th}}{2}$$

and so

$$\eta\% = \frac{E_{Th}/2}{E_{Th}} \times 100\% = \frac{1}{2} \times 100\% = \mathbf{50\%}$$

Since the efficiency will always be 50% under maximum power transfer conditions, there are situations in which the power *transfer* level may have to be reduced to attain a higher efficiency level. For instance, when large power levels are involved, high efficiency levels are extremely important. Can you imagine a power plant operating under conditions where only 50% of the power was distributed to the various loads?

Consider a change in load levels from $5\,\Omega$ to $10\,\Omega$. In Fig. 9.71, the power level has dropped only from $0.8\,\text{W}$ to $0.711\,\text{W}$ (an 11.125% drop), while the efficiency has increased to 66.7% (a 33.4% increase), as shown in Fig. 9.72. A balance point must be identified where the efficiency is sufficiently high without reducing the power to the load to too low a level.

The maximum power transfer theorem is frequently applied in the electronics and communications fields, where power levels are rela-

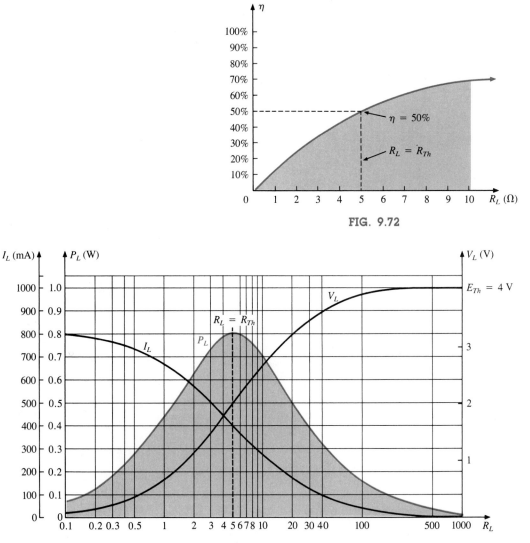

FIG. 9.72

FIG. 9.73

tively low and the efficiency criteria less critical than in the field of power transmission.

If P_L, V_L, and I_L are plotted on a semi-log graph for $E_{Th} = 4\,\text{V}$ and $R_{Th} = 5\,\Omega$, as shown in Fig. 9.73, the variation in levels for the wide range of resistance values becomes clear. On a semi-log graph, one axis is plotted on a log scale while the other employs a linear scale. Note that P_L reaches only one maximum (at $R_L = R_{Th}$), that V_L increases with increasing values of R_L as determined by the voltage divider rule, and that I_L drops with increasing levels of R_L as controlled by Ohm's law. One obvious advantage of log scales is that they permit a wide variation in the magnitude of a parameter (such as R_L in Fig. 9.73). Log scales will be described in detail in Chapter 20.

For any physical network, the value of E_{Th} can be determined experimentally by measuring the open-circuit voltage across the load terminals as shown in Fig. 9.74; $V_{aa'} = E_{Th}$. The value of R_{Th} can then be

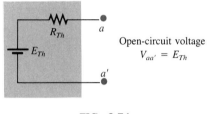

Open-circuit voltage
$$V_{aa'} = E_{Th}$$

FIG. 9.74

determined by completing the network with R_L, and varying R_L until the voltage appearing across the load is one-half the open-circuit value, or $V_L = E_{Th}/2$ (Fig. 9.75). When this condition is established, the voltage across the Thevenin and the load resistances are the same. Since the current through each is also the same, the resistance values of each are equal, and

$$R_L = \frac{E_{Th}/2}{I} = \frac{E_{oc}/2}{I(R_L = R_{Th})} = \frac{E_{oc}}{2I} = R_{Th} \qquad (9.3)$$

The power delivered to R_L under maximum power conditions $(R_L = R_{Th})$ is

$$I = \frac{E_{Th}}{R_{Th} + R_L} = \frac{E_{Th}}{2R_{Th}}$$

$$P_L = I^2 R_L = \left(\frac{E_{Th}}{2R_{Th}}\right)^2 R_{Th} = \frac{E_{Th}^2 R_{Th}}{4R_{Th}^2}$$

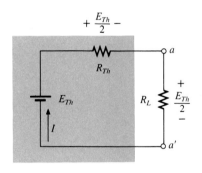

FIG. 9.75

and

$$\boxed{P_{L_{\text{max}}} = \frac{E_{Th}^2}{4R_{Th}}} \qquad \text{(watts, W)} \qquad (9.4)$$

For the Norton circuit of Fig. 9.70,

$$P_{L_{max}} = \frac{I_N^2 R_N}{4} \quad \text{(W)} \tag{9.5}$$

EXAMPLE 9.14. A dc generator, battery, and laboratory supply are connected to a resistive load R_L in Figs. 9.76(a), (b), and (c), respec-

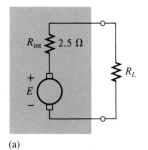

(a) (b) (c)

FIG. 9.76

tively. For each, determine the value of R_L for maximum power transfer to R_L.

Solution: For the dc generator,

$$R_L = \mathbf{2.5\ \Omega}$$

For the battery,

$$R_L = \mathbf{0.5\ \Omega}$$

For the laboratory supply,

$$R_L = \mathbf{40\ \Omega}$$

The results of the preceding example indicate that the following modified form of the statement of the maximum power transfer theorem is correct:

For loads connected directly to a dc voltage supply, maximum power will be delivered to the load when the load resistance is equal to the internal resistance of the source; that is, when

$$R_L = R_{int} \tag{9.6}$$

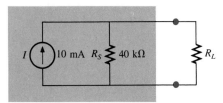

FIG. 9.77

EXAMPLE 9.15. Analysis of a transistor network resulted in the reduced configuration of Fig. 9.77. Determine the R_L necessary to transfer maximum power to R_L, and calculate the power to R_L under these conditions.

Solution: Eq. (9.2):

$$R_L = R_S = \mathbf{40\,k\Omega}$$

Eq. (9.5):

$$P_{L_{max}} = \frac{I_N^2 R_N}{4} = \frac{(10 \times 10^{-3})^2 (40\,k\Omega)}{4} = \mathbf{1\,W}$$

EXAMPLE 9.16. For the network of Fig. 9.78, determine the value of R for maximum power to R, and calculate the power delivered under these conditions.

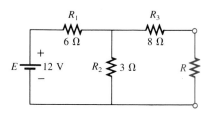

FIG. 9.78

Solution: See Fig. 9.79, and

$$R_{Th} = R_3 + R_1 \parallel R_2 = 8 + \frac{(6)(3)}{6+3} = 8 + 2$$

and

$$R = R_{Th} = \mathbf{10\,\Omega}$$

See Fig. 9.80, and

$$E_{Th} = \frac{R_2 E}{R_2 + R_1} = \frac{(3)(12)}{3+6} = \frac{36}{9} = \mathbf{4\,V}$$

and, by Eq. (9.4),

$$P_{L_{max}} = \frac{E_{Th}^2}{4R_{Th}} = \frac{(4)^2}{4(10)} = \mathbf{0.4\,W}$$

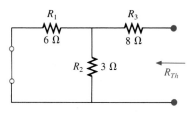

FIG. 9.79

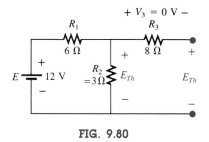

FIG. 9.80

EXAMPLE 9.17. Find the value of R_L in Fig. 9.81 for maximum power to R_L, and determine the maximum power.

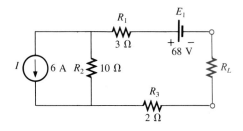

FIG. 9.81

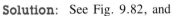

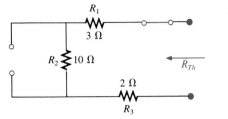

FIG. 9.82

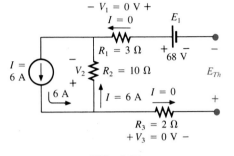

FIG. 9.83

Solution: See Fig. 9.82, and

$$R_{Th} = R_1 + R_2 + R_3 = 3 + 10 + 2 = 15\,\Omega$$

and

$$R_L = R_{Th} = \mathbf{15\,\Omega}$$

Note Fig. 9.83, where

$$V_1 = V_3 = 0\,V$$

and

$$V_2 = I_2 R_2 = IR_2 = (6)(10) = 60\,V$$

Applying Kirchhoff's voltage law,

$$\Sigma_{\circlearrowright} V = -V_2 - E_1 + E_{Th} = 0$$

and

$$E_{Th} = V_2 + E_1 = 60 + 68 = 128\,V$$

Thus,

$$P_{L_{max}} = \frac{E_{Th}^2}{4R_{Th}} = \frac{(128)^2}{4(15)} = \mathbf{273.07\,W}$$

9.6 COMPUTER ANALYSIS

Program 9.1 (Fig. 9.85) provides a complete Thevenin analysis of the network of Fig. 9.84 including a tabulation of power and efficiency versus load resistance.

The Thevenin resistance is determined by lines 200 and 210, and E_{Th} is determined using superposition on lines 220 through 260. E_1 on line 230 is the contribution to E_{Th} due to the voltage source E, and E_2 on line 250 is the contribution due to the current source I. E_{Th} is then determined on line 260. The results are printed out by lines 270 through 290. The details of the plotting routine will be left for your first computer course.

FIG. 9.84

```
10 REM ***** PROGRAM 9-1 *****
20 REM ********************************************
30 REM Program to tabulate changes in load levels for
40 REM a range of load values using Thevenin's theorem
50 REM ********************************************
60 REM
100 PRINT "For the network of Fig. 9.84"
110 PRINT "enter the following data:"
120 PRINT
130 INPUT "R1=";R1 :REM Enter 0 if resistor non-existant
140 INPUT "R2=";R2 :REM Enter 1E30 if resistor non-existant
150 INPUT "R3=";R3 :REM Enter 0 if resistor non-existant
160 INPUT "RL=";RL
170 INPUT "Supply voltage, E=":E
180 INPUT "and supply current, I=";I
190 PRINT
200 REM Determine Rth
210 RT=R3+R1*R2/(R1+R2)
```

Input Data (lines 130–180)

R_{Th} (lines 200–210)

```
           ┌ 220 REM Use superposition to determine Eth
             230 E1=R2*E/(R1+R2)
  E_Th        240 I2=R2*I/(R1+R2)
             250 E2=R1*R2*I/(R1+R2)
           └ 260 ET=E1+E2
  Output   ┌ 270 PRINT "Using Thevenin's theorem:"
  R_Th &     280 PRINT "Rth=";RT;"ohms"
  E_Th     └ 290 PRINT "and Eth=";ET;"volts"
             300 PRINT
  Table    ┌ 310 PRINT TAB(7);"RL";TAB(15);"IL";TAB(25);"VL";
  Heading  └ 320 PRINT TAB(35);"PL";TAB(45);"PD";TAB(55);"n%"
           ┌ 330 FOR RL=RT/4 TO 4*RT STEP RT/4
             340 IL=ET/(RT+RL)
  Calc.      350 VL=IL*RL
             360 PL=IL^2*RL
             370 PD=ET*IL
           └ 380 N=100*PL/PD
           ┌ 390 IF RL=RT THEN PRINT "Rth=";
  Output     400 PRINT TAB(5);RL;TAB(13);IL;TAB(23);VL;
  Control     410 PRINT TAB(33);PL;TAB(43);PD;TAB(53);N
           └ 420 NEXT RL
             430 END


  READY

  RUN


  For the network of Fig. 9.84
  enter the following data:

  R1=? 20

  R2=? 1E30

  R3=? 5

  RL=? 25

  Supply voltage, E=? -10

  and supply current, I=? 4


  Using Thevenin's theorem:
  Rth= 25 ohms
  and Eth= 70 volts
```

	RL	IL	VL	PL	PD	n%
	6.25	2.24	14	31.36	156.8	20
	12.5	1.8667	23.3333	43.5556	130.6667	33.3333
	18.75	1.6	30	48	112	42.8571
Rth=	25	1.4	35	49	98	50
	31.25	1.2444	38.8889	48.3951	87.1111	55.5556
	37.5	1.12	42	47.04	78.4	60
	43.75	1.0182	44.5455	45.3554	71.2727	63.6364
	50	.9333	46.6667	43.5556	65.3333	66.6667
	56.25	.8615	48.4615	41.7515	60.3077	69.2308
	62.5	.8	50	40	56	71.4286
	68.75	.7467	51.3333	38.3289	52.2667	73.3333
	75	.7	52.5	36.75	49	75
	81.25	.6588	53.5294	35.2664	46.1176	76.4706
	87.5	.6222	54.4444	33.8765	43.5556	77.7778
	93.75	.5895	55.2632	32.5762	41.2632	78.9474
	100	.56	56	31.36	39.2	80

```
  READY
```

FIG. 9.85
Program 9.1.

255

Note in the provided run that maximum power is delivered to the load when $R_L = R_{Th} = 25 \, \Omega$, and the use of a very large number, 1×10^{30}, to indicate the absence of a resistor R_2 in the network (to approximate an open circuit).

9.7 MILLMAN'S THEOREM

Through the application of Millman's theorem, any number of parallel voltage sources can be reduced to one. In Fig. 9.86, for example, the

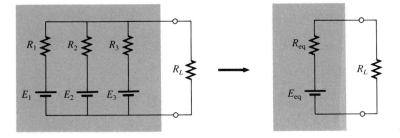

FIG. 9.86

three voltage sources can be reduced to one. This would permit finding the current through or voltage across R_L without having to apply a method such as mesh analysis, nodal analysis, superposition, and so on. The theorem can best be described by applying it to the network of Fig. 9.86. There are basically three steps included in its application.

Step 1: Convert all voltage sources to current sources as outlined in Section 8.3. This is performed in Fig. 9.87 for the network of Fig. 9.86.

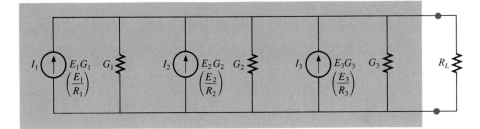

FIG. 9.87

Step 2: Combine parallel current sources as described in Section 8.4. The resulting network is shown in Fig. 9.88, where

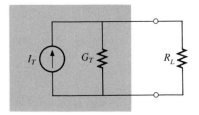

FIG. 9.88

$$I_T = I_1 + I_2 + I_3 \qquad G_T = G_1 + G_2 + G_3$$

Step 3: Convert the resulting current source to a voltage source, and the desired single-source network is obtained as shown in Fig. 9.89.

In general, Millman's theorem states that for any number of parallel voltage sources,

$$E_{eq} = \frac{I_T}{G_T} = \frac{\pm I_1 \pm I_2 \pm I_3 \pm \cdots \pm I_N}{G_1 + G_2 + G_3 + \cdots + G_N}$$

or

$$\boxed{E_{eq} = \frac{\pm E_1 G_1 \pm E_2 G_2 \pm E_3 G_3 \pm \cdots \pm E_N G_N}{G_1 + G_2 + G_3 + \cdots + G_N}} \qquad \textbf{(9.7)}$$

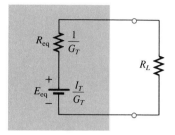

FIG. 9.89

The plus and minus signs appear in Eq. (9.7) to include those cases where the sources may not be supplying energy in the same direction. (Note Example 9.18.)

The equivalent resistance is

$$\boxed{R_{eq} = \frac{1}{G_T} = \frac{1}{G_1 + G_2 + G_3 + \cdots + G_N}} \qquad \textbf{(9.8)}$$

In terms of the resistance values,

$$\boxed{E_{eq} = \frac{\pm \dfrac{E_1}{R_1} \pm \dfrac{E_2}{R_2} \pm \dfrac{E_3}{R_3} \pm \cdots \pm \dfrac{E_N}{R_N}}{\dfrac{1}{R_1} + \dfrac{1}{R_2} + \dfrac{1}{R_3} + \cdots + \dfrac{1}{R_N}}} \qquad \textbf{(9.9)}$$

and

$$\boxed{R_{eq} = \frac{1}{\dfrac{1}{R_1} + \dfrac{1}{R_2} + \dfrac{1}{R_3} + \cdots + \dfrac{1}{R_N}}} \qquad \textbf{(9.10)}$$

The relatively few direct steps required may result in the student's applying each step rather than memorizing and employing Eqs. (9.7) through (9.10).

EXAMPLE 9.18. Using Millman's theorem, find the current through and voltage across the resistor R_L of Fig. 9.90.

Solution: By Eq. (9.9),

$$E_{eq} = \frac{+\dfrac{E_1}{R_1} - \dfrac{E_2}{R_2} + \dfrac{E_3}{R_3}}{\dfrac{1}{R_1} + \dfrac{1}{R_2} + \dfrac{1}{R_3}}$$

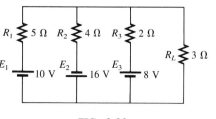

FIG. 9.90

The minus sign is used for E_2/R_2 because that supply has the opposite polarity of the other two. The chosen reference direction is therefore that of E_1 and E_3. The total conductance is unaffected by the direction, and

$$E_{eq} = \cfrac{+\dfrac{10}{5} - \dfrac{16}{4} + \dfrac{8}{2}}{\dfrac{1}{5} + \dfrac{1}{4} + \dfrac{1}{2}} = \frac{2 - 4 + 4}{0.2 + 0.25 + 0.5}$$

$$= \frac{2}{0.95} = \mathbf{2.105\ V}$$

with

$$R_{eq} = \cfrac{1}{\dfrac{1}{5} + \dfrac{1}{4} + \dfrac{1}{2}} = \frac{1}{0.95} = \mathbf{1.053\ \Omega}$$

The resultant source is shown in Fig. 9.91, and

$$I_L = \frac{2.105}{1.053 + 3} = \frac{2.105}{4.053} = \mathbf{0.519\ A}$$

with

$$V_L = I_L R_L = (0.519)(3) = \mathbf{1.557\ V}$$

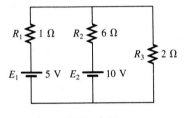

FIG. 9.91

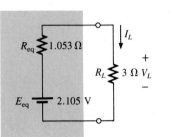

FIG. 9.92

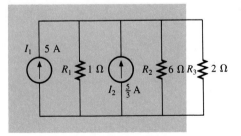

FIG. 9.93

EXAMPLE 9.19. Let us now consider the type of problem encountered in the introduction to mesh and nodal analysis in Chapter 8. Mesh analysis was applied to the network of Fig. 9.92 (Example 8.18). Let us now use Millman's theorem to find the current through the 2-Ω resistor and compare the results.

Solution:

a. Let us first apply each step and, in the (b) solution, Eq. (9.9). Converting sources yields Fig. 9.93. Combining sources and parallel conductance branches (Fig. 9.94) yields

$$I_T = I_1 + I_2 = 5 + \frac{5}{3} = \frac{15}{3} + \frac{5}{3} = \frac{20}{3}\ A$$

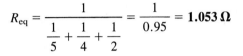

FIG. 9.94

$$G_T = G_1 + G_2 = 1 + \frac{1}{6} = \frac{6}{6} + \frac{1}{6} = \frac{7}{6} \, \text{S}$$

Converting the current source to a voltage source (Fig. 9.95), we obtain

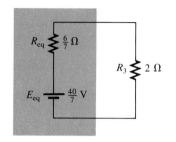

FIG. 9.95

$$E_{eq} = \frac{I_T}{G_T} = \frac{\dfrac{20}{3}}{\dfrac{7}{6}} = \frac{(6)(20)}{(3)(7)} = \frac{40}{7} \, \text{V}$$

and

$$R_{eq} = \frac{1}{G_T} = \frac{1}{\dfrac{7}{6}} = \frac{6}{7} \, \Omega$$

so that

$$I_{2\Omega} = \frac{E_{eq}}{R_{eq} + R_3} = \frac{\dfrac{40}{7}}{\dfrac{6}{7} + 2} = \frac{\dfrac{40}{7}}{\dfrac{6}{7} + \dfrac{14}{7}} = \frac{40}{20} = 2 \, \text{A}$$

which agrees with the result obtained in Example 8.18.
b. Let us now simply apply the proper equation, Eq. (9.9):

$$E_{eq} = \frac{+\dfrac{5}{1} + \dfrac{10}{6}}{\dfrac{1}{1} + \dfrac{1}{6}} = \frac{\dfrac{30}{6} + \dfrac{10}{6}}{\dfrac{6}{6} + \dfrac{1}{6}} = \frac{40}{7} \, \text{V}$$

and

$$R_{eq} = \frac{1}{\dfrac{1}{1} + \dfrac{1}{6}} = \frac{1}{\dfrac{6}{6} + \dfrac{1}{6}} = \frac{1}{\dfrac{7}{6}} = \frac{6}{7} \, \Omega$$

which are the same values obtained above.

The dual of Millman's theorem is the combining of series current sources. The dual of Fig. 9.86 is Fig. 9.96.

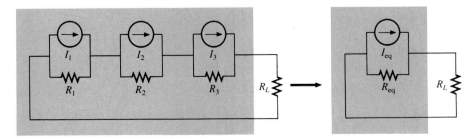

FIG. 9.96

It can be shown that I_{eq} and R_{eq}, as shown in Fig. 9.96, are given by

$$I_{eq} = \frac{\pm I_1 R_1 \pm I_2 R_2 \pm I_3 R_3}{R_1 + R_2 + R_3} \qquad \textbf{(9.11)}$$

and

$$R_{eq} = R_1 + R_2 + R_3 \qquad \textbf{(9.12)}$$

The derivation will appear as a problem at the end of the chapter.

9.8 SUBSTITUTION THEOREM

The substitution theorem states the following:

If the voltage across and current through any branch of a dc bilateral network are known, this branch can be replaced by any combination of elements that will maintain the same voltage across and current through the chosen branch.

More simply, the theorem states that for branch equivalence, the terminal voltage and current must be the same. Consider the circuit of Fig. 9.97 in which the voltage across and current through the branch *a-b* are determined. Through the use of the substitution theorem, a number of equivalent *a-a'* branches are shown in Fig. 9.98.

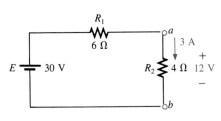

FIG. 9.97

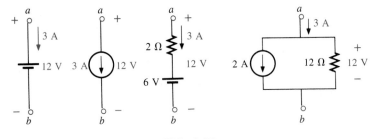

FIG. 9.98

Note that for each equivalent, the terminal voltage and current are the same. Also consider that the response of the remainder of the circuit of Fig. 9.97 is unchanged by substituting any one of the equivalent branches. As demonstrated by the single-source equivalents of Fig. 9.98, *a known potential difference and current in a network can be replaced by an ideal voltage source and current source, respectively.*

Understand that this theorem cannot be used to *solve* networks with two or more sources that are not in series or parallel. For it to be applied, a potential difference or current value must be known or found using one of the techniques discussed earlier. One application of the theorem is shown in Fig. 9.99. Note that in the figure the known potential difference V was replaced by a voltage source, permitting the isolation of the portion of the network including R_3, R_4, and R_5. Recall that this was basically the approach employed in the analysis of the ladder network as we worked our way back toward the terminal resistance R_5.

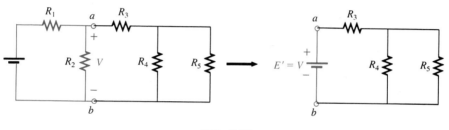

FIG. 9.99

The current source equivalence of the above is shown in Fig. 9.100, where a known current is replaced by an ideal current source, permitting the isolation of R_4 and R_5.

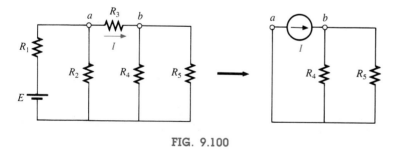

FIG. 9.100

You will also recall from the discussion of bridge networks that $V = 0$ and $I = 0$ were replaced by a short circuit and an open circuit, respectively. This substitution is a very specific application of the substitution theorem.

9.9 RECIPROCITY THEOREM

The reciprocity theorem is applicable only to single-source networks. It is, therefore, not a theorem employed in the analysis of multisource networks described thus far.

The theorem states the following:

The current I in any branch of a network, due to a single voltage source E anywhere else in the network, will equal the current through the branch in which the source was originally located if the source is placed in the branch in which the current I was originally measured.

In other words, the location of the voltage source and the resulting current may be interchanged without a change in current. The theorem requires that the polarity of the voltage source have the same correspondence with the direction of the branch current in each position. In the representative network of Fig. 9.101(a), the current *I* due to the voltage source *E* was determined. If the position of each is inter-

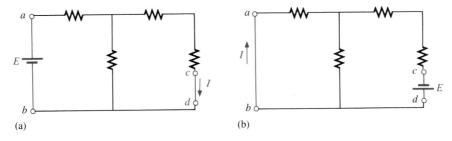

(a)   (b)

FIG. 9.101

changed as shown in Fig. 9.101(b), the current *I* will be the same value as indicated. To demonstrate the validity of this statement and the theorem, consider the network of Fig. 9.102, in which values for the elements of Fig. 9.101(a) have been assigned.

The total resistance is

FIG. 9.102

$$R_T = R_1 + R_2 \parallel (R_3 + R_4) = 12 + 6 \parallel (2 + 4)$$
$$= 12 + 6 \parallel 6 = 12 + 3 = 15 \, \Omega$$

and

$$I_T = \frac{E}{R_T} = \frac{45}{15} = 3 \, \text{A}$$

with

$$I = \frac{3}{2} = \textbf{1.5 A}$$

For the network of Fig. 9.103, which corresponds to that of Fig. 9.101(b), we find

$$R_T = R_4 + R_3 + R_1 \parallel R_2$$
$$= 4 + 2 + 12 \parallel 6 = 10 \, \Omega$$

and

$$I_T = \frac{E}{R_T} = \frac{45}{10} = 4.5 \, \text{A}$$

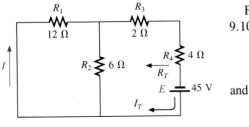

FIG. 9.103

so that

$$I = \frac{(6)(4.5)}{12 + 6} = \frac{4.5}{3} = \mathbf{1.5\,A}$$

which agrees with the above.

The uniqueness and power of such a theorem can best be demonstrated by considering a complex single-source network such as shown in Fig. 9.104.

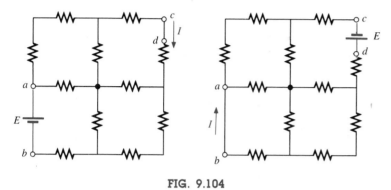

FIG. 9.104

PROBLEMS

Section 9.2

1. **a.** Using superposition, find the current through each resistor of the network of Fig. 9.105.
   b. Find the power delivered to R_1 for each source.
   c. Find the power delivered to R_1 using the total current through R_1.
   d. Does superposition apply to power effects? Explain.

2. Using superposition, find the current through the 10-Ω resistor for each of the networks of Fig. 9.106.

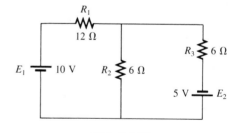

FIG. 9.105

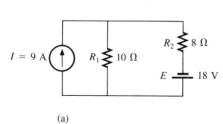

(a)

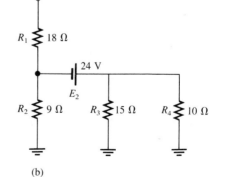

(b)

FIG. 9.106

*3. Using superposition, find the current through R_1 for each network of Fig. 9.107.

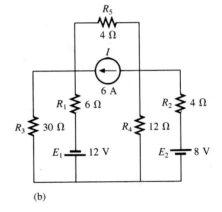

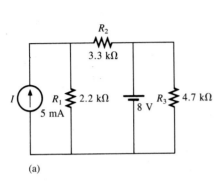

(a)   (b)

FIG. 9.107

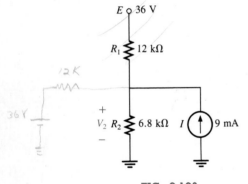

FIG. 9.108

4. Using superposition, find the voltage V_2 for the network of Fig. 9.108.

Section 9.3

5. a. Find the Thevenin equivalent circuit for the network external to the resistor R of Fig. 9.109.

  b. Find the current through R when R is 2, 30, and 100 Ω.

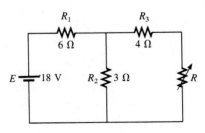

FIG. 9.109

6. a. Find the Thevenin equivalent circuit for the network external to the resistor R in each of the networks of Fig. 9.110.

b. Find the power delivered to R when R is $2\,\Omega$ and $100\,\Omega$.

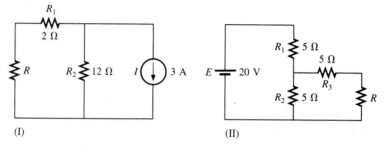

(I)   (II)

FIG. 9.110

7. Repeat Problem 6a for the networks of Fig. 9.111.

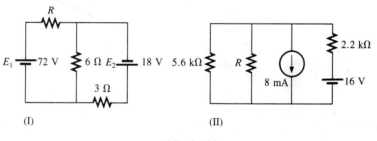

(I)   (II)

FIG. 9.111

*8. Repeat Problem 6 for the networks of Fig. 9.112.

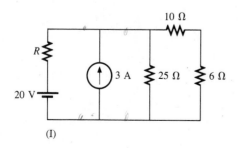
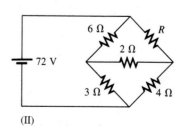

(I)   (II)

FIG. 9.112

*9. Find the Thevenin equivalent circuit for the portions of the networks of Fig. 9.113 external to points a and b.

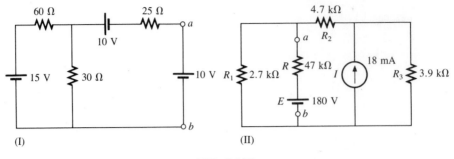

(I)

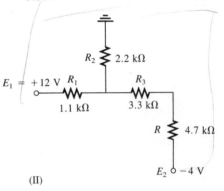

(II)

FIG. 9.113

*10. Determine the Thevenin equivalent circuit for the network external to the resistor R in both networks of Fig. 9.114.

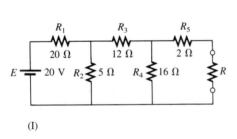

(I)

(II)

FIG. 9.114

Section 9.4

11. Find the Norton equivalent circuit for the network external to the resistor R in each network of Fig. 9.110 by
   a. following the procedure outlined in the text.
   b. converting the Thevenin equivalent circuit to a Norton equivalent circuit if available from a previous assignment.

12. Repeat Problem 11 for the networks of Fig. 9.111.

*13. Repeat Problem 11 for the networks of Fig. 9.112.

14. Find the Norton equivalent circuit for the portions of the networks of Fig. 9.113 external to branch a-b by
   a. following the procedure outlined in the text.
   b. converting the Thevenin equivalent circuit to a Norton equivalent circuit if available from a previous assignment.

*15. Repeat Problem 14 for the resistor R of Fig. 9.114.

16. Find the Norton equivalent circuit for the portions of the networks of Fig. 9.115 external to branch *a-b*.

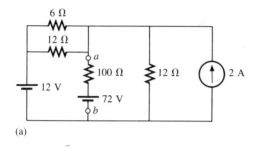

(a)

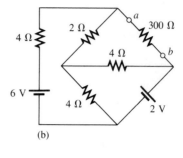

(b)

FIG. 9.115

Section 9.5

17. a. For each network of Fig. 9.110, find the value of R for maximum power to R.
   b. Determine the maximum power to R for each network.

18. Repeat Problem 17 for the networks of Fig. 9.111.

*19. Repeat Problem 17 for the networks of Fig. 9.112.

20. a. For the network of Fig. 9.116, determine the value of R for maximum power to R.
   b. Determine the maximum power to R.
   c. Plot a curve of power to R versus R for R equal to $\frac{1}{4}$, $\frac{1}{2}$, $\frac{3}{4}$, 1, $1\frac{1}{4}$, $1\frac{1}{2}$, $1\frac{3}{4}$, and 2 times the value obtained in part (a).

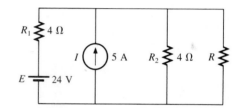

FIG. 9.116

*21. Find the resistance R_1 of Fig. 9.117 such that the resistor R_4 will receive maximum power. Think!

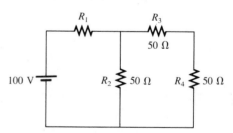

FIG. 9.117

*22. a. For the network of Fig. 9.118, determine the value R_2 for maximum power to R_4.
   b. Is there a general statement that can be made about situations such as those presented here and in Problem 21?

Section 9.6

23. Write a program to determine the current through the 10-Ω resistor of Fig. 9.106(a) (for any component values) using superposition.

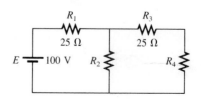

FIG. 9.118

24. Write a program to perform the analysis required for Problem 8, network (II) for any component values.

25. Write a program to perform the analysis of Problem 20 and tabulate the power to R for the values listed in part (c).

Section 9.7

26. Using Millman's theorem, find the current through and voltage across the resistor R_L of Fig. 9.119.

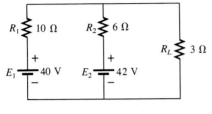

FIG. 9.119

27. Repeat Problem 26 for the network of Fig. 9.120.

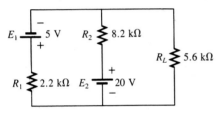

FIG. 9.120

28. Repeat Problem 26 for the network of Fig. 9.121.

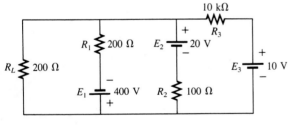

FIG. 9.121

29. Using the dual of Millman's theorem, find the current through and voltage across the resistor R_L of Fig. 9.122.

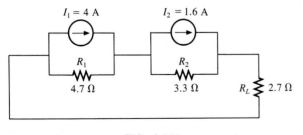

FIG. 9.122

*30. Repeat Problem 29 for the network of Fig. 9.123.

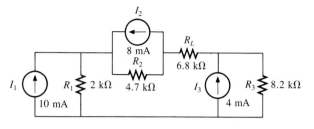

FIG. 9.123

Section 9.8

31. Using the substitution theorem, draw three equivalent branches for the branch *a-b* of the network of Fig. 9.124.

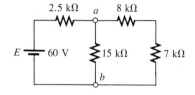

FIG. 9.124

32. Repeat Problem 31 for the network of Fig. 9.125.

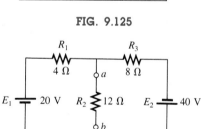

FIG. 9.125

*33. Repeat Problem 31 for the network of Fig. 9.126. Be careful!

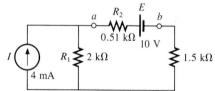

FIG. 9.126

Section 9.9

34. a. For the network of Fig. 9.127(a), determine the current *I*.
  b. Repeat part (a) for the network of Fig. 9.127(b).
  c. Is the reciprocity theorem satisfied?

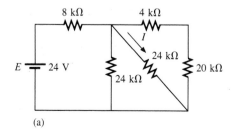
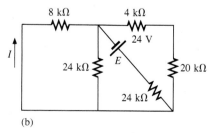

(a)   (b)

FIG. 9.127

35. Repeat Problem 34 for the networks of Fig. 9.128.

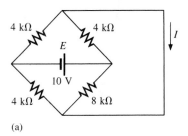

(a)

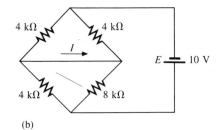

(b)

FIG. 9.128

36. a. Determine the voltage V for the network of Fig. 9.129(a).
 b. Repeat part (a) for the network of Fig. 9.129(b).
 c. Is the dual of the reciprocity theorem satisfied?

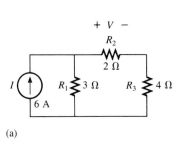

(a)

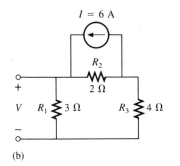

(b)

FIG. 9.129

GLOSSARY

Maximum power transfer theorem A theorem used to determine the load resistance necessary to insure maximum power transfer to the load.

Millman's theorem A method employing source conversions that will permit the determination of unknown variables in a multiloop network.

Norton's theorem A theorem that permits the reduction of any two-terminal linear dc network to one having a single current source and parallel resistor.

Reciprocity theorem A theorem that states that for single-source networks, the current in any branch of a network, due to a single voltage source in the network, will equal the current through the branch in which the source was originally located if the source is placed in the branch in which the current was originally measured.

Substitution theorem A theorem that states that if the voltage across and current through any branch of a dc bilateral network are known, the branch can be replaced by any combination of elements that will maintain the same voltage across and current through the chosen branch.

Superposition theorem A network theorem that permits considering the effects of each source independently. The resulting current and/or voltage is the algebraic sum of the currents and/or voltages developed by each source independently.

Thevenin's theorem A theorem that permits the reduction of any two-terminal linear dc network to one having a single voltage source and series resistor.

10
Capacitors

10.1 INTRODUCTION

Thus far, the only passive device appearing in the text has been the resistor. We will now consider two additional passive devices called the *capacitor* and the *inductor*, which are quite different from the resistor in purpose, operation, and construction.

Unlike the resistor, these elements display their total characteristics only when a change in voltage or current is made in the circuit in which they exist. In addition, if we consider the *ideal* situation, they do not dissipate energy like the resistor but store it in a form that can be returned to the circuit whenever required by the circuit design.

Proper treatment of each requires that we devote this entire chapter to the capacitor and Chapter 12 to the inductor. Since electromagnetic effects are a major consideration in the design of inductors, Chapter 11 on magnetic circuits will appear first.

10.2 THE ELECTRIC FIELD

Recall from Chapter 2 that a force of attraction or repulsion exists between two charged bodies. We shall now examine this phenomenon in greater detail by considering the electric field that exists in the region around any charged body. This electric field is represented by electric flux lines, which are drawn to indicate the strength of the electric field at any point around the charged body; that is, the denser the lines of flux, the stronger the electric field. In Fig. 10.1, the electric field strength is stronger at position a than at position b because the flux lines

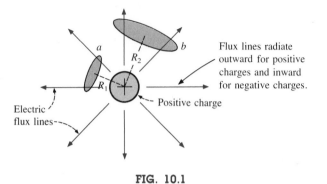

Flux lines radiate outward for positive charges and inward for negative charges.

Positive charge

Electric flux lines

FIG. 10.1

are denser at a than at b. The symbol for electric flux is the Greek letter ψ (psi). The flux per unit area (flux density) is represented by the capital letter D and is determined by

$$D = \frac{\psi}{A} \qquad \text{(flux/unit area)} \qquad \textbf{(10.1)}$$

The larger the charge Q in coulombs, the greater the number of flux lines extending or terminating per unit area, independent of the surrounding medium. Twice the charge will produce twice the flux per unit area. The two can therefore be equated:

$$\psi \equiv Q \qquad \text{(coulombs, C)} \qquad \textbf{(10.2)}$$

By definition, the *electric field strength* at a point is the force acting on a unit positive charge at that point; that is,

$$\mathscr{E} = \frac{F}{Q} \qquad \text{(newtons/coulomb, N/C)} \qquad \textbf{(10.3)}$$

The force exerted on a unit positive charge ($Q_2 = 1$), by a charge Q_1, r meters away, as determined by Coulomb's law is

$$F = \frac{kQ_1Q_2}{r^2} = \frac{kQ_1(1)}{r^2} = \frac{kQ_1}{r^2} \qquad (k = 9 \times 10^9)$$

Substituting this force F into Eq. (10.3) yields

$$\mathscr{E} = \frac{F}{Q_2} = \frac{kQ_1/r^2}{1}$$

$$\mathscr{E} = \frac{kQ_1}{r^2} \qquad \text{(N/C)} \qquad \textbf{(10.4)}$$

We can therefore conclude that the electric field strength at any point distance r from a point charge of Q coulombs is directly proportional to the magnitude of the charge and inversely proportional to the distance squared from the charge. The squared term in the denominator will result in a rapid decrease in the strength of the electric field with distance from the point charge. In Fig. 10.1, substituting distances R_1 and R_2 into Eq. (10.4) will verify our previous conclusion that the electric field strength is greater at a than at b.

Flux lines always extend from a positively charged to a negatively charged body, always extend or terminate perpendicular to the charged surfaces, and never intersect. For two charges of similar and opposite polarities, the flux distribution would appear as shown in Fig. 10.2.

The attraction and repulsion between charges can now be explained in terms of the electric field and its flux lines. In Fig. 10.2(a), the flux lines are not interlocked but tend to act as a buffer, preventing attraction and causing repulsion. Since the electric field strength is stronger (flux lines denser) for each charge the closer we are to the charge, the more we try to bring the two charges together, the stronger will be the force of repulsion between them. In Fig. 10.2(b), the flux lines extending from the positive charge are terminated at the negative charge. A basic law of physics states that electric flux lines always tend to be as short as possible. The two charges will therefore be drawn to each other. Again, the closer the two charges, the stronger the attraction between the two charges due to the increased field strengths.

10.3 CAPACITANCE

Up to this point we have considered only isolated positive and negative spherical charges, but the analysis can be extended to charged surfaces of any shape and size. In Fig. 10.3, for example, two parallel plates of a conducting material separated by an air gap have been connected through a switch and a resistor to a battery. If the parallel plates are initially uncharged and the switch is left open, no net positive or negative charge will exist on either plate. The instant the switch is closed, however, electrons are drawn from the upper plate through the resistor to the positive terminal of the battery. There will be a surge of current at first, limited in magnitude by the resistance present. The level of flow will then decline, as will be demonstrated in the sections to follow. This action creates a net positive charge on the top plate. Electrons are being repelled by the negative terminal through the lower conductor to the bottom plate at the same rate they are being drawn to the positive terminal. This transfer of electrons continues until the potential difference across the parallel plates is exactly equal to the battery voltage. The final result is a net positive charge on the top plate and a negative charge on the bottom plate, very similar in many respects to the two isolated charges of Fig. 10.2(b).

This element, constructed simply of two parallel conducting plates separated by an insulating material (in this case, air), is called a

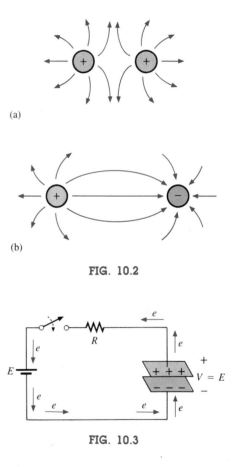

(a)

(b)

FIG. 10.2

FIG. 10.3

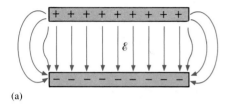

(a)

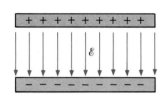

(b)

FIG. 10.4

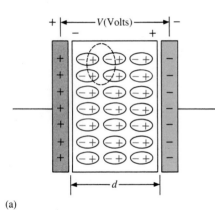

(a)

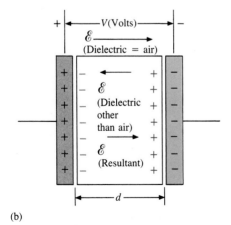

(b)

FIG. 10.5

capacitor. *Capacitance* is a measure of a capacitor's ability to store charge on its plates—in other words, its storage capacity. A capacitor has a capacitance of 1 farad if 1 coulomb of charge is deposited on the plates by a potential difference of 1 volt across the plates. The farad is named after Michael Faraday, a nineteenth-century English chemist and physicist. The farad, however, is generally too large a measure of capacitance for most practical applications, so the microfarad (10^{-6}) or picofarad (10^{-12}) is more commonly used. Expressed as an equation, the capacitance is determined by

$$C = \frac{Q}{V}$$

C = farads (F)
Q = coulombs (C)   **(10.5)**
V = volts (V)

Different capacitors for the same voltage across their plates will acquire greater or lesser amounts of charge on their plates. Hence the capacitors have a greater or lesser capacitance, respectively.

A cross-sectional view of the parallel plates is shown with the distribution of electric flux lines in Fig. 10.4(a). The number of flux lines per unit area (D) between the two plates is quite uniform. At the edges, the flux lines extend outside the common surface area of the plates, producing an effect known as *fringing*. This effect, which reduces the capacitance somewhat, can be neglected for most practical applications. For the analysis to follow, we will assume that all the flux lines leaving the positive plate will pass directly to the negative plate within the common surface area of the plates [Fig. 10.4(b)].

If a potential difference of V volts is applied across the two plates separated by a distance of d, the electric field strength between the plates is determined by

$$\mathscr{E} = \frac{V}{d} \qquad \text{(volts/meter, V/m)} \qquad \textbf{(10.6)}$$

The uniformity of the flux distribution in Fig. 10.4(b) also indicates that the electric field strength is the same at any point between the two plates.

Many values of capacitance can be obtained for the same set of parallel plates by the addition of certain insulating materials between the plates. In Fig. 10.5(a), an insulating material has been placed between a set of parallel plates having a potential difference of V volts across them.

Since the material is an insulator, the electrons within the insulator are unable to leave the parent atom and travel to the positive plate. The positive components (protons) and negative components (electrons) of each atom do shift, however [as shown in Fig. 10.5(a)], to form *dipoles*.

When the dipoles align themselves as shown in Fig. 10.5(a), the material is *polarized*. A close examination within this polarized material will indicate that the positive and negative components of adjoining dipoles are neutralizing the effects of each other [note the dashed area in Fig. 10.5(a)]. The layer of positive charge on one surface and the negative charge on the other are not neutralized, however, resulting in the establishment of an electric field within the insulator [$\mathscr{E}_{dielectric}$, Fig. 10.5(b)]. The net electric field between the plates ($\mathscr{E}_{resultant} = \mathscr{E}_{air} - \mathscr{E}_{dielectric}$) would therefore be reduced due to the insertion of the dielectric.

The purpose of the dielectric, therefore, is to create an electric field to oppose the electric field set up by free charges on the parallel plates. For this reason, the insulating material is referred to as a *dielectric, di* for *opposing* and *electric* for *electric field*.

In either case—with or without the dielectric—if the potential across the plates is kept constant and the distance between the plates is fixed, the net electric field within the plates must remain the same, as determined by the equation $\mathscr{E} = V/d$. We just ascertained, however, that the net electric field between the plates would decrease with insertion of the dielectric for a fixed amount of free charge on the plates. To compensate and keep the net electric field equal to the value determined by V and d, more charge must be deposited on the plates. [Look ahead to Eq. (10.11).] This additional charge for the same potential across the plates increases the capacitance, as determined by the following equation:

$$C\!\uparrow \;=\; \frac{Q\!\uparrow}{V}$$

For different dielectric materials between the same two parallel plates, different amounts of charge will be deposited on the plates. But $\psi \equiv Q$, so the dielectric is also determining the number of flux lines between the two plates and consequently the flux density ($D = \psi/A$) since A is fixed.

The ratio of the flux density to the electric field intensity in the dielectric is called the *permittivity* of the dielectric:

$$\boxed{\epsilon = \frac{D}{\mathscr{E}}} \qquad \text{(farads/meter, F/m)} \qquad \textbf{(10.7)}$$

It is a measure of how easily the dielectric will "permit" the establishment of flux lines within the dielectric. The greater its value, the greater the amount of charge deposited on the plates, and, consequently, the greater the flux density for a fixed area.

For a vacuum, the value of ϵ (denoted by ϵ_o) is 8.85×10^{-12} F/m. The ratio of the permittivity of any dielectric to that of a vacuum is called the *relative permittivity*, ϵ_r. In equation form,

$$\boxed{\epsilon_r = \frac{\epsilon}{\epsilon_o}} \qquad \textbf{(10.8)}$$

The value of ϵ for any material, therefore, is

$$\epsilon = \epsilon_r \epsilon_o$$

Note that ϵ_r is a dimensionless quantity. The relative permittivity, or *dielectric constant,* as it is often called, is given in Table 10.1 for various dielectric materials.

TABLE 10.1
Relative permittivity (dielectric constant) of various dielectrics.

Dielectric	ϵ_r (Average Values)
Vacuum	1.0
Air	1.0006
Teflon®	2.0
Paper, paraffined	2.5
Rubber	3.0
Transformer oil	4.0
Mica	5.0
Porcelain	6.0
Bakelite	7.0
Glass	7.5
Distilled water	80.0
Barium-strontium titanite (ceramic)	7500.0

Substituting for D and $\mathscr{E}$ in Eq. (10.7), we have

$$\epsilon = \frac{D}{\mathscr{E}} = \frac{\psi/A}{V/d} = \frac{Q/A}{V/d} = \frac{Qd}{VA}$$

But

$$C = \frac{Q}{V}$$

and therefore

$$\epsilon = \frac{Cd}{A}$$

and

$$\boxed{C = \frac{\epsilon A}{d}} \quad \text{(F)} \qquad \textbf{(10.9)}$$

or

$$\boxed{C = \epsilon_o \epsilon_r \frac{A}{d} = 8.85 \times 10^{-12} \epsilon_r \frac{A}{d}} \quad \text{(F)} \qquad \textbf{(10.10)}$$

where A is the area in square meters of the plates, d is the distance in meters between the plates, and ϵ_r is the relative permittivity. The capacitance, therefore, will be greater if the area of the plates is increased, or the distance between the plates is decreased, or the dielectric is changed so that ϵ_r is increased.

Solving for the distance d in Eq. (10.9), we have

$$d = \frac{\epsilon A}{C}$$

and substituting into Eq. (10.6) yields

$$\mathscr{E} = \frac{V}{d} = \frac{V}{\epsilon A/C} = \frac{CV}{\epsilon A}$$

But $Q = CV$, and therefore

$$\boxed{\mathscr{E} = \frac{Q}{\epsilon A}} \qquad (\text{V/m}) \qquad \textbf{(10.11)}$$

which gives the electric field intensity between the plates in terms of the permittivity ϵ, the charge Q, and the surface area A of the plates. The ratio

$$\frac{C = \epsilon A/d}{C_o = \epsilon_o A/d} = \frac{\epsilon}{\epsilon_o} = \epsilon_r$$

or

$$\boxed{C = \epsilon_r C_o} \qquad \textbf{(10.12)}$$

which, in words, states that for the same set of parallel plates, the capacitance using a dielectric of mica (or any other dielectric) is five (the relativity permittivity of the dielectric) times that obtained for a vacuum (or air, approximately) between the plates. This relationship between ϵ_r and the capacitances provides an excellent experimental method for finding the value of ϵ_r for various dielectrics.

EXAMPLE 10.1. Determine the capacitance of each capacitor on the right side of Fig. 10.6.

Solution:

a. $C = 3(5\,\mu\text{F}) = \textbf{15}\,\boldsymbol{\mu}\textbf{F}$

b. $C = \dfrac{1}{2}(0.1\,\mu\text{F}) = \textbf{0.05}\,\boldsymbol{\mu}\textbf{F}$

c. $C = 2.5(20\,\mu\text{F}) = \textbf{50}\,\boldsymbol{\mu}\textbf{F}$

d. $C = (5)\dfrac{4}{(1/8)}(1000\,\text{pF}) = (160)(1000\,\text{pF}) = \textbf{0.16}\,\boldsymbol{\mu}\textbf{F}$

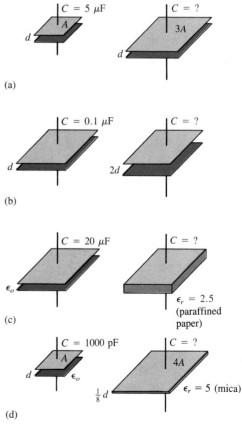

FIG. 10.6

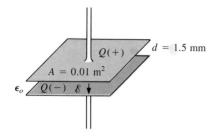

FIG. 10.7

EXAMPLE 10.2. For the capacitor of Fig. 10.7:
a. Determine the capacitance.
b. Determine the electric field strength between the plates if 450 V are applied across the plates.
c. Find the resulting charge on each plate.
Solutions:

a. $C_o = \dfrac{\epsilon_o A}{d} = \dfrac{(8.85 \times 10^{-12})(0.01)}{1.5 \times 10^{-3}} = 59.0 \times 10^{-12}$

$\qquad = \textbf{59 pF}$

b. $\mathscr{E} = \dfrac{V}{d} = \dfrac{450}{1.5 \times 10^{-3}}$

$\qquad \cong \textbf{300} \times \textbf{10}^3 \textbf{ V/m}$

c. $C = \dfrac{Q}{V}$

or

$Q = CV = (59.0 \times 10^{-12})(450)$
$\qquad = 26.550 \times 10^{-9}$
$\qquad = \textbf{26.55 nC}$

EXAMPLE 10.3. A sheet of mica 1.5 mm thick having the same area as the plates is inserted between the plates of Example 10.2.
a. Find the electric field strength between the plates.
b. Find the charge on each plate.
c. Find the capacitance.
Solutions:

a. $\mathscr{E}$ is fixed by

$$\mathscr{E} = \frac{V}{d} = \frac{450}{1.5 \times 10^3}$$

$$\cong \textbf{300} \times \textbf{10}^3 \textbf{ V/m}$$

b. $\mathscr{E} = \dfrac{Q}{\epsilon A}$

or

$Q = \epsilon \mathscr{E} A = \epsilon_r \epsilon_o \mathscr{E} A$
$\qquad = (5)(8.85 \times 10^{-12})(300 \times 10^3)(0.01)$
$\qquad = 132.75 \times 10^{-9} = \textbf{132.75 nC}$

(five times the amount for air between the plates)

c. $C = \epsilon_r C_o$
$\qquad = (5)(59 \times 10^{-12}) = \textbf{295 pF}$

10.4 DIELECTRIC STRENGTH

For every dielectric there is a potential that if applied across the dielectric will break the bonds within the dielectric and cause current to flow. The voltage required per unit length (electric field intensity) to establish conduction in a dielectric is an indication of its *dielectric strength* and is called the *breakdown voltage*. When breakdown occurs, the capacitor has characteristics very similar to those of a conductor. A typical example of breakdown is lightning, which occurs when the potential between the clouds and the earth is so high that charge can pass from one to the other through the atmosphere, which acts as the dielectric.

The average dielectric strengths for various dielectrics are tabulated in volts/mil in Table 10.2 (1 mil = 0.001 in.). The relative permittivity appears in parentheses to emphasize the importance of considering both factors in the design of capacitors. Take particular note of barium-strontium titanite and mica.

TABLE 10.2
Dielectric strength of some dielectric materials.

Dielectric	Dielectric Strength (Average Value), in Volts/Mil .00/	(ϵ_r)
Air	75	(1.006)
Barium-strontium titanite (ceramic)	75	(7500)
Porcelain	200	(6.0)
Transformer oil	400	(4.0)
Bakelite	400	(7.0)
Rubber	700	(3.0)
Paper, paraffined	1300	(2.5)
Teflon	1500	(2.0)
Glass	3000	(7.5)
Mica	5000	(5.0)

EXAMPLE 10.4. Find the maximum voltage that can be applied across a 0.2-μF capacitor having a plate area of 0.3 m^2. The dielectric is porcelain. Assume a linear relationship between the dielectric strength and the thickness of the dielectric.

Solution:

$$C = \frac{8.85\epsilon_r A}{10^{12}d}$$

or

$$d = \frac{8.85\epsilon_r A}{10^{12}C} = \frac{(8.85)(6)(0.3)}{(10^{12})(0.2 \times 10^{-6})} = 7.965 \times 10^{-5}$$

$$\cong \mathbf{79.65\ \mu m}$$

Converting millimeters to mils, we have

$$79.76 \, \mu m \left(\frac{10^{-6} \, m}{\mu m} \right) \left(\frac{39.371 \, in.}{m} \right) \left(\frac{1000 \, mils}{1 \, in.} \right) = 3.136 \, mils$$

Dielectric strength = 200 V/mil

Therefore,

$$\left(\frac{200 \, V}{mil} \right) (3.136 \, mils) = \mathbf{627.20 \, V}$$

10.5 LEAKAGE CURRENT

Up to this point, we have assumed that the flow of electrons will occur in a dielectric only when the breakdown voltage is reached. This is the ideal case. In actuality, there are free electrons in every dielectric. These free electrons are due in part to impurities in the dielectric and forces within the material itself.

When a voltage is applied across the plates of a capacitor, a leakage current due to the free electrons flows from one plate to the other. The current is usually so small, however, that it can be neglected for most practical applications. This effect is represented by a resistor in parallel with the capacitor, as shown in Fig. 10.8(a). Its value is usually in the order of 1000 megohms (MΩ). There are some capacitors, however, such as the electrolytic type, that have high leakage currents. When charged and then disconnected from the charging circuit, these capacitors lose their charge in a matter of seconds because of the flow of charge (leakage current) from one plate to the other [Fig. 10.8(b)].

(a)

(b)

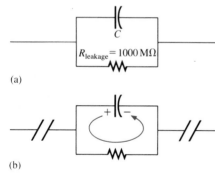

FIG. 10.8

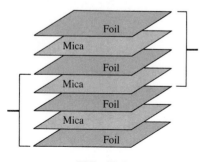

FIG. 10.9

10.6 TYPES OF CAPACITORS

Like resistors, all capacitors can be included under either of two general headings: fixed or variable. The symbol for a fixed capacitor is ╫ and for a variable capacitor ╫. The curved line represents the plate that is usually connected to the point of lower potential.

Many types of fixed capacitors are available today. Some of the most common are the mica, ceramic, electrolytic, tantalum, and polyester-film capacitors. The typical *mica capacitor* consists basically of mica sheets separated by sheets of metal foil. The plates are connected to two electrodes, as shown in Fig. 10.9. The total area is the area of one sheet times the number of dielectric sheets. The entire system is encased in a plastic insulating material as shown in Fig. 10.10(a). The mica capacitor exhibits excellent characteristics under stress of temperature variations and high voltage applications (its dielectric strength is 5000 V/mil). Its leakage current is also very small ($R_{leakage}$ about 1000 MΩ).

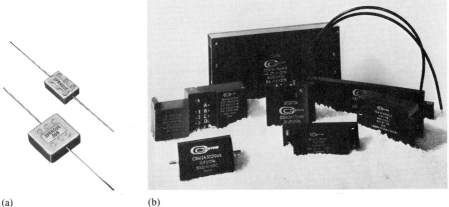

(a)　　　　　　(b)

FIG. 10.10
Mica capacitors (Courtesy of Custom Electronics Inc.)

Mica capacitors are typically between a few picofarads and 0.2 microfarad with voltages of 100 volts or more. The color code for the mica capacitors of Fig. 10.10(a) can be found in Appendix C.

A second type of mica capacitor appears in Fig. 10.10(b). Note in particular the cylindrical unit in the bottom left-hand corner of the figure. The ability to "roll" the mica to form the cylindrical shape is due to a process whereby the soluble contaminants in natural mica are removed, leaving a paperlike structure due to the cohesive forces in natural mica. It is commonly referred to as *reconstituted mica,* although the terminology does not mean "recycled" or "second-hand" mica. For some of the units in the photograph, different levels of capacitance are available between different sets of terminals.

The *ceramic capacitor* is made in many shapes and sizes, some of which are shown in Fig. 10.11. The basic construction, however, is

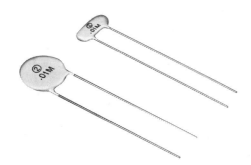

FIG. 10.11
Ceramic disc capacitors. (Courtesy of Sprague Electric Co.)

about the same for each, as shown in Fig. 10.12. A ceramic base is coated on two sides with a metal, such as copper or silver, to act as the two plates. The leads are then attached through electrodes to the plates.

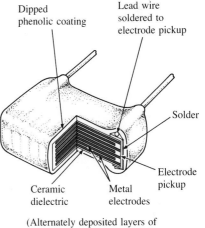

(Alternately deposited layers of ceramic dielectric material and metal electrodes fired into a single homogeneous block)

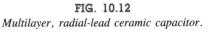

FIG. 10.12
Multilayer, radial-lead ceramic capacitor.

An insulating coating of ceramic or plastic is then applied over the plates and dielectric. Ceramic capacitors also have a very low leakage current (R_{leakage} about 1000 MΩ) and can be used in both dc and ac networks. They can be found in values ranging from a few picofarads to perhaps 2 microfarads, with very high working voltages such as 5000 volts or more.

In recent years there has been increasing interest in monolithic (single-structure) chip capacitors such as appearing in Fig. 10.13(a) due to their application on hybrid circuitry [networks using both discrete and integrated circuit (IC) components]. There has also been increasing use of microstrip (strip-line) circuitry such as appearing in Fig. 10.13(b). Note the small chips in this cutaway section. The L and H of Fig. 10.13(a) indicate the level of capacitance. For example, the letter H in black letters represents 16 units of capacitance (in picofarads), or 16 pF. If blue ink is used, a multiplier of 100 is applied, resulting in 1600 pF. Although the size is similar, the type of ceramic material controls the capacitance level.

The *electrolytic capacitor* is used most commonly in situations where capacitances of the order of one to several thousand microfarads are required. They are designed primarily for use in networks where only dc voltages will be applied across the capacitor. There are electrolytic capacitors available that can be used in ac circuits (for starting motors) and in cases where the polarity of the dc voltage will reverse across the capacitor for short periods of time.

The basic construction of the electrolytic capacitor consists of a roll of aluminum foil coated on one side with an aluminum oxide, the aluminum being the positive plate and the oxide the dielectric. A layer of paper or gauze saturated with an electrolyte is placed over the aluminum oxide on the positive plate. Another layer of aluminum without the oxide coating is then placed over this layer to assume the role of the negative plate. In most cases the negative plate is connected directly to the aluminum container, which then serves as the negative terminal for external connections. Because of the size of the roll of aluminum foil, the overall area of this capacitor is large; and due to the use of an oxide as the dielectric, the distance between the plates is extremely small. The negative terminal of the electrolytic capacitor is usually the one with no visible identification on the casing. The positive is usually indicated by such designs as +, △, □, and so on. Due to the polarity requirement, the symbol for an electrolytic will normally appear as ⊣⊢$^+$.

Associated with each electrolytic capacitor are the dc working voltage and the surge voltage. The *working voltage* is the voltage that can be applied across the capacitor for long periods of time without breakdown. The *surge voltage* is the maximum dc voltage that can be applied for a short period of time. Electrolytic capacitors are characterized as having low breakdown voltages and high leakage currents (R_{leakage} about 1 MΩ). Various types of electrolytic capacitors are shown in Fig. 10.14. They can be found in values extending from a few microfarads to several thousand microfarads and working voltages as high as 500 volts.

(a)

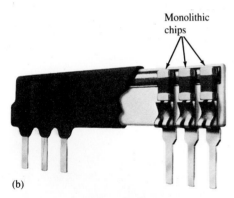

Monolithic chips

(b)

FIG. 10.13

Monolithic chip capacitors. (Courtesy of Vitramon, Inc.)

However, increased levels of voltage are normally associated with lower values of available capacitance.

There are fundamentally two types of *tantalum capacitors:* the *solid* and the *wet-slug.* In each case, tantalum powder of high purity is pressed into a rectangular or cylindrical shape as shown in Fig. 10.15. The anode (+) connection is then simply pressed into the resulting structures as shown in the figure. The resulting unit is then sintered (baked) in a vacuum at very high temperatures to establish a very porous material. The result is a structure with a very large surface area in a limited volume. Through immersion in an acid solution, a very thin manganese dioxide (MnO_2) coating is established on the large, porous surface area. An electrolyte is then added to establish contact between the surface area and the cathode, producing a solid tantalum capacitor. If an appropriate ''wet'' acid is introduced, it is called a *wet-slug* tantalum capacitor.

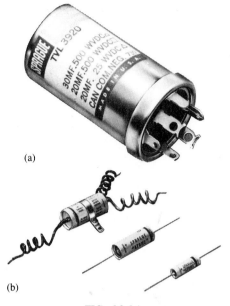

(a)

(b)

FIG. 10.14
Electrolytic capacitors. (Courtesy of Sprague Electric Co.)

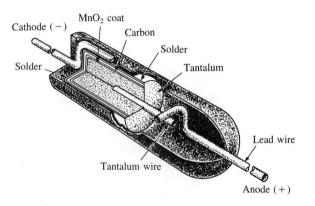

FIG. 10.15
Tantalum capacitor. (Courtesy of Union Carbide Corp.)

The last type of fixed capacitor to be introduced is the *polyester-film capacitor,* the basic construction of which is shown in Fig. 10.16. It consists simply of two metal foils separated by a strip of polyester material such as Mylar®. The outside layer of polyester is applied to act as an insulating jacket. Each metal foil is connected to a lead which extends either axially or radially from the capacitor. The rolled construction results in a large surface area, and the use of the plastic dielectric results in a very thin layer between the conducting surfaces.

Data such as capacitance and working voltage are printed on the outer wrapping if the polyester capacitor is large enough. Color coding is used on smaller devices (see Appendix C). A band (usually black) is sometimes printed near the lead that is connected to the outer metal foil. The lead nearest this band should always be connected to the point of lower potential. This capacitor can be used for both dc and ac networks. Its leakage resistance is of the order of 100 MΩ. A typical polyester

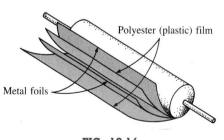

FIG. 10.16
Polyester-film capacitor.

FIG. 10.17

Orange Drop® tubular capacitor. (Courtesy of Sprague Electric Co.)

capacitor appears in Fig. 10.17. Polyester capacitors range in value from a few hundred picofarads to 10–20 microfarads, with working voltages as high as a few thousand volts.

The most common of the variable-type capacitors is shown in Fig. 10.18. The dielectric for each capacitor is air. The capacitance in Fig. 10.18(a) is changed by turning the shaft at one end to vary the common

(a)

(b)

FIG. 10.18

Variable air capacitors. (Part (a) courtesy of James Millen Manufacturing Co.; part (b) courtesy of Johnson Manufacturing Co.)

FIG. 10.19

Digital reading capacitance meter. (Courtesy of Global Specialties Corp.)

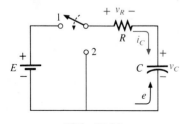

FIG. 10.20

area of the movable and fixed plates. The greater the common area, the larger the capacitance, as determined by Eq. (10.10). The capacitance of the trimmer capacitor in Fig. 10.18(b) is changed by turning the screw, which will vary the distance between the plates and thereby the capacitance. A digital reading capacitance meter appears in Fig. 10.19.

10.7 TRANSIENTS IN CAPACITIVE NETWORKS: CHARGING PHASE

Section 10.3 described how a capacitor acquires its charge. Let us now extend this discussion to include the potentials and current developed within the network of Fig. 10.20 following the closing of the switch (to position 1).

You will recall that the instant the switch is closed, electrons are drawn from the top plate and deposited on the bottom plate by the battery, resulting in a net positive charge on the top plate, and a negative charge on the bottom plate. The transfer of electrons is very rapid at first, slowing down as the potential across the capacitor approaches the applied voltage of the battery. When the voltage across the capacitor equals the battery voltage, the transfer of electrons will cease and the plates will have a net charge determined by $Q = CV_C = CE$.

Plots of the changing current and voltage appear in Figs. 10.21 and 10.22, respectively. When the switch is closed at $t = 0$ s, the current jumps to a value limited only by the resistance of the network and then decays to zero as the plates are charged. Note the rapid decay in current level revealing that the amount of charge deposited on the plates per unit time is rapidly decaying also. Since the voltage across the plates is directly related to the charge on the plates by $v_C = q/C$, the rapid rate with which charge is initially deposited on the plates will result in a rapid increase in v_C. Obviously, as the rate of flow of charge (I) decreases, the rate of change in voltage will follow suit. Eventually, the flow of charge will stop, the current I will be zero, and the voltage will cease to change in magnitude—the *charging phase* has passed. At this point the capacitor takes on the characteristics of an open circuit: a voltage drop across the plates without a flow of charge "between" the plates. As demonstrated in Fig. 10.23, the voltage across the capacitor

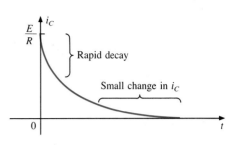

FIG. 10.21

i_C during the charging phase.

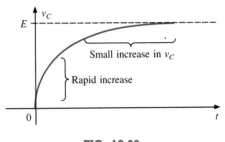

FIG. 10.22

v_C during the charging phase.

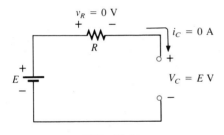

FIG. 10.23

Open-circuit equivalent for a capacitor.

is the source voltage since $i = i_C = i_R = 0$ A and $v_R = i_R R = (0)R = 0$ V. For all future analysis:

A capacitor can be replaced by an open-circuit equivalent once the charging phase in a dc network has passed.

Looking back at the instant the switch is closed, we can also surmise that a capacitor behaves like a short circuit the moment the switch is closed in a dc charging network, as shown in Fig. 10.24. The current $i = i_C = i_R = E/R$, and the voltage $v_C = E - v_R = E - i_R R = E - (E/R)R = E - E = 0$ V at $t = 0$ s.

Through the use of calculus, the following mathematical equation for the charging current i_C can be obtained:

$$i_C = \frac{E}{R}e^{-t/RC} \qquad \textbf{(10.13)}$$

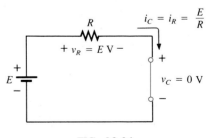

FIG. 10.24

Short-circuit equivalent for a capacitor.

The factor $e^{-t/RC}$ is an exponential function of the form e^{-x}, where $x = -t/RC$ and $e = 2.71828...$. A plot of e^{-x} for $x \geq 0$ appears in Fig.

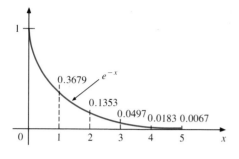

FIG. 10.25

The e^{-x} function ($x \geq 0$).

10.25. Exponentials are mathematical functions that all students of electrical, electronic, or computer systems must become very familiar with. They will appear throughout the analysis to follow in this course, and in succeeding courses.

Our current interest in the function e^{-x} is limited to values of x greater than zero, as noted by the curve of Fig. 10.21. All modern-day scientific calculators have the function e^x. To obtain e^{-x}, the sign of x must be changed using the sign key before the exponential function is keyed in. The magnitude of e^{-x} has been listed in Table 10.3 for a range of values of x. Note the rapidly decreasing magnitude of e^{-x} with increasing value of x.

TABLE 10.3

Selected values of e^{-x}.

$x = 0$	$e^{-x} = e^{-0} = \dfrac{1}{e^0} = \dfrac{1}{1} = 1$
$x = 1$	$e^{-1} = \dfrac{1}{e} = \dfrac{1}{2.71828...} = 0.3679$
$x = 2$	$e^{-2} = \dfrac{1}{e^2} = 0.1353$
$x = 5$	$e^{-5} = \dfrac{1}{e^5} = 0.00674$
$x = 10$	$e^{-10} = \dfrac{1}{e^{10}} = 0.0000454$
$x = 100$	$e^{-100} = \dfrac{1}{e^{100}} = 3.72 \times 10^{-44}$

The factor RC in Eq. (10.13) is called the *time constant* of the system and has the units of time as follows:

$$RC = \left(\frac{V}{I}\right)\left(\frac{Q}{V}\right) = \left(\frac{V}{Q/t}\right)\left(\frac{Q}{V}\right) = t$$

Its symbol is the Greek letter τ (tau), and its unit of measure is the second. Thus,

$$\boxed{\tau = RC} \qquad \text{(seconds, s)} \qquad \textbf{(10.14)}$$

If we substitute $\tau = RC$ into the exponential function $e^{-t/RC}$, we obtain $e^{-t/\tau}$. In one time constant, $e^{-t/\tau} = e^{-\tau/\tau} = e^{-1} = 0.3679$, or the function equals 36.79% of its maximum value of 1. At $t = 2\tau$, $e^{-t/\tau} = e^{-2\tau/\tau} = e^{-2} = 0.1353$, and the function has decayed to only 13.53% of its maximum value.

The magnitude of $e^{-t/\tau}$ and the percent change between time constants have been tabulated in Tables 10.4 and 10.5, respectively. Note that the current has dropped 63.2% ($100\% - 36.8\%$) in the first time constant but only 0.4% between the fifth and sixth time constants. The rate of change of i_C is therefore quite sensitive to the time constant

TABLE 10.4

i_C vs. τ (charging phase).

t	Magnitude	
0	100%	
1τ	36.8%	
2τ	13.5%	
3τ	5.0%	
4τ	1.8%	Less than
5τ	0.67%	1% of maximum
6τ	0.24%	

TABLE 10.5

Change in i_C between time constants.

$(0 \rightarrow 1)\tau$	63.2%	
$(1 \rightarrow 2)\tau$	23.3%	
$(2 \rightarrow 3)\tau$	8.6%	
$(3 \rightarrow 4)\tau$	3.0%	
$(4 \rightarrow 5)\tau$	1.2%	
$(5 \rightarrow 6)\tau$	0.4%	$\leftarrow$ Less than 1%

determined by the network parameters R and C. For this reason, the universal time constant chart of Fig. 10.26 is provided to permit a more accurate estimate of the value of the function e^{-x} for specific time intervals related to the time constant. The term *universal* is used because the axes are not scaled to specific values.

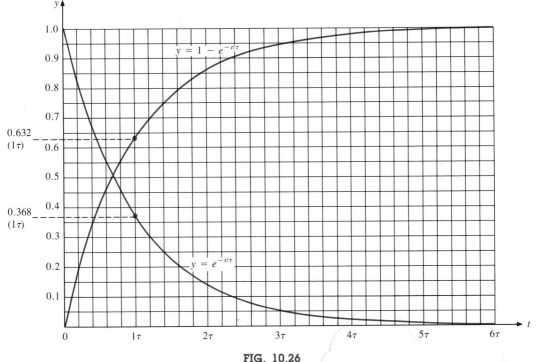

FIG. 10.26

Universal time constant chart.

Returning to Eq. (10.13), we find that the multiplying factor E/R is the maximum value the current i_C can attain, as shown in Fig. 10.21. Substituting $t = 0\,\text{s}$ into Eq. (10.13) yields

$$i_C = \frac{E}{R}e^{-t/RC} = \frac{E}{R}e^{-0} = \frac{E}{R}$$

verifying our earlier conclusion.

For increasing values of t, the magnitude of $e^{-t/\tau}$, and therefore the value of i_C, will decrease as shown in Fig. 10.27. Since the magnitude of i_C is less than 1% of its maximum after five time constants, we will assume for future analysis that:

The current i_C of a capacitive network is zero after five time constants of the charging phase have passed in a dc network.

Since C is usually found in microfarads or picofarads, the time constant $\tau = RC$ will never be greater than a few seconds unless R is very large.

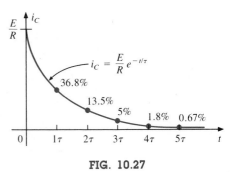

FIG. 10.27

i_C vs. t during the charging phase.

Let us now turn our attention to the charging voltage across the capacitor. Through further mathematical analysis, the following equation for the voltage across the capacitor can be determined:

$$v_C = E(1 - e^{-t/RC}) \qquad (10.15)$$

Note the presence of the same factor $e^{-t/RC}$ and the function $(1 - e^{-t/RC})$ appearing in Fig. 10.26. Since $e^{-t/\tau}$ is a decaying function, the factor $(1 - e^{-t/\tau})$ will grow toward a maximum value of 1 with time as shown in Fig. 10.26. In addition, since E is the multiplying factor, we can conclude that for all practical purposes the voltage v_C is E volts after five time constants of the charging phase. A plot of v_C versus t is provided in Fig. 10.28.

If we keep R constant and reduce C, the product RC will decrease, and the rise time of five time constants will decrease. The change in transient behavior of the voltage v_C is plotted in Fig. 10.29 for various values of C. The product RC will always have some numerical value, even though it may be very small in some cases. For this reason, *the voltage across a capacitor cannot change instantaneously*. In fact, the capacitance of a network is also a measure of how much it will oppose a change in voltage across the network. The larger the capacitance, the larger the time constant and the longer it takes to charge up to its final value (curve of C_3 in Fig. 10.29). A lesser capacitance would permit the voltage to build up more quickly since the time constant is less (curve of C_1 in Fig. 10.29).

The rate at which charge is deposited on the plates during the charging phase can be found by substituting the following for $v_C(t)$ in Eq. (10.15):

$$v_C(t) = \frac{q(t)}{C}$$

and

$$q(t) = Cv_C(t) = CE(1 - e^{-t/\tau}) \qquad \text{charging} \qquad (10.16)$$

indicating that the charging rate is very high during the first few time constants and less than 1% after five time constants.

The voltage across the resistor is determined by Ohm's law:

$$v_R(t) = i_R(t)R = Ri_C(t) = R\frac{E}{R}e^{-t/\tau}$$

or

$$v_R(t) = Ee^{-t/\tau} \qquad (10.17)$$

A plot of $v_R(t)$ appears in Fig. 10.30.

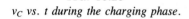

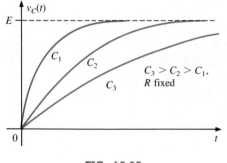

FIG. 10.28

v_C vs. t during the charging phase.

FIG. 10.29

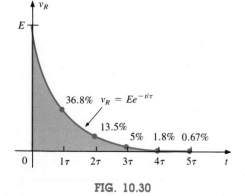

FIG. 10.30

EXAMPLE 10.5.

a. Find the mathematical expressions for the transient behavior of v_C, CAP Volt.
   CAP CUR i_C, and v_R for the circuit of Fig. 10.31 when the switch is moved to
   position 1. Plot the curves of v_C, i_C, and v_R.

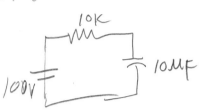

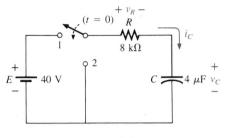

FIG. 10.31

b. How much time must pass before it can be assumed for all practical
   purposes that $i_C \cong 0$ A and $v_C \cong E$ volts?

Solutions:

a. $\tau = RC = (8 \times 10^3)(4 \times 10^{-6}) = 32 \times 10^{-3} = \mathbf{32\ ms}$

By Eq. (10.15),

$$v_C = E(1 - e^{-t/\tau}) = \mathbf{40(1 - e^{-t/(32 \times 10^{-3})})}$$

By Eq. (10.13),

$$i_C = \frac{E}{R}e^{-t/\tau} = \frac{40}{8\ k\Omega}e^{-t/(32 \times 10^{-3})}$$

$$= \mathbf{(5 \times 10^{-3})e^{-t/(32 \times 10^{-3})}}$$

By Eq. (10.17),

$$v_R = Ee^{-t/\tau} = \mathbf{40e^{-t/(32 \times 10^{-3})}}$$

The curves appear in Fig. 10.32.

b. $5\tau = 5(32\ ms) = \mathbf{160\ ms}$

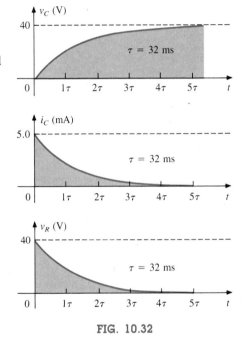

FIG. 10.32

Once the voltage across the capacitor has reached the input voltage
E, the capacitor is fully charged and will remain in this state if no
further changes are made in the circuit.

If the switch of Fig. 10.20 is opened, as shown in Fig. 10.33(a), the
capacitor will retain its charge for a period of time determined by its

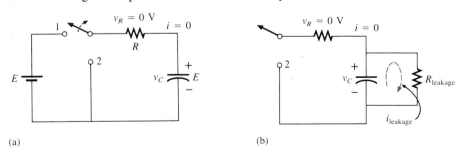

(a)   (b)

FIG. 10.33

leakage current. For capacitors such as the mica and ceramic, the leakage current ($i_{leakage} = v_C/R_{leakage}$) is very small, enabling the capacitor to retain its charge, and hence the potential difference across its plates, for a long period of time. For electrolytic capacitors, which have very high leakage currents, the capacitor will discharge more rapidly, as shown in Fig. 10.33(b). In any event, to insure that they are completely discharged, capacitors should be shorted by a lead or a screwdriver before they are handled.

10.8 DISCHARGE PHASE

The network of Fig. 10.20 is designed to both charge and discharge the capacitor. When the switch is placed in position 1, the capacitor will charge toward the supply voltage as described in the last section. At any point in the charging process, if the switch is moved to position 2, the capacitor will begin to discharge at a rate sensitive to the same time constant $\tau = RC$. The established voltage across the capacitor will create a flow of charge in the closed path that will eventually discharge the capacitor completely. In essence, the capacitor functions like a battery with a decreasing terminal voltage. Note in particular that the current i_C has reversed direction, changing the polarity of the voltage across R.

If the capacitor had charged to the full battery voltage as indicated in Fig. 10.34, the equation for the decaying voltage across the capacitor would be the following:

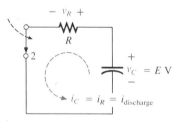

FIG. 10.34

$$v_C = Ee^{-t/RC}$$ *discharging* **(10.18)**

which employs the function e^{-x} and the same time constant used above. The resulting curve will have the same shape as the curve for i_C and v_R in the last section. During the discharge phase, the current i_C will also decrease with time as defined by the following equation:

$$i_C = \frac{E}{R}e^{-t/RC}$$ *discharging* **(10.19)**

The voltage $v_R = v_C$, and

$$v_R = Ee^{-t/RC}$$ *discharging* **(10.20)**

The complete discharge will occur, for all practical purposes, in five time constants. If the switch is moved between terminals 1 and 2 every five time constants, the waveshapes of Fig. 10.35 will result for v_C, i_C, and v_R. For each curve, the current direction and voltage polarities were defined by Fig. 10.20. Since the polarity of v_C is the same for both the charging and discharging phases, the entire curve lies above the axis. The current i_C reverses direction during the charging and discharging phases, producing a negative pulse for both the current and the voltage v_R. Note that the voltage v_C never changes magnitude instantaneously

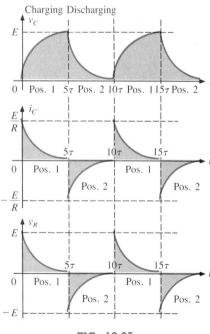

FIG. 10.35

but that the current i_C has the ability to change instantaneously as demonstrated by its vertical rises and drops to maximum values.

EXAMPLE 10.6. After v_C in Example 10.5 has reached its final value of 40 V, the switch is thrown into position 2 as shown in Fig. 10.36.

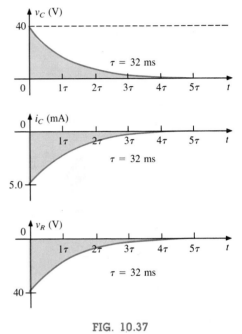

FIG. 10.36

Find the mathematical expressions for the transient behavior of v_C, i_C, and v_R after the closing of the switch. Plot the curves for v_C, i_C, and v_R using the defined directions and polarities of Fig. 10.31.
Solution:

$$\tau = 32 \text{ ms}$$

By Eq. (10.18),

$$v_C = Ee^{-t/\tau} = 40e^{-t/(32\times10^{-3})}$$

By Eq. (10.19),

$$i_C = \frac{E}{R}e^{-t/\tau} = (5\times10^{-3})e^{-t/(32\times10^{-3})}$$

By Eq. (10.20),

$$v_R = Ee^{-t/\tau} = 40e^{-t/(32\times10^{-3})}$$

The curves appear in Fig. 10.37.

FIG. 10.37

The above discussion and examples apply to situations in which the capacitor charges to the battery voltage. If the charging phase is disrupted before reaching the supply voltage, the capacitive voltage will be less, and the equation for the discharging voltage v_C will take on the form

$$v_C = V_i e^{-t/RC} \tag{10.21}$$

where V_i is the starting or *initial* voltage for the discharge phase. The equation for the decaying current is also modified by simply substituting V_i for E. That is,

$$i_C = \frac{V_i}{R}e^{-t/\tau} = I_i e^{-t/\tau} \tag{10.22}$$

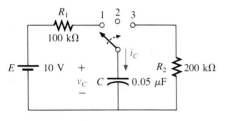

FIG. 10.38

Use of the above equations will be demonstrated in Examples 10.8 and 10.9.

EXAMPLE 10.7.

a. Find the mathematical expression for the transient behavior of the voltage across the capacitor of Fig. 10.38 if the switch is thrown into position 1 at $t = 0\,$s.
b. Repeat part (a) for i_C.
c. Find the mathematical expressions for the response of v_C and i_C if the switch is thrown into position 2 at 30 ms (assuming the leakage resistance of the capacitor is ∞ ohms).
d. Find the mathematical expressions for the voltage v_C and current i_C if the switch is thrown into position 3 at $t = 48\,$ms.
e. Plot the waveforms obtained in parts (a) through (d) on the same time axis for the voltage v_C and the current i_C using the defined polarity and current direction of Fig. 10.38.

Solutions:

a. Charging phase:

$$v_C = E(1 - e^{-t/\tau})$$
$$\tau = R_1C = (100 \times 10^3)(0.05 \times 10^{-6}) = 5 \times 10^{-3}$$
$$= 5\,\text{ms}$$
$$v_C = \mathbf{10(1 - e^{-t/(5\times10^{-3})})}$$

b. $i_C = \dfrac{E}{R_1}e^{-t/\tau}$

$$= \dfrac{10}{100 \times 10^3}e^{-t/(5\times10^{-3})}$$
$$i_C = \mathbf{(0.1 \times 10^{-3})e^{-t/(5\times10^{-3})}}$$

c. Storage phase:

$$v_C = E = \mathbf{10\,V}$$
$$i_C = \mathbf{0\,A}$$

d. Discharging phase:

$$v_C = Ee^{-t/\tau}$$
$$\tau = R_2C = (200 \times 10^3)(0.05 \times 10^{-6}) = 10 \times 10^{-3}$$
$$= 10\,\text{ms}$$
$$v_C = \mathbf{10e^{-t/(10\times10^{-3})}}$$
$$i_C = \dfrac{E}{R_2}e^{-t/\tau}$$
$$= \dfrac{10}{200 \times 10^3}e^{-t/(10\times10^{-3})}$$
$$i_C = \mathbf{(0.05 \times 10^{-3})e^{-t/(10\times10^{-3})}}$$

e. See Fig. 10.39.

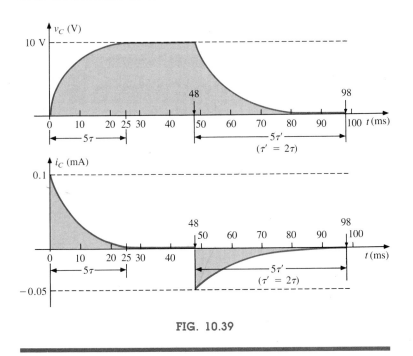

FIG. 10.39

10.9 INSTANTANEOUS VALUES

On occasion it will be necessary to determine the voltage or current at a particular instant of time. For example, if

$$v_C = 20(1 - e^{-t/(2\times10^{-3})})$$

the voltage v_C may be required at $t = 5$ ms, which does not correspond to a particular value of τ. Figure 10.26 reveals that $(1 - e^{-t/\tau})$ is approximately 0.93 at $t = 5$ ms $= 2.5\tau$, resulting in $v_C = 20(0.93) = 18.6$ V. Additional accuracy can be obtained simply by substituting $t = 5$ ms into the equation and solving for v_C using a calculator or table to determine $e^{-2.5}$. Thus,

$$\begin{aligned}
v_C &= 20(1 - e^{-5\text{ms}/2\text{ms}}) \\
&= 20(1 - e^{-2.5}) \\
&= 20(1 - 0.082) \\
&= 20(0.918) \\
&= \mathbf{18.36\ V}
\end{aligned}$$

The results are close, but accuracy beyond the tenths' place is suspect using Fig. 10.26. The above procedure can also be applied to any other equation introduced in this chapter for currents or other voltages.

There are also occasions when the time to reach a particular voltage or current is required. The procedure is complicated somewhat by the use of natural logs ($\log_e$, or ln), but today's calculators are equipped to handle the operation with ease. There are two forms that require some development. First, consider the following sequence:

$$v_C = E(1 - e^{-t/\tau})$$

$$\frac{v_C}{E} = 1 - e^{-t/\tau}$$

$$1 - \frac{v_C}{E} = e^{-t/\tau}$$

$$\log_e\left(1 - \frac{v_C}{E}\right) = \log_e e^{-t/\tau}$$

$$\log_e\left(1 - \frac{v_C}{E}\right) = -\frac{t}{\tau}$$

and

$$\boxed{t = -\tau \log_e\left(1 - \frac{v_C}{E}\right)} \qquad (10.23)$$

The second form is as follows:

$$v_C = Ee^{-t/\tau}$$

$$\frac{v_C}{E} = e^{-t/\tau}$$

$$\log_e \frac{v_C}{E} = \log_e e^{-t/\tau}$$

$$\log_e \frac{v_C}{E} = -\frac{t}{\tau}$$

and

$$\boxed{t = -\tau \log_e \frac{v_C}{E}} \qquad (10.24)$$

For $i_C = (E/R)e^{-t/\tau}$,

$$\boxed{t = -\tau \log_e\left(\frac{i_C R}{E}\right)} \qquad (10.25)$$

For example, suppose

$$v_C = 20(1 - e^{-t/(2 \times 10^{-3})})$$

and the time to reach 10 V is required. Substituting into Eq. (10.23), we have

$$t = (-2 \times 10^{-3}) \log_e\left(1 - \frac{10}{20}\right)$$

$$= (-2 \times 10^{-3}) \log_e 0.5$$
$$= (-2 \times 10^{-3})(-0.693)$$
$$= \mathbf{1.386\ ms} \qquad \text{ln} \;\; \text{key on calculator}$$

Using Fig. 10.26 we find at $(1 - e^{-t/\tau}) = v_C/E = 0.5$ that $t \cong 0.7\tau = 0.7(2 \text{ ms}) = 1.4 \text{ ms}$, which is relatively close to the above.

10.10 $\tau = R_{Th}C$

Occasions will arise in which the network does not have the simple series form of Fig. 10.20. It will then be necessary to first find the Thevenin equivalent circuit for the network external to the capacitive element. E_{Th} will then be the source voltage E of Eqs. (10.15) through (10.20) and R_{Th} the resistance R. The time constant is then $\tau = R_{Th}C$.

EXAMPLE 10.8. For the network of Fig. 10.40:

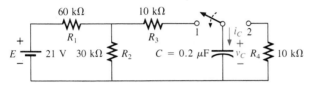

FIG. 10.40

a. Find the mathematical expression for the transient behavior of the voltage v_C and the current i_C following the closing of the switch (position 1 at $t = 0$ s).
b. Find the mathematical expression for the voltage v_C and current i_C as a function of time if the switch is thrown into position 2 at $t = 9$ ms.
c. Draw the resultant waveforms of parts (a) and (b) on the same time axis.

Solutions:

a. Applying Thevenin's theorem to the 0.2-μF capacitor, we obtain Fig. 10.41:

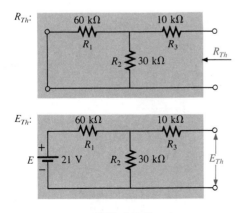

FIG. 10.41

$$R_{Th} = R_1 \parallel R_2 + R_3 = \frac{(60 \text{ k}\Omega)(30 \text{ k}\Omega)}{90 \text{ k}\Omega} + 10 \text{ k}\Omega$$

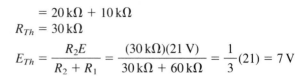

$$= 20\,k\Omega + 10\,k\Omega$$
$$R_{Th} = 30\,k\Omega$$

$$E_{Th} = \frac{R_2 E}{R_2 + R_1} = \frac{(30\,k\Omega)(21\,V)}{30\,k\Omega + 60\,k\Omega} = \frac{1}{3}(21) = 7\,V$$

The resultant Thevenin equivalent circuit with the capacitor replaced is shown in Fig. 10.42:

$$v_C(t) = E_{Th}(1 - e^{-t/\tau})$$
$$\tau = R_{Th}C = (30\,k\Omega)(0.2\,\mu F)$$
$$= (30 \times 10^3)(0.2 \times 10^{-6}) = 6 \times 10^{-3}$$
$$\tau = 6\,ms$$
$$v_C(t) = \mathbf{7(1 - e^{-t/(6 \times 10^{-3})})}$$

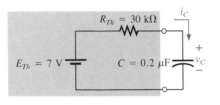

FIG. 10.42

and

$$i_C(t) = \frac{E_{Th}}{R}e^{-t/RC}$$

$$= \frac{7\,V}{30\,k\Omega}e^{-t/(6 \times 10^{-3})}$$

$$i_C(t) = \mathbf{(0.233 \times 10^{-3})e^{-t/(6 \times 10^{-3})}}$$

b. At $t = 9\,ms$,

$$v_C = E_{Th}(1 - e^{-t/\tau}) = 7(1 - e^{-(9 \times 10^{-3})/(6 \times 10^{-3})})$$
$$= 7(1 - e^{-1.5}) = 7(1 - 0.223)$$
$$v_C = 7(0.777) = 5.44\,V$$

$$i_C = \frac{E_{Th}}{R}e^{-t/\tau} = (0.233 \times 10^{-3})e^{-1.5}$$

$$= (0.233 \times 10^{-3})(0.223)$$
$$i_C = 0.052 \times 10^{-3} = 0.052\,mA$$

By Eq. (10.21),

$$v_C = V_i e^{-t/\tau'}$$

with

$$\tau' = R_4 C = (10 \times 10^3)(0.2 \times 10^{-6}) = 2 \times 10^{-3}$$
$$= 2\,ms$$

and

$$v_C = \mathbf{5.44 e^{-t/(2 \times 10^{-3})}}$$

By Eq. (10.22),

$$i_C = \frac{V_i}{R}e^{-t/\tau'} = I_i e^{-t/\tau'}$$

$$= \mathbf{(0.052 \times 10^{-3})e^{-t/(2 \times 10^{-3})}}$$

c. See Fig. 10.43.

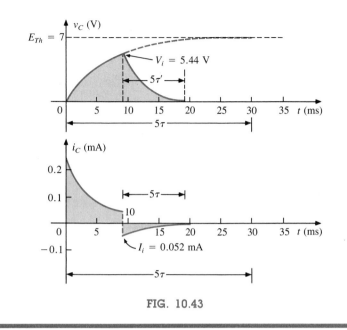

FIG. 10.43

EXAMPLE 10.9. The capacitor of Fig. 10.44 is initially charged to 80 V. Find the mathematical expression for v_C after the closing of the switch.

Solution: The network is redrawn in Fig. 10.45:

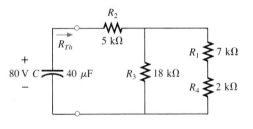

FIG. 10.45

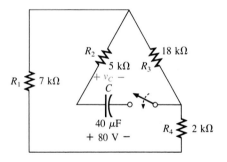

FIG. 10.44

$$R_{Th} = R_2 + R_3 \| (R_1 + R_4) = 5\,\text{k}\Omega + 18\,\text{k}\Omega \| (7\,\text{k}\Omega + 2\,\text{k}\Omega)$$
$$= 5\,\text{k}\Omega + 18\,\text{k}\Omega \| 9\,\text{k}\Omega = 5\,\text{k}\Omega + 6\,\text{k}\Omega = 11\,\text{k}\Omega$$
$$\tau = R_{Th}C = (11 \times 10^3)(40 \times 10^{-6})$$
$$= 440 \times 10^{-3} = 0.44\,\text{s}$$

and

$$v_C(t) = V_i e^{-t/\tau} = 80e^{-t/0.44}$$

EXAMPLE 10.10. For the network of Fig. 10.46, find the mathematical expression for the voltage v_C after the closing of the switch (at $t = 0$).

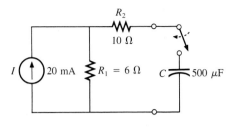

FIG. 10.46

Solution:

$$R_{Th} = R_1 + R_2 = 6 + 10 = 16\,\Omega$$
$$E_{Th} = V_1 + V_2 = IR_1 + 0$$
$$= (20 \times 10^{-3})(6) = 120 \times 10^{-3} = 0.12\,\text{V}$$

and

$$\tau = R_{Th}C = (16)(500 \times 10^{-6}) = 8\,\text{ms}$$

so that

$$v_C(t) = \mathbf{0.12(1 - }e^{-t/(8\times 10^{-3})}\mathbf{)}$$

10.11 THE CURRENT i_C

The current i_C associated with a capacitance C is related to the voltage across the capacitor by

$$i_C = C\frac{dv_C}{dt} \qquad \textbf{(10.26)}$$

where dv_C/dt is a measure of the change in v_C in a vanishingly small period of time. The function dv_C/dt is called the *derivative* of the voltage v_C with respect to time t.

If the voltage fails to change at a particular instant, then

$$dv_C = 0$$

and

$$i_C = C\frac{dv_C}{dt} = 0$$

In other words, if the voltage across a capacitor fails to change with time, the current i_C associated with the capacitor is zero. To take this a step further, the equation also states that the more rapid the change in voltage across the capacitor, the greater the resulting current.

In an effort to develop a clearer understanding of Eq. (10.26), let us calculate the average current associated with a capacitor for various voltages impressed across the capacitor. The average current is defined by the equation

$$i_{Cav} = C\frac{\Delta v_C}{\Delta t} \qquad \textbf{(10.27)}$$

where Δ indicates a finite (measurable) change in charge, voltage, or time. The instantaneous current can be derived from Eq. (10.27) by letting Δt become vanishingly small; that is,

$$i_{C\text{inst}} = \lim_{\Delta t \to 0} C\frac{\Delta v_C}{\Delta t} = C\frac{dv_C}{dt}$$

In the following example, the change in voltage Δv_C will be considered for each slope of the voltage waveform. If the voltage increases with time, the average current is the change in voltage divided by the change in time, with a positive sign. If the voltage decreases with time, the average current is again the change in voltage divided by the change in time, but with a negative sign.

EXAMPLE 10.11. Find the waveform for the average current if the voltage across a 2-μF capacitor is as shown in Fig. 10.47.

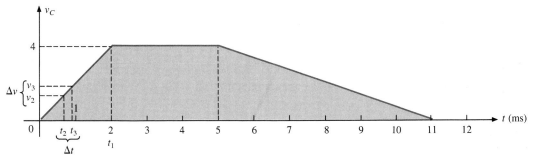

FIG. 10.47

Solution:

a. From 0 to 2 ms, the voltage increases linearly from 0 to 4 V, the change in voltage $\Delta v = 4 - 0 = 4$ (with a positive sign, since the voltage increases with time). The change in time $\Delta t = 2 - 0 = 2$ ms, and

$$i_{C\text{av}} = C\frac{\Delta v_C}{\Delta t} = (2 \times 10^{-6})\left(\frac{4}{2 \times 10^{-3}}\right)$$
$$= 4 \times 10^{-3} = 4\,\text{mA}$$

b. From 2 to 5 ms, the voltage remains constant at 4 V; the change in voltage $\Delta v = 0$. The change in time $\Delta t = 3$ ms, and

$$i_{C\text{av}} = C\frac{\Delta v_C}{\Delta t} = C\frac{0}{\Delta t} = 0$$

c. From 5 to 11 ms, the voltage decreases from 4 to 0 V. The change in voltage Δv is therefore $4 - 0 = 4$ V (with a negative sign, since the voltage is decreasing with time). The change in time $\Delta t = 11 - 5 = 6$ ms, and

$$i_{C\text{av}} = C\frac{\Delta v_C}{\Delta t} = -(2 \times 10^{-6})\left(\frac{4}{6 \times 10^{-3}}\right)$$
$$= -1.33 \times 10^{-3} = -1.33\,\text{mA}$$

d. From 11 ms on, the voltage remains constant at 0 and $\Delta v = 0$, so $i_{Cav} = 0$. The waveform for the average current for the impressed voltage is as shown in Fig. 10.48.

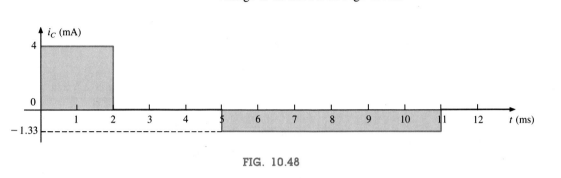

FIG. 10.48

In general, note in Example 10.11 that the steeper the slope, the greater the current, and when the voltage fails to change, the current is zero. In addition, the average value is the same as the instantaneous value at any point along the slope over which the average value was found. For example, if the interval Δt is reduced from $0 \rightarrow t_1$ to $t_2 - t_3$ as noted in Fig. 10.47, $\Delta v / \Delta t$ is still the same. In fact, no matter how small the interval Δt, the slope will be the same, and therefore the current i_{Cav} will be the same. If we consider the limit as $\Delta t \rightarrow 0$, the slope will still remain the same, and therefore $i_{Cav} = i_{Cinst}$ at any instant of time between 0 and t_1. The same can be said about any portion of the voltage waveform that has a constant slope.

An important point to be gained from this discussion is that it is not the magnitude of the voltage across a capacitor that determines the current but rather how quickly the voltage *changes* across the capacitor. An applied steady dc voltage of 10,000 V would (ideally) not create any flow of charge (current), but a change in voltage of 1 V in a very brief period of time could create a significant current.

The method described above is only for waveforms with straight-line (linear) segments. For nonlinear (curved) waveforms, a method of calculus (differentiation) must be employed.

10.12 CAPACITORS IN SERIES AND PARALLEL

Capacitors, like resistors, can be placed in series and parallel. Increasing levels of capacitance can be obtained by placing capacitors in parallel, while decreasing levels can be obtained by placing capacitors in series.

For capacitors in series, the charge is the same on each capacitor (Fig. 10.49):

$$Q_T = Q_1 = Q_2 = Q_3 \qquad \textbf{(10.28)}$$

Applying Kirchhoff's voltage law around the closed loop gives

$$E = V_1 + V_2 + V_3$$

However,

$$V = \frac{Q}{C}$$

so that

$$\frac{Q_T}{C_T} = \frac{Q_1}{C_1} + \frac{Q_2}{C_2} + \frac{Q_3}{C_3}$$

Using Eq. (10.28) and dividing both sides by Q yields

$$\frac{1}{C_T} = \frac{1}{C_1} + \frac{1}{C_2} + \frac{1}{C_3} \qquad \textbf{(10.29)}$$

which is similar to the manner in which we found the total resistance of a parallel resistive circuit. The total capacitance of two capacitors in series is

$$C_T = \frac{C_1 C_2}{C_1 + C_2} \qquad \textbf{(10.30)}$$

For capacitors in parallel as shown in Fig. 10.50, the voltage is the same across each capacitor, and the total charge is the sum of that on each capacitor:

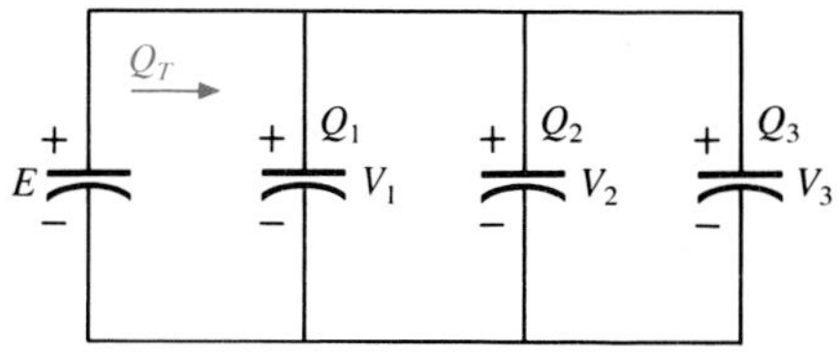

$$Q_T = Q_1 + Q_2 + Q_3 \qquad \textbf{(10.31)}$$

However,

$$Q = CV$$

Therefore,

$$C_T E = C_1 V_1 + C_2 V_2 + C_3 V_3$$

and

$$E = V_1 = V_2 = V_3$$

Thus,

$$C_T = C_1 + C_2 + C_3 \qquad \textbf{(10.32)}$$

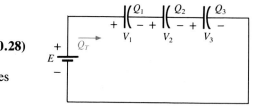

FIG. 10.49

FIG. 10.50

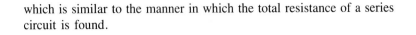

which is similar to the manner in which the total resistance of a series circuit is found.

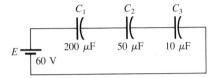

FIG. 10.51

EXAMPLE 10.12. For the circuit of Fig. 10.51:
a. Find the total capacitance.
b. Determine the charge on each plate.
c. Find the voltage across each capacitor.

Solutions:

a. $\dfrac{1}{C_T} = \dfrac{1}{C_1} + \dfrac{1}{C_2} + \dfrac{1}{C_3} = \dfrac{1}{200 \times 10^{-6}} + \dfrac{1}{50 \times 10^{-6}}$

$$+ \dfrac{1}{10 \times 10^{-6}}$$

$$= 0.005 \times 10^6 + 0.02 \times 10^6$$
$$+ 0.1 \times 10^6$$
$$= 0.125 \times 10^6$$

and

$$C_T = \dfrac{1}{0.125 \times 10^6} = \mathbf{8\ \mu F}$$

b. $Q_T = Q_1 = Q_2 = Q_3$
$Q_T = C_T E = (8 \times 10^{-6})(60) = \mathbf{480\ \mu C}$

c. $V_1 = \dfrac{Q_1}{C_1} = \dfrac{480 \times 10^{-6}}{200 \times 10^{-6}} = \mathbf{2.4\ V}$

$V_2 = \dfrac{Q_2}{C_2} = \dfrac{480 \times 10^{-6}}{50 \times 10^{-6}} = \mathbf{9.6\ V}$

$V_3 = \dfrac{Q_3}{C_3} = \dfrac{480 \times 10^{-6}}{10 \times 10^{-6}} = \mathbf{48.0\ V}$

and

$$E = V_1 + V_2 + V_3 = 2.4 + 9.6 + 48$$
$$= \mathbf{60\ V} \quad \text{(checks)}$$

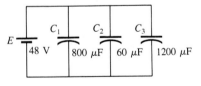

FIG. 10.52

EXAMPLE 10.13. For the network of Fig. 10.52:
a. Find the total capacitance.
b. Determine the charge on each plate.
c. Find the total charge.

Solutions:

a. $C_T = C_1 + C_2 + C_3 = 800\ \mu F + 60\ \mu F + 1200\ \mu F$
$$= \mathbf{2060\ \mu F}$$

b. $Q_1 = C_1 E = (800 \times 10^{-6})(48) = \mathbf{0.0384\ C}$
$Q_2 = C_2 E = (60 \times 10^{-6})(48) = \mathbf{0.00288\ C}$
$Q_3 = C_3 E = (1200 \times 10^{-6})(48) = \mathbf{0.0576\ C}$

c. $Q_T = Q_1 + Q_2 + Q_3 = 0.0384 + 0.00288 + 0.0576$
$$= \mathbf{0.09888\ C}$$

EXAMPLE 10.14. Find the voltage across and charge on each capacitor for the network of Fig. 10.53.

Solution:

$$C'_T = C_2 + C_3 = 4\,\mu F + 2\,\mu F = 6\,\mu F$$

$$C_T = \frac{C_1 C'_T}{C_1 + C'_T} = \frac{3 \times 6}{3 + 6} = 2\,\mu F$$

$$Q_T = C_T E = (2 \times 10^{-6})(120)$$
$$= \textbf{240}\,\boldsymbol{\mu}\textbf{C}$$

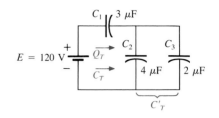

FIG. 10.53

An equivalent circuit (Fig. 10.54) has

$$Q_T = Q_1 = Q'_T$$

and therefore

$$Q_1 = \textbf{240}\,\boldsymbol{\mu}\textbf{C}$$

and

$$V_1 = \frac{Q_1}{C_1} = \frac{240 \times 10^{-6}}{3 \times 10^{-6}} = \textbf{80 V}$$

$$Q'_T = 240\,\mu C$$

and therefore

$$V'_T = \frac{Q'_T}{C'_T} = \frac{240 \times 10^{-6}}{6 \times 10^{-6}} = \textbf{40 V}$$

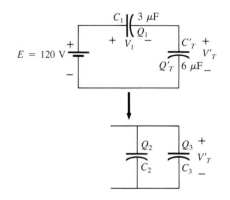

FIG. 10.54

and

$$Q_2 = C_2 V'_T = (4 \times 10^{-6})(40) = \textbf{160}\,\boldsymbol{\mu}\textbf{C}$$
$$Q_3 = C_3 V'_T = (2 \times 10^{-6})(40) = \textbf{80}\,\boldsymbol{\mu}\textbf{C}$$

EXAMPLE 10.15. Find the voltage across and charge on capacitor C_1 of Fig. 10.55 after it has charged up to its final value.

Solution: As previously discussed, the capacitor is effectively an open circuit for dc after charging up to its final value (Fig. 10.56).

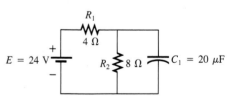

FIG. 10.55

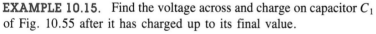

FIG. 10.56

Therefore,

$$V_C = \frac{(8)(24)}{4 + 8} = \textbf{16 V}$$

$$Q_1 = C_1 V_C = (20 \times 10^{-6})(16)$$
$$= \textbf{320}\,\boldsymbol{\mu}\textbf{C}$$

EXAMPLE 10.16. Find the voltage across and charge on each capacitor of the network of Fig. 10.57 after each has charged up to its final value.

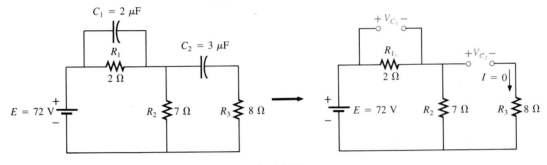

FIG. 10.57

Solution:

$$V_{C_2} = \frac{(7)(72)}{7 + 2} = \textbf{56 V}$$

$$V_{C_1} = \frac{(2)(72)}{2 + 7} = \textbf{16 V}$$

$$Q_1 = C_1 V_{C_1} = (2 \times 10^{-6})(16) = \textbf{32}\,\boldsymbol{\mu}\textbf{C}$$

$$Q_2 = C_2 V_{C_2} = (3 \times 10^{-6})(56) = \textbf{168}\,\boldsymbol{\mu}\textbf{C}$$

10.13 ENERGY STORED BY A CAPACITOR

The ideal capacitor does not dissipate any of the energy supplied to it. It stores the energy in the form of an electric field between the conducting surfaces. A plot of the voltage, current, and power to a capacitor during the charging phase is shown in Fig. 10.58. The power curve can be obtained by finding the product of the voltage and current at selected intervals of time and connecting the points obtained. (Note p_1 on the curve of Fig. 10.58.) The energy stored is represented by the shaded area under the power curve. Using calculus, we can determine the area under the curve:

$$W_C = \frac{1}{2}CE^2$$

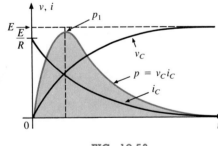

FIG. 10.58

In general,

$$\boxed{W_C = \frac{1}{2}CV^2} \qquad \text{(J)} \qquad \textbf{(10.33)}$$

where V is the steady-state voltage across the capacitor. In terms of Q and C,

$$W_C = \frac{1}{2}C\left(\frac{Q}{C}\right)^2$$

or

$$W_C = \frac{Q^2}{2C} \qquad \text{(J)} \qquad \qquad \textbf{(10.34)}$$

EXAMPLE 10.17. For the network of Fig. 10.57, determine the energy stored by each capacitor.

Solution: For C_1,

$$W_C = \frac{1}{2}CV^2$$

$$= \frac{1}{2}(2 \times 10^{-6})(16)^2 = (1 \times 10^{-6})(256)$$

$$= \textbf{256}\,\boldsymbol{\mu}\textbf{J}$$

For C_2,

$$W_C = \frac{1}{2}CV^2$$

$$= \frac{1}{2}(3 \times 10^{-6})(56)^2 = (1.5 \times 10^{-6})(3136)$$

$$= \textbf{4704}\,\boldsymbol{\mu}\textbf{J}$$

Due to the squared term, note the difference in energy stored due to a higher voltage.

10.14 STRAY CAPACITANCES

In addition to the capacitors discussed so far in this chapter, there are stray capacitances that exist not through design but simply because two conducting surfaces are relatively close to each other. Two conducting wires in the same network will have a capacitive effect between them, as shown in Fig. 10.59(a). In electronic circuits, capacitance levels exist between conducting surfaces of the transistor as shown in Fig. 10.59(b). In Chapter 12 we will discuss another element called the *inductor* which will have capacitive effects between the windings [Fig. 10.59(c)]. Stray capacitances can often lead to serious errors in system design if not considered carefully.

SKiP any Qwos. BETWEEN 10.1 - 10.4
IN TEXT.

PROBLEMS

Section 10.2

1. Find the electric field strength at a point 2 m from a charge of $4\,\mu$C.

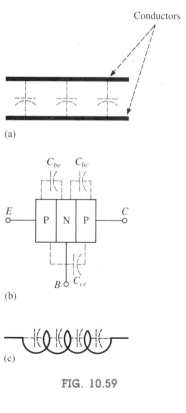

Conductors

(a)

C_{be} C_{bc}

E P | N | P C

B C_{ce}

(b)

(c)

FIG. 10.59

2. The electric field strength is 36 newtons/coulomb (N/C) at a point r meters from a charge of $0.064\ \mu$C. Find the distance r.

Section 10.3

3. Find the capacitance of a parallel plate capacitor if $1400\ \mu$C of charge are deposited on its plates when 20 V are applied across the plates.

4. How much charge is deposited on the plates of a 0.05-μF capacitor if 45 V are applied across the capacitor?

5. Find the electric field strength between the plates of a parallel plate capacitor if $100\ $mV are applied across the plates and the plates are 2 mm apart.

6. Repeat Problem 5 if the plates are separated by 4 mils.

7. A 4-μF parallel plate capacitor has $160\ \mu$C of charge on its plates. If the plates are 5 mm apart, find the electric field strength between the plates.

8. Find the capacitance of a parallel plate capacitor if the area of each plate is $0.075\ \text{m}^2$ and the distance between the plates is $1.77\ $mm. The dielectric is air.

9. Repeat Problem 8 if the dielectric is paraffin-coated paper.

10. Find the distance in mils between the plates of a 2-μF capacitor if the area of each plate is $0.09\ \text{m}^2$ and the dielectric is transformer oil.

11. The capacitance of a capacitor with a dielectric of air is $1200\ $pF. When a dielectric is inserted between the plates, the capacitance increases to $0.006\ \mu$F. Of what material is the dielectric made?

12. The plates of a parallel plate air capacitor are $0.2\ $mm apart, have an area of $0.08\ \text{m}^2$, and 200 V are applied across the plates.
    a. Determine the capacitance.
    b. Find the electric field intensity between the plates.
    c. Find the charge on each plate if the dielectric is air.

13. A sheet of Bakelite $0.2\ $mm thick having an area of $0.08\ \text{m}^2$ is inserted between the plates of Problem 12.
    a. Find the electric field strength between the plates.
    b. Determine the charge on each plate.
    c. Determine the capacitance.

Section 10.4

14. Find the maximum voltage ratings of the capacitors of Problems 12 and 13 assuming a linear relationship between the breakdown voltage and the thickness of the dielectric.

15. Find the maximum voltage that can be applied across a parallel plate capacitor of $0.006\ \mu$F. The area of one plate is $0.02\ \text{m}^2$ and the dielectric is mica. Assume a linear

relationship between the dielectric strength and the thickness of the dielectric.

16. Find the distance in millimeters between the plates of a parallel plate capacitor if the maximum voltage that can be applied across the capacitor is 1250 V. The dielectric is mica. Assume a linear relationship between the breakdown strength and the thickness of the dielectric.

Section 10.7

17. For the circuit of Fig. 10.60:
    a. Determine the time constant of the circuit.
    b. Write the mathematical equation for the voltage v_C following the closing of the switch.
    c. Determine the voltage v_C after one, three, and five time constants.
    d. Write the equations for the current i_C and the voltage v_R.
    e. Sketch the waveforms for v_C and i_C.

18. Repeat Problem 17 for $R = 1\,\text{M}\Omega$ and compare results.

19. Repeat Problem 17 for the network of Fig. 10.61.

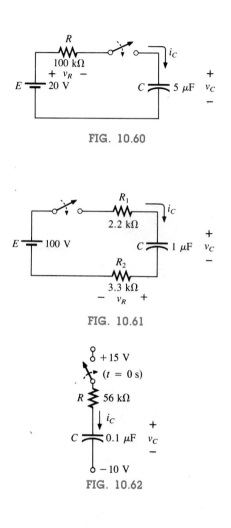

FIG. 10.60

FIG. 10.61

20. For the circuit of Fig. 10.62:
    a. Determine the time constant of the circuit.
    b. Write the mathematical expression for the voltage v_C following the closing of the switch.
    c. Write the mathematical expression for the current i_C following the closing of the switch.
    d. Sketch the waveforms of v_C and i_C.

FIG. 10.62

Section 10.8

21. For the circuit of Fig. 10.63:
    a. Determine the time constant of the circuit when the switch is thrown into position 1.
    b. Find the mathematical expression for the voltage across the capacitor after the switch is thrown into position 1.
    c. Determine the mathematical expression for the current following the closing of the switch (position 1).
    d. Determine the voltage v_C and the current i_C if the switch is thrown into position 2 at $t = 100\,\text{ms}$.

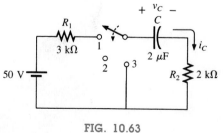

FIG. 10.63

e. Determine the mathematical expressions for the voltage v_C and the current i_C if the switch is thrown into position 3 at $t = 200$ ms.
f. Plot the waveforms of v_C and i_C for a period of time extending from $t = 0$ to $t = 300$ ms.

22. Repeat Problem 21 for a capacitance of $20 \mu F$.

23. For the network of Fig. 10.64:
   a. Find the mathematical expression for the voltage across the capacitor after the switch is thrown into position 1.
   b. Repeat part (a) for the current i_C.
   c. Find the mathematical expressions for the voltage v_C and current i_C if the switch is thrown into position 2 at a time equal to five time constants of the charging circuit.
   d. Plot the waveforms of v_C and i_C for a period of time extending from $t = 0$ to $t = 30 \mu s$.

24. The capacitor of Fig. 10.65 is initially charged to 40 V before the switch is closed. Write the expressions for the voltages v_C and v_R and the current i_C for the decay phase.

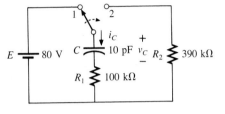

FIG. 10.64

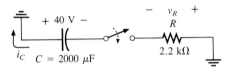

FIG. 10.65

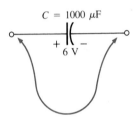

FIG. 10.66

25. The 1000-μF capacitor of Fig. 10.66 is charged to 6 V. To discharge the capacitor before further use, a wire with a resistance of $0.002 \,\Omega$ is placed across the capacitor.
   a. How long will it take to discharge the capacitor?
   b. What is the peak value of the current?
   c. Based on the result of part (b), is a spark expected when contact is made with both ends of the capacitor?

Section 10.9

26. Given the expression $v_C = 8(1 - e^{-t/(20 \times 10^{-6})})$:
   a. Determine v_C after five time constants.
   b. Determine v_C after 10 time constants.
   c. Determine v_C at $t = 5 \mu s$.

27. For the situation of Problem 25, determine when the discharge current is one-half its maximum value if contact is made at $t = 0$ s.

Section 10.10

28. For the circuit of Fig. 10.67:
   a. Find the mathematical expressions for the transient behavior of the voltage v_C and the current i_C following the closing of the switch.
   b. Sketch the waveforms of v_C and i_C.

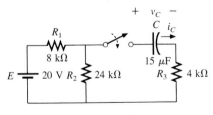

FIG. 10.67

*29. Repeat Problem 28 for the circuit of Fig. 10.68.

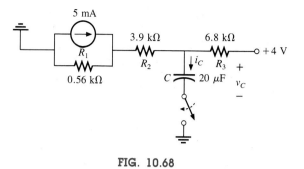

FIG. 10.68

Section 10.11

30. Find the waveform for the average current if the voltage across a 0.06-μF capacitor is as shown in Fig. 10.69.

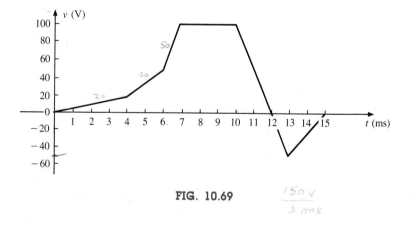

FIG. 10.69

$$\frac{150\ V}{3\ ms}$$

31. Repeat Problem 30 for the waveform of Fig. 10.70.

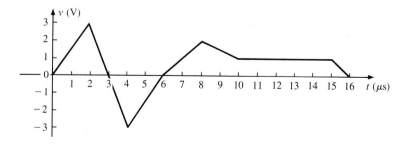

FIG. 10.70

32. Find the total capacitance C_T between points a and b of the circuits of Fig. 10.71.

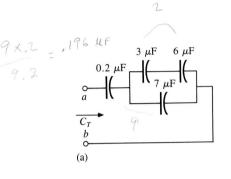

9×.2 = .196 μF
9.2

2

(a)

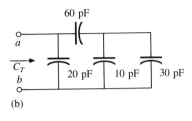

(b)

FIG. 10.71

33. Find the voltage across and charge on each capacitor for the circuits of Fig. 10.72.

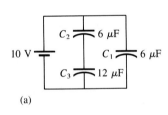

(a)

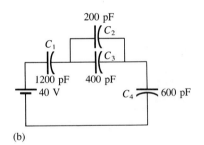

(b)

FIG. 10.72

34. For the circuits of Fig. 10.73, find the voltage across and charge on each capacitor after each capacitor has charged to its final value.

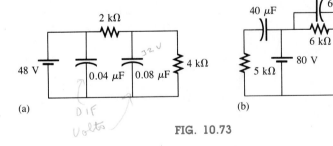

(a)

DIF
Volts

(b)

FIG. 10.73

Section 10.13

35. Find the energy stored by a 120-pF capacitor with 12 V across its plates.

36. If the energy stored by a 6-μF capacitor is 1200 J, find the charge Q on each plate of the capacitor.

*37. An electronic flashgun has a 1000-μF capacitor which is charged to 100 V.

   a. How much energy is stored by the capacitor?

   b. What is the charge on the capacitor?

   c. When the photographer takes a picture, the flash fires for 1/2000 s. What is the average current through the flashtube?

   d. Find the power delivered to the flashtube.

   e. After a picture is taken, the capacitor has to be recharged by a power supply which delivers a maximum current of 10 mA. How long will it take to charge the capacitor?

Computer Problems

38. Write a program to tabulate the voltage of Problem 26 for each time constant from one to 20 time constants.

39. Write a program to tabulate the average current of a capacitor given the waveform of the voltage across the capacitor as shown in Fig. 10.69 and the capacitance value. That is, list the level of current associated with each time interval encompassing a linear change in voltage.

40. Given three capacitors in any series-parallel arrangement, write a program to determine the total capacitance. That is, determine the total number of possibilities and ask the user to identify the configuration and provide the capacitor values. Then calculate the total capacitance.

GLOSSARY

Breakdown voltage Another term for *dielectric strength*, listed below.

Capacitance A measure of a capacitor's ability to store charge; measured in farads (F).

Capacitive time constant The product of resistance and capacitance that establishes the required time for the charging and discharging phases of a capacitive transient.

Capacitive transient The waveforms for the voltage and current of a capacitor that result during the charging and discharging phases.

Capacitor A fundamental electrical element having two conducting surfaces separated by an insulating material and having the capacity to store charge on its plates.

Coulomb's law An equation relating the force between two like or unlike charges.

Dielectric The insulating material between the plates of a capacitor that can have a pronounced effect on the charge stored on the plates of a capacitor.

Dielectric constant Another term for *relative permittivity*, listed below.

Dielectric strength An indication of the voltage required for unit length to establish conduction in a dielectric.

Electric field strength The force acting on a unit positive charge in the region of interest.

Electric flux lines Lines drawn to indicate the strength and direction of an electric field in a particular region.

Fringing An effect established by flux lines that do not pass directly from one conducting surface to another.

Leakage current The current that will result in the total discharge of a capacitor if the capacitor is disconnected from the charging network for a sufficient length of time.

Permittivity A measure of how well a dielectric will *permit* the establishment of flux lines within the dielectric.

Relative permittivity The permittivity of a material compared to that of air.

Stray capacitance Capacitances that exist not through design but simply because two conducting surfaces are relatively close to each other.

Surge voltage The maximum voltage that can be applied across the capacitor for very short periods of time.

Working voltage The voltage that can be applied across a capacitor for long periods of time without concern for dielectric breakdown.

11
Magnetic Circuits

11.1 INTRODUCTION

Magnetism plays an integral part in almost every electrical device used today in industry, research, or the home. Generators, motors, transformers, circuit breakers, televisions, computers, tape recorders, and telephones all employ magnetic effects to perform a variety of important tasks.

The compass, used by Chinese sailors as early as the second century A.D., relies on a *permanent magnet* for indicating direction. The permanent magnet is made of a material, such as steel or iron, that will remain magnetized for long periods of time without the need for an external source of energy.

In 1820, the Danish physicist Hans Christian Oersted discovered that the needle of a compass would deflect if brought near a current-carrying conductor. For the first time it was demonstrated that electricity and magnetism were related, and in the same year the French physicist André Marie Ampère performed experiments in this area and developed what is presently known as *Ampère's circuital law*. In subsequent years, men such as Michael Faraday, Karl Friedrich Gauss, and James Clerk Maxwell continued to experiment in this area and developed many of the basic concepts of *electromagnetism*—magnetic effects induced by the flow of charge, or current.

There is a great deal of similarity between the analyses of electric circuits and magnetic circuits. This will be demonstrated later in this chapter when we compare the basic equations and methods used to solve magnetic circuits with those used for electric circuits.

Difficulty in understanding methods used with magnetic circuits will often arise in simply learning to use the proper set of units, and not because of the equations themselves. The problem exists because three different systems of units are still being used in the industry. To the extent practical, the SI will be used throughout this chapter. For the CGS and English systems, a conversion table is provided in Appendix F.

11.2 MAGNETIC FIELDS

In the region surrounding a permanent magnet there exists a magnetic field, which can be represented by magnetic flux lines similar to electric flux lines. Magnetic flux lines, however, do not have origins or terminating points like electric flux lines but exist in continuous loops, as shown in Fig. 11.1. The symbol for magnetic flux is the Greek letter Φ (phi).

The magnetic flux lines radiate from the north pole to the south pole, returning to the north pole through the metallic bar. Note the equal spacing between the flux lines within the core and the symmetric distribution outside the magnetic material. These are additional properties of magnetic flux lines in homogeneous materials (that is, materials having uniform structure or composition throughout). It is also important to realize that the continuous magnetic flux line will strive to occupy as small an area as possible. This will result in magnetic flux lines of minimum length between the like poles, as shown in Fig. 11.2. The strength of a magnetic field in a particular region is directly related to the density of flux lines in that region. In Fig. 11.1, for example, the magnetic field strength at a is twice that at b since there are twice as many magnetic flux lines associated with the perpendicular plane at a than at b. Recall from childhood experiments how the strength of permanent magnets was always stronger near the poles.

If unlike poles of two permanent magnets are brought together, the magnets will attract, and the flux distribution will be as shown in Fig. 11.2. If like poles are brought together, the magnets will repel, and the flux distribution will be as shown in Fig. 11.3.

If a nonmagnetic material, such as glass or copper, is placed in the flux paths surrounding a permanent magnet, there will be an almost unnoticeable change in the flux distribution (Fig. 11.4). However, if a magnetic material, such as soft iron, is placed in the flux path, the flux

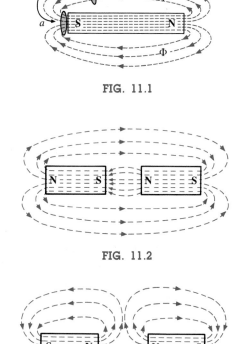

FIG. 11.1

FIG. 11.2

FIG. 11.3

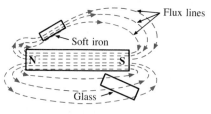

FIG. 11.4

lines will pass through the soft iron rather than the surrounding air because flux lines pass with greater ease through magnetic materials than through air. This principle is put to use in the shielding of sensitive electrical elements and instruments that can be affected by stray magnetic fields (Fig. 11.5).

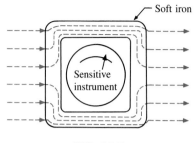

FIG. 11.5

As indicated in the introduction, a magnetic field (represented by concentric magnetic flux lines, as in Fig. 11.6) is present around every

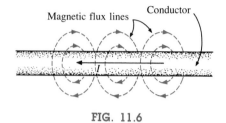

FIG. 11.6

wire that carries an electric current. The direction of the magnetic flux lines can be found simply by placing the thumb of the *right* hand in the direction of *conventional* current flow and noting the direction of the fingers. (This method is commonly called the *right-hand rule*.) If the conductor is wound in a single-turn coil (Fig. 11.7), the resulting flux will flow in a common direction through the center of the coil.

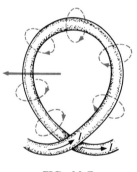

FIG. 11.7

A coil of more than one turn would produce a magnetic field that would exist in a continuous path through and around the coil (Fig. 11.8).

The flux distribution of the coil is quite similar to that of the permanent magnet. The flux lines leaving the coil from the left and entering to the right simulate a north and south pole, respectively. The principal difference between the two flux distributions is that the flux lines are more concentrated for the permanent magnet than for the coil. Also, since the strength of a magnetic field is determined by the density of the flux lines, the coil has a weaker field strength. The field strength of the coil can be effectively increased by placing certain materials, such as iron, steel, or cobalt, within the coil to increase the flux density within the coil. By increasing the field strength with the addition of the core, we have devised an *electromagnet* (Fig. 11.9) which, in addition to

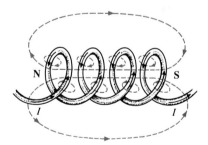

FIG. 11.8

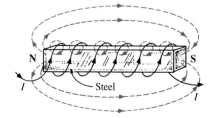

FIG. 11.9

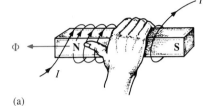

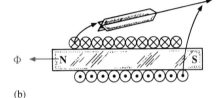

(a)

(b)

FIG. 11.10

having all the properties of a permanent magnet, also has a field strength that can be varied by changing one of the component values (current, turns, and so on). Of course, current must pass through the coil of the electromagnet in order for magnetic flux to be developed, whereas there is no need for the coil or current in the permanent magnet. The direction of flux lines can be determined for the electromagnet (or in any core with a wrapping of turns) by placing the fingers of the right hand in the direction of current flow around the core. The thumb will then point in the direction of the north pole of the induced magnetic flux. This is demonstrated in Fig. 11.10. A cross section of the same electromagnet was included in the figure to introduce the convention for directions perpendicular to the page. The cross and dot refer to the tail and head of the arrow, respectively.

Other areas of application for electromagnetic effects are shown in Fig. 11.11. The flux path for each is indicated in each figure.

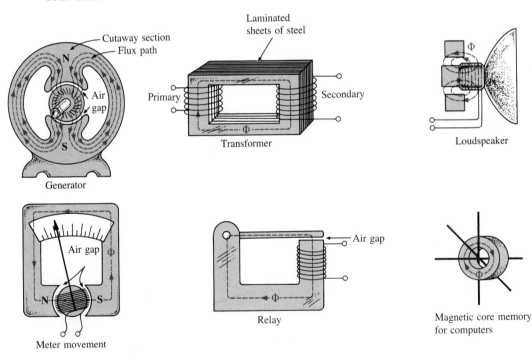

FIG. 11.11

11.3 FLUX DENSITY

In the SI system of units, magnetic flux is measured in webers (Wb) and has the symbol Φ. The number of flux lines per unit area is called the *flux density* and is denoted by the capital letter B. Its magnitude is determined by the following equation:

$$B = \frac{\Phi}{A}$$

B = teslas (T)
Φ = webers (Wb)
A = square meters (m^2)

(11.1)

where Φ is the number of flux lines passing through the area A (Fig. 11.12). The flux density at position a in Fig. 11.1 is twice that at b because twice as many flux lines are passing through the same area.

As noted in Eq. (11.1), magnetic flux density in the SI system of units is measured in *teslas,* for which the symbol is T. By definition,

$$1 \text{ tesla} = 1 \text{ Wb/m}^2$$

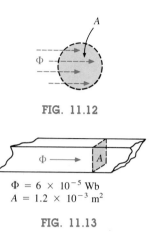

FIG. 11.12

EXAMPLE 11.1. For the core of Fig. 11.13, determine the flux density B in teslas.
Solution:

$$B = \frac{\Phi}{A} = \frac{6 \times 10^{-5}}{1.2 \times 10^{-3}} = \mathbf{5 \times 10^{-2}\, T}$$

$\Phi = 6 \times 10^{-5}$ Wb
$A = 1.2 \times 10^{-3}$ m^2

FIG. 11.13

EXAMPLE 11.2. In Fig. 11.13, if the flux density is 1.2 Wb/m^2 and the area is 0.25 in.^{2}, determine the flux through the core.
Solution: By Eq. (11.1),

$$\Phi = BA$$

However, converting 0.25 in.^{2} to metric units,

$$A = 0.25 \text{ in.}^2\left(\frac{1 \text{ m}}{39.37 \text{ in.}}\right)\left(\frac{1 \text{ m}}{39.37 \text{ in.}}\right) = 1.613 \times 10^{-4}\, \text{m}^2$$

and

$$\Phi = (1.2)(1.613 \times 10^{-4})$$
$$= \mathbf{1.936 \times 10^{-4}\, Wb}$$

An instrument designed to measure flux density in gauss (CGS system) appears in Fig. 11.14. Appendix F reveals that 1 Wb/m^2 = 10^4 gauss. The magnitude of the reading appearing on the face of the meter in Fig. 11.14 is therefore

$$1.964 \text{ gauss}\left(\frac{1 \text{ Wb/m}^2}{10^4 \text{ gauss}}\right) = 1.964 \times 10^{-4}\, \text{Wb/m}^2$$
$$= 1.964 \times 10^{-4}\, \text{T}$$

11.4 PERMEABILITY

If cores of different materials with the same physical dimensions are used in the electromagnet described in Section 11.2, the strength of the magnet will vary in accordance with the core used. This variation in

FIG. 11.14
Digital display gaussmeter. (Courtesy of LDJ Electronics, Inc.)

strength is due to the greater or lesser number of flux lines passing through the core. Materials in which flux lines can readily be set up are said to be *magnetic* and to have *high permeability*. The permeability (μ) of a material, therefore, is a measure of the ease with which magnetic flux lines can be established in the material. It is similar in many respects to conductivity in electric circuits. The permeability of free space μ_o (vacuum) is

$$\mu_o = 4\pi \times 10^{-7} \text{ weber/ampere-meter}$$

As indicated above, μ has the units of Wb/Am. Practically speaking, the permeability of all nonmagnetic materials, such as copper, aluminum, wood, glass, and air, is the same as that for free space. Materials that have permeabilities slightly less than that of free space are said to be *diamagnetic,* and those with permeabilities slightly greater than that of free space are said to be *paramagnetic*. Magnetic materials, such as iron, nickel, steel, cobalt, and alloys of these metals, have permeabilities hundreds and even thousands of times that of free space. Materials with these very high permeabilities are referred to as *ferromagnetic*.

The ratio of the permeability of a material to that of free space is called its *relative permeability;* that is,

$$\mu_r = \frac{\mu}{\mu_o} \tag{11.2}$$

In general, for ferromagnetic materials, $\mu_r \geq 100$, and for nonmagnetic materials, $\mu_r = 1$.

Since μ_r is a variable, dependent on other quantities of the magnetic circuit, values of μ_r are not tabulated. Methods of calculating μ_r from the data supplied by manufacturers will be considered in a later section.

11.5 RELUCTANCE

The resistance of a material to the flow of charge (current) is determined for electric circuits by the equation

$$R = \rho\frac{l}{A} \quad (\text{ohms, } \Omega)$$

The *reluctance* of a material to the setting up of magnetic flux lines in the material is determined by the following equation:

$$\mathcal{R} = \frac{l}{\mu A} \quad (\text{rels or At/Wb}) \tag{11.3}$$

where $\mathcal{R}$ is the reluctance, l is the length of the magnetic path, and A is its cross-sectional area. The t in the units At/Wb is the number of turns of the applied winding. More will be said about ampere-turns (At) in the next section. Note that the resistance and reluctance are inversely proportional to the area, indicating that an increase in area will result in a reduction in each and an *increase* in the desired result: current and flux. For an increase in length the opposite is true, and the desired effect is reduced. The reluctance, however, is inversely proportional to the permeability, while the resistance is directly proportional to the resistivity. The larger the μ or smaller the ρ, the smaller the reluctance and resistance, respectively. Obviously, therefore, materials with high permeability, such as the ferromagnetics, have very small reluctances and will result in an increased measure of flux through the core. There is no widely accepted unit for reluctance, although the *rel* and the At/Wb are usually applied.

11.6 OHM'S LAW FOR MAGNETIC CIRCUITS

Recall the equation

$$\text{Effect} = \frac{\text{cause}}{\text{opposition}}$$

appearing in Chapter 4 to introduce Ohm's law for electric circuits. For magnetic circuits, the effect desired is the flux Φ. The cause is the *magnetomotive force* (mmf) $\mathscr{F}$, which is the external force (or "pressure") required to set up the magnetic flux lines within the magnetic material. The opposition to the setting up of the flux Φ is the reluctance $\mathscr{R}$.

Substituting, we have

$$\Phi = \frac{\mathscr{F}}{\mathscr{R}} \tag{11.4}$$

The magnetomotive force $\mathscr{F}$ is proportional to the product of the number of turns around the core (in which the flux is to be established) and the current through the turns of wire (Fig. 11.15). In equation form,

$$\mathscr{F} = NI \qquad \text{(ampere-turns, At)} \tag{11.5}$$

The equation clearly indicates that an increase in the number of turns or the current through the wire will result in an increased "pressure" on the system to establish flux lines through the core.

Although there is a great deal of similarity between electric and magnetic circuits, one must continue to realize that the flux Φ is not a "flow" variable such as current in an electric circuit. Magnetic flux is established in the core through the alteration of the atomic structure of the core due to external pressure and is not a measure of the flow of some charged particles through the core.

FIG. 11.15

11.7 MAGNETIZING FORCE

The magnetomotive force per unit length is called the *magnetizing force* (*H*). In equation form,

$$H = \frac{\mathcal{F}}{l} \qquad \text{(At/m)} \qquad \textbf{(11.6)}$$

Substituting for the magnetomotive force will result in

$$H = \frac{NI}{l} \qquad \text{(At/m)} \qquad \textbf{(11.7)}$$

For the magnetic circuit of Fig. 11.16, if $NI = 40$ At and $l = 0.2$ m, then

$$H = \frac{NI}{l} = \frac{40}{0.2} = 200 \text{ At/m}$$

In words, the result indicates that there are 200 ampere-turns of "pressure" per meter to establish flux in the core.

Note in Fig. 11.16 that the direction of the flux Φ can be determined by placing the fingers of the right hand in the direction of current around the core and noting the direction of the thumb. It is interesting to realize that *the magnetizing force is independent of the type of core material—it is determined solely by the number of turns, the current, and the length of the core.*

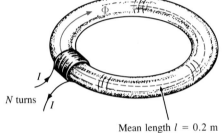

Mean length $l = 0.2$ m

FIG. 11.16

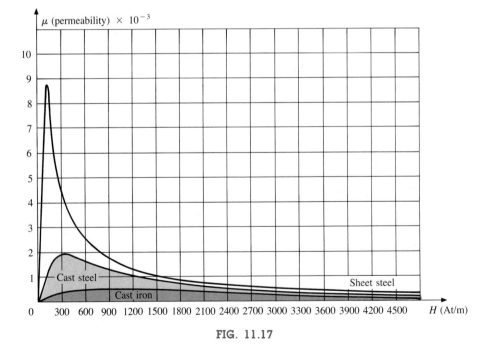

FIG. 11.17

The applied magnetizing force has a pronounced effect on the resulting permeability of a magnetic material. As the magnetizing force increases, the permeability rises to a maximum and then drops to a minimum, as shown in Fig. 11.17 for three commonly employed magnetic materials.

The flux density and the magnetizing force are related by the following equation:

$$B = \mu H$$ **(11.8)**

This equation indicates that for a particular magnetizing force, the greater the permeability, the greater will be the induced flux density.

11.8 HYSTERESIS

A curve of the flux density B versus the magnetizing force H of a material is of particular importance to the engineer. Curves of this type can usually be found in manuals and descriptive pamphlets and brochures published by manufacturers of magnetic materials. A typical B-H curve for a ferromagnetic material such as steel can be derived using the setup of Fig. 11.18.

The core is initially unmagnetized and the current $I = 0$. If the current I is increased to some value above zero, the magnetizing force H will increase to a value determined by

$$H\uparrow = \frac{NI\uparrow}{l}$$

The flux Φ and the flux density B ($B = \Phi/A$) will also increase with the current I (or H). If the material has no residual magnetism and the magnetizing force H is increased from zero to some value H_a, the B-H curve will follow the path shown in Fig. 11.19 between o and a. If the magnetizing force H is increased until saturation (H_S) occurs, the curve will continue as shown in the figure to point b. When saturation occurs, the flux density has, *for all practical purposes,* reached its maximum

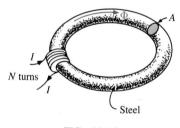

FIG. 11.18

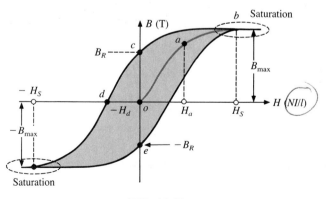

FIG. 11.19

value. Any further increase in current through the coil increasing $H = NI/l$ will result in a very small increase in flux density B.

If the magnetizing force is reduced to zero by letting I decrease to zero, the curve will follow the path of the curve between b and c. The flux density B_R, which remains when the magnetizing force is zero, is called the *residual flux density*. It is this residual flux density that makes it possible to create permanent magnets. If the coil is now removed from the core of Fig. 11.18, the core will still have the magnetic properties determined by the residual flux density, a measure of its "retentivity." If the current I is reversed, developing a magnetizing force, $-H$, the flux density B will decrease with increase in I. Eventually, the flux density will be zero when $-H_d$ (the portion of curve from c to d) is reached. The magnetizing force $-H_d$ required to "coerce" the flux density to reduce its level to zero is called the *coercive force*, a measure of the coercivity of the magnetic sample. As the force $-H$ is increased until saturation again occurs and is then reversed and brought back to zero, the path shown from d to e will result. If the magnetizing force is increased in the positive direction $(+H)$, the curve will trace the path shown from e to b. The entire curve represented by $bcded$ is called the *hysteresis curve* for the ferromagnetic material, from the Greek *hysterein*, meaning "to lag behind." The flux density B *lagged* behind the magnetizing force H during the entire plotting of the curve. When H was zero at c, B was not zero but had only begun to decline. Long after H had passed through zero and had become equal to $-H_d$ did the flux density B finally become equal to zero.

If the entire cycle is repeated, the curve obtained for the same core will be determined by the maximum H applied. Three hysteresis loops for the same material for maximum values of H less than the saturation value are shown in Fig. 11.20. In addition, the saturation curve is repeated for comparison purposes.

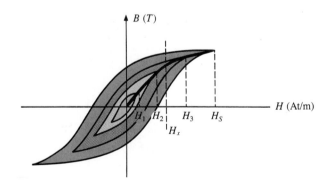

FIG. 11.20

Note from the various curves that for a particular value of H, say, H_x, the value of B can vary widely, as determined by the history of the core. In an effort to assign a particular value of B to each value of H, we compromise by connecting the tips of the hysteresis loops. The resulting curve, shown by the heavy, solid line in Fig. 11.20 and for various materials in Fig. 11.21, is called the *normal magnetization curve*. An expanded view of one region appears in Fig. 11.22.

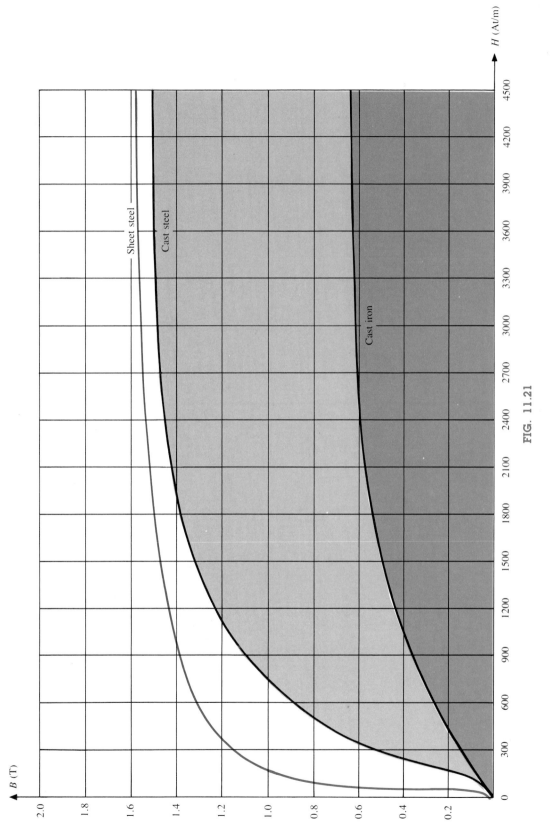

FIG. 11.21

323

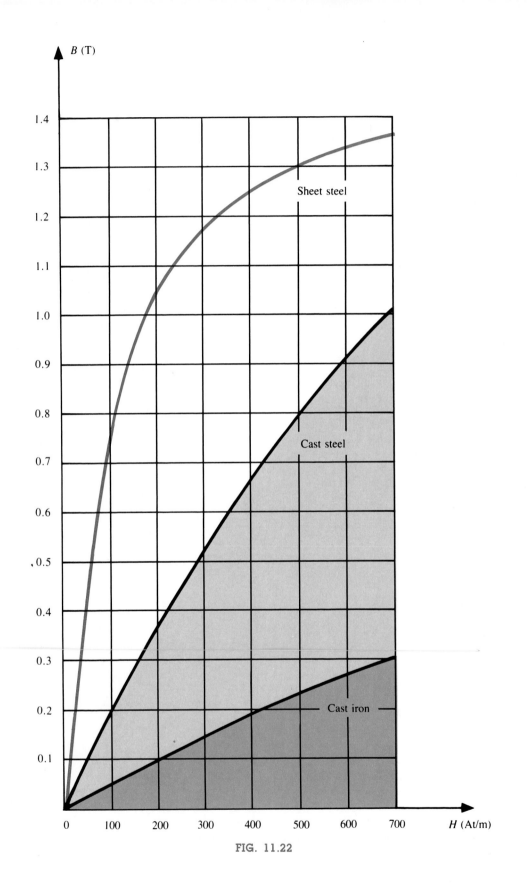

FIG. 11.22

324

A comparison of Figs. 11.17 and 11.21 shows that for the same value of H, the value of B is higher in Fig. 11.21 for the materials with the higher μ in Fig. 11.17. This is particularly obvious for low values of H. This correspondence between the two figures must exist, since $B = \mu H$. In fact, if in Fig. 11.21 we find μ for each value of H using the equation $\mu = B/H$, we will obtain the curves of Fig. 11.17.

An instrument that will provide a plot of the B-H curve for a magnetic sample appears in Fig. 11.23.

It is interesting to note that the hysteresis curves of Fig. 11.20 have a "point symmetry" about the origin. That is, the inverted pattern to the left of the vertical axis is the same as that appearing to the right of the vertical axis. In addition, you will find that a further application of the same magnetizing forces to the sample will result in the same plot. For a current I in $H = NI/l$ that will move between positive and negative maximums at a fixed rate, the same B-H curve will result during each cycle. Such will be the case when we examine ac (sinusoidal) networks in the later chapters. The reversal of the field (Φ) due to the changing current direction will result in a loss of energy that can best be described by first introducing the *domain theory of magnetism*.

Within each atom, the orbiting electrons (described in Chapter 2) are also spinning as they revolve around the nucleus. The atom, due to its spinning electrons, has a magnetic field associated with it. In non-magnetic materials, the net magnetic field is effectively zero, since the magnetic fields due to the atoms of the material oppose each other. In magnetic materials such as iron and steel, however, the magnetic fields of groups of atoms numbering in the order of 10^{12} are aligned, forming very small bar magnets. This group of magnetically aligned atoms is called a *domain*. Each domain is a separate entity; that is, each domain is independent of the surrounding domains. For an unmagnetized sample of magnetic material, these domains appear in a random manner, such as shown in Fig. 11.24(a). The net magnetic field in any one direction is zero.

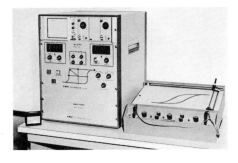

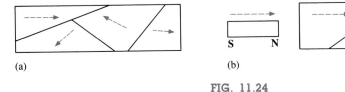

(a)   (b)

FIG. 11.24

When an external magnetizing force is applied, the domains that are nearly aligned with the applied field will grow at the expense of the less favorably oriented domains, such as shown in Fig. 11.24(b). Eventually, if a sufficiently strong field is applied, all of the domains will have the orientation of the applied magnetizing force, and any further increase in external field will not increase the strength of the magnetic flux through the core—a condition referred to as *saturation*. The elasticity of the above is evidenced by the fact that when the magnetizing force is removed, the alignment will be lost to some measure and the flux density will drop to B_R. In other words, the removal of the magnetizing force will result in the return of a number of misaligned domains within the core. The continued alignment of a number of the domains, however, accounts for our ability to create permanent magnets.

At a point just before saturation, the opposing unaligned domains are reduced to small cylinders of various shapes referred to as *bubbles*. These bubbles can be moved within the magnetic sample through the application of a *controlling* magnetic field. It is these magnetic bubbles that form the basis of the recently designed bubble memory system for computers.

11.9 AMPÈRE'S CIRCUITAL LAW

It was mentioned in the introduction to this chapter that there is a broad similarity between the analyses of electric and magnetic circuits. This has already been demonstrated to some extent for the quantities in Table 11.1.

TABLE 11.1

	Electric Circuits	Magnetic Circuits
Cause	E	$\mathcal{F}$
Effect	I	Φ
Opposition	R	$\mathcal{R}$

If we apply the "cause" analogy to Kirchhoff's voltage law ($\Sigma_\circlozenge V = 0$), we obtain the following:

$$\Sigma_\circlozenge \mathcal{F} = 0 \qquad \text{(for magnetic circuits)} \qquad \textbf{(11.9)}$$

which, in words, states that the algebraic sum of the rises and drops of the mmf around a closed loop of a magnetic circuit is equal to zero; that is, the sum of the mmf rises equals the sum of the mmf drops around a closed loop.

Equation (11.9) is referred to as *Ampère's circuital law*. When it is applied to magnetic circuits, sources of mmf are expressed by the equation

$$\mathcal{F} = NI \qquad \text{(At)} \qquad \textbf{(11.10)}$$

The equation for the mmf drop across a portion of a magnetic circuit can be found by applying the relationships listed in Table 11.1. That is, for electric circuits,

$$V = IR$$

resulting in the following for magnetic circuits:

$$\mathcal{F} = \Phi\mathcal{R} \qquad \text{(At)} \qquad \textbf{(11.11)}$$

where Φ is the flux passing through a section of the magnetic circuit and $\mathcal{R}$ is the reluctance of that section. The reluctance, however, is seldom calculated in the analysis of magnetic circuits. A more practical equation for the mmf drop is

$$\boxed{\mathscr{F} = Hl} \quad \text{(At)} \qquad \textbf{(11.12)}$$

as derived from Eq. (11.6), where H is the magnetizing force on a section of a magnetic circuit and l is the length of the section. As an example of Eq. (11.9), consider the magnetic circuit appearing in Fig. 11.25 constructed of three different ferromagnetic materials.

Applying Ampère's circuital law, we have

$$\Sigma_{\circlearrowleft} \mathscr{F} = 0$$

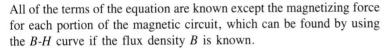

$$\underbrace{+NI}_{\text{rise}} - \underbrace{H_{ab}l_{ab}}_{\text{drop}} - \underbrace{H_{bc}l_{bc}}_{\text{drop}} - \underbrace{H_{ca}l_{ca}}_{\text{drop}} = 0$$

or

$$\underbrace{NI}_{\substack{\text{impressed}\\\text{mmf}}} = \underbrace{H_{ab}l_{ab} + H_{bc}l_{bc} + H_{ca}l_{ca}}_{\text{mmf drops}}$$

All of the terms of the equation are known except the magnetizing force for each portion of the magnetic circuit, which can be found by using the *B-H* curve if the flux density B is known.

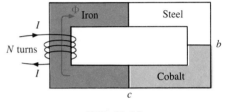

FIG. 11.25

11.10 THE FLUX Φ

If we continue to apply the relationships described in the previous section to Kirchhoff's current law, we will find that the sum of the fluxes entering a function is equal to the sum of the fluxes leaving a junction; that is, for the circuit of Fig. 11.26,

$$\Phi_a = \Phi_b + \Phi_c \qquad \text{(at junction } a\text{)}$$

or

$$\Phi_b + \Phi_c = \Phi_a \qquad \text{(at junction } b\text{)}$$

both of which are equivalent.

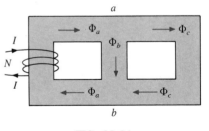

FIG. 11.26

11.11 SERIES MAGNETIC CIRCUITS: DETERMINING *NI*

We are now in a position to solve a few magnetic circuit problems, which are basically of two types. In one type, Φ is given, and the impressed mmf *NI* must be computed. This is the type of problem encountered in the design of motors, generators, and transformers. In the other type, *NI* is given, and the flux Φ of the magnetic circuit must be found. This type of problem is encountered primarily in the design of magnetic amplifiers and is more difficult since the approach is "hit or miss."

As indicated in earlier discussions, the value of μ will vary from point to point along the magnetization curve This eliminates the possibility of finding the reluctance of each "branch" or the "total reluctance" of a network as was done for electric circuits where ρ had a fixed value for any applied current or voltage. If the total reluctance could be determined, Φ could then be determined using the Ohm's law analogy for magnetic circuits.

For magnetic circuits, the level of B or H is determined from the other using the B-H curve, and μ is seldom calculated unless asked for.

An approach frequently employed in the analysis of magnetic circuits is the *table* method. Before a problem is analyzed in detail, a table is prepared listing in the extreme left-hand column the various sections of the magnetic circuit. The columns on the right are reserved for the quantities to be found for each section. In this way, the individual doing the problem can keep track of what is required to complete the problem and also of what the next step should be. After a few examples, the usefulness of this method should become clear.

This section will consider only *series* magnetic circuits in which the flux Φ is the same throughout. In each example, the magnitude of the magnetomotive force is to be determined.

EXAMPLE 11.3. For the series magnetic circuit of Fig. 11.27:

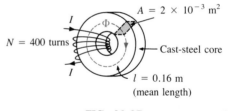

FIG. 11.27

a. Find the value of I required to develop a magnetic flux of $\Phi = 4 \times 10^{-4}$ Wb.

b. Determine μ and μ_r for the material under these conditions.

Solutions: The magnetic circuit can be represented by the system shown in Fig. 11.28(a). The electric circuit analogy is shown in Fig. 11.28(b). Analogies of this type can be very helpful in the solution of magnetic circuits. Table 11.2 is for part (a) of this problem. The table is fairly trivial for this example but it does define the quantities to be found.

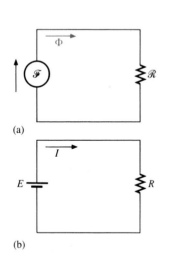

FIG. 11.28

(a) Magnetic circuit equivalent and (b) electric circuit analogy.

TABLE 11.2

Section	Φ (Wb)	A (m²)	B (T)	H (At/m)	l (m)	Hl (At)
One continuous section	4×10^{-4}	2×10^{-3}			0.16	

a. The flux density B is

$$B = \frac{\Phi}{A} = \frac{4 \times 10^{-4}}{2 \times 10^{-3}} = 2 \times 10^{-1} = 0.2 \, \text{T}$$

Using the *B-H* curves of Fig. 11.22, we can determine the magnetizing force H:

$$H \, (\text{cast steel}) = 110 \, \text{At/m}$$

Applying Ampère's circuital law yields

$$NI = Hl$$

and

$$I = \frac{Hl}{N} = \frac{(110)(0.16)}{400} = 0.044 = \mathbf{44 \, mA}$$

b. The permeability of the material can be found using Eq. (11.8):

$$\mu = \frac{B}{H} = \frac{0.2}{110} = \mathbf{1.818 \times 10^{-3} \, Wb/Am}$$

and the relative permeability is

$$\mu_r = \frac{\mu}{\mu_o}$$

$$= \frac{1.818 \times 10^{-3}}{4\pi \times 10^{-7}}$$

$$= \frac{1.818 \times 10^{-3}}{12.57 \times 10^{-7}}$$

$$= \mathbf{1446.3}$$

EXAMPLE 11.4. The electromagnet of Fig. 11.29 has picked up a section of cast iron. Determine the current I required to establish the indicated flux in the core.

Solution: To be able to use Figs. 11.21 and 11.22, the dimensions must first be converted to the metric system. However, since the area is the same throughout, we can determine the length for each material rather than work with the individual sections:

$$l_{efab} = 4 \, \text{in.} + 4 \, \text{in.} + 4 \, \text{in.} = 12 \, \text{in.}$$

$$l_{bcde} = 0.5 \, \text{in.} + 4 \, \text{in.} + 0.5 \, \text{in.} = 5 \, \text{in.}$$

$$12 \, \text{in.} \left(\frac{1 \, \text{m}}{39.37 \, \text{in.}} \right) = 304.8 \times 10^{-3} \, \text{m}$$

$$5 \, \text{in.} \left(\frac{1 \, \text{m}}{39.37 \, \text{in.}} \right) = 127 \times 10^{-3} \, \text{m}$$

$$1 \, \text{in.}^2 \left(\frac{1 \, \text{m}}{39.37 \, \text{in.}} \right) \left(\frac{1 \, \text{m}}{39.37 \, \text{in.}} \right) = 6.452 \times 10^{-4} \, \text{m}^2$$

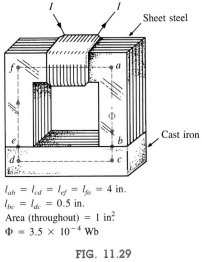

$$l_{ab} = l_{cd} = l_{ef} = l_{fa} = 4 \, \text{in.}$$
$$l_{bc} = l_{dc} = 0.5 \, \text{in.}$$
Area (throughout) = 1 in.²
$$\Phi = 3.5 \times 10^{-4} \, \text{Wb}$$

FIG. 11.29
Electromagnet for Example 11.4.

The information available from the specifications of the problem has been inserted in Table 11.3. When the problem has been completed,

TABLE 11.3

Section	Φ (Wb)	A (m²)	B (T)	H (At/m)	l (m)	Hl (At)
efab	3.5×10^{-4}	6.452×10^{-4}			304.8×10^{-3}	
bcde	3.5×10^{-4}	6.452×10^{-4}			127×10^{-3}	

each space will contain some information. Sufficient data to complete the problem can be found if we fill in each column from left to right. As the various quantities are calculated, they will be placed in a similar table found at the end of the example.

The flux density for each section is

$$B = \frac{\Phi}{A} = \frac{3.5 \times 10^{-4}}{6.452 \times 10^{-4}} = 0.542 \text{ T}$$

and the magnetizing force is

$$H \text{ (sheet steel, Fig. 11.22)} \cong 60 \text{ At/m}$$
$$H \text{ (cast iron, Fig. 11.21)} \cong 1600 \text{ At/m}$$

Note the extreme difference in magnetizing force for each material for the required flux density. In fact, when we apply Ampère's circuital law, we will find that the sheet steel section could be ignored with a minimal error in the solution.

Determining Hl for each section yields

$$H_{efab}l_{efab} = (60)(304.8 \times 10^{-3}) = 18.29 \text{ At}$$
$$H_{bcde}l_{bcde} = (1600)(127 \times 10^{-3}) = 203.2 \text{ At}$$

Inserting the above data in Table 11.3 will result in Table 11.4. The magnetic circuit equivalent and the electric circuit analogy for the system of Fig. 11.29 appear in Fig. 11.30.

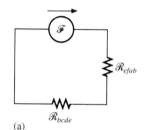

(a)

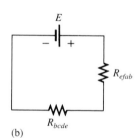

(b)

FIG. 11.30

(a) Magnetic circuit equivalent and (b) electric circuit analogy for the electromagnetic of Fig. 11.29.

TABLE 11.4

Section	Φ (Wb)	A (m²)	B (T)	H (At/m)	l (m)	Hl (At)
efab	3.5×10^{-4}	6.452×10^{-4}	0.542	60	304.8×10^{-3}	18.29
bcde	3.5×10^{-4}	6.452×10^{-4}	0.542	1600	127×10^{-3}	203.2

Applying Ampère's circuital law,

$$NI = H_{efab}l_{efab} + H_{bcde}l_{bcde}$$
$$= 18.29 + 203.2 = 221.49$$

and

$$50I = 221.49$$

or

$$I = \frac{221.49}{50} = \textbf{4.43 A}$$

EXAMPLE 11.5. Determine the secondary current I_2 for the transformer of Fig. 11.31 if the flux in the core is 1.5×10^{-5} Wb.

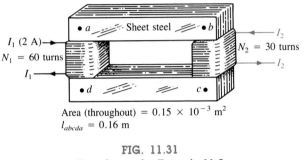

Area (throughout) $= 0.15 \times 10^{-3}$ m^2
$l_{abcda} = 0.16$ m

FIG. 11.31
Transformer for Example 11.5.

Solution: This is the first example with two magnetizing forces to consider. In the analogies of Fig. 11.32 you will note that the resulting flux of each is opposing, just as the two sources of voltage are opposing in the electric circuit analogy.

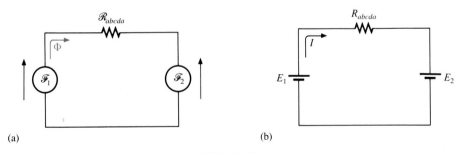

(a)   (b)

FIG. 11.32
(a) Magnetic circuit equivalent and (b) electric circuit analogy for the transformer of Fig. 11.31.

The structural data appear in Table 11.5.

TABLE 11.5

Section	Φ (Wb)	A (m^2)	B (T)	H (At/m)	l (m)	Hl (At)
abcda	1.5×10^{-5}	0.15×10^{-3}			0.16	

The flux density throughout is

$$B = \frac{\Phi}{A} = \frac{1.5 \times 10^{-5}}{0.15 \times 10^{-3}} = 10 \times 10^{-2} = 0.10 \text{ T}$$

and

$$H \text{ (from Fig. 11.22)} \cong \frac{1}{7}(100) = 14.29 \text{ At/m}$$

Applying Ampère's circuital law,

$$N_1 I_1 - N_2 I_2 = H_{abcda} l_{abcda}$$
$$(60)(2) - (30)(I_2) = (14.29)(0.16)$$
$$120 - 30 I_2 = 2.29$$

and

$$30 I_2 = 120 - 2.29$$

or

$$I_2 = \frac{117.71}{30} = \textbf{3.924 A}$$

For the analysis of most transformer systems, the equation $N_1 I_1 = N_2 I_2$ is employed. This would result in 4 A versus 3.924 A above. This difference is normally ignored, however, and the equation $N_1 I_1 = N_2 I_2$ considered exact.

Because of the nonlinearity of the *B-H* curve, *it is not possible to apply superposition to magnetic circuits;* that is, in the previous example, we cannot consider the effects of each source independently and then find the total effects by using superposition.

11.12 AIR GAPS

Before continuing with the illustrative examples, let us consider the effects an air gap has on a magnetic circuit. Note the presence of air gaps in the magnetic circuits of the motor and meter of Fig. 11.11. The spreading of the flux lines outside the common area of the core for the air gap in Fig. 11.33(a) is known as *fringing*. For our purposes, we shall neglect this effect and assume the flux distribution to be as in Fig. 11.33(b).

The flux density of the air gap in Fig. 11.33(b) is given by

$$\boxed{B_g = \frac{\Phi_g}{A_g}} \qquad \textbf{(11.13)}$$

where, for our purposes,

$$\Phi_g = \Phi_{\text{core}}$$

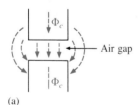
(a)

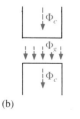
(b)

FIG. 11.33

and

$$A_g = A_{core}$$

For most practical applications, the permeability of air is taken to be equal to that of free space. The magnetizing force of the air gap is then determined by

$$H_g = \frac{B_g}{\mu_o}$$ **(11.14)**

and the mmf drop across the air gap is equal to $H_g l_g$. An equation for H_g is as follows:

$$H_g = \frac{B_g}{\mu_o} = \frac{B_g}{4\pi \times 10^{-7}}$$

and

$$H_g = (7.96 \times 10^5)B_g$$ (At/m)   **(11.15)**

EXAMPLE 11.6. Find the value of I required to establish a magnetic flux of $\Phi = 0.75 \times 10^{-4}$ Wb in the series magnetic circuit of Fig. 11.34.

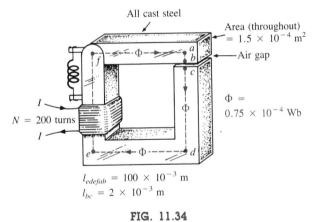

All cast steel

Area (throughout) = 1.5×10^{-4} m²

Air gap

$\Phi = 0.75 \times 10^{-4}$ Wb

I

$N = 200$ turns

I

$l_{edefab} = 100 \times 10^{-3}$ m
$l_{bc} = 2 \times 10^{-3}$ m

FIG. 11.34
Relay for Example 11.6.

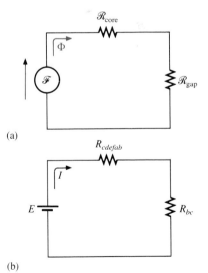

(a)

(b)

FIG. 11.35
(a) Magnetic circuit equivalent and (b) electric circuit analogy for the relay of Fig. 11.34.

Solution: An equivalent magnetic circuit and its electric circuit analogy are shown in Fig. 11.35.

The flux density for each section is

$$B = \frac{\Phi}{A} = \frac{0.75 \times 10^{-4}}{1.5 \times 10^{-4}} = 0.5 \text{ T}$$

From the *B-H* curves of Fig. 11.22,

$$H \text{ (cast steel)} \cong 280 \text{ At/m}$$

Applying Eq. (11.15),

$$H_g = (7.96 \times 10^5)B_g = (7.96 \times 10^5)(0.5) = 3.98 \times 10^5 \,\text{At/m}$$

The mmf drops are

$$H_{core}l_{core} = (280)(100 \times 10^{-3}) = 28 \,\text{At}$$
$$H_g l_g = (3.98 \times 10^5)(2 \times 10^{-3}) = 796 \,\text{At}$$

Applying Ampère's circuital law,

$$NI = H_{core}l_{core} + H_g l_g$$
$$= 28 + 796$$
$$200I = 824$$
$$I = \mathbf{4.12 \,A}$$

Note from the above that the air gap requires the biggest share (by far) of the impressed *NI* due to the fact that air is nonmagnetic.

11.13 SERIES-PARALLEL MAGNETIC CIRCUITS

As one might expect, the close analogies between electric and magnetic circuits will eventually lead to series-parallel magnetic circuits similar in many respects to those encountered in Chapter 7. In fact, the electric circuit analogy will prove helpful in defining the procedure to follow toward a solution.

EXAMPLE 11.7. Determine the current *I* required to establish a flux of 1.5×10^{-4} Wb in the section of the core indicated in Fig. 11.36.

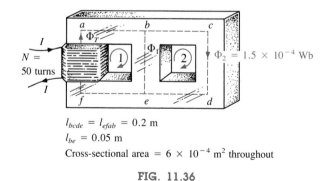

$$l_{bcde} = l_{efab} = 0.2 \,\text{m}$$
$$l_{be} = 0.05 \,\text{m}$$

Cross-sectional area $= 6 \times 10^{-4} \,\text{m}^2$ throughout

FIG. 11.36

Solution: The equivalent magnetic circuit and the electric circuit analogy appear in Fig. 11.37. We have

$$B_2 = \frac{\Phi_2}{A} = \frac{1.5 \times 10^{-4}}{6 \times 10^{-4}} = 0.25 \,\text{T}$$

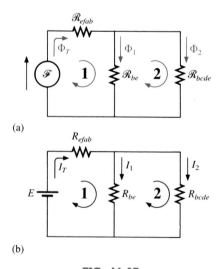

(a)

(b)

FIG. 11.37

(a) Magnetic circuit equivalent and (b) electric circuit analogy for the series-parallel system of Fig. 11.36.

From Fig. 11.22,

$$H_{bcde} \cong 30 \, \text{At/m}$$

Applying Ampère's circuital law around loop 2 of Figs. 11.36 and 11.37,

$$\Sigma_{\circlearrowleft} \mathscr{F} = 0$$
$$H_{be}l_{be} - H_{bcde}l_{bcde} = 0$$
$$H_{be}(0.05) - (30)(0.2) = 0$$
$$H_{be} = \frac{6}{0.05} = 120 \, \text{At}$$

From Fig. 11.22,

$$B_1 \cong 0.8 \, \text{T}$$

and

$$\Phi_1 = B_1 A = (0.8)(6 \times 10^{-4}) = 4.8 \times 10^{-4} \, \text{Wb}$$

The results are then entered in Table 11.6.

TABLE 11.6

Section	Φ (Wb)	A (m²)	B (T)	H (At/m)	l (m)	Hl (At)
bcde	1.5×10^{-4}	6×10^{-4}	0.25	30	0.2	6
be	4.8×10^{-4}	6×10^{-4}	0.8	120	0.05	6
efab		6×10^{-4}			0.2	

The table reveals that we must now turn our attention to section *efab:*

$$\Phi_T = \Phi_1 + \Phi_2 = 4.8 \times 10^{-4} + 1.5 \times 10^{-4}$$
$$= 6.3 \times 10^{-4} \, \text{Wb}$$
$$B = \frac{\Phi_T}{A} = \frac{6.3 \times 10^{-4}}{6 \times 10^{-4}}$$
$$= 1.05 \, \text{T}$$

From Fig. 10.21,

$$H_{efab} \cong 200 \, \text{At}$$

Applying Ampère's circuital law,

$$+NI - H_{efab}l_{efab} - H_{be}l_{be} = 0$$
$$NI = (200)(0.2) + (120)(0.05)$$
$$50I = 40 + 6$$
$$I = \frac{46}{50} = \mathbf{0.92 \, A}$$

To demonstrate that μ is sensitive to the magnetizing force, the permeability of each section is determined as follows. For section *bcde,*

$$\mu = \frac{B}{H} = \frac{0.25}{30} = 0.0083$$

and

$$\mu_r = \frac{\mu}{\mu_o} = \frac{0.0083}{12.57 \times 10^{-7}} = \mathbf{6603}$$

For section *be*,

$$\mu = \frac{B}{H} = \frac{0.8}{120} = 0.0067$$

and

$$\mu_r = \frac{\mu}{\mu_o} = \frac{0.0067}{12.57 \times 10^{-7}} = \mathbf{5330}$$

For section *efab*,

$$\mu = \frac{B}{H} = \frac{1.05}{200} = 0.00525$$

and

$$\mu_r = \frac{\mu}{\mu_o} = \frac{0.00525}{12.57 \times 10^{-7}} = \mathbf{4176.6}$$

11.14 DETERMINING Φ

The examples of this section are of the second type, where *NI* is given and the flux Φ must be found. This is a relatively straightforward problem if only one magnetic section is involved. Then

$$H = \frac{NI}{l} \qquad H \rightarrow B \ (B\text{-}H \text{ curve})$$

and

$$\Phi = BA$$

For magnetic circuits with more than one section, there is no set order of steps that will lead to an exact solution for every problem on the first attempt. In general, however, we proceed as follows. We must find the impressed mmf for a *calculated guess* of the flux Φ and then compare this with the specified value of mmf. We can then make adjustments on our guess to bring it closer to the actual value. For most applications, a value within $\pm 5\%$ of the actual Φ or specified *NI* is acceptable.

We can make a reasonable guess at the value of Φ if we realize that the maximum mmf drop appears across the material with the smallest permeability if the length and area of each material are the same. As shown in Example 11.6, if there is an air gap in the magnetic circuit, there will be a considerable drop in mmf across the gap. As a starting point for problems of this type, therefore, we shall assume that the total mmf (*NI*) is across the section with the lowest μ or greatest $\mathcal{R}$ (if the other physical dimensions are relatively similar). This assumption gives a value of Φ that will produce a calculated *NI* greater than the specified value. Then, after considering the results of our original assumption very carefully, we shall *cut* Φ and *NI* by introducing the effects (reluctance) of the other portions of the magnetic circuit and *try* the new solution. For obvious reasons, this approach is frequently called the *cut and try* method.

EXAMPLE 11.8. Calculate the magnetic flux Φ for the magnetic circuit of Fig. 11.38.

Solution: By Ampère's circuital law,

$$NI = H_{abcda}l_{abcda}$$

or

$$H_{abcda} = \frac{NI}{l_{abcda}} = \frac{(60)(5)}{0.3}$$
$$= \frac{300}{0.3} = 1000 \text{ At/m}$$

and

$$B_{abcda} \text{ (from Fig. 11.22)} \cong 0.38 \text{ T}$$

Since $B = \Phi/A$, we have

$$\Phi = BA = (0.38)(2 \times 10^{-4}) = \mathbf{0.760 \times 10^{-4} \, Wb}$$

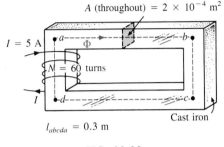

FIG. 11.38

EXAMPLE 11.9. Find the magnetic flux Φ for the series magnetic circuit of Fig. 11.39 for the specified impressed mmf.

Solution: Assuming that the total impressed mmf *NI* is across the air gap,

$$NI = H_g l_g$$

or

$$H_g = \frac{NI}{l_g} = \frac{400}{0.001} = 4 \times 10^5 \text{ At/m}$$

and

$$B_g = \mu_o H_g = (4\pi \times 10^{-7})(4 \times 10^5)$$
$$= 50.265 \times 10^{-2} \text{ T}$$

The flux

$$\Phi_g = \Phi_{\text{core}} = B_g A$$
$$= (50.265 \times 10^{-2})(0.003)$$
$$\Phi_{\text{core}} = 150.8 \times 10^{-5} \text{ Wb}$$

Using this value of Φ, we can find *NI*. The data are inserted in Table 11.7.

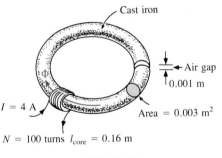

FIG. 11.39

TABLE 11.7

Section	Φ (Wb)	A (m^2)	B (T)	H (At/m)	l (m)	Hl (At)
Core	150.8×10^{-5}	0.003	0.50	1500 (B-H curve)	0.16	
Gap	150.8×10^{-5}	0.003	0.50	4×10^5	0.001	400

$$H_{core}l_{core} = (1500)(0.16) = 240 \text{ At}$$

Applying Ampère's circuital law results in

$$NI = H_{core}l_{core} + H_g l_g$$
$$= 240 + 400$$
$$NI = 640 > 400$$

Since we neglected the reluctance of all the magnetic paths but the air gap, the calculated value is greater than the specified value. We must therefore reduce this value by including the effect of these reluctances. Since approximately $(640 - 400)/640 = 140.8/640 \cong 37.5\%$ of our calculated value is above the desired value, let us reduce Φ by 37.5% and see how close we come to the impressed mmf of 400:

$$\Phi = (1 - 0.375)(150.8 \times 10^{-5})$$
$$= 94.25 \times 10^{-5} \text{ Wb}$$

See Table 11.8.

TABLE 11.8

Section	Φ (Wb)	A (m^2)	B (T)	H (At/m)	l (m)	Hl (At)
Core	94.25×10^{-5}	0.003			0.16	
Gap	94.25×10^{-5}	0.003			0.001	

$$B = \frac{\Phi}{A} = \frac{94.25 \times 10^{-5}}{0.003} = 31.42 \times 10^{-2} \cong 0.31 \text{ T}$$

$$H_g l_g = (7.96 \times 10^5) B_g l_g$$
$$= (7.96 \times 10^5)(0.31)(0.001)$$
$$\cong 246.76 \text{ At}$$

From B-H curves,

$$H_{core} \cong 730 \text{ At/m}$$

$$H_{core}l_{core} = (730)(0.16) = 116.8 \text{ At}$$

Applying Ampère's circuital law yields

$$NI = H_{core}l_{core} + H_g l_g$$
$$= 116.8 + 297.45$$
$$NI = \mathbf{414.25} > 400 \qquad \text{(but within } \pm 5\%$$
$$\text{and therefore acceptable)}$$

The solution is therefore

$$\Phi \cong \mathbf{94.25 \times 10^{-5} \, Wb}$$

PROBLEMS

Section 11.3

1. Using Appendix F, fill in the blanks in the following table. Indicate the units for each quantity.

	Φ	B
SI	5×10^{-4} Wb	8×10^{-4} T
CGS	——	——
English	——	——

2. Repeat Problem 1 for the following table if area $= 2$ in.^{2}:

	Φ	B
SI	——	——
CGS	60,000 maxwells	——
English	——	——

3. For the electromagnet of Fig. 11.40:
   a. Find the flux density in the core.
   b. Sketch the magnetic flux lines and indicate their direction.
   c. Indicate the north and south poles of the magnet.

Section 11.5

4. Which section of Fig. 11.41 [(a), (b), or (c)] has the largest reluctance to the setting up of flux lines through its longest dimension?

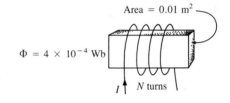

FIG. 11.40

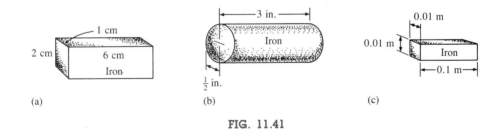

FIG. 11.41

Section 11.6

5. Find the reluctance of a magnetic circuit if a magnetic flux $\Phi = 4.2 \times 10^{-4}$ Wb is established by an impressed mmf of 400 At.

6. Repeat Problem 5 for $\Phi = 72,000$ maxwells and an impressed mmf of 120 gilberts.

Section 11.7

7. Find the magnetizing force H for Problem 5 in SI units if the magnetic circuit is 6 inches in length.

8. If a magnetizing force H of 600 At/m is applied to a magnetic circuit, a flux density B of 1200×10^{-4} Wb/m^2 is established. Find the permeability μ of a material that will produce twice the original flux density for the same magnetizing force.

Section 11.8

9. For the series magnetic circuit of Fig. 11.42, determine the current I necessary to establish the indicated flux.

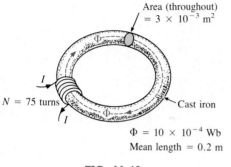

Area (throughout)
= 3×10^{-3} m^2

$N = 75$ turns

Cast iron

$\Phi = 10 \times 10^{-4}$ Wb
Mean length = 0.2 m

FIG. 11.42

10. Find the current necessary to establish a flux of $\Phi = 3 \times 10^{-4}$ Wb in the series magnetic circuit of Fig. 11.43.

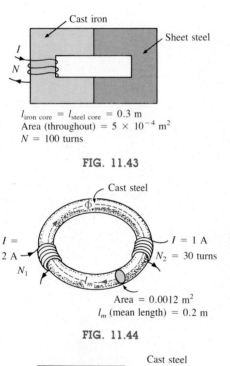

Cast iron

Sheet steel

$l_{\text{iron core}} = l_{\text{steel core}} = 0.3$ m
Area (throughout) = 5×10^{-4} m^2
$N = 100$ turns

FIG. 11.43

11. a. Find the number of turns N required to establish a flux $\Phi = 12 \times 10^{-4}$ Wb in the magnetic circuit of Fig. 11.44.
b. Find the permeability μ of the material.

Cast steel

$I = 2$ A

$I = 1$ A
$N_2 = 30$ turns

N_1

l_m

Area = 0.0012 m^2
l_m (mean length) = 0.2 m

FIG. 11.44

12. a. Find the mmf (NI) required to establish a flux $\Phi = 80,000$ lines in the magnetic circuit of Fig. 11.45.
b. Find the permeability of each material.

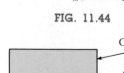

Cast steel

NI

Sheet steel

Uniform area
(throughout)
= 1 in^2

$l_{\text{cast steel}} = 5.5$ in.
$l_{\text{sheet steel}} = 0.5$ in.

FIG. 11.45

*13. For the series magnetic circuit of Fig. 11.46 with two impressed sources of magnetic "pressure," determine the current I. Each applied mmf establishes a flux pattern in the clockwise direction.

$\Phi = 0.8 \times 10^{-4}$ Wb

$l_{\text{cast steel}} = 5.5$ in.

$l_{\text{cast iron}} = 2.5$ in.

Cast steel

I $N_1 = 20$ turns I $N_2 = 30$ turns I

Cast iron

Area (throughout) $= 0.25$ in.^{2}

FIG. 11.46

Section 11.12

14. a. Find the current I required to establish a flux $\Phi = 2.4 \times 10^{-4}$ Wb in the magnetic circuit of Fig. 11.47.

b. Compare the mmf drop across the air gap to that across the rest of the magnetic circuit. Discuss your results using the value of μ for each material.

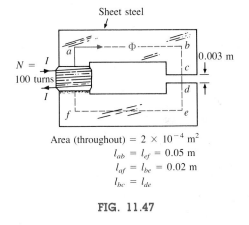

Sheet steel

$N = 100$ turns

0.003 m

Area (throughout) $= 2 \times 10^{-4}$ m^2

$l_{ab} = l_{ef} = 0.05$ m

$l_{af} = l_{be} = 0.02$ m

$l_{bc} = l_{de}$

FIG. 11.47

*15. The force carried by the plunger of the door chime of Fig. 11.48 is determined by

$$f = \frac{1}{2} NI \frac{d\phi}{dx} \quad \text{(newtons)}$$

where $d\phi/dx$ is the rate of change of flux linking the coil as the core is drawn into the coil. The greatest rate of change of flux will occur when the core is 1/4 to 3/4 the way through. In this region, if Φ changes from 0.5×10^{-4} Wb to 8×10^{-4} Wb, what is the force carried by the plunger?

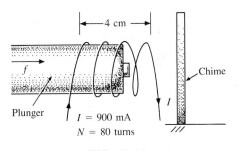

4 cm

Chime

Plunger

$I = 900$ mA

$N = 80$ turns

FIG. 11.48

Door chime.

16. Determine the current I_1 required to establish a flux of $\Phi = 2 \times 10^{-4}$ Wb in the magnetic circuit of Fig. 11.49.

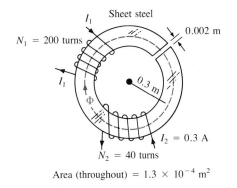

I_1 Sheet steel

$N_1 = 200$ turns

0.002 m

I_1

0.3 m

Φ

$I_2 = 0.3$ A

$N_2 = 40$ turns

Area (throughout) $= 1.3 \times 10^{-4}$ m^2

FIG. 11.49

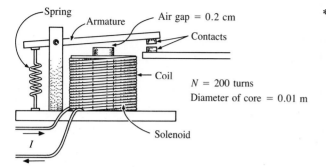

FIG. 11.50
Relay.

N = 200 turns
Diameter of core = 0.01 m

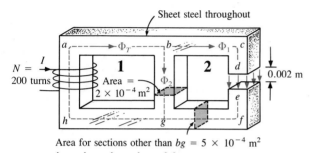

Area for sections other than bg = 5 × 10⁻⁴ m²
$l_{ab} = l_{bg} = l_{gh} = l_{ha}$ = 0.2 m
$l_{bc} = l_{fg}$ = 0.1 m, $l_{cd} = l_{ef}$ = 0.099 m

FIG. 11.51

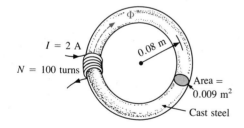

FIG. 11.52

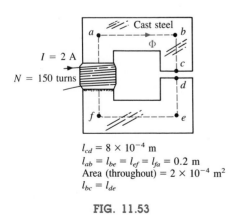

l_{cd} = 8 × 10⁻⁴ m
$l_{ab} = l_{be} = l_{ef} = l_{fa}$ = 0.2 m
Area (throughout) = 2 × 10⁻⁴ m²
$l_{bc} = l_{de}$

FIG. 11.53

*17. a. A flux of 0.2 × 10⁻⁴ Wb will establish sufficient attractive force for the armature of the relay of Fig. 11.50 to close the contacts. Determine the required current to establish this flux level if we assume the total mmf drop is across the air gap.

b. The force exerted on the armature is determined by the equation

$$F \text{ (newtons)} = \frac{1}{2} \cdot \frac{B_g^2 A}{\mu_o}$$

where B_g is the flux density within the air gap and A is the common area of the air gap. Find the force in newtons exerted when the flux Φ specified in part (a) is established.

*18. For the series-parallel magnetic circuit of Fig. 11.51, find the value of I required to establish a flux in the gap $\Phi_g = 2 \times 10^{-4}$ Wb.

Section 11.14

19. Find the magnetic flux Φ established in the series magnetic circuit of Fig. 11.52.

*20. Determine the magnetic flux Φ established in the series magnetic circuit of Fig. 11.53.

GLOSSARY

Ampère's circuital law A law establishing the fact that the algebraic sum of the rises and drops of the mmf around a closed loop of a magnetic circuit is equal to zero.

Diamagnetic materials Materials that have permeabilities slightly less than that of free space.

Domain A group of magnetically aligned atoms.

Electromagnetism Magnetic effects introduced by the flow of charge or current.

Ferromagnetic materials Materials having permeabilities hundreds and thousands of times greater than that of free space.

Flux density (B) A measure of the flux per unit area perpendicular to a magnetic flux path. It is measured in teslas (T) or webers per square meter (Wb/m^2).

Hysteresis The lagging effect between flux density of a material and the magnetizing force applied.

Magnetic flux lines Lines of a continuous nature that reveal the strength and direction of the magnetic field.

Magnetizing force (H) A measure of the magnetomotive force per unit length of a magnetic circuit.

Magnetomotive force ($\mathcal{F}$) The ''pressure'' required to establish magnetic flux in a ferromagnetic material. It is measured in ampere-turns (At).

Paramagnetic materials Materials that have permeabilities slightly greater than that of free space.

Permanent magnet A material such as steel or iron that will remain magnetized for long periods of time without the aid of external means.

Permeability (μ) A measure of the ease with which magnetic flux can be established in a material. It is measured in Wb/Am.

Relative permeability (μ_r) The ratio of the permeability of a material to that of free space.

Reluctance ($\mathcal{R}$) A quantity determined by the physical characteristics of a material that will provide an indication of the ''reluctance'' of that material to the setting up of magnetic flux lines in the material. It is measured in rels or At/Wb.

12

Inductors

12.1 INTRODUCTION

We have examined the resistor and the capacitor in detail. In this chapter we shall consider a third element, the *inductor,* which has a number of response characteristics similar in many respects to those of the capacitor. In fact, some sections of this chapter will proceed parallel to those for the capacitor to emphasize the similarity that exists between the two elements.

12.2 FARADAY'S LAW OF ELECTROMAGNETIC INDUCTION

If a conductor is moved through a magnetic field so that it cuts magnetic lines of flux, a voltage will be induced across the conductor, as shown in Fig. 12.1. The greater the number of flux lines cut per unit time (by increasing the speed with which the conductor passes through the field), or the stronger the magnetic field strength (for the same traversing speed), the greater will be the induced voltage across the conductor. If the conductor is held fixed and the magnetic field is moved so that its flux lines cut the conductor, the same effect will be produced.

If a coil of N turns is placed in the region of a changing flux, as in Fig. 12.2, a voltage will be induced across the coil as determined by *Faraday's law:*

$$e = N\frac{d\phi}{dt} \qquad \text{(volts, V)} \qquad \textbf{(12.1)}$$

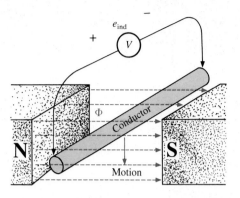

FIG. 12.1

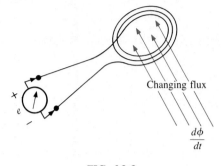

FIG. 12.2

where N represents the number of turns of the coil and $d\phi/dt$ is the instantaneous change in flux (in webers) linking the coil. The term *linking* refers to the flux within the turns of wire. The term *changing* simply indicates that either the strength of the field linking the coil changes in magnitude or the coil is moved through the field in such a way that the number of flux lines through the coil changes with time.

If the flux linking the coil ceases to change, such as when the coil simply sits still in a magnetic field of fixed strength, $d\phi/dt = 0$, and the induced voltage $e = N(d\phi/dt) = N(0) = 0$.

12.3 LENZ'S LAW

In Section 11.2 it was shown that the magnetic flux linking a coil of N turns with a current I has the distribution of Fig. 12.3.

If the current increases in magnitude, the flux linking the coil also increases. It was shown in Section 12.2, however, that a changing flux linking a coil induces a voltage across the coil. For this coil, therefore, an induced voltage is developed *across* the coil due to the change in current *through* the coil. The polarity of this induced voltage tends to establish a current in the coil which produces a flux that will oppose any change in the original flux. In other words, the induced effect (e_{ind}) is a result of the increasing current through the coil. However, the resulting induced voltage will tend to establish a current that will oppose the increasing change in current through the coil. Keep in mind that this is all occurring simultaneously. The instant the current begins to increase in magnitude, there will be an opposing effect trying to limit the change. It is "choking" the change in current through the coil. Hence, the term *choke* is often applied to the inductor or coil. In fact, we will find shortly that the current through a coil cannot change instantaneously. A period of time determined by the coil and the resistance of the circuit is required before the inductor discontinues its opposition to a momentary change in current. Recall a similar situation for the voltage across a capacitor in Chapter 10. The reaction above is true for increasing or decreasing levels of current through the coil. This effect is an example of a general principle known as *Lenz's law,* which states that *an induced effect is always such as to oppose the cause that produced it*.

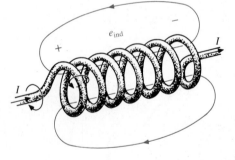

FIG. 12.3

12.4 SELF-INDUCTANCE

The ability of a coil to oppose any change in current is a measure of the *self-inductance L* of the coil. For brevity, the prefix *self* is usually dropped. Inductance is measured in henries (H), after the American physicist Joseph Henry.

Inductors are coils of various dimensions designed to introduce specified amounts of inductance into a circuit. The inductance of a coil varies directly with the magnetic properties of the coil. Ferromagnetic materials, therefore, are frequently employed to increase the inductance by increasing the flux linking the coil.

A close approximation, in terms of physical dimensions, for the inductance of the coils of Fig. 12.4 can be found using the following equation:

$$L = \frac{N^2 \mu A}{l} \qquad \text{(henries, H)} \qquad \textbf{(12.2)}$$

where N represents the number of turns, μ the permeability of the core (recall that μ is not a constant but depends on the level of B and H since $\mu = B/H$), A the area of the core in square meters, and l the mean length of the core in meters.

Substituting $\mu = \mu_r \mu_o$ into Eq. (12.2) yields

$$L = \frac{N^2 \mu_r \mu_o A}{l} = \mu_r \frac{N^2 \mu_o A}{l}$$

and

$$L = \mu_r L_o \qquad \textbf{(12.3)}$$

where L_o is the inductance of the coil with an air core. In other words, the inductance of a coil with a ferromagnetic core is the relative permeability of the core times the inductance achieved with an air core.

Equations for the inductance of coils different from those shown above can be found in reference handbooks. Most of the equations are more complex than those just described.

EXAMPLE 12.1. Find the inductance of the air-core coil of Fig. 12.5.
Solution:

$$\mu = \mu_r \mu_o = (1)(\mu_o) = \mu_o$$

$$A = \frac{\pi d^2}{4} = \frac{(3.1416)(4 \times 10^{-3})^2}{4} = 12.57 \times 10^{-6}\, \text{m}^2$$

$$L_o = \frac{N^2 \mu_o A}{l} = \frac{(100)^2 (4\pi \times 10^{-7})(12.57 \times 10^{-6})}{0.1}$$

$$= \frac{(10^4)(12.57)(12.57)(10^{-7})(10^{-6})}{0.1}$$

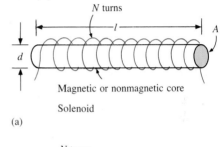

(a)

Magnetic or nonmagnetic core

Solenoid

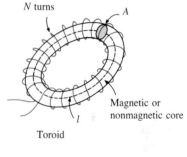

(b)

Magnetic or nonmagnetic core

Toroid

FIG. 12.4

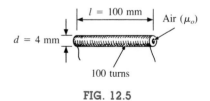

FIG. 12.5

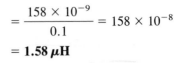

$$= \frac{158 \times 10^{-9}}{0.1} = 158 \times 10^{-8}$$

$$= \mathbf{1.58 \, \mu H}$$

Example 12.2. Repeat Example 12.1, but with an iron core and conditions such that $\mu_r = 2000$.

Solution: By Eq. (12.3),

$$L = \mu_r L_o = (2000)(1.58 \times 10^{-6}) = \mathbf{3.16 \, mH}$$

12.5 TYPES OF INDUCTORS

Associated with every inductor are a resistance equal to the resistance of the turns and a stray capacitance due to the capacitance between the turns of the coil. To include these effects, the equivalent circuit for the inductor is as shown in Fig. 12.6. However, for most applications considered in this text, the stray capacitance appearing in Fig. 12.6 can be ignored, resulting in the equivalent model of Fig. 12.7.

The resistance R_l can play an important role in the analysis of networks with inductive elements. For most applications, we have been able to treat the capacitor as an ideal element and maintain a high degree of accuracy. For the inductor, however, R_l must often be included in the analysis and can have a pronounced effect on the response of a system (see Chapter 20, Resonance). The level of R_l can extend from a few ohms to a few hundred ohms. Keep in mind that the longer or thinner the wire used in the construction of the inductor, the greater will be the dc resistance as determined by $R = \rho l / A$. Our initial analysis will treat the inductor as an ideal element. Once a general feeling for the response of the element is established, the effects of R_l will be included.

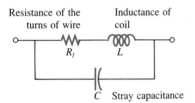

FIG. 12.6

Complete equivalent model for an inductor.

FIG. 12.7

Practical equivalent model for an inductor.

The primary function of the inductor, however, is to introduce inductance—not resistance or capacitance—into the network. For this reason, the symbols employed for inductance are as shown in Fig. 12.8.

All inductors, like capacitors, can be listed under two general headings: *fixed* and *variable*. The fixed air-core and iron-core inductors were described in the last section. The permeability-tuned variable coil has a ferromagnetic shaft that can be moved within the coil to vary the flux linkages of the coil and thereby its inductance. Several fixed and variable inductors appear in Fig. 12.9.

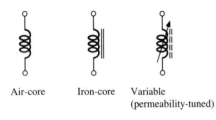

Air-core   Iron-core   Variable
(permeability-tuned)

FIG. 12.8

12.6 INDUCED VOLTAGE

The inductance of a coil is also a measure of the change in flux linking a coil due to a change in current through the coil; that is,

$$\boxed{L = N\frac{d\phi}{di}} \qquad \text{(H)} \qquad \textbf{(12.4)}$$

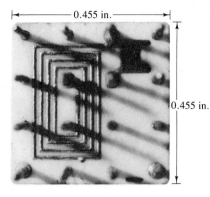

(a) Inductor and resistor on a module

(b) 1.0 H at 8 A, 8 kV working voltage

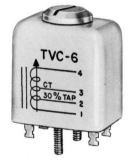

(c) Variable inductor, 0.2–2 H

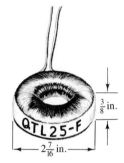

(d) Toroidal inductor, 25 mH. Typical
distributed capacitance 127 pF

(e) Molded inductors, 0.022–10,000 μH

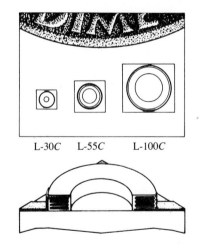

L-30C   L-55C   L-100C

(f) Microchip inductors. L-30C (30-mil outside diameter,
2–56 μH); L-55C (55-mil outside diameter, up to 250 μH);
and L-100C (100-mil outside diameter, up to 500 μH)

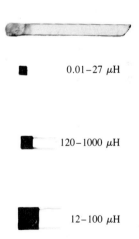

0.01–27 μH

120–1000 μH

12–100 μH

(g) Micro-i® chip inductors

FIG. 12.9

*Various types of inductors. (Part (a) courtesy of International Business
Machines Corp.; part (b) courtesy of Basler Electric Co.; part (c) courtesy
of United Transformer Corp.; part (d) courtesy of Microtan Company, Inc.;
part (e) courtesy of Delevan, Division of American Precision Industries,
Inc.; part (f) courtesy of Thinco Division, Hull Corp.; part (g) courtesy of
Delevan, Division of American Precision Industries, Inc.)*

where N is the number of turns, ϕ is the flux in webers, and i is the current through the coil. The equation states that the larger the inductance of a coil (with N fixed), the larger will be the instantaneous change in flux linking the coil due to an instantaneous change in current through the coil.

If we write Eq. (12.1) as

$$e_L = N\frac{d\phi}{dt} = \left(N\frac{d\phi}{di}\right)\left(\frac{di}{dt}\right)$$

and substitute Eq. (12.4), we then have

$$\boxed{e_L = L\frac{di}{dt}} \qquad \text{(V)} \qquad\qquad \textbf{(12.5)}$$

revealing that the magnitude of the voltage across an inductor is directly related to the inductance L and the instantaneous rate of change of current through the coil. Obviously, therefore, the greater the *rate* of change of current through the coil, the greater will be the induced voltage. This certainly agrees with our earlier discussion of Lenz's law.

When induced effects are employed in the generation of voltages such as available from dc or ac generators, the symbol e is appropriate for the induced voltage. However, in network analysis the voltage across an inductor will always have a polarity such as to oppose the source that produced it, and therefore the following notation will be used throughout the analysis to come:

$$\boxed{v_L = L\frac{di}{dt}} \qquad\qquad \textbf{(12.6)}$$

If the current through the coil fails to change at a particular instant, the induced voltage across the coil will be zero. For dc applications, after the transient effect has passed, $di/dt = 0$, and the induced voltage is

$$v_L = L\frac{di}{dt} = L(0) = 0\text{ V}$$

Recall that the equation for the current of a capacitor is the following:

$$i_C = C\frac{dv_C}{dt}$$

Note the similarity between this equation and Eq. (12.6). In fact, if we apply the duality $v \rightleftarrows i$ (that is, interchange the two) and $L \rightleftarrows C$ for capacitance and inductance, each equation can be derived from the other.

The average voltage across the coil is defined by the equation

$$\boxed{v_{L_{av}} = L\frac{\Delta i}{\Delta t}} \qquad \text{(V)} \qquad\qquad \textbf{(12.7)}$$

where Δ signifies finite change (a measurable change). Compare this to $i_C = C(\Delta v / \Delta t)$, and the meaning of Δ and application of this equation should be clarified from Chapter 10. An example follows.

EXAMPLE 12.3.   Find the waveform for the average voltage across the coil if the current through a 4-mH coil is as shown in Fig. 12.10.

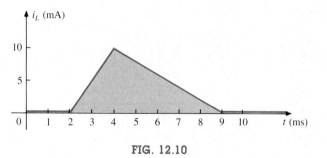

FIG. 12.10

Solution:

a. 0 to 2 ms: Since there is no change in current through the coil, there is no voltage induced across the coil; that is,

$$v_L = L\frac{\Delta i}{\Delta t} = L\frac{0}{\Delta t} = 0$$

b. 2 ms to 4 ms:

$$v_L = L\frac{\Delta i}{\Delta t} = (4 \times 10^{-3})\left(\frac{10 \times 10^{-3}}{2 \times 10^{-3}}\right) = 20 \times 10^{-3}$$

$$= 20\,\text{mV}$$

c. 4 ms to 9 ms:

$$v_L = L\frac{\Delta i}{\Delta t} = (-4 \times 10^{-3})\left(\frac{10 \times 10^{-3}}{5 \times 10^{-3}}\right) = -8 \times 10^{-3}$$

$$= -8\,\text{mV}$$

d. 9 ms to ∞:

$$v_L = L\frac{\Delta i}{\Delta t} = L\frac{0}{\Delta t} = 0$$

The waveform for the average voltage across the coil is shown in Fig. 12.11. Note from the curve that *the voltage across the coil is not*

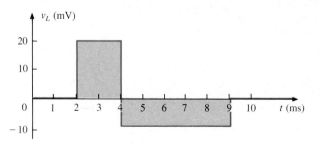

FIG. 12.11

determined by the magnitude of the change in current through the coil (Δi), *but by the* **rate** *of change of current through the coil* ($\Delta i/\Delta t$). A similar statement was made for the current of a capacitor due to a change in voltage across the capacitor.

A careful examination of Fig. 12.11 will also reveal that the area under the positive pulse from 2 ms to 4 ms equals the area under the negative pulse from 4 ms to 9 ms. In Section 12.13, we will find that the area under the curves represents the energy stored or released by the inductor. From 2 ms to 4 ms, the inductor is storing energy, while from 4 ms to 9 ms, the inductor is releasing the energy stored. For the full period zero to 10 ms, energy has simply been stored and released; there has been no dissipation as experienced for the resistive elements. Over a full cycle, both the ideal capacitor and inductor do not consume energy but simply store and release it in their respective forms.

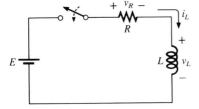

FIG. 12.12

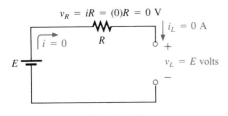

FIG. 12.13

Circuit of Fig. 12.12 the instant the switch is closed.

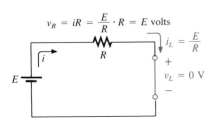

FIG. 12.14

Circuit of Fig. 12.12 under steady-state conditions.

12.7 *R-L* TRANSIENTS: STORAGE CYCLE

The changing voltages and current that result during the storing of energy in the form of a magnetic field by an inductor in a dc circuit can best be described using the network of Fig. 12.12. At the instant the switch is closed, the inductance of the coil will prevent an instantaneous change in current through the coil. The potential drop across the coil, v_L, will equal the impressed voltage E as determined by Kirchhoff's voltage law since $v_R = iR = (0)R = 0$ V. The current i_L will then build up from zero, establishing a voltage drop across the resistor and a corresponding drop in v_L. The current will continue to increase until the voltage across the inductor drops to zero volts and the full impressed voltage appears across the resistor. Initially, the current i_L increases quite rapidly, followed by a continually decreasing rate until it reaches its maximum value of E/R.

You will recall from the discussion of capacitors that a capacitor has a short-circuit equivalent when the switch is first closed and an open-circuit equivalent when steady-state conditions are established. The inductor assumes the opposite equivalents for each stage. The instant the switch of Fig. 12.12 is closed, the equivalent network will appear as shown in Fig. 12.13. Note the correspondence with the earlier comments regarding the levels of voltage and current. The inductor obviously meets all the requirements for an open-circuit equivalent—$v_L = E$ volts, $i_L = 0$ A.

When steady-state conditions have been established and the storage phase is complete, the "equivalent" network will appear as shown in Fig. 12.14. The network clearly reveals that:

An ideal inductor assumes a short-circuit equivalent in a dc network once steady-state conditions have been established.

Fortunately, the mathematical equations for the voltages and current for the storage phase are similar in many respects to those encountered for the *R-C* network. The experience gained with these equations in Chapter 10 will undoubtedly make the analysis of *R-L* networks somewhat easier to understand.

The equation for the current i_L during the storage phase is the following:

$$i_L = I_m(1 - e^{-t/\tau}) = \frac{E}{R}(1 - e^{-t/(L/R)}) \qquad \textbf{(12.8)}$$

Note the factor $(1 - e^{-t/\tau})$, which also appeared for the voltage v_C of a capacitor during the charging phase. A plot of the equation is given in Fig. 12.15, clearly indicating that the maximum steady-state value of i_L

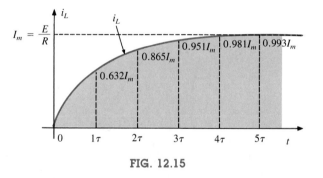

FIG. 12.15

is E/R, and that the rate of change in current decreases as time passes. The abscissa is scaled in time constants, with τ for inductive circuits defined by the following:

$$\tau = \frac{L}{R} \qquad \text{(seconds, s)} \qquad \textbf{(12.9)}$$

The fact that τ has the units of time can be verified by taking the equation for the induced voltage

$$v_L = L\frac{di}{dt}$$

and solving for *L*:

$$L = \frac{v_L}{di/dt}$$

which leads to the ratio

$$\tau = \frac{L}{R} = \frac{\dfrac{v_L}{di/dt}}{R} = \frac{v_L}{\dfrac{di}{dt}R} \Rightarrow \frac{V}{\dfrac{IR}{t}}$$

$$= \frac{\not{V}}{\dfrac{\not{V}}{t}} = t \text{ (seconds)}$$

Our experience with the factor $(1 - e^{-t/\tau})$ verifies the level of 63.2% after one time constant, 86.5% after two time constants, and so on. For convenience, Fig. 10.26 is repeated as Fig. 12.16 to evaluate the functions $(1 - e^{-t/\tau})$ and $e^{-t/\tau}$ at various values of τ.

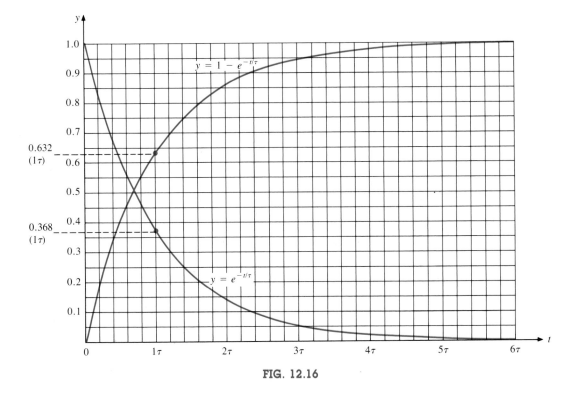

FIG. 12.16

If we keep R constant and increase L, the ratio L/R increases and the rise time increases. The change in transient behavior for the current i_L is plotted in Fig. 12.17 for various values of L. Note again the duality between these curves and those obtained for the R-C network in Fig. 10.29.

For most practical applications, we will assume that:

The storage phase has passed and steady-state conditions have been established once a period of time equal to five time constants has occurred.

In addition, since L/R will always have some numerical value even though it may be very small, the period 5τ will always be greater than zero, confirming the fact that *the current cannot change instantaneously in an inductive network.* In fact, the larger the inductance, the more the circuit will oppose a rapid buildup in current level.

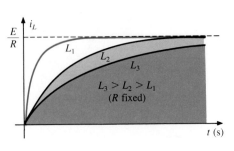

FIG. 12.17

Figures 12.13 and 12.14 clearly reveal that the voltage across the coil jumps to E volts when the switch is closed and decays to zero volts with time. The decay occurs in an exponential manner, and v_L during the storage phase can be described mathematically by the following equation:

$$v_L = Ee^{-t/\tau}$$ (12.10)

A plot of v_L appears in Fig. 12.18 with the time axis again divided into equal increments of τ. Obviously, the voltage v_L will decrease to

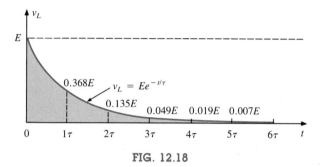

FIG. 12.18

zero volts at the same rate the current presses toward its maximum value.

In five time constants, $i_L = E/R$, $v_L = 0\,V$, and the inductor can be replaced by its short-circuit equivalent.

Since

$$v_R = i_R R = i_L R$$

then

$$v_R = \left[\frac{E}{R}(1 - e^{-t/\tau}) \right] R$$

and

$$v_R = E(1 - e^{-t/\tau})$$ (12.11)

and the curve for v_R will have the same shape as obtained for i_L.

EXAMPLE 12.4. Find the mathematical expressions for the transient behavior of i_L and v_L for the circuit of Fig. 12.19 after the closing of the switch. Sketch the resulting curves.

Solution:

$$\tau = \frac{L}{R_1} = \frac{4}{2\,k\Omega} = 2\,ms$$

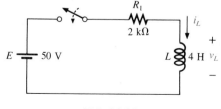

FIG. 12.19

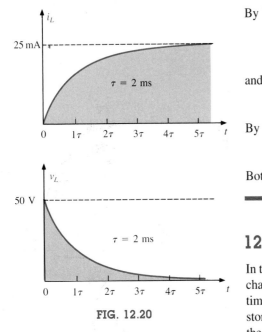

FIG. 12.20

By Eq. (12.8),

$$I_m = \frac{E}{R_1} = \frac{50}{2\,k\Omega} = 25 \times 10^{-3}\,\text{A} = 25\,\text{mA}$$

and

$$i_L(t) = (25 \times 10^{-3})(1 - e^{-t/(2 \times 10^{-3})})$$

By Eq. (12.10),

$$v_L(t) = 50e^{-t/(2 \times 10^{-3})}$$

Both waveforms appear in Fig. 12.20.

12.8 *R-L* TRANSIENTS: DECAY PHASE

In the analysis of *R-C* circuits, we found that the capacitor could hold its charge and store energy in the form of an electric field for a period of time determined by the leakage factors. In *R-L* circuits, the energy is stored in the form of a magnetic field established by the current through the coil. Unlike the capacitor, however, an isolated inductor cannot continue to store energy since the absence of a closed path would cause the current to drop to zero, releasing the energy stored in the form of a magnetic field. If the switch of Fig. 12.12 were opened quickly, a spark would probably occur across the contacts due to the rapid change in current from a maximum of *E/R* to zero amperes. The change in current *di/dt* of the equation $v_L = L(di/dt)$ would establish a high voltage v_L across the coil that would discharge across the points of the switch. This is the same mechanism as applied in the ignition system of a car to ignite the fuel in the cylinder. Some 25,000 volts are generated by the rapid decrease in ignition coil current that occurs when the switch in the system is opened. (In older systems, the "points" in the distributor served as the switch.) This inductive reaction is significant when you consider that the only independent source in a car is a 12-V battery.

If opening the switch to move it to another position will cause such a rapid discharge in stored energy, how can the decay phase of an *R-L* circuit be analyzed in much the same manner as for the *R-C* circuit? The solution is to use a network such as that appearing in Fig. 12.21. When the switch is closed, the voltage across the resistor R_2 is *E* volts and the *R-L* branch will respond in the same manner as described above, with the same waveforms and levels. A Thevenin network of *E* in parallel with R_2 would simply result in the source since R_2 would be shorted out by the short-circuit replacement of the voltage source *E* when the Thevenin resistance is determined.

After the storage phase has passed and steady-state conditions are established, the switch can be opened without the sparking effect or rapid discharge due to the resistor R_2, which provides a complete path for the current i_L. In fact, for clarity the discharge path is isolated in Fig. 12.22. The voltage v_L across the inductor will reverse polarity and have a magnitude determined by

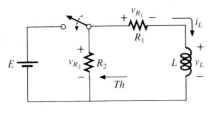

FIG. 12.21

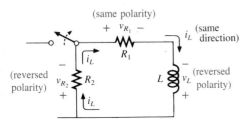

FIG. 12.22

Network of Fig. 12.21 the instant the switch is opened.

$$v_L = v_{R_1} + v_{R_2} \qquad (12.12)$$

Recall that the voltage across an inductor can change instantaneously but the current cannot. The result is that the current i_L must maintain the same direction and magnitude as shown in Fig. 12.22. Therefore, the instant after the switch is opened, i_L is still $I_m = E/R_1$, and

$$v_L = v_{R_1} + v_{R_2} = i_1 R_1 + i_2 R_2$$

$$= i_L(R_1 + R_2) = \frac{E}{R_1}(R_1 + R_2) = \left(\frac{R_1}{R_1} + \frac{R_2}{R_1}\right)E$$

and

$$v_L = \left(1 + \frac{R_2}{R_1}\right)E \qquad (12.13)$$

which is bigger than E volts by the ratio R_2/R_1. In other words, when the switch is opened, the voltage across the inductor will jump instantaneously from E to $[1 + (R_2/R_1)]E$ volts, with the reverse polarity.

As the inductor releases its stored energy, the voltage across the coil will decay to zero in the following manner:

$$v_L = V_i e^{-t/\tau'} \qquad (12.14)$$

with

$$V_i = \left(1 + \frac{R_2}{R_1}\right)E$$

and

$$\tau' = \frac{L}{R_T} = \frac{L}{R_1 + R_2}$$

The current will decay from a maximum of $I_m = E/R_1$ to zero, in the following manner:

$$i_L = I_m e^{-t/\tau'} \qquad (12.15)$$

with

$$I_m = \frac{E}{R_1} \quad \text{and} \quad \tau' = \frac{L}{R_1 + R_2}$$

The mathematical expression for the voltage across either resistor can then be determined using Ohm's law:

$$v_{R_1} = i_{R_1} R_1 = i_L R_1$$
$$= I_m e^{-t/\tau'} R_1$$
$$= \frac{E}{R_1} R_1 e^{-t/\tau'}$$

and

$$\boxed{v_{R_1} = Ee^{-t/\tau'}} \qquad (12.16)$$

The voltage v_{R_1} has the same polarity as during the storage phase since the current i_L has the same direction. The voltage v_{R_2} is expressed as follows:

$$\begin{aligned} v_{R_2} &= i_{R_2}R_2 = i_L R_2 \\ &= I_m e^{-t/\tau'} R_2 \\ &= \frac{E}{R_1} R_2 e^{-t/\tau'} \end{aligned}$$

and

$$\boxed{v_{R_2} = \frac{R_2}{R_1} Ee^{-t/\tau'}} \qquad (12.17)$$

with the polarity indicated in Fig. 12.22.

EXAMPLE 12.5. The resistor R_2 was added to the network of Fig. 12.19 as shown in Fig. 12.23.

a. Find the mathematical expressions for i_L, v_L, v_{R_1}, and v_{R_2} after the storage phase has been completed and the switch is opened.

b. Sketch the waveforms for each voltage and current for both phases covered by this example and Example 12.4 if five time constants pass between phases. Use the defined polarities of Fig. 12.21.

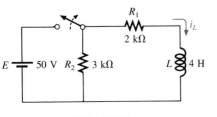

FIG. 12.23

Solutions:

a. $\tau' = \dfrac{L}{R_1 + R_2} = \dfrac{4}{2\,\text{k}\Omega + 3\,\text{k}\Omega} = \dfrac{4}{5 \times 10^3} = 0.8 \times 10^{-3}$

$ = 0.8\,\text{ms}$

By Eq. (12.14),

$$V_i = \left(1 + \frac{R_2}{R_1}\right)E = \left(1 + \frac{3\,\text{k}\Omega}{2\,\text{k}\Omega}\right)(50) = 125\,\text{V}$$

and

$$v_L = V_i e^{-t/\tau'} = \mathbf{125}e^{-t/(0.8 \times 10^{-3})}$$

By Eq. (12.15),

$$I_m = \frac{E}{R_1} = \frac{50}{2\,\text{k}\Omega} = 25\,\text{mA}$$

and

$$i_L = I_m e^{-t/\tau'} = \mathbf{(25 \times 10^{-3})}e^{-t/(0.8 \times 10^{-3})}$$

By Eq. (12.16),

$$v_{R_1} = Ee^{-t/\tau'} = \mathbf{50}e^{-t/(0.8 \times 10^{-3})}$$

By Eq. (12.17),

$$v_{R_2} = \frac{R_2}{R_1}Ee^{-t/\tau'} = \frac{3\text{ k}\Omega}{2\text{ k}\Omega}(50)e^{-t/\tau'} = \mathbf{75}e^{-t/(0.8\times10^{-3})}$$

b. See Fig. 12.24.

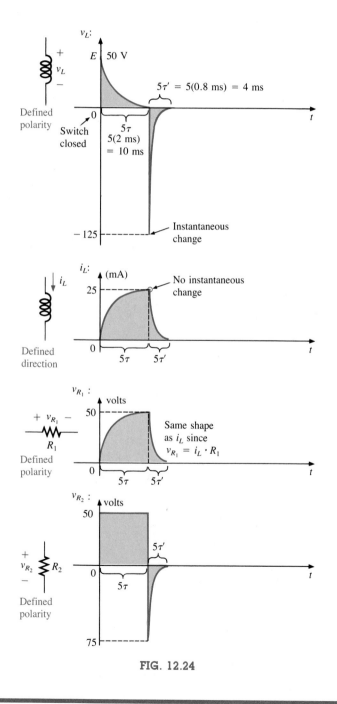

FIG. 12.24

In the analysis above, it was assumed that steady-state conditions were established during the charging phase and $I_m = E/R_1$, with $v_L = 0$ V. However, if the switch of Fig. 12.22 is opened before i_L reaches its maximum value, the equation for the decaying current of Fig. 12.22 must change to

$$i_L = I_i e^{-t/\tau'}$$ **(12.18)**

where I_i is the starting or *i*nitial current. Equation (12.14) would be modified as follows:

$$v_L = V_i e^{-t/\tau'}$$ **(12.19)**

with

$$V_i = I_i(R_1 + R_2)$$

12.9 INSTANTANEOUS VALUES

The development presented in Section 10.9 for capacitive networks can also be applied to *R-L* networks to determine instantaneous voltages, currents, and time. The instantaneous values of any voltage or current can be determined by simply inserting t into the equation and using a calculator or table to determine the magnitude of the exponential term.

The similarity between the equations $v_C = E(1 - e^{-t/\tau})$ and $i_L = I_m(1 - e^{-t/\tau})$ results in a derivation of the following for t which is identical to that used to obtain Eq. (10.21):

$$t = -\tau \log_e\left(1 - \frac{i_L}{I_m}\right)$$ **(12.20)**

For the other form, the equation $v_C = Ee^{-t/\tau}$ is a close match with $v_L = Ee^{-t/\tau}$, permitting a derivation similar to that employed for Eq. (10.22):

$$t = -\tau \log_e \frac{v_L}{E}$$ **(12.21)**

The similarities between the above and the equations in Chapter 10 should make the equation for t fairly easy to obtain.

12.10 $\tau = L/R_{Th}$

In Chapter 10 (Capacitors), we found that there are occasions when the circuit does not have the basic form of Fig. 12.12. The same is true for inductive networks. Again, it is necessary to find the Thevenin equivalent circuit before proceeding in the manner described in this chapter. Consider the following example.

EXAMPLE 12.6.   For the network of Fig. 12.25:
a. Find the mathematical expression for the transient behavior of the current i_L and the voltage v_L after the closing of the switch.
b. Draw the resultant waveform for each.

Solutions:
a. Applying Thevenin's theorem to the 80-mH inductor (Fig. 12.26) yields

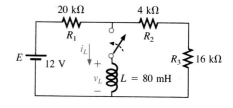

FIG. 12.25

$$R_{Th} = \frac{R}{N} = \frac{20\,k\Omega}{2} = 10\,k\Omega$$

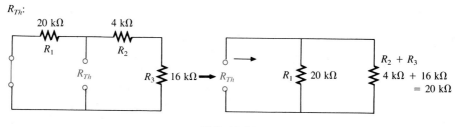

FIG. 12.26

Applying the voltage divider rule (Fig. 12.27),

$$E_{Th} = \frac{(R_2 + R_3)E}{R_1 + R_2 + R_3}$$

$$= \frac{(4\,k\Omega + 16\,k\Omega)(12)}{20\,k\Omega + 4\,k\Omega + 16\,k\Omega} = \frac{(20)(12)}{40} = 6\,V$$

The Thevenin equivalent circuit is shown in Fig. 12.28. Using Eq. (12.8),

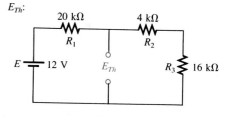

FIG. 12.27

$$i_L(t) = \frac{E_{Th}}{R}(1 - e^{-t/\tau})$$

$$\tau = \frac{L}{R_{Th}} = \frac{80 \times 10^{-3}}{10 \times 10^3} = 8 \times 10^{-6}\,s$$

$$I_m = \frac{E_{Th}}{R_{Th}} = \frac{6}{10 \times 10^3} = 0.6 \times 10^{-3}\,A$$

and

$$i_L(t) = (\mathbf{0.6 \times 10^{-3}})(\mathbf{1 - e^{-t/(8 \times 10^{-6})}})$$

Using Eq. (12.10),

$$v_L(t) = E_{Th}e^{-t/\tau}$$

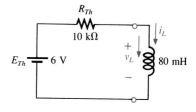

Thevenin equivalent circuit:

FIG. 12.28

so that

$$v_L(t) = \mathbf{6}e^{-t/(8 \times 10^{-6})}$$

b. See Fig. 12.29.

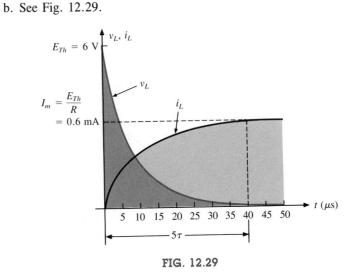

FIG. 12.29

12.11 INDUCTORS IN SERIES AND PARALLEL

Inductors, like resistors and capacitors, can be placed in series or parallel. Increasing levels of inductance can be obtained by placing inductors in series, while decreasing levels can be obtained by placing inductors in parallel.

For inductors in series, the total inductance is found in the same manner as the total resistance of resistors in series (Fig. 12.30):

FIG. 12.30

$$\boxed{L_T = L_1 + L_2 + L_3 + \cdots + L_N} \tag{12.22}$$

For inductors in parallel, the total inductance is found in the same manner as the total resistance of resistors in parallel (Fig. 12.31):

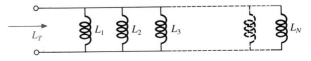

FIG. 12.31

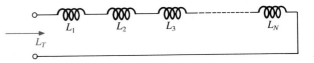

$$\frac{1}{L_T} = \frac{1}{L_1} + \frac{1}{L_2} + \frac{1}{L_3} + \cdots + \frac{1}{L_N} \qquad (12.23)$$

For two inductors in parallel,

$$L_T = \frac{L_1 L_2}{L_1 + L_2} \qquad (12.24)$$

12.12 *R-L* AND *R-L-C* CIRCUITS WITH dc INPUTS

We found in Section 12.7 that for all practical purposes, an inductor can be replaced by a short circuit in a dc circuit after a period of time greater than five time constants has passed. If in the following circuits we assume that all of the currents and voltages have reached their final values, the current through each inductor can be found by replacing each inductor by a short circuit. For the circuit of Fig. 12.32, for example,

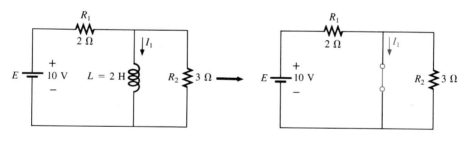

FIG. 12.32

$$I_1 = \frac{E}{R_1} = \frac{10}{2} = 5\,\text{A}$$

For the circuit of Fig. 12.33,

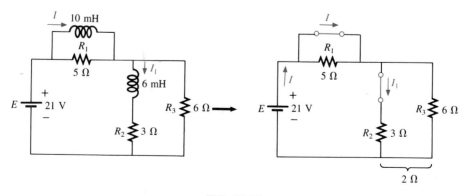

FIG. 12.33

$$I = \frac{E}{R_2 \| R_3} = \frac{21}{2} = 10.5 \, \text{A}$$

Applying the current divider rule,

$$I_1 = \frac{R_3 I}{R_3 + R_2} = \frac{(6)(10.5)}{6 + 3} = \frac{63}{9} = 7 \, \text{A}$$

In the following examples we will assume that the voltage across the capacitors and the current through the inductors have reached their final values. Under these conditions, the inductors can be replaced by short circuits, and the capacitors by open circuits.

EXAMPLE 12.7. Find the current I_L and the voltage V_C for the network of Fig. 12.34.

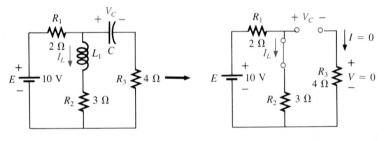

FIG. 12.34

Solution:

$$I_L = \frac{E}{R_1 + R_2} = \frac{10}{5} = 2 \, \text{A}$$

$$V_C = \frac{R_2 E}{R_2 + R_1} = \frac{(3)(10)}{3 + 2} = 6 \, \text{V}$$

EXAMPLE 12.8. Find the currents I_1 and I_2 and the voltages V_1 and V_2 for the network of Fig. 12.35.

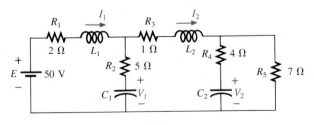

FIG. 12.35

Solution: Note Fig. 12.36:

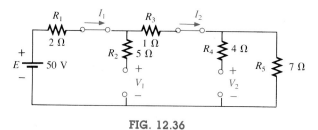

FIG. 12.36

$I_1 = I_2$

$$I_1 = \frac{E}{R_1 + R_3 + R_5} = \frac{50}{2 + 1 + 7} = \frac{50}{10} = \mathbf{5\ A}$$

$V_2 = I_2 R_5 = (5)(7) = \mathbf{35\ V}$

Applying the voltage divider rule,

$$V_1 = \frac{(R_3 + R_5)E}{R_1 + R_3 + R_5} = \frac{(1 + 7)(50)}{2 + 1 + 7} = \frac{(8)(50)}{10} = \mathbf{40\ V}$$

12.13 ENERGY STORED BY AN INDUCTOR

The ideal inductor, like the ideal capacitor, does not dissipate the electrical energy supplied to it. It stores the energy in the form of a magnetic field. A plot of the voltage, current, and power to an inductor is shown in Fig. 12.37 during the buildup of the magnetic field surrounding the

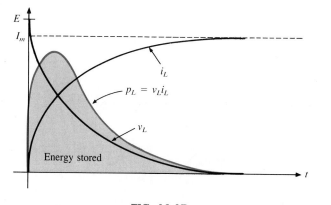

FIG. 12.37

inductor. The energy stored is represented by the shaded area under the power curve. Using calculus, we can show that the evaluation of the area under the curve yields

$$W_{\text{stored}} = \frac{1}{2}LI_m^2 \qquad \text{(joules, J)} \qquad \mathbf{(12.25)}$$

EXAMPLE 12.9. Find the energy stored by the inductor in the circuit of Fig. 12.38 when the current through it has reached its final value.

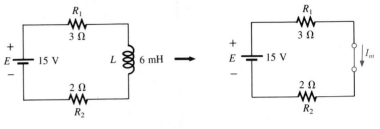

FIG. 12.38

Solution:

$$I_m = \frac{E}{R_1 + R_2} = \frac{15}{3 + 2} = \frac{15}{5} = 3 \text{ A}$$

$$W_{\text{stored}} = \frac{1}{2}LI_m^2 = \frac{1}{2}(6 \times 10^{-3})(3)^2 = \frac{54}{2} \times 10^{-3}$$

$$= 27 \times 10^{-3} \text{ J}$$

PROBLEMS

Section 12.2

1. If the flux linking a coil of 50 turns changes at a rate of 0.085 Wb/s, what is the induced voltage across the coil?

2. Determine the rate of change of flux linking a coil if 20 V are induced across a coil of 40 turns.

3. How many turns does a coil have if 42 mV are induced across the coil by a change of flux of 0.003 Wb/s?

Section 12.4

4. Find the inductance L in henries of the inductor of Fig. 12.39.

5. Repeat Problem 4 with $l = 4$ in. and $d = 0.25$ in.

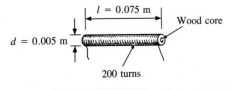

FIG. 12.39

6. **a.** Find the inductance L in henries of the inductor of Fig. 12.40.

   b. Repeat part (a) if a ferromagnetic core is added having a μ_r of 2000.

Section 12.6

7. Find the voltage induced across a coil of 5 H if the rate of change of current through the coil is
   a. 0.5 A/s
   b. 60 mA/s
   c. 0.04 A/ms

8. Find the induced voltage across a 50-mH inductor if the current through the coil changes at a rate of 0.1 mA/μs.

9. Find the waveform for the voltage induced across a 200-mH coil if the current through the coil is as shown in Fig. 12.41.

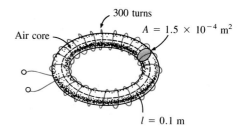

300 turns
$A = 1.5 \times 10^{-4}$ m^2

Air core

$l = 0.1$ m

FIG. 12.40

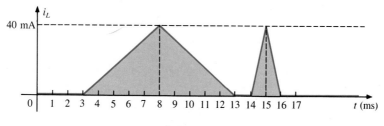

FIG. 12.41

10. Repeat Problem 9 for the waveform of Fig. 12.42.

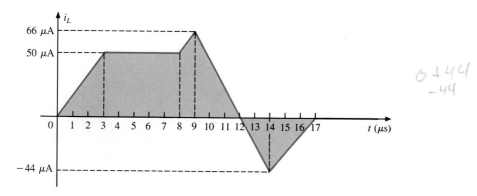

$0 + 44$
-44

FIG. 12.42

Section 12.7

11. For the circuit of Fig. 12.43:
   a. Determine the time constant.
   b. Write the mathematical expression for the current i_L after the switch is closed.
   c. Repeat part (b) for v_L and v_R.

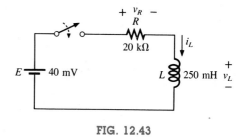

$+$ v_R $-$
R
20 kΩ
i_L

E ⎓ 40 mV

L 250 mH v_L $+$ $-$

FIG. 12.43

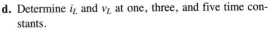

d. Determine i_L and v_L at one, three, and five time constants.
e. Sketch the waveforms of i_L, v_L, and v_R.

12. Repeat Problem 11 for the network of Fig. 12.44.

+ 12 V

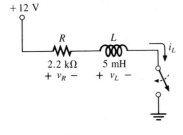

R   L
2.2 kΩ   5 mH
+ v_R −   + v_L −

i_L

FIG. 12.44

Section 12.8

13. For the network of Fig. 12.45:
   a. Determine the mathematical expressions for the current i_L and the voltage v_L when the switch is closed.
   b. Repeat part (a) if the switch is opened after a period of five time constants has passed.
   c. Sketch the waveforms of parts (a) and (b) on the same axis.

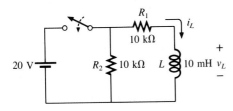

R_1
10 kΩ
i_L

20 V R_2 10 kΩ   L 10 mH v_L +   −

FIG. 12.45

14. a. Repeat Problem 13 for the network of Fig. 12.46.
   b. Sketch the waveform for the voltage across R_2 for the same period of time encompassed by i_L and v_L.

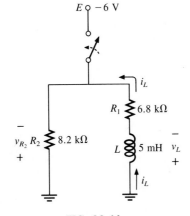

E − 6 V

i_L

R_1 6.8 kΩ

v_{R_2} R_2 8.2 kΩ

L 5 mH v_L

i_L

FIG. 12.46

15. For the network of Fig. 12.47:
   a. Determine the mathematical expressions for the current i_L and the voltage v_L following the closing of the switch.
   b. Repeat part (a) if the switch is opened at $t = 1$ μs.
   c. Sketch the waveforms of parts (a) and (b) on the same axis.

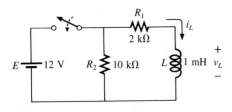

R_1
2 kΩ
i_L

E 12 V R_2 10 kΩ   L 1 mH v_L +   −

FIG. 12.47

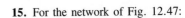

Sections 12.9 and 12.10

16. **a.** Determine the mathematical expressions for i_L and v_L following the closing of the switch in Fig. 12.48.

b. Determine i_L and v_L at $t = 100\,\mu s$.

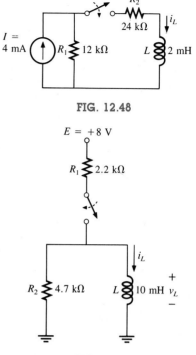

FIG. 12.48

17. **a.** Determine the mathematical expressions for i_L and v_L following the closing of the switch in Fig. 12.49.

b. Calculate i_L and v_L at $t = 10\,\mu s$.

c. Write the mathematical expressions for the current i_L and the voltage v_L if the switch is opened at $t = 10\,\mu s$.

d. Sketch the waveforms of i_L and v_L for parts (a) and (c).

Section 12.11

18. Find the total inductance of the circuits of Fig. 12.50.

FIG. 12.49

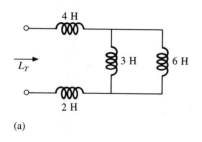

(a)

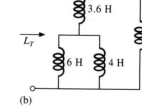

(b)

FIG. 12.50

19. Reduce the networks of Fig. 12.51 to the fewest elements.

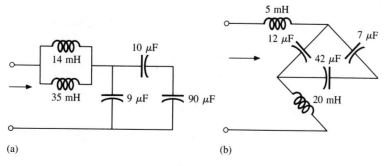

(a)

(b)

FIG. 12.51

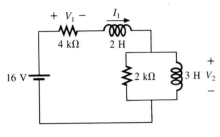

FIG. 12.52

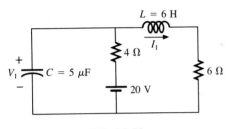

FIG. 12.53

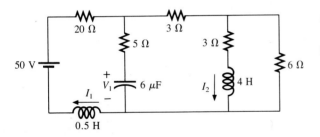

FIG. 12.54

Section 12.12

For Problems 20 through 22, assume that the voltage across each capacitor and the current through each inductor have reached their final values.

20. Find the voltages V_1 and V_2 and the current I_1 for the circuit of Fig. 12.52.

21. Find the current I_1 and the voltage V_1 for the circuit of Fig. 12.53.

22. Find the voltage V_1 and the current through each inductor in the circuit of Fig. 12.54.

Section 12.13

23. Find the energy stored in each inductor of Problem 20.

24. Find the energy stored in the capacitor and inductor of Problem 21.

25. Find the energy stored in each inductor of Problem 22.

Computer Problems

26. Write a program to provide a general solution for the circuit of Fig. 12.12. That is, given the network parameters, generate the equations for i_L, v_L, and v_R.

27. Write a program that will provide a general solution for the storage and decay phase of the network of Fig. 12.45. That is, given the network values, generate the equations for i_L and v_L for each phase. In this case, assume the storage phase has passed through five time constants before the decay phase begins.

28. Repeat Problem 27 but assume the storage phase was not completed, requiring that the instantaneous values of i_L and v_L be determined when the switch is opened.

GLOSSARY

Choke A term often applied to an inductor, due to the ability of an inductor to resist a change in current through it.

Faraday's law A law relating the voltage induced across a coil to the number of turns in the coil and the rate at which the flux linking the coil is changing.

Inductor A fundamental element of electrical systems constructed of numerous turns of wire around a ferromagnetic or air core.

Lenz's law A law stating that an induced effect is always such as to oppose the cause that produced it.

Self-inductance A measure of the ability of a coil to oppose any change in current through the coil and to store energy in the form of a magnetic field in the region surrounding the coil.

13

Sinusoidal Alternating Waveforms

13.1 INTRODUCTION

The analysis thus far has been limited to dc networks, networks in which the currents or voltages are fixed in magnitude except for transient effects. We will now turn our attention to the analysis of networks in which the magnitude of the source varies in a set manner. Of particular interest is the time-varying voltage that is commercially available in large quantities and is commonly called the *ac voltage*. (The letters *ac* are an abbreviation for *alternating current*.) To be absolutely rigorous, the terminology *ac voltage* or *ac current* is not sufficient to describe the type of signal we will be analyzing. Each waveform of Fig. 13.1 is an

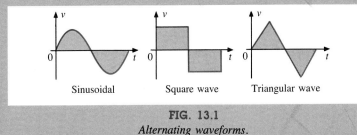

Sinusoidal   Square wave   Triangular wave

FIG. 13.1
Alternating waveforms.

alternating waveform available from commercial supplies. The term *alternating* indicates only that the waveform alternates between two prescribed levels in a set time sequence (Fig. 13.1). To be absolutely correct, the term *sinusoidal*, *square wave*, or *triangular* must also be applied. The pattern of particular interest is the *sinusoidal* ac voltage of Fig. 13.1. Since this type of signal is encountered in the vast majority of instances, the abbreviated phrases *ac voltage* and *ac current* are commonly applied without confusion. For the other patterns of Fig. 13.1, the descriptive term is always present, but frequently the *ac* abbreviation is dropped, resulting in the designation *square-wave* or *triangular* waveforms.

One of the important reasons for concentrating on the sinusoidal ac voltage is that it is the voltage generated by utilities throughout the world. Other reasons include its application throughout electrical, electronic, communication, and industrial systems. In addition, the chapters to follow will reveal that the waveform itself has a number of characteristics that will result in a unique response when it is applied to the basic electrical elements. The wide range of theorems and methods introduced for dc networks will also be applied to sinusoidal ac systems. Although the application of sinusoidal signals will raise the required math level, once the notation given in Chapter 14 is understood, most of the concepts introduced in the dc chapters can be applied to ac networks with a minimum of added difficulty.

The increasing number of computer systems used in the industrial community requires, at the very least, a brief introduction to the terminology employed with pulse waveforms and the response of some fundamental configurations to the application of such signals. Chapter 21 will serve such a purpose.

13.2 SINUSOIDAL ac VOLTAGE GENERATION

The characteristics of the sinusoidal voltage and current and their effect on the basic R, L, and C elements will be described in some detail in this chapter and those to follow. Of immediate interest is the generation of sinusoidal voltages.

The terminology *ac generator* or *alternator* should not be new to most technically oriented students. It is an electromechanical device capable of converting mechanical power to electrical power. As shown in the very basic ac generator of Fig. 13.2, it is constructed of two main components: the *rotor* (or armature, in this case) and the *stator*. As implied by the terminology, the rotor rotates within the framework of the stator, which is stationary. When the rotor is caused to rotate by some mechanical power such as is available from the forces of rushing water (dams) or steam-turbine engines, the conductors on the rotor will cut magnetic lines of force established by the poles of the stator, as shown in Fig. 13.2. The poles may be those of a permanent magnet or may result from the turns of wire around the ferromagnetic core of the pole through which a dc current is passed to establish the necessary magnetomotive force for the required flux density.

Dictated by Eq. (12.1), the length of conductor passing through the magnetic field will have a voltage induced across it as shown in Fig. 13.2. Note that the induced voltage across each conductor is additive, so the generated terminal voltage is the sum of the two induced voltages. Since the armature of Fig. 13.2 is rotating, and the output termi-

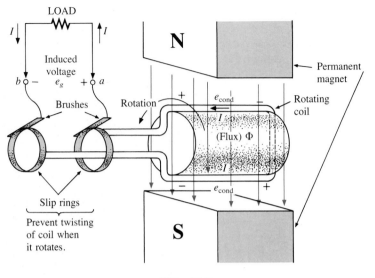

FIG. 13.2

nals a and b are connected to some fixed external load, there is the necessity for the induced *slip rings*. The slip rings are circular conducting surfaces that provide a path of conduction from the generated voltage to the load and prevent a twisting of the coil at a and b when the coil rotates. The induced voltage will have a polarity at terminals a and b and will develop a current I having the direction indicated in Fig. 13.2. Note that the direction of I is also the direction of increasing induced voltage within the generator.

A method will now be described for determining the direction of the resulting current or of the increasing induced voltage. For the generator, the thumb, forefinger, and middle finger are placed at right angles, as shown in Fig. 13.3. The thumb is placed in the direction of force or motion of a conductor, the index finger in the direction of the magnetic flux lines, and the middle finger in the direction of current flow resulting in the conductor if a load is attached. If a load is not attached, the middle finger indicates the direction of increasing induced voltage. The placement of the fingers is indicated in Fig. 13.3(a) for the top

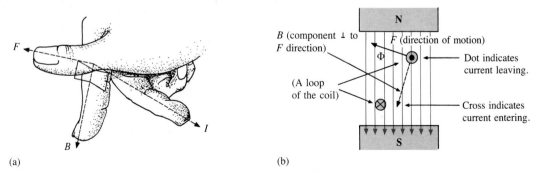

FIG. 13.3

(a) Right-hand rule; (b) current directions as determined using the right-hand rule for the indicated position of the rotating coil.

conductor of the rotor of Fig. 13.2 as it passes through the position indicated in Fig. 13.3(b). From this point on, we will assume that a load has been applied so that the current directions can be included using the dot (·)-cross (×) convention described in Chapter 11. Note that the resulting direction for the upper conductor is opposite that of the lower conductor. This is a necessary condition for the current I in the series configuration. The reversal in the direction of motion (the thumb) in this region will result in the opposite direction for I.

Let us now consider a few representative positions of the rotating coil and determine the relative magnitude and polarity of the generated voltage at these positions. At the instant the coil passes through position 1 in Fig. 13.4(a), there are no flux lines being cut, and the induced voltage is zero. As the coil moves from position 1 to position 2, indicated in Fig. 13.4(b), the number of flux lines cut per unit time will increase, resulting in an increased induced voltage across the coil. For position 2, the resulting current direction and polarity of terminals a and b are indicated as determined by the right-hand rule. At position 3, the number of flux lines being cut per unit time is a maximum, resulting in a maximum induced voltage. The polarities and current direction are the same as at position 2.

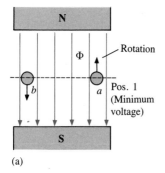
(a)

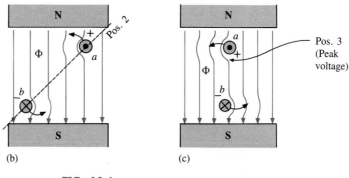
(b)   (c)

FIG. 13.4

As the coil continues to rotate toward position 4, indicated in Fig. 13.5(a), the polarity of the induced voltage and the current direction remain the same, as shown in the figure, although the induced voltage will drop due to the reduced number of flux lines cut per unit time. At

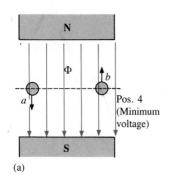
(a)

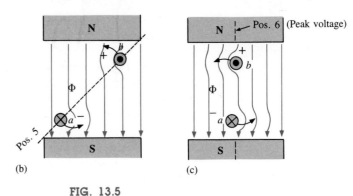
(b)   (c)

FIG. 13.5

position 4, the induced voltage is again zero, since the number of flux lines cut per unit time has dropped to zero. As the coil now turns toward position 5, the magnitude of the induced voltage will again increase, but note the change in polarity for terminals *a* and *b* and the reversal of current direction in each conductor. The similarities between the coil positions of positions 2 and 5 [Fig. 13.5(b)], and of 3 and 6 [Fig. 13.5(c)], indicate that the magnitude of the induced voltage is the same although the polarity of *a-b* has reversed.

A continuous plot of the induced voltage e_g appears in Fig. 13.6. The polarities of the induced voltage are shown for terminals *a* and *b* to the left of the vertical axis.

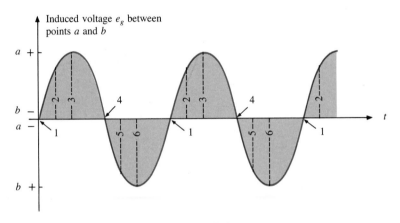

FIG. 13.6
Sinusoidal waveform.

Take a moment to relate the various positions to the resulting waveform of Fig. 13.6. This waveform will become very familiar in the discussions to follow. Note some of its obvious characteristics. As shown in the figure, if the coil is allowed to continue rotating, the generated voltage will repeat itself in equal intervals of time. Note also that the pattern is exactly the same below the axis as it is above, and that it changes continually with time (the horizontal axis). At the risk of being repetitious, let us again state that the waveform of Fig. 13.6 is the appearance of a *sinusoidal ac voltage*.

The function generator of Fig. 13.7, which employs semiconductor electronic components, will provide sinusoidal, square-wave, and triangular signals over a wide frequency range determined by the dial setting and the chosen frequency range.

If a sinusoidal signal with a frequency of 1000 Hz and a peak value of 10 V were required, the sinusoidal function switch would first be depressed as shown in the figure. Next, the dial would be set to 1 and the 1k range switch depressed as shown. The output frequency is the product of the dial position and the chosen range setting. The amplitude control would adjust the output until an ac voltmeter or oscilloscope indicated an output with a peak value of 10 V. The same frequency could also have been set by choosing a dial position of 10 and pressing the 100 range switch.

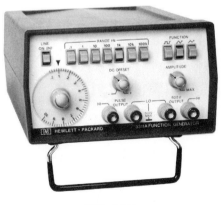

FIG. 13.7
Function generator. (Courtesy of Hewlett Packard Co.)

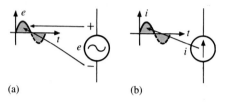

(a)   (b)

FIG. 13.8

(a) Sinusoidal ac voltage sources; (b) sinusoidal ac current sources.

13.3 DEFINED POLARITIES AND DIRECTION

In the following analysis, we will find it necessary to establish a set of polarities for the sinusoidal ac voltage and a direction for the sinusoidal ac current. In each case, the polarity and current direction will be for an instant of time in the positive portion of the sinusoidal waveform. This is shown in Fig. 13.8 with the symbols for the sinusoidal ac voltage and current. A lowercase letter is employed for each to indicate that the quantity is time dependent; that is, its magnitude will change with time. The need for defining polarities and current direction will become quite obvious when we consider multisource ac networks. Note in the last sentence the absence of the term *sinusoidal* before the phrase *ac networks*. This will occur to an increasing degree as we progress; *sinusoidal* is to be understood unless otherwise indicated.

13.4 DEFINITIONS

The sinusoidal waveform of Fig. 13.9 with its additional notation will now be used as a model in defining a few basic terms. These terms can, however, be applied to any alternating waveform.

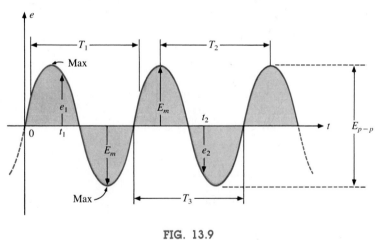

FIG. 13.9

Sinusoidal voltage.

Waveform: The path traced by a quantity, such as the voltage in Fig. 13.9, plotted as a function of some variable such as time (as above), position, degrees, radians, temperature, and so on.

Instantaneous value: The magnitude of a waveform at any instant of time; denoted by lowercase letters (e_1, e_2).

Amplitude, or *peak value:* The maximum value of a waveform; denoted by uppercase letters (E_m for sources of voltage and V_m for the voltage across common loads).

Peak-to-peak value: Denoted by $E_{p\text{-}p}$ or $V_{p\text{-}p}$, the full voltage between positive and negative peaks of the waveform, that is, the sum of the magnitude of the positive and negative peaks.

Periodic waveform: A waveform that continually repeats itself after the same time interval. The waveform of Fig. 13.9 is a periodic waveform.

Period (T): The time interval between successive repetitions of a periodic waveform; the period $T_1 = T_2 = T_3$ in Fig. 13.9, so long as successive *similar points* of the periodic waveform are used in determining T.

Cycle: The portion of a waveform contained in *one period* of time. The cycles within T_1, T_2, and T_3 of Fig. 13.9 may appear different in Fig. 13.10, but they are all bounded by one period of time and therefore satisfy the definition of a cycle.

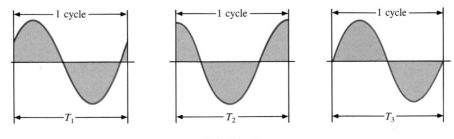

FIG. 13.10

Frequency (f): The number of cycles that occur in 1 second. The frequency of the waveform of Fig. 13.11(a) is 1 cycle per second, and for Fig. 13.11(b), $2\frac{1}{2}$ cycles per second. If a waveform of similar shape

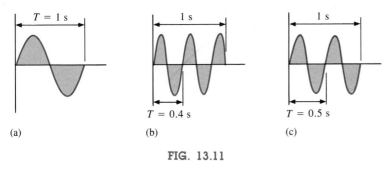

FIG. 13.11

had a period of 0.5 second [Fig. 13.11(c)], the frequency would be 2 cycles per second.

The unit of measure for frequency is the *hertz* (Hz), where

$$\boxed{1 \text{ hertz (Hz)} = 1 \text{ cycle per second (c/s)}} \qquad \textbf{(13.1)}$$

The unit hertz is derived from the surname of Heinrich Rudolph Hertz, who did original research in the area of alternating currents and voltages and their effect on the basic R, L, and C elements. The frequency standard for North America is 60 Hz, while for Europe it is predominantly 50 Hz.

Since the frequency is inversely related to the period—that is, as one increases the other decreases by an equal amount—the two can be related by the following equation:

$$\boxed{f = \frac{1}{T}} \qquad \begin{array}{l} f = \text{Hz} \\ T = \text{seconds (s)} \end{array} \qquad \textbf{(13.2)}$$

or

$$\boxed{T = \frac{1}{f}} \qquad\qquad\qquad \textbf{(13.3)}$$

EXAMPLE 13.1. Find the period of a periodic waveform with a frequency of
a. 60 Hz
b. 1000 Hz
Solutions:

a. $T = \dfrac{1}{f} = \dfrac{1}{60} = 0.01667\,\text{s}$ or **16.67 ms**

   (a recurring value since 60 Hz is so prevalent)

b. $T = \dfrac{1}{f} = \dfrac{1}{1000} = 10^{-3}\,\text{s} = \textbf{1 ms}$

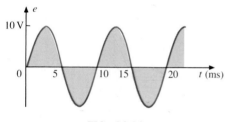

FIG. 13.12

EXAMPLE 13.2. Determine the frequency of the waveform of Fig. 13.12.

Solution: From the figure, $T = 10\,\text{ms}$, and

$$f = \frac{1}{T} = \frac{1}{10 \times 10^{-3}} = \textbf{100 Hz}$$

EXAMPLE 13.3. The oscilloscope is an instrument that will display alternating waveforms such as those described above. A sinusoidal pattern appears on the oscilloscope of Fig. 13.13 with the indicated scale

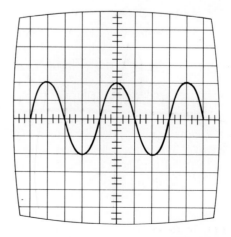

Vertical sensitivity = 0.1 V/cm

Horizontal sensitivity = 50 μs/cm

FIG. 13.13

settings. Determine the period, frequency, and peak value of the wave-form.

Solution: One cycle spans 4 cm. The period is

$$T = 4(50\,\mu\text{s}) = \mathbf{200\,\mu s}$$

and the frequency is

$$f = \frac{1}{T} = \frac{1}{200 \times 10^{-6}} = \mathbf{5\,kHz}$$

The vertical height above the horizontal axis encompasses 2 cm. There-fore,

$$V_m = 2(0.1) = \mathbf{0.2\,V}$$

13.5 THE SINE WAVE

The terms defined in the previous section can be applied to any type of periodic waveform, whether smooth or discontinuous. The sinusoidal waveform is of particular importance, however, since it lends itself readily to the mathematics and the physical phenomena associated with electric circuits. Consider the power of the following statement:

The sine wave is the only alternating waveform whose appearance is unaffected by the response characteristics of R, L, and C elements.

In other words, if the voltage across a resistor, coil, or capacitor is sinusoidal in nature, the resulting current for each will also have sinus-oidal characteristics. If a square wave or a triangular wave were ap-plied, such would not be the case. It must be pointed out that the above statement is also applicable to the cosine wave since the waves differ only by a 90° shift on the horizontal axis, as shown in Fig. 13.14.

The unit of measurement for the horizontal axis of Fig. 13.14 is the *degree*. A second unit of measurement frequently used is the *radian* (rad). It is defined by a quadrant of a circle such as in Fig. 13.15 where the distance subtended on the circumference equals the radius of the circle.

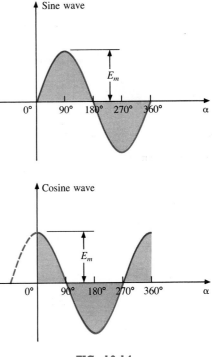

FIG. 13.14

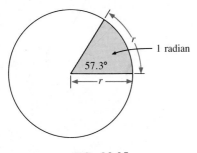

FIG. 13.15
Defining the radian.

If we define x as the number of intervals of r (the radius) around the circumference of the circle, then

$$C = 2\pi r = x \cdot r$$

and we find

$$x = 2\pi$$

Therefore, there are 2π radians around a 360° circle, as shown in Fig. 13.16, and

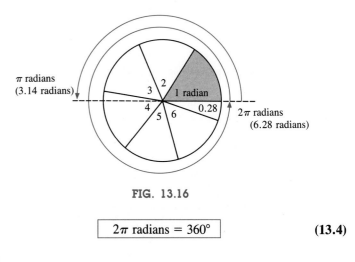

FIG. 13.16

$$\boxed{2\pi \text{ radians} = 360°} \tag{13.4}$$

with

$$\boxed{1 \text{ radian} \cong 57.3°} \tag{13.5}$$

A number of electrical formulas contain a multiplier of π. This is one reason it is sometimes preferable to measure angles in radians rather than in degrees.

The quantity π has been determined to an extended number of places primarily in an attempt to see if a repetitive sequence of numbers appears. It does not. A sampling of the effort appears below:

$$\pi = 3.14159\ 26535\ 89793\ 23846\ 26433\ \ldots$$

For our purposes, the following approximation will be used:

$$\boxed{\pi = 3.14} \tag{13.6}$$

For 180° and 360°, the two units of measurement are related as shown in Fig. 13.16. The conversion equations between the two are the following:

$$\boxed{\text{Radians} = \left(\frac{\pi}{180°}\right) \times (\text{degrees})} \tag{13.7}$$

$$\text{Degrees} = \left(\frac{180°}{\pi}\right) \times \text{(radians)} \qquad \textbf{(13.8)}$$

Applying these equations, we find

$$\textbf{90°:} \quad \text{Radians} = \frac{\pi}{180°}(90°) = \frac{\pi}{2}\textbf{ rad}$$

$$\textbf{30°:} \quad \text{Radians} = \frac{\pi}{180°}(30°) = \frac{\pi}{6}\textbf{ rad}$$

$$\frac{\pi}{3}: \quad \text{Degrees} = \frac{180°}{\pi}\left(\frac{\pi}{3}\right) = \textbf{60°}$$

$$\frac{3\pi}{2}: \quad \text{Degrees} = \frac{180°}{\pi}\left(\frac{3\pi}{2}\right) = \textbf{270°}$$

Using the radian as the unit of measurement for the abscissa, we would obtain a sine wave as shown in Fig. 13.17.

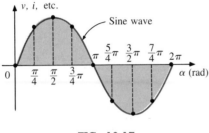

FIG. 13.17

It is of particular interest that the sinusoidal waveform can be derived from the length of the *vertical projection* of a radius vector rotating in a uniform circular motion about a fixed point. Starting as shown in Fig. 13.18(a) and plotting the amplitude (above and below zero) on the coordinates drawn to the right [Figs. 13.18(b) through (i)], we will trace a complete sinusoidal waveform after the radius vector has completed a 360° rotation about the center.

The velocity with which the radius vector rotates about the center, called the *angular velocity,* can be determined from the following equation:

$$\text{Angular velocity} = \frac{\text{distance (degrees or radians)}}{\text{time (seconds)}} \qquad \textbf{(13.9)}$$

Substituting into Eq. (13.9) and assigning the Greek letter omega (ω) to the angular velocity, we have

$$\omega = \frac{\alpha}{t} \qquad \textbf{(13.10)}$$

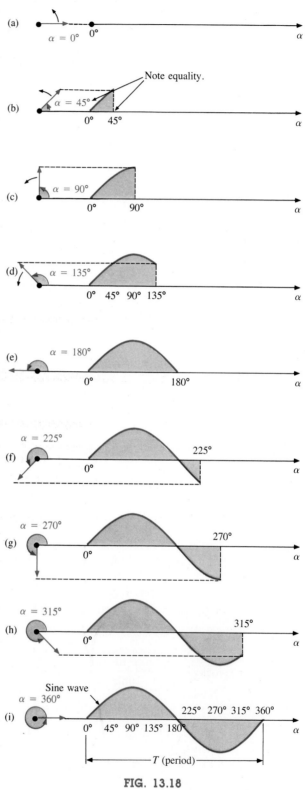

(a) $\alpha = 0°$ 0° α

Note equality.

(b) $\alpha = 45°$ 0°   45° α

(c) $\alpha = 90°$ 0°   90° α

(d) $\alpha = 135°$ 0°   45°   90°   135° α

(e) $\alpha = 180°$ 0°   180° α

(f) $\alpha = 225°$ 0°   225° α

(g) $\alpha = 270°$ 0°   270° α

(h) $\alpha = 315°$ 0°   315° α

Sine wave

(i) $\alpha = 360°$ 0°   45°   90°   135°   180°   225°   270°   315°   360° α

T (period)

FIG. 13.18

and

$$\boxed{\alpha = \omega t} \qquad\qquad (13.11)$$

Since ω is typically provided in radians per second, the angle α obtained using Eq. (13.11) is usually in radians. If α is required in degrees, Eq. (13.8) must be applied. The importance of remembering the above will become obvious in the examples to follow.

In Fig. 13.18, the time required to complete one revolution is equal to the period (T) of the sinusoidal waveform of Fig. 13.18(i). The radians subtended in this time interval are 2π. Substituting, we have

$$\boxed{\omega = \frac{2\pi}{T}} \qquad \text{(rad/s)} \qquad (13.12)$$

In words, this equation states that the smaller the period of the sinusoidal waveform of Fig. 13.18(i), or the smaller the time interval before one complete cycle is generated, the greater must be the angular velocity of the rotating radius vector. Certainly this statement agrees with what we have learned thus far. We can now go one step further and apply the fact that the frequency of the generated waveform is inversely related to the period of the waveform; that is, $f = 1/T$. Thus,

$$\boxed{\omega = 2\pi f} \qquad \text{(rad/s)} \qquad (13.13)$$

This equation states that the higher the frequency of the generated sinusoidal waveform, the higher must be the angular velocity. Equations (13.12) and (13.13) are verified somewhat by Fig. 13.19, where for the same radius vector, $\omega = 100$ rad/s and 500 rad/s.

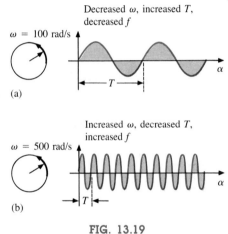

(a)

(b)

FIG. 13.19

EXAMPLE 13.4. Determine the angular velocity of a sine wave having a frequency of 60 Hz.
Solution:

$$\omega = 2\pi f = (6.28)(60) \cong \textbf{377 rad/s}$$

(a recurring value due to 60-Hz predominance)

EXAMPLE 13.5. Determine the frequency and period of the sine wave of Fig. 13.19(b).
Solution: Since $\omega = 2\pi/T$,

$$T = \frac{2\pi}{\omega} = \frac{2\pi}{500} = \frac{6.28}{500} = \textbf{12.56 ms}$$

and

$$f = \frac{1}{T} = \frac{1}{12.56 \times 10^{-3}} = \mathbf{79.62\,Hz}$$

EXAMPLE 13.6. Given $\omega = 200$ rad/s, determine how long it will take the sinusoidal waveform to pass through an angle of 90°.

Solution: Eq. (13.11): $\alpha = \omega t$, and

$$t = \frac{\alpha}{\omega}$$

However, α must be substituted as $\pi/2$ (= 90°) since ω is in radians per second:

$$t = \frac{\alpha}{\omega} = \frac{\pi/2}{200} = \frac{\pi}{400} = \frac{3.14}{400} = \mathbf{7.85\,ms}$$

EXAMPLE 13.7. Find the angle a sinusoidal waveform of 60 Hz will pass through in a period of 5 ms.

Solution: Eq. (13.11): $\alpha = \omega t$, or

$$\alpha = 2\pi f t = (6.28)(60)(5 \times 10^{-3}) = \mathbf{1.884\,rad}$$

If not careful, one might be tempted to interpret the answer as 1.884°. However,

$$\alpha\ (°) = \frac{180°}{\pi}(1.884) = \mathbf{108°}$$

13.6 GENERAL FORMAT FOR THE SINUSOIDAL VOLTAGE OR CURRENT

The basic mathematical format for the sinusoidal waveform is

$$\boxed{A_m \sin\ \alpha} \qquad \textbf{(13.14)}$$

where A_m is the peak value of the waveform and α is the unit of measure for the horizontal axis as shown in Fig. 13.20.

The equation $\alpha = \omega t$ states that the angle α through which the rotating vector of Fig. 13.18 will pass is determined by the angular velocity of the rotating vector and the length of time the vector rotates. For example, for a particular angular velocity (fixed ω), the longer the radius vector is permitted to rotate (that is, the greater the value of t), the

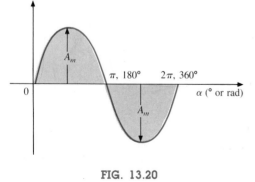

FIG. 13.20

greater will be the number of degrees or radians through which the vector will pass. Relating this statement to the sinusoidal waveform, for a particular angular velocity, the longer the time, the greater the number of cycles shown. For a fixed time interval, the greater the angular velocity, the greater the number of cycles generated.

Due to Eq. (13.11), the general format of a sine wave can also be written

$$\boxed{A_m \sin \omega t} \qquad\qquad \textbf{(13.15)}$$

with ωt as the horizontal unit of measure.

For electrical quantities such as current and voltage, the general format is

$$i = I_m \sin \omega t = I_m \sin \alpha$$
$$e = E_m \sin \omega t = E_m \sin \alpha$$

where the capital letters with the subscript m represent the amplitude and the lowercase letters i and e represent the instantaneous value of current or voltage, respectively, at any time t.

EXAMPLE 13.8. Given $e = 5 \sin \alpha$, determine e at $\alpha = 40°$ and $\alpha = 0.8\pi$.
Solution: For $\alpha = 40°$,

$$e = 5 \sin 40° = 5(0.6428) = \textbf{3.214 V}$$

For $\alpha = 0.8\pi$,

$$\alpha \ (°) = \frac{180°}{\pi}(0.8\pi) = 144°$$

and

$$e = 5 \sin 144° = 5(0.5878) = \textbf{2.939 V}$$

The angle at which a particular voltage level is attained can be determined by rearranging the equation

$$e = E_m \sin \alpha$$

in the following manner:

$$\sin \alpha = \frac{e}{E_m}$$

which can be written

$$\boxed{\alpha = \sin^{-1} \frac{e}{E_m}} \qquad\qquad \textbf{(13.16)}$$

Similarly, for a particular current level,

$$\alpha = \sin^{-1} \frac{i}{I_m} \tag{13.17}$$

The function $\sin^{-1}$ is available on all scientific calculators.

EXAMPLE 13.9.

a. Determine the angle at which the magnitude of the sinusoidal function $v = 10 \sin 377t$ is 4 V.
b. Determine the time at which the magnitude is attained.

Solutions:

a. Eq. (13.16):

$$\alpha_1 = \sin^{-1} \frac{e}{E_m} = \sin^{-1} \frac{4}{10} = \sin^{-1} 0.4 = \mathbf{23.578°}$$

However, Fig. 13.21 reveals that the magnitude of 4 V (positive) will be attained at two points between 0° and 180°. The second intersection is determined by

$$\alpha_2 = 180° - 23.578° = \mathbf{156.422°}$$

In general, therefore, keep in mind that Eqs. (13.16) and (13.17) will provide an angle with a magnitude between 0° and 90°.

b. Eq. (13.11): $\alpha = \omega t$, and so $t = \alpha/\omega$. However, α must be in radians. Thus,

$$\alpha \text{ (rad)} = \frac{\pi}{180°}(23.578°) = 0.411 \text{ rad}$$

and

$$t_1 = \frac{\alpha}{\omega} = \frac{0.411}{377} = \mathbf{1.09 \, ms}$$

For the second intersection,

$$T = \frac{2\pi}{\omega} = \frac{6.28}{377} = 16.66 \text{ ms}$$

and

$$\frac{T}{2} = \frac{16.66 \text{ ms}}{2} = 8.33 \text{ ms}$$

with

$$t_2 = \frac{T}{2} - t_1 = (8.33 - 1.09) \text{ ms} = \mathbf{7.24 \, ms}$$

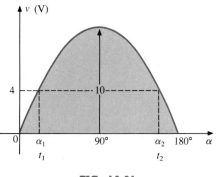

FIG. 13.21

The sine wave can also be plotted against *time* on the horizontal axis. The time period for each interval can be determined from $t = \alpha/\omega$, but the most direct route is simply to find the period T from $T = 1/f$ and break it up into the required intervals. This latter technique will be demonstrated in Example 13.10.

Before reviewing the example, take special note of the relative simplicity of the mathematical equation that can represent a sinusoidal waveform. Any alternating waveform whose characteristics differ from those of the sine wave cannot be represented by a single term, but may require two, four, six, or perhaps an infinite number of terms to be represented accurately. A further description of nonsinusoidal waveforms can be found in Chapter 23.

EXAMPLE 13.10.   Sketch $e = 10 \sin 314t$ with the abscissa
a. angle (α) in degrees.
b. angle (α) in radians.
c. time (t) in seconds.
Solutions:
a. See Fig. 13.22. (Note that no calculations are required.)

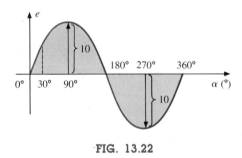

FIG. 13.22

b. See Fig. 13.23. (Once the relationship between degrees and radians is understood, there is again no need for calculations.)

c. $360°$: $T = \dfrac{2\pi}{\omega} = \dfrac{6.28}{314} = 20\,\text{ms}$

   $180°$: $\dfrac{T}{2} = \dfrac{20}{2} \times 10^{-3} = 10\,\text{ms}$

   $90°$: $\dfrac{T}{4} = \dfrac{20}{4} \times 10^{-3} = 5\,\text{ms}$

   $30°$: $\dfrac{T}{12} = \dfrac{20}{12} \times 10^{-3} = 1.67\,\text{ms}$

See Fig. 13.24.

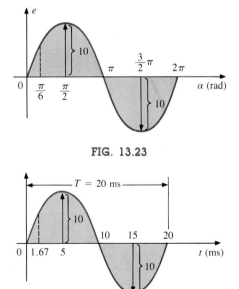

FIG. 13.23

FIG. 13.24

EXAMPLE 13.11. Given $i = 6 \times 10^{-3} \sin 1000t$, determine i at $t = 2\,\text{ms}$.

Solution:

$$\alpha = \omega t = 1000t = 1000(2 \times 10^{-3}) = 2\,\text{rad}$$

$$\alpha\,(°) = \frac{180°}{\pi}(2) = 114.59°$$

$$i = 6 \times 10^{-3} \sin 114.59° = (6 \times 10^{-3})(0.9093) = \mathbf{5.46\,mA}$$

13.7 PHASE RELATIONS

Thus far, we have considered only sine waves that have maxima at $\pi/2$ and $3\pi/2$, with a zero value at 0, π, and 2π, as shown in Fig. 13.23. If the waveform is shifted to the right or left of $0°$, the expression becomes

$$\boxed{A_m \sin(\omega t \pm \theta)} \tag{13.18}$$

where θ is the angle in degrees or radians that the waveform has been shifted.

If the waveform passes through the horizontal axis with a *positive-going* (increasing with time) slope *before* $0°$, as shown in Fig. 13.25, the expression is

$$\boxed{A_m \sin(\omega t + \theta)} \tag{13.19}$$

At $\omega t = \alpha = 0°$, the magnitude is determined by $A_m \sin \theta$. If the waveform passes through the horizontal axis with a positive-going slope *after* $0°$, as shown in Fig. 13.26, the expression is

$$\boxed{A_m \sin(\omega t - \theta)} \tag{13.20}$$

And at $\omega t = \alpha = 0°$, the magnitude is $A_m \sin(-\theta)$, which by a trigonometric identity is $-A_m \sin \theta$.

If the waveform crosses the horizontal axis with a positive-going slope $90°$ $(\pi/2)$ sooner, as shown in Fig. 13.27, it is called a *cosine wave*. That is,

$$\boxed{\sin(\omega t + 90°) = \sin\left(\omega t + \frac{\pi}{2}\right) = \cos \omega t} \tag{13.21}$$

or

$$\boxed{\sin \omega t = \cos(\omega t - 90°) = \cos\left(\omega t - \frac{\pi}{2}\right)} \tag{13.22}$$

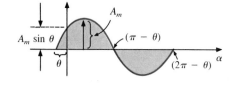

FIG. 13.25

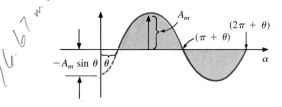

FIG. 13.26

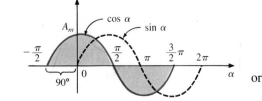

FIG. 13.27

The terms *lead* and *lag* are used to indicate the relationship between two sinusoidal waveforms of the *same frequency* plotted on the same set of axes. In Fig. 13.27, the cosine curve is said to *lead* the sine curve by 90°, and the sine curve is said to *lag* the cosine curve by 90°. The 90° is referred to as the phase angle between the two waveforms. In language commonly applied, the waveforms are *out of phase* by 90°. Note that the phase angle between the two waveforms is measured between those two points on the horizontal axis through which each passes with the *same slope*. If both waveforms cross the axis at the same point with the same slope, they are *in phase*.

A few additional geometric relations that may prove useful in applications involving sines or cosines in phase relationships are the following:

$$
\begin{array}{|l|}
\hline
\sin(-\alpha) = -\sin\alpha \\
\cos(-\alpha) = \cos\alpha \\
-\sin(\alpha) = \sin(\alpha \pm 180°) \\
-\cos(\alpha) = \cos(\alpha \pm 180°) \\
\hline
\end{array}
\qquad \textbf{(13.23)}
$$

If a sinusoidal expression should appear as

$$ e = -E_m \sin \omega t $$

the negative sign is associated with the sine portion of the expression, not the peak value E_m. In other words, the expression, if not for convenience, would be written

$$ e = E_m(-\sin \omega t) $$

Since

$$ -\sin \omega t = \sin(\omega t \pm 180°) $$

the expression can also be written

$$ e = E_m \sin(\omega t \pm 180°) $$

revealing that a negative sign can be replaced by a 180° change in phase angle (+ or −). That is,

$$
\begin{aligned}
e = -E_m \sin \omega t &= E_m \sin(\omega t + 180°) \\
&= E_m \sin(\omega t - 180°)
\end{aligned}
$$

A plot of each will clearly show their equivalence. There are, therefore, two correct mathematical representations for the functions.

The *phase relationship* between two waveforms indicates which one leads or lags, and by how many degrees or radians.

EXAMPLE 13.12. What is the phase relationship between the sinusoidal waveforms of each of the following sets?
a. $v = 10 \sin(\omega t + 30°)$
    $i = 5 \sin(\omega t + 70°)$

b. $i = 15 \sin(\omega t + 60°)$
    $v = 10 \sin(\omega t - 20°)$
c. $i = 2 \cos(\omega t + 10°)$
    $v = 3 \sin(\omega t - 10°)$
d. $i = -\sin(\omega t + 30°)$
    $v = 2 \sin(\omega t + 10°)$
e. $i = -2 \cos(\omega t - 60°)$
    $v = 3 \sin(\omega t - 150°)$

Solutions:

a. See Fig. 13.28.
    ***i* leads *v* by 40°, or *v* lags *i* by 40°.**

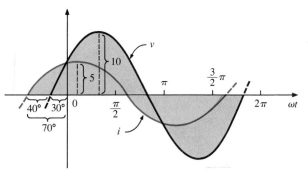

FIG. 13.28

b. See Fig. 13.29.
    ***i* leads *v* by 80°, or *v* lags *i* by 80°.**

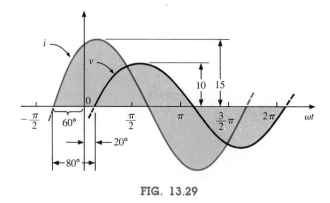

FIG. 13.29

c. See Fig. 13.30.

$$i = 2 \cos(\omega t + 10°) = 2 \sin(\omega t + 10° + 90°)$$
$$= 2 \sin(\omega t + 100°)$$

***i* leads *v* by 110°, or *v* lags *i* by 110°.**

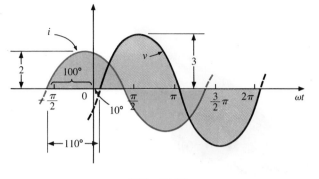

FIG. 13.30

d. See Fig. 13.31.

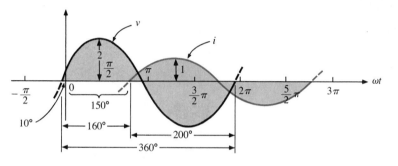

FIG. 13.31

$$-\sin(\omega t + 30°) = \sin(\omega t + 30° \overset{\text{Note}}{-} 180°)$$
$$= \sin(\omega t - 150°)$$

v leads i by 160°, or i lags v by 160°.

Or using

$$-\sin(\omega t + 30°) = \sin(\omega t + 30° \overset{\text{Note}}{+} 180°)$$
$$= \sin(\omega t + 210°)$$

i leads v by 200°, or v lags i by 200°.

e. See Fig. 13.32.

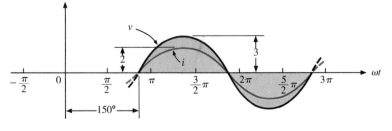

FIG. 13.32

$$i = -2 \cos(\omega t - 60°) = 2 \cos(\omega t - 60° \angle 180°) \quad \text{By choice}$$
$$= 2 \cos(\omega t - 240°)$$

However,

$$\cos \alpha = \sin(\alpha + 90°)$$

so that

$$2 \cos(\omega t - 240°) = 2 \sin(\omega t - 240° + 90°)$$
$$= 2 \sin(\omega t - 150°)$$

v and *i* are in phase.

13.8 AVERAGE VALUE

After traveling a considerable distance by car, some drivers like to calculate their average speed for the entire trip. This is usually done by dividing the miles traveled by the hours required to drive that distance. For example, if a person traveled 180 mi in 5 h, his average speed was 180/5 or 36 mi/h. This same distance may have been traveled at various speeds for various intervals of time, as shown in Fig. 13.33.

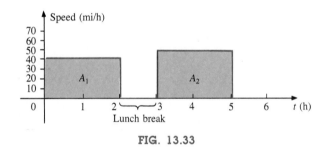

FIG. 13.33

By finding the total area under the curve for the 5 h and then dividing the area by 5 h (the total time for the trip), we obtain the same result of 36 mi/h; that is,

$$\text{Average speed} = \frac{\text{area under curve}}{\text{length of curve}} \quad \textbf{(13.24)}$$

$$= \frac{A_1 + A_2}{5}$$

$$= \frac{(40)(2) + (50)(2)}{5}$$

$$= \frac{80}{5}$$

$$= \textbf{36 mi/h}$$

Equation (13.24) can be extended to include any variable quantity, such as current or voltage, if we let G denote the average value, as follows:

$$G \text{ (average value)} = \frac{\text{algebraic sum of areas}}{\text{length of curve}} \qquad \textbf{(13.25)}$$

The algebraic sum of the areas must be determined, since some area contributions will be from below the horizontal axis. Areas above the axis will be assigned a positive sign, and those below a negative sign. A positive average value will then be above the axis, and a negative value below.

The average value of *any* current or voltage is the value indicated on a dc meter. In other words, over a complete cycle, the average value is the equivalent dc value. In the analysis of electronic circuits to be considered in a later course, both dc and ac sources of voltage will be applied to the same network. It will then be necessary to know or determine the dc (or average value) and ac components of the voltage or current in various parts of the system.

EXAMPLE 13.13. Find the average values of the following waveforms over one full cycle:
a. Fig. 13.34.
b. Fig. 13.35.

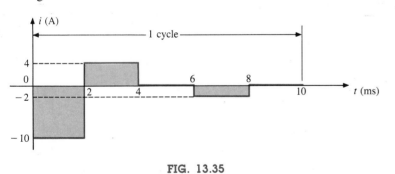

FIG. 13.35

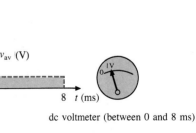

FIG. 13.34

Solutions:

a. $G = \dfrac{+(3)(4) - (1)(4)}{8} = \dfrac{12 - 4}{8} = \textbf{1 V}$

Note Fig. 13.36.

b. $G = \dfrac{-(10)(2) + (4)(2) - (2)(2)}{10} = \dfrac{-20 + 8 - 4}{10}$

$\qquad = -\dfrac{16}{10} = \textbf{-1.6 A}$

Note Fig. 13.37.

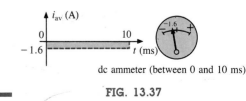

FIG. 13.36

dc voltmeter (between 0 and 8 ms)

dc ammeter (between 0 and 10 ms)

FIG. 13.37

We found the areas under the curves in the preceding example by using a simple geometric formula. If we should encounter a sine wave or any other unusual shape, however, we must find the area by some other means. We can obtain a good approximation of the area by attempting to reproduce the original wave shape using a number of small rectangles or other familiar shapes, the area of which we already know through simple geometric formulas. For example, *the actual area of the positive (or negative) pulse of a sine wave is* $2A_m$. Approximating this waveform by two triangles (Fig. 13.38), we obtain (using *area* = 1/2 *base* × *height* for the area of a triangle) a rough idea of the actual area:

$$\text{Area (shaded)} = 2\left(\frac{1}{2}bh\right) = 2\left[\left(\frac{1}{2}\right)\overbrace{\left(\frac{\pi}{2}\right)}^{b}\overbrace{(A_m)}^{h}\right] = \frac{\pi}{2}A_m$$

$$\cong 1.58A_m$$

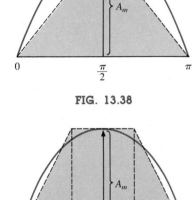

FIG. 13.38

A closer approximation might be a rectangle with two similar triangles (Fig. 13.39):

$$\text{Area} = A_m\frac{\pi}{3} + 2\left(\frac{1}{2}bh\right) = A_m\frac{\pi}{3} + \frac{\pi}{3}A_m = \frac{2}{3}\pi A_m$$

$$= 2.094A_m$$

which is certainly close to the actual area. If an infinite number of forms were used, an exact answer of $2A_m$ could be obtained. For irregular waveforms, this method can be especially useful if data such as the average value are desired.

FIG. 13.39

The procedure of calculus that gives the exact solution $2A_m$ is known as *integration*. Integration is presented here only to make the method recognizable to the reader; it is not necessary to be proficient in its use to continue with this text. It is a useful mathematical tool, however, and should be learned. Finding the area under the positive pulse of a sine wave using integration, we have

$$\text{Area} = \int_0^\pi A_m \sin \alpha \, d\alpha$$

where ∫ is the sign of integration, π and 0 are the limits of integration, $A_m \sin \alpha$ is the function to be integrated, and $d\alpha$ indicates that we are integrating with respect to α.

Integrating, we obtain

$$\text{Area} = A_m[-\cos \alpha]_0^\pi$$
$$= -A_m(\cos \pi - \cos 0°)$$
$$= -A_m[-1 - (+1)] = -A_m(-2)$$

$$\boxed{\text{Area} = 2A_m}$$

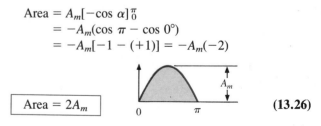

(13.26)

Since we know the area under the positive (or negative) pulse, we can easily determine the average value of the positive (or negative) region of a sine wave pulse by applying Eq. (13.25):

$$G = \frac{2A_m}{\pi}$$

and

$$\boxed{G = 0.637A_m}$$

(13.27)

For the waveform of Fig. 13.40,

$$G = \frac{(2A_m/2)}{\pi/2} = \frac{2A_m}{\pi} \qquad \text{(average the same as for a full pulse)}$$

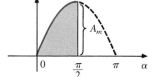

FIG. 13.40

EXAMPLE 13.14. Find the average value of the following waveforms over one full cycle:
a. Fig. 13.41.

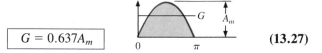

FIG. 13.41

b. Fig. 13.42.
c. Fig 13.43. For this waveform, simply indicate whether the average value is positive and what its approximate value is.

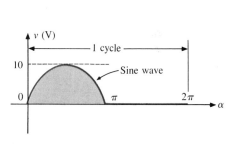

FIG. 13.42

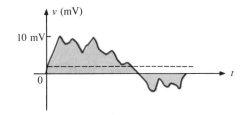

FIG. 13.43

Solutions:

a. $G = \dfrac{+2A_m - 2A_m}{2\pi} = \mathbf{0}$

The average value of a sine wave (or cosine wave) is zero, which should be obvious from the appearance of the waveform over one full cycle.

b. $G = \dfrac{2A_m + 0}{2\pi} = \dfrac{2(10)}{2\pi} \cong \mathbf{3.18\ V}$

c. From the appearance of the waveform, the average value is positive and in the vicinity of 2 mV. Occasionally, judgments of this type will have to be made.

13.9 EFFECTIVE VALUES

This section will begin to relate dc and ac quantities with respect to the power delivered to a load. It will help us determine the amplitude of a sinusoidal ac current required to deliver the same power as a particular dc current. The question frequently arises, How is it possible for a sinusoidal ac quantity to deliver a net power if, over a full cycle, the net current in any one direction is zero (average value = 0)? It would almost appear that the power delivered during the positive portion of the sinusoidal waveform is withdrawn during the negative portion, and since the two are equal in magnitude, the net power delivered is zero. However, understand that *irrespective of direction,* current of any magnitude through a resistor will deliver power *to that resistor.* In other words, during the positive or negative portions of a sinusoidal ac current, power is being delivered at *each instant of time* to the resistor. The power delivered at each instant will, of course, vary with the magnitude of the sinusoidal ac current, but there will be a net flow during either the positive or negative pulses with a net flow over the full cycle. The net power flow will equal twice that delivered by either the positive or negative regions of the sinusoidal quantity.

A fixed relationship between ac and dc voltages and currents can be derived from the experimental setup shown in Fig. 13.44. A resistor in a water bath is connected by switches to a dc and an ac supply. If switch 1 is closed, a dc current I, determined by the resistance R and battery voltage E, will be established through the resistor R. The temperature reached by the water is determined by the dc power dissipated in the form of heat by the resistor.

If switch 2 is closed and switch 1 left open, the ac current through the resistor will have a peak value of I_m. The temperature reached by the

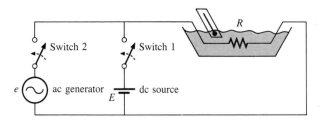

FIG. 13.44

water is now determined by the ac power dissipated in the form of heat by the resistor. The ac input is varied until the temperature is the same as that reached with the dc input. When this is accomplished, the average electrical power delivered to the resistor R by the ac source is the same as that delivered by the dc source.

The power delivered by the ac supply at any instant of time is

$$P_{ac} = (i_{ac})^2 R = (I_m \sin \omega t)^2 R = (I_m^2 \sin^2 \omega t)R$$

but

$$\sin^2 \omega t = \frac{1}{2}(1 - \cos 2\omega t) \qquad \text{(trigonometric identity)}$$

Therefore,

$$P_{ac} = I_m^2 \left[\frac{1}{2}(1 - \cos 2\omega t) \right] R$$

and

$$\boxed{P_{ac} = \frac{I_m^2 R}{2} - \frac{I_m^2 R}{2} \cos 2\omega t} \qquad \textbf{(13.28)}$$

The *average power* delivered by the ac source is just the first term, since the average value of a cosine wave is zero even though the wave may have twice the frequency of the original input current waveform. Equating the average power delivered by the ac generator to that delivered by the dc source,

$$P_{av(ac)} = P_{dc}$$
$$\frac{I_m^2 R}{2} = I_{dc}^2 R$$

and

$$I_m = \sqrt{2} I_{dc}$$

or

$$I_{dc} = \frac{I_m}{\sqrt{2}} = 0.707 I_m$$

which, in words, states that *the equivalent dc value of a sinusoidal current or voltage is $1/\sqrt{2}$ or 0.707 of its maximum value. The equivalent dc value is called the effective value of the sinusoidal quantity.*

In summary,

$$\boxed{I_{eq\ dc} = I_{eff} = 0.707 I_m} \qquad \textbf{(13.29)}$$

or

$$\boxed{I_m = \sqrt{2} I_{eff} = 1.414 I_{eff}} \qquad \textbf{(13.30)}$$

and

$$E_{\text{eff}} = 0.707 E_m \tag{13.31}$$

or

$$E_m = \sqrt{2} E_{\text{eff}} = 1.414 E_{\text{eff}} \tag{13.32}$$

As a simple numerical example, it would require an ac current with a peak value of $\sqrt{2}(10) = 14.14$ A to deliver the same power to the resistor in Fig. 13.44 as a dc current of 10 A. The effective value of any quantity plotted as a function of time can be found by using the following equation derived from the experiment just described:

$$I_{\text{eff}} = \sqrt{\frac{\int_0^T i(t)^2 \, dt}{T}} \tag{13.33}$$

or

$$I_{\text{eff}} = \sqrt{\frac{\text{area}[i(t)^2]}{T}} \tag{13.34}$$

which, in words, states that to find the effective value, the function $i(t)$ must first be squared. After $i(t)$ is squared, the area under the curve is found by integration. It is then divided by T, the length of the cycle or period of the waveform, to obtain the average or *mean* value of the squared waveform. The final step is to take the *square root* of the mean value. This procedure gives us another designation for the effective value, the *root-mean-square* (rms) value.

EXAMPLE 13.15.   Find the effective values of the sinusoidal waveforms in each part of Fig. 13.45.

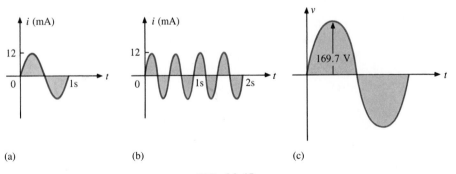

FIG. 13.45

Solution: For part (a), $I_{\text{eff}} = 0.707(12 \times 10^{-3}) = \textbf{8.484 mA.}$ For part (b), again $I_{\text{eff}} = \textbf{8.484 mA.}$ Note that frequency did not change the effective value in (b) above as compared to (a). For part (c), $V_{eff} = 0.707(169.73) \cong$ **120 V,** as available from a home outlet.

EXAMPLE 13.16. Find the effective or rms value of the waveform of Fig. 13.46.

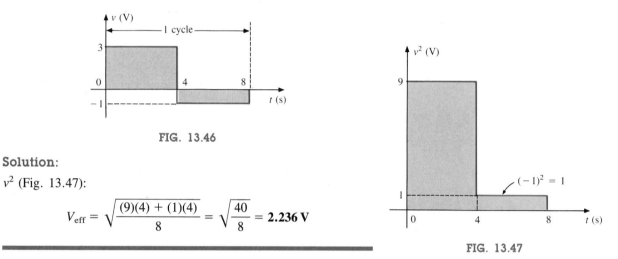

FIG. 13.46

Solution:

v^2 (Fig. 13.47):

$$V_{eff} = \sqrt{\frac{(9)(4) + (1)(4)}{8}} = \sqrt{\frac{40}{8}} = \mathbf{2.236\ V}$$

FIG. 13.47

EXAMPLE 13.17. Calculate the effective value of the voltage of Fig. 13.48.

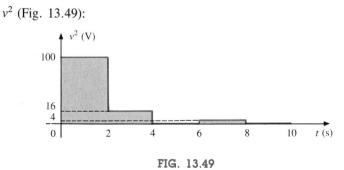

FIG. 13.48

Solution:

v^2 (Fig. 13.49):

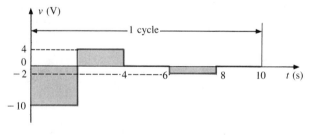

FIG. 13.49

$$V_{\text{eff}} = \sqrt{\frac{(100)(2) + (16)(2) + (4)(2)}{10}} = \sqrt{\frac{240}{10}}$$

$$= \textbf{4.899 V}$$

EXAMPLE 13.18. Determine the average and effective values of the square wave of Fig. 13.50.

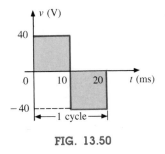

FIG. 13.50

Solution: By inspection, the average value is zero.

v^2 (Fig. 13.51):

$$V_{\text{eff}} = \sqrt{\frac{(1600)(10 \times 10^{-3}) + (1600)(10 \times 10^{-3})}{20 \times 10^{-3}}}$$

$$= \sqrt{\frac{32{,}000 \times 10^{-3}}{20 \times 10^{-3}}} = \sqrt{1600}$$

$$V_{\text{eff}} = \textbf{40 V}$$

(the maximum value of the waveform of Fig. 13.50)

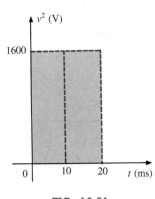

FIG. 13.51

The waveforms appearing in these examples are the same as those used in the examples on the average value. It might prove interesting to compare the effective and average values of these waveforms.

The effective values of sinusoidal quantities such as voltage or current will be represented by E and I. These symbols are the same as those used for dc voltages and currents. To avoid confusion, the peak value of a waveform will always have a subscript m associated with it: $I_m \sin \omega t$. *Caution:* When finding the effective value of the positive pulse of a sine wave, note that the squared area is *not* simply $(2A_m)^2 = 4A_m^2$; it must be found by a completely new integration. This will always be the case for any waveform that is not rectangular.

13.10 ac METERS AND INSTRUMENTS

The d'Arsonval movement employed in dc meters can also be used to measure sinusoidal voltages and currents if the *bridge rectifier* of Fig. 13.52 is placed between the signal to be measured and the average reading movement.

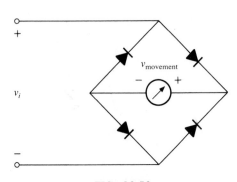

FIG. 13.52
Full-wave bridge rectifier.

The bridge rectifier composed of four diodes (electronic switches) will convert the input signal of zero average value to one having an average value sensitive to the peak value of the input signal. The conversion process is well described in most basic electronics texts. Fundamentally, conduction is permitted through the diodes in such a manner as to convert the sinusoidal input of Fig. 13.53(a) to one having the appearance of Fig. 13.53(b). The negative portion of the input has been effectively "flipped over" by the bridge configuration. The resulting waveform of Fig. 13.53(b) is called a *full-wave rectified waveform*.

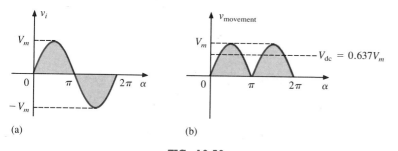

FIG. 13.53

(*a*) *Sinusoidal input; (b) full-wave rectified signal.*

The zero average value of Fig. 13.53(a) has been replaced by a pattern having an average value determined by

$$G = \frac{2V_m + 2V_m}{2\pi} = \frac{4V_m}{2\pi} = \frac{2V_m}{\pi} = 0.637V_m$$

The movement of the pointer will therefore be directly related to the peak value of the signal by the factor 0.637.

Forming the ratio between the rms and dc levels will result in

$$\frac{V_{rms}}{V_{dc}} = \frac{0.707V_m}{0.637V_m} \cong 1.11$$

revealing that the scale indication is 1.11 times the dc level measured by the movement. That is,

$$\boxed{\text{Meter indication} = 1.11(\text{dc or average value})}_{full\text{-}wave} \qquad \textbf{(13.35)}$$

Some ac meters use a half-wave rectifier arrangement that results in the waveform of Fig. 13.54, which has half the average value of Fig. 13.53(b) over one full cycle. The result is

$$\boxed{\text{Meter indication} = 2.22(\text{dc or average value})}_{half\text{-}wave} \qquad \textbf{(13.36)}$$

A second movement, called the electrodynamometer movement (Fig. 13.55), can measure both ac and dc quantities without a change in internal circuitry. The movement can, in fact, read the effective value of any periodic or nonperiodic waveform, because a reversal in current direction reverses the fields of both the stationary and the movable coils, so the deflection of the pointer is always up-scale.

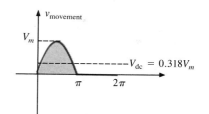

FIG. 13.54

Half-wave rectified signal.

FIG. 13.55

Electrodynamometer movement. (Courtesy of Weston Instruments, Inc.)

The VOM, introduced in Chapter 2, can be used to measure both dc and ac voltages using a d'Arsonval movement and the proper switching networks. That is, when the meter is used for dc measurements, the dial setting will establish the proper series resistance for the chosen scale and permit the appropriate dc level to pass directly to the movement. For ac measurements, the dial setting will introduce a network that employs a full- or half-wave rectifier to establish a dc level. As discussed above, each setting is properly calibrated to indicate the desired quantity on the face of the instrument.

EXAMPLE 13.19. Determine the reading of each meter for each situation of Fig. 13.56.

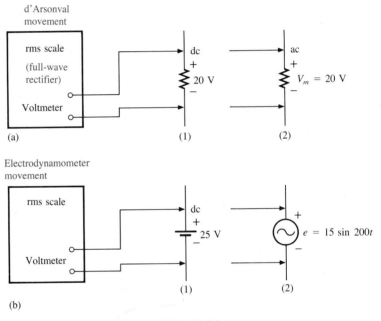

FIG. 13.56

Solution: For part (a), situation (1): By Eq. (13.35),

$$\text{Meter indication} = 1.11(20) = \mathbf{22.2\ V}$$

For part (a), situation (2):

$$V_{\text{rms}} = 0.707V_m = 0.707(20) = \mathbf{14.14\ V}$$

For part (b), situation (1):

$$V_{\text{rms}} = V_{\text{dc}} = \mathbf{25\ V}$$

For part (b), situation (2):

$$V_{\text{rms}} = 0.707V_m = 0.707(15) \cong \mathbf{10.6\ V}$$

Most DMMs employ a full-wave rectification system to convert the input ac signal to one with an average value. In fact, for the DMM of Fig. 2.25, the same scale factor of Eq. (13.35) is employed. That is, the average value is scaled up by a factor of 1.11 to obtain the rms value. In the digital meters, however, there are no moving parts such as in the d'Arsonval or electrodynamometer movements to display the signal level. Rather, the average value is sensed by a multiprocessor integrated circuit (IC), which in turn determines which digits should appear on the digital display.

Digital meters can also be used to measure nonsinusoidal signals, but the scale factor of each input waveform must first be known (normally provided by the manufacturer in the operator's manual). For instance, the scale factor for an average responding DMM on the ac rms scale will produce an indication for a square-wave input that is 1.11 times the peak value. For a triangular input, the response is 0.555 times the peak value. Obviously, for a sine wave input the response is 0.707 times the peak value.

For any instrument, it is always good practice to read (if only briefly) the operator's manual if it appears you will use the instrument on a regular basis.

For frequency measurements, the frequency counter of Fig. 13.57 provides a digital readout of the frequency or period of waveforms having a frequency range from 5 Hz to 80 MHz. In a period average mode, it can average the cycle time over 10, 100, or 1000 cycles. It has an input impedance of 1 MΩ and an internal rechargeable battery for portability. Note the high degree of accuracy available from the six-digit display.

The Amp-Clamp® of Fig. 13.58 is an instrument that can measure alternating current in the ampere range without having to open the circuit. The loop is opened by squeezing the ''trigger''; then it is placed

FIG. 13.57
Frequency counter. (Courtesy of Tektronix, Inc.)

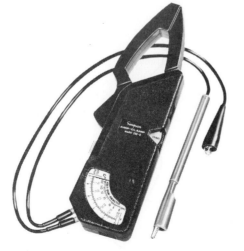

FIG. 13.58
Amp-Clamp®. (Courtesy of Simpson Instruments, Inc.)

around the current-carrying conductor. Through transformer action, the level of current in rms units will appear on the appropriate scale. The accuracy of this instrument is ±3% of full scale at 60 Hz, and its scales have maximum values ranging from 6 A to 300 A. The addition of two leads as indicated in the figure permits its use as both a voltmeter and an ohmmeter.

One of the most versatile and important instruments in the electronics industry is the oscilloscope. It provides a display of the waveform on a cathode-ray tube to permit the detection of irregularities and the determination quantities such as magnitude, frequency, period, dc component, and so on. The unit of Fig. 13.59 is particularly interesting for two reasons: It is portable (working off internal batteries), and it is very small and lightweight. It weighs only 3.5 lb and is approximately 3″ × 5″ × 10″ in size. It has an input impedance of 1 MΩ and a time base that can be set for 5 μs to 500 ms per horizontal division. The vertical scale can be set to sensitivities extending from 1 mV to 50 V per division. This oscilloscope can also display two signals (dual trace) at the same time for magnitude and phase comparisons.

FIG. 13.59

Miniscope. (*Courtesy of Tektronix, Inc.*)

PROBLEMS

Section 13.4

1. For the cycles of the periodic waveform shown in Fig. 13.60:
   a. Find the period T.
   b. How many cycles are shown?
   c. What is the frequency?
   *d. Determine the positive amplitude and peak-to-peak value (think!).

2. Repeat Problem 1 for the periodic waveform of Fig. 13.61.

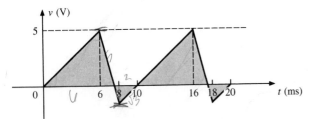

FIG. 13.60

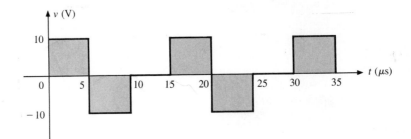

FIG. 13.61

3. Determine the period and frequency of the sawtooth waveform of Fig. 13.62.

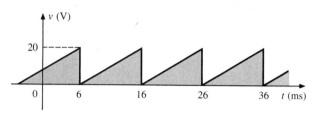

FIG. 13.62

4. Find the period of a periodic waveform whose frequency is

  a. 25 Hz   **b.** 35 MHz

  c. 55 kHz   **d.** 1 Hz

5. Find the frequency of a repeating waveform whose period is

  a. 1/60 s   **b.** 0.01 s

  c. 34 ms   **d.** 25 μs

6. Find the period of a sinusoidal waveform that completes 80 cycles in 24 ms.

7. If a periodic waveform has a frequency of 20 Hz, how long (in seconds) will it take to complete 5 cycles?

8. What is the frequency of a periodic waveform that completes 42 cycles in 6 s?

9. Sketch a periodic square wave like that appearing in Fig. 13.61 with a frequency of 20,000 Hz and a peak value of 10 mV.

Section 13.5

10. Convert the following degrees to radians:

  a. 45°   **b.** 60°

  c. 120°   **d.** 270°

  e. 178°   **f.** 221°

11. Convert the following radians to degrees:

  a. $\pi/4$ **b.** $\pi/6$

  c. $\frac{1}{10}\pi$ **d.** $\frac{7}{6}\pi$

  e. 3π **f.** 0.55π

12. Find the angular velocity of a waveform with a period of

  a. 2 s   **b.** 0.3 ms

  c. 4 μs   **d.** 1/25 s

13. Find the angular velocity of a waveform with a frequency of

  a. 50 Hz   **b.** 600 Hz

  c. 2 kHz   **d.** 0.004 MHz

14. Find the frequency and period of sine waves having an angular velocity of

  a. 754 rad/s   **b.** 8.4 rad/s

  c. 6000 rad/s   **d.** 1/16 rad/s

15. Given $f = 60\,\text{Hz}$, determine how long it will take the sinusoidal waveform to pass through an angle of $45°$.

16. If a sinusoidal waveform passes through an angle of $30°$ in 5 ms, determine the angular velocity of the waveform.

Section 13.6

17. Find the amplitude and frequency of the following waves:
  a. 20 sin 377*t*   **b.** 5 sin 754*t*
  c. 10^6 sin 10,000*t*   **d.** 0.001 sin 942*t*
  e. −7.6 sin 43.6*t*   **f.** 1/42 sin 6.28*t*

18. Sketch 5 sin 754*t* with the abscissa
  a. angle in degrees.
  b. angle in radians.
  c. time in seconds.

19. Sketch 10^6 sin 10,000*t* with the abscissa
  a. angle in degrees.
  b. angle in radians.
  c. time in seconds.

20. Sketch −7.6 sin 43.6*t* with the abscissa
  a. angle in degrees.
  b. angle in radians.
  c. time in seconds.

21. If $e = 300$ sin 157*t*, how long (in seconds) does it take this waveform to complete 1/2 cycle?

22. Given $i = 0.5$ sin α, determine i at $\alpha = 72°$.

23. Given $v = 20$ sin α, determine v at $\alpha = 1.2\pi$.

24. Given $v = 30 \times 10^{-3}$ sin α, determine the angles at which v will be 6 mV.

*25. If $v = 40\,\text{V}$ at $\alpha = 30°$ and $t = 1\,\text{ms}$, determine the mathematical expression for the sinusoidal voltage.

Section 13.7

26. Sketch sin(377*t* + 60°) with the abscissa
  a. angle in degrees.
  b. angle in radians.
  c. time in seconds.

27. Sketch the following waveforms:
  a. 50 sin(ωt + 0°)   **b.** −20 sin(ωt + 2°)
  c. 5 sin(ωt + 60°)   **d.** 4 cos ωt
  e. 2 cos(ωt + 10°)   **f.** −5 cos(ωt + 20°)

28. Find the phase relationship between the waveforms of each set:
  a. $v = 4$ sin(ωt + 50°)
     $i = 6$ sin(ωt + 40°)
  b. $v = 25$ sin(ωt − 80°)
     $i = 5 \times 10^{-3}$ sin(ωt − 10°)
  c. $v = 0.2$ sin(ωt − 60°)
     $i = 0.1$ sin(ωt + 20°)
  d. $v = 200$ sin(ωt − 210°)
     $i = 25$ sin(ωt − 60°)

*29. Repeat Problem 28 for the following sets:

a. $v = 2 \cos(\omega t - 30°)$ **b.** $v = -1 \sin(\omega t + 20°)$
    $i = 5 \sin(\omega t + 60°)$ $i = 10 \sin(\omega t - 70°)$

c. $v = -4 \cos(\omega t + 90°)$
    $i = -2 \sin(\omega t + 10°)$

30. Write the analytical expression for the waveforms of Fig. 13.63 with the phase angle in degrees.

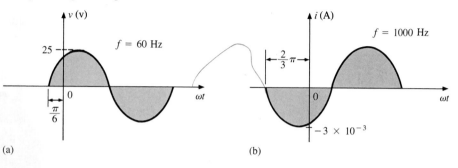

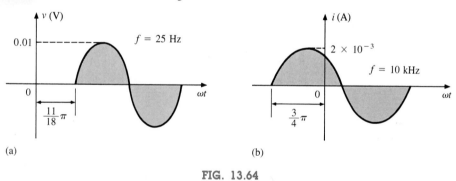

FIG. 13.63

31. Repeat Problem 30 for the waveforms of Fig. 13.64.

FIG. 13.64

Section 13.8

32. Find the average value of the periodic waveforms of Fig. 13.65 over one full cycle.

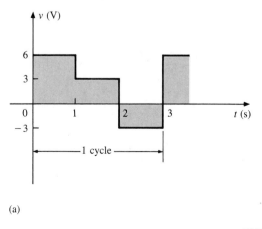

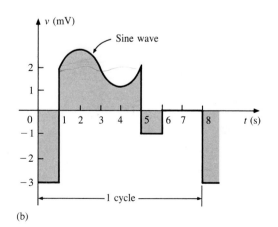

FIG. 13.65

33. Repeat Problem 32 for the waveforms of Fig. 13.66.

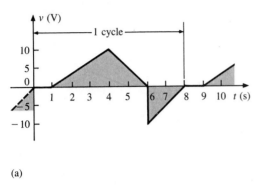

(a)

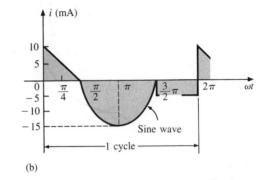

(b)

FIG. 13.66

34. a. By the method of approximation, using familiar geometric shapes, find the area under the curve of Fig. 13.67 from zero to 10 s. Compare your solution with the actual area of 5 volt-seconds (V-s).
b. Find the average value of the waveform from zero to 10 s.

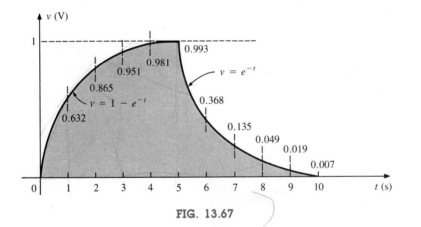

FIG. 13.67

Section 13.9

35. Find the effective values of the following sinusoidal waveforms:
a. $v = 20 \sin 754t$
b. $v = 7.07 \sin 377t$
c. $i = 0.006 \sin(400t + 20°)$
d. $i = 16 \times 10^{-3} \sin(377t - 10°)$

36. Write the sinusoidal expressions for voltages and currents having the following effective values at a frequency of 60 Hz with zero phase shift:
a. 1.414 V   **b.** 70.7 V
c. 0.06 A   **d.** 24 μA

37. Find the effective value of the periodic waveform of Fig. 13.68 over one full cycle.

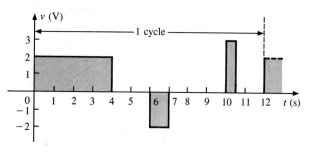

FIG. 13.68

38. Repeat Problem 37 for the waveform of Fig. 13.69.

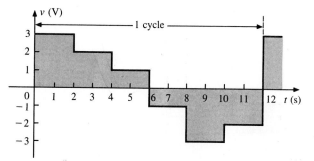

FIG. 13.69

39. What are the average and effective values of the square wave of Fig. 13.70?

40. What are the average and effective values of the waveform of Fig. 13.61?

41. What is the average value of the waveform of Fig. 13.62?

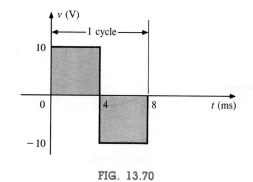

FIG. 13.70

Section 13.10

42. Determine the reading of the meter for each situation of Fig. 13.71.

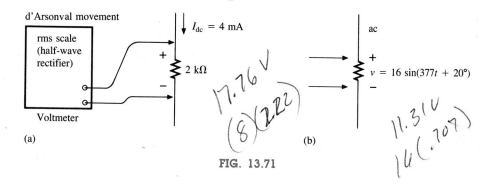

FIG. 13.71

Computer Problems

43. Given a sinusoidal function, write a program to determine the effective value, frequency, and period.

44. Given two sinusoidal functions, write a program to determine the phase shift between the two waveforms, and indicate which is leading or lagging.

45. Given an alternating pulse waveform, write a program to determine the average and effective value of the waveform over one complete cycle.

GLOSSARY

Alternating waveform   A waveform that oscillates above and below a defined reference level.

Amp-Clamp®   A clamp-type instrument that will permit noninvasive current measurements and that can be used as a conventional voltmeter or ohmmeter.

Angular velocity   The velocity with which a radius vector projecting a sinusoidal function rotates about its center.

Average value   The level of a waveform defined by the condition that the area enclosed by the curve above this level is exactly equal to the area enclosed by the curve below this level.

Cycle   A portion of a waveform contained in one period of time.

Effective value   The equivalent dc value of any alternating voltage or current.

Electrodynamometer meters   Instruments that can measure both ac and dc quantities without a change in internal circuitry.

Frequency (f)   The number of cycles of a periodic waveform that occur in one second.

Frequency counter   An instrument that will provide a digital display of the frequency or period of a periodic time-varying signal.

Instantaneous value   The magnitude of a waveform at any instant of time, denoted by lowercase letters.

Oscilloscope   An instrument that will display, through the use of a cathode-ray tube, the characteristics of a time-varying signal.

Peak-to-peak value   The magnitude of the total swing of a signal from positive to negative peaks. The sum of the absolute values of the positive and negative peak values.

Peak value   The maximum value of a waveform, denoted by uppercase letters.

Period (T)   The time interval between successive repetitions of a periodic waveform.

Periodic waveform   A waveform that continually repeats itself after a defined time interval.

Phase relationship   An indication of which of two waveforms leads or lags the other, and by how many degrees or radians.

Radian   A unit of measure used to define a particular segment of a circle. One radian is approximately equal to 57.3°; 2π radians are equal to 360°.

Rectifier-type ac meter   An instrument calibrated to indicate the effective value of a current or voltage through the use of a rectifier network and d'Arsonval-type movement.

rms value   The root-mean-square or effective value of a waveform.

Sinusoidal ac waveform   An alternating waveform of unique characteristics that oscillates with equal amplitude above and below a given axis.

VOM   A multimeter with the capability to measure resistance and both ac and dc levels of current and voltage.

Waveform   The path traced by a quantity, plotted as a function of some variable such as position, time, degrees, temperature, and so on.

14

The Basic Elements and Phasors

14.1 INTRODUCTION

The response of the basic R, L, and C elements to a sinusoidal voltage and current will be examined in this chapter with special note of how frequency will affect the "opposing" characteristic of each element. Phasor notation will then be introduced to establish a method of analysis that permits a direct correspondence with a number of the methods, theorems, and concepts introduced in the dc chapters.

14.2 THE DERIVATIVE

It is fundamental to the understanding of the response of the basic R, L, and C elements to a sinusoidal signal that the concept of the *derivative* be examined in some detail. It will not be necessary that you become proficient in the mathematical technique but simply that you understand the impact of a relationship defined by a derivative.

Recall from Section 10.11 that the derivative dx/dt is defined as the rate of change of x with respect to time. If x fails to change at a particular instant, $dx = 0$, and the derivative is zero. For the sinusoidal waveform, dx/dt is zero only at the positive and negative peaks ($\omega t = \pi/2$ and $\frac{3}{2}\pi$ in Fig. 14.1), since x fails to change at these instants of time. The derivative dx/dt is actually the slope of the graph at any instant of time.

A close examination of the sinusoidal waveform will also indicate that the greatest change in x will occur at the instants $\omega t = 0$, π, and

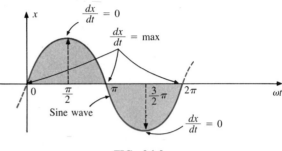

FIG. 14.1

2π. The derivative is therefore a maximum at these points. At 0 and 2π, x increases at its greatest rate, and the derivative is given a positive sign since x increases with time. At π, dx/dt decreases at the same rate as it increases at 0 and 2π, but the derivative is given a negative sign since x decreases with time. Since the rate of change at 0, π, and 2π is the same, the magnitude of the derivative at these points is the same also. For various values of ωt between these maxima and minima, the derivative will exist and have values from the minimum to the maximum inclusive. A plot of the derivative in Fig. 14.2 shows that the derivative of a sine wave is a cosine wave.

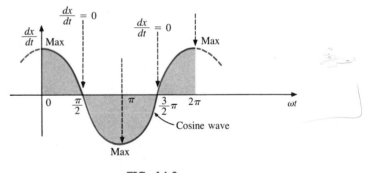

FIG. 14.2

The peak value of the cosine wave is directly related to the frequency of the original waveform. The higher the frequency, the steeper the slope at the abscissa, and the greater the value of dx/dt, as shown in Fig. 14.3.

Note in Fig. 14.3 that even though both waveforms have the same peak value, the sinusoidal function with the higher frequency produces the larger peak value for the derivative. In addition, note that the derivative has the same period and frequency as the original sinusoidal waveform.

For a sinusoidal voltage

$$e(t) = E_m \sin(\omega t \pm \theta)$$

the derivative can be found directly by differentiation (calculus) to produce the following:

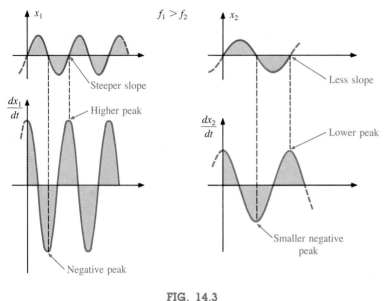

$f_1 > f_2$

FIG. 14.3

$$\frac{d}{dt} e(t) = \omega E_m \cos(\omega t \pm \theta)$$

$$= 2\pi f E_m \cos(\omega t \pm \theta)$$

(14.1)

clearly substantiating that the derivative of a sinusoidal function is a cosine wave and that the peak value is directly related to the frequency.

By similar means, if

$$e(t) = E_m \cos(\omega t \pm \theta)$$

then

$$\frac{d}{dt} e(t) = -\omega E_m \sin(\omega t \pm \theta)$$

$$= -2\pi f E_m \sin(\omega t \pm \theta)$$

(14.2)

14.3 RESPONSE OF BASIC R, L, AND C ELEMENTS TO A SINUSOIDAL VOLTAGE OR CURRENT

Using Ohm's law and the basic equations for the capacitor and inductor, we can now apply the sinusoidal voltage or current to the basic R, L, and C elements.

Resistor

For power-line frequencies and frequencies up to a few kilohertz, resistance is, for all practical purposes, unaffected by the frequency of the

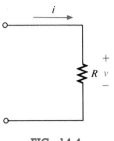

FIG. 14.4

applied sinusoidal voltage or current. For this frequency region, the resistor R of Fig. 14.4 can be treated as a constant, and Ohm's law applied as follows. For $v = V_m \sin \omega t$,

$$i = \frac{v}{R} = \frac{V_m \sin \omega t}{R} = \frac{V_m}{R} \sin \omega t = I_m \sin \omega t$$

where

$$\boxed{I_m = \frac{V_m}{R}} \qquad\qquad \textbf{(14.3)}$$

In addition, for a given i,

$$v = iR = (I_m \sin \omega t)R = I_m R \sin \omega t = V_m \sin \omega t$$

where

$$\boxed{V_m = I_m R} \qquad\qquad \textbf{(14.4)}$$

A plot of v and i in Fig. 14.5 reveals that for a purely resistive element, the voltage across and the current through the element are *in phase*.

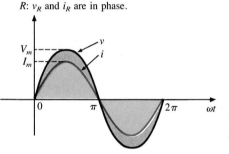

R: v_R and i_R are in phase.

FIG. 14.5

Inductor

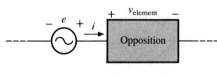

FIG. 14.6

For the series configuration of Fig. 14.6, the voltage appearing across the boxed-in element opposes the source e and thereby reduces the magnitude of the current i. The magnitude of the voltage across the element is directly related to the opposition of the element to the flow of charge, or current i. For a resistive element, we have found that the opposition is its resistance and that v_{element} and i are determined by $v_{\text{element}} = iR$.

For the inductor, we found in Chapter 12 that the voltage across an inductor is directly related to the rate of change of current through the coil. Consequently, the higher the frequency, the greater will be the rate of change of current through the coil, and the greater the magnitude of the voltage. In addition, we found in the same chapter that the inductance of a coil will determine the rate of change of flux linking a coil for a particular change in current through the coil. The higher the inductance, the greater the rate of change of the flux linkages, and the greater the resulting voltage across the coil.

The inductive voltage, therefore, is directly related to the frequency (or, more specifically, the angular velocity of the sinusoidal ac current through the coil) and the inductance of the coil. For increasing values of ω and L in Fig. 14.7, the magnitude of v_L will increase as described above.

Utilizing the similarities between Figs. 14.6 and 14.7, we find that increasing levels of v_L are directly related to increasing levels of opposi-

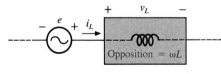

FIG. 14.7

tion in Fig. 14.6. Since v_L will increase with both ω and L, the opposition of an inductive element is defined as their product in Fig. 14.7.

We will now verify some of the preceding conclusions using a more mathematical approach and then define a few important quantities to be employed in the sections and chapters to follow.

For the inductor of Fig. 14.8, we know that

$$v_L = L \frac{di_L}{dt}$$

and, applying differentiation,

$$\frac{di_L}{dt} = \frac{d}{dt} (I_m \sin \omega t) = \omega I_m \cos \omega t$$

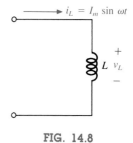

FIG. 14.8

Therefore,

$$v_L = L \frac{di_L}{dt} = L(\omega I_m \cos \omega t) = \omega L I_m \cos \omega t$$

or

$$v_L = V_m \sin(\omega t + 90°)$$

where

$$V_m = \omega L I_m$$

Note that the peak value of v_L is directly related to ω and L as predicted in the discussion above.

A plot of v_L and i_L in Fig. 14.9 reveals that v_L *leads* i_L by 90°, or i_L *lags* v_L by 90°.

If

$$i_L = I_m \sin(\omega t \pm \theta)$$

then

$$v_L = \omega L I_m \sin(\omega t \pm \theta + 90°)$$

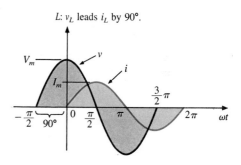

FIG. 14.9

The opposition to current developed by an inductor in a sinusoidal ac network can be found by applying Eq. (4.1):

$$\text{Effect} = \frac{\text{cause}}{\text{opposition}}$$

which for our purposes can be written

$$\text{Opposition} = \frac{\text{cause}}{\text{effect}}$$

Substituting values, we have

$$\text{Opposition} = \frac{V_m}{I_m} = \frac{\omega L I_m}{I_m} = \omega L$$

which agrees with the results obtained earlier.

The quantity ωL, called the *reactance* (from the word *reaction*) of an inductor, is symbolically represented by X_L and is measured in ohms; that is,

$$\boxed{X_L = \omega L} \qquad \text{(ohms, } \Omega\text{)} \qquad \textbf{(14.5)}$$

Inductive reactance is the opposition to the flow of current, which results in the continual interchange of energy between the source and the magnetic field of the inductor. In other words, reactance, unlike resistance (which dissipates energy in the form of heat), does not dissipate electrical energy.

Capacitor

Let us now return to the series configuration of Fig. 14.6 and insert the capacitor as the element of interest. For the capacitor, however, we will determine i for a particular voltage across the element. When this approach reaches its conclusion, the relationship between the voltage and current will be known, and the opposing voltage (v_{element}) can be determined for any sinusoidal current i.

Our investigation of the inductor revealed that the inductive voltage across a coil opposes the instantaneous change in current through the coil. For capacitive networks, the voltage across the capacitor is limited by the rate at which charge can be deposited on, or released by, the plates of the capacitor during the charging and discharging phases, respectively. In other words, an instantaneous change in voltage across a capacitor is opposed by the fact that there is an element of time required to deposit charge on (or release charge from) the plates of a capacitor, and $V = Q/C$.

Since capacitance is a measure of the rate at which a capacitor will store charge on its plates *for a particular change in voltage across the capacitor, the greater the value of capacitance, the greater will be the resulting capacitive current*. In addition, the fundamental equation relating the voltage across a capacitor to the current of a capacitor [$i = C (dv/dt)$] indicates that *for a particular capacitance, the greater the rate of change of voltage across the capacitor, the greater the capacitive current*. Certainly, an increase in frequency corresponds to an increase in the rate of change of voltage across the capacitor and to an increase in the current of the capacitor.

The current of a capacitor is therefore directly related to the frequency (or, again more specifically, the angular velocity) and the capacitance of the capacitor. An increase in either quantity will result in an increase in the current of the capacitor. For the basic configuration of Fig. 14.10, however, we are interested in determining the opposition of the capacitor as related to the resistance of a resistor and ωL for the inductor. Since an increase in current corresponds to a decrease in opposition, and i_C is proportional to ω and C, the opposition of a capacitor is directly related to the reciprocal of ωC, or $1/\omega C$, as shown in Fig. 14.10. In other words, the higher the angular velocity (or frequency) and capacitance, the less the opposition to the current i_C or the lower the

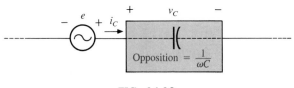

FIG. 14.10

countervoltage of the capacitor (v_C), which limits the current i_C as indicated by Fig. 14.6.

We will now verify, as we did for the inductor, some of these conclusions using a more mathematical approach. Certain important quantities used repeatedly in the following analysis will then be defined.

For the capacitor of Fig. 14.11, we know that

$$i_C = C \frac{dv_C}{dt}$$

and that

$$\frac{dv_C}{dt} = \frac{d}{dt}(V_m \sin \omega t) = \omega V_m \cos \omega t$$

Therefore,

$$i_C = C \frac{dv_C}{dt} = C(\omega V_m \cos \omega t) = \omega C V_m \cos \omega t$$

or

$$i_C = I_m \sin(\omega t + 90°)$$

where

$$I_m = \omega C V_m$$

Note that the peak value of i_C is directly related to ω and C, as predicted in the discussion above.

A plot of v_C and i_C in Fig. 14.12 reveals that i_C *leads* v_C by 90°, or v_C *lags* i_C by 90°.

If

$$v_C = V_m \sin(\omega t \pm \theta)$$

then

$$i_C = \omega C V_m \sin(\omega t \pm \theta + 90°)$$

Applying

$$\text{Opposition} = \frac{\text{cause}}{\text{effect}}$$

and substituting values, we obtain

$$\text{Opposition} = \frac{V_m}{I_m} = \frac{V_m}{\omega C V_m} = \frac{1}{\omega C}$$

which agrees with the results obtained above.

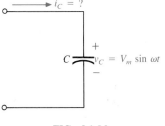

FIG. 14.11

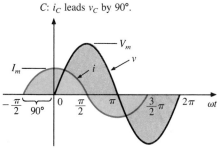

C: i_C leads v_C by 90°.

FIG. 14.12

The quantity $1/\omega C$, called the *reactance* of a capacitor, is symbolically represented by X_C and is measured in ohms; that is,

$$\boxed{X_C = \frac{1}{\omega C}} \qquad (\Omega) \qquad \textbf{(14.6)}$$

Capacitive reactance is the opposition to the flow of charge, which results in the continual interchange of energy between the source and the electric field of the capacitor. Like the inductor, the capacitor does *not* dissipate energy in any form (ignoring the effects of the leakage resistance).

In the circuits just considered, the current was given in the inductive circuit, and the voltage in the capacitive circuit. This was done to avoid the use of integration in finding the unknown quantities. In the inductive circuit,

$$v_L = L\,\frac{di_L}{dt}$$

but

$$\boxed{i_L = \frac{1}{L}\int v_L\,dt} \qquad \textbf{(14.7)}$$

In the capacitive circuit,

$$i_C = C\,\frac{dv_C}{dt}$$

but

$$\boxed{v_C = \frac{1}{C}\int i_C\,dt} \qquad \textbf{(14.8)}$$

Shortly, we shall consider a method of analyzing ac circuits that will permit us to solve for an unknown quantity with sinusoidal input without having to use direct integration or differentiation.*

It is possible to determine whether a circuit with one or more elements is predominantly capacitive or inductive by noting the phase relationship between the input voltage and current. *If the current leads the voltage, the circuit is predominantly capacitive, and if the voltage leads the current, it is predominantly inductive.*

Since we now have an equation for the reactance of an inductor or capacitor, we do not need to use derivatives or integration in the examples to be considered. Simply applying Ohm's law, $I_m = E_m/X_L$ (or

---

*A mnemonic phrase sometimes used to remember the phase relationship between the voltage and current of a coil and capacitor is "*ELI* the *ICE* man." Note that the *L* (inductor) has the *EL* before the *I* (*e* leads *i* by 90°), and the *C* (capacitor) has the *I* before the *E* (*i* leads *e* by 90°).

X_C), and keeping in mind the phase relationship between the voltage and current for each element, will be sufficient to complete the examples.

EXAMPLE 14.1. The voltage across a resistor is indicated. Find the sinusoidal expression for the current if the resistor is 10 Ω. Sketch the v and i curves with the angle ωt as the abscissa.

a. $v = 100 \sin 377t$
b. $v = 25 \sin(377t + 60°)$

Solutions:

a. $i = \dfrac{v}{R} = \dfrac{100}{10} \sin 377t$

and

$i = \mathbf{10 \sin 377}t$

The curves are sketched in Fig. 14.13.

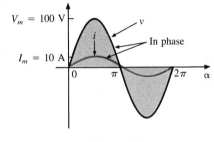

FIG. 14.13

b. $i = \dfrac{v}{R} = \dfrac{25}{10} \sin(377t + 60°)$

and

$i = \mathbf{2.5 \sin(377}t + \mathbf{60°)}$

The curves are sketched in Fig. 14.14.

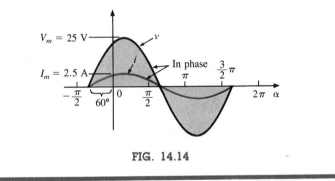

FIG. 14.14

EXAMPLE 14.2. The current through a 5-Ω resistor is given. Find the sinusoidal expression for the voltage across the resistor for $i = 40 \sin(377t + 30°)$.

Solution:

$$v = iR = (40)(5) \sin(377t + 30°)$$

and

$$v = \mathbf{200 \sin(377}t + \mathbf{30°)}$$

EXAMPLE 14.3. The current through a 0.1-H coil is given. Find the sinusoidal expression for the voltage across the coil. Sketch the v and i curves.

a. $i = 10 \sin 377t$

b. $i = 7 \sin(377t - 70°)$

Solutions:

a. $X_L = \omega L = 37.7\,\Omega$

$V_m = I_m X_L = (10)(37.7) = 377\text{ V}$

and we know that for a coil, v leads i by 90°. Therefore,

$v = \mathbf{377\ sin(377\mathit{t} + 90°)}$

The curves are sketched in Fig. 14.15.

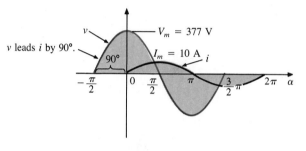

FIG. 14.15

b. $X_L = 37.7\,\Omega$

$V_m = I_m X_L = (7)(37.7) = 263.9\text{ V}$

and we know that for a coil, v leads i by 90°. Therefore,

$v = 263.9 \sin(377t - 70° + 90°)$

and

$v = \mathbf{263.9\ sin(377\mathit{t} + 20°)}$

The curves are sketched in Fig. 14.16.

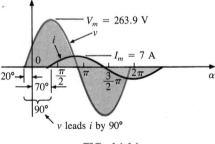

FIG. 14.16

EXAMPLE 14.4. The voltage across a 0.5-H coil is given. What is the sinusoidal expression for the current?

$$v = 100 \sin 20t$$

Solution:

$$X_L = \omega L = (20)(0.5) = 10\,\Omega$$

$$I_m = \frac{V_m}{X_L} = \frac{100}{10} = 10\text{ A}$$

and we know that i lags v by 90°. Therefore,

$$i = \mathbf{10\ sin(20\mathit{t} - 90°)}$$

EXAMPLE 14.5. The voltage across a 1-μF capacitor is given. What is the sinusoidal expression for the current? Sketch the v and i curves.

$$v = 30 \sin 400t$$

Solution:

$$X_C = \frac{1}{\omega C} = \frac{1}{(400)(1 \times 10^{-6})} = \frac{10^6}{400} = 2500 \,\Omega$$

$$I_m = \frac{V_m}{X_C} = \frac{30}{2500} = 0.0120 \text{ A} = 12.0 \text{ mA}$$

and we know that for a capacitor, i leads v by 90°. Therefore,

$$i = \mathbf{12 \times 10^{-3} \sin(400}t\mathbf{ + 90°)}$$

The curves are sketched in Fig. 14.17.

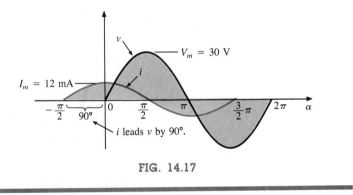

FIG. 14.17

EXAMPLE 14.6. The current through a 100-μF capacitor is given. Find the sinusoidal expression for the voltage across the capacitor.

$$i = 40 \sin(500t + 60°)$$

Solution:

$$X_C = \frac{1}{\omega C} = \frac{1}{(500)(100 \times 10^{-6})} = \frac{10^6}{5 \times 10^4} = \frac{10^2}{5} = 20 \,\Omega$$

$$V_m = I_m X_C = (40)(20) = 800 \text{ V}$$

and we know that for a capacitor, v lags i by 90°. Therefore,

$$v = 800 \sin(500t + 60° - 90°)$$

and

$$v = \mathbf{800 \sin(500}t\mathbf{ - 30°)}$$

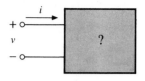

FIG. 14.18

EXAMPLE 14.7. For the following pairs of voltages and currents, indicate whether the element involved is a capacitor, inductor, or resistor, and determine the value of C, L, or R, if sufficient data are given (Fig. 14.18):

a. $v = 100 \sin(\omega t + 40°)$
   $i = 20 \sin(\omega t + 40°)$

b. $v = 1000 \sin(377t + 10°)$
   $i = 5 \sin(377t - 80°)$

c. $v = 500 \sin(157t + 30°)$
   $i = 1 \sin(157t + 120°)$

d. $v = 50 \cos(\omega t + 20°)$
   $i = 5 \sin(\omega t + 110°)$

Solutions:

a. Since v and i are *in phase*, the element is a *resistor*, and

$$R = \frac{V_m}{I_m} = \frac{100}{20} = \mathbf{5\,\Omega}$$

b. Since v *leads* i by 90°, the element is an *inductor*, and

$$X_L = \frac{V_m}{I_m} = \frac{1000}{5} = 200\,\Omega$$

so that

$$X_L = \omega L = 200\,\Omega$$

or

$$L = \frac{200}{\omega} = \frac{200}{377} = \mathbf{0.531\,H}$$

c. Since i *leads* v by 90°, the element is a *capacitor*, and

$$X_C = \frac{V_m}{I_m} = \frac{500}{1} = 500\,\Omega$$

so that

$$X_C = \frac{1}{\omega C} = 500\,\Omega$$

or

$$C = \frac{1}{\omega 500} = \frac{1}{(157)(500)} = \mathbf{12.74\,\mu F}$$

d. $v = 50 \cos(\omega t + 20°) = 50 \sin(\omega t + 20° + 90°)$
   $= 50 \sin(\omega t + 110°)$

Since v and i are *in phase*, the element is a *resistor*, and

$$R = \frac{V_m}{I_m} = \frac{50}{5} = \mathbf{10\,\Omega}$$

For dc circuits, the frequency is zero, since the currents and voltages have constant magnitudes. The reactance of the coil for dc is therefore

$$X_L = 2\pi f L = 2\pi 0 L = 0\,\Omega$$

hence the short-circuit representation for the inductor in dc circuits (Chapter 12). At very high frequencies, $X_L\uparrow\; = 2\pi f\uparrow L$ is very large, and for some practical applications the inductor can be replaced by an open circuit.

The capacitor can be replaced by an open circuit in dc circuits since $f = 0$, and

$$X_C = \frac{1}{2\pi f C} = \frac{1}{2\pi 0 C} = \infty\,\Omega$$

once again substantiating our previous action (Chapter 10). At very high frequencies, for finite capacitances,

$$X_C\downarrow\; = \frac{1}{2\pi f\uparrow C}$$

is very small, and for some practical applications the capacitor can be replaced by a short circuit.

14.4 FREQUENCY RESPONSE OF THE BASIC ELEMENTS

The analysis of Section 14.3 was limited to a particular applied frequency. What is the effect of varying the frequency on the level of opposition offered by a resistive, inductive, or capacitive element? We are aware from the last section that the inductive reactance increases with frequency while the capacitive reactance decreases. However, what is the pattern to this increase or decrease in opposition? Since applied signals may have frequencies extending from a few hertz to megahertz, it is important to be aware of the effect of frequency on the opposition level.

R

For the frequency range of interest for this text, it will be assumed that the resistance of a resistor remains constant for the frequency range of interest. The result is the resistance-versus-frequency plot of Fig. 14.19.

L

The equation

$$X_L = \omega L = 2\pi f L = 2\pi L f$$

is directly related to the straight-line equation

$$y = mx + b$$

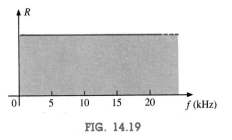

FIG. 14.19

with a slope (m) of $2\pi L$ and a y-intercept (b) of zero. X_L is the y variable and f is the x variable.

Quite obviously, the larger the inductance, the greater the slope for the same frequency range, as shown in Fig. 14.20. Keep in mind, as reemphasized by Fig. 14.20, that the opposition of an inductor at very low frequencies approaches that of a short circuit, while at high frequencies the reactance approaches that of an open circuit.

For the capacitor, the reactance equation

$$X_C = \frac{1}{2\pi f C}$$

can be written

$$X_C f = \frac{1}{2\pi C}$$

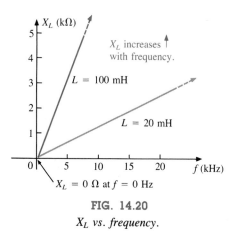

FIG. 14.20

X_L vs. frequency.

which matches the basic format of a parabola,

$$yx = k$$

with the constant $k = 1/2\pi C$, $y = X_C$, and $x = f$.

At $f = 0$ Hz, the reactance of the capacitor is so large, as shown in Fig. 14.21, that it can be replaced by an open-circuit equivalent. As the frequency increases, the reactance decreases, until eventually a short-circuit equivalent would be appropriate. Note how an increase in capacitance causes the reactance to drop off more rapidly with frequency.

EXAMPLE 14.8. At what frequency will the reactance of a 200-mH inductor match the resistance level of a 5-kΩ resistor?

Solution: The resistance remains constant at 5 kΩ for the frequency range of the inductor. Therefore,

$$R = 5000 = X_L = 2\pi f L = 2\pi L f$$
$$= (6.28)(200 \times 10^{-3})f = 1.256f$$

and

$$f = \frac{5000}{1.256} \cong \mathbf{3.98\ kHz}$$

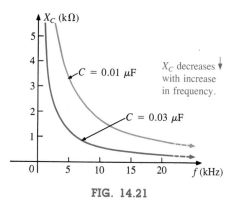

FIG. 14.21

EXAMPLE 14.9. At what frequency will an inductor of 5 mH have the same reactance as a capacitor of 0.1 μF?

Solution:

$$X_L = X_C$$
$$2\pi f L = \frac{1}{2\pi f C}$$
$$f^2 = \frac{1}{4\pi^2 LC}$$

and

$$f = \frac{1}{2\pi\sqrt{LC}} = \frac{1}{6.28\sqrt{(5 \times 10^{-3})(0.1 \times 10^{-6})}}$$

$$= \frac{1}{6.28\sqrt{5 \times 10^{-10}}} = \frac{1}{(6.28)(2.236 \times 10^{-5})}$$

$$f = \frac{10^5}{14.04} \cong \mathbf{7.12\,kHz}$$

14.5 AVERAGE POWER AND POWER FACTOR

The instantaneous power to the load of Fig. 14.22 is

$$p = vi$$

If we consider the general case where

$$v = V_m \sin(\omega t + \beta) \quad \text{and} \quad i = I_m \sin(\omega t + \psi)$$

then

$$p = vi = V_m \sin(\omega t + \beta)I_m \sin(\omega t + \psi)$$
$$= V_m I_m \sin(\omega t + \beta) \sin(\omega t + \psi)$$

Using the trigonometric identity

$$\sin A \sin B = \frac{\cos(A - B) - \cos(A + B)}{2}$$

the function $\sin(\omega t + \beta) \sin(\omega t + \psi)$ becomes

$$\sin(\omega t + \beta) \sin(\omega t + \psi)$$
$$= \frac{\cos[(\omega t + \beta) - (\omega t + \psi)] - \cos[(\omega t + \beta) + (\omega t + \psi)]}{2}$$
$$= \frac{\cos(\beta - \psi) - \cos(2\omega t + \beta + \psi)}{2}$$

so that

$$p = \left[\frac{V_m I_m}{2} \cos(\beta - \psi)\right] - \left[\frac{V_m I_m}{2} \cos(2\omega t + \beta + \psi)\right]$$

A plot of v, i, and p on the same set of axes is shown in Fig. 14.23.

Note that the second factor in the preceding equation is a cosine wave with an amplitude of $V_m I_m/2$, and a frequency twice that of the voltage or current. The average value of this term is zero, producing no net transfer of energy in any one direction.

The first term in the preceding equation, however, has a constant magnitude (no time dependence) and therefore provides some net

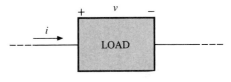

FIG. 14.22

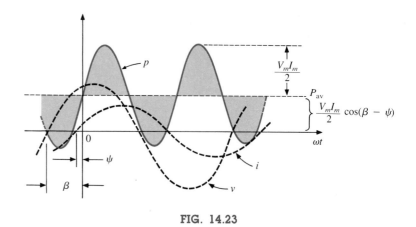

FIG. 14.23

transfer of energy. This term is referred to as the *average power,* the reason for which is obvious from Fig. 14.23. The average power, or *real* power as it is sometimes called, is the power delivered to and dissipated by the load. It corresponds to the power calculations performed for dc networks. The angle $(\beta - \psi)$ is the phase angle between v and i. *Since* $\cos(-\alpha) = \cos \alpha$, *the magnitude of average power is independent of whether v leads i or i leads v.* Defining θ as equal to $|\beta - \psi|$, where $|-|$ indicates that only the magnitude is important and the sign is immaterial, we have

$$P = \frac{V_m I_m}{2} \cos \theta \qquad \text{(watts, W)} \qquad \textbf{(14.9)}$$

where P is the average power in watts. This equation can also be written

$$P = \left(\frac{V_m}{\sqrt{2}}\right)\left(\frac{I_m}{\sqrt{2}}\right) \cos \theta$$

Or, since

$$V_{\text{eff}} = \frac{V_m}{\sqrt{2}} \quad \text{and} \quad I_{\text{eff}} = \frac{I_m}{\sqrt{2}}$$

Eq. (14.9) becomes

$$P = V_{\text{eff}} I_{\text{eff}} \cos \theta \qquad \textbf{(14.10)}$$

Let us now apply Eqs. (14.9) and (14.10) to the basic R, L, and C elements.

Resistor

In a purely resistive circuit, since v and i are in phase, $|\beta - \psi| = \theta = 0°$, and $\cos \theta = \cos 0° = 1$, so that

$$\boxed{P = \frac{V_m I_m}{2} = V_{\text{eff}} I_{\text{eff}}} \quad \text{(W)} \qquad \textbf{(14.11)}$$

Or, since

$$I_{\text{eff}} = \frac{V_{\text{eff}}}{R}$$

then

$$\boxed{P = \frac{V_{\text{eff}}^2}{R} = I_{\text{eff}}^2 R} \quad \text{(W)} \qquad \textbf{(14.12)}$$

Inductor

In a purely inductive circuit, since v leads i by 90°, $|\beta - \psi| = \theta = 90°$. Therefore,

$$P = \frac{V_m I_m}{2} \cos 90° = \frac{V_m I_m}{2}(0) = \mathbf{0}$$

The average power or power dissipated by the ideal inductor (no associated resistance) is always zero.

Capacitor

In a purely capacitive circuit, since i leads v by 90°, $|\beta - \psi| = \theta = |-90°| = 90°$. Therefore,

$$P = \frac{V_m I_m}{2} \cos(90°) = \frac{V_m I_m}{2}(0) = \mathbf{0}$$

The average power or power dissipated by the ideal capacitor (no associated resistance) is always zero.

EXAMPLE 14.10. Find the average power dissipated in a circuit whose input current and voltage are the following:

$$i = 5 \sin(\omega t + 40°)$$
$$v = 10 \sin(\omega t + 40°)$$

Solution: Since v and i are in phase, the circuit appears at the input terminals to be purely resistive. Therefore,

$$P = \frac{V_m I_m}{2} = \frac{(10)(5)}{2} = \mathbf{25\,W}$$

or

$$R = \frac{V_m}{I_m} = \frac{10}{5} = 2\,\Omega$$

and

$$P = \frac{V_{eff}^2}{R} = \frac{[(0.707)(10)]^2}{2} = 25\ \text{W}$$

or

$$P = I_{eff}^2 R = [(0.707)(5)]^2(2) = 25\ \text{W}$$

For the following examples, the circuit consists of a combination of resistances and reactances producing phase angles between the input current and voltage different from 0° or 90°.

EXAMPLE 14.11. Determine the average power delivered to networks having the following input voltage and current:
a. $v = 100\ \sin(\omega t + 40°)$
   $i = 20\ \sin(\omega t + 70°)$
b. $v = 150\ \sin(\omega t - 70°)$
   $i = 3\ \sin(\omega t - 50°)$
Solutions:
a. $V_m = 100,\ \beta = 40°$
   $I_m = 20,\ \psi = 70°$
   $\theta = |\beta - \psi| = |40° - 70°| = |-30°| = 30°$

and

$$P = \frac{V_m I_m}{2}\cos\theta = \frac{(100)(20)}{2}\cos(30°) = (1000)(0.866)$$
$$= \mathbf{866\ W}$$

b. $V_m = 150,\ \beta = -70°$
   $I_m = 3,\ \psi = -50°$
   $\theta = |\beta - \psi| = |-70° - (-50°)|$
   $= |-70° + 50°| = |-20°| = 20°$

and

$$P = \frac{V_m I_m}{2}\cos\theta = \frac{(150)(3)}{2}\cos(20°) = (225)(0.9397)$$
$$= \mathbf{211.43\ W}$$

Power Factor

The frequency and the elements of the parallel network of Fig. 14.24(a) were chosen such that

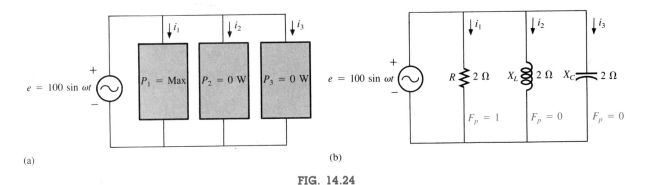

(a)

(b)

FIG. 14.24

$$i_1 = 50 \sin \omega t$$
$$i_2 = 50 \sin(\omega t - 90°)$$
$$i_3 = 50 \sin(\omega t + 90°)$$

Therefore, for each parallel branch, the peak values (or effective values) of the voltage and current are the same. Yet, as indicated in Fig. 14.24(a), the power to two of the branches is zero, and a maximum to the third. In our power equation, the only factor that accounts for this variation is cos θ, related to the phase relationship between v and i. This factor, cos θ, is called the *power factor* and is symbolically represented by F_p; that is,

$$\boxed{\text{Power factor} = F_p = \cos \theta} \qquad (14.13)$$

The more reactive a load, the lower the power factor, and the smaller the average power delivered. The more resistive the load, the higher the power factor, and the greater the real power delivered. The elements within each load of Fig. 14.24(a) appear in Fig. 14.24(b). Note the power factor for each element in the figure. Low power factors are usually avoided, since a high current would be required to deliver any appreciable power. This higher current demand produces higher heating losses and, consequently, the system operates at a lower efficiency.

In terms of the average power and the terminal voltage and current,

$$\boxed{F_p = \cos \theta = \frac{P}{V_{\text{eff}} I_{\text{eff}}}} \qquad (14.14)$$

The terms *leading* and *lagging* are often written in conjunction with the power factor. *They are defined by the current through the load.* If the current leads the voltage across a load, the load has a leading power factor. If the current lags the voltage across the load, the load has a lagging power factor. In other words, *capacitive networks have leading power factors, and inductive networks have lagging power factors.*

EXAMPLE 14.12. Determine the power factors of the following loads, and indicate whether they are leading or lagging:

a. Fig. 14.25

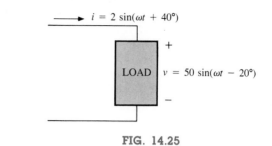

FIG. 14.25

b. Fig. 14.26

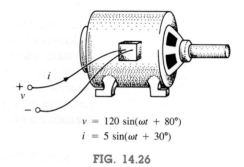

$v = 120 \sin(\omega t + 80°)$
$i = 5 \sin(\omega t + 30°)$

FIG. 14.26

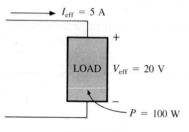

FIG. 14.27

c. Fig. 14.27

Solutions:

a. $F_p = \cos \theta = \cos 60° = \textbf{0.5 leading}$

b. $F_p = \cos \theta = \cos 50° = \textbf{0.6428 lagging}$

c. $F_p = \cos \theta = \dfrac{P}{V_{\text{eff}} I_{\text{eff}}} = \dfrac{100}{(20)(5)} = \dfrac{100}{100} = \textbf{1}$

The load is resistive, and F_p is neither leading nor lagging.

14.6 COMPLEX NUMBERS

In our analysis of dc networks, we found it necessary to determine the algebraic sum of voltages and currents. Since the same will also be true for ac networks, the question arises, How do we determine the algebraic sum of two or more voltages (or currents) that are varying sinusoidally? Although one solution would be to find the algebraic sum on a point-to-point basis (as shown in Section 14.12), this would be a long and tedious process in which accuracy would be directly related to the scale employed.

It is the purpose of this chapter to introduce a system of *complex numbers* which, when related to the sinusoidal ac waveform, will result in a technique for finding the algebraic sum of sinusoidal waveforms that is quick, direct, and accurate. In the following chapters, the technique will be extended to permit the analysis of sinusoidal ac networks in a manner very similar to that applied to dc networks. The methods and theorems as described for dc networks can then be applied to sinusoidal ac networks with little difficulty.

A *complex number* represents a point in a two-dimensional plane located with reference to two distinct axes. This point can also determine a radius vector drawn from the origin to the point. The horizontal axis is called the *real* axis, while the vertical axis is called the *imaginary* axis. Both are labeled in Fig. 14.28. For reasons that will be obvious later, the real axis is sometimes called the *resistance* axis, and the imaginary axis, the *reactance* axis. Every number from zero to $\pm\infty$ can be represented by some point along the real axis. Prior to the development of this system of complex numbers, it was believed that any number not on the real axis would not exist—hence the term *imaginary* for the vertical axis.

In the complex plane, the horizontal or real axis represents all positive numbers to the right of the imaginary axis and all negative numbers to the left of the imaginary axis. All positive imaginary numbers are represented above the real axis, and all negative imaginary numbers, below the real axis. The symbol j (or sometimes i) is used to denote an imaginary number.

There are two forms used to represent a complex number: *rectangular* and *polar*. Each can represent a point in the plane or a radius vector drawn from the origin to that point.

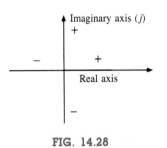

FIG. 14.28

14.7 RECTANGULAR FORM

The format for the rectangular form is

$$\boxed{C = A + jB}$$ (14.15)

as shown in Fig. 14.29.

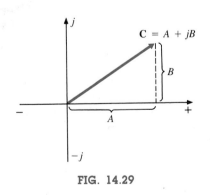

FIG. 14.29

EXAMPLE 14.13.   Sketch the following complex numbers in the complex plane:

a. $\mathbf{C} = 3 + j4$
b. $\mathbf{C} = 0 - j6$
c. $\mathbf{C} = -10 - j20$

Solutions:

a. See Fig. 14.30.

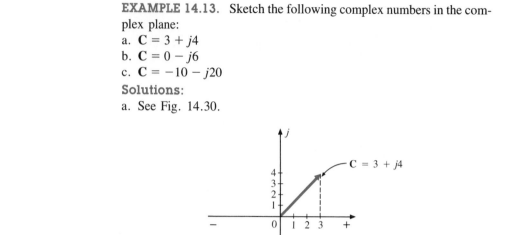

FIG. 14.30

b. See Fig. 14.31.
c. See Fig. 14.32.

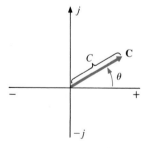

FIG. 14.31

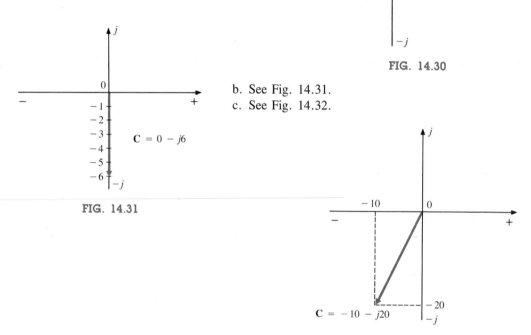

FIG. 14.32

14.8 POLAR FORM

The format for the polar form is

$$\boxed{\mathbf{C} = C \angle \theta} \qquad \textbf{(14.16)}$$

where C indicates magnitude only and θ is always measured counterclockwise (CCW) from the *positive real axis*, as shown in Fig. 14.33.

FIG. 14.33

A negative sign has the effect shown in Fig. 14.34:

$$-\mathbf{C} = -C \, \angle\theta = C \, \angle\theta \pm \pi \qquad\qquad \textbf{(14.17)}$$

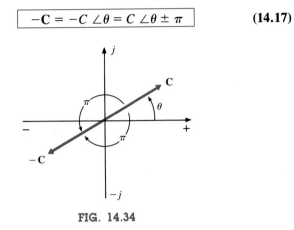

FIG. 14.34

EXAMPLE 14.14. Sketch the following complex numbers in the complex plane:

a. $\mathbf{C} = 5 \, \angle 30°$
b. $\mathbf{C} = 7 \, \angle 120°$
c. $\mathbf{C} = -4.2 \, \angle 60°$

Solutions:

a. See Fig. 14.35.

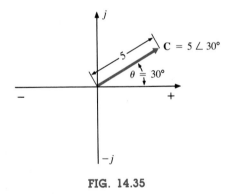

FIG. 14.35

b. See Fig. 14.36.

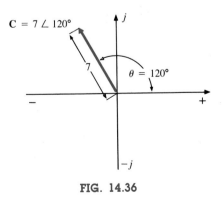

FIG. 14.36

c. See Fig. 14.37.

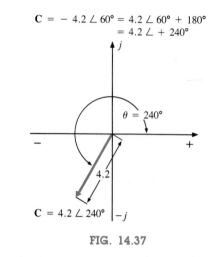

$$\mathbf{C} = -4.2 \angle 60° = 4.2 \angle 60° + 180°$$
$$= 4.2 \angle +240°$$

FIG. 14.37

14.9 CONVERSION BETWEEN FORMS

The two forms are related by the following equations.

Rectangular to Polar

$$C = \sqrt{A^2 + B^2} \qquad (14.18)$$

$$\theta = \tan^{-1}\frac{B}{A} \qquad (14.19)$$

Note Fig. 14.38.

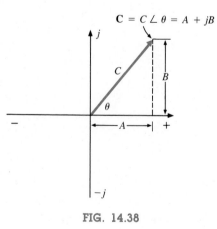

FIG. 14.38

Polar to Rectangular

$$A = C \cos \theta \qquad \textbf{(14.20)}$$

$$B = C \sin \theta \qquad \textbf{(14.21)}$$

EXAMPLE 14.15. Convert the following from rectangular to polar form:

$$\mathbf{C} = 3 + j4 \qquad (\text{Fig. 14.39})$$

Solution:

$$C = \sqrt{(3)^2 + (4)^2} = \sqrt{25} = 5$$

$$\theta = \tan^{-1}\left(\frac{4}{3}\right) = 53.13°$$

and

$$\mathbf{C} = 5 \angle 53.13°$$

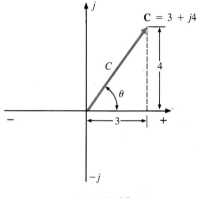

FIG. 14.39

EXAMPLE 14.16. Convert the following from polar to rectangular form:

$$\mathbf{C} = 10 \angle 45° \qquad (\text{Fig. 14.40})$$

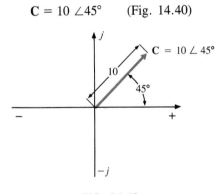

FIG. 14.40

Solution:

$$A = 10 \cos 45° = (10)(0.707) = 7.07$$
$$B = 10 \sin 45° = (10)(0.707) = 7.07$$

and

$$\mathbf{C} = 7.07 + j7.07$$

If the complex number should appear in the second, third, or fourth quadrant, simply convert it in that quadrant, and carefully determine the proper angle to be associated with the magnitude of the vector.

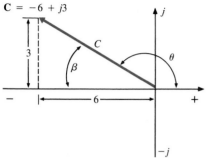

FIG. 14.41

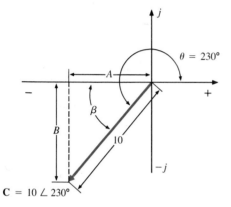

FIG. 14.42

EXAMPLE 14.17. Convert the following from rectangular to polar form:

$$C = -6 + j3 \qquad \text{(Fig. 14.41)}$$

Solution:

$$C = \sqrt{(6)^2 + (3)^2} = \sqrt{45} = 6.71$$

$$\beta = \tan^{-1}\left(\frac{3}{6}\right) = 26.57°$$

$$\theta = 180 - 26.57° = 153.43°$$

and

$$\mathbf{C} = \mathbf{6.71} \, \angle \mathbf{153.43°}$$

EXAMPLE 14.18. Convert the following from polar to rectangular form:

$$C = 10 \, \angle 230° \qquad \text{(Fig. 14.42)}$$

Solution:

$$A = C \cos \beta = 10 \cos(230° - 180°) = 10 \cos 50°$$
$$= (10)(0.6428) = 6.428$$
$$B = C \sin \beta = 10 \sin 50° = (10)(0.7660) = 7.660$$

and

$$\mathbf{C} = \mathbf{-6.428 - j7.660}$$

14.10 MATHEMATICAL OPERATIONS WITH COMPLEX NUMBERS

Complex numbers lend themselves readily to the basic mathematical operations of addition, subtraction, multiplication, and division. A few basic rules and definitions must be understood before considering these operations.

Let us first examine the symbol j associated with imaginary numbers. By definition,

$$\boxed{j = \sqrt{-1}} \qquad \text{(14.22)}$$

Thus,

$$\boxed{j^2 = -1} \qquad \text{(14.23)}$$

and

$$j^3 = j^2 j = -1j = -j$$

with

$$j^4 = j^2 j^2 = (-1)(-1) = +1$$
$$j^5 = j$$

and so on. Further,

$$\frac{1}{j} = (1)\left(\frac{1}{j}\right) = \left(\frac{j}{j}\right)\left(\frac{1}{j}\right) = \frac{j}{j^2} = \frac{j}{-1}$$

and

$$\boxed{\frac{1}{j} = -j} \qquad\qquad \textbf{(14.24)}$$

Complex Conjugate

The *conjugate* or *complex conjugate* of a complex number can be found by simply changing the sign of the imaginary part in the rectangular form or by using the negative of the angle of the polar form. For example, the conjugate of

$$\mathbf{C} = 2 + j3$$

is

$$2 - j3$$

as shown in Fig. 14.43. The conjugate of

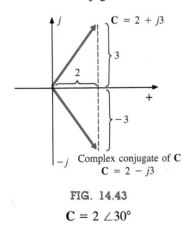

FIG. 14.43

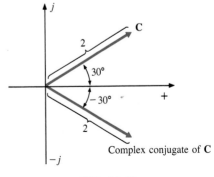

FIG. 14.44

$$\mathbf{C} = 2 \angle 30°$$

is

$$2 \angle -30°$$

as shown in Fig. 14.44.

Reciprocal

The *reciprocal* of a complex number is 1 divided by the complex number. For example, the reciprocal of

$$\mathbf{C} = A + jB$$

is

$$\frac{1}{A + jB}$$

and of $C \angle\theta$,

$$\frac{1}{C \angle\theta}$$

We are now prepared to consider the four basic operations of *addition, subtraction, multiplication,* and *division* with complex numbers.

Addition

To add two or more complex numbers, simply add the real and imaginary parts separately. For example, if

$$\mathbf{C}_1 = \pm A_1 \pm jB_1 \quad \text{and} \quad \mathbf{C}_2 = \pm A_2 \pm jB_2$$

then

$$\boxed{\mathbf{C}_1 + \mathbf{C}_2 = (\pm A_1 \pm A_2) + j(\pm B_1 \pm B_2)} \qquad \textbf{(14.25)}$$

There is really no need to memorize the equation. Simply set one above the other and consider the real and imaginary parts separately, as shown in Example 14.19.

EXAMPLE 14.19.
a. Add $\mathbf{C}_1 = 2 + j4$ and $\mathbf{C}_2 = 3 + j1$.
b. Add $\mathbf{C}_1 = 3 + j6$ and $\mathbf{C}_2 = -6 + j3$.
Solutions:
a. By Eq. (14.25),

$$\mathbf{C}_1 + \mathbf{C}_2 = (2 + 3) + j(4 + 1) = \mathbf{5 + j5}$$

Note Fig. 14.45. An alternative method is

$$
\begin{array}{c}
2 + j4 \\
\underline{3 + j1} \\
\downarrow \quad\quad \downarrow \\
\mathbf{5 + j5}
\end{array}
$$

b. By Eq. (14.25),

$$\mathbf{C}_1 + \mathbf{C}_2 = (3 - 6) + j(6 + 3) = \mathbf{-3 + j9}$$

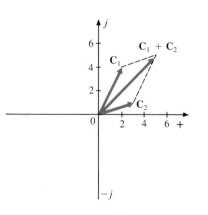

FIG. 14.45

Note Fig. 14.46. An alternative method is

$$3 + j6$$
$$\underline{- 6 + j3}$$
$$\downarrow \qquad \downarrow$$
$$\mathbf{-3 + j9}$$

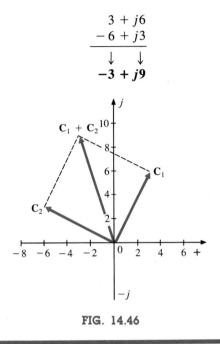

FIG. 14.46

Subtraction

In subtraction, the real and imaginary parts are again considered separately. For example, if

$$\mathbf{C}_1 = \pm A_1 \pm jB_1 \quad \text{and} \quad \mathbf{C}_2 = \pm A_2 \pm jB_2$$

then

$$\boxed{\mathbf{C}_1 - \mathbf{C}_2 = [\pm A_2 - (\pm A_2)] + j[\pm B_1 - (\pm B_2)]} \quad \textbf{(14.26)}$$

Again, there is no need to memorize the equation if the alternative method of Example 14.20 is employed.

EXAMPLE 14.20.
a. Subtract $\mathbf{C}_2 = 1 + j4$ from $\mathbf{C}_1 = 4 + j6$.
b. Subtract $\mathbf{C}_2 = -2 + j5$ from $\mathbf{C}_1 = +3 + j3$.
Solutions:
a. By Eq. (14.26),

$$\mathbf{C}_1 - \mathbf{C}_2 = (4 - 1) + j(6 - 4) = \mathbf{3 + j2}$$

Note Fig. 14.47. An alternative method is

$$4 + j6$$
$$\underline{-(1 + j4)}$$
$$\downarrow \qquad \downarrow$$
$$\mathbf{3 + j2}$$

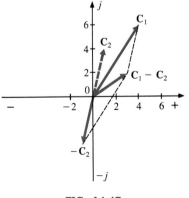

FIG. 14.47

b. By Eq. (14.26),

$$C_1 - C_2 = [3 - (-2)] + j(3 - 5) = 5 - j2$$

Note Fig. 14.48. An alternative method is

$$\begin{array}{c} 3 + j3 \\ -(-2 + j5) \\ \hline \downarrow \quad \downarrow \\ \mathbf{5 - j2} \end{array}$$

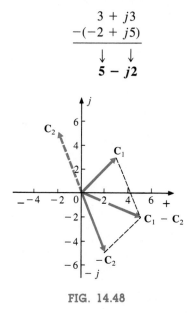

FIG. 14.48

Addition or subtraction cannot be performed in polar form unless the complex numbers have the same angle θ or differ only by multiples of 180°.

EXAMPLE 14.21.

$$2 \angle 45° + 3 \angle 45° = \mathbf{5} \angle \mathbf{45°}$$

Note Fig. 14.49. Or

$$2 \angle 0° - 4 \angle 180° = \mathbf{6} \angle \mathbf{0°}$$

Note Fig. 14.50.

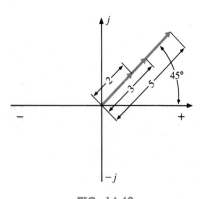

FIG. 14.49

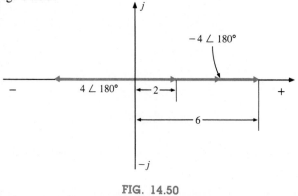

FIG. 14.50

Multiplication

To multiply two complex numbers in *rectangular* form, multiply the real and imaginary parts of one in turn by the real and imaginary parts of the other. For example, if

$$\mathbf{C}_1 = A_1 + jB_1 \quad \text{and} \quad \mathbf{C}_2 = A_2 + jB_2$$

then

$$
\begin{aligned}
\mathbf{C}_1 \cdot \mathbf{C}_2 = \quad & A_1 + jB_1 \\
& \underline{A_2 + jB_2} \\
& A_1A_2 + jB_1A_2 \\
& \underline{\quad + jA_1B_2 + j^2B_1B_2} \\
& A_1A_2 + j(B_1A_2 + A_1B_2) + B_1B_2(-1)
\end{aligned}
$$

and

$$\boxed{\mathbf{C}_1 \cdot \mathbf{C}_2 = (A_1A_2 - B_1B_2) + j(B_1A_2 + A_1B_2)} \quad \textbf{(14.27)}$$

In Example 14.22(b), we obtain a solution without resorting to memorizing Eq. (14.27). Simply carry along the j factor when multiplying each part of one vector with the real and imaginary parts of the other.

EXAMPLE 14.22.
a. Find $\mathbf{C}_1 \cdot \mathbf{C}_2$ if

$$\mathbf{C}_1 = 2 + j3 \quad \text{and} \quad \mathbf{C}_2 = 5 + j10$$

b. Find $\mathbf{C}_1 \cdot \mathbf{C}_2$ if

$$\mathbf{C}_1 = -2 - j3 \quad \text{and} \quad \mathbf{C}_2 = +4 - j6$$

Solutions:
a. Using the format above, we have

$$
\begin{aligned}
\mathbf{C}_1 \cdot \mathbf{C}_2 &= [(2)(5) - (3)(10)] + j[(3)(5) + (2)(10)] \\
&= -20 + j35
\end{aligned}
$$

b. Without using the format, we obtain

$$
\begin{aligned}
& -2 - j3 \\
& \underline{+4 - j6} \\
& -8 - j12 \\
& \underline{\quad + j12 + j^2 18} \\
& -8 + j(-12 + 12) - 18
\end{aligned}
$$

and

$$\mathbf{C}_1 \cdot \mathbf{C}_2 = -26 = 26 \angle 180°$$

In *polar* form, the magnitudes are multiplied and the angles added algebraically. For example, for

$$\mathbf{C}_1 = C_1 \angle \theta_1 \quad \text{and} \quad \mathbf{C}_2 = C_2 \angle \theta_2$$

we write

$$\boxed{\mathbf{C}_1 \cdot \mathbf{C}_2 = C_1 C_2 \,\underline{/\theta_1 + \theta_2}} \qquad (14.28)$$

EXAMPLE 14.23.
a. Find $\mathbf{C}_1 \cdot \mathbf{C}_2$ if

$$\mathbf{C}_1 = 5 \angle 20° \quad \text{and} \quad \mathbf{C}_2 = 10 \angle 30°$$

b. Find $\mathbf{C}_1 \cdot \mathbf{C}_2$ if

$$\mathbf{C}_1 = 2 \angle -40° \quad \text{and} \quad \mathbf{C}_2 = 7 \angle +120°$$

Solutions:
a. $\mathbf{C}_1 \cdot \mathbf{C}_2 = (5)(10) \,\underline{/20° + 30°} = \mathbf{50} \angle \mathbf{50°}$
b. $\mathbf{C}_1 \cdot \mathbf{C}_2 = (2)(7) \,\underline{/-40° + 120°} = \mathbf{14} \angle \mathbf{+80°}$

To multiply a complex number in rectangular form by a real number requires that both the real part and the imaginary part be multiplied by the real number. For example,

$$(10)(2 + j3) = 20 + j30$$

and

$$50 \angle 0°(0 + j6) = j300 = 300 \angle 90°$$

Division

To divide two complex numbers in *rectangular* form, multiply the numerator and denominator by the conjugate of the denominator and the resulting real and imaginary parts collected. That is, if

$$\mathbf{C}_1 = A_1 + jB_1 \quad \text{and} \quad \mathbf{C}_2 = A_2 + jB_2$$

then

$$\frac{\mathbf{C}_1}{\mathbf{C}_2} = \frac{(A_1 + jB_1)(A_2 - jB_2)}{(A_2 + jB_2)(A_2 - jB_2)}$$
$$= \frac{(A_1 A_2 + B_1 B_2) + j(A_2 B_1 - A_1 B_2)}{A_2^2 + B_2^2}$$

and

$$\boxed{\frac{\mathbf{C}_1}{\mathbf{C}_2} = \frac{A_1 A_2 + B_1 B_2}{A_2^2 + B_2^2} + j\frac{A_2 B_1 - A_1 B_2}{A_2^2 + B_2^2}} \qquad (14.29)$$

The equation does not have to be memorized if the steps above used to obtain it are employed. That is, first multiply the numerator by the complex conjugate of the denominator and separate the real and imaginary terms. Then divide each term by the real number obtained by multiplying the denominator by its conjugate.

EXAMPLE 14.24.
a. Find $\mathbf{C}_1/\mathbf{C}_2$ if

$$\mathbf{C}_1 = 1 + j4 \quad \text{and} \quad \mathbf{C}_2 = 4 + j5$$

b. Find $\mathbf{C}_1/\mathbf{C}_2$ if

$$\mathbf{C}_1 = -4 - j8 \quad \text{and} \quad \mathbf{C}_2 = +6 - j1$$

Solutions:
a. By Eq. (14.29),

$$\frac{\mathbf{C}_1}{\mathbf{C}_2} = \frac{(1)(4) + (4)(5)}{4^2 + 5^2} + j\,\frac{(4)(4) - (1)(5)}{4^2 + 5^2}$$

$$= \frac{24}{41} + \frac{j11}{41} \cong \mathbf{0.585 + j0.268}$$

b. Using an alternative method, we obtain

$$
\begin{array}{r}
-4 - j8 \\
+6 + j1 \\
\hline
-24 - j48 \\
- j4 - j^2 8 \\
\hline
-24 - j52 + 8 = -16 - j52
\end{array}
$$

$$
\begin{array}{r}
+6 - j1 \\
+6 + j1 \\
\hline
36 + j6 \\
- j6 - j^2 1 \\
\hline
36 + 0 + 1 = 37
\end{array}
$$

and

$$\frac{\mathbf{C}_1}{\mathbf{C}_2} = \frac{-16}{37} - \frac{j52}{37} = \mathbf{-0.432 - j1.405}$$

To divide a complex number in rectangular form by a real number, both the real part and the imaginary part must be divided by the real number. For example,

$$\frac{8 + j10}{2} = 4 + j5$$

and

$$\frac{6.8 - j0}{2} = 3.4 - j0 = 3.4 \angle 0°$$

In *polar* form, division is accomplished by simply dividing the magnitude of the numerator by the magnitude of the denominator and subtracting the angle of the denominator from that of the numerator. That is, for

$$\mathbf{C}_1 = C_1 \angle\theta_1 \quad \text{and} \quad \mathbf{C}_2 = C_2 \angle\theta_2$$

we write

$$\boxed{\frac{\mathbf{C}_1}{\mathbf{C}_2} = \frac{C_1}{C_2} \angle\theta_1 - \theta_2} \tag{14.30}$$

EXAMPLE 14.25.

a. Find $\mathbf{C}_1/\mathbf{C}_2$ if

$$\mathbf{C}_1 = 15 \angle 10° \quad \text{and} \quad \mathbf{C}_2 = 2 \angle 7°$$

b. Find $\mathbf{C}_1/\mathbf{C}_2$ if

$$\mathbf{C}_1 = 8 \angle 120° \quad \text{and} \quad \mathbf{C}_2 = 16 \angle -50°$$

Solutions:

a. $\dfrac{\mathbf{C}_1}{\mathbf{C}_2} = \dfrac{15}{2} \angle 10° - 7° = \mathbf{7.5} \angle \mathbf{3°}$

b. $\dfrac{\mathbf{C}_1}{\mathbf{C}_2} = \dfrac{8}{16} \angle 120° - (-50°) = \mathbf{0.5} \angle \mathbf{170°}$

We obtain the *reciprocal* in the rectangular form by multiplying the numerator and denominator by the complex conjugate of the denominator:

$$\frac{1}{A + jB} = \left(\frac{1}{A + jB}\right)\left(\frac{A - jB}{A - jB}\right) = \frac{A - jB}{A^2 + B^2}$$

and

$$\boxed{\frac{1}{A + jB} = \frac{A}{A^2 + B^2} - j\frac{B}{A^2 + B^2}} \tag{14.31}$$

In the polar form, the reciprocal is

$$\boxed{\frac{1}{C \angle\theta} = \frac{1}{C} \angle -\theta} \tag{14.32}$$

Some concluding examples using the four basic operations follow.

EXAMPLE 14.26. Perform the following operations, leaving the answer in polar or rectangular form:

a. $\dfrac{(2 + j3) + (4 + j6)}{(7 + j7) - (3 - j3)} = \dfrac{(2 + 4) + j(3 + 6)}{(7 - 3) + j(7 + 3)}$

$= \dfrac{(6 + j9)(4 - j10)}{(4 + j10)(4 - j10)}$

$= \dfrac{[(6)(4) + (9)(10)] + j[(4)(9) - (6)(10)]}{4^2 + 10^2}$

$= \dfrac{114 - j24}{116} = \mathbf{0.983 - j0.207}$

b. $\dfrac{(50 \angle 30°)(5 + j5)}{10 \angle -20°} = \dfrac{(50 \angle 30°)(7.07 \angle 45°)}{10 \angle -20°} = \dfrac{353.5 \angle 75°}{10 \angle -20°}$

$= 35.35\ \underline{/75° - (-20°)} = \mathbf{35.35 \angle 95°}$

c. $\dfrac{(2 \angle 20°)^2(3 + j4)}{8 - j6} = \dfrac{(2 \angle 20°)(2 \angle 20°)(5 \angle 53.13°)}{10 \angle -36.87°}$

$= \dfrac{(4 \angle 40°)(5 \angle 53.13°)}{10 \angle -36.87°} = \dfrac{20 \angle 93.13°}{10 \angle -36.87°}$

$= 2\ \underline{/93.13° - (-36.87°)} = \mathbf{2.0 \angle 130°}$

d. $3 \angle 27° - 6 \angle -40° = (2.673 + j1.362) - (4.596 - j3.857)$
   $= (2.673 - 4.596) + j(1.362 + 3.857)$
   $= \mathbf{-1.923 + j5.219}$

14.11 TECHNIQUES OF CONVERSION

For many years the technologist was dependent on tables, charts, and the slide rule to perform conversions between complex numbers. Today, calculators such as that shown in Fig. 14.51 are available that can perform the conversion to eight-place accuracy. In fact, this particular calculator is programmed to perform this particular operation by your pressing just a few buttons. The →R and →P refer to the rectangular and polar forms, respectively.

You will appreciate the speed and accuracy of calculator conversion after using other techniques of conversion. There are inexpensive calculators that have the necessary functions to perform the conversions using Eqs. (14.18) through (14.21). For those who seriously expect to stay in this field, a calculator with the proper functions would be a wise investment.

14.12 PHASORS

As noted earlier in this chapter, the addition of sinusoidal voltages and currents will frequently be required in the analysis of ac circuits. One indicated method of performing this operation is to place both sinusoi-

FIG. 14.51
Scientific calculator. (Courtesy of Hewlett Packard Co.)

dal waveforms on the same set of axes and add algebraically the magnitudes of each at every point along the abscissa, as shown for $c = a + b$ in Fig. 14.52. This, however, can be a long and tedious process with

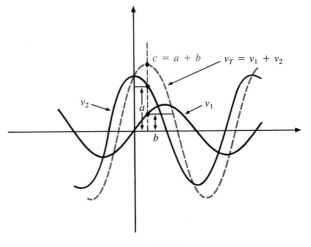

FIG. 14.52

limited accuracy. A shorter method uses the rotating radius vector shown in Fig. 13.18. This *radius vector,* having a *constant magnitude* (length) with *one end fixed at the origin,* is called a *phasor* when applied to electric circuits. During its rotational development of the sine wave, the phasor will, at the instant $t = 0$, have the positions shown in Fig. 14.53(a) for each waveform in Fig. 14.53(b).

Note in Fig. 14.53(b) that v_1 passes through the horizontal axis at $t = 0$, requiring that the radius vector in Fig. 14.53(a) be on the hori-

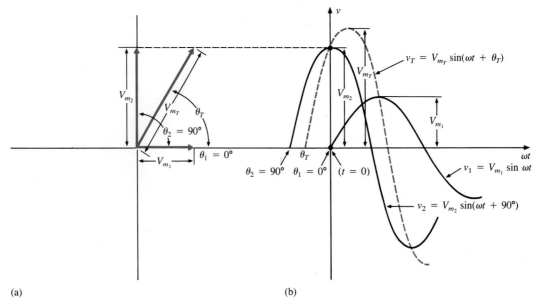

(a)

(b)

FIG. 14.53

zontal axis. Its length in Fig. 14.53(a) is equal to the peak value of the sinusoid as required by the radius vector of Fig. 13.18. The other sinusoid (actually a cosine wave) has passed through 90° of its rotation by the time $t = 0$ is reached and therefore has its maximum vertical projection as shown in Fig. 14.53(a). (Recall that 90° separates the minimum and maximum values of any sinusoidal function.) Since the vertical projection is a maximum, the peak value of the sinusoid that it will generate is also attained at $t = 0$, as shown in Fig. 14.53(b). Note also that $v_T = v_2$ at $t = 0$ since $v_1 = 0$ at this instant.

It can be shown [see Fig. 14.53(a)] using the vector algebra described in Section 14.10 that

$$V_{m_1} \angle 0° + V_{m_2} \angle 90° = V_{m_T} \angle \theta_T$$

In other words, if we convert v_1 and v_2 to the phasor form using

$$v = V_m \sin(\omega t \pm \theta) \Rightarrow V_m \angle \pm \theta$$

and add them using vector algebra, we can find the phasor form for v_T with very little difficulty. It can then be converted to the time domain and plotted on the same set of axes as shown in Fig. 14.53(b). Figure 14.53(a), showing the magnitudes and relative positions of the various phasors, is called a *phasor diagram*. It is actually a "snapshot" of the phasors representing the sinusoidal waveforms at $t = 0$.

In the future, therefore, if the addition of two sinusoids is required, they should first be converted to the phasor domain and the sum found using complex algebra. The result can then be converted to the time domain if required.

As an example, consider the case of

$$v_1 = 5 \sin \omega t$$
$$v_2 = 10 \sin(\omega t + 90°)$$

in Figs. 14.52 and 14.53. In the phasor domain,

$$\mathbf{V}_1 = 5 \angle 0°$$
$$\mathbf{V}_2 = 10 \angle 90°$$
$$\mathbf{V}_T = \mathbf{V}_1 + \mathbf{V}_2 = 5 + j10 = 11.180 \angle 63.43° = V_{m_T} \angle \theta_T$$

In other words,

$$V_{m_T} = 11.180 \text{ V}$$
$$\theta_T = 63.43°$$
$$v_T = 11.180 \sin(\omega t + 63.43°)$$

as verified by Figs. 14.52 and 14.53.

The case of two sinusoidal functions having phase angles different from 0° and 90° appears in Fig. 14.54. Note again that the vertical height of the functions in Fig. 14.54(b) is determined by the rotational positions of the radius vectors in Fig. 14.54(a).

Since the effective, rather than the peak, values are used almost exclusively in the analysis of ac circuits, the phasor will now be redefined for the purposes of practicality and uniformity as having a

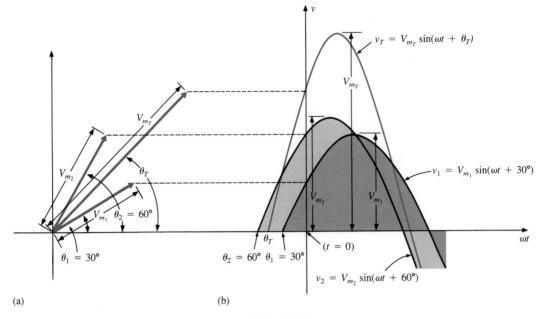

FIG. 14.54

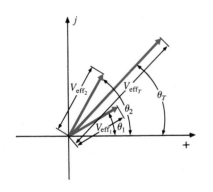

FIG. 14.55

magnitude equal to the *effective value* of the sine wave it represents. The angle associated with the phasor will remain as previously described—the phase angle. The phasor diagram will therefore be as shown in Fig. 14.55, replacing that of Fig. 14.54(a).

In general, for all of the analyses to follow, the phasor form of a sinusoidal voltage or current will be

$$\mathbf{V} = V \angle \theta \quad \text{and} \quad \mathbf{I} = I \angle \theta$$

where V and I are effective values and θ is the phase angle. It should be pointed out that in phasor notation, the sine wave is always the reference, and the frequency is not represented. Phasor algebra for sinusoidal quantities is applicable only for waveforms having the *same frequency*, so that it can be carried along without any special notation.

EXAMPLE 14.27. Convert the following from the time to the phasor domain:

Time Domain	Phasor Domain
a. $\sqrt{2}(50) \sin \omega t$	**50 $\angle 0°$**
b. $69.6 \sin(\omega t + 72°)$	$(0.707)(69.6) \angle 72° = \mathbf{49.21 \angle 72°}$
c. $45 \cos \omega t$	$(0.707)(45) \angle 90° = \mathbf{31.82 \angle 90°}$

EXAMPLE 14.28. Write the sinusoidal expression for the following phasors if the frequency is 60 Hz:

Phasor Domain	Time Domain
a. $\mathbf{I} = 10 \angle 30°$	$i = \sqrt{2}(10)\sin(2\pi 60t + 30°)$ and $i = \mathbf{14.14\ sin(377}t + \mathbf{30°)}$
b. $\mathbf{V} = 115 \angle -70°$	$v = \sqrt{2}(115)\sin(377t - 70°)$ and $v = \mathbf{162.6\ sin(377}t - \mathbf{70°)}$

EXAMPLE 14.29. Find the input voltage of the circuit of Fig. 14.56 if

$$v_a = 50\sin(377t + 30°) \atop v_b = 30\sin(377t + 60°) \Big\} \quad f = 60\,Hz$$

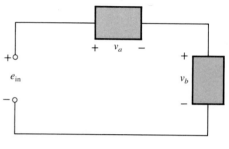

FIG. 14.56

Solution: Applying Kirchhoff's voltage law, we have

$$e_{in} = v_a + v_b$$

Converting from the time to the phasor domain yields

$$v_a = 50\sin(377t + 30°) \Rightarrow \mathbf{V}_a = 35.35 \angle 30°$$
$$v_b = 30\sin(377t + 60°) \Rightarrow \mathbf{V}_b = 21.21 \angle 60°$$

Converting from polar to rectangular form for addition yields

$$\mathbf{V}_a = 35.35 \angle 30° = 30.61 + j17.68$$
$$\mathbf{V}_b = 21.21 \angle 60° = 10.61 + j18.37$$

Then

$$\mathbf{E}_{in} = \mathbf{V}_a + \mathbf{V}_b = (30.61 + j17.68) + (10.61 + j18.37)$$
$$= 41.22 + j36.05$$

Converting from rectangular to polar form, we have

$$\mathbf{E}_{in} = 41.22 + j36.05 = 54.76 \angle 41.17$$

Converting from the phasor to the time domain, we obtain

$$\mathbf{E}_{in} = 54.76 \angle 41.17° \Rightarrow e_{in} = \sqrt{2}(54.76)\sin(377t + 41.17°)$$

and

$$e_{in} = \mathbf{77.43\ sin(377}t + \mathbf{41.17°)}$$

A plot of the three waveforms is shown in Fig. 14.57. Note that at each instant of time, the sum of the two waveforms does in fact add up to e_{in}. At $t = 0$ ($\omega t = 0$), e_{in} is the sum of the two positive values, while at a value of ωt almost midway between $\pi/2$ and π, the sum of the positive value of v_a and the negative value of v_b results in $e_{in} = 0$.

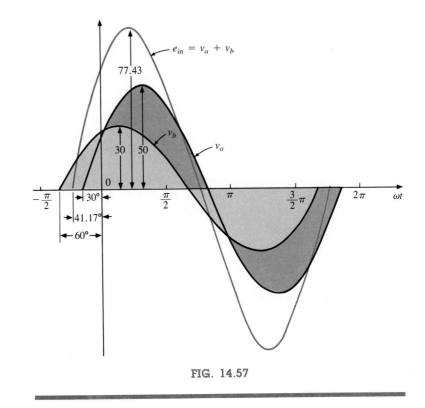

FIG. 14.57

EXAMPLE 14.30. Determine the current i_2 for the network of Fig. 14.58.

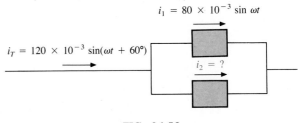

$i_1 = 80 \times 10^{-3} \sin \omega t$

$i_T = 120 \times 10^{-3} \sin(\omega t + 60°)$

$i_2 = ?$

FIG. 14.58

Solution: Applying Kirchhoff's current law, we obtain

$$i_T = i_1 + i_2$$

or

$$i_2 = i_T - i_1$$

Converting from the time to the phasor domain yields

$$i_T = 120 \times 10^{-3} \sin(\omega t + 60°) \Rightarrow 84.84 \times 10^{-3} \angle 60°$$
$$i_1 = 80 \times 10^{-3} \sin \omega t \Rightarrow 56.56 \times 10^{-3} \angle 0°$$

Converting from polar to rectangular form for subtraction yields

$$\mathbf{I}_T = 84.84 \times 10^{-3} \angle 60° = 42.42 \times 10^{-3} + j73.47 \times 10^{-3}$$
$$\mathbf{I}_1 = 56.56 \times 10^{-3} \angle 0° = 56.56 \times 10^{-3} + j0$$

Then

$$\mathbf{I}_2 = \mathbf{I}_T - \mathbf{I}_1$$
$$= (42.42 \times 10^{-3} + j73.47 \times 10^{-3}) - (56.56 \times 10^{-3} + j0)$$

and

$$\mathbf{I}_2 = -14.14 \times 10^{-3} + j73.47 \times 10^{-3}$$

Converting from rectangular to polar form, we have

$$\mathbf{I}_2 = 74.82 \times 10^{-3} \angle 100.89°$$

Converting from the phasor to the time domain, we have

$$\mathbf{I}_2 = 74.82 \times 10^{-3} \angle 100.89° \Rightarrow$$
$$i_2 = \sqrt{2}(74.82 \times 10^{-3}) \sin(\omega t + 100.89°)$$

and

$$\mathbf{i_2 = 105.8 \times 10^{-3} \sin(\omega t + 100.89°)}$$

A plot of the three waveforms appears in Fig. 14.59. The waveforms clearly indicate that $i_T = i_1 + i_2$.

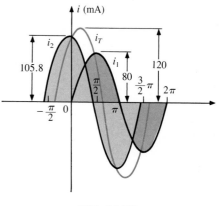

FIG. 14.59

14.13 COMPUTER METHODS: CONVERSION ROUTINE

A program designed to perform a conversion from polar to rectangular form or rectangular to polar form appears in Fig. 14.60 as Program 14.1. The input parameters of the rectangular form are entered on line

```
10 REM ***** PROGRAM 14-1 *****
20 REM ****************************************
30 REM Program to perform selected conversions
40 REM ****************************************
50 REM
100 PRINT
110 PRINT "Enter (1) for rectangular to polar conversion"
120 PRINT "      (2) for polar to rectangular conversion"
130 PRINT TAB(20);
140 INPUT "Choice=";C :REM C is choice 1 or 2
150 IF C<0 OR C>2 THEN GOTO 110
160 ON C GOSUB 200,300
170 PRINT:INPUT "More(YES or NO)";A$
180 IF A$="YES" THEN GOTO 100
190 END
200 REM Use rectangular to polar conversion module
210 PRINT:PRINT:PRINT "Enter rectangular data:"
220 INPUT "X=";X :INPUT "Y=";Y
230 GOSUB 2000
240 PRINT:PRINT "Polar form is";Z;"at an angle of";TH;"degrees"
250 RETURN
```

Input-
(Rect.)

Output-
(Polar)

```
           300 REM Use polar to rectangular conversion
Input   ┌ 310 PRINT:PRINT "Enter polar data:":PRINT:INPUT "Z=";Z
(Polar) └ 320 INPUT "Angle(degrees), TH=";TH
           330 GOSUB 2100
Output  ┌ 340 PRINT:PRINT "Rectangular form is";X;
(Rect.) │ 350 IF Y>=0 THEN PRINT "+j";Y
        └ 360 IF Y<0 THEN PRINT "-j";ABS(Y)
           370 RETURN
        ┌ 2000 REM Module to convert from rectangular to polar form.
        │ 2010 REM Enter with X, Y - Return with Z, TH(eta)
        │ 2020 Z=SQR(X^2+Y^2)
Rect.   │ 2030 IF X<0 THEN TH=(180/3.14159)*ATN(Y/X)
  ↓     │ 2040 IF X<0 THEN TH=180*SGN(Y)+(180/3.14159)*ATN(Y/X)
Polar   │ 2050 IF X=0 THEN TH=90*SGN(Y)
        │ 2060 IF Y=0 THEN IF X<0 THEN TH=180
        └ 2070 RETURN
        ┌ 2100 REM Module to convert from polar to rectangular form.
Polar   │ 2110 REM Enter with Z, TH(eta) - return with X, Y
  ↓     │ 2120 X=Z*COS(TH*3.14159/180)
Rect.   │ 2130 Y=Z*SIN(TH*3.14159/180)
        └ 2140 RETURN


    READY


    Enter (1) for rectangular to polar conversion
          (2) for polar to rectangular conversion
                    Choice=? 2


    Enter polar data:

    Z=? 5

    Angle(degrees), TH=? -53.13


    Rectangular form is 3 -j 4

    More(YES or NO)? YES


    Enter (1) for rectangular to polar conversion
          (2) for polar to rectangular conversion
                    Choice=? 1



    Enter rectangular data:
    X=? -10

    Y=? 20


    Polar form is 22.3607 at an angle of 116.565 degrees

    More(YES or NO)? YES


    Enter (1) for rectangular to polar conversion
          (2) for polar to rectangular conversion
                    Choice=? 2


    Enter polar data:

    Z=? 12

    Angle(degrees), TH=? 35
```

```
Rectangular form is 9.8298 +j 6.8829

More(YES or NO)? NO


READY
```

FIG. 14.60
Program 14.1.

220, and the polar form on lines 310 and 320. Line 240 outputs the polar form, and lines 340 through 360 the rectangular form.

The rectangular-to-polar conversion routine appears on lines 2000 through 2070, while the polar-to-rectangular conversion appears on lines 2100 through 2140. Note on line 2020 the equation for the magnitude of the polar form $Z = \sqrt{X^2 + Y^2}$ and the testing of X and Y to determine the correct value of θ on lines 2030 through 2060. Lines 2120 and 2130 determine X and Y using the equations $X = Z \cos \theta$ and $Y = Z \sin \theta$, respectively. Note the need to convert the input angle in degrees (TH) to radians before the BASIC language can act on the SIN and COS functions.

PROBLEMS

Section 14.2

1. Plot the following waveform versus time showing one clear complete cycle. Then determine the derivative of the waveform using Eq. (14.1), and sketch one complete cycle of the derivative directly under the original waveform. Compare the magnitude of the derivative at various points versus the slope of the original sinusoidal function.

$$v = 1 \sin 3.14t$$

2. Repeat Problem 1 for the following sinusoidal function and compare results. In particular, determine the frequency of the waveforms of Problems 1 and 2 and compare the magnitude of the derivative.

$$v = 1 \sin 15.7t$$

3. What is the derivative of each of the following sinusoidal expressions?
   a. $10 \sin 377t$ **b.** $0.6 \cos 754t$
   c. $0.05 \cos(157t - 10°)$ **d.** $25 \cos(20t - 150°)$

4. The voltage across a 5-Ω resistor is as indicated. Find the sinusoidal expression for the current. In addition, sketch the v and i curves with the abscissa in radians.
   a. $150 \sin 377t$ **b.** $30 \sin(377t + 20°)$
   c. $40 \cos(\omega t + 10°)$ **d.** $-80 \sin(\omega t + 40°)$

5. The current through a 7-kΩ resistor is as indicated. Find the sinusoidal expression for the voltage. In addition, sketch the v and i curves with the abscissa in radians.
   a. 0.03 sin 754t
   b. 2×10^{-3} sin($400t - 120°$)
   c. 6×10^{-6} cos($\omega t - 2°$)
   d. -0.004 cos($\omega t - 90°$)

6. Determine the inductive reactance (in ohms) of a 2-H coil for
   a. dc
   and for the following frequencies:
   b. 25 Hz   c. 60 Hz
   d. 2000 Hz   e. 100,000 Hz

7. Determine the inductance of a coil that has a reactance of
   a. 20 Ω at $f = 2$ Hz.
   b. 1000 Ω at $f = 60$ Hz.
   c. 5280 Ω at $f = 1000$ Hz.

8. Determine the frequency at which a 10-H inductance has the following inductive reactances:
   a. 50 Ω   b. 3770 Ω
   c. 15.7 kΩ   d. 243 Ω

9. The current through a 20-Ω inductive reactance is given. What is the sinusoidal expression for the voltage? Sketch the v and i curves with the abscissa in radians.
   a. $i = 5$ sin ωt b. $i = 0.4$ sin($\omega t + 60°$)
   c. $i = -6$ sin($\omega t - 30°$)   d. $i = 3$ cos($\omega t + 10°$)

10. The current through a 0.1-H coil is given. What is the sinusoidal expression for the voltage?
    a. 30 sin 30t
    b. 0.006 sin 377t
    c. 5×10^{-6} sin($400t + 20°$)
    d. -4 cos($20t - 70°$)

11. The voltage across a 50-Ω inductive reactance is given. What is the sinusoidal expression for the current? Sketch the v and i curves with the abscissa in radians.
    a. 50 sin ωt b. 30 sin($\omega t + 20°$)
    c. 40 cos($\omega t + 10°$)   d. -80 sin($377t + 40°$)

12. The voltage across a 0.2-H coil is given. What is the sinusoidal expression for the current?
    a. 1.5 sin 60t
    b. 0.016 sin($t + 4°$)
    c. -4.8 sin($0.05t + 50°$)
    d. 9×10^{-3} cos($377t + 360°$)

13. Determine the capacitive reactance (in ohms) of a 5-μF capacitor for
    a. dc
    and for the following frequencies:
    b. 60 Hz   c. 120 Hz
    d. 1800 Hz   e. 24,000 Hz

14. Determine the capacitance in microfarads if a capacitor has a reactance of
    a. $250 \, \Omega$ at $f = 60 \, \text{Hz}$.
    b. $55 \, \Omega$ at $f = 312 \, \text{Hz}$.
    c. $10 \, \Omega$ at $f = 25 \, \text{Hz}$.

15. Determine the frequency at which a 50-μF capacitor has the following capacitive reactances:
    a. $342 \, \Omega$ b. $684 \, \Omega$
    c. $171 \, \Omega$ d. $2000 \, \Omega$

16. The voltage across a 2.5-Ω capacitive reactance is given. What is the sinusoidal expression for the current? Sketch the v and i curves with the abscissa in radians.
    a. $100 \sin \omega t$ b. $0.4 \sin(\omega t + 20°)$
    c. $8 \cos(\omega t + 10°)$ d. $-70 \sin(\omega t + 40°)$

17. The voltage across a 1-μF capacitor is given. What is the sinusoidal expression for the current?
    a. $30 \sin 200t$ b. $90 \sin 377t$
    c. $-120 \sin(374t + 30°)$ d. $70 \cos(800t - 20°)$

18. The current through a 10-Ω capacitive reactance is given. Write the sinusoidal expression for the voltage. Sketch the v and i curves with the abscissa in radians.
    a. $i = 50 \sin \omega t$ b. $i = 40 \sin(\omega t + 60°)$
    c. $i = -6 \sin(\omega t - 30°)$ d. $i = 3 \cos(\omega t + 10°)$

19. The current through a 0.5-μF capacitor is given. What is the sinusoidal expression for the voltage?
    a. $0.20 \sin 300t$ b. $0.007 \sin 377t$
    c. $0.048 \cos 754t$ d. $0.08 \sin(1600t - 80°)$

*20. For the following pairs of voltages and currents, indicate whether the element involved is a capacitor, inductor, or resistor, and the value of C, L, or R if sufficient data are given:
    a. $v = 550 \sin(377t + 40°)$
       $i = 11 \sin(377t - 50°)$
    b. $v = 36 \sin(754t + 80°)$
       $i = 4 \sin(754t + 170°)$
    c. $v = 10.5 \sin(\omega t + 13°)$
       $i = 1.5 \sin(\omega t + 13°)$

*21. Repeat Problem 20 for the following pairs of voltages and currents:
    a. $v = 2000 \sin \omega t$
       $i = 5 \cos \omega t$
    b. $v = 80 \sin(157t + 150°)$
       $i = 2 \sin(157t + 60°)$
    c. $v = 35 \sin(\omega t - 20°)$
       $i = 7 \cos(\omega t - 110°)$

Section 14.4

22. Plot X_L versus frequency for a 5-mH coil using a frequency range of zero to 100 kHz on a linear scale.

23. Plot X_C versus frequency for a 1-μF capacitor using a frequency range of zero to 10 kHz on a linear scale.

24. At what frequency will the reactance of a 1-μF capacitor equal the resistance of a 2-kΩ resistor?

25. The reactance of a coil equals the resistance of a 10-kΩ resistor at a frequency of 5 kHz. Determine the inductance of the coil.

26. Determine the frequency at which a 1-μF capacitor and a 10-mH inductor will have the same reactance.

27. Determine the capacitance required to establish a capacitive reactance that will match that of a 2-mH coil at a frequency of 50 kHz.

Section 14.5

28. Find the average power loss in watts for each set of Problem 20.

29. Find the average power loss in watts for each set of Problem 21.

*30. Find the average power loss and power factor for each of the circuits whose input current and voltage are as follows:
   a. $v = 60 \sin(\omega t + 30°)$
      $i = 15 \sin(\omega t + 60°)$
   b. $v = -50 \sin(\omega t - 20°)$
      $i = -2 \sin(\omega t + 40°)$
   c. $v = 50 \sin(\omega t + 80°)$
      $i = 3 \cos(\omega t + 20°)$
   d. $v = 75 \sin(\omega t - 5°)$
      $i = 0.08 \sin(\omega t - 35°)$

31. If the current through and voltage across an element are $i = 8 \sin(\omega t + 40°)$ and $v = 48 \sin(\omega t + 40°)$, respectively, compute the power by I^2R, $(V_m I_m/2) \cos \theta$, and $VI \cos \theta$, and compare answers.

32. A circuit dissipates 100 W (average power) at 150 V (effective input voltage) and 2 A (effective input current). What is the power factor? Repeat if the power is 0 W; 300 W.

*33. The power factor of a circuit is 0.5 lagging. The power delivered in watts is 500. If the input voltage is $50 \sin(\omega t + 10°)$, find the sinusoidal expression for the input current.

34. In Fig. 14.61, $e = 30 \sin(377t + 20°)$.
   a. What is the sinusoidal expression for the current?
   b. Find the power loss in the circuit.
   c. How long (in seconds) does it take the current to complete 6 cycles?

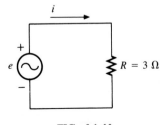

FIG. 14.61

35. In Fig. 14.62, $e = 100 \sin(157t + 30°)$.
  a. Find the sinusoidal expression for i.
  b. Find the value of the inductance L.
  c. Find the average power loss by the inductor.

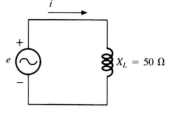

FIG. 14.62

36. In Fig. 14.63, $i = 3 \sin(377t - 20°)$.
  a. Find the sinusoidal expression for e.
  b. Find the value of the capacitance C in microfarads.
  c. Find the average power loss in the capacitor.

Section 14.9

37. Convert the following from rectangular to polar form:
  a. $4 + j3$ **b.** $2 + j2$
  c. $3.5 + j16$ **d.** $100 + j800$
  e. $1000 + j400$ **f.** $0.001 + j0.0065$
  g. $7.6 - j9$ **h.** $-8 + j4$
  i. $-15 - j60$ **j.** $+78 - j65$
  k. $-2400 + j3600$
  l. $5 \times 10^{-3} - j25 \times 10^{-3}$

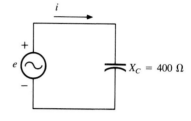

FIG. 14.63

38. Convert the following from polar to rectangular form:
  a. $6 \angle 30°$ **b.** $40 \angle 80°$
  c. $7400 \angle 70°$ **d.** $4 \times 10^{-4} \angle 8°$
  e. $0.04 \angle 80°$ **f.** $0.0093 \angle 23°$
  g. $65 \angle 150°$ **h.** $1.2 \angle 135°$
  i. $500 \angle 200°$ **j.** $6320 \angle -35°$
  k. $7.52 \angle -125°$ **l.** $0.008 \angle 310°$

39. Convert the following from rectangular to polar form:
  a. $1 + j15$ **b.** $60 + j5$
  c. $0.01 + j0.3$ **d.** $100 - j2000$
  e. $-5.6 + j86$ **f.** $-2.7 - j38.6$

40. Convert the following from polar to rectangular form:
  a. $13 \angle 5°$ **b.** $160 \angle 87°$
  c. $7 \times 10^{-6} \angle 2°$ **d.** $8.7 \angle 177°$
  e. $76 \angle -4°$ **f.** $396 \angle +265°$

Section 14.10

Perform the following operations.

41. Addition and subtraction (express your answers in rectangular form):
  a. $(4.2 + j6.8) + (7.6 + j0.2)$
  b. $(142 + j7) + (9.8 + j42) + (0.1 + j0.9)$
  c. $(4 \times 10^{-6} + j76) + (7.2 \times 10^{-7} - j5)$
  d. $(9.8 + j6.2) - (4.6 + j4.6)$
  e. $(167 + j243) - (-42.3 - j68)$
  f. $(-36.0 + j78) - (-4 - j6) + (10.8 - j72)$
  g. $6 \angle 20° + 8 \angle 80°$
  h. $42 \angle 45° + 62 \angle 60° - 70 \angle 120°$

42. Multiplication [express your answers in rectangular form for parts (i) through (1), and in polar form for parts (m) through (p)]:

   i. $(2 + j3)(6 + j8)$
   j. $(7.8 + j1)(4 + j2)(7 + j6)$
   k. $(0.002 + j0.006)(-2 + j2)$
   l. $(400 - j200)(-0.01 - j0.5)(-1 + j3)$
   m. $(2 \angle 60°)(4 \angle 22°)$
   n. $(6.9 \angle 8°)(7.2 \angle -72°)$
   o. $0.002 \angle 120°)(0.5 \angle 200°)(40 \angle -60°)$
   p. $(540 \angle -20°)(-5 \angle 180°)(6.2 \angle 0°)$

43. Division (express your answers in polar form):

   q. $(42 \angle 10°)/(7 \angle 60°)$
   r. $(0.006 \angle 120°)/(30 \angle -20°)$
   s. $(4360 \angle -20°)/(40 \angle 210°)$
   t. $(650 \angle -80°)/(8.5 \angle 360°)$
   u. $(8 + j8)/(2 + j2)$
   v. $(8 + j42)/(-6 + j60)$
   w. $(0.05 + j0.25)/(8 - j60)$
   x. $(-4.5 - j6)/(0.1 - j0.4)$

44. Perform the following operations (express your answers in rectangular form):

   a. $\dfrac{(4 + j3) + (6 - j8)}{(3 + j3) - (2 + j3)}$

   b. $\dfrac{8 \angle 60°}{(2 \angle 0°) + (100 + j100)}$

   c. $\dfrac{(6 \angle 20°)(120 \angle -40°)(3 + j4)}{2 \angle -30°}$

   d. $\dfrac{(0.4 \angle 60°)^2(300 \angle 40°)}{3 + j9}$

   e. $\dfrac{(150 \angle 2°)(4 \times 10^{-6} \angle 88°)}{(0.002 \angle 10°)^2(4 \angle 30°)}$

Section 14.12

45. Express the following in phasor form:
   a. $\sqrt{2}(100) \sin(\omega t + 30°)$
   b. $\sqrt{2}(0.25) \sin(157t - 40°)$
   c. $100 \sin(\omega t - 90°)$
   d. $42 \sin(377t + 0°)$
   e. $6 \times 10^{-6} \cos \omega t$
   f. $3.6 \times 10^{-6} \cos(754t - 20°)$

46. Express the following phasor currents and voltages as sine waves if the frequency is 60 Hz:
   a. $\mathbf{I} = 40 \angle 20°$ **b.** $\mathbf{V} = 120 \angle 0°$
   c. $\mathbf{I} = 8 \times 10^{-3} \angle 120°$ **d.** $\mathbf{V} = 5 \angle 90°$
   e. $\mathbf{I} = 1200 \angle -120°$ **f.** $\mathbf{V} = \dfrac{6000}{\sqrt{2}} \angle -180°$

47. For the system of Fig. 14.64, find the sinusoidal expression for the unknown voltage v_a if

$$e_{in} = 60 \sin(377t + 20°)$$
$$v_b = 20 \sin 377t$$

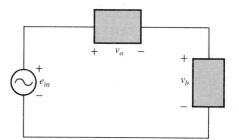

FIG. 14.64

48. For the system of Fig. 14.65, find the sinusoidal expression for the unknown current i_1 if

$$i_T = 20 \times 10^{-6} \sin(\omega t + 90°)$$
$$i_2 = 6 \times 10^{-6} \sin(\omega t - 60°)$$

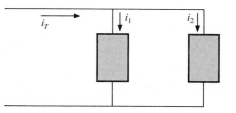

FIG. 14.65

49. Find the sinusoidal expression for the voltage v_c for the system of Fig. 14.66 if

$$e_{in} = 120 \sin(\omega t + 30°)$$
$$v_a = 60 \sin \omega t$$
$$v_b = 30 \sin \omega t$$

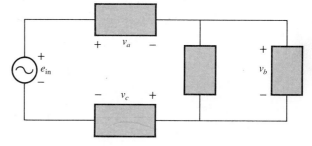

FIG. 14.66

50. Find the sinusoidal expression for the current i_T for the system of Fig. 14.67 if

$$i_1 = 6 \times 10^{-3} \sin(377t + 180°)$$
$$i_2 = 8 \times 10^{-3} \sin 377t$$
$$i_3 = 2i_2$$

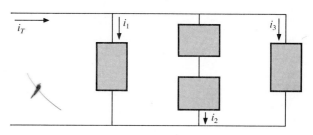

Computer Problems

FIG. 14.67

51. Given a sinusoidal function, write a program to print out the derivative.

52. Given the sinusoidal expression for the current, determine the expression for the voltage across a resistor, capacitor, or inductor, depending on the element involved. In other words, the program will ask which element is to be investigated and will then request the pertinent data to obtain the mathematical expression for the sinusoidal voltage.

53. Write a program to tabulate the reactance versus frequency for an inductor or capacitor for a specified frequency range.

54. Given the sinusoidal expression for the voltage and current of a load, write a program to determine the average power and power factor.

55. Given two sinusoidal functions, write a program to convert each to the phasor domain, add the two, and print out the sum in the phasor and time domains.

GLOSSARY

Average or real power The power delivered to and dissipated by the load over a full cycle.

Complex conjugate A complex number defined by simply changing the sign of an imaginary component of a complex number in the rectangular form.

Complex number A number that represents a point in a two-dimensional plane located with reference to two distinct axes. It defines a vector drawn from the origin to that point.

Derivative The instantaneous rate of change of a function with respect to time or another variable.

Leading and lagging power factors An indication of whether a network is primarily capacitive or inductive in nature. Leading power factors are associated with capacitive networks, and lagging power factors with inductive networks.

Phasor A radius vector that has a constant magnitude at a fixed angle from the positive real axis and that represents a sinusoidal voltage or current in the vector domain.

Phasor diagram A "snapshot" of the phasors that represent a number of sinusoidal waveforms at $t = 0$.

Polar form A method of defining a point in a complex plane that includes a single magnitude to represent the distance from the origin, and an angle to reflect the counterclockwise distance from the positive real axis.

Power factor (F_p) An indication of how reactive or resistive an electrical system is. The higher the power factor, the greater the resistive component.

Reactance The opposition of an inductor or capacitor to the flow of charge that results in the continual exchange of energy between the circuit and magnetic field of an inductor or the electric field of a capacitor.

Reciprocal A format defined by 1 divided by the complex number.

Rectangular form A method of defining a point in a complex plane that includes the magnitude of the real component and the magnitude of the imaginary component, the latter component being defined by an associated letter j.

15

Series and Parallel ac Circuits

15.1 INTRODUCTION

In this chapter, phasor algebra will be used to develop a quick, direct method for solving both the series and the parallel ac circuits. The close relationship that exists between this method for solving for unknown quantities and the approach used for dc circuits will become apparent after a few simple examples are considered. Once this association is established, many of the rules (current divider rule, voltage divider rule, and so on) for dc circuits can be readily applied to ac circuits.

SERIES ac CIRCUITS

15.2 IMPEDANCE AND THE PHASOR DIAGRAM

In Chapter 14, we found, for the purely resistive circuit of Fig. 15.1, that v and i were in phase, and the magnitude

$$I_m = \frac{V_m}{R} \quad \text{or} \quad V_m = I_m R$$

In phasor form,

$$v = V_m \sin \omega t \Rightarrow \mathbf{V} = V \angle 0°$$

where $V = 0.707 V_m$.

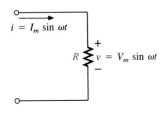

FIG. 15.1

Applying Ohm's law and using phasor algebra, we have

$$\mathbf{I} = \frac{V \angle 0°}{R \angle \theta_R} = \frac{V}{R} \underline{/0° - \theta_R}$$

Since i and v are in phase, the angle associated with i also must be $0°$. To satisfy this condition, θ_R must equal $0°$. Substituting $\theta_R = 0°$, we find

$$\mathbf{I} = \frac{V \angle 0°}{R \angle 0°} = \frac{V}{R} \underline{/0° - 0°} = \frac{V}{R} \angle 0°$$

so that in the time domain,

$$i = \sqrt{2}\left(\frac{V}{R}\right) \sin \omega t$$

The complex number in the denominator of the above equation,

$$\boxed{\mathbf{R} = R \angle 0°} \tag{15.1}$$

does not represent a sinusoidal function in the phasor domain even though it has the same format. It is a radius vector in the complex plane that has a fixed magnitude R at an angle of $0°$ (the positive real axis). The relative advantages of associating $\angle 0°$ with purely resistive elements is demonstrated in the following examples. You will find that it is no longer necessary to keep in mind that v and i are in phase. This fact was included when we associated an angle of $0°$ with R.

EXAMPLE 15.1. Using phasor algebra, find the current i for the circuit of Fig. 15.2. Sketch the waveforms of v and i.

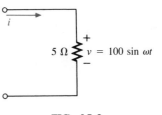

FIG. 15.2

Solution: Note Fig. 15.3:

$$v = 100 \sin \omega t \Rightarrow \text{phasor form } \mathbf{V} = 70.7 \angle 0°$$

$$\mathbf{I} = \frac{\mathbf{V}}{\mathbf{R}} = \frac{70.7 \angle 0°}{5 \angle 0°} = 14.14 \angle 0°$$

and

$$i = \sqrt{2}(14.14) \sin \omega t = \mathbf{20 \sin \omega t}$$

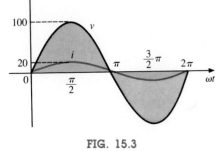

FIG. 15.3

EXAMPLE 15.2. Using phasor algebra, find the voltage v for the circuit of Fig. 15.4. Sketch the waveforms of v and i.

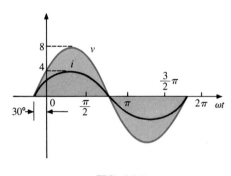

$$i = 4 \sin(\omega t + 30°)$$

$2\ \Omega$

FIG. 15.4

Solution: Note Fig. 15.5:

$$i = 4 \sin(\omega t + 30°) \Rightarrow \text{phasor form } \mathbf{I} = 2.828 \angle 30°$$

$$\mathbf{V} = \mathbf{I}R = (2.828 \angle 30°)(2 \angle 0°) = 5.656 \angle 30°$$

and

$$v = \sqrt{2}(5.656) \sin(\omega t + 30°) = \mathbf{8.0 \sin(\omega t + 30°)}$$

FIG. 15.5

It is often helpful in the analysis of networks to have a *phasor diagram,* which shows at a glance the *magnitudes* and *phase relations* between the various quantities within the network. For example, the phasor diagrams of the circuits considered in the preceding examples would be as shown in Fig. 15.6. In both cases, it is immediately obvious that v and i are in phase, since they both have the same phase angle.

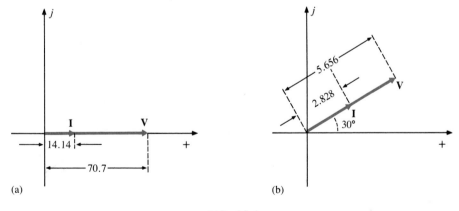

FIG. 15.6

For the pure inductor of Fig. 15.7, it was learned in Chapter 13 that the voltage leads the current by $90°$, and that the reactance of the coil X_L is determined by ωL.

$$v = V_m \sin \omega t \Rightarrow \text{phasor form } \mathbf{V} = V \angle 0°$$

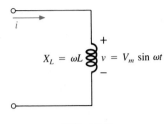

FIG. 15.7

By Ohm's law,

$$\mathbf{I} = \frac{V \angle 0°}{X_L \angle \theta_L} = \frac{V}{X_L} \underline{/0° - \theta_L}$$

Since v leads i by 90°, i must have an angle of $-90°$ associated with it. To satisfy this condition, θ_L must equal $+90°$. Substituting $\theta_L = 90°$, we obtain

$$\mathbf{I} = \frac{V \angle 0°}{X_L \angle 90°} = \frac{V}{X_L} \underline{/0° - 90°} = \frac{V}{X_L} \angle -90°$$

so that in the time domain,

$$i = \sqrt{2}\left(\frac{V}{X_L}\right) \sin(\omega t - 90°)$$

The complex number in the denominator of the preceding equation,

$$\boxed{\mathbf{X}_L = X_L \angle 90°} \qquad (15.2)$$

does not represent a sinusoidal function in the phasor domain even though it has the same format. It is a radius vector in the complex plane that has a fixed magnitude X_L at an angle of 90°.

EXAMPLE 15.3. Using phasor algebra, find the current i for the circuit of Fig. 15.8. Sketch the v and i curves.

Solution: Note Fig. 15.9:

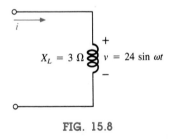

FIG. 15.8

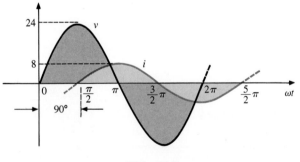

FIG. 15.9

$v = 24 \sin \omega t \Rightarrow$ phasor form $\mathbf{V} = 16.968 \angle 0°$

$$\mathbf{I} = \frac{\mathbf{V}}{\mathbf{X}_L} = \frac{16.968 \angle 0°}{3 \angle 90°} = 5.656 \angle -90°$$

and

$$i = \sqrt{2}(5.656) \sin(\omega t - 90°) = \mathbf{8.0 \sin(\omega t - 90°)}$$

$Q = \dfrac{14.0}{.05}$

EXAMPLE 15.4. Using phasor algebra, find the voltage v for the circuit of Fig. 15.10. Sketch the v and i curves.
Solution: Note Fig. 15.11:

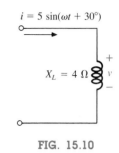

$i = 5 \sin(\omega t + 30°)$

$X_L = 4\ \Omega$ v

FIG. 15.10

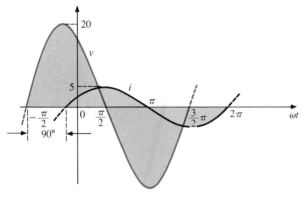

FIG. 15.11

$$i = 5 \sin(\omega t + 30°) \Rightarrow \text{phasor form } \mathbf{I} = 3.535 \angle 30°$$
$$\mathbf{V} = \mathbf{I}X_L = (3.535 \angle 30°)(4 \angle +90°) = 14.140 \angle 120°$$

and

$$v = \sqrt{2}(14.140) \sin(\omega t + 120°) = \mathbf{20\ sin(\omega t + 120°)}$$

The phasor diagrams for the two circuits of the preceding examples are shown in Fig. 15.12. Both indicate quite clearly that the voltage leads the current by 90°.

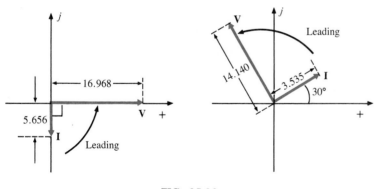

FIG. 15.12

For the pure capacitor of Fig. 15.13, it was learned in Chapter 13 that the current leads the voltage by 90°, and that the reactance of the capacitor X_C is determined by $1/\omega C$.

$$v = V_m \sin \omega t \Rightarrow \text{phasor form } \mathbf{V} = V \angle 0°$$

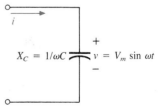

$X_C = 1/\omega C$ $v = V_m \sin \omega t$

FIG. 15.13

Applying Ohm's law and using phasor algebra, we find

$$\mathbf{I} = \frac{V \angle 0°}{X_C \angle \theta_C} = \frac{V}{X_C} \underline{/0° - \theta_C}$$

Since we know i leads v by 90°, i must have an angle of $+90°$ associated with it. To satisfy this condition, θ_C must equal $-90°$. Substituting $\theta_C = -90°$ yields

$$\mathbf{I} = \frac{V \angle 0°}{X_C \angle -90°} = \frac{V}{X_C} \underline{/0° - (-90°)} = \frac{V}{X_C} \angle 90°$$

so, in the time domain,

$$i = \sqrt{2}\left(\frac{V}{X_C}\right) \sin(\omega t + 90°)$$

Once more, the complex number in the denominator of the above equation,

$$\boxed{\mathbf{X_C} = X_C \angle -90°} \qquad (15.3)$$

does not represent a sinusoidal function in the phasor domain even though it has the same format. It is a radius vector in the complex plane that has a fixed magnitude X_C at an angle of $-90°$.

EXAMPLE 15.5. Using phasor algebra, find the current i in the circuit of Fig. 15.14. Sketch the v and i curves.

Solution: Note Fig. 15.15:

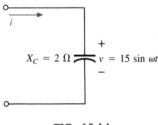

FIG. 15.14

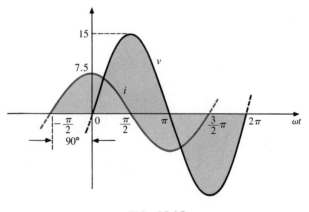

FIG. 15.15

$$v = 15 \sin \omega t \Rightarrow \text{phasor notation } \mathbf{V} = 10.605 \angle 0°$$

$$\mathbf{I} = \frac{\mathbf{V}}{\mathbf{X_C}} = \frac{10.605 \angle 0°}{2 \angle -90°} = 5.303 \angle 90°$$

and

$$i = \sqrt{2}(5.303) \sin(\omega t + 90°) = \mathbf{7.5 \sin(\omega t + 90°)}$$

EXAMPLE 15.6. Using phasor algebra, find the voltage v in the circuit of Fig. 15.16. Sketch the v and i curves.
Solution: Note Fig. 15.17:

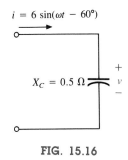

FIG. 15.16

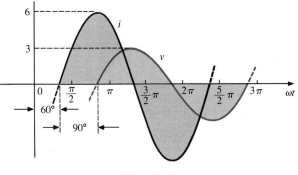

FIG. 15.17

$$i = 6 \sin(\omega t - 60°) \Rightarrow \text{phasor notation } \mathbf{I} = 4.242 \ \angle -60°$$
$$\mathbf{V} = \mathbf{I}X_C = (4.242 \ \angle -60°)(0.5 \ \angle -90°) = 2.121 \ \angle -150°$$

and

$$v = \sqrt{2}(2.121) \sin(\omega t - 150°) = \mathbf{3.0 \sin(\omega t - 150°)}$$

The phasor diagrams for the two circuits of the preceding examples are shown in Fig. 15.18. Both indicate quite clearly that the current i leads the voltage v by 90°.

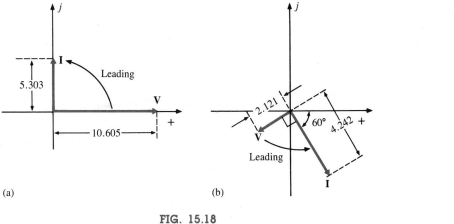

(a)   (b)

FIG. 15.18

A plot of resistance, inductive reactance, and capacitive reactance appears in Fig. 15.19. For any network, the resistance will *always* appear on the positive real axis, the inductive reactance on the positive imaginary axis, and the capacitive reactance on the negative imaginary axis.

Any *one or combination* of these elements in an ac circuit is called the *impedance* of the circuit. It is a measure of how much the circuit will

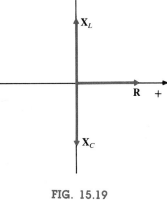

FIG. 15.19

impede, or hinder, the flow of current through it. The diagram of Fig. 15.19 is referred to as an *impedance diagram.* The symbol for impedance is Z.

For the individual elements,

$$\text{Resistance: } \mathbf{Z} = \mathbf{R} = R \angle 0° = R + j0 \qquad \textbf{(15.4)}$$

$$\text{Inductive reactance: } \mathbf{Z} = \mathbf{X}_L$$
$$= X_L \angle 90° = 0 + jX_L \qquad \textbf{(15.5)}$$

$$\text{Capacitive reactance: } \mathbf{Z} = \mathbf{X}_C$$
$$= X_C \angle -90° = 0 - jX_C \qquad \textbf{(15.6)}$$

15.3 SERIES CONFIGURATION

The overall properties of series ac circuits (Fig. 15.20) are the same as those for dc circuits; that is, *the current is the same through series*

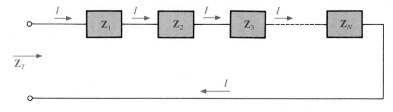

FIG. 15.20

elements, and the total impedance of a system is the sum of the individual impedances:

$$\mathbf{Z}_T = \mathbf{Z}_1 + \mathbf{Z}_2 + \mathbf{Z}_3 + \cdots + \mathbf{Z}_N \qquad \textbf{(15.7)}$$

EXAMPLE 15.7. Draw the impedance diagram for the circuit of Fig. 15.21 and find the total impedance.

Solution: As indicated by Fig. 15.22, the input impedance can be found graphically from the impedance diagram by properly scaling the

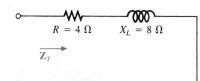

FIG. 15.21

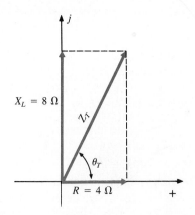

FIG. 15.22

real and imaginary axes and finding the length of the resultant vector Z_T and angle θ_T. Or, by using vector algebra, we obtain

$$\begin{aligned} \mathbf{Z}_T &= \mathbf{Z}_1 + \mathbf{Z}_2 \\ &= R \angle 0° + X_L \angle 90° \\ &= R + jX_L = 4 + j8 \\ \mathbf{Z}_T &= \mathbf{8.944} \angle \mathbf{63.43°} \end{aligned}$$

EXAMPLE 15.8. Determine the input impedance to the series network of Fig. 15.23. Draw the impedance diagram.

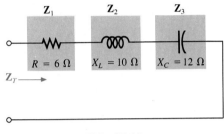

$$\mathbf{Z}_1 \qquad \mathbf{Z}_2 \qquad \mathbf{Z}_3$$

$$R = 6\ \Omega \qquad X_L = 10\ \Omega \qquad X_C = 12\ \Omega$$

$$Z_T \longrightarrow$$

FIG. 15.23

Solution:

$$\begin{aligned} \mathbf{Z}_T &= \mathbf{Z}_1 + \mathbf{Z}_2 + \mathbf{Z}_3 \\ &= R \angle 0° + X_L \angle 90° + X_C \angle -90° \\ &= R + jX_L - jX_C \\ &= R + j(X_L - X_C) = 6 + j(10 - 12) = 6 - j2 \\ \mathbf{Z}_T &= \mathbf{6.325} \angle \mathbf{-18.43°} \end{aligned}$$

The impedance diagram appears in Fig. 15.24. Note that in this example, series inductive and capacitive reactances are in direct opposition. For the circuit of Fig. 15.23, if the inductive reactance were equal to the capacitive reactance, the input impedance would be purely resistive. We will have more to say about this particular condition in a later chapter.

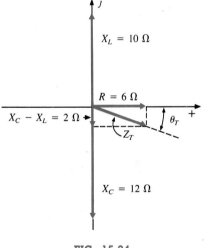

FIG. 15.24

As mentioned in Chapter 13, it can be determined whether a circuit is predominantly inductive or capacitive by noting the phase relationship between the input current and voltage. The term to be applied can also be determined by noting the angle θ_T associated with the total impedance Z_T of a circuit. If θ_T is in the first quadrant, or $0° < \theta_T < 90°$, the circuit is predominantly inductive, and if θ_T is in the fourth quadrant, or $-90° < \theta_T < 0°$, the circuit is predominantly capacitive. If $\theta_T = 0°$, the circuit is resistive.

In many of the circuits to be considered, $3 + j4 = 5 \angle 53.13°$ and $4 + j3 = 5 \angle 36.87°$ will be used quite frequently to insure that the approach is as clear as possible and not lost in mathematical complexity.

Let us now examine the *R-L*, *R-C*, and *R-L-C* series networks. Their basic nature dictates that they be examined in some detail. Numerical values were assigned to make the description as informative as possible.

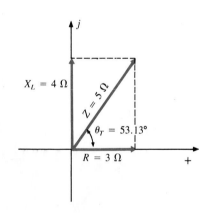

FIG. 15.25

R-L (Fig. 15.25)

Phasor notation:

$$e = 141.4 \sin \omega t \Rightarrow \mathbf{E} = 100 \angle 0°$$

Note Fig. 15.26.

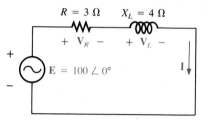

FIG. 15.26

$\mathbf{Z}_T$:

$$\mathbf{Z}_T = \mathbf{Z}_1 + \mathbf{Z}_2 = 3 \angle 0° + 4 \angle 90° = 3 + j4$$

and

$$\mathbf{Z}_T = \mathbf{5} \angle \mathbf{53.13°}$$

Impedance diagram: As shown in Fig. 15.27.

$\mathbf{I}$:

$$\mathbf{I} = \frac{\mathbf{E}}{\mathbf{Z}_T} = \frac{100 \angle 0°}{5 \angle 53.13°} = \mathbf{20} \angle \mathbf{-53.13°}$$

$\mathbf{V}_R$, $\mathbf{V}_L$:

Kirchhoff's voltage law:

$$\Sigma_{\circlearrowright} \mathbf{V} = \mathbf{E} - \mathbf{V}_R - \mathbf{V}_L = 0$$

or

$$\mathbf{E} = \mathbf{V}_R + \mathbf{V}_L$$

In rectangular form,

$$\mathbf{V}_R = 60 \angle -53.13° = 36 - j48$$
$$\mathbf{V}_L = 80 \angle +36.87° = 64 + j48$$

and

$$\mathbf{E} = \mathbf{V}_R + \mathbf{V}_L = (36 - j48) + (64 + j48) = 100 + j0$$
$$= 100 \angle 0°$$

as applied.

FIG. 15.27

Phasor diagram: Note that for the phasor diagram of Fig. 15.28, **I** is in phase with the voltage across the resistor and lags the voltage across the inductor by 90°.

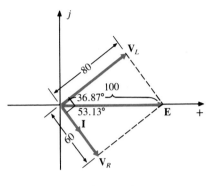

FIG. 15.28

Power: The total power in watts delivered to the circuit is

$$P_T = EI \cos \theta_T$$
$$= (100)(20) \cos 53.13° = (2000)(0.6)$$
$$= \mathbf{1200 \ W}$$

where E and I are effective values and θ_T is the phase angle between E and I, or

$$P_T = I^2R$$
$$= (20)^2(3) = (400)(3)$$
$$= \mathbf{1200 \ W}$$

where I is the effective value, or, finally,

$$P_T = P_R + P_L = V_R I \cos \theta_R + V_L I \cos \theta_L$$
$$= (60)(20) \cos 0° + (80)(20) \cos 90°$$
$$= 1200 + 0$$
$$= \mathbf{1200 \ W}$$

where θ_R is the phase angle between V_R and I, and θ_L is the phase angle between V_L and I.

Power factor: The power factor F_p of the circuit is cos 53.13° = **0.6 lagging,** where 53.13° is the phase angle between **E** and **I.**

If we write the basic power equation $P = EI \cos \theta$ in the following form:

$$\cos \theta = \frac{P}{EI}$$

where E and I are the input quantities and P is the power delivered to the network, and then perform the following substitutions from the basic series ac circuit:

$$\cos \theta = \frac{P}{EI} = \frac{I^2R}{EI} = \frac{IR}{E} = \frac{R}{E/I} = \frac{R}{Z_T}$$

we find

$$\boxed{F_p = \cos \theta_T = \frac{R}{Z_T}} \qquad \textbf{(15.8)}$$

Reference to Fig. 15.27 also indicates that θ is the impedance angle θ_T as written in Eq. (15.8). In other words, *the impedance angle θ_T is also the phase angle between the input voltage and current for a series ac circuit.* To determine the power factor, it is necessary only to form the ratio of the total resistance to the magnitude of the input impedance.

For the case at hand,

$$F_p = \cos\theta = \frac{R}{Z_T} = \frac{3}{5} = \textbf{0.6 lagging}$$

as found above.

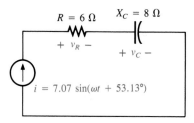

FIG. 15.29

R-C (Fig. 15.29)

Phasor notation:

$$i = 7.07\sin(\omega t + 53.13°) \Rightarrow \textbf{I} = 5 \angle 53.13°$$

Note Fig. 15.30.

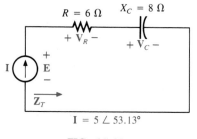

FIG. 15.30

$\textbf{Z}_T$:

$$\textbf{Z}_T = \textbf{Z}_1 + \textbf{Z}_2 = 6\angle 0° + 8\angle -90° = 6 - j8$$

and

$$\textbf{Z}_T = \textbf{10}\angle \textbf{-53.13°}$$

Impedance diagram: As shown in Fig. 15.31.

$\textbf{E}$:

$$\textbf{E} = \textbf{IZ}_T = (5\angle 53.13°)(10\angle -53.13°) = \textbf{50}\angle\textbf{0°}$$

$\textbf{V}_R$, $\textbf{V}_C$:

$$\textbf{V}_R = \textbf{IR} = (5\angle 53.13°)(6\angle 0°) = \textbf{30}\angle\textbf{53.13°}$$
$$\textbf{V}_C = \textbf{IX}_C = (5\angle 53.13°)(8\angle -90°) = \textbf{40}\angle\textbf{-36.87°}$$

Kirchhoff's voltage law:

$$\Sigma_\circ \textbf{V} = \textbf{E} - \textbf{V}_R - \textbf{V}_C = 0$$

or

$$\textbf{E} = \textbf{V}_R + \textbf{V}_C$$

which can be verified by vector algebra as demonstrated for the *R-L* circuit.

Phasor diagram: Note on the phasor diagram of Fig. 15.32 that the current **I** is in phase with the voltage across the resistor and leads the voltage across the capacitor by 90°.

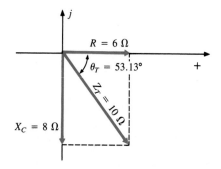

FIG. 15.31

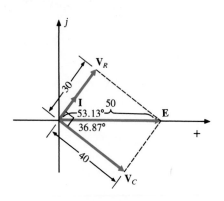

FIG. 15.32

Time domain: In the time domain,

$$e = \sqrt{2}(50) \sin \omega t = \mathbf{70.70 \sin \omega t}$$
$$v_R = \sqrt{2}(30) \sin(\omega t + 53.13°) = \mathbf{42.42 \sin(\omega t + 53.13°)}$$
$$v_C = \sqrt{2}(40) \sin(\omega t - 36.87°) = \mathbf{56.56 \sin(\omega t - 36.87°)}$$

A plot of all of the voltages and the current of the circuit appears in Fig. 15.33. Note again that i and v_R are in phase and that v_C lags i by 90°.

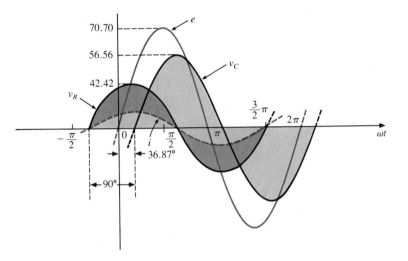

FIG. 15.33

Power: The total power in watts delivered to the circuit is

$$P_T = EI \cos \theta_T = (50)(5) \cos 53.13°$$
$$= (250)(0.6) = \mathbf{150\ W}$$

or

$$P_T = I^2R = (5)^2(6) = (25)(6)$$
$$= \mathbf{150\ W}$$

or, finally,

$$P_T = P_R + P_C = V_RI \cos \theta_R + V_CI \cos \theta_C$$
$$= (30)(5) \cos 0° + (40)(5) \cos 90°$$
$$= 150 + 0$$
$$= \mathbf{150\ W}$$

Power factor: The power factor of the circuit is

$$F_p = \cos \theta = \cos 53.13° = \mathbf{0.6\ leading}$$

Using Eq. (15.8), we obtain

$$F_p = \cos \theta = \frac{R}{Z_T} = \frac{6}{10}$$
$$= \mathbf{0.6\ leading}$$

as determined above.

R-L-C (Fig. 15.34)

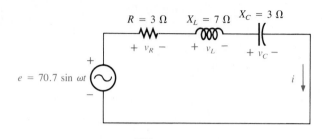

FIG. 15.34

Phasor notation: As shown in Fig. 15.35.

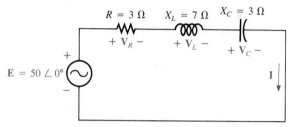

FIG. 15.35

$\mathbf{Z}_T$:

$$\mathbf{Z}_T = \mathbf{Z}_1 + \mathbf{Z}_2 + \mathbf{Z}_3 = R\ \angle 0° + X_L\ \angle 90° + X_C\ \angle -90°$$
$$= 3 + j7 - j3 = 3 + j4$$

and

$$\mathbf{Z}_T = 5\ \angle \mathbf{53.13°}$$

Impedance diagram: As shown in Fig. 15.36.

$\mathbf{I}$:

$$\mathbf{I} = \frac{\mathbf{E}}{\mathbf{Z}_T} = \frac{50\ \angle 0°}{5\ \angle 53.13°} = \mathbf{10\ \angle -53.13°}$$

$\mathbf{V}_R, \mathbf{V}_L, \mathbf{V}_C$:

$$\mathbf{V}_R = \mathbf{I}R = (10\ \angle -53.13°)(3\ \angle 0°) = \mathbf{30\ \angle -53.13°}$$
$$\mathbf{V}_L = \mathbf{I}X_L = (10\ \angle -53.13°)(7\ \angle 90°) = \mathbf{70\ \angle 36.87°}$$
$$\mathbf{V}_C = \mathbf{I}X_C = (10\ \angle -53.13°)(3\ \angle -90°) = \mathbf{30\ \angle -143.13°}$$

Kirchhoff's voltage law:

$$\Sigma_\circlearrowleft\mathbf{V} = \mathbf{E} - \mathbf{V}_R - \mathbf{V}_L - \mathbf{V}_C = 0$$

or

$$\mathbf{E} = \mathbf{V}_R + \mathbf{V}_L + \mathbf{V}_C$$

which can also be verified through vector algebra.

FIG. 15.36

Phasor diagram: The phasor diagram of Fig. 15.37 indicates that the current **I** is in phase with the voltage across the resistor, lags the voltage across the inductor by 90°, and leads the voltage across the capacitor by 90°.

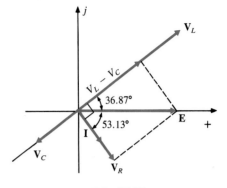

FIG. 15.37

Time domain:

$$i = \sqrt{2}(10) \sin(\omega t - 53.13°) = \mathbf{14.14\ sin(\omega t - 53.13°)}$$
$$v_R = \sqrt{2}(30) \sin(\omega t - 53.13°) = \mathbf{42.42\ sin(\omega t - 53.13°)}$$
$$v_L = \sqrt{2}(70) \sin(\omega t + 36.87°) = \mathbf{98.98\ sin(\omega t + 36.87°)}$$
$$v_C = \sqrt{2}(30) \sin(\omega t - 143.13°) = \mathbf{42.42\ sin(\omega t - 143.13°)}$$

A plot of all the voltages and the current of the circuit appears in Fig. 15.38.

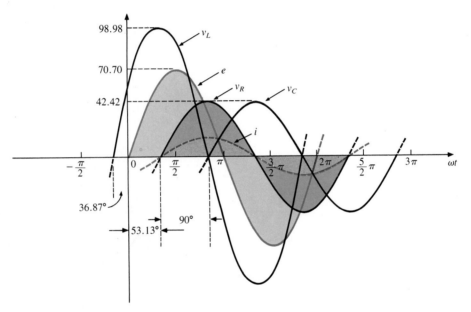

FIG. 15.38

Power: The total power in watts delivered to the circuit is

$$P_T = EI \cos \theta_T = (50)(10) \cos 53.13° = (500)(0.6) = \mathbf{300\ W}$$

or

$$P_T = I^2R = (10)^2(3) = (100)(3) = \mathbf{300\ W}$$

or

$$P_T = P_R + P_L + P_C$$
$$= V_RI \cos \theta_R + V_LI \cos \theta_L + V_CI \cos \theta_C$$
$$= (30)(10) \cos 0° + (70)(10) \cos 90° + (30)(10) \cos 90°$$
$$= (30)(10) + 0 + 0 = \mathbf{300\ W}$$

Power factor: The power factor of the circuit is

$$F_p = \cos \theta_T = \cos 53.13° = \mathbf{0.6\ lagging}$$

Using Eq. (15.8), we obtain

$$F_p = \cos \theta = \frac{R}{Z_T} = \frac{3}{5} = \textbf{0.6 lagging}$$

15.4 VOLTAGE DIVIDER RULE

The basic format for the voltage divider rule in ac circuits is exactly the same as that for dc circuits:

$$\boxed{\mathbf{V}_x = \frac{\mathbf{Z}_x \mathbf{E}}{\mathbf{Z}_T}} \tag{15.9}$$

where $\mathbf{V}_x$ is the voltage across one or more elements in series that have total impedance $\mathbf{Z}_x$, $\mathbf{E}$ is the total voltage appearing across the series circuit, and $\mathbf{Z}_T$ is the total impedance of the series circuit.

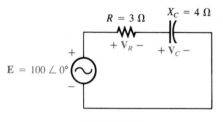

FIG. 15.39

EXAMPLE 15.9. Using the voltage divider rule, find the voltage across each element of the circuit of Fig. 15.39.
Solution:

$$\mathbf{V}_C = \frac{\mathbf{X}_C \mathbf{E}}{\mathbf{X}_C + \mathbf{R}} = \frac{(4 \angle -90°)(100 \angle 0°)}{4 \angle -90° + 3 \angle 0°} = \frac{400 \angle -90°}{3 - j4}$$

$$= \frac{400 \angle -90°}{5 \angle -53.13°} = \textbf{80} \angle -\textbf{36.87°}$$

$$\mathbf{V}_R = \frac{\mathbf{R} \mathbf{E}}{\mathbf{X}_C + \mathbf{R}} = \frac{(3 \angle 0°)(100 \angle 0°)}{5 \angle -53.13°} = \frac{300 \angle 0°}{5 \angle -53.13°}$$

$$= \textbf{60} \angle +\textbf{53.13°}$$

EXAMPLE 15.10. Using the voltage divider rule, find the unknown voltages $\mathbf{V}_R$, $\mathbf{V}_L$, $\mathbf{V}_C$, and $\mathbf{V}_1$ for the circuit of Fig. 15.40.

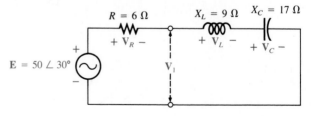

FIG. 15.40

Solution:

$$\mathbf{V}_R = \frac{\mathbf{R} \mathbf{E}}{\mathbf{R} + \mathbf{X}_L + \mathbf{X}_C} = \frac{(6 \angle 0°)(50 \angle 30°)}{6 \angle 0° + 9 \angle 90° + 17 \angle -90°}$$

$$= \frac{300 \angle 30°}{6 + j9 - j17} = \frac{300 \angle 30°}{6 - j8}$$

$$= \frac{300 \angle 30°}{10 \angle -53.13°} = \mathbf{30 \angle 83.13°}$$

$$\mathbf{V}_L = \frac{\mathbf{X}_L\mathbf{E}}{\mathbf{Z}_T} = \frac{(9 \angle 90°)(50 \angle 30°)}{10 \angle -53.13°} = \frac{450 \angle 120°}{10 \angle -53.13°}$$

$$= \mathbf{45 \angle 173.13°}$$

$$\mathbf{V}_C = \frac{\mathbf{X}_C\mathbf{E}}{\mathbf{Z}_T} = \frac{(17 \angle -90°)(50 \angle 30°)}{10 \angle -53°} = \frac{850 \angle -60°}{10 \angle -53°}$$

$$= \mathbf{85 \angle -6.87°}$$

$$\mathbf{V}_1 = \frac{(\mathbf{X}_L + \mathbf{X}_C)\mathbf{E}}{\mathbf{Z}_T} = \frac{(9 \angle 90° + 17 \angle -90°)(50 \angle 30°)}{10 \angle -53.13°}$$

$$= \frac{(8 \angle -90°)(50 \angle 30°)}{10 \angle -53.13°}$$

$$= \frac{400 \angle -60°}{10 \angle -53.13°} = \mathbf{40 \angle -6.87°}$$

EXAMPLE 15.11. For the circuit of Fig. 15.41:

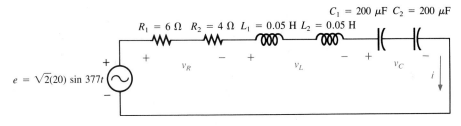

$C_1 = 200 \ \mu\text{F} \quad C_2 = 200 \ \mu\text{F}$
$R_1 = 6 \ \Omega \quad R_2 = 4 \ \Omega \quad L_1 = 0.05 \ \text{H} \quad L_2 = 0.05 \ \text{H}$

$e = \sqrt{2}(20) \sin 377t$

FIG. 15.41

a. Calculate i, v_R, v_L, and v_C in phasor form.
b. Calculate the total power factor.
c. Calculate the average power delivered to the circuit.
d. Draw the phasor diagram.
e. Obtain the phasor sum of $\mathbf{V}_R$, $\mathbf{V}_L$, and $\mathbf{V}_C$, and show that it equals the input voltage $\mathbf{E}$.
f. Find $\mathbf{V}_R$ and $\mathbf{V}_C$ using the voltage divider rule.

Solutions:

a. Combining common elements and finding the reactance of the inductor and capacitor, we obtain

$$R_T = 6 + 4 = 10 \ \Omega$$

$$L_T = 0.05 + 0.05 = 0.1 \ \text{H}$$

$$C_T = \frac{200}{2} = 100 \ \mu\text{F}$$

$$X_L = \omega L = (377)(0.1) = 37.70 \,\Omega$$

$$X_C = \frac{1}{\omega C} = \frac{1}{(377)(100 \times 10^{-6})} = \frac{10^6}{37,700} = 26.53 \,\Omega$$

Redrawing the circuit using phasor notation results in Fig. 15.42.

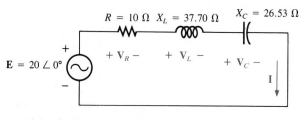

FIG. 15.42

For the circuit of Fig. 15.42,

$$\mathbf{Z}_T = R \angle 0° + X_L \angle 90° + X_C \angle -90°$$
$$= 10 + j37.70 - j26.53$$
$$= 10 + j11.17 = \mathbf{15 \angle 48.16°}$$

The current **I** is

$$\mathbf{I} = \frac{\mathbf{E}}{\mathbf{Z}_T} = \frac{20 \angle 0°}{15 \angle 48.16°} = \mathbf{1.33 \angle -48.16°}$$

The voltage across the resistor, inductor, and capacitor can be found using Ohm's law:

$$\mathbf{V}_R = \mathbf{IR} = (1.33 \angle -48.16°)(10 \angle 0°)$$
$$= \mathbf{13.30 \angle -48.16°}$$

$$\mathbf{V}_L = \mathbf{I}X_L = (1.33 \angle -48.16°)(37.70 \angle 90°)$$
$$= \mathbf{50.14 \angle 41.84°}$$

$$\mathbf{V}_C = \mathbf{I}X_C = (1.33 \angle -48.16°)(26.53 \angle -90°)$$
$$= \mathbf{35.28 \angle -138.16°}$$

b. The total power factor is determined by the angle between the applied voltage **E,** and the resulting current **I** is 48.16°:

$$F_p = \cos \theta = \cos 48.16° = \mathbf{0.667 \ lagging}$$

or

$$F_p = \cos \theta = \frac{R}{Z_T} = \frac{10}{15} = \mathbf{0.667 \ lagging}$$

c. The total power in watts delivered to the circuit is

$$P_T = EI \cos \theta = (20)(1.33)(0.667) = \mathbf{17.74 \ W}$$

d. The phasor diagram appears in Fig. 15.43.

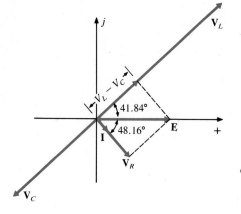

FIG. 15.43

e. The phasor sum of $\mathbf{V}_R$, $\mathbf{V}_L$, and $\mathbf{V}_C$ is

$$\mathbf{E} = \mathbf{V}_R + \mathbf{V}_L + \mathbf{V}_C$$
$$= 13.30 \angle -48.16° + 50.14 \angle 41.84° + 35.28 \angle -138.16°$$
$$\mathbf{E} = 13.30 \angle -48.16° + 14.86 \angle 41.84°$$

Therefore,

$$E = \sqrt{(13.30)^2 + (14.86)^2} = \mathbf{20}$$

and

$$\theta_E = \mathbf{0°} \qquad \text{(from phasor diagram)}$$

and

$$\mathbf{E} = 20 \angle 0°$$

f. $\mathbf{V}_R = \dfrac{\mathbf{R}\mathbf{E}}{\mathbf{Z}_T} = \dfrac{(10 \angle 0°)(20 \angle 0°)}{15 \angle 48.16°} = \dfrac{200 \angle 0°}{15 \angle 48.16°}$

$$= \mathbf{13.3} \angle -\mathbf{48.16°}$$

$\mathbf{V}_C = \dfrac{\mathbf{X}_C\mathbf{E}}{\mathbf{Z}_T} = \dfrac{(26.5 \angle -90°)(20 \angle 0°)}{15 \angle 48.16°} = \dfrac{530.6 \angle -90°}{15 \angle 48.16°}$

$$= \mathbf{35.37} \angle -\mathbf{138.16°}$$

15.5 FREQUENCY RESPONSE OF THE R-C NETWORK

Thus far, the analysis of series circuits has been limited to a particular frequency. We will now examine the effect of frequency on the response of an *R-C* series configuration such as that in Fig. 15.44.

The amplitude of the source is fixed at 14.14 V but the frequency range of analysis will extend from zero to 20 kHz.

Applying the voltage divider rule to determine the voltage across the capacitor in phasor form yields

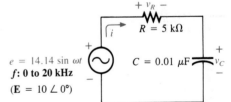

FIG. 15.44

$$\mathbf{V}_C = \frac{\mathbf{X}_C\mathbf{E}}{\mathbf{R} + \mathbf{X}_C}$$

$$= \frac{(X_C \angle -90°)(E \angle 0°)}{R - jX_C}$$

$$= \frac{X_C E \angle -90°}{R - jX_C}$$

$$= \frac{X_C E \angle -90°}{\sqrt{R^2 + X_C^2} \, \underline{/-\tan^{-1} X_C/R}}$$

or

$$= \frac{X_C E}{\sqrt{R^2 + X_C^2}} \, \underline{/ -90° + \tan^{-1} (X_C/R)}$$

The magnitude of $\mathbf{V}_C$ is therefore determined by

$$V_C = \frac{X_C E}{\sqrt{R^2 + X_C^2}} \qquad \textbf{(15.10)}$$

and the phase angle θ by which $\mathbf{V}_C$ lags $\mathbf{E}$ is given by

$$\theta = \left| -90° + \tan^{-1} \frac{X_C}{R} \right| \qquad \textbf{(15.11)}$$

To determine the frequency response, X_C must be calculated for each frequency of interest and inserted into Eqs. (15.10) and (15.11).

To begin our analysis, it makes good sense to consider the case of $f = 0\,\text{Hz}$ (dc conditions).

$f = 0\,\text{Hz}$:

$$X_C = \frac{1}{2\pi f C} = \frac{1}{0} \Rightarrow \text{very large value}$$

Applying the open-circuit equivalent for the capacitor based on the above calculation will result in the following:

$$\mathbf{V}_C = \mathbf{E} = 10 \angle 0°$$

If we apply Eq. (15.10), we find

$$X_C^2 \gg R^2$$

and

$$\sqrt{R^2 + X_C^2} \cong \sqrt{X_C^2} = X_C$$

and

$$V_C = \frac{X_C E}{\sqrt{R^2 + X_C^2}} = \frac{X_C E}{X_C} = E$$

with

$$\theta = \left| -90° + \tan^{-1}(\text{very large value}) \right|$$
$$= \left| -90° + 90° \right|$$
$$= 0°$$

verifying the above conclusions.

$f = 1\,\text{kHz}$:

$$X_C = \frac{1}{2\pi f C} = \frac{1}{(6.28)(1 \times 10^3)(0.01 \times 10^{-6})} \cong \textbf{15.9}\,\text{k}\Omega$$
$$\sqrt{R^2 + X_C^2} = \sqrt{(5\,\text{k}\Omega)^2 + (15.9\,\text{k}\Omega)^2} \cong 16.67\,\text{k}\Omega$$

and

$$V_C = \frac{X_C E}{\sqrt{R^2 + X_C^2}} = \frac{(15.9\,\text{k}\Omega)(10)}{16.67\,\text{k}\Omega} = \textbf{9.53}\,\textbf{V}$$

with

$$\theta = \left| -90° + \tan^{-1} \frac{X_C}{R} \right|$$

$$= \left| -90° + \tan^{-1} \frac{15.9\,k\Omega}{5\,k\Omega} \right|$$

$$= |-90° + 72.54°| = |-17.46°|$$

$$= \mathbf{17.46°}$$

and

$$\mathbf{V}_C = \mathbf{9.53} \angle \mathbf{-17.46°}$$

As expected, the high reactance of the capacitor at low frequencies has resulted in the major part of the applied voltage appearing across the capacitor.

If we plot the phasor diagrams for $f = 0$ Hz and $f = 1$ kHz, as shown in Fig. 15.45, we find that $\mathbf{V}_C$ is beginning a clockwise rotation with increase in frequency that will increase the angle θ and decrease the phase angle between $\mathbf{I}$ and $\mathbf{E}$. Recall that for a purely capacitive netork, $\mathbf{I}$ leads $\mathbf{E}$ by 90°. As the frequency increases, therefore, the capacitive reactance is decreasing, and eventually $R \gg X_C$ with $\theta = 90°$, and the angle between $\mathbf{I}$ and $\mathbf{E}$ will approach 0°. Keep in mind as we proceed through the other frequencies that θ is the phase angle between $\mathbf{V}_C$ and $\mathbf{E}$ and that the magnitude of the angle by which $\mathbf{I}$ leads $\mathbf{E}$ is determined by

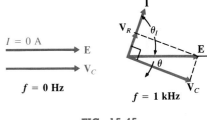

FIG. 15.45

$$\boxed{|\theta_I| = 90° - |\theta|} \qquad (15.12)$$

$f = \mathbf{5\,kHz}$:

$$X_C = \frac{1}{2\pi f C} = \frac{1}{(6.28)(5 \times 10^3)(0.01 \times 10^{-6})} \cong \mathbf{3.2\,k\Omega}$$

Note the dramatic drop in X_C from 1 kHz to 5 kHz. In fact, X_C is now less than the resistance R of the network, and the phase angle determined by $\tan^{-1}(X_C/R)$ must be less than 45°. Here,

$$V_C = \frac{X_C E}{\sqrt{R^2 + X_C^2}} = \frac{(3.2\,k\Omega)(10)}{\sqrt{(5\,k\Omega)^2 + (3.2\,k\Omega)^2}} = \mathbf{5.39\,V}$$

with

$$\theta = \left| -90° + \tan^{-1} \frac{X_C}{R} \right|$$

$$= \left| -90° + \tan^{-1} \frac{3.2\,k\Omega}{5\,k\Omega} \right|$$

$$= |-90° + 32.62°| = |-57.38°|$$

$$= \mathbf{57.38°}$$

$f = \mathbf{10\,kHz}$: $X_C \cong 1.6\,k\Omega$, $V_C = \mathbf{3.05\,V}$, $\theta = \mathbf{72.26°}$

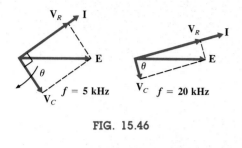

FIG. 15.46

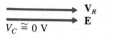

f = **very high frequencies**

FIG. 15.47

$f = $ **15 kHz:** $X_C \cong 1.06 \, k\Omega$, $V_C = $ **2.07 V**, $\theta = $ **78.03°**

$f = $ **20 kHz:** $X_C \cong 0.8 \, k\Omega$, $V_C = $ **1.58 V**, $\theta = $ **80.95°**

The phasor diagrams for $f = 5$ kHz and $f = 20$ kHz appear in Fig. 15.46 to show the continuing rotation of the $\mathbf{V}_C$ vector.

Note also from Figs. 15.45 and 15.46 that the vector $\mathbf{V}_R$ and the current $\mathbf{I}$ have grown in magnitude with reduction in the capacitive reactance. Eventually, at very high frequencies X_C will approach zero ohms and the short-circuit equivalent can be applied, resulting in $V_C = 0$ V and $\theta = 90°$, producing the phasor diagram of Fig. 15.47. The network is then purely resistive and the phase angle between $\mathbf{I}$ and $\mathbf{E}$ is zero degrees and V_R and I are their maximum values.

A plot of V_C versus frequency appears in Fig. 15.48, and θ versus frequency in Fig. 15.49.

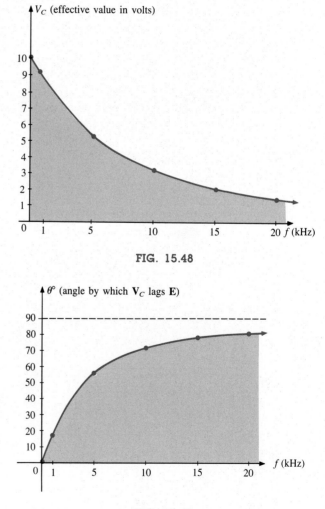

FIG. 15.48

FIG. 15.49

A plot of V_R versus frequency would approach E volts from 0 volts with increase in frequency, but remember $V_R \neq E - V_C$ due to vector relationship. The phase angle between **I** and **E** could be plotted directly from Fig. 15.49 using Eq. (15.12).

In Chapter 20, the analysis of this section will be extended to a much wider frequency range using a log axis for frequency. It will be demonstrated that an R-C circuit such as that in Fig. 15.44 can be used as a filter to determine which frequencies will have the greatest impact on the stage to follow. From our current analysis, it is obvious that any network connected across the capacitor will receive the greatest potential level at low frequencies and be effectively ''shorted out'' at very high frequencies.

The analysis of a series R-L circuit would proceed in much the same manner, except that X_L and V_L would increase with frequency and the angle between **I** and **E** would approach 90° (voltage leading the current) rather than 0°. If V_L were plotted versus frequency, $\mathbf{V}_L$ would approach **E,** and X_L would eventually attain a level at which the open-circuit equivalent would be appropriate.

PARALLEL ac CIRCUITS

15.6 ADMITTANCE AND SUSCEPTANCE

The discussion for parallel ac circuits will be very similar to that for dc circuits. In dc circuits, *conductance* (G) was defined as equal to $1/R$. The total conductance of a parallel circuit was then found by adding the conductance of each branch. The total resistance R_T is simply $1/G_T$.

In ac circuits, we define *admittance* (Y) as equal to $1/Z$. The unit of measure for admittance as defined by the SI system is *siemens,* which has the symbol S. Admittance is a measure of how well an ac circuit will *admit,* or allow, current to flow in the circuit. The larger its value, therefore, the heavier the current flow for the same applied potential. The total admittance of a circuit can also be found by finding the sum of the parallel admittances. The total impedance Z_T of the circuit is then $1/Y_T$; that is, for the network of Fig. 15.50,

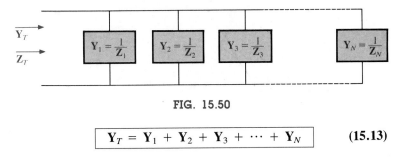

FIG. 15.50

$$\mathbf{Y}_T = \mathbf{Y}_1 + \mathbf{Y}_2 + \mathbf{Y}_3 + \cdots + \mathbf{Y}_N \qquad (15.13)$$

or, since $\mathbf{Z} = 1/\mathbf{Y}$,

$$\frac{1}{\mathbf{Z}_T} = \frac{1}{\mathbf{Z}_1} + \frac{1}{\mathbf{Z}_2} + \frac{1}{\mathbf{Z}_3} + \cdots + \frac{1}{\mathbf{Z}_N} \qquad \textbf{(15.14)}$$

For two impedances in parallel,

$$\frac{1}{\mathbf{Z}_T} = \frac{1}{\mathbf{Z}_1} + \frac{1}{\mathbf{Z}_2}$$

If the manipulations used in Chapter 6 to find the total resistance of two parallel resistors are now applied, the following similar equation will result:

$$\mathbf{Z}_T = \frac{\mathbf{Z}_1\mathbf{Z}_2}{\mathbf{Z}_1 + \mathbf{Z}_2} \qquad \textbf{(15.15)}$$

For three parallel impedances,

$$\mathbf{Z}_T = \frac{\mathbf{Z}_1\mathbf{Z}_2\mathbf{Z}_3}{\mathbf{Z}_1\mathbf{Z}_2 + \mathbf{Z}_2\mathbf{Z}_3 + \mathbf{Z}_1\mathbf{Z}_3} \qquad \textbf{(15.16)}$$

As pointed out in the introduction to this section, conductance is the reciprocal of resistance, and

$$\mathbf{Y} = \frac{1}{\mathbf{R}} = \frac{1}{R \angle 0°} = G \angle 0°$$

so that

$$\mathbf{G} = G \angle 0° \qquad \textbf{(15.17)}$$

The reciprocal of reactance $(1/X)$ is called *susceptance* and is a measure of how *susceptible* an element is to the passage of current through it. Susceptance is also measured in *siemens*, and is represented by the capital letter B.

For the inductor,

$$\mathbf{Y} = \frac{1}{\mathbf{X}_L} = \frac{1}{X_L \angle 90°} = \frac{1}{\omega L \angle 90°} = \frac{1}{\omega L} \angle -90°$$

Defining

$$B_L = \frac{1}{X_L} \qquad \text{(siemens, S)} \qquad \textbf{(15.18)}$$

we have

$$\mathbf{B}_L = B_L \angle -90° \qquad \textbf{(15.19)}$$

Note that for inductance, an increase in frequency or inductance will result in a decrease in susceptance or, correspondingly, in admittance.

For the capacitor,

$$Y = \frac{1}{X_C} = \frac{1}{X_C \angle -90°} = \frac{1}{1/\omega C \angle -90°} = \omega C \angle 90°$$

Defining

$$\boxed{B_C = \frac{1}{X_C}} \quad \text{(S)} \qquad \textbf{(15.20)}$$

we have

$$\boxed{\mathbf{B}_C = B_C \angle 90°} \qquad \textbf{(15.21)}$$

For the capacitor, therefore, an increase in frequency or capacitance will result in an increase in its susceptibility.

In summary, for parallel circuits,

Resistance:

$$\mathbf{Y} = \frac{1}{\mathbf{R}} = \mathbf{G} = G \angle 0° = G + j0 \qquad \textbf{(15.22)}$$

Inductance:

$$\mathbf{Y} = \frac{1}{\mathbf{X}_L} = \mathbf{B}_L = B_L \angle -90° = 0 - jB_L \qquad \textbf{(15.23)}$$

Capacitance:

$$\mathbf{Y} = \frac{1}{\mathbf{X}_C} = \mathbf{B}_C = B_C \angle 90° = 0 + jB_C \qquad \textbf{(15.24)}$$

For parallel ac circuits, the *admittance diagram* is used with the three admittances represented as shown in Fig. 15.51.

Note in Fig. 15.51 that the conductance (like resistance) is on the positive real axis, while inductive and capacitive susceptances are still in direct opposition on the imaginary axis.

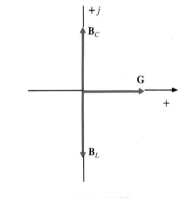

FIG. 15.51

EXAMPLE 15.12. For the network of Fig. 15.52:
a. Find the admittance of each parallel branch.
b. Determine the input admittance.
c. Calculate the input impedance.
d. Draw the admittance diagram.

Solutions:

a. $\mathbf{Y}_1 = \mathbf{G} = G \angle 0° = \dfrac{1}{R} \angle 0° = \dfrac{1}{20} \angle 0°$

   $= 0.05 \angle 0° = 0.05 + j0$

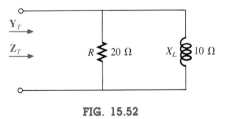

FIG. 15.52

$$\mathbf{Y}_2 = \mathbf{B}_L = B_L \angle -90° = \frac{1}{X_L} \angle -90° = \frac{1}{10} \angle -90°$$

$$= 0.1 \angle -90° = 0 - j0.1$$

b. $\mathbf{Y}_T = \mathbf{Y}_1 + \mathbf{Y}_2 = (0.05 + j0) + (0 - j0.1)$

$$= 0.05 - j0.1 = G - jB_L$$

c. $\mathbf{Z}_T = \dfrac{1}{\mathbf{Y}_T} = \dfrac{1}{0.05 - j0.1} = \dfrac{1}{0.112 \angle -63.43°}$

$$= 8.93 \angle 63.43°$$

or Eq. (15.15):

$$\mathbf{Z}_T = \frac{\mathbf{Z}_1 \mathbf{Z}_2}{\mathbf{Z}_1 + \mathbf{Z}_2} = \frac{(20 \angle 0°)(10 \angle 90°)}{20 + j10}$$

$$= \frac{200 \angle 90°}{22.36 \angle 26.57°} = 8.93 \angle 63.43°$$

d. The admittance diagram appears in Fig. 15.53.

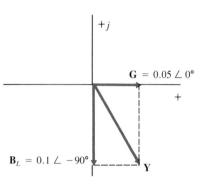

FIG. 15.53

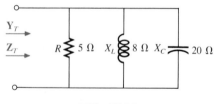

FIG. 15.54

EXAMPLE 15.13. Repeat Example 15.12 for the parallel network of Fig. 15.54.

Solutions:

a. $\mathbf{Y}_1 = \mathbf{G} = G \angle 0° = \dfrac{1}{R} \angle 0° = \dfrac{1}{5} \angle 0°$

$$= 0.2 \angle 0° = 0.2 + j0$$

$\mathbf{Y}_2 = \mathbf{B}_L = B_L \angle -90° = \dfrac{1}{X_L} \angle -90° = \dfrac{1}{8} \angle -90°$

$$= 0.125 \angle -90° = 0 - j0.125$$

$\mathbf{Y}_3 = \mathbf{B}_C = B_C \angle 90° = \dfrac{1}{X_C} \angle 90° = \dfrac{1}{20} \angle 90°$

$$= 0.050 \angle +90° = 0 + j0.050$$

b. $\mathbf{Y}_T = \mathbf{Y}_1 + \mathbf{Y}_2 + \mathbf{Y}_3$

$$= (0.2 + j0) + (0 - j0.125) + (0 + j0.050)$$

$$= 0.2 - j0.075 = 0.2136 \angle -20.56°$$

c. $\mathbf{Z}_T = \dfrac{1}{0.2136 \angle -20.56°} = 4.68 \angle 20.56°$

or

$$\mathbf{Z}_T = \frac{\mathbf{Z}_1 \mathbf{Z}_2 \mathbf{Z}_3}{\mathbf{Z}_1 \mathbf{Z}_2 + \mathbf{Z}_2 \mathbf{Z}_3 + \mathbf{Z}_1 \mathbf{Z}_3}$$

$$= \frac{(5 \angle 0°)(8 \angle 90°)(20 \angle -90°)}{(5 \angle 0°)(8 \angle 90°) + (8 \angle 90°)(20 \angle -90°)}$$
$$+ (5 \angle 0°)(20 \angle -90°)$$

$$= \frac{800 \angle 0°}{40 \angle 90° + 160 \angle 0° + 100 \angle -90°}$$

$$= \frac{800}{160 + j40 - j100} = \frac{800}{160 - j60}$$

$$= \frac{800}{170.88 \angle -20.56°}$$

$$= \mathbf{4.68 \angle 20.56°}$$

d. The admittance diagram appears in Fig. 15.55.

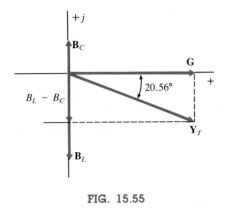

FIG. 15.55

On many occasions, the inverse relationship $\mathbf{Y}_T = 1/\mathbf{Z}_T$ or $\mathbf{Z}_T = 1/\mathbf{Y}_T$ will require that we divide the number 1 by a complex number having a real and an imaginary part. This division, if not performed in the polar form, requires that we multiply the numerator and denominator by the conjugate of the denominator, as follows:

$$\mathbf{Y}_T = \frac{1}{\mathbf{Z}_T} = \frac{1}{4 + j6} = \frac{1}{4 + j6} \frac{(4 - j6)}{(4 - j6)} = \frac{4 - j6}{4^2 + 6^2}$$

and

$$\mathbf{Y}_T = \frac{4}{52} - j\frac{6}{52}$$

To avoid this laborious task each time we want to find the reciprocal of a complex number in rectangular form, a format can be developed using the following complex number, which is symbolic of any impedance or admittance in the first or fourth quadrant:

$$\frac{1}{a_1 \pm jb_1} = \left(\frac{1}{a_1 \pm jb_1}\right)\left(\frac{a_1 \mp jb_1}{a_1 \mp jb_1}\right) = \frac{a_1 \mp jb_1}{a_1^2 + b_1^2}$$

or

$$\boxed{\frac{1}{a_1 \pm jb_1} = \frac{a_1}{a_1^2 + b_1^2} \mp j\frac{b_1}{a_1^2 + b_1^2}} \qquad \textbf{(15.25)}$$

Note that the denominator is simply the sum of the squares of each term. The sign is inverted between the real and imaginary parts. A few examples will develop some familiarity with the use of this equation.

EXAMPLE 15.14. Find the admittance of each set of series elements in Fig. 15.56.

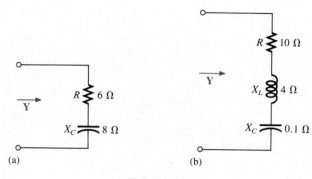

(a)   (b)

FIG. 15.56

Solutions:

a. $\mathbf{Z} = 6 - j8$

$$\mathbf{Y} = \frac{1}{6 - j8} = \frac{6}{(6)^2 + (8)^2} + j\frac{8}{(6)^2 + (8)^2} = \frac{6}{100} + j\frac{8}{100}$$

b. $\mathbf{Z} = 10 + j4 + (-j0.1) = 10 + j3.9$

$$\mathbf{Y} = \frac{1}{\mathbf{Z}} = \frac{1}{10 + j3.9} = \frac{10}{(10)^2 + (3.9)^2} - j\frac{3.9}{(10)^2 + (3.9)^2}$$

$$= \frac{10}{115.21} - j\frac{3.9}{115.21} = \mathbf{0.087 - j0.034}$$

To determine whether a series ac circuit is predominantly capacitive or inductive, it is necessary only to note whether the sign of the imaginary part of the total impedance is positive or negative. For parallel circuits, a negative sign in front of the imaginary part of the total admittance indicates that the circuit is predominantly inductive, and a positive sign indicates that it is predominantly capacitive.

15.7 R-L, R-C, AND R-L-C PARALLEL ac NETWORKS

R-L (Fig. 15.57)

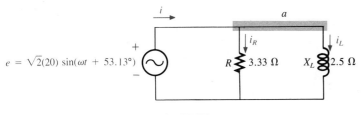

$e = \sqrt{2}(20)\sin(\omega t + 53.13°)$

FIG. 15.57

Phasor notation: As shown in Fig. 15.58.

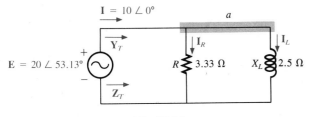

FIG. 15.58

Y_T, Z_T:

$$\mathbf{Y}_T = \mathbf{Y}_1 + \mathbf{Y}_2 = \mathbf{G} + \mathbf{B}_L = \frac{1}{3.33}\angle 0° + \frac{1}{2.5}\angle -90°$$

$$= 0.3\angle 0° + 0.4\angle -90° = 0.3 - j0.4$$
$$= \mathbf{0.5\angle -53.13°}$$

$$\mathbf{Z}_T = \frac{1}{\mathbf{Y}_T} = \frac{1}{0.5\angle -53.13°} = \mathbf{2\angle 53.13°}$$

Admittance diagram: As shown in Fig. 15.59.

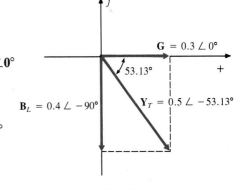

FIG. 15.59

I:

$$\mathbf{I} = \frac{\mathbf{E}}{\mathbf{Z}_T} = \mathbf{EY}_T = (20\angle 53.13°)(0.5\angle -53.13°) = \mathbf{10\angle 0°}$$

I_R, I_L:

$$\mathbf{I}_R = \frac{\mathbf{E}}{\mathbf{R}} = \mathbf{EG} = (20\angle 53.13°)(0.3\angle 0°) = \mathbf{6\angle 53.13°}$$

$$\mathbf{I}_L = \frac{\mathbf{E}}{\mathbf{X}_L} = \mathbf{EB}_L = (20\angle 53.13°)(0.4\angle -90°)$$
$$= \mathbf{8\angle -36.87°}$$

Kirchhoff's current law: At node a,

$$\mathbf{I} - \mathbf{I}_R - \mathbf{I}_L = 0$$

or

$$\mathbf{I} = \mathbf{I}_R + \mathbf{I}_L$$
$$10\angle 0° = 6\angle 53.13° + 8\angle -36.87°$$
$$10\angle 0° = (3.60 + j4.80) + (6.40 - j4.80) = 10 + j0$$

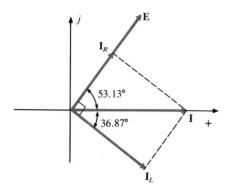

FIG. 15.60

and

$$10 \angle 0° = 10 \angle 0° \quad \text{(checks)}$$

Phasor diagram: The phasor diagram of Fig. 15.60 indicates that the applied voltage **E** is in phase with the current $\mathbf{I}_R$ and leads the current $\mathbf{I}_L$ by 90°.

Power: The total power in watts delivered to the circuit is

$$\begin{aligned} P_T &= EI \cos \theta_T \\ &= (20)(10) \cos 53.13° = (220)(0.6) \\ &= \textbf{120 W} \end{aligned}$$

or

$$P_T = I^2R = \frac{V_R^2}{R} = V_R^2 G = (20)^2(0.3) = \textbf{120 W}$$

or, finally,

$$\begin{aligned} P_T &= P_R + P_L = EI_R \cos \theta_R + EI_L \cos \theta_L \\ &= (20)(6) \cos 0° + (20)(8) \cos 90° = 120 + 0 \\ &= \textbf{120 W} \end{aligned}$$

Power factor: The power factor of the circuit is

$$F_p = \cos \theta_T = \cos 53.13° = \textbf{0.6 lagging}$$

or, through an analysis similar to that employed for a series ac circuit,

$$\cos \theta_T = \frac{P}{EI} = \frac{E^2/R}{EI} = \frac{EG}{I} = \frac{G}{I/V} = \frac{G}{Y_T}$$

and

$$\boxed{F_p = \cos \theta_T = \frac{G}{Y_T}} \qquad \textbf{(15.26)}$$

where G and Y_T are the magnitudes of the total conductance and admittance of the parallel network. For this case,

$$F_p = \cos \theta_T = \frac{0.3}{0.5} = \textbf{0.6 lagging}$$

Impedance approach: The current **I** can also be found by first finding the total impedance of the network:

$$\begin{aligned} \mathbf{Z}_T &= \frac{\mathbf{Z}_1\mathbf{Z}_2}{\mathbf{Z}_1 + \mathbf{Z}_2} = \frac{(3.33 \angle 0°)(2.5 \angle 90°)}{3.33 \angle 0° + 2.5 \angle 90°} \\ &= \frac{8.325 \angle 90°}{4.164 \angle 36.87°} = \textbf{2} \angle\textbf{53.13°} \end{aligned}$$

And then, using Ohm's law, we obtain

$$\mathbf{I} = \frac{\mathbf{E}}{\mathbf{Z}_T} = \frac{20 \angle 53.13°}{2 \angle 53.13°} = \textbf{10} \angle\textbf{0°}$$

R-C (Fig. 15.61)

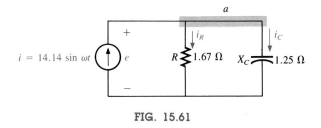

FIG. 15.61

Phasor notation: As shown in Fig. 15.62.

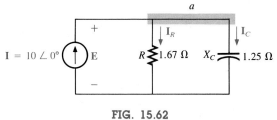

FIG. 15.62

Y_T, Z_T:

$$\mathbf{Y}_T = \mathbf{Y}_1 + \mathbf{Y}_2 = \mathbf{G} + \mathbf{B}_C = \frac{1}{1.67} \angle 0° + \frac{1}{1.25} \angle 90°$$

$$= 0.6 \angle 0° + 0.8 \angle 90° = 0.6 + j0.8 = \mathbf{1.0 \angle 53.13°}$$

$$\mathbf{Z}_T = \frac{1}{\mathbf{Y}_T} = \frac{1}{1.0 \angle 53.13°} = \mathbf{1 \angle -53.13°}$$

Admittance diagram: As shown in Fig. 15.63.

E:

$$\mathbf{E} = \mathbf{I}\mathbf{Z}_T = \frac{\mathbf{I}}{\mathbf{Y}_T} = \frac{10 \angle 0°}{1 \angle 53.13°} = \mathbf{10 \angle -53.13°}$$

I_R, I_C:

$$\mathbf{I}_R = \mathbf{E}\mathbf{G} = (10 \angle -53.13°)(0.6 \angle 0°) = \mathbf{6 \angle -53.13°}$$
$$\mathbf{I}_C = \mathbf{E}\mathbf{B}_C = (10 \angle -53.13°)(0.8 \angle 90°) = \mathbf{8 \angle 36.87°}$$

Kirchhoff's current law: At node *a*,

$$\mathbf{I} - \mathbf{I}_R - \mathbf{I}_C = 0$$

or

$$\mathbf{I} = \mathbf{I}_R + \mathbf{I}_C$$

which can also be verified (as for the *R-L* network) through vector algebra.

Phasor diagram: The phasor diagram of Fig. 15.64 indicates that **E** is in phase with the current through the resistor $\mathbf{I}_R$ and lags the capacitive current $\mathbf{I}_C$ by 90°.

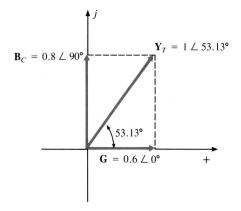

FIG. 15.63

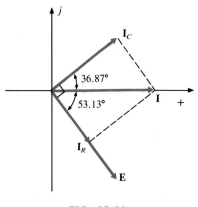

FIG. 15.64

Time domain:

$$e = \sqrt{2}(10) \sin(\omega t - 53.13°) = \mathbf{14.14 \sin(\omega t - 53.13°)}$$
$$i_R = \sqrt{2}(6) \sin(\omega t - 53.13°) = \mathbf{8.48 \sin(\omega t - 53.13°)}$$
$$i_C = \sqrt{2}(8) \sin(\omega t + 36.87°) = \mathbf{11.31 \sin(\omega t + 36.87°)}$$

A plot of all of the currents and the voltage appears in Fig. 15.65. Note that e and i_R are in phase and e lags i_C by 90°.

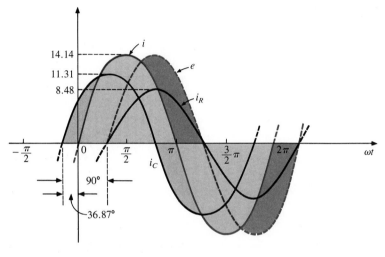

FIG. 15.65

Power:

$$P_T = EI \cos \theta = (10)(10) \cos 53.13° = (10)^2(0.6)$$
$$= \mathbf{60\,W}$$

or

$$P_T = E^2G = (10)^2(0.6) = \mathbf{60\,W}$$

or, finally,

$$P_T = P_R + P_C = EI_R \cos \theta_R + EI_C \cos \theta_C$$
$$= (10)(6) \cos 0° + (10)(8) \cos 90°$$
$$= \mathbf{60\,W}$$

Power factor: The power factor of the circuit is

$$F_p = \cos 53.13° = \mathbf{0.6\ leading}$$

Using Eq. (15.26), we have

$$F_p = \cos \theta_T = \frac{G}{Y_T} = \frac{0.6}{1.0} = \mathbf{0.6\ leading}$$

Impedance approach: The voltage **E** can also be found by first finding the total impedance of the circuit:

$$\mathbf{Z}_T = \frac{\mathbf{Z}_1 \mathbf{Z}_2}{\mathbf{Z}_1 + \mathbf{Z}_2} = \frac{(1.67 \angle 0°)(1.25 \angle -90°)}{1.67 \angle 0° + 1.25 \angle -90°}$$

$$= \frac{2.09 \angle -90°}{2.09 \angle -36.81°} = \mathbf{1 \angle -53.19°}$$

and then, using Ohm's law, we find

$$\mathbf{E} = \mathbf{IZ}_T = (10 \angle 0°)(1 \angle -53.19°) = \mathbf{10 \angle -53.19°}$$

R-L-C (Fig. 15.66)

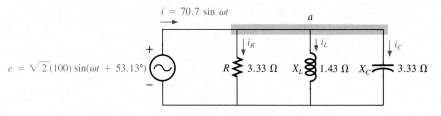

FIG. 15.66

Phasor notation: As shown in Fig. 15.67.

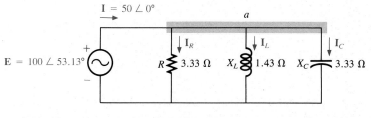

FIG. 15.67

$\mathbf{Y}_T$, $\mathbf{Z}_T$:

$$\mathbf{Y}_T = \mathbf{Y}_1 + \mathbf{Y}_2 + \mathbf{Y}_3 = \mathbf{G} + \mathbf{B}_L + \mathbf{B}_C$$

$$= \frac{1}{3.33} \angle 0° + \frac{1}{1.43} \angle -90° + \frac{1}{3.33} \angle 90°$$

$$= 0.3 \angle 0° + 0.7 \angle -90° + 0.3 \angle 90°$$

$$= 0.3 - j0.7 + j0.3$$

$$= 0.3 - j0.4 = \mathbf{0.5 \angle -53.13°}$$

$$\mathbf{Z}_T = \frac{1}{\mathbf{Y}_T} = \frac{1}{0.5 \angle -53.13°} = \mathbf{2 \angle 53.13°}$$

Admittance diagram: As shown in Fig. 15.68.

I:

$$\mathbf{I} = \frac{\mathbf{E}}{\mathbf{Z}_T} = \mathbf{EY}_T = (100 \angle 53.13°)(0.5 \angle -53.13°) = \mathbf{50 \angle 0°}$$

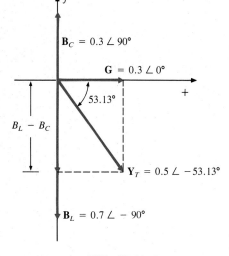

FIG. 15.68

$\mathbf{I}_R$, $\mathbf{I}_L$, $\mathbf{I}_C$:

$$\mathbf{I}_R = \mathbf{EG} = (100 \angle 53.13°)(0.3 \angle 0°) = \mathbf{30 \angle 53.13°}$$
$$\mathbf{I}_L = \mathbf{EB}_L = (100 \angle 53.13°)(0.7 \angle -90°) = \mathbf{70 \angle -36.87°}$$
$$\mathbf{I}_C = \mathbf{EB}_C = (100 \angle 53.13°)(0.3 \angle +90°) = \mathbf{30 \angle 143.13°}$$

Kirchhoff's current law: At node *a*,

$$\mathbf{I} - \mathbf{I}_R - \mathbf{I}_L - \mathbf{I}_C = 0$$

or

$$\mathbf{I} = \mathbf{I}_R + \mathbf{I}_L + \mathbf{I}_C$$

Phasor diagram: The phasor diagram of Fig. 15.69 indicates that the impressed voltage **E** is in phase with the current $\mathbf{I}_R$ through the resistor, leads the current $\mathbf{I}_L$ through the inductor by 90°, and lags the current $\mathbf{I}_C$ of the capacitor by 90°.

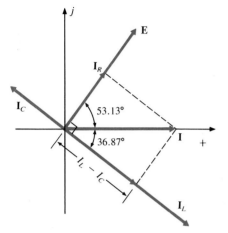

FIG. 15.69

Time domain:

$$i = \sqrt{2}(50) \sin \omega t = \mathbf{70.70 \sin \omega t}$$
$$i_R = \sqrt{2}(30) \sin(\omega t + 53.13°) = \mathbf{42.42 \sin(\omega t + 53.13°)}$$
$$i_L = \sqrt{2}(70) \sin(\omega t - 36.87°) = \mathbf{98.98 \sin(\omega t - 36.87°)}$$
$$i_C = \sqrt{2}(30) \sin(\omega t + 143.13°) = \mathbf{42.42 \sin(\omega t + 143.13°)}$$

A plot of all of the currents and the impressed voltage appears in Fig. 15.70.

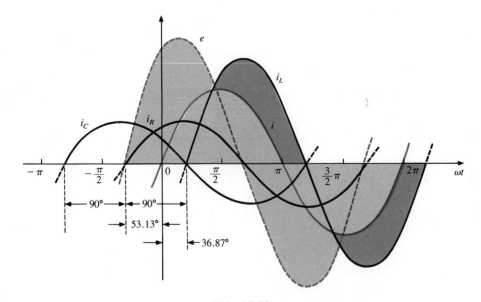

FIG. 15.70

Power: The total power in watts delivered to the circuit is

$$P_T = EI \cos \theta = (100)(50) \cos 53.13° = (5000)(0.6)$$
$$= \mathbf{3000 \ W}$$

or

$$P_T = E^2G = (100)^2(0.3) = \textbf{3000 W}$$

or, finally,

$$
\begin{aligned}
P_T &= P_R + P_L + P_C \\
&= EI_R \cos \theta_R + EI_L \cos \theta_L + EI_C \cos \theta_C \\
&= (100)(30) \cos 0° + (100)(70) \cos 90° \\
&\quad + (100)(30) \cos 90° \\
&= 3000 + 0 + 0 \\
&= \textbf{3000 W}
\end{aligned}
$$

Power factor: The power factor of the circuit is

$$F_p = \cos \theta_T = \cos 53.13° = \textbf{0.6 lagging}$$

Using Eq. (15.26), we obtain

$$F_p = \cos \theta_T = \frac{G}{Y_T} = \frac{0.3}{0.5} = \textbf{0.6 lagging}$$

Impedance approach: The input current **I** can also be determined by first finding the total impedance in the following manner:

$$\mathbf{Z}_T = \frac{\mathbf{Z}_1\mathbf{Z}_2\mathbf{Z}_3}{\mathbf{Z}_1\mathbf{Z}_2 + \mathbf{Z}_2\mathbf{Z}_3 + \mathbf{Z}_1\mathbf{Z}_3} = \mathbf{2 \angle 53.13°}$$

and, applying Ohm's law, we obtain

$$\mathbf{I} = \frac{\mathbf{E}}{\mathbf{Z}_T} = \frac{100 \angle 53.13°}{2 \angle 53.13°} = \mathbf{50 \angle 0°}$$

15.8 CURRENT DIVIDER RULE

The basic format for the current divider rule in ac circuits is exactly the same as that for dc circuits; that is, for two parallel branches with impedances $\mathbf{Z}_1$ and $\mathbf{Z}_2$ as shown in Fig. 15.71,

$$\mathbf{I}_1 = \frac{\mathbf{Z}_2\mathbf{I}_T}{\mathbf{Z}_1 + \mathbf{Z}_2} \quad \text{or} \quad \mathbf{I}_2 = \frac{\mathbf{Z}_1\mathbf{I}_T}{\mathbf{Z}_1 + \mathbf{Z}_2} \qquad \textbf{(15.27)}$$

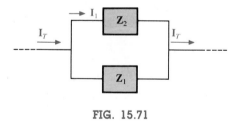

FIG. 15.71

EXAMPLE 15.15. Using the current divider rule, find the current through each impedance of Fig. 15.72.

Solution:

$$
\begin{aligned}
\mathbf{I}_R &= \frac{\mathbf{X}_L\mathbf{I}_T}{\mathbf{R} + \mathbf{X}_L} = \frac{(4 \angle 90°)(20 \angle 0°)}{3 \angle 0° + 4 \angle 90°} = \frac{80 \angle 90°}{5 \angle 53.13°} \\
&= \mathbf{16 \angle 36.87°}
\end{aligned}
$$

$$
\begin{aligned}
\mathbf{I}_L &= \frac{\mathbf{R}\mathbf{I}_T}{\mathbf{R} + \mathbf{X}_L} = \frac{(3 \angle 0°)(20 \angle 0°)}{5 \angle 53.13°} = \frac{60 \angle 0°}{5 \angle 53.13°} \\
&= \mathbf{12 \angle -53.13°}
\end{aligned}
$$

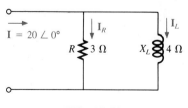

FIG. 15.72

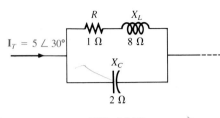

FIG. 15.73

EXAMPLE 15.16. Using the current divider rule, find the current through each parallel branch of Fig. 15.73.

Solution:

$$\mathbf{I}_{R\text{-}L} = \frac{\mathbf{X}_C\mathbf{I}_T}{\mathbf{X}_C + \mathbf{Z}_{R\text{-}L}} = \frac{(2\angle-90°)(5\angle30°)}{-j2 + 1 + j8} = \frac{10\angle-60°}{1 + j6}$$

$$= \frac{10\angle-60°}{6.083\angle80.54°} \cong \mathbf{1.644}\angle-\mathbf{140.54°}$$

$$\mathbf{I}_C = \frac{\mathbf{Z}_{R\text{-}L}\mathbf{I}_T}{\mathbf{Z}_{R\text{-}L} + \mathbf{X}_C} = \frac{(1 + j8)(5\angle30°)}{6.08\angle80.54°}$$

$$= \frac{(8.06\angle82.87°)(5\angle30°)}{6.08\angle80.54°} = \frac{40.30\angle112.87°}{6.083\angle80.54°}$$

$$= \mathbf{6.625}\angle\mathbf{32.33°}$$

15.9 EQUIVALENT CIRCUITS

In a series ac circuit, the total impedance of two or more elements in series is often equivalent to an impedance that can be achieved with fewer elements of different values, the elements and their values being determined by the frequency applied. This is also true for parallel circuits. For the circuit of Fig. 15.74(a),

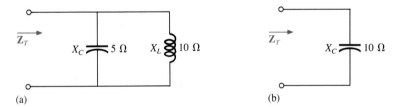

(a)   (b)

FIG. 15.74

$$\mathbf{Z}_T = \frac{\mathbf{X}_C\mathbf{X}_L}{\mathbf{X}_C + \mathbf{X}_L} = \frac{(5\angle-90°)(10\angle90°)}{5\angle-90° + 10\angle90°} = \frac{50\angle0°}{5\angle90°}$$

$$= 10\angle-90°$$

The total impedance at the frequency applied is equivalent to a capacitor with a reactance of $10\,\Omega$, as shown in Fig. 15.74(b). Always keep in mind that this equivalence is true only at the applied frequency. If the frequency changes, the reactance of each element changes, and the equivalent circuit will change—perhaps from capacitive to inductive in the above example.

Another interesting development appears if the impedance of a parallel circuit, such as the one of Fig. 15.75(a), is found in rectangular coordinates. In this case,

$$\mathbf{Z}_T = \frac{\mathbf{X}_L\mathbf{R}}{\mathbf{X}_L + \mathbf{R}} = \frac{(4\angle90°)(3\angle0°)}{4\angle90° + 3\angle0°}$$

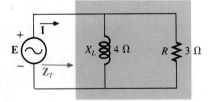

(a)

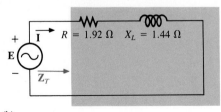

(b)

FIG. 15.75

$$= \frac{12 \angle 90°}{5 \angle 53.13°} = 2.40 \angle 36.87°$$

$$= 1.920 + j1.440$$

which is the impedance of a series circuit with a resistor of 1.92 Ω and an inductive reactance of 1.44 Ω, as shown in Fig. 15.75(b).

The current **I** will be the same in each circuit of Fig. 15.74 or Fig. 15.75 if the same input voltage **E** is applied. For a parallel circuit of one resistive element and one reactive element, the series circuit with the same input impedance will always be composed of one resistive and one reactive element. The impedance of each element of the series circuit will be different from that of the parallel circuit, but the reactive elements will always be of the same type; that is, an *R-L* circuit and an *R-C* parallel circuit will have an equivalent *R-L* and *R-C* series circuit, respectively. The same is true when converting from a series to a parallel circuit. In the discussion to follow, keep in mind that *the term* equivalent *refers only to the fact that for the same applied potential, the same impedance and input current will result.*

To formulate the equivalence between the series and parallel circuits, the equivalent series circuit for a resistor and reactance in parallel can be found by determining the total impedance of the circuit in rectangular form; that is, for the circuit of Fig. 15.76(a),

$$\mathbf{Y}_p = \frac{1}{R_p} + \frac{1}{\pm jX_p}$$

and

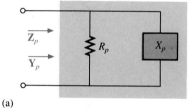

(a)

$$\mathbf{Z}_p = \frac{1}{\mathbf{Y}_p} = \frac{1}{(1/R_p) \mp j(1/X_p)}$$

$$= \frac{1/R_p}{(1/R_p)^2 + (1/X_p)^2} \pm j\frac{1/X_p}{(1/R_p)^2 + (1/X_p)^2}$$

Multiplying the numerator and denominator of each term by $R_p^2 X_p^2$ results in

$$\mathbf{Z}_p = \frac{R_p X_p^2}{X_p^2 + R_p^2} \pm j\frac{R_p^2 X_p}{X_p^2 + R_p^2}$$

$$= R_s \pm jX_s \quad [\text{Fig. 15.76(b)}]$$

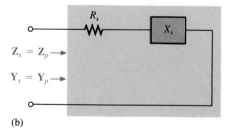

(b)

FIG. 15.76

and

$$\boxed{R_s = \frac{R_p X_p^2}{X_p^2 + R_p^2}} \qquad \textbf{(15.28)}$$

with

$$\boxed{X_s = \frac{R_p^2 X_p}{X_p^2 + R_p^2}} \qquad \textbf{(15.29)}$$

For the network of Fig. 15.75,

$$R_s = \frac{R_p X_p^2}{X_p^2 + R_p^2} = \frac{(3)(4)^2}{(4)^2 + (3)^2} = \frac{48}{25} = \mathbf{1.920\,\Omega}$$

and

$$X_s = \frac{R_p^2 X_p}{X_p^2 + R_p^2} = \frac{(3)^2(4)}{25} = \frac{36}{25} = \mathbf{1.440\,\Omega}$$

which agrees with the previous result.

The equivalent parallel circuit for a circuit with a resistor and reactance in series can be found by simply finding the total admittance of the system in rectangular form; that is, for the circuit of Fig. 15.76(b),

$$\mathbf{Z}_s = R_s \pm jX_s$$

$$\mathbf{Y}_s = \frac{1}{\mathbf{Z}_s} = \frac{1}{R_s \pm jX_s} = \frac{R_s}{R_s^2 + X_s^2} \mp j\frac{X_s}{R_s^2 + X_s^2}$$

$$= G_p \mp jB_p = \frac{1}{R_p} \mp j\frac{1}{X_p} \qquad [\text{Fig. 15.76(a)}]$$

or

$$\boxed{R_p = \frac{R_s^2 + X_s^2}{R_s}} \qquad (15.30)$$

with

$$\boxed{X_p = \frac{R_s^2 + X_s^2}{X_s}} \qquad (15.31)$$

For the above example,

$$R_p = \frac{R_s^2 + X_s^2}{R_s} = \frac{(1.92)^2 + (1.44)^2}{1.92} = \frac{5.76}{1.92} = \mathbf{3.0\,\Omega}$$

and

$$X_p = \frac{R_s^2 + X_s^2}{X_s} = \frac{5.76}{1.44} = \mathbf{4.0\,\Omega}$$

as shown in Fig. 15.75(a).

EXAMPLE 15.17. Determine the series equivalent circuit for the network of Fig. 15.77.

Solution:

$$R_p = 8\,\text{k}\Omega$$

$$X_p\ (\text{resultant}) = |X_L - X_C| = |9\,\text{k}\Omega - 4\,\text{k}\Omega|$$
$$= 5\,\text{k}\Omega$$

and

$$R_s = \frac{R_p X_p^2}{X_p^2 + R_p^2} = \frac{(8\,k\Omega)(5\,k\Omega)^2}{(5\,k\Omega)^2 + (8\,k\Omega)^2} = \frac{200\,k\Omega}{89} = \textbf{2.247 k}\boldsymbol{\Omega}$$

with

$$X_s = \frac{R_p^2 X_p}{X_p^2 + R_p^2} = \frac{(8\,k\Omega)^2(5\,k\Omega)}{89\,k\Omega} = \frac{320\,k\Omega}{89}$$
$$= \textbf{3.596 k}\boldsymbol{\Omega} \qquad \textbf{(inductive)}$$

The equivalent series circuit appears in Fig. 15.78.

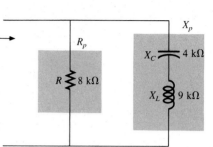

FIG. 15.77

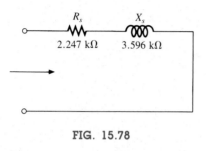

FIG. 15.78

EXAMPLE 15.18. For the network of Fig. 15.79:

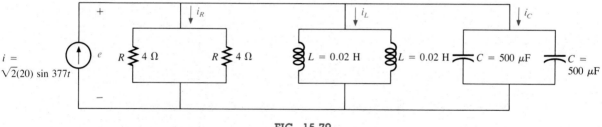

FIG. 15.79

a. Compute e, i_R, i_L, and i_C in phasor form.
b. Compute the total power factor.
c. Compute the total power delivered to the network.
d. Draw the phasor diagram.
e. Obtain the phasor sum of $\mathbf{I}_R$, $\mathbf{I}_L$, and $\mathbf{I}_C$, and show that it equals $\mathbf{I}$.
f. Compute the impedance of the parallel combination of $\mathbf{X}_L$ and $\mathbf{X}_C$, and then find $\mathbf{I}_R$ by the current divider rule.
g. Determine the equivalent series circuit as far as the total impedance and current $\mathbf{I}$ are concerned.

Solutions:

a. Combining common elements and finding the reactance of the inductor and capacitor, we obtain

$$R_T = \frac{R}{2} = \frac{4}{2} = 2\,\Omega$$

$$L_T = \frac{0.02}{2} = 0.01\,\text{H}$$

$$C_T = 500\,\mu\text{F} + 500\,\mu\text{F} = 1000\,\mu\text{F}$$

$$X_L = \omega L = (377)(0.01) = 3.77\,\Omega$$

$$X_C = \frac{1}{\omega C} = \frac{1}{(377)(10^3 \times 10^{-6})} = 2.65\,\Omega$$

The network is redrawn in Fig. 15.80 with phasor notation. The total admittance is

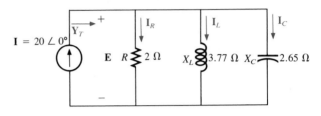

FIG. 15.80

$$\mathbf{Y}_T = \mathbf{Y}_1 + \mathbf{Y}_2 + \mathbf{Y}_3 = \mathbf{G} + \mathbf{B}_L + \mathbf{B}_C$$

$$= \frac{1}{2}\,\angle 0° + \frac{1}{3.77}\,\angle -90° + \frac{1}{2.65}\,\angle +90°$$

$$= 0.5\,\angle 0° + 0.265\,\angle -90° + 0.377\,\angle +90°$$

$$= 0.5 - j0.265 + j0.377$$

$$= 0.5 + j0.112 = \mathbf{0.512\,\angle 12.63°}$$

The input voltage is

$$\mathbf{E} = \frac{\mathbf{I}}{\mathbf{Y}_T} = \frac{20\,\angle 0°}{0.512\,\angle 12.63°}$$

$$= \mathbf{39.06\,\angle -12.63°}$$

The current through the resistor, inductor, and capacitor can be found using Ohm's law as follows:

$$\mathbf{I}_R = \mathbf{EG} = (39.06\,\angle -12.63°)(0.5\,\angle 0°)$$

$$= \mathbf{19.53\,\angle -12.63°}$$

$$\mathbf{I}_L = \mathbf{EB}_L = (39.06\,\angle -12.63°)(0.265\,\angle -90°)$$

$$= \mathbf{10.35\,\angle -102.63°}$$

$$\mathbf{I}_C = \mathbf{EB}_C = (39.06\,\angle -12.63°)(0.377\,\angle +90°)$$

$$= \mathbf{14.73\,\angle +77.37°}$$

b. The total power factor is

$$F_p = \cos\theta = \frac{G}{Y_T} = \frac{0.5}{0.512} = \mathbf{0.977}$$

c. The total power in watts delivered to the circuit is

$$P_T = I_R^2 R = E^2 G = (39.06)^2(0.5)$$

$$= \mathbf{762.84\,W}$$

d. The phasor diagram is shown in Fig. 15.81.

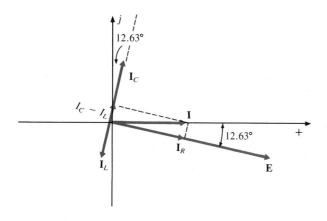

FIG. 15.81

e. The phasor sum of $\mathbf{I}_R$, $\mathbf{I}_L$, and $\mathbf{I}_C$ is

$$\mathbf{I} = \mathbf{I}_R + \mathbf{I}_L + \mathbf{I}_C$$
$$= 19.53 \angle -12.63° + 10.35 \angle -102.63°$$
$$+ \ 14.73 \angle +77.37°$$
$$= 19.53 \angle -12.63° + 4.38 \angle +77.37°$$
$$I_T = \sqrt{(19.53)^2 + (4.38)^2} = 20$$

and θ_T (from phasor diagram) $= 0°$. Therefore,

$$\mathbf{I}_T = \mathbf{20 \angle 0°}$$

which agrees with the original input.

f. The impedance of the parallel combination of $\mathbf{X}_L$ and $\mathbf{X}_C$ is

$$\mathbf{Z}_{T_1} = \frac{\mathbf{X}_L \mathbf{X}_C}{\mathbf{X}_L + \mathbf{X}_C} = \frac{(3.77 \angle 90°)(2.65 \angle -90°)}{3.77 \angle 90° + 2.65 \angle -90°}$$
$$= \frac{10 \angle 0°}{1.12 \angle 90°}$$
$$= 8.93 \angle -90°$$

The current $\mathbf{I}_R$, using the current divider rule, is

$$\mathbf{I}_R = \frac{\mathbf{Z}_{T_1} \mathbf{I}}{\mathbf{Z}_{T_1} + \mathbf{R}} = \frac{(8.93 \angle -90°)(20 \angle 0°)}{8.93 \angle -90° + 2 \angle 0°}$$
$$= \frac{178.60 \angle -90°}{9.15 \angle -77.37°} = \mathbf{19.52 \angle -12.63°}$$

g. $\mathbf{Z}_T = \dfrac{1}{\mathbf{Y}_T} = \dfrac{1}{0.512 \angle 12.63°} = \mathbf{1.95 \angle -12.63°}$

which, in rectangular form, is $1.90 - j0.427$, and

$$C = \frac{1}{\omega X_C} = \frac{1}{(377)(0.427)} = 0.006212 \, \text{F} = \mathbf{6212 \, \mu F}$$

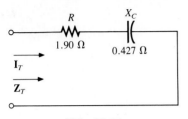

FIG. 15.82

The series circuit appears in Fig. 15.82. Since Z_{T_1} is available from part (f) above, we can apply Eqs. (15.28) and (15.29). That is, $Z_{T_1} = X_p$, and

$$R_s = \frac{R_p X_p^2}{X_p^2 + R_p^2} = \frac{(2)(8.93)^2}{(8.93)^2 + (2)^2} = \frac{159.49}{83.74} = \mathbf{1.90\ \Omega}$$

with

$$X_s = \frac{R_p^2 X_p}{X_p^2 + R_p^2} = \frac{(2)^2(8.93)}{83.74} = \frac{35.72}{83.74} = \mathbf{0.427\ \Omega}$$

as obtained above.

PROBLEMS

Section 15.2

1. Express the impedances of Fig. 15.83 in both polar and rectangular form.

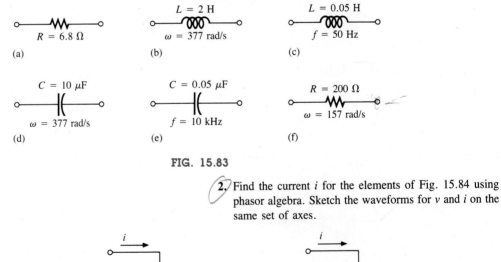

FIG. 15.83

2. Find the current i for the elements of Fig. 15.84 using phasor algebra. Sketch the waveforms for v and i on the same set of axes.

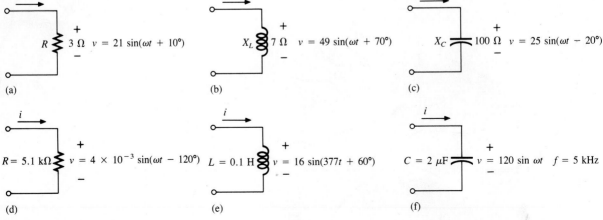

FIG. 15.84

3. Find the voltage v for the elements of Fig. 15.85 using phasor algebra. Sketch the waveforms of v and i on the same set of axes.

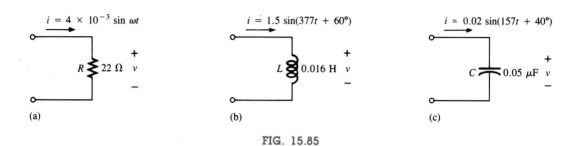

$i = 4 \times 10^{-3} \sin \omega t$

$i = 1.5 \sin(377t + 60°)$

$i = 0.02 \sin(157t + 40°)$

R 22 Ω v

L 0.016 H v

C 0.05 µF v

(a)   (b)   (c)

FIG. 15.85

Section 15.3

4. Calculate the total impedance of the circuits of Fig. 15.86. Express your answer in rectangular and polar form, and draw the impedance diagram.

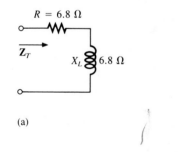

$R = 6.8 \ \Omega$

$\mathbf{Z}_T$

X_L 6.8 Ω

(a)

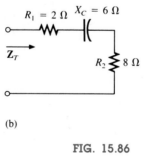

$R_1 = 2 \ \Omega \quad X_C = 6 \ \Omega$

$\mathbf{Z}_T$

R_2 8 Ω

(b)

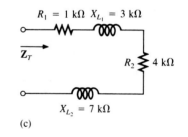

$R_1 = 1 \ \text{k}\Omega \quad X_{L_1} = 3 \ \text{k}\Omega$

$\mathbf{Z}_T$

R_2 4 kΩ

$X_{L_2} = 7 \ \text{k}\Omega$

(c)

FIG. 15.86

5. Repeat Problem 4 for the circuits of Fig. 15.87.

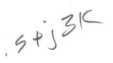

$.5 + j \ 3K$

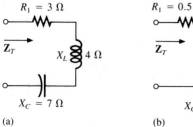

$R_1 = 3 \ \Omega$

$\mathbf{Z}_T$

X_L 4 Ω

$X_C = 7 \ \Omega$

(a)

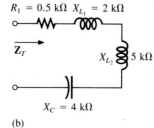

$R_1 = 0.5 \ \text{k}\Omega \quad X_{L_1} = 2 \ \text{k}\Omega$

$\mathbf{Z}_T$

X_{L_2} 5 kΩ

$X_C = 4 \ \text{k}\Omega$

(b)

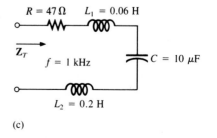

$R = 47 \ \Omega \qquad L_1 = 0.06 \ \text{H}$

$\mathbf{Z}_T$

$f = 1 \ \text{kHz}$

$C = 10 \ \mu\text{F}$

$L_2 = 0.2 \ \text{H}$

(c)

FIG. 15.87

6. Find the type and impedance in ohms of the series circuit elements that must be in the closed container of Fig. 15.88 in order for the indicated voltages and currents to exist at the input terminals. (Find the simplest series circuit that will satisfy the indicated conditions.)

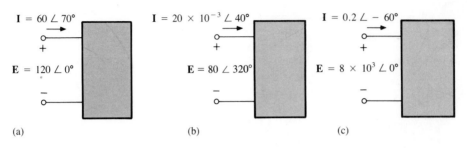

$\mathbf{I} = 60 \angle 70°$

$\mathbf{E} = 120 \angle 0°$

(a)

$\mathbf{I} = 20 \times 10^{-3} \angle 40°$

$\mathbf{E} = 80 \angle 320°$

(b)

$\mathbf{I} = 0.2 \angle -60°$

$\mathbf{E} = 8 \times 10^3 \angle 0°$

(c)

FIG. 15.88

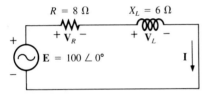

$R = 8 \, \Omega$ $X_L = 6 \, \Omega$

$+ \mathbf{V}_R -$ $+ \mathbf{V}_L -$

$\mathbf{E} = 100 \angle 0°$ $\mathbf{I}$

FIG. 15.89

7. For the circuit of Fig. 15.89:
   a Find the total impedance $\mathbf{Z}_T$ in polar form.
   b. Draw the impedance diagram.
   c. Find the current $\mathbf{I}$ and the voltages $\mathbf{V}_R$ and $\mathbf{V}_L$ in phasor form.
   d. Draw the phasor diagram of the voltages $\mathbf{E}$, $\mathbf{V}_R$, and $\mathbf{V}_L$, and the current $\mathbf{I}$.
   e. Verify Kirchhoff's voltage law around the closed loop.
   f. Find the average power delivered to the circuit.
   g. Find the power factor of the circuit and indicate whether it is leading or lagging.
   h. Find the sinusoidal expressions for the voltages and current if the frequency is 60 Hz.
   i. Plot the waveforms for the voltages and current on the same set of axes.

8. Repeat Problem 7 for the circuit of Fig. 15.90, replacing $\mathbf{V}_L$ by $\mathbf{V}_C$ in parts (c) and (d).

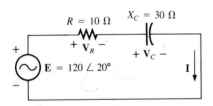

$R = 10 \, \Omega$ $X_C = 30 \, \Omega$

$+ \mathbf{V}_R -$ $+ \mathbf{V}_C -$

$\mathbf{E} = 120 \angle 20°$ $\mathbf{I}$

FIG. 15.90

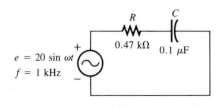

R C

$0.47 \, \text{k}\Omega$ $0.1 \, \mu\text{F}$

$e = 20 \sin \omega t$
$f = 1 \, \text{kHz}$

FIG. 15.91

9. Given the network of Fig. 15.91:
   a. Determine $\mathbf{Z}_T$.
   b. Find $\mathbf{I}$.
   c. Calculate $\mathbf{V}_R$ and $\mathbf{V}_L$.
   d. Find P and F_p.

10. For the circuit of Fig. 15.92:
   a. Find the total impedance Z_T in polar form.
   b. Draw the impedance diagram.
   c. Find the value of C in microfarads and L in henries.
   d. Find the current i and the voltages v_R, v_L, and v_C in phasor form.
   e. Draw the phasor diagram of the voltages **E**, **V**$_R$, **V**$_L$, and **V**$_C$, and the current **I**.
   f. Verify Kirchhoff's voltage law around the closed loop.
   g. Find the average power delivered to the circuit.
   h. Find the power factor of the circuit and indicate whether it is leading or lagging.
   i. Find the sinusoidal expressions for the voltages and current.
   j. Plot the waveforms for the voltages and current on the same set of axes.

11. Repeat Problem 10 for the circuit of Fig. 15.93.

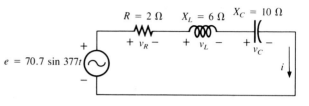

FIG. 15.92

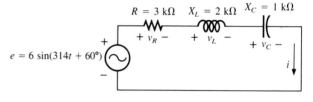

FIG. 15.93

Section 15.4

12. Calculate the voltages **V**$_1$ and **V**$_2$ for the circuit of Fig. 15.94 in phasor form using the voltage divider rule.

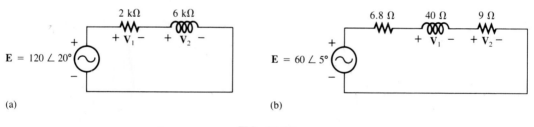

(a)   (b)

FIG. 15.94

13. Repeat Problem 12 for the circuits of Fig. 15.95.

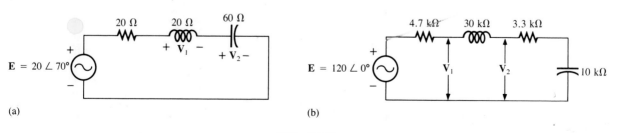

(a)   (b)

FIG. 15.95

*14. For the circuit of Fig. 15.96:
  a. Determine i, v_R, and v_C in phasor form.
  b. Calculate the total power factor and indicate whether it is leading or lagging.
  c. Calculate the average power delivered to the circuit.
  d. Draw the impedance diagram.
  e. Draw the phasor diagram of the voltages **E**, **V**$_R$, and **V**$_C$, and the current **I**.
  f. Find the voltages **V**$_R$ and **V**$_C$ using the voltage divider rule, and compare with part (a) above.
  g. Draw the equivalent series circuit of the above as far as the total impedance and the current i are concerned.

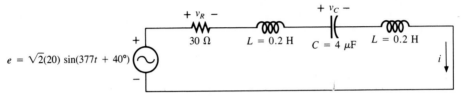

$+ v_R -$ $+ v_C -$

$30\ \Omega$ $L = 0.2\ H$ $C = 4\ \mu F$ $L = 0.2\ H$

$e = \sqrt{2}(20)\ \sin(377t + 40°)$ i

FIG. 15.96

*15. Repeat Problem 14 if the capacitance is changed to 1000 μF.

16. An electrical load has a power factor of 0.8 lagging. It dissipates 8 kW at a voltage of 200 V. Calculate the impedance of this load in rectangular coordinates.

*17. Find the series element or elements that must be in the enclosed container of Fig. 15.97 to satisfy the following conditions:
  a. Average power to circuit = 300 W.
  b. Circuit has a lagging power factor.

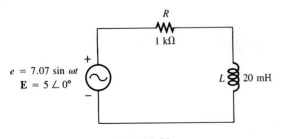

$\mathbf{I} = 3 \angle \theta$

$2\ \Omega$

$\mathbf{E} = 120 \angle 0°$

FIG. 15.97

Section 15.5

18. For the circuit of Fig. 15.98:
  a. Plot V_L versus frequency for a frequency range of zero to 20 kHz.
  b. Plot θ_L versus frequency for the same frequency range as in part (a).
  c. Plot V_R versus frequency for the frequency range of part (a).

R

$1\ k\Omega$

$e = 7.07\ \sin \omega t$
$\mathbf{E} = 5 \angle 0°$

L ⬚ 20 mH

FIG. 15.98

19. For the circuit of Fig. 15.99:

    a. Plot V_C versus frequency for a frequency range of zero to 10 kHz.

    b. Plot θ_C versus frequency for the same frequency range as in part (a).

    c. Plot V_R versus frequency for the frequency range of part (a).

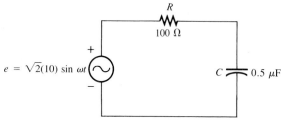

FIG. 15.99

Section 15.6

20. Find the total admittance and impedance of the circuits of Fig. 15.100. Identify the values of conductance and susceptance, and draw the admittance diagram.

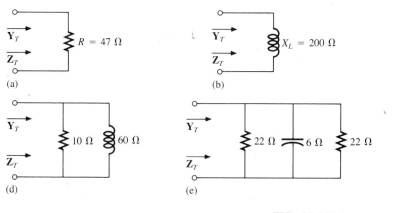
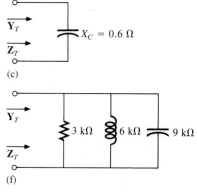

FIG. 15.100

21. Repeat Problem 20 for the networks of Fig. 15.101.

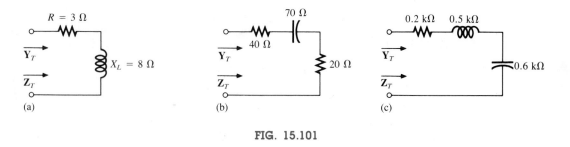

FIG. 15.101

22. Repeat Problem 6 for the parallel circuit elements that must be in the closed container for the same voltage and current to exist at the input terminals. (Find the simplest parallel circuit that will satisfy the conditions indicated.)

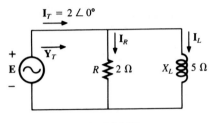

$I_T = 2 \angle 0°$

FIG. 15.102

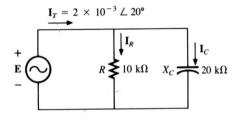

$I_T = 2 \times 10^{-3} \angle 20°$

FIG. 15.103

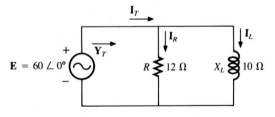

I_T

FIG. 15.104

Section 15.7

23. For the circuit of Fig. 15.102:
    a. Find the total admittance $\mathbf{Y}_T$ in polar form.
    b. Draw the admittance diagram.
    c. Find the voltage $\mathbf{E}$ and the currents $\mathbf{I}_R$ and $\mathbf{I}_L$ in phasor form.
    d. Draw the phasor diagram of the currents $\mathbf{I}_T$, $\mathbf{I}_R$, and $\mathbf{I}_L$, and the voltage $\mathbf{E}$.
    e. Verify Kirchhoff's current law at one node.
    f. Find the average power delivered to the circuit.
    g. Find the power factor of the circuit and indicate whether it is leading or lagging.
    h. Find the sinusoidal expressions for the currents and voltage if the frequency is 60 Hz.
    i. Plot the waveforms for the currents and voltage on the same set of axes.

24. Repeat Problem 23 for the circuit of Fig. 15.103, replacing $\mathbf{I}_L$ by $\mathbf{I}_C$ in parts (c) and (d).

25. Repeat Problem 23 for the circuit of Fig. 15.104, replacing $\mathbf{E}$ by $\mathbf{I}_T$ in part (c).

26. For the circuit of Fig. 15.105:
    a. Find the total admittance $\mathbf{Y}_T$ in polar form.
    b. Draw the admittance diagram.
    c. Find the value of C in microfarads and L in henries.
    d. Find the voltage e and currents i_R, i_L, and i_C in phasor form.
    e. Draw the phasor diagram of the currents $\mathbf{I}_T$, $\mathbf{I}_R$, $\mathbf{I}_L$, and $\mathbf{I}_C$, and the voltage $\mathbf{E}$.
    f. Verify Kirchhoff's current law at one node.
    g. Find the average power delivered to the circuit.
    h. Find the power factor of the circuit and indicate whether it is leading or lagging.
    i. Find the sinusoidal expressions for the currents and voltage.
    j. Plot the waveforms for the currents and voltage on the same set of axes.

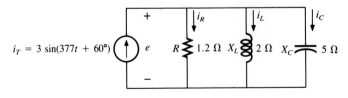

$i_T = 3 \sin(377t + 60°)$

FIG. 15.105

27. Repeat Problem 26 for the circuit of Fig. 15.106.

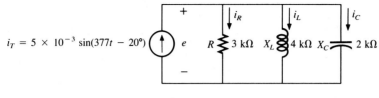

$i_T = 5 \times 10^{-3} \sin(377t - 20°)$

e $R \gtrless 3 \text{ k}\Omega$ $X_L \gtrless 4 \text{ k}\Omega$ $X_C \rightrightarrows 2 \text{ k}\Omega$

FIG. 15.106

28. Repeat Problem 26 for the circuit of Fig. 15.107, replacing e by i_T in part (d).

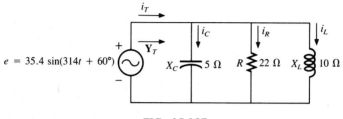

$e = 35.4 \sin(314t + 60°)$

$X_C \rightrightarrows 5 \Omega$ $R \gtrless 22 \Omega$ $X_L \gtrless 10 \Omega$

FIG. 15.107

Section 15.8

29. Calculate the currents $\mathbf{I}_1$ and $\mathbf{I}_2$ of Fig. 15.108 in phasor form using the current divider rule.

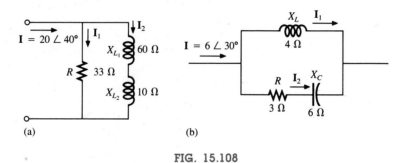

$\mathbf{I} = 20 \angle 40°$ $X_{L_1} \gtrless 60 \Omega$ $R \gtrless 33 \Omega$ $X_{L_2} \gtrless 10 \Omega$

$\mathbf{I} = 6 \angle 30°$ X_L $\mathbf{I}_1$ 4Ω R $\mathbf{I}_2$ X_C 3Ω 6Ω

(a)   (b)

FIG. 15.108

Section 15.9

30. For the series circuits of Fig. 15.109, find a parallel circuit that will have the same total impedance ($\mathbf{Z}_T$).

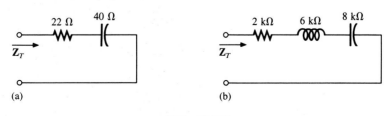

22Ω 40Ω $\mathbf{Z}_T$

$2 \text{ k}\Omega$ $6 \text{ k}\Omega$ $8 \text{ k}\Omega$ $\mathbf{Z}_T$

(a)   (b)

FIG. 15.109

31. For the parallel circuits of Fig. 15.110, find a series circuit that will have the same total impedance.

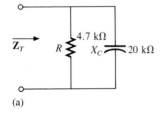

(a)

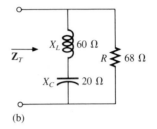

(b)

FIG. 15.110

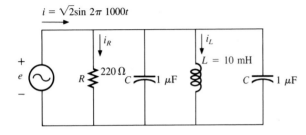

FIG. 15.111

32. For the network of Fig. 15.111:
  a. Calculate e, i_R, and i_L in phasor form.
  b. Calculate the total power factor and indicate whether it is leading or lagging.
  c. Calculate the average power delivered to the circuit.
  d. Draw the admittance diagram.
  e. Draw the phasor diagram of the currents $\mathbf{I}_T$, $\mathbf{I}_R$, and $\mathbf{I}_L$, and the voltage $\mathbf{E}$.
  f. Find the current $\mathbf{I}_C$ for each capacitor using only Kirchhoff's current law.
  g. Find the series circuit of one resistive and reactive element that will have the same impedance as the original circuit.

*33. Repeat Problem 32 if the inductance is changed to 1 H.

34. Find the element or elements that must be in the closed container of Fig. 15.112 to satisfy the following conditions. (Find the simplest parallel circuit that will satisfy the indicated conditions.)
  a. Average power to the circuit = 3000 W.
  b. Circuit has a lagging power factor.

Computer Problems

35. Write a program to generate the sinusoidal expression for the current of a resistor, inductor, or capacitor given the value of R, L, or C and the applied voltage in sinusoidal form.

36. Given the impedance of each element in rectangular form, write a program to determine the total impedance in rectangular form of any number of series elements.

37. Given two phasors in polar form in the first quadrant, write a program to generate the sum of the two phasors in polar form.

38. Given a series or parallel network as shown in Fig. 15.76, write a program to generate the parallel or series equivalent.

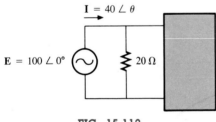

FIG. 15.112

GLOSSARY

Admittance A measure of how easily a network will "admit" the passage of current through that system. It is measured in siemens, abbreviated S, and is represented by the capital letter Y.

Admittance diagram A vector display that clearly depicts the magnitude of the admittance of the conductance, capacitive susceptance, and inductive susceptance, and the magnitude and angle of the total admittance of the system.

Current divider rule A method by which the current through either of two parallel branches can be determined in an ac network without first finding the voltage across the parallel branches.

Equivalent circuits For every series ac network there is a parallel ac network (and vice versa) that will be "equivalent" in the sense that the input current and impedance are the same.

Impedance diagram A vector display that clearly depicts the magnitude of the impedance of the resistive, reactive, and capacitive components of a network, and the magnitude and angle of the total impedance of the system.

Parallel ac circuits A connection of elements in an ac network in which all of the elements have two points in common. The voltage is the same across each element.

Phasor diagram A vector display that provides at a glance the magnitude and phase relationships among the various voltages and currents of a network.

Series ac configuration A connection of elements in an ac network in which no two impedances have more than one terminal in common and the current is the same through each element.

Susceptance A measure of how "susceptible" an element is to the passage of current through it. It is measured in siemens, abbreviated S, and is represented by the capital letter B.

Voltage divider rule A method through which the voltage across one element of a series of elements in an ac network can be determined without first having to find the current through the elements.

16

Series-Parallel ac Networks

16.1 INTRODUCTION

In this chapter, we shall utilize the fundamental concepts of the previous chapter to develop a technique for solving series-parallel ac networks. A brief review of Chapter 7 may be helpful before considering these networks, since the approach here will be quite similar to that undertaken earlier. The circuits to be discussed will have only one source of energy, either potential or current. Networks with two or more sources will be considered in Chapters 17 and 18, using methods previously described for dc circuits.

In general, when working with series-parallel ac networks, consider the following approach:

1. Redraw the network employing block impedances to combine obvious series and parallel elements to reduce the network to one that clearly reveals the fundamental structure of the system.
2. Study the problem and make a brief mental sketch of the overall approach you plan to use. Doing this may result in time- and energy-saving shortcuts.
3. After the overall approach has been determined, consider each branch involved in your method independently before tying them together in series-parallel combinations.
4. When you have arrived at a solution, check to see that it is reasonable by considering the magnitudes of the energy source and the elements in the circuit. If it does not seem reasonable, either solve the network using another approach, or check over your work very carefully.

16.2 ILLUSTRATIVE EXAMPLES

EXAMPLE 16.1.   For the network of Fig. 16.1:

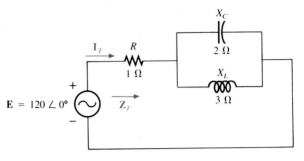

FIG. 16.1

a. Calculate $\mathbf{Z}_T$.
b. Determine $\mathbf{I}_T$.
c. Find $\mathbf{I}_C$.
d. Calculate $\mathbf{V}_R$ and $\mathbf{V}_C$.
e. Compute the power delivered.
f. Find F_p of the network.

For convenience, the network is redrawn in Fig. 16.2. It is good practice when working with complex networks to represent impedances in the manner indicated. When the unknown quantity is found in terms of these subscripted impedances, the numerical values can then be substituted to find the magnitude of the unknown. This will usually save time and prevent calculation errors.

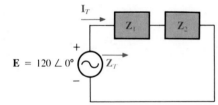

FIG. 16.2

Solutions:

a. $\mathbf{Z}_1 = R\ \angle 0° = 1\ \angle 0°$

$$\mathbf{Z}_2 = \mathbf{X}_C \parallel \mathbf{X}_L = \frac{\mathbf{X}_C \mathbf{X}_L}{\mathbf{X}_C + \mathbf{X}_L} = \frac{(2\ \angle -90°)(3\ \angle 90°)}{-j2 + j3}$$

$$= \frac{6\ \angle 0°}{j1} = \frac{6\ \angle 0°}{1\ \angle 90°}$$

$$= 6\ \angle -90°$$

and

$$\mathbf{Z}_T = \mathbf{Z}_1 + \mathbf{Z}_2 = 1 - j6 = \mathbf{6.08}\ \angle \mathbf{-80.54°}$$

b. $\mathbf{I}_T = \dfrac{\mathbf{E}}{\mathbf{Z}_T} = \dfrac{120\ \angle 0°}{6.08\ \angle -80.5°} = \mathbf{19.74}\ \angle \mathbf{80.54°}$

c. By the current divider rule,

$$\mathbf{I}_C = \frac{\mathbf{X}_L \mathbf{I}_T}{\mathbf{X}_L + \mathbf{X}_C} = \frac{(3\ \angle 90°)(19.74\ \angle 80.54°)}{1\ \angle 90°}$$

$$= \frac{59.22\ \angle 170.54°}{1\ \angle 90°} = \mathbf{59.22}\ \angle \mathbf{80.54°}$$

d. $\mathbf{V}_R = \mathbf{I}_T\mathbf{Z}_1 = (19.74 \angle 80.54°)(1 \angle 0°) = \mathbf{19.74} \angle \mathbf{80.54°}$
$\mathbf{V}_C = \mathbf{I}_T\mathbf{Z}_2 = (19.74 \angle 80.54°)(6 \angle -90°) = \mathbf{118.44} \angle \mathbf{-9.46°}$

e. $P_{del} = I_T^2 R = (19.74)^2(1) = \mathbf{389.67\ W}$

f. $F_p = \cos \theta = \cos 80.54° = \mathbf{0.164\ leading,}$ indicating a very reactive network. That is, the closer F_p is to zero, the more reactive and less resistive is the network.

EXAMPLE 16.2. For the network of Fig. 16.3:

a. If **I** is 50 ∠30°, calculate $\mathbf{I}_1$ using the current divider rule.
b. Repeat part (a) for $\mathbf{I}_2$.
c. Verify Kirchhoff's current law at one node.

Solutions:

a. Redrawing the circuit as in Fig. 16.4, we have

$$\mathbf{Z}_1 = R + jX_L = 3 + j4 = 5 \angle 53.13°$$
$$\mathbf{Z}_2 = -jX_C = -j8 = 8 \angle -90°$$

Using the current divider rule yields

$$\mathbf{I}_1 = \frac{\mathbf{Z}_2\mathbf{I}}{\mathbf{Z}_2 + \mathbf{Z}_1} = \frac{(8 \angle -90°)(50 \angle 30°)}{(-j8) + (3 + j4)} = \frac{400 \angle -60°}{3 - j4}$$

$$= \frac{400 \angle -60°}{5 \angle -53.13°} = \mathbf{80} \angle \mathbf{-6.87°}$$

b. $\mathbf{I}_2 = \dfrac{\mathbf{Z}_1\mathbf{I}}{\mathbf{Z}_2 + \mathbf{Z}_1} = \dfrac{(5 \angle 53.13°)(50 \angle 30°)}{5 \angle -53.13°} = \dfrac{250 \angle 83.13°}{5 \angle -53.13°}$

$$= \mathbf{50} \angle \mathbf{136.26°}$$

c. $\mathbf{I} = \mathbf{I}_1 + \mathbf{I}_2$
$50 \angle 30° = 80 \angle -6.87° + 50 \angle 136.26°$
$\quad\quad\quad = (79.43 - j9.57) + (-36.12 + j34.57)$
$\quad\quad\quad = 43.31 + j25.0$
$50 \angle 30° = 50 \angle 30°$ (checks)

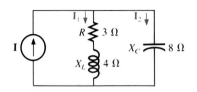

FIG. 16.3

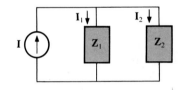

FIG. 16.4

EXAMPLE 16.3. For the network of Fig. 16.5:

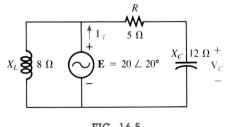

FIG. 16.5

a. Calculate the voltage $\mathbf{V}_C$ using the voltage divider rule.
b. Calculate the current $\mathbf{I}_T$.

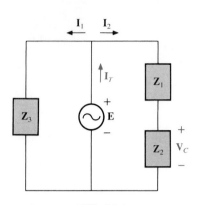

FIG. 16.6

Solutions:

a. The network is redrawn as shown in Fig. 16.6, with

$$\mathbf{Z}_1 = 5 = 5 \angle 0°$$
$$\mathbf{Z}_2 = -j12 = 12 \angle -90°$$
$$\mathbf{Z}_3 = +j8 = 8 \angle 90°$$

Since $\mathbf{V}_C$ is desired, we will not combine R and X_C into a single block impedance. Note also how Fig. 16.6 clearly reveals that $\mathbf{E}$ is the total voltage across the series combination of $\mathbf{Z}_1$ and $\mathbf{Z}_2$, permitting the use of the voltage divider rule to calculate $\mathbf{V}_C$:

$$\mathbf{V}_C = \frac{\mathbf{Z}_2 \mathbf{E}}{\mathbf{Z}_1 + \mathbf{Z}_2} = \frac{(12 \angle -90°)(20 \angle 20°)}{5 - j12} = \frac{240 \angle -70°}{13 \angle -67.38°}$$
$$= \mathbf{18.46 \angle -2.62°}$$

b. $\mathbf{I}_1 = \dfrac{\mathbf{E}}{\mathbf{Z}_3} = \dfrac{20 \angle 20°}{8 \angle 90°} = 2.5 \angle -70°$

$\mathbf{I}_2 = \dfrac{\mathbf{E}}{\mathbf{Z}_1 + \mathbf{Z}_2} = \dfrac{20 \angle 20°}{13 \angle -67.38°} = 1.54 \angle 87.38°$

and

$$\mathbf{I}_T = \mathbf{I}_1 + \mathbf{I}_2$$
$$= 2.5 \angle -70° + 1.54 \angle 87.38°$$
$$= (0.86 - j2.35) + (0.07 + j1.54)$$
$$\mathbf{I}_T = 0.93 - j0.81 = \mathbf{1.23 \angle -41.05°}$$

EXAMPLE 16.4. For Fig. 16.7:

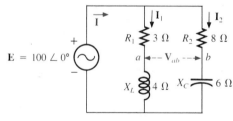

FIG. 16.7

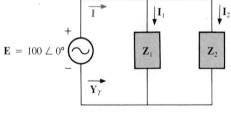

FIG. 16.8

a. Calculate the current **I**.
b. Find the voltage $\mathbf{V}_{ab}$.

Solutions:

a. Redrawing the circuit as in Fig. 16.8, we obtain

$$\mathbf{Z}_1 = R_1 + jX_L = 3 + j4 = 5 \angle 53.13°$$
$$\mathbf{Z}_2 = R_2 - jX_C = 8 - j6 = 10 \angle -36.87°$$

$$\mathbf{Y}_T = \mathbf{Y}_1 + \mathbf{Y}_2$$

$$= \frac{1}{\mathbf{Z}_1} + \frac{1}{\mathbf{Z}_2}$$

$$= \frac{1}{5 \angle 53.13°} + \frac{1}{10 \angle -36.87°}$$

$$= 0.2 \angle -53.13° + 0.1 \angle 36.87°$$

$$= (0.12 - j0.16) + (0.08 + j0.06)$$

$$\mathbf{Y}_T = 0.2 - j0.1 = \mathbf{0.224 \angle -26.57°}$$

By Ohm's law,

$$\mathbf{I} = \frac{\mathbf{E}}{\mathbf{Z}_T} = \mathbf{EY}_T = (100 \angle 0°)(0.224 \angle -26.57°)$$

$$= \mathbf{22.4 \angle -26.57°}$$

Using another approach, we find the total impedance:

$$\mathbf{Z}_T = \frac{\mathbf{Z}_1 \mathbf{Z}_2}{\mathbf{Z}_1 + \mathbf{Z}_2} = \frac{(5 \angle 53.13°)(10 \angle -36.87°)}{(3 + j4) + (8 - j6)}$$

$$= \frac{50 \angle 16.26°}{11 - j2} = \frac{50 \angle 16.26°}{11.18 \angle -10.30°}$$

$$= \mathbf{4.47 \angle 26.56°}$$

and

$$\mathbf{I} = \frac{\mathbf{E}}{\mathbf{Z}_T} = \frac{100 \angle 0°}{4.47 \angle 26.30°} = \mathbf{22.37 \angle -26.30°}$$

b. By Ohm's law,

$$\mathbf{I}_1 = \frac{\mathbf{E}}{\mathbf{Z}_1} = \frac{100 \angle 0°}{5 \angle 53.13°} = \mathbf{20 \angle -53.13°}$$

$$\mathbf{I}_2 = \frac{\mathbf{E}}{\mathbf{Z}_2} = \frac{100 \angle 0°}{10 \angle -36.87°} = \mathbf{10 \angle 36.87°}$$

Returning to Fig. 16.7, we have

$$\mathbf{V}_{R_1} = \mathbf{I}_1 \mathbf{R}_1 = (20 \angle -53.13°)(3 \angle 0°) = \mathbf{60 \angle -53.13°}$$

$$\mathbf{V}_{R_2} = \mathbf{I}_2 \mathbf{R}_2 = (10 \angle +36.87°)(8 \angle 0°) = \mathbf{80 \angle +36.87°}$$

Instead of using the two steps just shown, $\mathbf{V}_{R_1}$ or $\mathbf{V}_{R_2}$ could have been determined in one step using the voltage divider rule:

$$\mathbf{V}_{R_1} = \frac{(3 \angle 0°)(100 \angle 0°)}{3 \angle 0° + 4 \angle 90°} = \frac{300 \angle 0°}{5 \angle 53.13°} = \mathbf{60 \angle -53.13°}$$

In order to find $\mathbf{V}_{ab}$, Kirchhoff's voltage law must be applied around the loop (Fig. 16.9) consisting of the 3-Ω and 8-Ω resistors. By Kirchhoff's voltage law,

$$\mathbf{V}_{ab} + \mathbf{V}_{R_1} - \mathbf{V}_{R_2} = 0$$

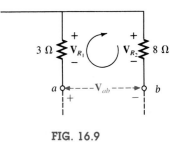

FIG. 16.9

or

$$\begin{aligned}
\mathbf{V}_{ab} &= \mathbf{V}_{R_2} - \mathbf{V}_{R_1} \\
&= 80\ \angle 36.87° - 60\ \angle -53.13° \\
&= (64 + j48) - (36 - j48) \\
&= 28 + j96 \\
\mathbf{V}_{ab} &= \mathbf{100}\ \angle \mathbf{73.74°}
\end{aligned}$$

EXAMPLE 16.5. For the network of Fig. 16.10:

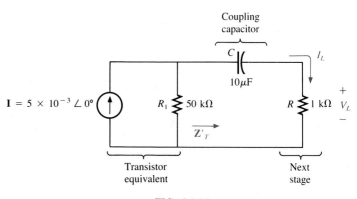

FIG. 16.10

a. Find $\mathbf{Z}'_T$ and compare it to $R_1 = 50\,\text{k}\Omega$ at $f = 0.1\,\text{kHz}$ and $f = 20\,\text{kHz}$.
b. Determine $\mathbf{I}_L$ at both frequencies and compare it to $\mathbf{I}$, keeping in mind the results of part (a).
c. Calculate $\mathbf{V}_L$ at each frequency.

(The type of network found in Fig. 16.10 will appear frequently in the analysis of transistor networks. The coupling capacitor is designed to be an open circuit for dc and as low an impedance as possible for the frequencies of interest to insure that $\mathbf{V}_L$ is a maximum value. The frequencies chosen in the example represent the low and high ends of the audio range.)

Solutions:

a. $f = 0.1\,\text{kHz} = 100\,\text{Hz}$:

$$X_C = \frac{1}{2\pi f C} = \frac{1}{(6.28)(100)(10 \times 10^{-6})} = 159.24\,\Omega$$

$$\mathbf{Z}'_T = R - jX_C = 1000 - j159.24 = \mathbf{1012.6}\ \angle \mathbf{-9.05°}$$

$f = 20\,\text{kHz}$:

$$X_C = \frac{1}{2\pi f C} = \frac{1}{(6.28)(20 \times 10^3)(10 \times 10^{-6})} = 0.796\,\Omega$$

$$\mathbf{Z}'_T = R - jX_C = 1000 - j0.796 = 1000\ \angle -0.046°$$
$$\cong \mathbf{1000}\ \angle \mathbf{0°} = R$$

Note the dramatic change in X_C with frequency. Obviously, the higher the frequency, the better the short-circuit approximation for X_C for ac conditions.

b. $f = 100$ Hz:

$$\mathbf{I}_L = \frac{R_1\mathbf{I}}{R_1 + R - jX_C} = \frac{(50 \times 10^3 \angle 0°)(5 \times 10^{-3} \angle 0°)}{50{,}000 + 1000 - j159.24}$$

$$= \frac{250 \angle 0°}{51{,}000 - j159.24} \cong \frac{250 \angle 0°}{51{,}000 \angle 0°} = \mathbf{4.9 \times 10^{-3} \angle 0°}$$

$f = 20$ kHz:

$$\mathbf{I}_L = \frac{R_1\mathbf{I}}{R_1 + R - jX_C} = \frac{(50 \times 10^3 \angle 0°)(5 \times 10^{-3} \angle 0°)}{50{,}000 + 1000 + j0.796}$$

$$= \frac{250 \angle 0°}{51{,}000 \angle 0°} = \mathbf{4.9 \times 10^{-3} \angle 0°}$$

Note that at both frequencies, the effect of X_C on $\mathbf{I}_L$ is negligible.

c. Since $\mathbf{I}_L = 4.9 \times 10^{-3} \angle 0°$ at both frequencies,

$$\mathbf{V}_L = \mathbf{I}_L\mathbf{R} = (4.9 \times 10^{-3} \angle 0°)(1 \times 10^3 \angle 0°) = \mathbf{4.9\ V}$$

Ideal conditions require that $R_1 = \infty\ \Omega$ (open circuit) and $X_C = 0\ \Omega$, resulting in

$$\mathbf{V}_L = \mathbf{I}_L\mathbf{R} = (5 \times 10^{-3} \angle 0°)(1 \times 10^3 \angle 0°) = \mathbf{5\ V}$$

The percent difference is

$$\frac{5 - 4.9}{5} \times 100\% = \mathbf{2\%}$$

EXAMPLE 16.6. For the network of Fig. 16.11:

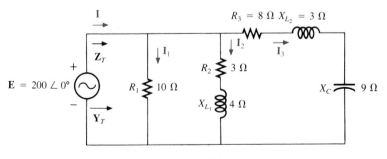

FIG. 16.11

a. Compute **I**.
b. Find $\mathbf{I}_1$, $\mathbf{I}_2$, and $\mathbf{I}_3$.
c. Verify Kirchhoff's current law by showing that

$$\mathbf{I} = \mathbf{I}_1 + \mathbf{I}_2 + \mathbf{I}_3$$

d. Find the total impedance of the circuit.

Solutions:

a. Redrawing the circuit as in Fig. 16.12, we obtain

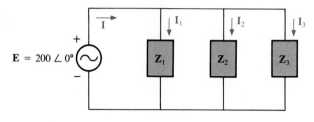

FIG. 16.12

$$\mathbf{Z}_1 = R_1 = 10 \angle 0°$$
$$\mathbf{Z}_2 = R_2 + jX_{L_1} = 3 + j4$$
$$\mathbf{Z}_3 = R_3 + jX_{L_2} - jX_C = 8 + j3 - j9 = 8 - j6$$

The total admittance is

$$\mathbf{Y}_T = \mathbf{Y}_1 + \mathbf{Y}_2 + \mathbf{Y}_3$$
$$= \frac{1}{\mathbf{Z}_1} + \frac{1}{\mathbf{Z}_2} + \frac{1}{\mathbf{Z}_3} = \frac{1}{10} + \frac{1}{3+j4} + \frac{1}{8-j6}$$

However,

$$\frac{1}{\mathbf{Z}_2} = \frac{1}{3+j4} = \frac{3}{(3)^2 + (4)^2} - j\frac{4}{(3)^2 + (4)^2} = \frac{3}{25} - j\frac{4}{25}$$

and

$$\frac{1}{\mathbf{Z}_3} = \frac{1}{8-j6} = \frac{8}{(8)^2 + (6)^2} + j\frac{6}{(8)^2 + (6)^2} = \frac{8}{100} + j\frac{6}{100}$$

Therefore,

$$\mathbf{Y}_1 = \frac{1}{10} + j0 = \frac{10}{100} + j0$$

$$\mathbf{Y}_2 = \frac{3}{25} - j\frac{4}{25} = \frac{12}{100} - j\frac{16}{100}$$

$$\mathbf{Y}_3 = \frac{8}{100} + j\frac{6}{100} = \frac{8}{100} + j\frac{6}{100}$$

$$\mathbf{Y}_T = \mathbf{Y}_1 + \mathbf{Y}_2 + \mathbf{Y}_3 = \frac{30}{100} - j\frac{10}{100}$$

$$\mathbf{I} = \mathbf{E}\mathbf{Y}_T = 200 \angle 0° \left(\frac{30}{100} - j\frac{10}{100} \right) = \mathbf{60 - j20}$$

b. Since the voltage is the same across parallel branches,

$$\mathbf{I}_1 = \frac{\mathbf{E}}{\mathbf{Z}_1} = \frac{200 \angle 0°}{10 \angle 0°} = \mathbf{20 \angle 0°}$$

$$\mathbf{I}_2 = \frac{\mathbf{E}}{\mathbf{Z}_2} = \frac{200 \angle 0°}{5 \angle 53.13°} = \mathbf{40 \angle -53.13°}$$

$$\mathbf{I}_3 = \frac{\mathbf{E}}{\mathbf{Z}_3} = \frac{200 \angle 0°}{10 \angle -36.87°} = \mathbf{20 \angle +36.87°}$$

c. $\mathbf{I} = \mathbf{I}_1 + \mathbf{I}_2 + \mathbf{I}_3$
$60 - j20 = 20 \angle 0° + 40 \angle -53.13° + 20 \angle +36.87°$
$\qquad\quad = (20 + j0) + (24 - j32) + (16 + j12)$
$60 - j20 = 60 - j20 \qquad \text{(checks)}$

d. $\mathbf{Z}_T = \dfrac{1}{\mathbf{Y}_T} = \dfrac{1}{0.3 - j0.1}$

$\qquad = \dfrac{0.3}{(0.3)^2 + (0.1)^2} + j\dfrac{0.1}{(0.3)^2 + (0.1)^2}$

$\qquad = \dfrac{0.3}{0.1} + j\dfrac{0.1}{0.1} = \mathbf{3 + j}$

EXAMPLE 16.7. For the network of Fig. 16.13:

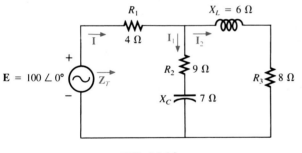

FIG. 16.13

a. Calculate the total impedance $\mathbf{Z}_T$.
b. Compute $\mathbf{I}$.
c. Find the total power factor.
d. Calculate $\mathbf{I}_1$ and $\mathbf{I}_2$.
e. Find the average power delivered to the circuit.

Solutions:

a. Redrawing the circuit as in Fig. 16.14, we have

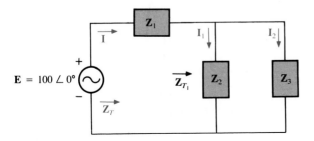

FIG. 16.14

$$\mathbf{Z}_1 = R_1 = 4 \angle 0°$$
$$\mathbf{Z}_2 = R_2 - jX_C = 9 - j7 = 11.40 \angle -37.87°$$
$$\mathbf{Z}_3 = R_3 + jX_L = 8 + j6 = 10 \angle +36.87°$$

The total impedance is

$$\mathbf{Z}_T = \mathbf{Z}_1 + \mathbf{Z}_{T_1}$$

$$= \mathbf{Z}_1 + \frac{\mathbf{Z}_2 \mathbf{Z}_3}{\mathbf{Z}_2 + \mathbf{Z}_3}$$

$$= 4 + \frac{(11.4 \angle -37.87°)(10 \angle 36.87°)}{(9 - j7) + (8 + j6)}$$

$$= 4 + \frac{114 \angle -1.00°}{17.03 \angle -3.37°}$$

$$= 4 + 6.69 \angle 2.37°$$

$$= 4 + 6.68 + j0.28$$

$$\mathbf{Z}_T = 10.68 + j0.28 = \mathbf{10.68} \angle \mathbf{1.5°}$$

b. $\mathbf{I} = \dfrac{\mathbf{E}}{\mathbf{Z}_T} = \dfrac{100 \angle 0°}{10.68 \angle 1.5°} = \mathbf{9.36} \angle \mathbf{-1.5°}$

c. $F_p = \cos \theta_T = \dfrac{R}{Z_T} = \dfrac{10.68}{10.68} = 1$

(interesting, considering the complexity of the network)

d. $\mathbf{I}_2 = \dfrac{\mathbf{Z}_2 \mathbf{I}}{\mathbf{Z}_2 + \mathbf{Z}_3} = \dfrac{(11.40 \angle -37.87°)(9.36 \angle -1.5°)}{(9 - j7) + (8 + j6)}$

$$= \frac{106.7 \angle -39.37°}{17 - j1} = \frac{106.7 \angle -39.37°}{17.03 \angle -3.37°}$$

$$\mathbf{I}_2 = \mathbf{6.27} \angle \mathbf{-36°}$$

Applying Kirchhoff's current law yields

$$\mathbf{I} = \mathbf{I}_1 + \mathbf{I}_2$$

or

$$\mathbf{I}_1 = \mathbf{I} - \mathbf{I}_2$$
$$= (9.36 \angle -1.5°) - (6.27 \angle -36°)$$
$$= (9.36 - j0.25) - (5.07 - j3.69)$$
$$\mathbf{I}_1 = 4.29 + j3.44 = \mathbf{5.5} \angle \mathbf{38.72°}$$

e. $P_T = EI \cos \theta_T$
$$= (100)(9.36) \cos 1.5°$$
$$= (936)(1)$$
$$P_T = \mathbf{936 \ W}$$

16.3 LADDER NETWORKS

Ladder networks were discussed in some detail in Chapter 7. This section will simply apply the first method described in Section 7.3 to the general sinusoidal ac ladder network of Fig. 16.15. The current I_6 is desired.

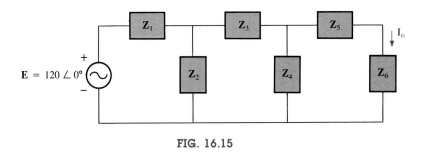

FIG. 16.15

Impedances Z_T, Z'_T, and Z''_T and currents I_1 and I_3 are defined in Fig. 16.16:

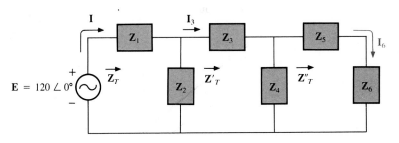

FIG. 16.16

$$Z''_T = Z_5 + Z_6$$

and

$$Z'_T = Z_3 + Z_4 \parallel Z''_T$$

with

$$Z_T = Z_1 + Z_2 \parallel Z'_T$$

Then

$$I = \frac{E}{Z_T}$$

and

$$I_3 = \frac{Z_2 I}{Z_2 + Z'_T}$$

with

$$I_6 = \frac{Z_4 I_3}{Z_4 + Z''_T}$$

PROBLEMS

Section 16.2

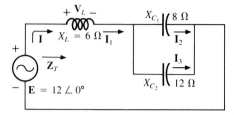

FIG. 16.17

1. For the series-parallel network of Fig. 16.17:
   a. Calculate Z_T.
   b. Determine **I**.
   c. Determine I_1.
   d. Find I_2 and I_3.
   e. Find V_L.

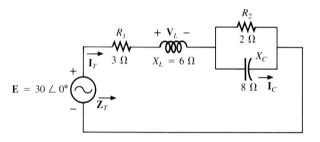

FIG. 16.18

2. For the network of Fig. 16.18:
   a. Find the total impedance Z_T.
   b. Determine the current I_T.
   c. Calculate I_C using the current divider rule.
   d. Calculate V_L using the voltage divider rule.

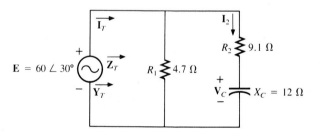

FIG. 16.19

3. For the network of Fig. 16.19:
   a. Find the total impedance Z_T and the total admittance Y_T.
   b. Find the current I_T.
   c. Calculate I_2 using the current divider rule.
   d. Calculate V_C.
   e. Calculate the average power delivered to the network.

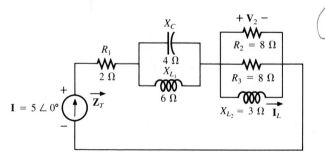

FIG. 16.20

4. For the network of Fig. 16.20:
   a. Find the total impedance Z_T.
   b. Calculate the voltage V_2 and the current I_L.
   c. Find the power factor of the network.

5. For the network of Fig. 16.21:
  a. Find the current **I**.
  b. Find the voltage $\mathbf{V}_C$.
  c. Find the average power delivered to the network.

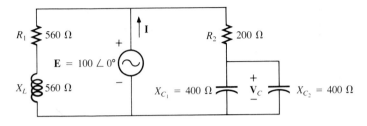

FIG. 16.21

*6. For the network of Fig. 16.22:
  a. Find the current $\mathbf{I}_1$.
  b. Calculate the voltage $\mathbf{V}_C$ using the voltage divider rule.
  c. Find the voltage $\mathbf{V}_{ab}$.

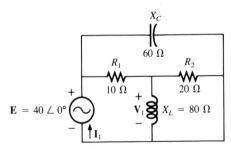

FIG. 16.22

*7. For the network of Fig. 16.23:
  a. Find the current $\mathbf{I}_1$.
  b. Find the voltage $\mathbf{V}_1$.
  c. Calculate the average power delivered to the network.

8. For the network of Fig. 16.24:
  a. Find the total impedance $\mathbf{Z}_T$ and the admittance $\mathbf{Y}_T$.
  b. Find the currents $\mathbf{I}_1$, $\mathbf{I}_2$, and $\mathbf{I}_3$.
  c. Verify Kirchhoff's current law by showing that $\mathbf{I}_T = \mathbf{I}_1 + \mathbf{I}_2 + \mathbf{I}_3$.
  d. Find the power factor of the network and indicate whether it is leading or lagging.

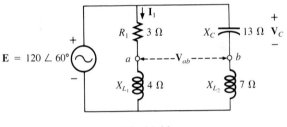

FIG. 16.23

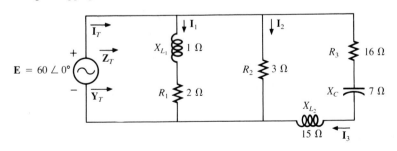

FIG. 16.24

*9. For the network of Fig. 16.25:
  a. Find the total admittance $\mathbf{Y}_T$.
  b. Find the voltages $\mathbf{V}_1$ and $\mathbf{V}_2$.
  c. Find the current $\mathbf{I}_3$.

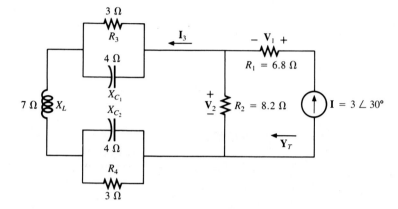

FIG. 16.25

*10. For the network of Fig. 16.26:
  a. Find the total impedance $\mathbf{Z}_T$ and the admittance $\mathbf{Y}_T$.
  b. Find the current i_T in phasor form.
  c. Find the currents i_1 and i_2 in phasor form.
  d. Find the voltages v_1 and v_{ab} in phasor form.
  e. Find the average power delivered to the network.
  f. Find the power factor of the network and indicate whether it is leading or lagging.

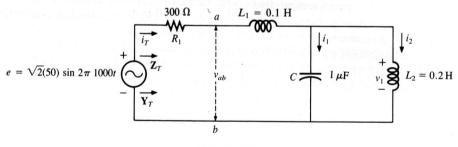

FIG. 16.26

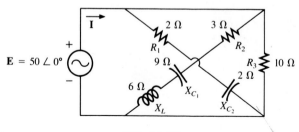

*11. Find the current $\mathbf{I}$ for the network of Fig. 16.27.

FIG. 16.27

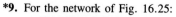

Section 16.3

12. Find the current I_5 for the network of Fig. 16.28. Note the effect of one reactive element on the resulting calculations.

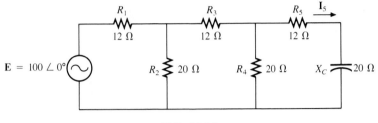

FIG. 16.28

13. Find the average power delivered to R_4 in Fig. 16.29.

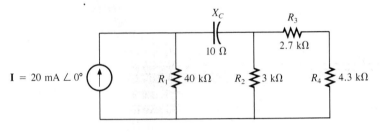

FIG. 16.29

14. Find the current I_1 for the network of Fig. 16.30.

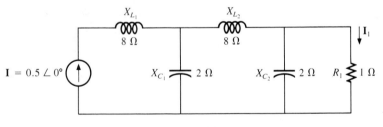

FIG. 16.30

Computer Problems

15. Write a program to provide a general solution to Problem 1. That is, given the reactance of each element, generate a solution for parts (a) through (e).

16. Given the network of Fig. 16.19, write a program to generate a solution for parts (a) and (b) of Problem 2. Use the values given.

17. Generate a program to obtain a general solution for the network of Fig. 16.21 for the questions asked in parts (a) through (c) of Problem 2. That is, given the resistance and reactance of the elements, determine the requested current, voltage, and power.

GLOSSARY

Ladder network A repetitive combination of series and parallel branches that has the appearance of a ladder.

Series-parallel ac network A combination of series and parallel branches in the same network configuration. Each branch may contain any number of elements whose impedance is dependent on the applied frequency.

17

Methods of Analysis and Selected Topics (ac)

17.1 INTRODUCTION

For networks with two or more sources that are not in series or parallel, the methods described in the last two chapters cannot be applied. Rather, such methods as mesh analysis or nodal analysis must be employed. Since these methods were discussed in detail for dc circuits in Chapter 8, this chapter will consider the variations required to apply these methods to ac circuits.

The branch-current method will not be discussed again, since it falls within the framework of mesh analysis. In addition to the methods mentioned above, the bridge network and Δ-Y, Y-Δ conversions will also be discussed for ac circuits.

Before we examine these topics, however, we must consider the subject of independent and controlled sources.

17.2 INDEPENDENT VERSUS DEPENDENT (CONTROLLED) SOURCES

In the previous chapters, each source appearing in the analysis of dc or ac networks was an *independent source* such as E and I (or **E** and **I**) in Fig. 17.1. The term *independent* indicates that the magnitude of the source is *independent of the network* to which it is applied and that it displays its terminal characteristics even if completely isolated.

A *dependent* or *controlled source* is one whose magnitude is determined (or controlled) by a current or voltage of the system in which it

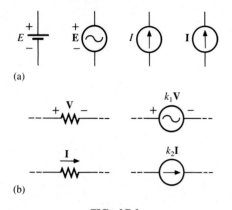

(a)

(b)

FIG. 17.1

Sources. (a) Independent; (b) controlled or dependent.

appears. The magnitude of the voltage source k_1V in Fig. 17.1 is determined by the voltage V appearing across the resistive element and the constant k_1. The magnitude of the current source k_2I is determined by the current through another resistive element and the constant k_2.

Possible combinations for controlled sources are indicated in Fig. 17.2. Note that the magnitude of current sources or voltage sources can be controlled by a voltage and a current, respectively. Unlike with the independent source, isolation such that V or $I = 0$ in Fig. 17.2 will result in the short-circuit or open-circuit equivalent as indicated in Fig. 17.2. Note that the type of representation under these conditions is controlled by whether it is a current source or a voltage source, not by the controlling agent (V or I).

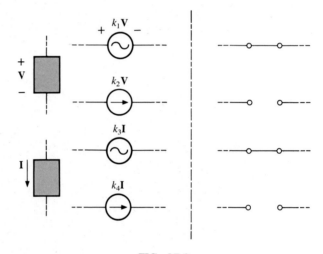

FIG. 17.2

Conditions of $V = 0$ and $I = 0$ for a controlled source.

One reason for our interest in controlled sources stems from electronic components such as the transistor, which employs a dependent source in its *equivalent circuit*. An equivalent circuit is a combination of elements, active (sources) and passive (R, L, C), properly chosen, that best represent the terminal characteristics and response of the device under a particular set of operating conditions. A few examples in this chapter will indicate the basic appearance of these equivalent circuits.

17.3 SOURCE CONVERSIONS

When applying the methods to be discussed, it may be necessary to convert a current source to a voltage source, or a voltage source to a current source. This can be accomplished in much the same manner as for dc circuits, except now we shall be dealing with phasors and impedances instead of just real numbers and resistors.

In general, the format for converting from one to the other is as shown in Fig. 17.3.

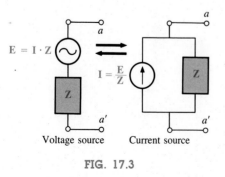

Voltage source   Current source

FIG. 17.3

EXAMPLE 17.1. Convert the voltage source of Fig. 17.4(a) to a current source.

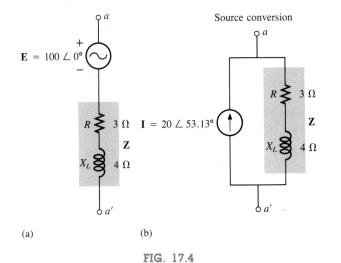

Source conversion

(a)   (b)

FIG. 17.4

Solution:

$$\mathbf{I} = \frac{\mathbf{E}}{\mathbf{Z}} = \frac{100 \angle 0°}{5 \angle 53.13°}$$

$$= \mathbf{20 \angle -53.13°} \qquad [\text{Fig. 17.4(b)}]$$

EXAMPLE 17.2. Convert the current source of Fig. 17.5(a) to a voltage source.

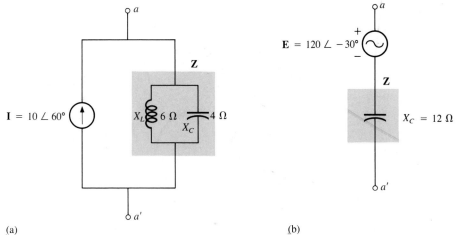

(a)   (b)

FIG. 17.5

Solution:

$$\mathbf{Z} = \frac{(4 \angle -90°)(6 \angle 90°)}{-j4 + j6} = \frac{24 \angle 0°}{2 \angle 90°}$$

$$= 12 \angle -90° \qquad [\text{Fig. 17.5(b)}]$$

$$\mathbf{E} = \mathbf{IZ} = (10 \angle 60°)(12 \angle -90°)$$

$$= 120 \angle -30° \qquad [\text{Fig. 17.5(b)}]$$

For dependent sources, the direct conversion of Fig. 17.3 can be applied if the controlling variable (**V** or **I** in Fig. 17.2) is not determined by a portion of the network to which the conversion is to be applied. For example, in Figs. 17.6 and 17.7, **V** and **I**, respectively, are controlled by an external portion of the network. Conversions of the other kind, where **V** and **I** are controlled by a portion of the network to be converted, will be considered in Sections 18.3 and 18.4.

EXAMPLE 17.3. Convert the voltage source of Fig. 17.6(a) to a current source.

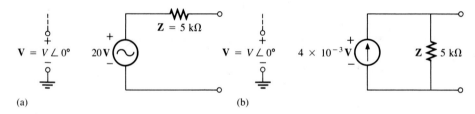

(a)   (b)

FIG. 17.6

Solution:

$$\mathbf{I} = \frac{\mathbf{E}}{\mathbf{Z}} = \frac{20\mathbf{V} \angle 0°}{5 \times 10^3 \angle 0°}$$

$$= 4 \times 10^{-3}\mathbf{V} \angle 0° \qquad [\text{Fig. 17.6(b)}]$$

EXAMPLE 17.4. Convert the current source of Fig. 17.7(a) to a voltage source.

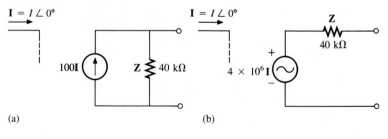

(a)   (b)

FIG. 17.7

Solution:

$$\mathbf{E} = \mathbf{IZ} = (100\mathbf{I} \angle 0°)(40 \times 10^3 \angle 0°)$$
$$= 4 \times 10^6 \mathbf{I} \angle 0° \quad [\text{Fig. } 17.7(\text{b})]$$

17.4 MESH ANALYSIS (FORMAT APPROACH)

The first method to be considered is mesh analysis using the format approach, introduced in Section 8.9. The steps for applying this method are repeated here with changes for its use in ac circuits:

1. Assign a loop current to each independent closed loop (as in the previous section) in a *clockwise* direction.
2. The number of required equations is equal to the number of chosen independent closed loops. Column 1 of each equation is formed by simply summing the impedance values of those impedances through which the loop current of interest passes and multiplying the result by that loop current.
3. We must now consider the mutual terms which are always subtracted from the terms in the first column. It is possible to have more than one mutual term if the loop current of interest has an element in common with more than one other loop current. Each mutual term is the product of the mutual impedance and the other loop current passing through the same element.
4. The column to the right of the equality sign is the algebraic sum of the voltage sources through which the loop current of interest passes. Positive signs are assigned to those sources of voltage having a polarity such that the loop current passes from the negative to the positive terminal. A negative sign is assigned to those potentials for which the reverse is true.
5. Solve resulting simultaneous equations for the desired loop currents.

Note that the only change in the above as compared to its appearance in Chapter 8 (dc circuits) was to replace the word *resistor* by the word *impedance*. This and the use of phasors instead of real numbers will be the only major changes in applying these methods to ac circuits.

The technique is applied as above for all networks with independent sources or networks with dependent sources where the controlling variable is not a part of the network under investigation. If the controlling variable is part of the network being examined, additional care must be taken when applying the above steps.

EXAMPLE 17.5. Using mesh analysis, find the current $\mathbf{I}_1$ in Fig. 17.8.

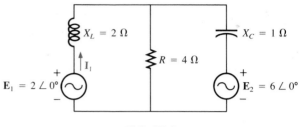

FIG. 17.8

Solution: When applying these methods to ac circuits, it is good practice to represent the resistors and reactances (or combinations thereof) by subscripted impedances. When the total solution is found in terms of these subscripted impedances, the numerical values can be substituted to find the unknown quantities.

The network is redrawn in Fig. 17.9 with subscripted impedances:

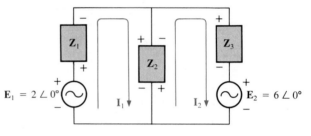

FIG. 17.9

$$\mathbf{Z}_1 = +jX_L = +j2$$
$$\mathbf{Z}_2 = R = 4$$
$$\mathbf{Z}_3 = -jX_C = -j$$

Step 1 is as indicated in Fig. 17.9.
Steps 2 to 4:

$$\mathbf{I}_1(\mathbf{Z}_1 + \mathbf{Z}_2) - \mathbf{I}_2\mathbf{Z}_2 = \mathbf{E}_1$$
$$\mathbf{I}_2(\mathbf{Z}_2 + \mathbf{Z}_3) - \mathbf{I}_1\mathbf{Z}_2 = -\mathbf{E}_2$$

which are rewritten as

$$\mathbf{I}_1(\mathbf{Z}_1 + \mathbf{Z}_2) - \mathbf{I}_2\mathbf{Z}_2 \qquad = \mathbf{E}_1$$
$$-\mathbf{I}_1\mathbf{Z}_2 \qquad + \mathbf{I}_2(\mathbf{Z}_2 + \mathbf{Z}_3) = -\mathbf{E}_2$$

Step 5: Using determinants, we obtain

$$\mathbf{I}_1 = \frac{\begin{vmatrix} \mathbf{E}_1 & -\mathbf{Z}_2 \\ -\mathbf{E}_2 & \mathbf{Z}_2 + \mathbf{Z}_3 \end{vmatrix}}{\begin{vmatrix} \mathbf{Z}_1 + \mathbf{Z}_2 & -\mathbf{Z}_2 \\ -\mathbf{Z}_2 & \mathbf{Z}_2 + \mathbf{Z}_3 \end{vmatrix}}$$

$$= \frac{\mathbf{E}_1(\mathbf{Z}_2 + \mathbf{Z}_3) - \mathbf{E}_2(\mathbf{Z}_2)}{(\mathbf{Z}_1 + \mathbf{Z}_2)(\mathbf{Z}_2 + \mathbf{Z}_3) - (\mathbf{Z}_2)^2}$$

$$= \frac{(E_1 - E_2)Z_2 + E_1Z_3}{Z_1Z_2 + Z_1Z_3 + Z_2Z_3}$$

Substituting numerical values yields

$$I_1 = \frac{(2 - 6)(4) + (2)(-j)}{(+j2)(4) + (+j2)(-j) + (4)(-j)} = \frac{-16 - j2}{j8 - j^2 2 - j4}$$

$$= \frac{-16 - j2}{2 + j4} = \frac{16.12 \angle -172.87°}{4.47 \angle 63.43°}$$

$$= 3.61 \angle -236.30° \quad \text{or} \quad 3.61 \angle 123.70°$$

Therefore,

$$I_1 = 3.61 \angle 123.70°$$

EXAMPLE 17.6. Using mesh analysis, find the current I_2 in Fig. 17.10.

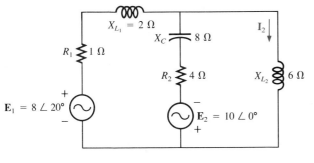

FIG. 17.10

Solution: The network is redrawn in Fig. 17.11:

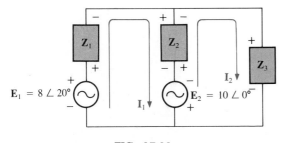

FIG. 17.11

$$Z_1 = R_1 + jX_{L_1} = 1 + j2$$
$$Z_2 = R_2 - jX_C = 4 - j8$$
$$Z_3 = +jX_{L_2} = +j6$$

Note the reduction in complexity of the problem with the substitution of the subscripted impedances.

Step 1 is as indicated in Fig. 17.11.

Steps 2 to 4:

$$\mathbf{I}_1(\mathbf{Z}_1 + \mathbf{Z}_2) - \mathbf{I}_2\mathbf{Z}_2 = \mathbf{E}_1 + \mathbf{E}_2$$
$$\mathbf{I}_2(\mathbf{Z}_2 + \mathbf{Z}_3) - \mathbf{I}_1\mathbf{Z}_2 = -\mathbf{E}_2$$

which are rewritten as

$$\mathbf{I}_1(\mathbf{Z}_1 + \mathbf{Z}_2) - \mathbf{I}_2\mathbf{Z}_2 \qquad = \mathbf{E}_1 + \mathbf{E}_2$$
$$-\mathbf{I}_1\mathbf{Z}_2 \qquad + \mathbf{I}_2(\mathbf{Z}_2 + \mathbf{Z}_3) = -\mathbf{E}_2$$

Step 5: Using determinants, we have

$$\mathbf{I}_2 = \frac{\begin{vmatrix} \mathbf{Z}_1 + \mathbf{Z}_2 & \mathbf{E}_1 + \mathbf{E}_2 \\ -\mathbf{Z}_2 & -\mathbf{E}_2 \end{vmatrix}}{\begin{vmatrix} \mathbf{Z}_1 + \mathbf{Z}_2 & -\mathbf{Z}_2 \\ -\mathbf{Z}_2 & \mathbf{Z}_2 + \mathbf{Z}_3 \end{vmatrix}}$$

$$= \frac{-(\mathbf{Z}_1 + \mathbf{Z}_2)\mathbf{E}_2 + \mathbf{Z}_2(\mathbf{E}_1 + \mathbf{E}_2)}{(\mathbf{Z}_1 + \mathbf{Z}_2)(\mathbf{Z}_2 + \mathbf{Z}_3) - \mathbf{Z}_2^2}$$

$$= \frac{-\mathbf{Z}_1\mathbf{E}_2 + \mathbf{Z}_2\mathbf{E}_1}{\mathbf{Z}_1\mathbf{Z}_2 + \mathbf{Z}_1\mathbf{Z}_3 + \mathbf{Z}_2\mathbf{Z}_3}$$

Substituting numerical values yields

$$\mathbf{I}_2 = \frac{-(1 + j2)(10 \angle 0°) + (4 - j8)(8 \angle 20°)}{(1 + j2)(4 - j8) + (1 + j2)(+j6) + (4 - j8)(+j6)}$$

$$= \frac{-(10 + j20) + (4 - j8)(7.52 + j2.74)}{20 + (j6 - 12) + (j24 + 48)}$$

$$= \frac{-(10 + j20) + (52.0 - j49.20)}{56 + j30} = \frac{+42.0 - j69.20}{56 + j30}$$

$$= \frac{80.95 \angle -58.74°}{63.53 \angle 28.18°} = 1.27 \angle -86.92°$$

Therefore,

$$\mathbf{I}_2 = 1.27 \angle -86.92°$$

EXAMPLE 17.7. Write the mesh equations for the network of Fig. 17.12. Do not solve.

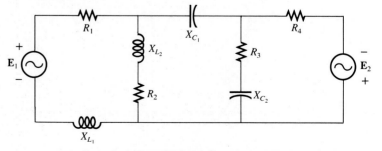

FIG. 17.12

Solution: The network is redrawn in Fig. 17.13. Again note the reduced complexity and increased clarity by use of subscripted impedances:

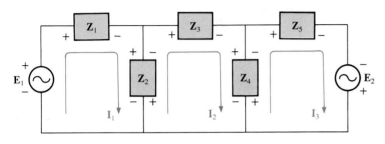

FIG. 17.13

$$\mathbf{Z}_1 = R_1 + jX_{L_1}$$
$$\mathbf{Z}_2 = R_2 + jX_{L_2}$$
$$\mathbf{Z}_3 = jX_{C_1}$$
$$\mathbf{Z}_4 = R_3 - jX_{C_2}$$
$$\mathbf{Z}_5 = R_4$$

and

$$\mathbf{I}_1(\mathbf{Z}_1 + \mathbf{Z}_2) - \mathbf{I}_2\mathbf{Z}_2 = \mathbf{E}_1$$
$$\mathbf{I}_2(\mathbf{Z}_2 + \mathbf{Z}_3 + \mathbf{Z}_4) - \mathbf{I}_1\mathbf{Z}_2 - \mathbf{I}_3\mathbf{Z}_4 = 0$$
$$\mathbf{I}_3(\mathbf{Z}_4 + \mathbf{Z}_5) - \mathbf{I}_2\mathbf{Z}_4 = \mathbf{E}_2$$

or

$$\mathbf{I}_1(\mathbf{Z}_1 + \mathbf{Z}_2) - \mathbf{I}_2(\mathbf{Z}_2) \qquad + 0 \qquad = \mathbf{E}_1$$
$$\mathbf{I}_1(\mathbf{Z}_2) \quad - \mathbf{I}_2(\mathbf{Z}_2 + \mathbf{Z}_3 + \mathbf{Z}_4) + \mathbf{I}_3(\mathbf{Z}_4) \quad = 0$$
$$0 \qquad - \mathbf{I}_2(\mathbf{Z}_4) \qquad + \mathbf{I}_3(\mathbf{Z}_4 + \mathbf{Z}_5) = \mathbf{E}_2$$

EXAMPLE 17.8. Using the format approach, write the mesh equations for the network of Fig. 17.14.

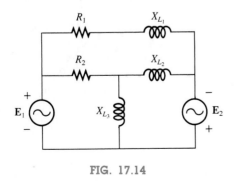

FIG. 17.14

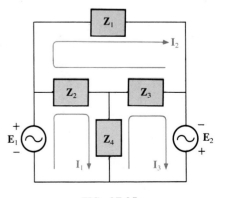

FIG. 17.15

Solution: The network is redrawn as shown in Fig. 17.15, where

$$\mathbf{Z}_1 = R_1 + jX_{L_1}$$

$$\mathbf{Z}_2 = R_2$$
$$\mathbf{Z}_3 = jX_{L_2}$$
$$\mathbf{Z}_4 = jX_{L_3}$$

and

$$\mathbf{I}_1(\mathbf{Z}_2 + \mathbf{Z}_4) - \mathbf{I}_2\mathbf{Z}_2 - \mathbf{I}_3\mathbf{Z}_4 = \mathbf{E}_1$$
$$\mathbf{I}_2(\mathbf{Z}_1 + \mathbf{Z}_2 + \mathbf{Z}_3) - \mathbf{I}_1\mathbf{Z}_2 - \mathbf{I}_3\mathbf{Z}_3 = 0$$
$$\mathbf{I}_3(\mathbf{Z}_3 + \mathbf{Z}_4) - \mathbf{I}_2\mathbf{Z}_3 - \mathbf{I}_1\mathbf{Z}_4 = \mathbf{E}_2$$

or

$$
\begin{array}{llll}
\mathbf{I}_1(\mathbf{Z}_2 + \mathbf{Z}_4) & -\ \mathbf{I}_2\mathbf{Z}_2 & -\ \mathbf{I}_3\mathbf{Z}_4 & = \mathbf{E}_1 \\
-\mathbf{I}_1\mathbf{Z}_2 & +\ \mathbf{I}_2(\mathbf{Z}_1 + \mathbf{Z}_2 + \mathbf{Z}_3) & -\ \mathbf{I}_3\mathbf{Z}_3 & = 0 \\
-\mathbf{I}_1\mathbf{Z}_4 & -\ \mathbf{I}_2\mathbf{Z}_3 & +\ \mathbf{I}_3(\mathbf{Z}_3 + \mathbf{Z}_4) & = \mathbf{E}_2
\end{array}
$$

Note the symmetry *about* the diagonal axis. That is, note the location of $-\mathbf{Z}_2$, $-\mathbf{Z}_4$, and $-\mathbf{Z}_3$ off the diagonal.

17.5 NODAL ANALYSIS (FORMAT APPROACH)

Before examining the application of this method to ac circuits, the student should review the section on nodal analysis in Chapter 8, since we shall repeat only the final conclusions as they apply to ac circuits. Recall that these conclusions made the writing of the nodal equations quite direct and in a form convenient for the use of determinants. For sinusoidal ac networks, the procedure is the following:

1. Choose a reference node and assign a subscripted voltage label to the $(N - 1)$ remaining independent nodes of the network.
2. The number of equations required for a complete solution is equal to the number of subscripted voltages $(N - 1)$. Column 1 of each equation is formed by summing the admittances tied to the node of interest and multiplying the result by that subscripted nodal voltage.
3. The mutual terms are always subtracted from the terms of the first column. It is possible to have more than one mutual term if the nodal voltage of interest has an element in common with more than one other nodal voltage. Each mutual term is the product of the mutual admittance and the other nodal voltage tied to that admittance.
4. The column to the right of the equality sign is the algebraic sum of the current sources tied to the node of interest. A current source is assigned a positive sign if it supplies current to a node, and a negative sign if it draws current from the node.

5. Solve resulting simultaneous equations for the desired nodal voltages. The comments offered for mesh analysis regarding independent and dependent sources apply here also.

EXAMPLE 17.9. Using nodal analysis, find the voltage across the 4-Ω resistor in Fig. 17.16.

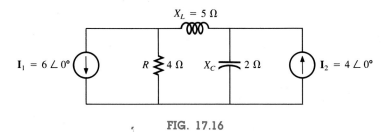

FIG. 17.16

Solution: Choosing nodes (Fig. 17.17) and writing the nodal equations, we have

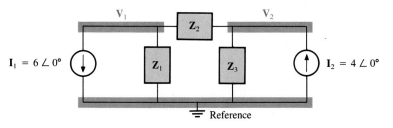

FIG. 17.17

$$\mathbf{Z}_1 = R = 4 \qquad \mathbf{Z}_2 = jX_L = j5 \qquad \mathbf{Z}_3 = -jX_C = -j2$$
$$\mathbf{V}_1(\mathbf{Y}_1 + \mathbf{Y}_2) - \mathbf{V}_2(\mathbf{Y}_2) = -\mathbf{I}_1$$
$$\underline{\mathbf{V}_2(\mathbf{Y}_3 + \mathbf{Y}_2) - \mathbf{V}_1(\mathbf{Y}_2) = +\mathbf{I}_2}$$

or

$$\mathbf{V}_1(\mathbf{Y}_1 - \mathbf{Y}_2) - \mathbf{V}_2(\mathbf{Y}_2) \qquad\quad = -\mathbf{I}_1$$
$$\overline{-\mathbf{V}_1(\mathbf{Y}_2) \qquad\quad + \mathbf{V}_2(\mathbf{Y}_3 + \mathbf{Y}_2) = +\mathbf{I}_2}$$

$$\mathbf{Y}_1 = \frac{1}{\mathbf{Z}_1} \qquad \mathbf{Y}_2 = \frac{1}{\mathbf{Z}_2} \qquad \mathbf{Y}_3 = \frac{1}{\mathbf{Z}_3}$$

Using determinants yields

$$\mathbf{V}_1 = \frac{\begin{vmatrix} -\mathbf{I}_1 & -\mathbf{Y}_2 \\ +\mathbf{I}_2 & \mathbf{Y}_3 + \mathbf{Y}_2 \end{vmatrix}}{\begin{vmatrix} \mathbf{Y}_1 + \mathbf{Y}_2 & -\mathbf{Y}_2 \\ -\mathbf{Y}_2 & \mathbf{Y}_3 + \mathbf{Y}_2 \end{vmatrix}}$$

$$= \frac{-(\mathbf{Y}_3 + \mathbf{Y}_2)\mathbf{I}_1 + \mathbf{I}_2\mathbf{Y}_2}{(\mathbf{Y}_1 + \mathbf{Y}_2)(\mathbf{Y}_3 + \mathbf{Y}_2) - \mathbf{Y}_2^2}$$

$$= \frac{-(\mathbf{Y}_3 + \mathbf{Y}_2)\mathbf{I}_1 + \mathbf{I}_2\mathbf{Y}_2}{\mathbf{Y}_1\mathbf{Y}_3 + \mathbf{Y}_2\mathbf{Y}_3 + \mathbf{Y}_1\mathbf{Y}_2}$$

Substituting numerical values, we have

$$\mathbf{V}_1 = \frac{-[(1/-j2) + (1/j5)]6 \angle 0° + 4 \angle 0°(1/j5)}{(1/4)(1/-j2) + (1/j5)(1/-j2) + (1/4)(1/j5)}$$

$$= \frac{-(+j0.5 - j0.2)6 \angle 0° + 4 \angle 0°(-j0.2)}{(1/-j8) + (1/10) + (1/j20)}$$

$$= \frac{(-0.3 \angle 90°)(6 \angle 0°) + (4 \angle 0°)(0.2 \angle -90°)}{j0.125 + 0.1 - j0.05}$$

$$= \frac{-1.8 \angle 90° + 0.8 \angle -90°}{0.1 + j0.075}$$

$$= \frac{2.6 \angle -90°}{0.125 \angle 36.87°}$$

$$= \mathbf{20.80 \angle -126.87°}$$

EXAMPLE 17.10. Write the nodal equations for the network of Fig. 17.18. In this case, a voltage source appears in the network.

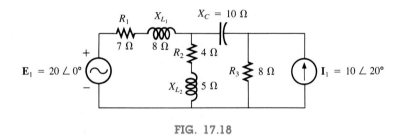

FIG. 17.18

Solution: The circuit is redrawn in Fig. 17.19, where

$$\mathbf{Z}_1 = R_1 + jX_{L_1} = 7 + j8$$
$$\mathbf{Z}_2 = R_2 + jX_{L_2} = 4 + j5$$
$$\mathbf{Z}_3 = -jX_C = -j10$$
$$\mathbf{Z}_4 = R_3 = 8$$

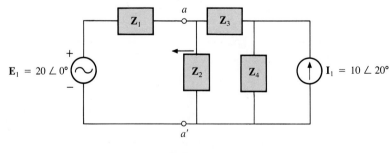

FIG. 17.19

Converting the voltage source to a current source and choosing nodes, we obtain Fig. 17.20. Note the "neat" appearance of the network using the subscripted impedances. Working directly with Fig. 17.18 would be difficult and would probably produce errors.

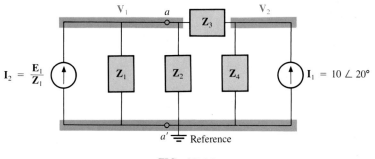

FIG. 17.20

Write the nodal equations:

$$V_1(Y_1 + Y_2 + Y_3) - V_2(Y_3) = +I_2$$
$$V_2(Y_3 + Y_4) - V_1(Y_3) = +I_1$$

$$Y_1 = \frac{1}{Z_1} \qquad Y_2 = \frac{1}{Z_2} \qquad Y_3 = \frac{1}{Z_3} \qquad Y_4 = \frac{1}{Z_4}$$

which are rewritten as

$$V_1(Y_1 + Y_2 + Y_3) - V_2(Y_3) = +I_2$$
$$-V_1(Y_3) + V_2(Y_3 + Y_4) = +I_1$$

$$Y_1 = \frac{1}{7 + j8} \qquad Y_2 = \frac{1}{4 + j5} \qquad Y_3 = \frac{1}{-j10} \qquad Y_4 = \frac{1}{8}$$

$$I_2 = \frac{20 \angle 0°}{7 + j8} \qquad I_1 = 10 \angle 20°$$

EXAMPLE 17.11. Write the nodal equations for the network of Fig. 17.21.

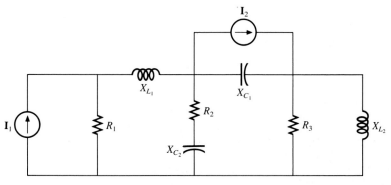

FIG. 17.21

Solution: Choose nodes (Fig. 17.22):

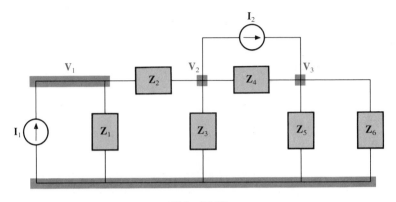

FIG. 17.22

$$\mathbf{Z}_1 = R_1 \qquad \mathbf{Z}_2 = jX_{L_1} \qquad \mathbf{Z}_3 = R_2 - jX_{C_2}$$
$$\mathbf{Z}_4 = -jX_{C_1} \qquad \mathbf{Z}_5 = R_3 \qquad \mathbf{Z}_6 = jX_{L_2}$$

and write nodal equations:

$$\mathbf{V}_1(\mathbf{Y}_1 + \mathbf{Y}_2) - \mathbf{V}_2(\mathbf{Y}_2) = +\mathbf{I}_1$$
$$\mathbf{V}_2(\mathbf{Y}_2 + \mathbf{Y}_3 + \mathbf{Y}_4) - \mathbf{V}_1(\mathbf{Y}_2) - \mathbf{V}_3(\mathbf{Y}_4) = -\mathbf{I}_2$$
$$\mathbf{V}_3(\mathbf{Y}_4 + \mathbf{Y}_5 + \mathbf{Y}_6) - \mathbf{V}_2(\mathbf{Y}_4) = +\mathbf{I}_2$$

which are rewritten as

$$
\begin{array}{llll}
\mathbf{V}_1(\mathbf{Y}_1 + \mathbf{Y}_2) - \mathbf{V}_2(\mathbf{Y}_2) & & + 0 & = +\mathbf{I}_1 \\
-\mathbf{V}_1(\mathbf{Y}_2) & + \mathbf{V}_2(\mathbf{Y}_2 + \mathbf{Y}_3 + \mathbf{Y}_4) - \mathbf{V}_3(\mathbf{Y}_4) & & = -\mathbf{I}_2 \\
0 & - \mathbf{V}_2(\mathbf{Y}_4) & + \mathbf{V}_3(\mathbf{Y}_4 + \mathbf{Y}_5 + \mathbf{Y}_6) & = +\mathbf{I}_2
\end{array}
$$

$$\mathbf{Y}_1 = \frac{1}{R_1} \qquad \mathbf{Y}_2 = \frac{1}{jX_{L_1}} \qquad \mathbf{Y}_3 = \frac{1}{R_2 - jX_{C_2}}$$

$$\mathbf{Y}_4 = \frac{1}{-jX_{C_1}} \qquad \mathbf{Y}_5 = \frac{1}{R_3} \qquad \mathbf{Y}_6 = \frac{1}{jX_{L_2}}$$

Note the symmetry about the diagonal for this example and those preceding it in this section.

EXAMPLE 17.12. Apply nodal analysis to the network of Fig. 17.23. Determine the voltage $\mathbf{V}_L$.

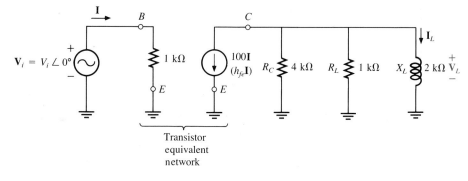

FIG. 17.23

Solution: In this case there is no need for a source conversion. The network is redrawn in Fig. 17.24 with the chosen node voltage and subscripted impedances.

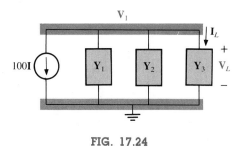

FIG. 17.24

Apply the format approach:

$$\mathbf{Y}_1 = \frac{1}{\mathbf{Z}_1} = \frac{1}{4 \times 10^3} = 0.25 \times 10^{-3} \angle 0° = \mathbf{G}_1$$

$$\mathbf{Y}_2 = \frac{1}{\mathbf{Z}_2} = \frac{1}{1 \times 10^3} = 1.0 \times 10^{-3} \angle 0° = \mathbf{G}_2$$

$$\mathbf{Y}_3 = \frac{1}{\mathbf{Z}_3} = \frac{1}{2 \times 10^3 \angle 90°} = 0.5 \times 10^{-3} \angle -90°$$

$$= -j0.5 \times 10^{-3} = \mathbf{B}_L$$

$$\mathbf{V}_1: (\mathbf{Y}_1 + \mathbf{Y}_2 + \mathbf{Y}_3)\mathbf{V}_1 = -100\mathbf{I}$$

and

$$\mathbf{V}_1 = \frac{-100\mathbf{I}}{\mathbf{Y}_1 + \mathbf{Y}_2 + \mathbf{Y}_3}$$

$$= \frac{-100\mathbf{I}}{(0.25 \times 10^{-3}) + (1 \times 10^{-3}) + (-j0.5 \times 10^{-3})}$$

$$= \frac{-100 \times 10^3\mathbf{I}}{1.25 - j0.5} = \frac{-100 \times 10^3\mathbf{I}}{1.3463 \angle -21.80°}$$

$$= -74.28 \times 10^3\mathbf{I} \angle 21.80°$$

$$= -74.28 \times 10^3 \left(\frac{\mathbf{V}_i}{1 \, k\Omega} \right) \angle 21.80°$$

$$\mathbf{V}_1 = \mathbf{V}_L = -74.28 \mathbf{V}_i \angle \mathbf{21.80°}$$

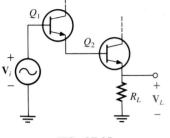

FIG. 17.25

EXAMPLE 17.13. The transistor configuration of Fig. 17.25 will result in a network very similar in appearance to Fig. 17.26 when the equivalent circuits are substituted. The quantities h_1 and h_2 are characteristic constants of the transistors. Determine $\mathbf{V}_L$ for the network of Fig. 17.26. The resistance values were chosen for clarity and are not typical values. In this case, one of the controlling variables is part of the network to be analyzed. Care must be exercised when applying the method.

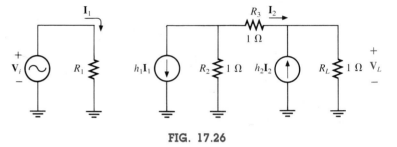

FIG. 17.26

Solution: The network is redrawn in Fig. 17.27. Note that the controlling variable is defined by a separate network:

$$\mathbf{I}_2 = \frac{\mathbf{V}_1 - \mathbf{V}_2}{R_3} = \frac{\mathbf{V}_1 - \mathbf{V}_2}{1} = \mathbf{V}_1 - \mathbf{V}_2$$

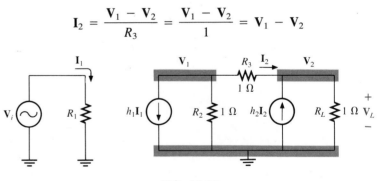

FIG. 17.27

and

$$\mathbf{I}_1 = \frac{\mathbf{V}_i}{R_1} = \frac{\mathbf{V}_i}{1} = \mathbf{V}_i$$

Applying the format approach, we have

$$\mathbf{V}_1: \mathbf{V}_1(1 + 1) - (1)\mathbf{V}_2 = -h_1\mathbf{I}_1$$
$$\mathbf{V}_2: \mathbf{V}_2(1 + 1) - (1)\mathbf{V}_1 = h_2\mathbf{I}_2$$

and

$$2\mathbf{V}_1 - \mathbf{V}_2 = -h_1\mathbf{V}_i$$
$$2\mathbf{V}_2 - \mathbf{V}_1 = h_2(\mathbf{V}_1 - \mathbf{V}_2) = h_2\mathbf{V}_1 - h_2\mathbf{V}_2$$

so that
and

$$2\mathbf{V}_1 - \mathbf{V}_2 = -h_1\mathbf{V}_i$$
$$-\mathbf{V}_1 - h_2\mathbf{V}_1 + 2\mathbf{V}_2 + h_2\mathbf{V}_2 = 0$$

Or
with

$$2\mathbf{V}_1 - \mathbf{V}_2 = -h_1\mathbf{V}_i$$
$$-(1 + h_2)\mathbf{V}_1 + (2 + h_2)\mathbf{V}_2 = 0$$

and

$$\mathbf{V}_2 = \mathbf{V}_L = \frac{\begin{vmatrix} 2 & -h_1\mathbf{V}_i \\ -(1 + h_2) & 0 \end{vmatrix}}{\begin{vmatrix} 2 & -1 \\ -(1 + h_2) & (2 + h_2) \end{vmatrix}}$$

$$= \frac{-h_1(1 + h_2)\mathbf{V}_i}{2(2 + h_2) - (1 + h_2)}$$

$$= \frac{-h_1(1 + h_2)\mathbf{V}_i}{4 + 2h_2 - 1 - h_2}$$

so that

$$\mathbf{V}_L = \left[-\frac{\mathbf{h}_1(\mathbf{1} + \mathbf{h}_2)}{\mathbf{3} + \mathbf{h}_2}\right]\mathbf{V}_i$$

For $h_1 = h_2 = 100$ (typical),

$$\mathbf{V}_L = \frac{(-100)(101)}{3 + 100}\mathbf{V}_i \cong -98\mathbf{V}_i$$

17.6 BRIDGE NETWORKS (ac)

The basic bridge configuration was discussed in some detail in Section 8.12 for dc networks. We now continue to examine bridge networks by considering those that have reactive components and a sinusoidal ac voltage or current applied.

We will first analyze various familiar forms of the bridge network using mesh analysis and nodal analysis (the format approach). The balance conditions will be investigated throughout the section.

Apply *mesh analysis* to the network of Fig. 17.28. The network is redrawn in Fig. 17.29, where

$$\mathbf{Z}_1 = \frac{1}{\mathbf{Y}_1} = \frac{1}{G_1 + jB_C} = \frac{G_1}{G_1^2 + B_C^2} - j\frac{B_C}{G_1^2 + B_C^2}$$

$$\mathbf{Z}_2 = R_2 \qquad \mathbf{Z}_3 = R_3 \qquad \mathbf{Z}_4 = R_4 + jX_L \qquad \mathbf{Z}_5 = R_5$$

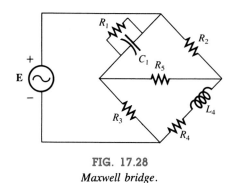

FIG. 17.28
Maxwell bridge.

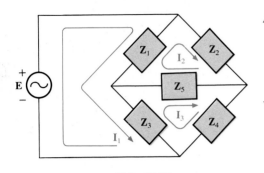

FIG. 17.29

Apply the format approach:

$$(\mathbf{Z}_1 + \mathbf{Z}_3)\mathbf{I}_1 - (\mathbf{Z}_1)\mathbf{I}_2 - (\mathbf{Z}_3)\mathbf{I}_3 = \mathbf{E}$$
$$(\mathbf{Z}_1 + \mathbf{Z}_2 + \mathbf{Z}_5)\mathbf{I}_2 - (\mathbf{Z}_1)\mathbf{I}_1 - (\mathbf{Z}_5)\mathbf{I}_3 = 0$$
$$(\mathbf{Z}_3 + \mathbf{Z}_4 + \mathbf{Z}_5)\mathbf{I}_3 - (\mathbf{Z}_3)\mathbf{I}_1 - (\mathbf{Z}_5)\mathbf{I}_2 = 0$$

which are rewritten as

$$\mathbf{I}_1(\mathbf{Z}_1 + \mathbf{Z}_3) - \mathbf{I}_2\mathbf{Z}_1 \qquad\qquad - \mathbf{I}_3\mathbf{Z}_3 \qquad\qquad = \mathbf{E}$$
$$-\mathbf{I}_1\mathbf{Z}_1 + \mathbf{I}_2(\mathbf{Z}_1 + \mathbf{Z}_2 + \mathbf{Z}_5) - \mathbf{I}_3\mathbf{Z}_5 = 0$$
$$-\mathbf{I}_1\mathbf{Z}_3 \qquad\qquad - \mathbf{I}_2\mathbf{Z}_5 \qquad\qquad + \mathbf{I}_3(\mathbf{Z}_3 + \mathbf{Z}_4 + \mathbf{Z}_5) = 0$$

Note the symmetry about the diagonal of the above equations. For balance, $\mathbf{I}_{\mathbf{Z}_5} = 0\,\text{A}$, and

$$\mathbf{I}_{\mathbf{Z}_5} = \mathbf{I}_2 - \mathbf{I}_3 = 0$$

From the above equations,

$$\mathbf{I}_2 = \frac{\begin{vmatrix} \mathbf{Z}_1 + \mathbf{Z}_3 & \mathbf{E} & -\mathbf{Z}_3 \\ -\mathbf{Z}_1 & 0 & -\mathbf{Z}_5 \\ -\mathbf{Z}_3 & 0 & (\mathbf{Z}_3 + \mathbf{Z}_4 + \mathbf{Z}_5) \end{vmatrix}}{\begin{vmatrix} \mathbf{Z}_1 + \mathbf{Z}_3 & -\mathbf{Z}_1 & -\mathbf{Z}_3 \\ -\mathbf{Z}_1 & (\mathbf{Z}_1 + \mathbf{Z}_2 + \mathbf{Z}_3) & -\mathbf{Z}_5 \\ -\mathbf{Z}_3 & -\mathbf{Z}_5 & (\mathbf{Z}_3 + \mathbf{Z}_4 + \mathbf{Z}_5) \end{vmatrix}}$$

$$= \frac{\mathbf{E}(\mathbf{Z}_1\mathbf{Z}_3 + \mathbf{Z}_1\mathbf{Z}_4 + \mathbf{Z}_1\mathbf{Z}_5 + \mathbf{Z}_3\mathbf{Z}_5)}{\Delta}$$

where Δ signifies the determinant of the denominator (or coefficients). Similarly,

$$\mathbf{I}_3 = \frac{\mathbf{E}(\mathbf{Z}_1\mathbf{Z}_3 + \mathbf{Z}_3\mathbf{Z}_2 + \mathbf{Z}_1\mathbf{Z}_5 + \mathbf{Z}_3\mathbf{Z}_5)}{\Delta}$$

and

$$\mathbf{I}_{\mathbf{Z}_5} = \mathbf{I}_2 - \mathbf{I}_3 = \frac{\mathbf{E}(\mathbf{Z}_1\mathbf{Z}_4 - \mathbf{Z}_3\mathbf{Z}_2)}{\Delta}$$

For $\mathbf{I}_{\mathbf{Z}_5} = 0$, the following must be satisfied (for a finite Δ not equal to zero):

$$\boxed{\mathbf{Z}_1\mathbf{Z}_4 = \mathbf{Z}_3\mathbf{Z}_2} \qquad \mathbf{I}_{\mathbf{Z}_5} = 0 \qquad (17.1)$$

This condition will be analyzed in greater depth later in this section.

Applying *nodal analysis* to the network of Fig. 17.30 will result in the configuration of Fig. 17.31, where

$$\mathbf{Y}_1 = \frac{1}{\mathbf{Z}_1} = \frac{1}{R_1 - jX_C} \qquad \mathbf{Y}_2 = \frac{1}{\mathbf{Z}_2} = \frac{1}{R_2}$$

$$\mathbf{Y}_3 = \frac{1}{\mathbf{Z}_3} = \frac{1}{R_3} \qquad \mathbf{Y}_4 = \frac{1}{\mathbf{Z}_4} = \frac{1}{R_4 + jX_L} \qquad \mathbf{Y}_5 = \frac{1}{R_5}$$

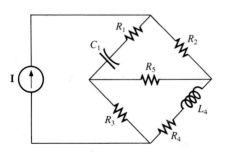

FIG. 17.30
Hay bridge.

and

$$(\mathbf{Y}_1 + \mathbf{Y}_2)\mathbf{V}_1 - (\mathbf{Y}_1)\mathbf{V}_2 - (\mathbf{Y}_2)\mathbf{V}_3 = \mathbf{I}$$
$$(\mathbf{Y}_1 + \mathbf{Y}_3 + \mathbf{Y}_5)\mathbf{V}_2 - (\mathbf{Y}_1)\mathbf{V}_1 - (\mathbf{Y}_5)\mathbf{V}_3 = 0$$
$$(\mathbf{Y}_2 + \mathbf{Y}_4 + \mathbf{Y}_5)\mathbf{V}_3 - (\mathbf{Y}_2)\mathbf{V}_1 - (\mathbf{Y}_5)\mathbf{V}_2 = 0$$

which are rewritten as

$$\mathbf{V}_1(\mathbf{Y}_1 + \mathbf{Y}_2) - \mathbf{V}_2\mathbf{Y}_1 \qquad\qquad - \mathbf{V}_3\mathbf{Y}_2 \qquad\qquad = \mathbf{I}$$
$$-\mathbf{V}_1\mathbf{Y}_1 \qquad + \mathbf{V}_2(\mathbf{Y}_1 + \mathbf{Y}_3 + \mathbf{Y}_5) - \mathbf{V}_3\mathbf{Y}_5 \qquad = 0$$
$$-\mathbf{V}_1\mathbf{Y}_2 \qquad\qquad - \mathbf{V}_2\mathbf{Y}_5 \qquad + \mathbf{V}_3(\mathbf{Y}_2 + \mathbf{Y}_4 + \mathbf{Y}_5) = 0$$

Again, note the symmetry about the diagonal axis. For balance, $\mathbf{V}_{\mathbf{Z}_5} = 0\,\text{V}$, and

$$\mathbf{V}_{\mathbf{Z}_5} = \mathbf{V}_2 - \mathbf{V}_3 = 0$$

From the above equations,

$$\mathbf{V}_2 = \frac{\begin{vmatrix} \mathbf{Y}_1 + \mathbf{Y}_2 & \mathbf{I} & -\mathbf{Y}_2 \\ -\mathbf{Y}_1 & 0 & -\mathbf{Y}_5 \\ -\mathbf{Y}_2 & 0 & (\mathbf{Y}_2 + \mathbf{Y}_4 + \mathbf{Y}_5) \end{vmatrix}}{\begin{vmatrix} \mathbf{Y}_1 + \mathbf{Y}_2 & -\mathbf{Y}_1 & -\mathbf{Y}_2 \\ -\mathbf{Y}_1 & (\mathbf{Y}_1 + \mathbf{Y}_3 + \mathbf{Y}_5) & -\mathbf{Y}_5 \\ -\mathbf{Y}_2 & -\mathbf{Y}_5 & (\mathbf{Y}_2 + \mathbf{Y}_4 + \mathbf{Y}_5) \end{vmatrix}}$$

$$= \frac{\mathbf{I}(\mathbf{Y}_1\mathbf{Y}_3 + \mathbf{Y}_1\mathbf{Y}_4 + \mathbf{Y}_1\mathbf{Y}_5 + \mathbf{Y}_3\mathbf{Y}_5)}{\Delta}$$

Similarly,

$$\mathbf{V}_3 = \frac{\mathbf{I}(\mathbf{Y}_1\mathbf{Y}_3 + \mathbf{Y}_3\mathbf{Y}_2 + \mathbf{Y}_1\mathbf{Y}_5 + \mathbf{Y}_3\mathbf{Y}_5)}{\Delta}$$

Note the similarities between the above equations and those obtained for mesh analysis. Then

$$\mathbf{V}_{\mathbf{Z}_5} = \mathbf{V}_2 - \mathbf{V}_3 = \frac{\mathbf{I}(\mathbf{Y}_1\mathbf{Y}_4 - \mathbf{Y}_3\mathbf{Y}_2)}{\Delta}$$

For $\mathbf{V}_{\mathbf{Z}_5} = 0$, the following must be satisfied for a finite Δ not equal to zero:

$$\boxed{\mathbf{Y}_1\mathbf{Y}_4 = \mathbf{Y}_3\mathbf{Y}_2} \qquad \mathbf{V}_{\mathbf{Z}_5} = 0 \qquad (17.2)$$

However, substituting $\mathbf{Y}_1 = 1/\mathbf{Z}_1$, $\mathbf{Y}_2 = 1/\mathbf{Z}_2$, $\mathbf{Y}_3 = 1/\mathbf{Z}_3$, and $\mathbf{Y}_4 = 1/\mathbf{Z}_4$, we have

$$\frac{1}{\mathbf{Z}_1\mathbf{Z}_4} = \frac{1}{\mathbf{Z}_3\mathbf{Z}_2}$$

or

$$\boxed{\mathbf{Z}_1\mathbf{Z}_4 = \mathbf{Z}_3\mathbf{Z}_2} \qquad \mathbf{V}_{\mathbf{Z}_5} = 0$$

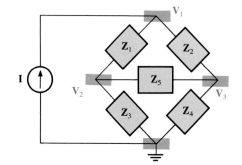

FIG. 17.31

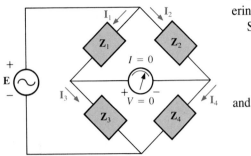

FIG. 17.32

corresponding with Eq. (17.1) obtained earlier.

Let us now investigate the balance criteria in more detail by considering the network of Fig. 17.32, where it is specified that $\mathbf{I}, \mathbf{V} = 0$.

Since $\mathbf{I} = 0$,

$$\boxed{\mathbf{I}_1 = \mathbf{I}_3} \qquad (17.3a)$$

and

$$\boxed{\mathbf{I}_2 = \mathbf{I}_4} \qquad (17.3b)$$

In addition, for $\mathbf{V} = 0$,

$$\boxed{\mathbf{I}_1\mathbf{Z}_1 = \mathbf{I}_2\mathbf{Z}_2} \qquad (17.3c)$$

and

$$\boxed{\mathbf{I}_3\mathbf{Z}_3 = \mathbf{I}_4\mathbf{Z}_4} \qquad (17.3d)$$

Substituting the current relations above into Eq. (17.3d), we have

$$\mathbf{I}_1\mathbf{Z}_3 = \mathbf{I}_2\mathbf{Z}_4$$

and

$$\mathbf{I}_2 = \frac{\mathbf{Z}_3}{\mathbf{Z}_4}\mathbf{I}_1$$

Substituting this relationship for $\mathbf{I}_2$ into Eq. (17.3c) yields

$$\mathbf{I}_1\mathbf{Z}_1 = \left(\frac{\mathbf{Z}_3}{\mathbf{Z}_4}\mathbf{I}_1\right)\mathbf{Z}_2$$

and

$$\mathbf{Z}_1\mathbf{Z}_4 = \mathbf{Z}_2\mathbf{Z}_3$$

as obtained above. Rearranging, we have

$$\boxed{\frac{\mathbf{Z}_1}{\mathbf{Z}_3} = \frac{\mathbf{Z}_2}{\mathbf{Z}_4}} \qquad (17.4)$$

corresponding with Eq. (8.7) for dc resistive networks.

For the network of Fig. 17.30, which is referred to as a *Hay bridge* when $\mathbf{Z}_5$ is replaced by a sensitive galvanometer,

$$\mathbf{Z}_1 = R_1 - jX_C$$
$$\mathbf{Z}_2 = R_2$$
$$\mathbf{Z}_3 = R_3$$
$$\mathbf{Z}_4 = R_4 + jX_L$$

This particular network is used for measuring the resistance and inductance of coils in which the resistance is a small fraction of the reactance X_L.

Substitute into Eq. (17.4) in the following form:

$$\mathbf{Z}_2\mathbf{Z}_3 = \mathbf{Z}_4\mathbf{Z}_1$$
$$R_2R_3 = (R_4 + jX_L)(R_1 - jX_C)$$

or

$$R_2R_3 = R_1R_4 + j(R_1X_L - R_4X_C) + X_CX_L$$

so that

$$R_2R_3 + j0 = (R_1R_4 + X_CX_L) + j(R_1X_L - R_4X_C)$$

In order for the equations to be equal, *the real and imaginary parts must be equal*. Therefore, for a balanced Hay bridge,

$$\boxed{R_2R_3 = R_1R_4 + X_CX_L} \qquad \textbf{(17.5a)}$$

and

$$\boxed{0 = R_1X_L - R_4X_C} \qquad \textbf{(17.5b)}$$

or substituting

$$X_L = \omega L$$
$$X_C = \frac{1}{\omega C}$$

we have

$$X_CX_L = \left(\frac{1}{\omega C}\right)(\omega L) = \frac{L}{C}$$

and

$$R_2R_3 = R_1R_4 + \frac{L}{C}$$

with

$$R_1\omega L = \frac{R_4}{\omega C}$$

Solving for R_4 in the last equation yields

$$R_4 = \omega^2 LCR_1$$

and substituting into the previous equation, we have

$$R_2R_3 = R_1(\omega^2 LCR_1) + \frac{L}{C}$$

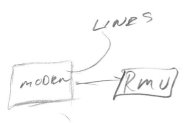

Multiply through by C and factor:

$$CR_2R_3 = L(\omega^2C^2R_1^2 + 1)$$

and

$$\boxed{L = \frac{CR_2R_3}{1 + \omega^2C^2R_1^2}} \qquad \textbf{(17.6a)}$$

with further algebra yielding

$$\boxed{R_4 = \frac{\omega^2C^2R_1R_2R_3}{1 + \omega^2C^2R_1^2}} \qquad \textbf{(17.6b)}$$

Equations (17.5) and (17.6) are the balance conditions for the Hay bridge. Note that each is frequency dependent. For different frequencies, the resistive and capacitive elements must vary for a particular coil to achieve balance. For a coil placed in the Hay bridge as shown in Fig. 17.31, the resistance and inductance of the coil can be determined by Eqs. (17.6a) and (17.6b) when balance is achieved.

The bridge of Fig. 17.28 is referred to as a *Maxwell bridge* when $\mathbf{Z}_5$ is replaced by a sensitive galvanometer. This setup is used for inductance measurements when the resistance of the coil is large enough not to require a Hay bridge.

Application of Eq. (17.4) will yield the following results for the inductance and resistance of the inserted coil:

$$\boxed{L = CR_2R_3} \qquad \textbf{(17.7)}$$

$$\boxed{R_4 = \frac{R_2R_3}{R_1}} \qquad \textbf{(17.8)}$$

The derivation of these equations is quite similar to that employed for the Hay bridge. Keep in mind that the real and imaginary parts must be equal.

One remaining popular bridge is the *capacitance comparison bridge* of Fig. 17.33. An unknown capacitance and its associated resistance can be determined using this bridge. Application of Eq. (17.4) will yield the following results:

$$\boxed{C_4 = C_3\frac{R_1}{R_2}} \qquad \textbf{(17.9)}$$

$$\boxed{R_4 = \frac{R_2R_3}{R_1}} \qquad \textbf{(17.10)}$$

The derivation of these equations will appear as a problem at the end of the chapter.

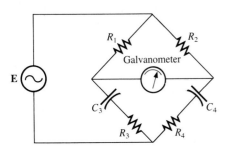

FIG. 17.33

Capacitance comparison bridge.

17.7 Δ-Y, Y-Δ CONVERSIONS

The Δ-Y, Y-Δ (or π-T, T-π as defined in Section 8.13) conversions for ac circuits will not be derived here since the development corresponds exactly with that for dc circuits. Taking the Δ-Y configuration shown in Fig. 17.34, we find the general equations for the impedances of the Y in terms of those for the Δ:

$$\mathbf{Z}_1 = \frac{\mathbf{Z}_A \mathbf{Z}_C}{\mathbf{Z}_A + \mathbf{Z}_B + \mathbf{Z}_C} \qquad (17.11)$$

$$\mathbf{Z}_2 = \frac{\mathbf{Z}_B \mathbf{Z}_C}{\mathbf{Z}_A + \mathbf{Z}_B + \mathbf{Z}_C} \qquad (17.12)$$

$$\mathbf{Z}_3 = \frac{\mathbf{Z}_A \mathbf{Z}_B}{\mathbf{Z}_A + \mathbf{Z}_B + \mathbf{Z}_C} \qquad (17.13)$$

For the impedances of the Δ in terms of those for the Y, the equations are

$$\mathbf{Z}_A = \frac{\mathbf{Z}_1 \mathbf{Z}_2 + \mathbf{Z}_1 \mathbf{Z}_3 + \mathbf{Z}_2 \mathbf{Z}_3}{\mathbf{Z}_2} \qquad (17.14)$$

$$\mathbf{Z}_B = \frac{\mathbf{Z}_1 \mathbf{Z}_2 + \mathbf{Z}_1 \mathbf{Z}_3 + \mathbf{Z}_2 \mathbf{Z}_3}{\mathbf{Z}_1} \qquad (17.15)$$

$$\mathbf{Z}_C = \frac{\mathbf{Z}_1 \mathbf{Z}_2 + \mathbf{Z}_1 \mathbf{Z}_3 + \mathbf{Z}_2 \mathbf{Z}_3}{\mathbf{Z}_3} \qquad (17.16)$$

Note that each impedance of the Y is equal to the product of the impedances in the two closest branches of the Δ, divided by the sum of the impedances in the Δ; and the value of each impedance of the Δ is equal to the sum of the possible product combinations of the impedances of the Y, divided by the impedances of the Y farthest from the impedance to be determined.

Drawn in different forms (Fig. 17.35), they are also referred to as the T and π configurations.

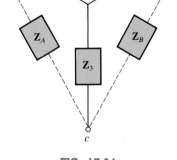

FIG. 17.34
Δ-Y configuration.

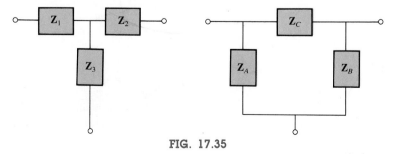

FIG. 17.35

In the study of dc networks, we found that if all of the resistors of the Δ or Y were the same, the conversion from one to the other could be accomplished using the equation

$$R_\Delta = 3R_Y \quad \text{or} \quad R_Y = \frac{R_\Delta}{3}$$

For ac networks,

$$\boxed{\mathbf{Z}_\Delta = 3\mathbf{Z}_Y \quad \text{or} \quad \mathbf{Z}_Y = \frac{\mathbf{Z}_\Delta}{3}} \qquad \textbf{(17.17)}$$

Be careful when using this simplified form. It is not sufficient for all the impedances of the Δ or Y to be of the same magnitude: *The angle associated with each must also be the same.*

EXAMPLE 17.14. Find the total impedance $\mathbf{Z}_T$ of the network of Fig. 17.36.

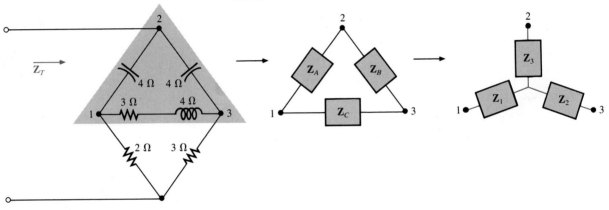

FIG. 17.36

Solution:

$$\mathbf{Z}_A = -j4 \qquad \mathbf{Z}_B = -j4 \qquad \mathbf{Z}_C = 3 + j4$$

$$\mathbf{Z}_1 = \frac{\mathbf{Z}_A\mathbf{Z}_C}{\mathbf{Z}_A + \mathbf{Z}_B + \mathbf{Z}_C} = \frac{(-j4)(3 + j4)}{-j4 - j4 + 3 + j4}$$

$$= \frac{(4\angle-90°)(5\angle53.13°)}{3 - j4} = \frac{20\angle-36.87°}{5\angle-53.13°}$$

$$= 4\angle16.13° = 3.84 + j1.11$$

$$\mathbf{Z}_2 = \frac{\mathbf{Z}_B\mathbf{Z}_C}{\mathbf{Z}_A + \mathbf{Z}_B + \mathbf{Z}_C} = \frac{(-j4)(3 + j4)}{5\angle-53.13°}$$

$$= 4\angle16.13° = 3.84 + j1.11$$

Recall from the study of dc circuits that if two branches of the Y or Δ were the same, the corresponding Δ or Y, respectively, would also have two similar branches. In this example, $\mathbf{Z}_A = \mathbf{Z}_B$. Therefore, $\mathbf{Z}_1 = \mathbf{Z}_2$, and

$$\mathbf{Z}_3 = \frac{\mathbf{Z}_A\mathbf{Z}_B}{\mathbf{Z}_A + \mathbf{Z}_B + \mathbf{Z}_C} = \frac{(-j4)(-j4)}{5 \angle -53.13°}$$

$$= \frac{16 \angle -180°}{5 \angle -53.13°} = 3.2 \angle -126.87° = -1.92 - j2.56$$

Replace the Δ by the Y (Fig. 17.37):

$$\mathbf{Z}_1 = 3.84 + j1.11 \qquad \mathbf{Z}_2 = 3.84 + j1.11$$
$$\mathbf{Z}_3 = -1.92 - j2.56 \qquad \mathbf{Z}_4 = 2 \qquad \mathbf{Z}_5 = 3$$

Impedances $\mathbf{Z}_1$ and $\mathbf{Z}_4$ are in series:

$$\mathbf{Z}_{T_1} = \mathbf{Z}_1 + \mathbf{Z}_4 = 3.84 + j1.11 + 2 = 5.84 + j1.11$$
$$= 5.94 \angle 10.76°$$

Impedances $\mathbf{Z}_2$ and $\mathbf{Z}_5$ are in series:

$$\mathbf{Z}_{T_2} = \mathbf{Z}_2 + \mathbf{Z}_5 = 3.84 + j1.11 + 3 = 6.84 + j1.11$$
$$= 6.93 \angle 9.22°$$

Impedances $\mathbf{Z}_{T_1}$ and $\mathbf{Z}_{T_2}$ are in parallel:

$$\mathbf{Z}_{T_3} = \frac{\mathbf{Z}_{T_1}\mathbf{Z}_{T_2}}{\mathbf{Z}_{T_1} + \mathbf{Z}_{T_2}} = \frac{(5.94 \angle 10.76°)(6.93 \angle 9.22°)}{5.84 + j1.11 + 6.84 + j1.11}$$

$$= \frac{41.16 \angle 19.98°}{12.68 + j2.22} = \frac{41.16 \angle 19.98°}{12.87 \angle 9.93°} = 3.198 \angle 10.05°$$

$$= 3.15 + j0.56$$

Impedances $\mathbf{Z}_3$ and $\mathbf{Z}_{T_3}$ are in series. Therefore,

$$\mathbf{Z}_T = \mathbf{Z}_3 + \mathbf{Z}_{T_3} = -1.92 - j2.56 + 3.15 + j0.56$$
$$= 1.23 - j2.0 = \mathbf{2.35 \angle -58.41°}$$

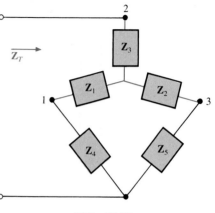

FIG. 17.37

EXAMPLE 17.15. Using both the Δ-Y and Y-Δ transformations, find the total impedance $\mathbf{Z}_T$ for the network of Fig. 17.38.

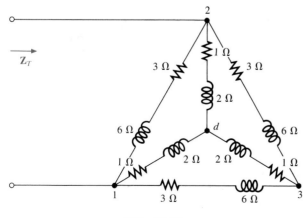

FIG. 17.38

Solution: *Using the Δ-Y transformation,* we obtain Fig. 17.39. In this case, since both systems are balanced (same impedance in each branch), the center point *d'* of the transformed Δ will be the same as the point *d* of the original Y:

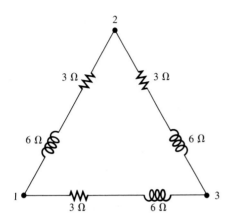
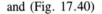
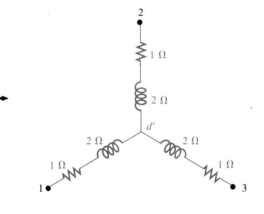

FIG. 17.39

$$\mathbf{Z}_Y = \frac{\mathbf{Z}_\Delta}{3} = \frac{3 + j6}{3} = 1 + j2$$

and (Fig. 17.40)

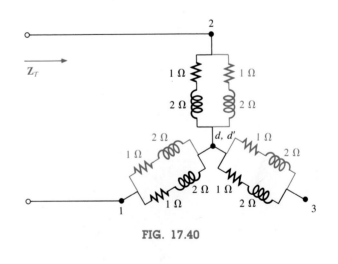

FIG. 17.40

$$\mathbf{Z}_T = 2\left(\frac{1 + j2}{2}\right) = \mathbf{1 + j2}$$

Using the Y-Δ transformation (Fig. 17.41), we obtain

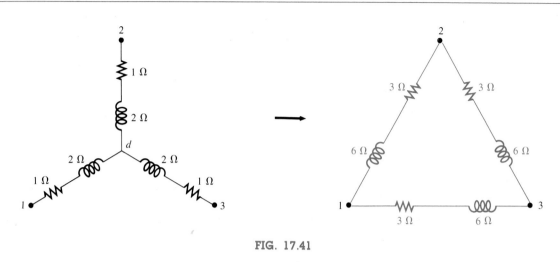

FIG. 17.41

$$\mathbf{Z}_\Delta = 3\mathbf{Z}_Y = 3(1 + j2) = 3 + j6$$

Each resulting parallel combination in Fig. 17.42 will have the following impedance:

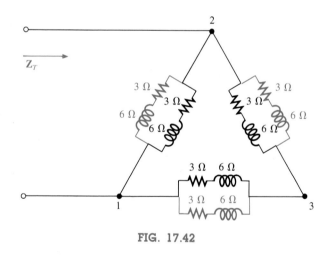

FIG. 17.42

$$\mathbf{Z}' = \frac{3 + j6}{2} = 1.5 + j3$$

and

$$\mathbf{Z}_T = \frac{\mathbf{Z}'(2\mathbf{Z}')}{\mathbf{Z}' + 2\mathbf{Z}'} = \frac{2(\mathbf{Z}')^2}{3\mathbf{Z}'} = \frac{2\mathbf{Z}'}{3}$$

$$= \frac{2(1.5 + j3)}{3} = \mathbf{1 + j2}$$

which compares with the above result.

PROBLEMS

Section 17.2

1. Discuss, in your own words, the difference between a controlled and an independent source.

Section 17.3

2. Convert the voltage sources of Fig. 17.43 to current sources.

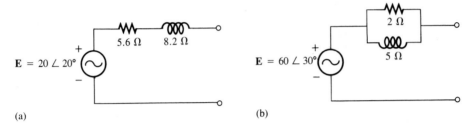

(a)   (b)

FIG. 17.43

3. Convert the current sources of Fig. 17.44 to voltage sources.

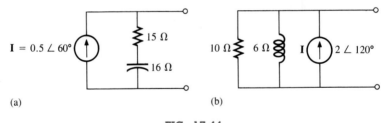

(a)   (b)

FIG. 17.44

4. Convert the voltage source of Fig. 17.45(a) to a current source and the current source of Fig. 17.45(b) to a voltage source.

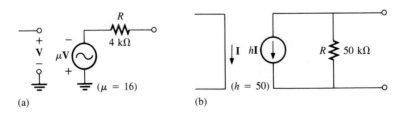

(a)   (b)

FIG. 17.45

Section 17.4

5. Write the mesh equations for the networks of Fig. 17.46.
Determine the current through the resistor R_1.

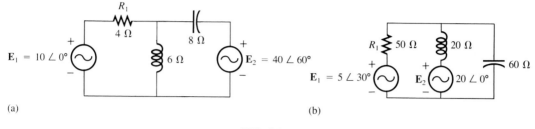

(a)

(b)

FIG. 17.46

6. Repeat Problem 5 for the networks of Fig. 17.47.

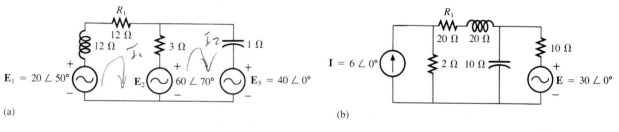

(a)

(b)

FIG. 17.47

*7. Repeat Problem 5 for the networks of Fig. 17.48.

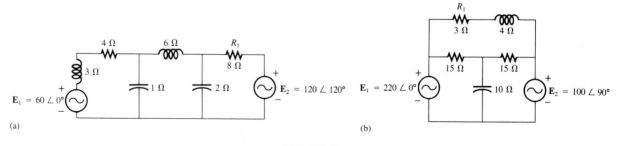

(a)

(b)

FIG. 17.48

*8. Repeat Problem 5 for the networks of Fig. 17.49.

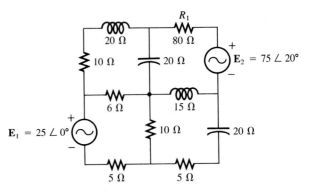

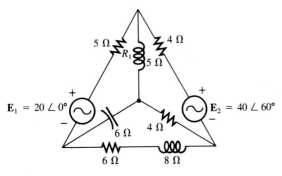

FIG. 17.49

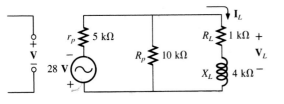

FIG. 17.50

9. Using the mesh-analysis format approach, determine the current I_L (in terms of **V**) for the network of Fig. 17.50.

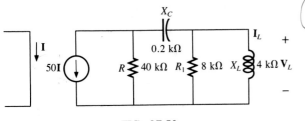

FIG. 17.51

*10. Using the mesh-analysis format approach, determine the current I_L (in terms of **I**) for the network of Fig. 17.51.

Section 17.5

11. Using the format approach, write the nodal equations for each network of Fig. 17.52. Is symmetry present? Determine the nodal voltages.

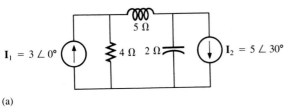

(a)

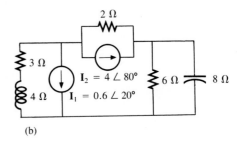

(b)

FIG. 17.52

*12. Repeat Problem 11 for the networks of Fig. 17.53.

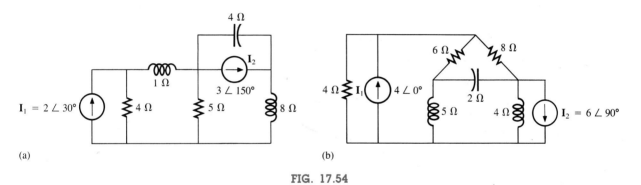

(a)

(b)

FIG. 17.53

13. Repeat Problem 11 for the networks of Fig. 17.54.

(a)

(b)

FIG. 17.54

14. Using the nodal-analysis approach, repeat Problem 9 (determine V_L in terms of V).

*15. Using the nodal-analysis approach, repeat Problem 10 (determine V_L in terms of I).

*16. For the network of Fig. 17.55:
   a. Write the nodal equations.
   b. Is symmetry present? If not, why?
   c. Determine V_L in terms of E_i.

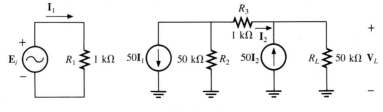

FIG. 17.55

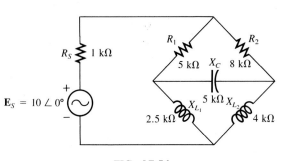

FIG. 17.56

FIG. 17.57

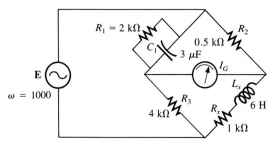

FIG. 17.58

FIG. 17.59

Section 17.6

17. For the bridge network of Fig. 17.56:
   a. Is the bridge balanced?
   b. Using mesh analysis, determine the current through the capacitive reactance.
   c. Using nodal analysis, determine the voltage across the capacitive reactance.

18. Repeat Problem 17 for the bridge of Fig. 17.57.

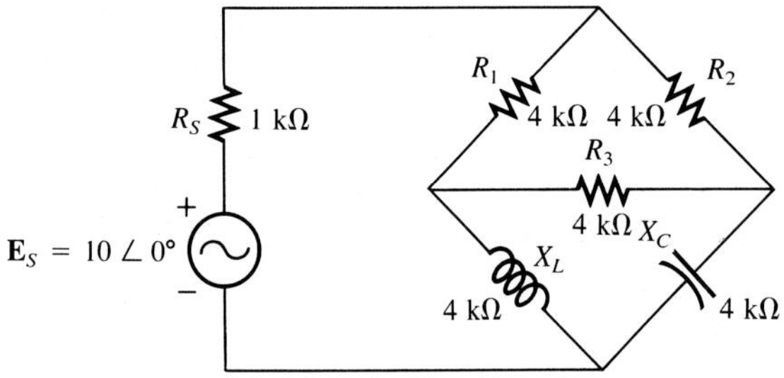

19. The Hay bridge of Fig. 17.58 is balanced. Using Eq. (17.1), determine the unknown inductance L_x and resistance R_x.

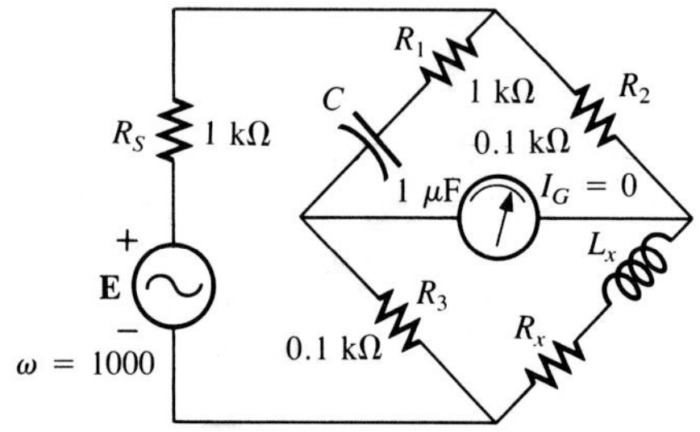

20. Determine whether the Maxwell bridge of Fig. 17.59 is balanced ($\omega = 1000$).

21. Derive the balance equations (17.9) and (17.10) for the capacitance comparison bridge.

22. Determine the balance equations for the inductance bridge of Fig. 17.60.

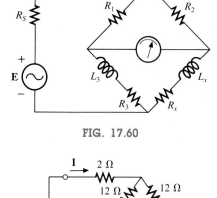

FIG. 17.60

Section 17.7

23. Using the Δ-Y or Y-Δ conversion, determine the current **I** for the networks of Fig. 17.61.

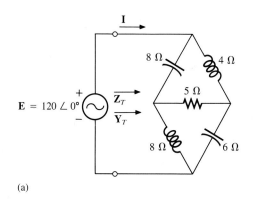

(a)

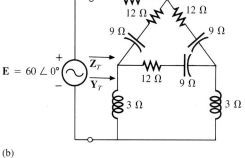

(b)

FIG. 17.61

24. Repeat Problem 23 for the networks of Fig. 17.62. (**E** = 100 ∠0° in each case.)

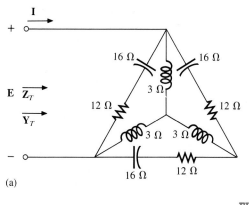

(a)

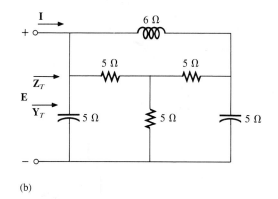

(b)

FIG. 17.62

Computer Problems

25. Write a computer program that will provide a general solution for the network of Fig. 17.8. That is, given the reactance of each element and the parameters of the source voltages, generate a solution in phasor form for both mesh currents.

26. Repeat Problem 25 for the nodal voltages of Fig. 17.16.

27. Given a bridge composed of series impedances in each branch, write a program to test the balance condition as defined by Eq. (17.4).

GLOSSARY

Bridge network A network configuration having the appearance of a diamond in which no two branches are in series or parallel.

Capacitance comparison bridge A bridge configuration having a galvanometer in the bridge arm that is used to determine an unknown capacitance and associated resistance.

Delta (Δ) configuration A network configuration having the appearance of the capital Greek letter delta.

Dependent (controlled) source A source whose magnitude and/or phase angle is determined (controlled) by a current or voltage of the system in which it appears.

Hay bridge A bridge configuration used for measuring the resistance and inductance of coils in those cases where the resistance is a small fraction of the reactance of the coil.

Independent source A source whose magnitude is independent of the network to which it is applied. It displays its terminal characteristics even if completely isolated.

Maxwell bridge A bridge configuration used for inductance measurements when the resistance of the coil is large enough not to require a Hay bridge.

Mesh analysis A method through which the loop (or mesh) currents of a network can be determined. The branch currents of the network can then be determined directly from the loop currents.

Nodal analysis A method through which the node voltages of a network can be determined. The voltage across each element can then be determined through application of Kirchhoff's voltage law.

Source conversion The changing of a voltage source to a current source, or vice versa, which will result in the same terminal behavior of the source. In other words, the external network is unaware of the change in sources.

Wye (Y) configuration A network configuration having the appearance of the capital letter Y.

18

Network Theorems (ac)

18.1 INTRODUCTION

This chapter will parallel Chapter 9, which dealt with network theorems as applied to dc networks. It would be time well spent to review each theorem in Chapter 9 before beginning this chapter, as many of the comments offered there will not be repeated.

Due to the need for developing confidence in the application of the various theorems to networks with controlled (dependent) sources, some sections have been divided into two parts: independent sources and dependent sources.

Theorems to be considered in detail include the superposition theorem, Thevenin and Norton theorems, and the maximum power theorem. The substitution and reciprocity theorems and Millman's theorem are not discussed in detail here, since a review of Chapter 9 will enable you to apply them to sinusoidal ac networks with little difficulty.

18.2 SUPERPOSITION THEOREM

You will recall from Chapter 9 that the superposition theorem eliminated the need for solving simultaneous linear equations by considering the effects of each source independently. To consider the effects of each source, we had to remove the remaining sources. This was accomplished by setting voltage sources to zero (short-circuit representation) and current sources to zero (open-circuit representation). The current through, or voltage across, a portion of the network produced by each source was then added algebraically to find the total solution for the current or voltage.

The only variation in applying this method to ac networks with independent sources is that we will now be working with impedances and phasors instead of just resistors and real numbers.

The superposition theorem is not applicable to power effects in ac networks, since we are still dealing with a nonlinear relationship. It can be applied to networks with sources of different frequencies only if the total response for *each* frequency is found independently and the results are expanded in a nonsinusoidal expression as appearing in Chapter 23.

We will first consider networks with only independent sources. The analysis is then very similar to that for dc networks.

Independent Sources

EXAMPLE 18.1. Using the superposition theorem, find the current **I** through the 4-Ω reactance (X_{L_2}) in Fig. 18.1.

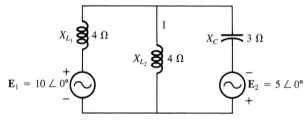

FIG. 18.1

Solution: In the redrawn circuit (Fig. 18.2),

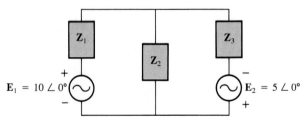

FIG. 18.2

$$\mathbf{Z}_1 = +jX_{L_1} = j4$$
$$\mathbf{Z}_2 = +jX_{L_2} = j4$$
$$\mathbf{Z}_3 = -jX_C = -j3$$

Considering the effects of the voltage source $\mathbf{E}_1$ (Fig. 18.3), we have

$$\mathbf{Z}_{2\|3} = \frac{\mathbf{Z}_2\mathbf{Z}_3}{\mathbf{Z}_2 + \mathbf{Z}_3} = \frac{(j4)(-j3)}{j4 - j3} = \frac{12}{j} = -j12$$
$$= 12 \angle -90°$$

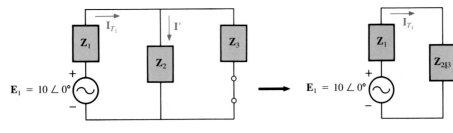

FIG. 18.3

$$\mathbf{I}_{T_1} = \frac{\mathbf{E}_1}{\mathbf{Z}_{2\|3} + \mathbf{Z}_1} = \frac{10 \angle 0°}{-j12 + j4} = \frac{10 \angle 0°}{8 \angle -90°}$$
$$= 1.25 \angle 90°$$

and

$$\mathbf{I'} = \frac{\mathbf{Z}_3 \mathbf{I}_{T_1}}{\mathbf{Z}_2 + \mathbf{Z}_3} \qquad \text{(current divider rule)}$$
$$= \frac{(-j3)(j1.25)}{j4 - j3} = \frac{3.75}{j} = 3.75 \angle -90°$$

Considering the effects of the voltage source $\mathbf{E}_2$ (Fig. 18.4), we have

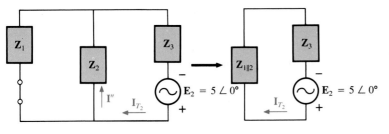

FIG. 18.4

$$\mathbf{Z}_{1\|2} = \frac{\mathbf{Z}_1}{N} = \frac{j4}{2} = j2$$
$$\mathbf{I}_{T_2} = \frac{\mathbf{E}_2}{\mathbf{Z}_{1\|2} + \mathbf{Z}_3} = \frac{5 \angle 0°}{j2 - j3} = \frac{5 \angle 0°}{1 \angle -90°} = 5 \angle 90°$$

and

$$\mathbf{I''} = \frac{\mathbf{I}_{T_2}}{2} = 2.5 \angle 90°$$

The total current through the 4-Ω reactance X_{L_2} (Fig. 18.5) is

$$\mathbf{I} = \mathbf{I'} - \mathbf{I''}$$
$$= 3.75 \angle -90° - 2.50 \angle 90° = -j3.75 - j2.50$$
$$= -j6.25$$
$$\mathbf{I} = \mathbf{6.25} \angle -\mathbf{90°}$$

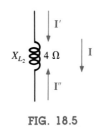

FIG. 18.5

EXAMPLE 18.2. Using superposition, find the current **I** through the 6-Ω resistor in Fig. 18.6.

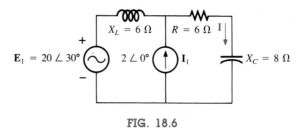

FIG. 18.6

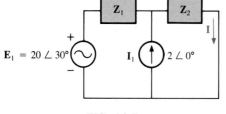

FIG. 18.7

Solution: In the redrawn circuit (Fig. 18.7),

$$\mathbf{Z}_1 = j6 \qquad \mathbf{Z}_2 = 6 - j8$$

Consider the effects of the current source (Fig. 18.8). Applying the current divider rule, we have

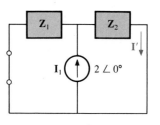

FIG. 18.8

$$\mathbf{I}' = \frac{\mathbf{Z}_1 \mathbf{I}_1}{\mathbf{Z}_1 + \mathbf{Z}_2} = \frac{j6(2)}{j6 + 6 - j8} = \frac{j12}{6 - j2}$$

$$= \frac{12 \angle 90°}{6.32 \angle -18.43°}$$

$$\mathbf{I}' = 1.9 \angle 108.43°$$

Consider the effects of the voltage source (Fig. 18.9). Applying Ohm's law gives us

$$\mathbf{I}'' = \frac{\mathbf{E}_1}{\mathbf{Z}_T} = \frac{\mathbf{E}_1}{\mathbf{Z}_1 + \mathbf{Z}_2} = \frac{20 \angle 30°}{6.32 \angle -18.43°}$$

$$= 3.16 \angle 48.43°$$

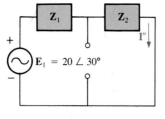

FIG. 18.9

The total current through the 6-Ω resistor (Fig. 18.10) is

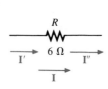

FIG. 18.10

$$\mathbf{I} = \mathbf{I'} + \mathbf{I''}$$
$$= 1.9 \ \angle 108.43° + 3.16 \ \angle 48.43°$$
$$= (-0.60 + j1.80) + (2.10 + j2.36)$$
$$= 1.50 + j4.16$$
$$\mathbf{I} = \mathbf{4.42 \ \angle 70.2°}$$

EXAMPLE 18.3. Using superposition, find the voltage across the 6-Ω resistor in Fig. 18.6. Check the results against $\mathbf{V}_{6\Omega} = \mathbf{I}(6)$, where $\mathbf{I}$ is the current found through the 6-Ω resistor in the previous example.

Solution: For the current source,

$$\mathbf{V'}_{6\Omega} = \mathbf{I'}(6) = (1.9 \ \angle 108.43°)(6) = 11.4 \ \angle 108.43°$$

For the voltage source,

$$\mathbf{V''}_{6\Omega} = \mathbf{I''}(6) = (3.16 \ \angle 48.43°)(6) = 18.96 \ \angle 48.43°$$

The total voltage across the 6-Ω resistor (Fig. 18.11) is

$$\mathbf{V}_{6\Omega} = \mathbf{V'}_{6\Omega} + \mathbf{V''}_{6\Omega}$$
$$= 11.4 \ \angle 108.43° + 18.96 \ \angle 48.43°$$
$$= (-3.60 + j10.82) + (12.58 + j14.18)$$
$$= 8.98 + j25.0$$
$$\mathbf{V}_{6\Omega} = \mathbf{26.5 \ \angle 70.2°}$$

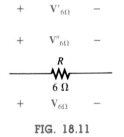

FIG. 18.11

Checking the result, we have

$$\mathbf{V}_{6\Omega} = \mathbf{I}(6) = (4.42 \ \angle 70.2°)(6) = \mathbf{26.5 \ \angle 70.2°} \qquad \text{(checks)}$$

Dependent Sources

For dependent sources in which the controlling variable is not determined by the network to which the superposition theorem is to be applied, the application of the theorem is basically the same as for independent sources. The solution obtained will simply be in terms of the controlling variables.

EXAMPLE 18.4. Using the superposition theorem, determine the current $\mathbf{I}_2$ for the network of Fig. 18.12. The quantities μ and h are constants.

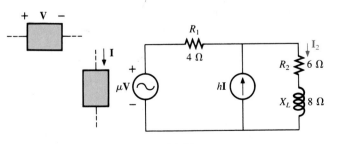

FIG. 18.12

Solution: With a portion of the system redrawn (Fig. 18.13),

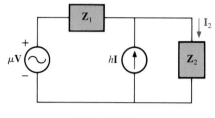

FIG. 18.13

$$\mathbf{Z}_1 = R_1 = 4 \qquad \mathbf{Z}_2 = R_2 + jX_L = 6 + j8$$

For the voltage source (Fig. 18.14),

$$\mathbf{I'} = \frac{\mu\mathbf{V}}{\mathbf{Z}_1 + \mathbf{Z}_2} = \frac{\mu\mathbf{V}}{4 + 6 + j8} = \frac{\mu\mathbf{V}}{10 + j8}$$

$$= \frac{\mu\mathbf{V}}{12.8 \angle 38.66°} = 0.078\ \mu\mathbf{V} \angle -38.66°$$

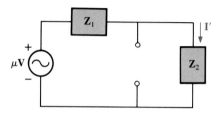

FIG. 18.14

For the current source (Fig. 18.15),

$$\mathbf{I''} = \frac{\mathbf{Z}_1(h\mathbf{I})}{\mathbf{Z}_1 + \mathbf{Z}_2} = \frac{4(h\mathbf{I})}{12.8 \angle 38.66°} = 4(0.078)h\mathbf{I} \angle -38.66°$$

$$= 0.312h\mathbf{I} \angle -38.66°$$

The current $\mathbf{I}_2$ is

$$\mathbf{I}_2 = \mathbf{I'} + \mathbf{I''}$$
$$= \mathbf{0.078}\ \mu\mathbf{V} \angle \mathbf{-38.66°} + \mathbf{0.312}h\mathbf{I} \angle \mathbf{-38.66°}$$

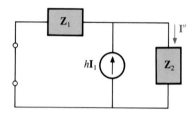

FIG. 18.15

For $\mathbf{V} = 10 \angle 0°$, $\mathbf{I} = 20 \times 10^{-3} \angle 0°$, $\mu = 20$, $h = 100$,

$$\mathbf{I}_2 = 0.078(20)(10) \angle -38.66°$$
$$+ 0.312(100)(20 \times 10^{-3}) \angle -38.66°$$
$$= 15.60 \angle -38.66° + 0.62 \angle -38.66°$$
$$\mathbf{I}_2 = \mathbf{16.22} \angle \mathbf{-38.66°}$$

For dependent sources in which the controlling variable is determined by the network to which the theorem is to be applied, the dependent source cannot be set to zero unless the controlling variable is also zero. For networks containing dependent sources such as indicated in Example 18.4 and dependent sources of the type just introduced above, the superposition theorem is applied for each independent source and each dependent source not having a controlling variable in the portions of the network under investigation. It must be reemphasized that dependent sources are not sources of energy in the sense that if all independent sources are removed from a system, all currents and voltages must be zero.

EXAMPLE 18.5. Determine the current $\mathbf{I}_L$ through the resistor R_L of Fig. 18.16.

Solution: Note that the controlling variable $\mathbf{V}$ is determined by the network to be analyzed. From the above discussions, it is understood that the dependent source cannot be set to zero unless $\mathbf{V}$ is zero. If we set $\mathbf{I}$ to zero, the network lacks a source of voltage, and $\mathbf{V} = 0$ with $\mu\mathbf{V} = 0$. The resulting $\mathbf{I}_L$ under this condition is zero. Obviously, therefore, the network must be analyzed as it appears in Fig. 18.16, with the result that neither source can be eliminated, as is normally done using the superposition theorem.

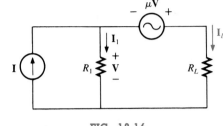

FIG. 18.16

Applying Kirchhoff's voltage law, we have

$$\mathbf{V}_L = \mathbf{V} + \mu\mathbf{V} = (1 + \mu)\mathbf{V}$$

and

$$\mathbf{I}_L = \frac{(1 + \mu)\mathbf{V}}{R_L}$$

The result, however, must be found in terms of $\mathbf{I}$ since $\mathbf{V}$ and $\mu\mathbf{V}$ are only dependent variables.

Applying Kirchhoff's current law gives us

$$\mathbf{I} = \mathbf{I}_1 + \mathbf{I}_L = \frac{\mathbf{V}}{R_1} + \frac{(1 + \mu)\mathbf{V}}{R_L}$$

and

$$\mathbf{I} = \mathbf{V}\left(\frac{1}{R_1} + \frac{1 + \mu}{R_L}\right)$$

or

$$\mathbf{V} = \frac{\mathbf{I}}{(1/R_1) + [(1 + \mu)/R_L]}$$

Substituting into the above yields

$$\mathbf{I}_L = \frac{(1 + \mu)\mathbf{V}}{R_L} = \frac{(1 + \mu)}{R_L}\left(\frac{\mathbf{I}}{(1/R_1) + [(1 + \mu)/R_L]}\right)$$

Therefore,

$$\mathbf{I}_L = \frac{(1 + \mu)R_1\mathbf{I}}{R_L + (1 + \mu)R_1}$$

18.3 THEVENIN'S THEOREM

Thevenin's theorem, as stated for sinusoidal ac circuits, is changed only to include the term *impedance* instead of *resistance;* that is, *any two-terminal linear ac network can be replaced by an equivalent circuit*

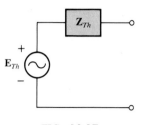

FIG. 18.17

consisting of a voltage source and an impedance in series as shown in Fig. 18.17.

Since the reactances of a circuit are frequency dependent, the Thevenin circuit found for a particular network is applicable only at *one* frequency.

The steps required to apply this method to dc circuits are repeated here with changes for sinusoidal ac circuits. As before, the only change is the replacement of the term *resistance* by *impedance*. Again, dependent and independent sources will be treated separately.

Independent Sources

1. Remove that portion of the network across which the Thevenin equivalent circuit is to be found.
2. Mark ($\bigcirc$, $\bullet$, and so on) the terminals of the remaining two-terminal network.
3. Calculate $\mathbf{Z}_{Th}$ by first setting all voltage and current sources to zero (short circuit and open circuit, respectively) and then finding the resulting impedance between the two marked terminals.
4. Calculate $\mathbf{E}_{Th}$ by first replacing the voltage and current sources and then finding the open-circuit voltage between the marked terminals.
5. Draw the Thevenin equivalent circuit with the portion of the circuit previously removed replaced between the terminals of the Thevenin equivalent circuit.

EXAMPLE 18.6. Find the Thevenin equivalent circuit for the network external to resistor R in Fig. 18.18.

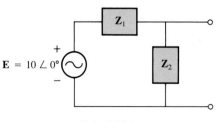

FIG. 18.18

Solution:
Steps 1 and 2 (Fig. 18.19):

$$\mathbf{Z}_1 = jX_L = j8 \qquad \mathbf{Z}_2 = -jX_C = -j2$$

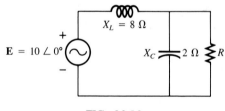

FIG. 18.19

Step 3 (Fig. 18.20):

$$\dot{Z}_{Th} = \frac{Z_1 Z_2}{Z_1 + Z_2} = \frac{(j8)(-j2)}{j8 - j2} = \frac{-j^2 16}{j6} = \frac{16}{6 \angle 90°}$$

$$= \mathbf{2.67 \angle -90°}$$

Step 4 (Fig. 18.21):

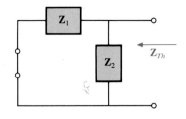

FIG. 18.20

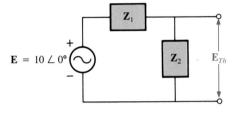

FIG. 18.21

$$\mathbf{E}_{Th} = \frac{Z_2 \mathbf{E}}{Z_1 + Z_2} \qquad \text{(voltage divider rule)}$$

$$= \frac{(-j2)(10)}{j8 - j2} = \frac{-j20}{j6} = \mathbf{3.33 \angle -180°}$$

Step 5: The Thevenin equivalent circuit is shown in Fig. 18.22.

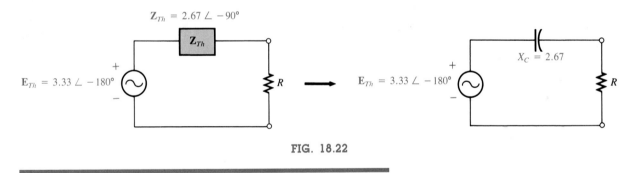

FIG. 18.22

EXAMPLE 18.7. Find the Thevenin equivalent circuit for the network external to branch *a-a'* in Fig. 18.23.

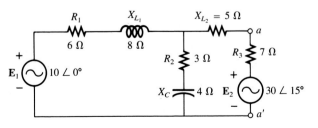

FIG. 18.23

Solution:

Steps 1 and 2 (Fig. 18.24): Note the reduced complexity with subscripted impedances:

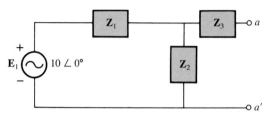

FIG. 18.24

$$\mathbf{Z}_1 = R_1 + jX_{L_1} = 6 + j8$$
$$\mathbf{Z}_2 = R_2 - jX_C = 3 - j4$$
$$\mathbf{Z}_3 = +jX_{L_2} = j5$$

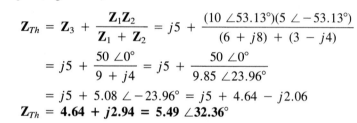

FIG. 18.25

Step 3 (Fig. 18.25):

$$\mathbf{Z}_{Th} = \mathbf{Z}_3 + \frac{\mathbf{Z}_1\mathbf{Z}_2}{\mathbf{Z}_1 + \mathbf{Z}_2} = j5 + \frac{(10 \angle 53.13°)(5 \angle -53.13°)}{(6 + j8) + (3 - j4)}$$

$$= j5 + \frac{50 \angle 0°}{9 + j4} = j5 + \frac{50 \angle 0°}{9.85 \angle 23.96°}$$

$$= j5 + 5.08 \angle -23.96° = j5 + 4.64 - j2.06$$

$$\mathbf{Z}_{Th} = \mathbf{4.64 + j2.94 = 5.49 \angle 32.36°}$$

Step 4 (Fig. 18.26): Since a-a' is an open circuit, $\mathbf{I}_{Z_3} = 0$. Then $\mathbf{E}_{Th}$ is the voltage drop across $\mathbf{Z}_2$:

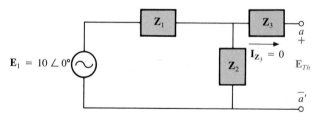

FIG. 18.26

$$\mathbf{E}_{Th} = \frac{\mathbf{Z}_2\mathbf{E}}{\mathbf{Z}_2 + \mathbf{Z}_1} \qquad \text{(voltage divider rule)}$$

$$= \frac{(5 \angle -53.13°)(10 \angle 0°)}{9.85 \angle 23.96°}$$

$$\mathbf{E}_{Th} = \frac{50 \angle -53.13°}{9.85 \angle 23.96°} = \mathbf{5.08 \angle -77.09°}$$

Step 5: The Thevenin equivalent circuit is shown in Fig. 18.27.

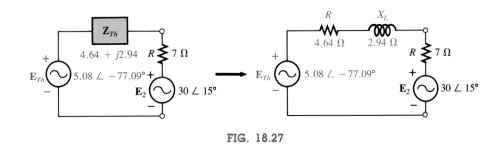

FIG. 18.27

Dependent Sources

For dependent sources with a controlling variable not in the network under investigation, the procedure indicated above can be applied. However, for dependent sources of the other type, where the controlling variable is part of the network to which the theorem is to be applied, another approach must be employed. The necessity for a different approach will be demonstrated in an example to follow. The method is not limited to dependent sources of the latter type. It can also be applied to any dc or sinusoidal ac network. However, for networks of independent sources, the method of application employed in Chapter 9 and the first portion of this section is generally more direct, with usual savings in time and errors.

The new approach to Thevenin's theorem can best be introduced at this stage in the development by considering the Thevenin equivalent circuit of Fig. 18.28(a). As indicated in Fig. 18.28(b), the open-circuit terminal voltage ($\mathbf{E}_{oc}$) of the Thevenin equivalent circuit is the Thevenin equivalent voltage. That is,

$$\mathbf{E}_{oc} = \mathbf{E}_{Th} \qquad \text{(18.1)}$$

If the external terminals are short circuited as in Fig. 18.28(c), the resulting short-circuit current is determined by

$$\mathbf{I}_{sc} = \frac{\mathbf{E}_{Th}}{\mathbf{Z}_{Th}} \qquad \text{(18.2)}$$

or, rearranged,

$$\mathbf{Z}_{Th} = \frac{\mathbf{E}_{Th}}{\mathbf{I}_{sc}}$$

and

$$\mathbf{Z}_{Th} = \frac{\mathbf{E}_{oc}}{\mathbf{I}_{sc}} \qquad \text{(18.3)}$$

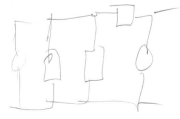

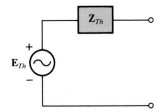

(a)

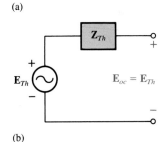

(b)

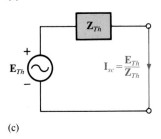

(c)

FIG. 18.28

Equations (18.1) and (18.3) indicate that for any linear bilateral dc or ac network with or without dependent sources of any type, if the open-circuit terminal voltage of a portion of a network can be determined along with the short-circuit current between the same two terminals, the Thevenin equivalent circuit is effectively known. A few examples will make the method quite clear. The advantage of the method, which was stressed earlier in this section for independent sources, should now be more obvious. The current $\mathbf{I}_{sc}$, which is necessary to find $\mathbf{Z}_{Th}$, is in general more difficult to obtain since all of the sources are present.

There is a third approach to the Thevenin equivalent circuit that is also useful from a practical viewpoint. The Thevenin voltage is found as in the two previous methods. However, the Thevenin impedance is obtained by applying a source of voltage to the terminals of interest and determining the source current as indicated in Fig. 18.29. For this method, the source voltage of the original network is set to zero. The Thevenin impedance is then determined by the following equation:

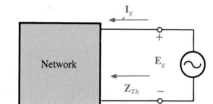

FIG. 18.29
Determining $\mathbf{Z}_{Th}$.

$$\mathbf{Z}_{Th} = \frac{\mathbf{E}_g}{\mathbf{I}_g} \qquad (18.4)$$

Note that for each technique, $\mathbf{E}_{Th} = \mathbf{E}_{oc}$, but the Thevenin impedance is found in different ways.

The first two examples include network configurations frequently encountered in the analysis of electronic networks. They have a dependent source with an *external* controlling variable, permitting the use of any one of the three techniques described in this section. In fact, each will be applied to the network of each example to demonstrate its validity.

EXAMPLE 18.8.   Using each of the three techniques described in this section, determine the Thevenin equivalent circuit for the network of Fig. 18.30.

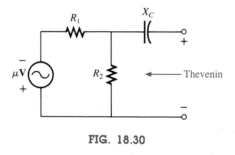

FIG. 18.30

Solution: Since for each approach the Thevenin voltage is found in exactly the same manner, it will be determined first. From Fig. 18.30, where $\mathbf{I}_{X_C} = 0$,

Due to the polarity for **V** and
defined terminal polarities

$$\mathbf{V}_{R_1} = \mathbf{E}_{Th} = \mathbf{E}_{oc} = \frac{R_2(\mu\mathbf{V})}{R_1 + R_2} = -\frac{\mu R_2\mathbf{V}}{R_1 + R_2}$$

The following three methods for determining the Thevenin impedance appear in the order in which they were introduced in this section.

Method 1 (Fig. 18.31):

$$\mathbf{Z}_{Th} = R_1 \| R_2 - jX_C$$

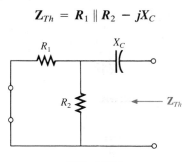

FIG. 18.31

Method 2 (Fig. 18.32): Converting the voltage source to a current

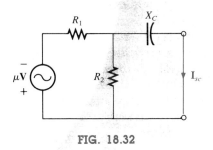

FIG. 18.32

source (Fig. 18.33), we have (current divider rule)

$$\mathbf{I}_{sc} = \frac{-(R_1 \| R_2)\dfrac{\mu\mathbf{V}}{R_1}}{(R_1 \| R_2) - jX_C} = \frac{-\dfrac{R_1R_2}{R_1 + R_2}\left(\dfrac{\mu\mathbf{V}}{R_1}\right)}{(R_1 \| R_2) - jX_C}$$

$$= \frac{\dfrac{-\mu R_2\mathbf{V}}{R_1 + R_2}}{(R_1 \| R_2) - jX_C}$$

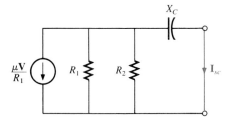

FIG. 18.33

and

$$\mathbf{Z}_{Th} = \frac{\mathbf{E}_{oc}}{\mathbf{I}_{sc}} = \frac{\dfrac{-\mu R_2\mathbf{V}}{R_1 + R_2}}{\dfrac{-\mu R_2\mathbf{V}}{\dfrac{R_1 + R_2}{(R_1 \| R_2) - jX_C}}} = \frac{1}{\dfrac{1}{(R_1 \| R_2) - jX_C}}$$

$$= R_1 \| R_2 - jX_C$$

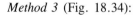

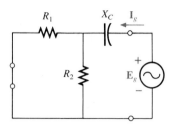

FIG. 18.34

Method 3 (Fig. 18.34):

$$\mathbf{I}_g = \frac{\mathbf{E}_g}{(R_1 \parallel R_2) - jX_C}$$

and

$$\mathbf{Z}_{Th} = \frac{\mathbf{E}_g}{\mathbf{I}_g} = R_1 \parallel R_2 - jX_C$$

In each case, the Thevenin impedance is the same. The resulting Thevenin equivalent circuit is shown in Fig. 18.35.

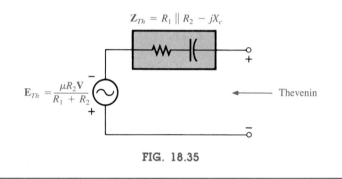

FIG. 18.35

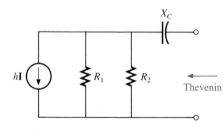

FIG. 18.36

EXAMPLE 18.9. Repeat Example 18.8 for the network of Fig. 18.36.
Solution: From Fig. 18.36, $\mathbf{E}_{Th}$ is

$$\mathbf{E}_{Th} = \mathbf{E}_{oc} = -h\mathbf{I}(R_1 \parallel R_2) = \frac{hR_1R_2\mathbf{I}}{R_1 + R_2}$$

Method 1 (Fig. 18.37):

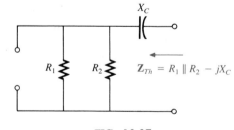

FIG. 18.37

$$\mathbf{Z}_{Th} = R_1 \parallel R_2 - jX_C$$

Note the similarity between this solution and that obtained for the previous example.

Method 2 (Fig. 18.38):

$$\mathbf{I}_{sc} = \frac{-(R_1 \parallel R_2)h\mathbf{I}}{(R_1 \parallel R_2) - jX_C}$$

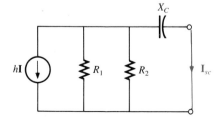

FIG. 18.38

and

$$Z_{Th} = \frac{E_{oc}}{I_{sc}} = \frac{-hI(R_1 \parallel R_2)}{\dfrac{-(R_1 \parallel R_2)hI}{(R_1 \parallel R_2) - jX_C}} = R_1 \parallel R_2 - jX_C$$

Method 3 (Fig. 18.39):

$$I_g = \frac{E_g}{(R_1 \parallel R_2) - jX_C}$$

and

$$Z_{Th} = \frac{E_g}{I_g} = R_1 \parallel R_2 - jX_C$$

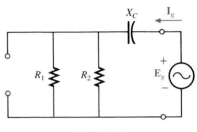

FIG. 18.39

The following example has a dependent source that will not permit the use of the method described in the beginning of this section for independent sources. All three methods will be applied, however, so that the results can be compared.

EXAMPLE 18.10. For the network of Fig. 18.40 (introduced in Example 18.5), determine the Thevenin equivalent circuit between the indicated terminals using each method described in this section. Compare your results.

Solution: First, using Kirchhoff's voltage law, E_{Th} (which is the same for each method) is written

$$E_{Th} = V + \mu V = (1 + \mu)V$$

However,

$$V = IR_1$$

so

$$E_{Th} = (1 + \mu)IR_1$$

Z_{Th}:

Method 1 (Fig. 18.41): Since $I = 0$, V and $\mu V = 0$, and

$$\bcancel{Z_{Th} = R_1} \quad \text{(incorrect)}$$

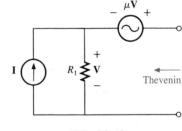

FIG. 18.40

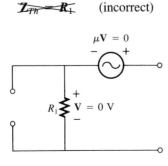

FIG. 18.41

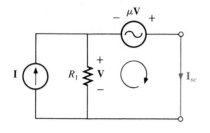

FIG. 18.42

Method 2 (Fig. 18.42): Kirchhoff's voltage law around the indicated loop gives us

$$\mathbf{V} + \mu\mathbf{V} = 0$$

and

$$\mathbf{V}(1 + \mu) = 0$$

Since μ is a positive constant, the above equation can be satisfied only when $\mathbf{V} = 0$. Substitution of this result into Fig. 18.42 will yield the configuration of Fig. 18.43, and

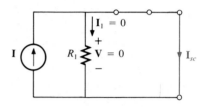

FIG. 18.43

$$\mathbf{I}_{sc} = \mathbf{I}$$

with

$$\mathbf{Z}_{Th} = \frac{\mathbf{E}_{oc}}{\mathbf{I}_{sc}} = \frac{(1 + \mu)\mathbf{I}R_1}{\mathbf{I}} = (1 + \mu)R_1 \qquad \text{(correct)}$$

Method 3 (Fig. 18.44):

$$\mathbf{E}_g = \mathbf{V} + \mu\mathbf{V} = (1 + \mu)\mathbf{V}$$

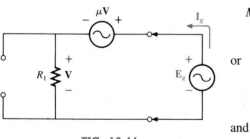

FIG. 18.44

or

$$\mathbf{V} = \frac{\mathbf{E}_g}{1 + \mu}$$

and

$$\mathbf{I}_g = \frac{\mathbf{V}}{R_1} = \frac{\mathbf{E}_g}{(1 + \mu)R_1}$$

and

$$\mathbf{Z}_{Th} = \frac{\mathbf{E}_g}{\mathbf{I}_g} = (1 + \mu)R_1 \qquad \text{(correct)}$$

The Thevenin equivalent circuit appears in Fig. 18.45, and

$$\mathbf{I}_L = \frac{(1 + \mu)R_1\mathbf{I}}{R_L + (1 + \mu)R_1}$$

which compares with the result of Example 18.5.

$\mathbf{E}_{Th} = (1 + \mu)\mathbf{I}R_1$

FIG. 18.45

The network of Fig. 18.46 is the basic configuration of the transistor equivalent circuit applied most frequently today. Needless to say, it is necessary to know its characteristics and be adept in its use. Note that there is a controlled voltage and current source, each controlled by variables in the configuration.

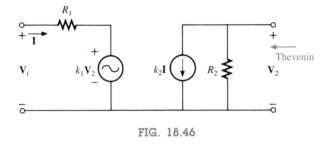

FIG. 18.46

EXAMPLE 18.11. Determine the Thevenin equivalent circuit for the indicated terminals of the network of Fig. 18.46.

Solution: Apply the second method introduced in this section.

$\mathbf{E}_{Th}$:

$$\mathbf{E}_{oc} = \mathbf{V}_2$$

$$\mathbf{I} = \frac{\mathbf{V}_i - k_1\mathbf{V}_2}{R_1} = \frac{\mathbf{V}_i - k_1\mathbf{E}_{oc}}{R_1}$$

and

$$\mathbf{E}_{oc} = -k_2IR_2 = -k_2R_2\left(\frac{\mathbf{V}_i - k_1\mathbf{E}_{oc}}{R_1}\right)$$

$$= \frac{-k_2R_2\mathbf{V}_i}{R_1} + \frac{k_1k_2R_2\mathbf{E}_{oc}}{R_1}$$

or

$$\mathbf{E}_{oc}\left(1 - \frac{k_1k_2R_2}{R_1}\right) = \frac{-k_2R_2\mathbf{V}_i}{R_1}$$

and

$$\mathbf{E}_{oc}\left(\frac{R_1 - k_1k_2R_2}{R_1}\right) = \frac{-k_2R_2\mathbf{V}_i}{R_1}$$

so

$$\boxed{\mathbf{E}_{oc} = \frac{-k_2R_2\mathbf{V}_i}{R_1 - k_1k_2R_2} = \mathbf{E}_{Th}}$$ **(18.5)**

$\mathbf{I}_{sc}$:

For the network of Fig. 18.47, where

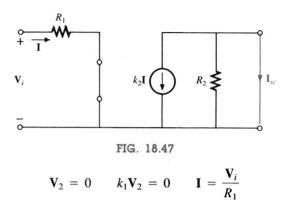

FIG. 18.47

$$V_2 = 0 \qquad k_1 V_2 = 0 \qquad I = \frac{V_i}{R_1}$$

and

$$I_{sc} = -k_2 I = \frac{-k_2 V_i}{R_1}$$

so

$$Z_{Th} = \frac{E_{oc}}{I_{sc}} = \frac{\dfrac{-k_2 R_2 V_i}{R_1 - k_1 k_2 R_2}}{\dfrac{-k_2 V_i}{R_1}} = \frac{R_1 R_2}{R_1 - k_1 k_2 R_2}$$

and

$$\boxed{Z_{Th} = \frac{R_2}{1 - \dfrac{k_1 k_2 R_2}{R_1}}} \qquad\qquad \textbf{(18.6)}$$

Frequently, the approximation $k_1 \cong 0$ is applied. Then, the Thevenin voltage and impedance are

$$\boxed{E_{Th} = \frac{-k_2 R_2 V_i}{R_1}} \qquad k_1 = 0 \qquad\qquad \textbf{(18.7)}$$

$$\boxed{Z_{Th} = R_2} \qquad k_1 = 0 \qquad\qquad \textbf{(18.8)}$$

Apply $Z_{Th} = E_g / I_g$ to the network of Fig. 18.48, where

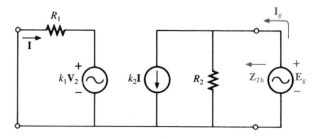

FIG. 18.48

$$I = \frac{-k_1 \mathbf{V}_2}{R_1}$$

But

$$\mathbf{V}_2 = \mathbf{E}_g$$

so

$$I = \frac{-k_1 \mathbf{E}_g}{R_1}$$

Applying Kirchhoff's current law, we have

$$\mathbf{I}_g = k_2 \mathbf{I} + \frac{\mathbf{E}_g}{R_2} = k_2 \left(-\frac{k_1 \mathbf{E}_g}{R_1} \right) + \frac{\mathbf{E}_g}{R_2}$$

$$= \mathbf{E}_g \left(\frac{1}{R_2} - \frac{k_1 k_2}{R_1} \right)$$

and

$$\frac{\mathbf{I}_g}{\mathbf{E}_g} = \frac{R_1 - k_1 k_2 R_2}{R_1 R_2}$$

or

$$\mathbf{Z}_{Th} = \frac{\mathbf{E}_g}{\mathbf{I}_g} = \frac{R_1 R_2}{R_1 - k_1 k_2 R_2}$$

as obtained above.

The last two methods presented in this section were applied only to networks in which the magnitudes of the controlled sources were dependent on a variable within the network for which the Thevenin equivalent circuit was to be obtained. Understand that both of those methods can also be applied to any dc or sinusoidal ac network containing only independent sources or dependent sources of the other kind.

18.4 NORTON'S THEOREM

The three methods described for Thevenin's theorem will each be altered to permit their use with Norton's theorem. Since the Thevenin and Norton impedances are the same for a particular network, certain portions of the discussion will be quite similar to those encountered in the previous section. We will first consider independent sources and the approach developed in Chapter 9, followed by dependent sources and the new techniques developed for Thevenin's theorem.

You will recall from Chapter 9 that Norton's theorem allows us to replace any two-terminal linear bilateral ac network by an equivalent circuit consisting of a current source and impedance, as in Fig. 18.49.

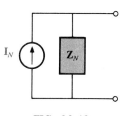

FIG. 18.49

The Norton equivalent circuit, like the Thevenin equivalent circuit, is applicable at only one frequency since the reactances are frequency dependent.

Independent Sources

The procedure outlined below to find the Norton equivalent of a sinusoidal ac network is changed (from that in Chapter 9) in only one respect: to replace the term *impedance* with the term *resistance*.

1. Remove that portion of the network across which the Norton equivalent circuit is to be found.
2. Mark ($\bigcirc$, $\bullet$, and so on) the terminals of the remaining two-terminal network.
3. Calculate $\mathbf{Z}_N$ by first setting all voltage and current sources to zero (short circuit and open circuit, respectively) and then finding the resulting impedance between the two marked terminals.
4. Calculate $\mathbf{I}_N$ by first replacing the voltage and current sources and then finding the short-circuit current between the marked terminals.
5. Draw the Norton equivalent circuit with the portion of the circuit previously removed replaced between the terminals of the Norton equivalent circuit.

The Norton and Thevenin equivalent circuits can be found from each other by using the source transformation shown in Fig. 18.50. The source transformation is applicable for any Thevenin or Norton equivalent circuit determined from a network with any combination of independent or dependent sources.

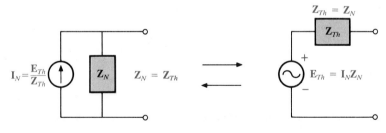

FIG. 18.50

EXAMPLE 18.12. Determine the Norton equivalent circuit for the network external to the 6-Ω resistor of Fig. 18.51.

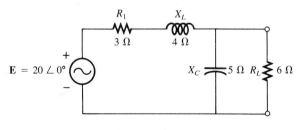

FIG. 18.51

Solution:
Steps 1 and 2 (Fig. 18.52):

$$\mathbf{Z}_1 = R_1 + jX_L = 3 + j4 = 5 \angle 53.13°$$
$$\mathbf{Z}_2 = -jX_C = -j5$$

Step 3 (Fig. 18.53):

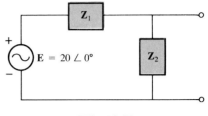

FIG. 18.52

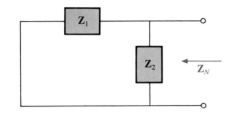

FIG. 18.53

$$\mathbf{Z}_N = \frac{\mathbf{Z}_1\mathbf{Z}_2}{\mathbf{Z}_1 + \mathbf{Z}_2} = \frac{(5 \angle 53.13°)(5 \angle -90°)}{3 + j4 - j5} = \frac{25 \angle -36.87°}{3 - j1}$$

$$= \frac{25 \angle -36.87°}{3.16 \angle -18.43°} = 7.91 \angle -18.44° = \mathbf{7.50 - j2.50}$$

Step 4 (Fig. 18.54):

$$\mathbf{I}_N = \mathbf{I}_1 = \frac{\mathbf{E}}{\mathbf{Z}_1} = \frac{20 \angle 0°}{5 \angle 53.13°} = \mathbf{4 \angle -53.13°}$$

Step 5: The Norton equivalent circuit is shown in Fig. 18.55.

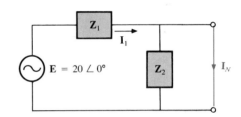

FIG. 18.54

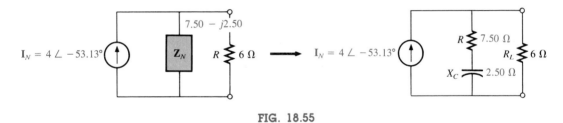

FIG. 18.55

EXAMPLE 18.13. Find the Norton equivalent circuit for the network external to the 7-Ω capacitive reactance in Fig. 18.56.

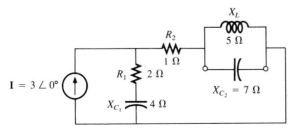

FIG. 18.56

Solution:

Steps 1 and 2 (Fig. 18.57):

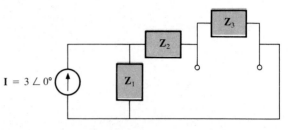

FIG. 18.57

$$\mathbf{Z}_1 = R_1 - jX_{C_1} = 2 - j4$$
$$\mathbf{Z}_2 = R_2 = 1$$
$$\mathbf{Z}_3 = +jX_L = j5$$

Step 3 (Fig. 18.58):

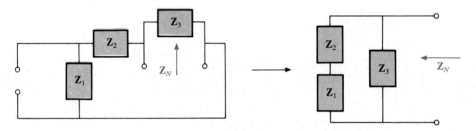

FIG. 18.58

$$\mathbf{Z}_N = \frac{\mathbf{Z}_3(\mathbf{Z}_1 + \mathbf{Z}_2)}{\mathbf{Z}_3 + (\mathbf{Z}_1 + \mathbf{Z}_2)}$$

$$\mathbf{Z}_1 + \mathbf{Z}_2 = 2 - j4 + 1 = 3 - j4 = 5 \angle -53.13°$$

$$\mathbf{Z}_N = \frac{(5 \angle 90°)(5 \angle -53.13°)}{j5 + 3 - j4} = \frac{25 \angle 36.87°}{3 + j1}$$

$$= \frac{25 \angle 36.87°}{3.16 \angle +18.43°}$$

$$\mathbf{Z}_N = 7.91 \angle 18.44° = \mathbf{7.50 + j2.50}$$

Step 4 (Fig. 18.59):

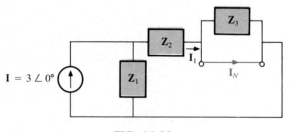

FIG. 18.59

$$\mathbf{I}_N = \mathbf{I}_1 = \frac{\mathbf{Z}_1\mathbf{I}}{\mathbf{Z}_1 + \mathbf{Z}_2} \qquad \text{(current divider rule)}$$

$$= \frac{(2 - j4)(3)}{3 - j4} = \frac{6 - j12}{5 \angle -53.13°} = \frac{13.4 \angle -63.43°}{5 \angle -53.13°}$$

$$\mathbf{I}_N = \mathbf{2.68} \angle \mathbf{-10.3°}$$

Step 5: The Norton equivalent circuit is shown in Fig. 18.60.

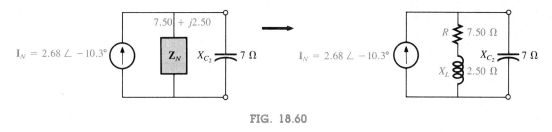

FIG. 18.60

EXAMPLE 18.14. Find the Thevenin equivalent circuit for the network external to the 7-Ω capacitive reactance in Fig. 18.56.

Solution: Using the conversion between sources (Fig. 18.61), we obtain

$$\mathbf{Z}_{Th} = \mathbf{Z}_N = \mathbf{7.50} + \mathbf{j2.50}$$
$$\mathbf{E}_{Th} = \mathbf{I}_N\mathbf{Z}_N = (2.68 \angle -10.3°)(7.91 \angle 18.44°)$$
$$= \mathbf{21.2} \angle \mathbf{8.14°}$$

The Thevenin equivalent circuit is shown in Fig. 18.62.

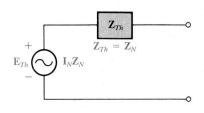

FIG. 18.61

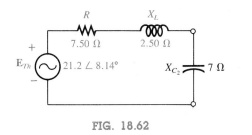

FIG. 18.62

Dependent Sources

As stated for Thevenin's theorem, dependent sources in which the controlling variable is not determined by the network for which the Norton equivalent circuit is to be found do not alter the procedure outlined above.

For dependent sources of the other kind, one of the following procedures must be applied. Both of these procedures can also be applied to networks with any combination of independent sources and dependent sources not controlled by the network under investigation.

The Norton equivalent circuit appears in Fig. 18.63(a). In Fig. 18.63(b), we find that

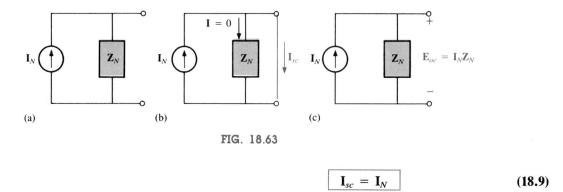

(a)   (b)   (c)

FIG. 18.63

$$\boxed{\mathbf{I}_{sc} = \mathbf{I}_N} \qquad\qquad (18.9)$$

and in Fig. 18.63(c) that

$$\mathbf{E}_{oc} = \mathbf{I}_N\mathbf{Z}_N$$

Or, rearranging, we have

$$\mathbf{Z}_N = \frac{\mathbf{E}_{oc}}{\mathbf{I}_N}$$

and

$$\boxed{\mathbf{Z}_N = \frac{\mathbf{E}_{oc}}{\mathbf{I}_{sc}}} \qquad\qquad (18.10)$$

The Norton impedance can also be determined by applying a source of voltage $\mathbf{E}_g$ to the terminals of interest and finding the resulting $\mathbf{I}_g$, as shown in Fig. 18.64. All independent sources and dependent sources not controlled by a variable in the network of interest are set to zero, and

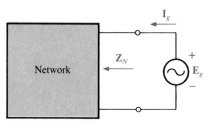

FIG. 18.64

$$\boxed{\mathbf{Z}_N = \frac{\mathbf{E}_g}{\mathbf{I}_g}} \qquad\qquad (18.11)$$

For this latter approach, the Norton current is still determined by the short-circuit current.

EXAMPLE 18.15. Using each method described for dependent sources, find the Norton equivalent circuit for the network of Fig. 18.65.

Solution:

I_N:

For each method, I_N is determined in the same manner. From Fig. 18.66, using Kirchhoff's current law, we have

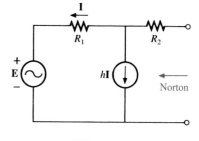

FIG. 18.65

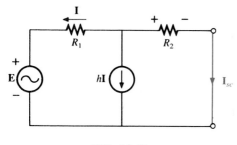

FIG. 18.66

$$0 = I + hI + I_{sc}$$

or

$$I_{sc} = -(1 + h)I$$

Applying Kirchhoff's voltage law gives us

$$E + IR_1 - I_{sc}R_2 = 0$$

and

$$IR_1 = I_{sc}R_2 - E$$

or

$$I = \frac{I_{sc}R_2 - E}{R_1}$$

so

$$I_{sc} = -(1 + h)I = -(1 + h)\left(\frac{I_{sc}R_2 - E}{R_1}\right)$$

or

$$R_1 I_{sc} = -(1 + h)I_{sc}R_2 + (1 + h)E$$
$$I_{sc}[R_1 + (1 + h)R_2] = (1 + h)E$$
$$I_{sc} = \frac{(1 + h)E}{R_1 + (1 + h)R_2} = I_N$$

Z_N:

Method 1: E_{oc} is determined from the network of Fig. 18.67. By Kirchhoff's current law,

$$0 = I + hI \quad \text{or} \quad I(h + 1) = 0$$

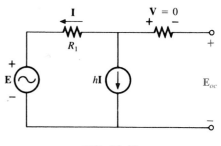

FIG. 18.67

For h, a positive constant $\mathbf{I}$ must equal zero to satisfy the above. Therefore,

$$\mathbf{I} = 0 \quad \text{and} \quad h\mathbf{I} = 0$$

and

$$\mathbf{E}_{oc} = \mathbf{E}$$

with

$$\mathbf{Z}_N = \frac{\mathbf{E}_{oc}}{\mathbf{I}_{sc}} = \frac{\mathbf{E}}{\dfrac{(1 + h)\mathbf{E}}{R_1 + (1 + h)R_2}} = \frac{R_1 + (1 + h)R_2}{(1 + h)}$$

Method 2: Note Fig. 18.68. By Kirchhoff's current law,

$$\mathbf{I}_g = \mathbf{I} + h\mathbf{I} = (1 + h)\mathbf{I}$$

By Kirchhoff's voltage law,

$$\mathbf{E}_g - \mathbf{I}_g R_2 - \mathbf{I}R_1 = 0$$

or

$$\mathbf{I} = \frac{\mathbf{E}_g - \mathbf{I}_g R_2}{R_1}$$

Substituting, we have

$$\mathbf{I}_g = (1 + h)\mathbf{I} = (1 + h)\left(\frac{\mathbf{E}_g - \mathbf{I}_g R_2}{R_1}\right)$$

and

$$\mathbf{I}_g R_1 = (1 + h)\mathbf{E}_g - (1 + h)\mathbf{I}_g R_2$$

so

$$\mathbf{E}_g(1 + h) = \mathbf{I}_g[R_1 + (1 + h)R_2]$$

or

$$\mathbf{Z}_N = \frac{\mathbf{E}_g}{\mathbf{I}_g} = \frac{R_1 + (1 + h)R_2}{1 + h}$$

which agrees with the above.

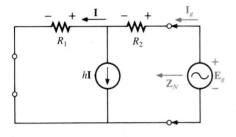

FIG. 18.68

EXAMPLE 18.16. Find the Norton equivalent circuit for the network configuration of Fig. 18.46.

Solution: By source conversion,

$$\mathbf{I}_N = \frac{\mathbf{E}_{Th}}{\mathbf{Z}_{Th}} = \frac{\dfrac{-k_2 R_2 \mathbf{V}_i}{R_1 - k_1 k_2 R_2}}{\dfrac{R_1 R_2}{R_1 - k_1 k_2 R_2}}$$

and

$$\boxed{\mathbf{I}_N = \frac{-k_2\mathbf{V}_i}{R_1}} \qquad \textbf{(18.12)}$$

which is $\mathbf{I}_{sc}$ as determined in that example, and

$$\boxed{\mathbf{Z}_N = \mathbf{Z}_{Th} = \frac{R_2}{1 - \dfrac{k_1 k_2 R_2}{R_1}}} \qquad \textbf{(18.13)}$$

For $k_1 \cong 0$, we have

$$\boxed{\mathbf{I}_N = \frac{-k_2\mathbf{V}_i}{R_1}} \qquad k_1 = 0 \qquad \textbf{(18.14)}$$

$$\boxed{\mathbf{Z}_N = R_2} \qquad k_1 = 0 \qquad \textbf{(18.15)}$$

18.5 MAXIMUM POWER TRANSFER THEOREM

When applied to ac circuits, the maximum power transfer theorem states that *maximum power will be delivered to a load when the load impedance is the conjugate of the Thevenin impedance across its terminals.* That is, for Fig. 18.69, for maximum power transfer to the load,

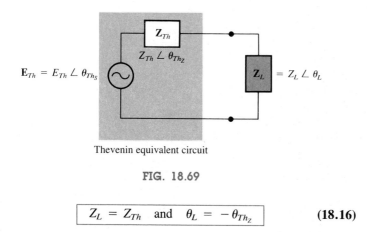

Thevenin equivalent circuit

FIG. 18.69

$$\boxed{Z_L = Z_{Th} \quad \text{and} \quad \theta_L = -\theta_{Th_z}} \qquad \textbf{(18.16)}$$

or, in rectangular form,

$$\boxed{R_L = R_{Th} \quad \text{and} \quad \pm jX_L = \mp jX_{Th}} \qquad \textbf{(18.17)}$$

The conditions just mentioned will make the total impedance of the circuit appear purely resistive, as indicated in Fig. 18.70:

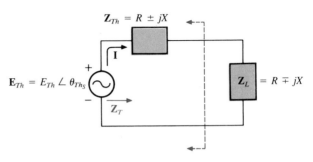

FIG. 18.70

$$\mathbf{Z}_T = (R \pm jX) + (R \mp jX)$$

and

$$\boxed{\mathbf{Z}_T = 2R} \qquad (18.18)$$

Since the circuit is purely resistive, the power factor of the circuit under maximum power conditions is 1. That is,

$$\boxed{F_p = 1} \qquad \text{(maximum power transfer)} \qquad (18.19)$$

The magnitude of the current **I** of Fig. 18.70 is

$$I = \frac{E_{Th}}{Z_T} = \frac{E_{Th}}{2R}$$

The maximum power to the load is

$$P_{\max} = I^2 R = \left(\frac{E_{Th}}{2R}\right)^2 R$$

and

$$\boxed{P_{\max} = \frac{E_{Th}^2}{4R}} \qquad (18.20)$$

EXAMPLE 18.17. Find the load impedance in Fig. 18.71 for maximum power to the load, and find the maximum power.

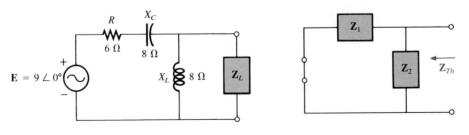

FIG. 18.71

Solution:

$$\mathbf{Z}_1 = R - jX_C = 6 - j8 = 10 \angle -53.13°$$

$$\mathbf{Z}_2 = +jX_L = j8$$

$$\mathbf{Z}_{Th} = \frac{\mathbf{Z}_1\mathbf{Z}_2}{\mathbf{Z}_1 + \mathbf{Z}_2} = \frac{(10 \angle -53.13°)(8 \angle 90°)}{6 - j8 + j8} = \frac{80 \angle 36.87°}{6 \angle 0°}$$

$$= 13.33 \angle 36.87° = 10.66 + j8$$

and

$$\mathbf{Z}_L = 13.3 \angle -36.87° = \mathbf{10.66 - j8}$$

In order to find the maximum power, we must first find $\mathbf{E}_{Th}$ (Fig. 18.72), as follows:

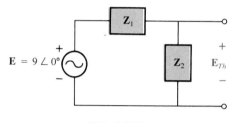

FIG. 18.72

$$\mathbf{E}_{Th} = \frac{\mathbf{Z}_2\mathbf{E}}{\mathbf{Z}_2 + \mathbf{Z}_1} \qquad \text{(voltage divider rule)}$$

$$= \frac{(8 \angle 90°)(9 \angle 0°)}{j8 + 6 - j8} = \frac{72 \angle 90°}{6 \angle 0°} = 12 \angle 90°$$

Then

$$P_{\max} = \frac{E_{Th}^2}{4R} = \frac{(12)^2}{4(10.66)} = \frac{144}{42.64} = \mathbf{3.38\ W}$$

EXAMPLE 18.18. Find the load impedance in Fig. 18.73 for maximum power to the load, and find the maximum power.

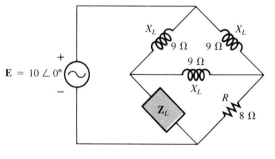

FIG. 18.73

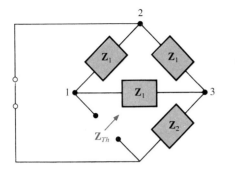

FIG. 18.74

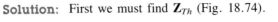

Solution: First we must find $\mathbf{Z}_{Th}$ (Fig. 18.74).

$$\mathbf{Z}_1 = +jX_L = j9 \qquad \mathbf{Z}_2 = R = 8$$

Converting from a Δ to a Y (Fig. 18.75), we have

$$\mathbf{Z}'_1 = \frac{\mathbf{Z}_1}{3} = j3 \qquad \mathbf{Z}_2 = 8$$

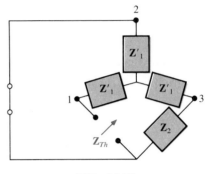

FIG. 18.75

FIG. 18.76

The redrawn circuit (Fig. 18.76) shows

$$\mathbf{Z}_{Th} = \mathbf{Z}'_1 + \frac{\mathbf{Z}'_1(\mathbf{Z}'_1 + \mathbf{Z}_2)}{\mathbf{Z}'_1 + (\mathbf{Z}'_1 + \mathbf{Z}_2)}$$

$$= j3 + \frac{3\angle 90°(j3 + 8)}{j6 + 8}$$

$$= j3 + \frac{(3\angle 90°)(8.54\angle 20.56°)}{10\angle 36.87°}$$

$$= j3 + \frac{25.62\angle 110.56°}{10\angle 36.87°} = j3 + 2.56\angle 73.69°$$

$$= j3 + 0.72 + j2.46$$

$$\mathbf{Z}_{Th} = 0.72 + j5.46$$

and

$$\mathbf{Z}_L = \mathbf{0.72} - j\mathbf{5.46}$$

For $\mathbf{E}_{Th}$, use the modified circuit of Fig. 18.77 with the voltage source replaced in its original position. Since $I_1 = 0$, $\mathbf{E}_{Th}$ is the voltage across the series impedance of $\mathbf{Z}'_1$ and $\mathbf{Z}_2$. Using the voltage divider rule gives us

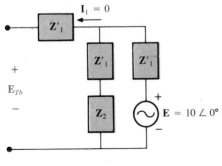

FIG. 18.77

$$\mathbf{E}_{Th} = \frac{(\mathbf{Z}'_1 + \mathbf{Z}_2)\mathbf{E}}{\mathbf{Z}'_1 + \mathbf{Z}_2 + \mathbf{Z}'_1} = \frac{(j3 + 8)(10\angle 0°)}{8 + j6}$$

$$= \frac{(8.54\angle 20.56°)(10\angle 0°)}{10\angle 36.87°}$$

$$\mathbf{E}_{Th} = 8.54\angle -16.31°$$

and

$$P_{max} = \frac{E_{Th}^2}{4R} = \frac{(8.54)^2}{4(0.72)} = \frac{72.93}{2.88}$$
$$= \textbf{25.32 W}$$

18.6 SUBSTITUTION, RECIPROCITY, AND MILLMAN'S THEOREMS

As indicated in the introduction to this chapter, the substitution and reciprocity theorems and Millman's theorem will not be considered here in detail. A careful review of Chapter 9 will enable you to apply these theorems to sinusoidal ac networks with little difficulty. A number of problems in the use of these theorems appear in the problem section.

PROBLEMS

Section 18.2

1. Using superposition, for each network of Fig. 18.78, determine the current through the inductance X_L.

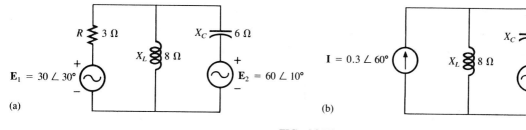

FIG. 18.78

*2. Repeat Problem 1 for the networks of Fig. 18.79.

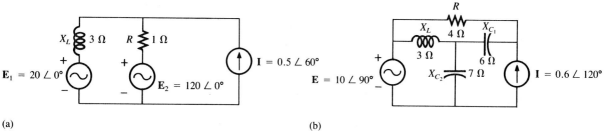

FIG. 18.79

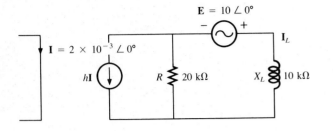

FIG. 18.80

3. Using superposition, for the network of Fig. 18.80, determine the current I_L ($h = 100$).

4. Using superposition, for the network of Fig. 18.81, determine the voltage V_L ($\mu = 20$).

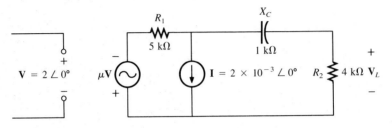

FIG. 18.81

*5. Using superposition, determine the current I_L for the network of Fig. 18.82 ($\mu = 20$; $h = 100$).

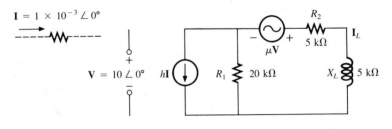

FIG. 18.82

*6. Determine V_L for the network of Fig. 18.83 ($h = 50$).

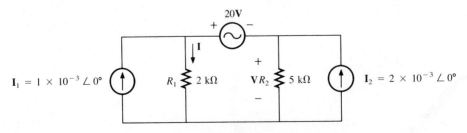

FIG. 18.83

*7. Calculate the current I for the network of Fig. 18.84.

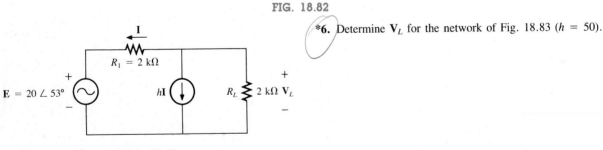

FIG. 18.84

Section 18.3

8. Find the Thevenin equivalent circuit for the portions of the networks of Fig. 18.85 external to the elements between points a and b.

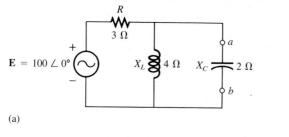

(a)

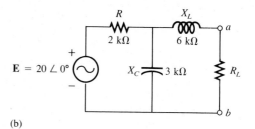

(b)

FIG. 18.85

*9. Repeat Problem 8 for the networks of Fig. 18.86.

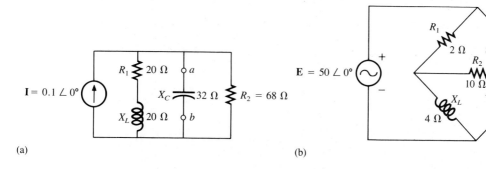

(a)

(b)

FIG. 18.86

*10. Repeat Problem 8 for the networks of Fig. 18.87.

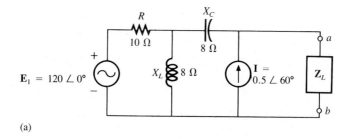

(a)

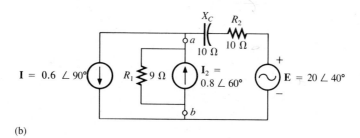

(b)

FIG. 18.87

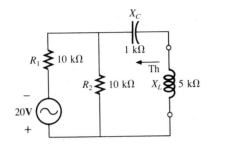

FIG. 18.88

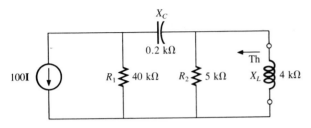

FIG. 18.89

11. Determine the Thevenin equivalent circuit for the network external to the 5-kΩ inductive reactance of Fig. 18.88 (in terms of **V**).

12. Determine the Thevenin equivalent circuit for the network external to the 4-kΩ inductive reactance of Fig. 18.89 (in terms of **I**).

13. Find the Thevenin equivalent circuit for the network external to the 10-kΩ inductive reactance of Fig. 18.80.

14. Determine the Thevenin equivalent circuit for the network external to the 4-kΩ resistor of Fig. 18.81.

*15. Find the Thevenin equivalent circuit for the network external to the 5-kΩ inductive reactance of Fig. 18.82.

*16. Determine the Thevenin equivalent circuit for the network external to the 2-kΩ resistor of Fig. 18.83.

*17. Find the Thevenin equivalent circuit for the network external to the resistor R_1 of Fig. 18.84.

Section 18.4

18. Find the Norton equivalent circuit for the portions of the networks of Fig. 18.85 external to the elements between points a and b.

19. Repeat Problem 18 for the networks of Fig. 18.86.

*20. Repeat Problem 18 for the networks of Fig. 18.87.

*21. Repeat Problem 18 for the networks of Fig. 18.90.

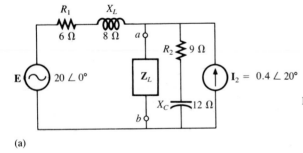

(a)

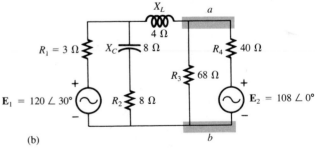

(b)

FIG. 18.90

22. Determine the Norton equivalent circuit for the network external to the 5-kΩ inductive reactance of Fig. 18.88.

23. Determine the Norton equivalent circuit for the network external to the 4-kΩ inductive reactance of Fig. 18.89.

24. Find the Norton equivalent circuit for the network external to the 4-kΩ resistor of Fig. 18.81.

*25. Find the Norton equivalent circuit for the network external to the 5-kΩ inductive reactance of Fig. 18.82.

*26. For the network of Fig. 18.91, find the Norton equivalent circuit for the network external to the 2-kΩ resistor.

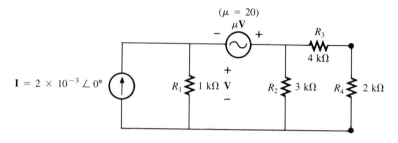

FIG. 18.91

*27. Find the Norton equivalent circuit for the network external to the $\mathbf{I}_1$ current source of Fig. 18.84.

Section 18.5

28. Find the load impedance $\mathbf{Z}_L$ for the networks of Fig. 18.92 for maximum power to the load, and find the maximum power to the load.

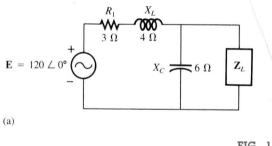
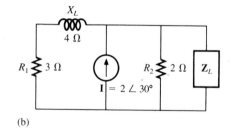

(a)   (b)

FIG. 18.92

*29. Repeat Problem 28 for the networks of Fig. 18.93.

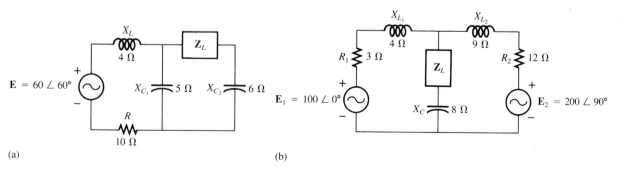

(a)   (b)

FIG. 18.93

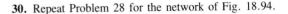

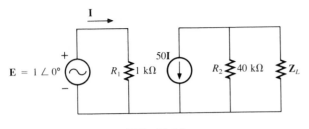

FIG. 18.94

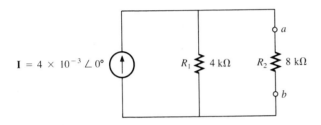

FIG. 18.95

30. Repeat Problem 28 for the network of Fig. 18.94.

Section 18.6

31. For the network of Fig. 18.95, determine two equivalent branches through the substitution theorem for the branch a-b.

32. a. For the network of Fig. 18.96(a), find the current **I**.
    b. Repeat part (a) for the network of Fig. 18.96(b).
    c. Do the results of parts (a) and (b) compare?

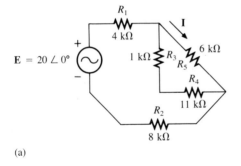

(a)

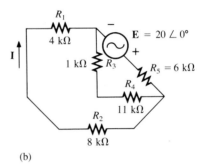

(b)

FIG. 18.96

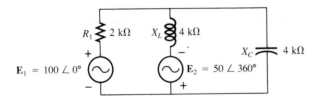

FIG. 18.97

33. Using Millman's theorem, determine the current through the 4-kΩ capacitive reactance of Fig. 18.97.

Computer Problems

34. Given the network of Fig. 18.1, write a program to determine a general solution for the current **I** using superposition. That is, given the reactance of the same network elements, determine **I** for voltage sources of any magnitude but the same angle.

35. Given the network of Fig. 18.18, write a program to determine the Thevenin voltage and impedance for any level of reactance for each element and any magnitude of voltage for the voltage source. The angle of the voltage source should remain at zero degrees.

36. Given the configuration of Fig. 18.98, demonstrate that
maximum power is delivered to the load when $X_C = X_L$
by tabulating the power to the load for X_C varying from
0.1 kΩ to 2 kΩ in increments of 0.1 kΩ.

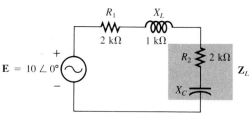

FIG. 18.98

GLOSSARY

Maximum power transfer theorem A theorem used to determine the load impedance necessary to insure maximum
power to the load.

Millman's theorem A method employing voltage-to-current
source conversions which will permit the determination of
unknown variables in a multiloop network.

Norton's theorem A theorem that permits the reduction of
any two-terminal linear ac network to one having a single
current source and parallel impedance. The resulting configuration can then be employed to determine a particular
current or voltage in the original network or to examine the
effects of a specific portion of the network on a particular
variable.

Reciprocity theorem A theorem stating that for single-source
networks, the magnitude of the current in any branch of a
network, due to a single voltage source anywhere else in the
network, will equal the magnitude of the current through the
branch in which the source was originally located if the
source is placed in the branch in which the current was
originally measured.

Substitution theorem A theorem stating that if the voltage
across and current through any branch of an ac bilateral
network are known, the branch can be replaced by any combination of elements that will maintain the same voltage
across and current through the chosen branch.

Superposition theorem A method of network analysis that
permits considering the effects of each source independently. The resulting current and/or voltage is the phasor
sum of the currents and/or voltages developed by each
source independently.

Thevenin's theorem A theorem that permits the reduction of
any two-terminal linear ac network to one having a single
voltage source and series impedance. The resulting configuration can then be employed to determine a particular current or voltage in the original network or to examine the
effects of a specific portion of the network on a particular
variable.

19
Power (ac)

P q
S

19.1 INTRODUCTION

The discussion of power in Chapter 14 included only the average power delivered to an ac network. We will now examine the total power equation in a slightly different form and introduce two additional types of power: *apparent* and *reactive*.

Let us define, for the configuration of Fig. 19.1,

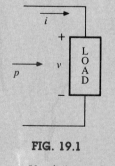

FIG. 19.1

$$v = V_m \sin(\omega t + \theta)$$

and

$$i = I_m \sin \omega t$$

where θ is the phase angle by which v leads i, and since

$$\mathbf{Z} = \frac{V \angle \theta}{I \angle 0°} = \frac{V}{I} \angle \theta = Z \angle \theta$$

it is also the angle associated with the total impedance of the load of Fig. 19.1.

The power delivered to the load of Fig. 19.1 at any instant of time is determined by

$$p = vi$$

Substituting into the above, we have

$$p = V_m I_m \sin \omega t \sin(\omega t + \theta)$$

If we now apply a number of trigonometric identities, we will find that the power equation can also be written

$$p = VI \cos \theta(1 - \cos 2\omega t) + VI \sin \theta(\sin 2\omega t) \qquad \textbf{(19.1)}$$

where V and I are effective values.

It would appear initially that nothing has been gained by putting the equation in this form. However, the usefulness of the form of Eq. (19.1) will be demonstrated in the following sections. The derivation of Eq. (19.1) from the initial form will appear as an assignment at the end of the chapter.

If Eq. (19.1) is expanded to the form

$$p = \underbrace{VI \cos \theta}_{\text{Average}} - \underbrace{VI \cos \theta}_{\text{Peak}} \underbrace{\cos 2\omega t}_{2x} + \underbrace{VI \sin \theta}_{\text{Peak}} \underbrace{\sin 2\omega t}_{2x}$$

there are two obvious points that can be made. First, the average power still appears as an isolated term that is time independent. Second, both terms that follow vary at a frequency twice that of the applied voltage or current with peak values having a very similar format.

In an effort to insure completeness and order in presentation, each basic element (R, L, and C) will first be treated separately.

19.2 RESISTIVE CIRCUIT

For a purely resistive circuit (such as that in Fig. 19.2), v and i are in phase, and $\theta = 0°$. Substituting $\theta = 0°$ into Eq. (19.1), we obtain

$$p_R = VI \cos(0°)(1 - \cos 2\omega t) + VI \sin(0°) \sin 2\omega t$$
$$= VI(1 - \cos 2\omega t) + 0$$

or

$$p_R = VI - VI \cos 2\omega t \qquad \textbf{(19.2)}$$

where VI is the average or dc term and $-VI \cos 2\omega t$ is a negative cosine wave with twice the frequency of either input quantity (v or i) and a peak value of VI.

Plotting the waveform for p_R (Fig. 19.3), we see that

$$T_1 = \text{period of input quantities}$$
$$T_2 = \text{period of power curve } p_R$$

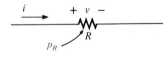

FIG. 19.2

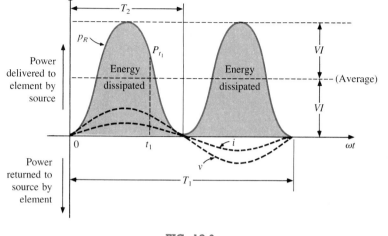

FIG. 19.3

Note that in Fig. 19.3 the power curve passes through two cycles about its average value of VI for each cycle of either v or i ($T_1 = 2T_2$ or $f_2 = 2f_1$). Consider also that since the peak and average values of the power curve are the same, the curve is always above the horizontal axis. This indicates that *the total power delivered to a resistor will be dissipated*. The power returned to the source is represented by the portion of the curve below the axis, which in this case is zero. The power dissipated at any instant of time t_1 by the resistor can be found by substituting t_1 into Eq. (19.2). The average power from Eq. (19.2), or Fig. 19.3, is VI; or, as a summary,

$$P = VI = \frac{V_m I_m}{2} = I^2 R = \frac{V^2}{R} \qquad \text{(watts, W)} \qquad \textbf{(19.3)}$$

as derived in Chapter 14.

The energy dissipated by the resistor (W_R) *over one full cycle of the power curve* (Fig. 19.3) can be found using the following equation:

$$W_R = \int_0^{T_2} p_R \, dt \qquad \text{(joules, J)}$$

$$= \text{area under the power curve from } 0 \text{ to } T_2 \text{ (period of } p_R)$$

$$\textbf{(19.4)}$$

The area under the curve = (average value) × (length of the curve), and

$$W_R = (VI) \times (T_2)$$

or

$$\boxed{W_R = VIT_2} \qquad \text{(J)} \qquad \textbf{(19.5)}$$

or, since $T_2 = 1/f_2$, where f_2 is the frequency of the p_R curve,

$$W_R = \frac{VI}{f_2} \qquad \text{(J)} \qquad \textbf{(19.6)}$$

19.3 APPARENT POWER

FIG. 19.4

From our analysis of dc networks (and resistive elements above), it would seem *apparent* that the power delivered to the load of Fig. 19.4 is simply determined by the product of the applied voltage and current, with no concern for the components of the load. That is, $P = VI$. However, we found in Chapter 14 that the power factor (cos θ) of the load will have a pronounced effect on the power dissipated, less pronounced for more reactive loads. Although the product of the voltage and current is not always the power delivered, it is a power rating of significant usefulness in the description and analysis of sinusoidal ac networks and in the maximum rating of a number of electrical components and systems. It is called the *apparent power* and is represented symbolically by S.* Since it is simply the product of voltage and current, its units are *volt-amperes,* for which the abbreviation is VA. Its magnitude is determined by

$$S = VI \qquad \text{(VA)} \qquad \textbf{(19.7)}$$

or, since

$$V = IZ \quad \text{and} \quad I = \frac{V}{Z}$$

then

$$S = I^2Z \qquad \text{(VA)} \qquad \textbf{(19.8)}$$

and

$$S = \frac{V^2}{Z} \qquad \text{(VA)} \qquad \textbf{(19.9)}$$

The average power to a system is defined by the equation

$$P = VI \cos \theta$$

However,

$$S = VI$$

---

*Prior to 1968, the symbol for apparent power was the more descriptive P_a.

Therefore,

$$\boxed{P = S \cos \theta} \qquad \text{(W)} \qquad\qquad \textbf{(19.10)}$$

or the power factor of a system F_p is

$$\boxed{F_p = \cos \theta = \frac{P}{S}} \qquad\qquad \textbf{(19.11)}$$

The power factor of a circuit, therefore, is the ratio of the average power to the apparent power. For a purely resistive circuit, we have

$$P = VI = S$$

and

$$F_p = \cos \theta = \frac{P}{S} = 1$$

In general, electric equipment is rated in volt-amperes (VA) or in kilovolt-amperes (kVA), not in watts. By knowing the volt-ampere rating and the rated voltage of a device, we can readily determine the *maximum* current rating. For example, a device rated at 10 kVA at 200 V has a maximum current rating of $I = 10,000/200 = 50$ A when operated under rated conditions. The volt-ampere rating of a piece of equipment is equal to the wattage rating only when the F_p is 1. It is therefore a maximum power dissipation rating. This condition exists only when the total impedance of a system $Z \angle \theta$ is such that $\theta = 0°$.

The exact current demand of a device, when used under normal operating conditions, could be determined if the wattage rating and power factor were given instead of the volt-ampere rating. However, the power factor is sometimes not available, or it may vary with the load.

The reason for rating electrical equipment in kilovolt-amperes rather than in kilowatts is obvious from the configuration of Fig. 19.5. The

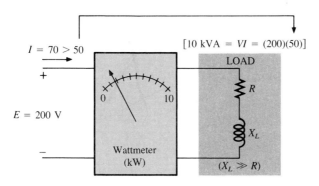

FIG. 19.5

load has an apparent power rating of 10 kVA and a current rating of 50 A at the applied voltage, 200 V. As indicated, the current demand is above the rated value and could damage the load element, yet the reading on the wattmeter is very low since the load is highly reactive. In other words, the wattmeter reading is an indication not of the current drawn but simply of the watts dissipated. Theoretically, if the load were purely reactive, the wattmeter reading could be zero, and the device burning up due to the high current demand.

19.4 INDUCTIVE CIRCUIT AND REACTIVE POWER

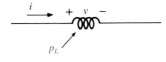

FIG. 19.6

For a purely inductive circuit (such as that in Fig. 19.6), v leads i by 90°. Therefore, in Eq. (19.1), $\theta = 90°$. Substituting $\theta = 90°$ into Eq. (19.1) yields

$$p_L = VI \cos(90°)(1 - \cos 2\omega t) + VI \sin(90°)(\sin 2\omega t)$$
$$= 0 + VI \sin 2\omega t$$

or

$$\boxed{p_L = VI \sin 2\omega t} \qquad \textbf{(19.12)}$$

where $VI \sin 2\omega t$ is a sine wave with twice the frequency of either input quantity (v or i) and a peak value of VI. Note the absence of an average or constant term in the equation.

Plotting the waveform for p_L (Fig. 19.7), we obtain

$$T_1 = \text{period of either input quantity}$$
$$T_2 = \text{period of } p_L \text{ curve}$$

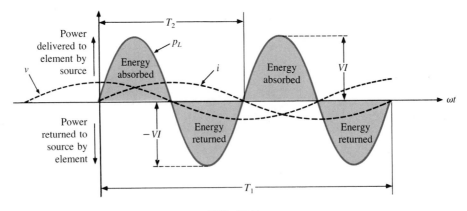

FIG. 19.7

Note that over one full cycle of p_L (T_2), the area above the horizontal axis in Fig. 19.7 is exactly equal to that below the axis. This indicates that over a full cycle of p_L, the power delivered by the source to the inductor is exactly equal to that returned to the source by the inductor. *The net flow of power to the pure inductor is therefore zero over a full cycle, and no energy is lost in the transaction.* The power absorbed or returned by the inductor at any instant of time t_1 can be found simply by substituting t_1 into Eq. (19.12). The peak value of the curve *VI* is defined as the *reactive power* associated with a pure inductor.

In general, the reactive power associated with any circuit is defined to be *VI* sin θ, a factor appearing in the second term of Eq. (19.1). Note that it is the peak value of that term of the total power equation that produces no net transfer of energy. The symbol for reactive power is Q, and its unit of measure is the *volt-ampere reactive* (VAR).* The Q is derived from the quadrature (90°) relationship between the various powers to be discussed in detail in a later section. Therefore,

$$\boxed{Q = VI \sin \theta} \qquad \text{(VAR)} \qquad \textbf{(19.13)}$$

where θ is the phase angle between V and I.

For the inductor,

$$\boxed{Q_L = VI} \qquad \text{(VAR)} \qquad \textbf{(19.14)}$$

or, since $V = IX_L$ or $I = V/X_L$,

$$\boxed{Q_L = I^2 X_L} \qquad \text{(VAR)} \qquad \textbf{(19.15)}$$

or

$$\boxed{Q_L = \frac{V^2}{X_L}} \qquad \text{(VAR)} \qquad \textbf{(19.16)}$$

The apparent power associated with an inductor is $S = VI$, and the average power is $P = 0$, as noted in Fig. 19.7. The power factor is therefore

$$F_p = \cos \theta = \frac{P}{S} = \frac{0}{VI} = 0$$

If the average power is zero, and the energy supplied is returned within one cycle, why is reactive power of any significance? The reason is not obvious but can be explained using the curve of Fig. 19.7. At every instant of time along the power curve that the curve is above the axis (positive), energy must be supplied to the inductor even though it will be returned during the negative portion of the cycle. This power requirement during the positive portion of the cycle requires that the generating plant provide this energy during that interval. Therefore, the effect of reactive elements such as the inductor can be to raise the power

---

*Prior to 1968, the symbol for reactive power was the more descriptive P_q.

requirement of the generating plant even though the reactive power is not dissipated but simply "borrowed." The increased power demand during these intervals is a cost factor that must be passed on to the industrial consumer. In fact, most larger users of electrical energy pay for the apparent power demand rather than the watts dissipated since the volt-amperes used are sensitive to the reactive power requirement (see Section 19.6). In other words, the closer the power factor of an industrial outfit is to 1, the more efficient is the plant's operation, since it is limiting its use of "borrowed" power.

The energy stored by the inductor during the positive portion of the cycle (Fig. 19.7) is equal to that returned during the negative portion and can be determined using the integral

$$W_L = \int_{T_2/2}^{T_2} p_L \, dt \quad \text{(J)}$$

$$= \text{area under the power curve from } T_2/2 \text{ to } T_2$$

Recall from Chapter 14 that the average value of the positive portion of a sinusoid equals 2(peak value/π), and that the area under any curve equals (average value) $\times$ (length of the curve):

$$W_L = \left(\frac{2VI}{\pi}\right) \times \left(\frac{T_2}{2}\right)$$

or

$$W_L = \frac{VIT_2}{\pi} \quad \text{(J)} \qquad \textbf{(19.17)}$$

or, since $T_2 = 1/f_2$, where f_2 is the frequency of the p_L curve, we have

$$W_L = \frac{VI}{\pi f_2} \quad \text{(J)} \qquad \textbf{(19.18)}$$

Since the frequency f_2 of the power curve is twice that of the input quantity, if we substitute the frequency f_1 of the input voltage or current, Eq. (19.18) becomes

$$W_L = \frac{VI}{\pi(2f_1)}$$

or

$$W_L = \frac{VI}{\omega_1} \quad \text{(J)} \qquad \textbf{(19.19)}$$

However,

$$V = IX_L = I\omega_1 L$$

so

$$W_L = \frac{(I\omega_1 L)I}{\omega_1}$$

or

$$\boxed{W_L = LI^2} \qquad \text{(J)} \qquad\qquad \textbf{(19.20)}$$

19.5 CAPACITIVE CIRCUIT

For a purely capacitive circuit (such as that in Fig. 19.8), i leads v by 90°. Therefore, in Eq. (19.1), $\theta = -90°$. Substituting $\theta = -90°$ into Eq. (19.1), we obtain

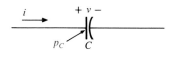

FIG. 19.8

$$p_C = VI\cos(-90°)(1 - \cos 2\omega t) + VI\sin(-90°)(\sin 2\omega t)$$
$$= 0 - VI\sin 2\omega t$$

or

$$\boxed{p_C = -VI\sin 2\omega t} \qquad\qquad \textbf{(19.21)}$$

where $-VI\sin 2\omega t$ is a negative sine wave with twice the frequency of either input (v or i) and a peak value of VI. Again, note the absence of an average or constant term.

Plotting the waveform for p_C (Fig. 19.9) gives us

$$T_1 = \text{period of either input quantity}$$
$$T_2 = \text{period of } p_C \text{ curve}$$

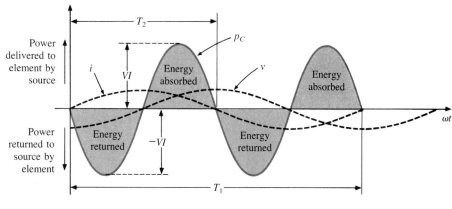

FIG. 19.9

Note that the same situation exists here for the p_C curve as existed for the p_L curve. The power delivered by the source to the capacitor is exactly equal to that returned to the source by the capacitor over one full cycle. *The net flow of power to the pure capacitor is therefore zero over a full cycle, and no energy is lost in the transaction.* The power absorbed or returned by the capacitor at any instant of time t_1 can be found by substituting t_1 into Eq. (19.21).

The reactive power associated with the capacitor is equal to the peak value of the p_C curve, as follows:

$$Q_C = VI \qquad \text{(VAR)} \qquad \textbf{(19.22)}$$

but, since $V = IX_C$ and $I = V/X_C$, the reactive power to the capacitor can also be written

$$Q_C = I^2 X_C \qquad \text{(VAR)} \qquad \textbf{(19.23)}$$

and

$$Q_C = \frac{V^2}{X_C} \qquad \text{(VAR)} \qquad \textbf{(19.24)}$$

The apparent power associated with the capacitor is

$$S = VI \qquad \text{(VA)} \qquad \textbf{(19.25)}$$

and the average power is $P = 0$, as noted from Eq. (19.21) or Fig. 19.9. The power factor is therefore

$$F_p = \cos \theta = \frac{P}{S} = \frac{0}{VI} = 0$$

The energy stored by the capacitor during the positive portion of the cycle (Fig. 19.9) is equal to that returned during the negative portion and can be determined using the integral

$$W_C = \int_0^{T_2/2} p_C \, dt \qquad \text{(J)}$$

$$= \text{area under the power curve from 0 to } T_2/2$$

Proceeding in a manner similar to that used for the inductor, we can show that

$$W_C = \frac{VIT_2}{\pi} \qquad \text{(J)} \qquad \textbf{(19.26)}$$

or, since $T_2 = 1/f_2$, where f_2 is the frequency of the p_C curve,

$$W_C = \frac{VI}{\pi f_2} \qquad \text{(J)} \qquad \textbf{(19.27)}$$

In terms of the frequency f_1 of the input quantities v and i,

$$W_C = \frac{VI}{\omega_1} \qquad \text{(J)}$$

(19.28)

and

$$W_C = CV^2 \qquad \text{(J)}$$

(19.29)

19.6 THE POWER TRIANGLE

The three quantities—average power, apparent power, and reactive power—can be related in the vector domain by

$$\boxed{S = P + Q}$$

(19.30)

with

$$\mathbf{P} = P \angle 0° \qquad \mathbf{Q}_L = Q_L \angle 90° \qquad \mathbf{Q}_C = Q_C \angle -90°$$

For an inductive load, the *phasor power* **S,** as it is often called, is defined by

$$\mathbf{S} = P + jQ_L$$

as shown in Fig. 19.10.

The 90° shift in Q_L from P is the source of another term for reactive power: *quadrature power*.

For a capacitive load, the phasor power **S** is defined by

$$\mathbf{S} = P - jQ_C$$

as shown in Fig. 19.11.

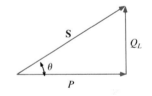

FIG. 19.10

Power diagram for inductive loads.

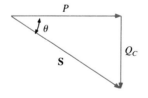

FIG. 19.11

Power diagram for capacitive loads.

If a network has both capacitive and inductive elements, the reactive component of the power triangle will be determined by the *difference* between the reactive power delivered to each. If $Q_L > Q_C$, the resultant power triangle will be similar to Fig. 19.10. If $Q_C > Q_L$, the resultant power triangle will be similar to Fig. 19.11.

That the total reactive power is the difference between the reactive powers of the inductive and capacitive elements can be demonstrated by considering Eqs. (19.12) and (19.21). Using these equations, the reactive power delivered to each reactive element has been plotted for a series *L-C* circuit on the same set of axes in Fig. 19.12. The reactive elements were chosen such that $X_L > X_C$. Note that the power curve for each is exactly 180° out of phase. The curve for the resultant reactive power is therefore determined by the algebraic resultant of the two at each instant of time. Since the reactive power is defined as the peak value, the reactive component of the power triangle is as indicated in the figure: $I^2(X_L - X_C)$.

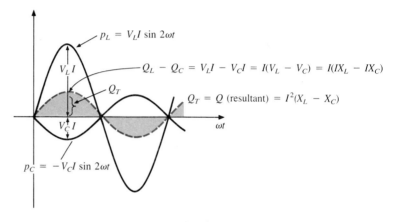

FIG. 19.12

An additional verification can be derived by first considering the impedance diagram of a series *R-L-C* circuit (Fig. 19.13). If we multiply each radius vector by the current squared (I^2), we obtain the results shown in Fig. 19.14, which is the power triangle for a predominantly inductive circuit.

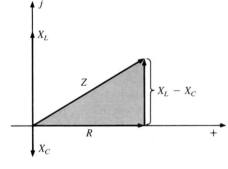

FIG. 19.13

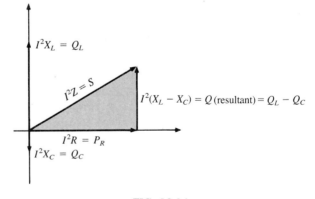

FIG. 19.14

Since the reactive power and average power are always angled 90° to each other, the three powers are related by the Pythagorean theorem; that is,

$$S^2 = P^2 + Q^2$$

(19.31)

Therefore, the third power can always be found if the other two are known.

It is particularly interesting that the equation

$$\boxed{\mathbf{S} = \mathbf{VI}*}$$ **(19.32)**

will provide the vector form of the apparent power of a system. Here, $\mathbf{V}$ is the voltage across the system and $\mathbf{I}*$ is the complex conjugate of the current.

Consider, for example, the simple R-L circuit of Fig. 19.15, where

$$\mathbf{I} = \frac{\mathbf{V}}{\mathbf{Z}_T} = \frac{10\ \angle 0°}{3 + j4} = \frac{10\ \angle 0°}{5\ \angle 53.13°} = 2\ \angle -53.13°$$

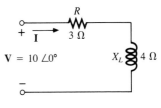

FIG. 19.15

The real power (the term *real* being derived from the positive real axis of the complex plane) is

$$P = I^2 R = (2)^2(3) = 12\ \text{W}$$

and the reactive power is

$$Q_L = I^2 X_L = (2)^2(4) = 16\ \text{VAR}$$

with

$$\mathbf{S} = P + jQ_L = 12 + j16 = 20\ \angle 53.13°$$

as shown in Fig. 19.16. Applying Eq. (19.32) yields

$$\mathbf{S} = \mathbf{VI}* = (10\ \angle 0°)(2\ \angle +53.13°) = 20\ \angle 53.13°$$

as obtained above.

The angle θ associated with $\mathbf{S}$ and appearing in Figs. 19.10, 19.11, and 19.16 is the power-factor angle of the network. Since

$$P = VI \cos \theta$$

or

$$P = S \cos \theta$$

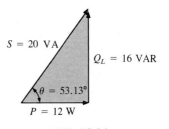

FIG. 19.16

then

$$\boxed{F_p = \cos \theta = \frac{P}{S}}$$ **(19.33)**

19.7 THE TOTAL P, Q, AND S

The total number of watts, volt-amperes reactive, volt-amperes, and the power factor of any system can be found using the following procedure:

1. Find the real power and reactive power for each branch of the circuit.

2. The total real power of the system (P_T) is then the sum of the average power delivered to each branch.
3. The total reactive power (Q_T) is the difference between the reactive power of the inductive loads and that of the capacitive loads.
4. The total apparent power is $S_T = \sqrt{P_T^2 + Q_T^2}$.
5. The total power factor is P_T/S_T.

There are two important points in the above tabulation. First, the total apparent power must be determined from the total average and reactive powers and *cannot* be determined from the apparent powers of each branch. Second, and more important, it is *not necessary* to consider the series-parallel arrangement of branches. In other words, the total real, reactive, or apparent power is independent of whether the loads are in series, parallel, or series-parallel. The following examples will demonstrate the relative ease with which all the quantities of interest can be found.

EXAMPLE 19.1. Find the total number of watts, volt-amperes reactive, volt-amperes, and the power factor F_p of the network in Fig. 19.17. Draw the power triangle and find the current in phasor form.

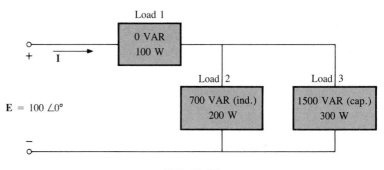

FIG. 19.17

Solution: Use a table:

Load	W	VAR	VA
1	100	0	100
2	200	700 (ind.)	$\sqrt{(200)^2 + (700)^2} = 728.0$
3	300	1500 (cap.)	$\sqrt{(300)^2 + (1500)^2} = 1529.71$
	$P_T = \mathbf{600}$	$Q_T = \mathbf{800}$ **(cap.)**	$S_T = \sqrt{(600)^2 + (800)^2} = \mathbf{1000}$
	Total power dissipated	Resultant reactive power of network	(Note that $S_T \neq$ sum of each branch: $1000 \neq 100 + 728 + 1529.71$)

$$F_p = \frac{P_T}{S_T} = \frac{600}{1000} = \mathbf{0.6 \ leading \ (cap.)}$$

The power triangle is shown in Fig. 19.18.

Since $S_T = VI = 1000$, $I = 1000/100 = 10$ A; and since θ of $\cos \theta = F_p$ is the angle between the input voltage and current,

$$\mathbf{I} = 10 \angle +53.13°$$

The plus sign is associated with the phase angle since the circuit is predominantly capacitive.

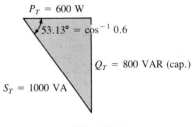

$P_T = 600$ W

$53.13° = \cos^{-1} 0.6$

$Q_T = 800$ VAR (cap.)

$S_T = 1000$ VA

FIG. 19.18

EXAMPLE 19.2.
a. Find the total number of watts, volt-amperes reactive, volt-amperes, and the power factor F_p for the network of Fig. 19.19.

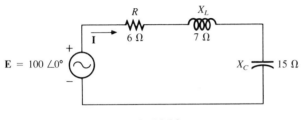

R

X_L

$6\ \Omega$ $7\ \Omega$

$\mathbf{I}$

$+$

$\mathbf{E} = 100 \angle 0°$

$X_C \rightleftharpoons 15\ \Omega$

$-$

FIG. 19.19

b. Sketch the power triangle.
c. Find the energy dissipated by the resistor over one full cycle of the input voltage if the frequency of the input quantities is 60 Hz.
d. Find the energy stored in, or returned by, the capacitor or inductor over 1/2 cycle of the power curve for each if the frequency of the input quantities is 60 Hz.

Solutions:

a. $\mathbf{I} = \dfrac{\mathbf{E}}{\mathbf{Z}_T} = \dfrac{100 \angle 0°}{6 + j7 - j15} = \dfrac{100 \angle 0°}{10 \angle -53.13°} = 10 \angle 53.13°$

$\mathbf{V}_R = \mathbf{IR} = (10 \angle 53.13°)(6 \angle 0°) = 60 \angle 53.13°$
$\mathbf{V}_L = \mathbf{IX}_L = (10 \angle 53.13°)(7 \angle 90°) = 70 \angle 143.13°$
$\mathbf{V}_C = \mathbf{IX}_C = (10 \angle 53.13°)(15 \angle -90°) = 150 \angle -36.87°$

$P_T = EI \cos \theta = (100)(10) \cos 53.13° = \mathbf{600\ W}$

$\quad = I^2R = (10)^2(6) = \mathbf{600\ W}$

$\quad = \dfrac{V_R^2}{R} = \dfrac{(60)^2}{6} = \mathbf{600\ W}$

$S_T = EI = (100)(10) = \mathbf{1000\ VA}$

$\quad = I^2Z_T = (10)^2(10) = \mathbf{1000\ VA}$

$\quad = \dfrac{E^2}{Z_T} = \dfrac{(100)^2}{10} = \mathbf{1000\ VA}$

$Q_T = EI \sin \theta = (100)(10) \sin 53.13° = \mathbf{800\ VAR}$

$\quad = Q_C - Q_L$

$$= I^2(X_C - X_L) = (100)(15 - 7) = \textbf{800 VAR}$$

$$Q_T = \frac{V_C^2}{X_C} - \frac{V_L^2}{X_L} = \frac{(150)^2}{15} - \frac{(70)^2}{7} = 1500 - 700 = \textbf{800 VAR}$$

$$F_p = \frac{P_T}{S_T} = \frac{600}{1000} = \textbf{0.6 leading (cap.)}$$

b. The power triangle is as shown in Fig. 19.20.

c. $W_R = 2\left(\dfrac{V_R I}{f_2}\right) = 2\left(\dfrac{V_R I}{2f_1}\right) = \dfrac{V_R I}{f_1} = \dfrac{(60)(10)}{60} = \textbf{10 J}$

d. $W_L = \dfrac{V_L I}{2\pi f_1} = \dfrac{(70)(10)}{(6.28)(60)} = \dfrac{700}{377} = \textbf{1.86 J}$

$W_C = \dfrac{V_C I}{2\pi f_1} = \dfrac{(150)(10)}{377} = \dfrac{1500}{377} = \textbf{3.98 J}$

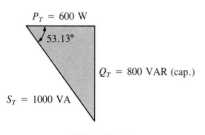

$P_T = 600$ W

53.13°

$Q_T = 800$ VAR (cap.)

$S_T = 1000$ VA

FIG. 19.20

EXAMPLE 19.3. For the network of Fig. 19.21:

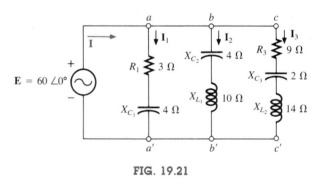

FIG. 19.21

a. Find the average power, apparent power, reactive power, and F_p for each branch.
b. Find the total number of watts, volt-amperes reactive, volt-amperes, and the power factor F_p of the circuit. Sketch the power triangle.
c. Find the current **I.**

Solutions:

a. For branch $a\text{-}a'$:

$$\mathbf{I}_1 = \frac{\mathbf{E}}{\mathbf{Z}_{a\text{-}a'}} = \frac{60 \angle 0°}{3 - j4} = \frac{60 \angle 0°}{5 \angle 53.13°} = 12 \angle +53.13°$$

$$P = I_1^2 R_1 = (12)^2(3) = (144)(3) = \textbf{432 W}$$

$$S = EI_1 = (60)(12) = \textbf{720 VA}$$

$$Q = I_1^2 X_{C_1} = (12)^2(4) = \textbf{576 VAR}$$

$$F_p = \frac{P}{S} = \frac{432}{720} = \textbf{0.6 leading (cap.)}$$

For branch b-b':

$$I_2 = \frac{E}{Z_{b-b'}} = \frac{60\angle 0°}{j10 - j4} = \frac{60\angle 0°}{6\angle 90°} = 10\angle -90°$$
$$P = I_2^2 R = I_2^2(0) = \textbf{0 W}$$
$$S = EI_2 = (60)(10) = \textbf{600 VA}$$
$$Q_T = I_2^2(X_{L_1} - X_{C_2}) = (10)^2(6) = \textbf{600 VAR}$$
$$F_p = \frac{P}{S} = \frac{0}{600} = \textbf{0}$$

For branch c-c':

$$I_3 = \frac{E}{Z_{c-c'}} = \frac{60\angle 0°}{9 - j2 + j14} = \frac{60\angle 0°}{9 + j12} = \frac{60\angle 0°}{15\angle 53.13°}$$
$$= 4\angle -53.13°$$
$$P = I_3^2 R_3 = (4)^2(9) = \textbf{144 W}$$
$$S = EI_3 = (60)(4) = \textbf{240 VA}$$
$$Q = I_3^2(X_{L_2} - X_{C_3}) = (4)^2(12) = \textbf{192 VAR}$$
$$F_p = \cos\theta = \frac{P}{S} = \frac{144}{240} = \textbf{0.6 lagging (ind.)}$$

b. The total system can now be represented as shown in Fig. 19.22.

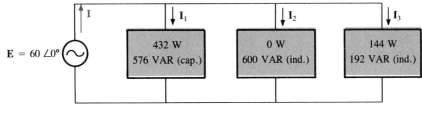

FIG. 19.22

Using a table,
$$S_T = \sqrt{(576)^2 + (216)^2} = \textbf{615.17 VA}$$

Branch	W	VAR
a-a'	432	576 (cap.)
b-b'	0	600 (ind.)
c-c'	144	192 (ind.)
	$P_T = \textbf{576 W}$	$Q_T = \textbf{216 (ind.)}$

$$F_{p(T)} = \cos\theta = \frac{P_T}{S_T} = \frac{576}{615.17} = \textbf{0.9363 lagging (ind.)}$$
$$\theta = \cos^{-1} 0.9363 = 20.56°$$

The power triangle is shown in Fig. 19.23.

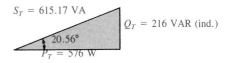

FIG. 19.23

c. $S_T = EI = 615.17$ VA. Therefore,

$$I = \frac{615}{60} = 10.25 \text{ A}$$

The circuit is inductive. Thus, **I** lags **E** by 20°, and

$$\mathbf{I} = \mathbf{10.25} \angle \mathbf{-20.56°}$$

EXAMPLE 19.4. An electrical device is rated 5 kVA, 100 V at a 0.6 power-factor lag. What is the impedance of the device in rectangular coordinates?

Solution:

$$S = EI = 5000 \text{ VA}$$

Therefore,

$$I = \frac{5000}{100} = 50 \text{ A}$$

For $F_p = 0.6$, we have

$$\theta = \cos^{-1} 0.6 = 53.13°$$

Since the power factor is lagging, the circuit is predominantly inductive, and **I** lags **E.** Or, for $\mathbf{E} = 100 \angle 0°$,

$$\mathbf{I} = 50 \angle -53.13°$$

However,

$$\mathbf{Z}_T = \frac{\mathbf{E}}{\mathbf{I}} = \frac{100 \angle 0°}{50 \angle -53.13°} = 2 \angle 53.13° = \mathbf{1.2 + j1.6}$$

which is the impedance of the circuit of Fig. 19.24.

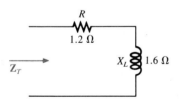

FIG. 19.24

19.8 POWER-FACTOR CORRECTION

The design of any power transmission system is very sensitive to the magnitude of the current in the lines as determined by the applied loads. Increased currents result in increased power losses (by a squared factor since $P = I^2R$) in the transmission lines due to the resistance of the lines. Heavier currents also require larger conductors, increasing the amount of copper needed for the system.

Every effort must therefore be made to keep current levels at a minimum. Since the line voltage of a transmission system is fixed, the apparent power is directly related to the current level. In turn, the smaller the net apparent power, the smaller the current drawn from the supply. Minimum current is therefore drawn from a supply when $S = P$ and $Q_T = 0$. Note the effect of decreasing levels of Q_T on the length (and magnitude) of S in Fig. 19.25 for the same real power. Note also that the power-factor angle approaches zero degrees and F_p approaches 1, revealing that the network is appearing more and more resistive at the input terminals.

The process of introducing reactive elements to bring the power factor closer to unity is called *power-factor correction*. Since most loads are inductive, the process normally involves introducing elements with capacitive terminal characteristics having the sole purpose of improving the power factor.

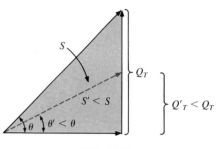

FIG. 19.25

EXAMPLE 19.5. A 5-hp motor with a 0.6 lagging power factor and an efficiency of 92% is connected to a 208-V, 60-Hz supply. What level of capacitance in parallel with the motor will raise the power factor of the combined system to unity?

Solution: First we establish the power triangle for the 5-hp motor. Since 1 hp = 746 W,

$$P_o = 5 \text{ hp} = 5(746) = 3730 \text{ W}$$

and

$$P_i \text{ (drawn from the line)} = \frac{P_o}{\eta} = \frac{3730}{0.92} = 4054.35 \text{ W}$$

Also,

$$F_p = \cos \theta = 0.6$$

and

$$\theta = \cos^{-1} 0.6 = 53.13°$$

Then we use

$$\tan \theta = \frac{Q_L}{P_i}$$

to obtain

$$Q_L = P_i \tan \theta = 4054.35 \tan 53.13°$$
$$= (4054.35)(1.333) = 5404.45 \text{ VAR}$$

and

$$S = \sqrt{P_i^2 + Q_L^2} = \sqrt{(4054.35)^2 + (5404.45)^2} = 6756.17 \text{ VA}$$

The power triangle appears in Fig. 19.26.

A net unity power-factor level is established by introducing a capacitive reactive power level of 5404.45 VAR to balance Q_L. Since

$$Q_C = \frac{V^2}{X_C}$$

then

$$X_C = \frac{V^2}{Q_C}$$

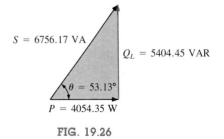

FIG. 19.26

and

$$X_C = \frac{(208)^2}{5404.45} = 8 \, \Omega$$

and

$$C = \frac{1}{2\pi f X_C} \qquad \text{(from } X_C = 1/2\pi fC)$$

$$= \frac{1}{(6.28)(60)(8)} = \mathbf{332 \, \mu F}$$

At $0.6 F_p$,

$$S = VI = 6756.17 \, \text{VA}$$

and

$$I = \frac{S}{V} = \frac{6756.17}{208} = 32.48 \, \text{A}$$

At unity F_p,

$$S = VI = 4054.35 \, \text{VA}$$

and

$$I = \frac{S}{V} = \frac{4054.35}{208} = 19.49 \, \text{A}$$

producing a 40% reduction in supply current.

EXAMPLE 19.6. A small industrial plant has a 10-kW heating load and a 20-kVA inductive load due to a bank of induction motors. The heating elements are considered purely resistive ($F_p = 1$) and the induction motors have a lagging power factor of 0.7. If the supply is 1000 V at 60 Hz, determine the capacitive element required to raise the power factor to 0.95.

Solution: For the induction motors,

$$S = VI = 20 \, \text{kVA}$$
$$P = VI \cos \theta = (20 \times 10^3)(0.7) = 14 \times 10^3 \, \text{W}$$
$$\theta = \cos^{-1} 0.7 \cong 45.6°$$

and

$$Q_L = VI \sin \theta = (20 \times 10^3)(0.714) = 14.28 \times 10^3 \, \text{VAR}$$

The power triangle for the total system appears in Fig. 19.27. Note the addition of real powers and the increased value of S:

$$S_T = \sqrt{(24 \, \text{kW})^2 + (14.28 \, \text{kVAR})^2} = 27.93 \, \text{kVA}$$

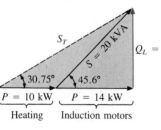

$Q_L = 14.28$ kVAR

FIG. 19.27

with

$$I = \frac{S_T}{V} = \frac{27.93 \, \text{kVA}}{1000} = 27.93 \, \text{A}$$

A power factor of 0.95 results in an angle between S and P of

$$\theta = \cos^{-1} 0.95 = 18.19°$$

changing the power triangle to the following (Fig. 19.28):

$$\tan \theta = \frac{Q'_L}{P_T} \Rightarrow Q'_L = P_T \tan \theta = (24 \times 10^3)(\tan 18.19°)$$

$$= (24 \times 10^3)(0.329) = 7.9 \, \text{kVAR}$$

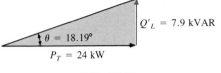

$Q'_L = 7.9$ kVAR

$\theta = 18.19°$

$P_T = 24$ kW

FIG. 19.28

The inductive reactive power must therefore be reduced by

$$Q_L - Q'_L = 14.28 \, \text{kVAR} - 7.9 \, \text{kVAR} = 6.38 \, \text{kVAR}$$

Therefore, $Q_C = 6.38 \, \text{kVAR}$, and using

$$Q_C = \frac{V^2}{X_C}$$

we obtain

$$X_C = \frac{V^2}{Q_C} = \frac{(10^3)^2}{6.38 \times 10^3} = 156.74 \, \Omega$$

and

$$C = \frac{1}{2\pi f X_C} = \frac{1}{(6.28)(60)(156.74)} = \mathbf{16.93 \, \mu F}$$

19.9 THE WATTMETER

The wattmeter, as the name suggests, is an instrument designed to read the power to an element or network. It employs an electrodynamometer-type movement or solid-state electronic system to measure the power in a dc or ac network. It can, in fact, be used to measure the wattage of any circuit with a periodic or nonperiodic input.

In an electrodynamometer movement, a moving coil rotates in a magnetic field produced by the current of a stationary coil. The fluxes of the stationary and movable coils interact to develop a torque on the pointer connected to the movable coil. In the wattmeter configuration (Fig. 19.29), the current in the stationary coils is the line current, while the current in the moving coil is derived from the line voltage. The instrument then indicates power in watts on a linear scale. A typical wattmeter using an electrodynamometer movement appears in Fig. 19.30.

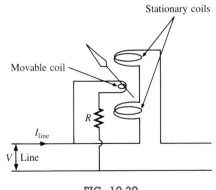

Stationary coils

Movable coil

R

I_{line}

V Line

FIG. 19.29

Potential terminals   Current terminals

FIG. 19.30

*Wattmeter. (Courtesy of Electrical
Instrument Service, Inc.)*

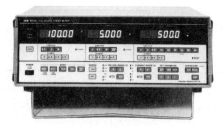

FIG. 19.31

*Digital wattmeter. (Courtesy of Yokogawa
Corporation of America)*

The digital display wattmeter of Fig. 19.31 employs a sophisticated electronic package to sense the voltage and current levels and through the use of an analog-to-digital conversion unit display the proper digits on the display.

For an up-scale deflection, the electrodynamometer wattmeter is connected as shown in Fig. 19.32. Some electrodynamometer wattmeters will always give a wattage reading that is higher than that actually delivered to the load. They are high by the amount of power consumed by the potential coil (V_{pc}^2/R_{pc}). This correction is important and should be considered with every set of data. Many wattmeters are designed to compensate for this correction, and therefore they eliminate the need for any other adjustment in the reading. The wattmeter is always connected with the potential terminals in parallel, and the current terminals in series, with the portion of the network to which the power is being measured.

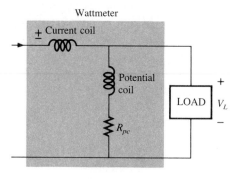

FIG. 19.32

The power delivered to R_1 in Fig. 19.33 can be found by connecting the electrodynamometer wattmeter as shown in Fig. 19.33(a). To find the power delivered to the total network, it should be connected as

shown in Fig. 19.33(b). The connections for the digital meter are fundamentally the same as for the electrodynamometer meter.

When using the electrodynamometer wattmeter, the operator must
take care not to exceed the current, voltage, or wattage rating. The
product of the voltage and current ratings may or may not equal the
wattage rating. In the high-power-factor wattmeter, the product of the
voltage and current ratings is usually equal to the wattage rating, or at
least 80% of it. For a low-power-factor wattmeter, the product of the
current and voltage ratings is much greater than the wattage rating. For
obvious reasons, the low-power-factor meter is used only in circuits
with low power factors (total impedance highly reactive). Typical ratings for high-power-factor (HPF) and low-power-factor (LPF) meters
are shown in Table 19.1. Meters of both high and low power factors
have an accuracy of 0.5% to 1% of full scale.

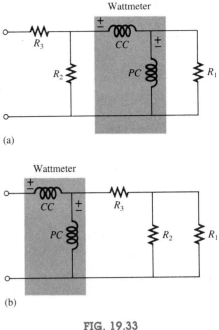

FIG. 19.33

TABLE 19.1

Meter	Current Ratings	Voltage Ratings	Wattage Ratings
HPF	2.5 A	150 V	1500/750/375
	5.0 A	300 V	
LPF	2.5 A	150 V	300/150/75
	5.0 A	300 V	

19.10 EFFECTIVE RESISTANCE

The resistance of a conductor as determined by the equation $R = \rho(l/A)$
is often called the *dc, ohmic,* or *geometric* resistance. It is a constant
quantity determined only by the material used and its physical dimensions. In ac circuits, the actual resistance of a conductor (called the
effective resistance) differs from the dc resistance because of the varying
currents and voltages which introduce effects not present in dc circuits.

These effects include radiation losses, skin effect, eddy currents, and
hysteresis losses. The first two effects apply to any network, while the
latter two are concerned with the additional losses introduced by the
presence of ferromagnetic materials in a changing magnetic field.

The effective resistance of an ac circuit cannot be measured by ratio
V/I, since this is now the impedance of a circuit that may have both
resistance and reactance. The effective resistance can be found, however, by using the power equation $P = I^2R$, where

$$R_{\text{eff}} = \frac{P}{I^2} \qquad \textbf{(19.34)}$$

A wattmeter and an ammeter are therefore necessary for measuring the
effective resistance of an ac circuit.

Let us now examine the various losses in greater detail. The radiation
loss is the loss of energy in the form of electromagnetic waves during

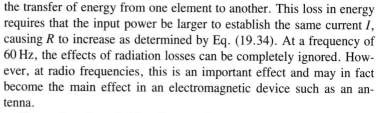
the transfer of energy from one element to another. This loss in energy requires that the input power be larger to establish the same current I, causing R to increase as determined by Eq. (19.34). At a frequency of 60 Hz, the effects of radiation losses can be completely ignored. However, at radio frequencies, this is an important effect and may in fact become the main effect in an electromagnetic device such as an antenna.

The explanation of skin effect requires the use of some basic concepts previously described. It will be remembered from Chapter 11 that a magnetic field exists around every current-carrying conductor (Fig. 19.34). Since the amount of charge flowing in ac circuits changes with time, the magnetic field surrounding the moving charge (current) also changes. Recall also that a wire placed in a changing magnetic field will have an induced voltage across its terminals as determined by Faraday's law $e = N (d\phi/dt)$. The higher the frequency of the changing flux as determined by an alternating current, the greater the induced voltage will be.

For a conductor carrying alternating current, the changing magnetic field surrounding the wire links the wire itself, thus developing within the wire an induced voltage that opposes the original flow of charge or current. These effects are more pronounced at the center of the conductor than at the surface since the center is linked by the changing flux inside the wire as well as that outside the wire. As the frequency of the applied signal increases, the flux linking the wire will change at a greater rate. An increase in frequency will therefore increase the counter induced voltage at the center of the wire to the point where the current will, for all practical purposes, flow on the surface of the conductor. At 60 Hz, the effects of skin effect are almost noticeable. However, at radio frequencies, the skin effect is so pronounced that large conductors are frequently made hollow since the center part is relatively ineffective. The skin effect, therefore, reduces the effective area through which the current can flow, and causes the resistance of the conductor, given by the equation $R\uparrow = \rho(l/A\downarrow)$, to increase.

As mentioned earlier, hysteresis and eddy current losses will appear when a ferromagnetic material is placed in the region of a changing magnetic field. To describe eddy current losses in greater detail, we will consider the effects of an alternating current passing through a coil wrapped around a ferromagnetic core. As the alternating current passes through the coil, it will develop a changing magnetic flux Φ linking both the coil and the core which will develop an induced voltage within the core as determined by Faraday's law. This induced voltage and the geometric resistance of the core $R_C = \rho(l/A)$ cause currents to be developed within the core, $i_{\text{core}} = (e_{\text{ind}}/R_C)$, called *eddy currents*. The currents flow in circular paths, as shown in Fig. 19.35, changing direction with the applied ac potential.

The eddy current losses are determined by

$$P_{\text{eddy}} = i_{\text{eddy}}^2 R_{\text{core}}$$

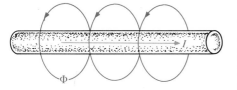

FIG. 19.34

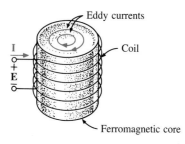

FIG. 19.35

The magnitude of these losses is determined primarily by the type of core used. If the core is nonferromagnetic—and has a high resistivity like wood or air—the eddy current losses can be neglected. In terms of the frequency of the applied signal and the magnetic field strength produced, the eddy current loss is proportional to the square of the frequency times the square of the magnetic field strength:

$$P_{eddy} \propto f^2 B^2$$

Eddy current losses can be reduced if the core is constructed of thin, laminated sheets of ferromagnetic material insulated from one another and aligned parallel to the magnetic flux. Such construction reduces the magnitude of the eddy currents by placing more resistance in their path.

Hysteresis losses were described in Section 11.8. You will recall that in terms of the frequency of the applied signal and the magnetic field strength produced, the hysteresis loss is proportional to the frequency to the 1st power times the magnetic field strength to the nth power:

$$P_{hys} \propto f^1 B^n$$

where n can vary from 1.4 to 2.6, depending on the material under consideration.

Hysteresis losses can be effectively reduced by the injection of small amounts of silicon into the magnetic core, constituting some 2% or 3% of the total composition of the core. This must be done carefully, however, as too much silicon makes the core brittle and difficult to machine into the shape desired.

EXAMPLE 19.7.

a. An air-core coil is connected to a 120-V, 60-Hz source as shown in Fig. 19.36. The current is found to be 5 A, and a wattmeter reading of 75 W is observed. Find the effective resistance and the inductance of the coil.

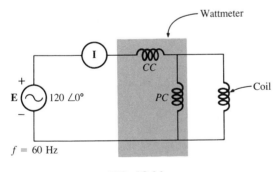

FIG. 19.36

b. A brass core is then inserted in the coil, and the ammeter reads 4 A and the wattmeter reads 80 W. Calculate the effective resistance of the core. To what do you attribute the increase in value over that of part (a)?

c. If a solid iron core is inserted in the coil, the current is found to be 2 A, and the wattmeter reads 52 W. Calculate the resistance and the inductance of the coil. Compare these values to those of part (a), and account for the changes.

Solutions:

a. $R = \dfrac{P}{I^2} = \dfrac{75}{(5)^2} = \mathbf{3\,\Omega}$

$Z_T = \dfrac{E}{I} = \dfrac{120}{5} = 24\,\Omega$

$X_L = \sqrt{Z_T^2 - R^2} = \sqrt{(24)^2 - (3)^2} = 23.81\,\Omega$

and

$R_L = 2\pi f L$

or

$L = \dfrac{X_L}{2\pi f} = \dfrac{23.81}{377} = \mathbf{63.16\,mH}$

b. $R = \dfrac{P}{I^2} = \dfrac{80}{(4)^2} = \dfrac{80}{16} = \mathbf{5\,\Omega}$

The brass core has less reluctance than the air core. Therefore a greater magnetic flux density B will be created in it. Since $P_{eddy} \propto f^2 B^2$, and $P_{hys} \propto f^1 B^n$, as the flux density increases, the core losses and the effective resistance increase.

c. $R = \dfrac{P}{I^2} = \dfrac{52}{(2)^2} = \dfrac{52}{4} = \mathbf{13\,\Omega}$

$Z_T = \dfrac{E}{I} = \dfrac{120}{2} = 60\,\Omega$

$X_L = \sqrt{Z_T^2 - R^2} = \sqrt{(60)^2 - (13)^2} = 58.57\,\Omega$

$L = \dfrac{X_L}{2\pi f} = \dfrac{58.57}{377} = \mathbf{155.36\,mH}$

The iron core has less reluctance than the air or brass cores. Therefore a greater magnetic flux density B will be developed in the core. Again, since $P_{eddy} \propto f^2 B^2$, and $P_{hys} \propto f^1 B^n$, the increased flux density will cause the core losses and the effective resistance to increase.

Since the inductance L is related to the change in flux by the equation $L = N\,(d\phi/di)$, the inductance will be greater for the iron core because the changing flux linking the core will increase.

19.11 COMPUTER ANALYSIS

Program 19.1 of Fig. 19.37 demonstrates how the computer can calculate the total real, reactive, and apparent power, and the power factor and supply current of the system.

```
        10 REM *****   PROGRAM 19-1   *****
        20 REM *******************************************
        30 REM This program calculates the total real,
        40 REM reactive and apparent power of a network
        50 REM with five individual loads.
        60 REM *******************************************
        70 REM
       100 DIM P(5),Q(5),S(5)
       110 PRINT "This program calculates the total real,"
       120 PRINT "reactive and apparent power of a network"
       130 PRINT "with five individual loads."
       140 PRINT
       150 PRINT "Input the following data:"
       160 PRINT "(use negative sign for capacitive vars)"
       170 PRINT
       180 INPUT "E=";E
       190 INPUT "at an angle=";EA
       200 PRINT
       210 FOR I=1 TO 5
       220 PRINT "For";I;"   ";
       230 INPUT "P(watts)=";P(I)
       240 PRINT TAB(8);
       250 INPUT "Q(vars)=";Q(I)
       260 PT=PT+P(I)
       270 QT=QT+Q(I)
       280 NEXT I
       290 PRINT
       300 PRINT "The apparent power associated with each load"
       310 PRINT "is the following:"
       320 PRINT
       330 FOR I=1 TO 5
       340 S(I)=SQR(P(I)^2+Q(I)^2)
       350 PRINT "S";I;"=";S(I)
       360 NEXT I
       370 ST=SQR(PT^2+QT^2)
       380 PRINT:PRINT
       390 PRINT "Total real power, PT=";PT;"watts"
       400 PRINT
       410 PRINT "Total reactive power, QT=";QT;"vars"
       420 PRINT
       430 PRINT "Total apparent power, ST=";ST;"VA"
       440 FP=PT/ST
       450 TH=57.296*ATN(ABS(QT)/PT)
       460 IF QT>0 THEN IA=EA-TH
       470 IF QT<0 THEN IA=EA+TH
       480 PRINT
       490 PRINT "Power factor angle=";TH;"degrees"
       500 PRINT
       510 PRINT "Power factor=";FP;
       520 IF QT>0 THEN PRINT "(lagging)"
       530 IF QT<0 THEN PRINT "(leading)"
       540 I=ST/E
       550 PRINT
       560 PRINT "Input current:";I;"at an angle of";IA;"degrees"
       570 END
```

Labels in left margin:
- Input — lines 180–190
- Input — lines 210–250
- P,Q — lines 260–270
- S — lines 300–370
- Power Output — lines 390–430
- F_p — lines 440–530
- I_T — lines 540–560

```
READY

RUN


This program calculates the total real,
reactive and apparent power of a network
with five individual loads.

Input the following data:
(use negative sign for capacitive vars)

E=? 50

at an angle=? 60
```

```
For  1    P(watts)=? 200

          Q(vars)=? 100

For  2    P(watts)=? 200

          Q(vars)=? 100

For  3    P(watts)=? 100

          Q(vars)=? -200

For  4    P(watts)=? 100

          Q(vars)=? -200

For  5    P(watts)=? 0

          Q(vars)=? 0


The apparent power associated with each load
is the following:

S 1 = 224
S 2 = 224
S 3 = 224
S 4 = 224
S 5 = 0


Total real power, PT= 600 watts

Total reactive power, QT=-200 vars

Total apparent power, ST= 632.456 VA

Power factor angle= 18.435 degrees

Power factor= .949 (leading)

Input current: 12.649 at an angle of 78.435 degrees
```

FIG. 19.37
Program 19.1.

The program is limited to five individual loads as defined by lines 110 through 130. Lines 260 and 270 determine the total real and reactive power using a loop routine that begins on line 210 and ends on line 280. The apparent power for each load is determined on line 340, and the total apparent power on line 370. The results are printed out by lines 390 through 430, and the power factor is determined by lines 440 through 530. The input current is then determined by lines 540 through 560.

A run of the program for four loads is provided with parameter values that permit a relatively easy check of the program through review of the results.

PROBLEMS

Section 19.7

1. For the network of Fig. 19.38:
  a. Find the average power delivered to each element.
  b. Find the reactive power for each element.
  c. Find the apparent power for each element.
  d. Find the total number of watts, volt-amperes reactive, volt-amperes, and the power factor F_p of the circuit.
  e. Sketch the power triangle.
  f. Find the energy dissipated by the resistor over one full cycle of the input voltage.
  g. Find the energy stored or returned by the capacitor and the inductor over 1/2 cycle of the power curve for each.

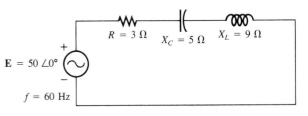

FIG. 19.38

2. For the system of Fig. 19.39:
  a. Find the total number of watts, volt-amperes reactive, volt-amperes, and the power factor F_p.
  b. Draw the power triangle.
  c. Find the current $\mathbf{I}_T$.

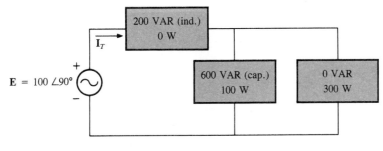

FIG. 19.39

3. Repeat Problem 2 for the system of Fig. 19.40.

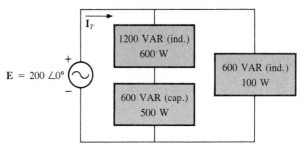

FIG. 19.40

4. Repeat Problem 2 for the system of Fig. 19.41.

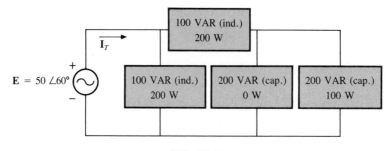

FIG. 19.41

5. For the circuit of Fig. 19.42:
   a. Find the average, reactive, and apparent power for the 20-Ω resistor.
   b. Repeat part (a) for the 10-Ω inductive reactance.
   c. Find the total number of watts, volt-amperes reactive, volt-amperes, and the power factor F_p.
   d. Find the current $\mathbf{I}_T$.

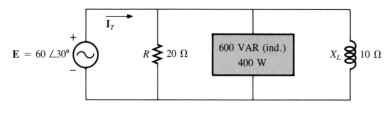

FIG. 19.42

6. Repeat Problem 1 for the circuit of Fig. 19.43.

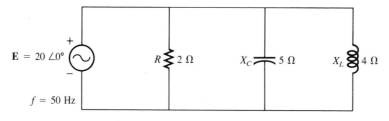

FIG. 19.43

7. Repeat Problem 1 for the circuit of Fig. 19.44.

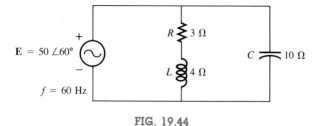

FIG. 19.44

*8. Repeat Problem 1 for the circuit of Fig. 19.45.

9. An electrical system is rated 10 kVA, 200 V at a 0.5 leading power factor.
   a. Determine the impedance of the system in rectangular coordinates.
   b. Find the average power delivered to the system.

10. Repeat Problem 9 for an electrical system rated 5 kVA, 120 V at a 0.8 lagging power factor.

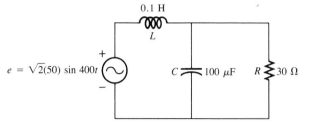

FIG. 19.45

*11. For the system of Fig. 19.46:
   a. Find the total number of watts, volt-amperes reactive, volt-amperes, and F_p.
   b. Find the current $\mathbf{I}_T$.
   c. Draw the power triangle.
   d. Find the type elements and their impedance in ohms within each electrical box. (Assume that all elements within the boxes are in series.)
   e. Verify that the result of part (b) is correct by finding the current $\mathbf{I}_T$ using only the input voltage $\mathbf{E}$ and the results of part (d). Compare the value of $\mathbf{I}_T$ with that obtained for part (b).

*12. For the system of Fig. 19.47:
   a. Find the total number of watts, volt-amperes reactive, volt-amperes, and F_p.
   b. Find the current $\mathbf{I}$.
   c. Find the type elements and their impedance in each box. (Assume that the elements within each box are in series.)

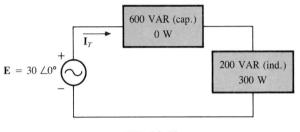

FIG. 19.46

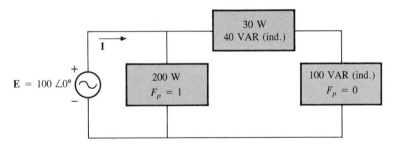

FIG. 19.47

13. For the circuit of Fig. 19.48:
   a. Find the total number of watts, volt-amperes reactive, volt-amperes, and F_p.
   b. Find the voltage $\mathbf{E}$.
   c. Find the type elements and their impedance in each box. (Assume that the elements within each box are in series.)

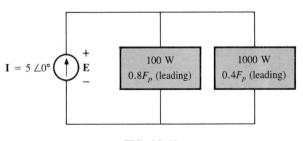

FIG. 19.48

*14. Repeat Problem 11 for the system of Fig. 19.49.

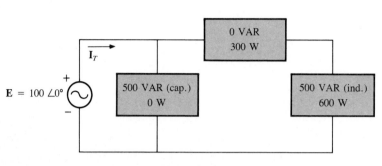

FIG. 19.49

Section 19.8

15. The lighting and motor loads of a small factory establish a 10-kVA power demand at a 0.7 lagging power factor on a 208-V, 60-Hz supply. Determine what level of capacitance in parallel with the load will raise the power factor level to
   a. unity
   b. 0.9

16. The load on a 1200-V, 60-Hz supply is 5 kW (resistive), 8 kVAR (inductive), and 2 kVAR (capacitive).
   a. Determine the F_p of the combined loads.
   b. Find the total kilovolt-amperes.
   c. Find the current drawn from the supply.
   d. Calculate the capacitance necessary to establish a unity power factor.
   e. Find the current drawn from the supply at unity power factor.

17. The loading of a factory on a 1000-V, 60-Hz system includes:

20-kW heating (unity power factor)

10-kW (P_i) induction motors (0.7 lagging power factor)

5-kW lighting (0.85 lagging power factor)

   a. Determine the total kilovolt-amperes.
   b. Find the total F_p.
   c. Derive the net reactive power.
   d. Find the current drawn from the supply.
   e. Calculate the capacitive contribution necessary to establish unity power factor for the total load.
   f. Calculate the current drawn from the supply under unity power-factor conditions.

Section 19.9

18. a. A wattmeter is connected with its current coil as shown in Fig. 19.50 and the potential coil across points f-g. What does the wattmeter read?

b. Repeat part (a) with the potential coil (PC) across a-b, b-c, a-c, a-d, c-d, d-e, and f-e.

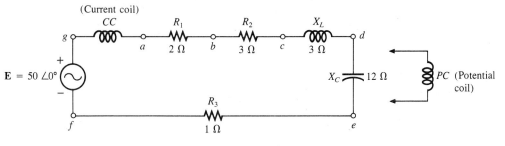

FIG. 19.50

19. The voltage source of Fig. 19.51 delivers 660 VA at 120 V with a supply current that lags the voltage by a power factor of 0.6.

a. Determine the voltmeter, ammeter, and wattmeter readings.

b. Find the load impedance in rectangular form.

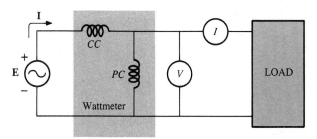

FIG. 19.51

Section 19.10

20. a. An air-core coil is connected to a 200-V, 60-Hz source. The current is found to be 4 A, and a wattmeter reading of 80 W is observed. Find the effective resistance and the inductance of the coil.

b. A brass core is inserted in the coil. The ammeter reads 3 A and the wattmeter reads 90 W. Calculate the effective resistance of the core. Explain the increase over the value of part (a).

c. If a solid iron core is inserted in the coil, the current is found to be 2 A, and the wattmeter reads 60 W. Calculate the resistance and inductance of the coil. Compare these values to the values of part (a), and account for the changes.

21. a. The inductance of an air-core coil is 0.08 H, and the effective resistance is 4 Ω when a 60-V, 50-Hz source is connected across the coil. Find the current passing through the coil and the reading of a wattmeter across the coil.

b. If a brass core is inserted in the coil, the effective resistance increases to 7 Ω, and the wattmeter reads 30 W. Find the current passing through the coil and the inductance of the coil.

c. If a solid iron core is inserted in the coil, the effective resistance of the coil increases to 10 Ω, and the current decreases to 1.7 A. Find the wattmeter reading and the inductance of the coil.

22. Write a program that provides a general solution for the network of Fig. 19.19. That is, given the resistance or reactance of each element and the source voltage at zero degrees, calculate the real, reactive, and apparent power of the system.

23. Write a program that will demonstrate the effect of increasing reactive power on the power factor of a system. Tabulate the real power, reactive power, and power factor of the system for a fixed real power and a reactive power that starts at 10% of the real power and continues through to five times the real power in increments of 10% of the real power.

24. Write a program that will provide a general solution for exercises like that defined by Example 19.5.

GLOSSARY

Apparent power   The power delivered to a load without consideration of the effects of a power-factor angle of the load. It is determined solely by the product of the terminal voltage and current of the load.

Average (real) power   The delivered power that is dissipated in the form of heat by a network or system.

Eddy currents   Small, circular currents in a paramagnetic core, causing an increase in the power losses and the effective resistance of the material.

Effective resistance   The resistance value that includes the effects of radiation losses, skin effect, eddy currents, and hysteresis losses.

Hysteresis losses   Losses in a magnetic material introduced by changes in the direction of the magnetic flux within the material.

Power-factor correction   The addition of reactive components (typically capacitive) to establish a system power factor closer to unity.

Radiation losses   The loss of energy in the form of electromagnetic waves during the transfer of energy from one element to another.

Reactive power   The power associated with reactive elements that provides a measure of the energy associated with setting up the magnetic and electric fields of inductive and capacitive elements, respectively.

Skin effect   At high frequencies, a counter induced voltage builds up at the center of a conductor resulting in an increased flow near the surface (skin) of the conductor and a great reduction near the center. As a result, resistance increases as determined by the basic equation for resistance in terms of the geometric shape of the conductor.

20
Resonance

20.1 INTRODUCTION

This chapter will introduce the very important resonant (or tuned) circuit, which is fundamental to the operation of a wide variety of electrical and electronic systems in use today. The resonant circuit is a combination of R, L, and C elements having a frequency response characteristic as shown in Fig. 20.1. Note in the figure that the response

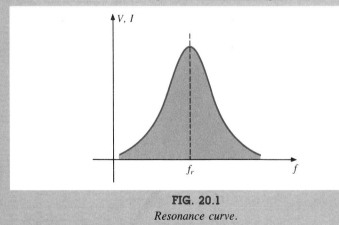

FIG. 20.1
Resonance curve.

is a maximum for the frequency f_r, decreasing to the right and left of this frequency. In other words, the resonant circuit selects a range of frequencies for which the response will be near or equal to the maximum. The frequencies to the far left or right are, for all practical purposes, nullified with respect to their effect on the system's response. The radio or television receiver has a response curve for each broadcast station of the type indicated in Fig. 20.1. When the receiver is set (or tuned) to a particular station, it is set on or near the frequency f_r of Fig. 20.1. Stations transmitting at frequencies to the far right or left of this resonant frequency are not carried through with significant power to affect the program of interest. The tuning process (setting the dial to f_r) as described above is the reason for the terminology *tuned circuit*. When the response is a maximum, the circuit is said to be in a state of *resonance*, with f_r as the *resonant frequency*.

The concept of resonance is not limited to electrical or electronic systems. If mechanical impulses are applied to a mechanical system at the proper frequency, the system will enter a state of resonance in which sustained vibrations of very large amplitude will develop. The frequency at which this occurs is called the *natural frequency* of the system. The classic example of this effect was the Tacoma Narrows Bridge built in 1940 over Puget Sound in Washington State. It had a suspended span of 2800 feet. Four months after the bridge was completed, a 42-mi/h pulsating gale set the bridge into oscillations at its natural frequency. The amplitude of the oscillations increased to the point where the main span broke up and fell into the water below. It has since been replaced by the new Tacoma Narrows Bridge, completed in 1950.

The resonant electrical circuit *must* have both inductance and capacitance. In addition, resistance will always be present due either to the lack of ideal elements or to the control offered on the shape of the resonance curve. When resonance occurs due to the application of the proper frequency (f_r), the energy absorbed at any instant by one reactive element is exactly equal to that released by another reactive element within the system. In other words, energy pulsates from one reactive element to the other. Therefore, once the system has reached a state of resonance, it requires no further reactive power since it is self-sustaining. The total apparent power is then simply equal to the average power dissipated by the resistive elements. *The average power absorbed by the system will also be a maximum at resonance,* just as the transfer of energy to the mechanical system above was a maximum at the natural frequency.

There are two types of resonant circuits: *series* and *parallel*. Each will be considered in some detail in this chapter.

SERIES RESONANCE

20.2 SERIES RESONANT CIRCUIT

The basic circuit configuration for the series resonant circuit appears in Fig. 20.2. The resistance R_l is the internal resistance of the coil. The resistance R_s is the source resistance and any other resistance added in series to affect the shape of the resonance curve.

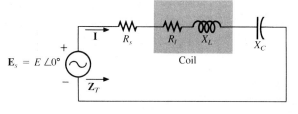

FIG. 20.2
Series resonant circuit.

Defining $R = R_s + R_l$, the total impedance of this network at any frequency is determined by

$$\mathbf{Z}_T = R + jX_L - jX_C = R + j(X_L - X_C)$$

Series resonance will occur when

$$\boxed{X_L = X_C} \qquad \textbf{(20.1)}$$

which removes the reactive component from the total impedance equation. The total impedance at resonance is then simply

$$\boxed{\mathbf{Z}_{T_s} = R} \qquad \textbf{(20.2)}$$

representing the minimum value of $\mathbf{Z}_T$ at any frequency. The subscript s will be employed to indicate series resonant conditions.

The resonant frequency can be determined in terms of the inductance and capacitance by examining the defining equation for resonance [Eq. (20.1)]:

$$X_L = X_C$$

Substituting yields

$$\omega L = \frac{1}{\omega C} \quad \text{and} \quad \omega^2 = \frac{1}{LC}$$

and

$$\boxed{\omega_s = \frac{1}{\sqrt{LC}}} \qquad \textbf{(20.3)}$$

or

$$\boxed{f_s = \frac{1}{2\pi\sqrt{LC}}} \qquad \begin{array}{l} f = \text{hertz (Hz)} \\ L = \text{H} \\ C = \text{farads (F)} \end{array} \qquad \textbf{(20.4)}$$

The current through the circuit at resonance is

$$\mathbf{I} = \frac{E \angle 0°}{R \angle 0°} = \frac{E}{R} \angle 0°$$

which you will note is the maximum current for the circuit of Fig. 20.2 for an applied voltage **E,** since $\mathbf{Z}_T$ is a minimum value. Consider also that *the input voltage and current are in phase at resonance*.

Since the current is the same through the capacitor and inductor, the voltage across each is equal in magnitude but 180° out of phase at resonance:

$$\left.\begin{array}{l} \mathbf{V}_L = \mathbf{IX}_L = (I \angle 0°)(X_L \angle 90°) = IX_L \angle 90° \\ \mathbf{V}_C = \mathbf{IX}_C = (I \angle 0°)(X_C \angle -90°) = IX_C \angle -90° \end{array}\right\} \begin{array}{l} 180° \\ \text{out of} \\ \text{phase} \end{array}$$

and, since $X_L = X_C$,

$$\boxed{V_{L_s} = V_{C_s}} \tag{20.5}$$

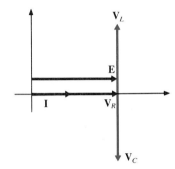

FIG. 20.3

Phasor diagram for the series resonant circuit at resonance.

Figure 20.3, a phasor diagram of the voltages and current, clearly indicates that the voltage across the resistor at resonance is the input voltage.

The average power to the resistor at resonance is equal to I^2R, and the reactive power to the capacitor and inductor are I^2X_C and I^2X_L, respectively.

The power triangle at resonance (Fig. 20.4) shows that the total apparent power is equal to the average power dissipated by the resistor, since $Q_L = Q_C$. The power factor of the circuit at resonance is

$$F_p = \cos \theta = \frac{P}{S} = 1$$

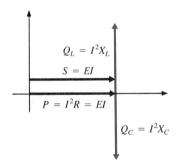

FIG. 20.4

Power triangle for the series resonant circuit at resonance.

Plotting the power curves of each element on the same set of axes (Fig. 20.5), we note that even though the total reactive power at any instant is

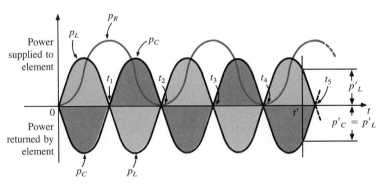

FIG. 20.5

Power curves at resonance for the series resonant circuit.

equal to zero ($t = t'$), energy is still being absorbed and released by the inductor and capacitor at resonance.

A closer examination reveals that the energy absorbed by the inductor from time 0 to t_1 is the same as the energy being released by the capacitor from 0 to t_1. The reverse occurs from t_1 to t_2, and so on. Therefore, the total apparent power continues to be equal to the average power, even though the inductor and capacitor are absorbing and releasing energy. This condition occurs only at resonance. The slightest change in frequency introduces a reactive component into the power triangle which will increase the apparent power of the system above the average power dissipation, and resonance will no longer exist.

20.3 THE QUALITY FACTOR (Q)

The *quality factor Q* of a series resonant circuit is defined as the ratio of the reactive power of either the inductor or the capacitor to the average power of the resistor at resonance; that is,

$$Q_s = \frac{\text{reactive power}}{\text{average power}} \qquad \textbf{(20.6)}$$

The quality factor is also an indication of how much energy is placed in storage (continual transfer from one reactive element to the other) as compared to that dissipated. The lower the level of dissipation for the same reactive power, the larger the Q_s factor and the more concentrated and intense the region of resonance.

Substituting for an inductive reactance in Eq. (20.6) at resonance gives us

$$Q_s = \frac{I^2 X_L}{I^2 R}$$

and

$$Q_s = \frac{X_L}{R} = \frac{\omega_s L}{R} \qquad \textbf{(20.7)}$$

or, for the capacitive reactance,

$$Q_s = \frac{I^2 X_C}{I^2 R}$$

and

$$Q_s = \frac{X_C}{R} = \frac{1}{\omega_s C R} \qquad \textbf{(20.8)}$$

If the resistance R is just the resistance of the coil (R_l), we can speak of the Q of the coil, where

$$Q_{coil} = Q = \frac{X_L}{R_l} \quad \Bigg|_{R = R_l}$$

$$\text{(20.9)}$$

Since the quality factor of a coil is typically the information provided by manufacturers of inductors, it is given the symbol Q without an associated subscript. It would appear from Eq. (20.9) that Q will increase linearly with frequency. That is, if the frequency doubles, then Q will also increase by a factor of 2. This is approximately true for the low range to the midrange of frequencies such as shown for the coil of Fig. 20.6. Unfortunately, however, as the frequency increases, the effective resistance of the coil will also increase and the resulting Q will decrease. In addition, the capacitive effects between the windings will increase, reducing the net inductance of the coil, further reducing the Q of the coil. For this reason, the Q of a coil must be specified at a particular frequency (usually at the maximum). For wide frequency applications, a plot of Q versus frequency is often provided. The maximum Q for most commercially available coils approaches 100.

If we substitute into Eq. (20.7) the fact that

$$\omega_s = 2\pi f_s$$

and

$$f_s = \frac{1}{2\pi\sqrt{LC}}$$

then

$$Q_s = \frac{\omega_s L}{R} = \frac{2\pi f_s L}{R} = \frac{2\pi}{R}\left(\frac{1}{2\pi\sqrt{LC}}\right)L$$

$$= \frac{L}{R}\left(\frac{1}{\sqrt{LC}}\right) = \left(\frac{\sqrt{L}}{\sqrt{L}}\right)\frac{L}{R\sqrt{LC}}$$

and

$$Q_s = \frac{1}{R}\sqrt{\frac{L}{C}}$$

$$\text{(20.10)}$$

For series resonant circuits used in communication systems, Q_s is usually greater than 1. By applying the voltage divider rule to the circuit of Fig. 20.2, we obtain

$$V_L = \frac{X_L E}{Z_T} = \frac{X_L E}{R} \quad \text{(at resonance)}$$

and

$$V_{L_s} = Q_s E$$

$$\text{(20.11)}$$

(a)

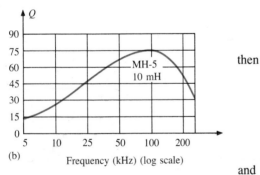

(b)

Frequency (kHz) (log scale)

FIG. 20.6

Q vs. frequency for a TRW/UTC 10-mH coil. (Courtesy of United Transformer Corp.)

or

$$V_C = \frac{X_C E}{Z_T} = \frac{X_C E}{R}$$

and

$$\boxed{V_{C_s} = Q_s E} \qquad (20.12)$$

Since Q_s is usually greater than 1, the voltage across the capacitor or inductor of a series resonant circuit is usually greater than the input voltage. In fact, in many cases the Q_s is so high that careful design and handling (including adequate insulation) are mandatory with respect to the voltage across the capacitor and inductor.

In the circuit of Fig. 20.7, for example, which is in the state of resonance,

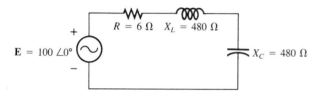

FIG. 20.7
High-Q series resonant circuit.

$$Q_s = \frac{X_L}{R} = \frac{480}{6} = 80$$

and

$$V_L = V_C = Q_s E = (80)(100) = 8000 \text{ V}$$

which is certainly a potential to be handled with great care.

20.4 Z_T VERSUS FREQUENCY

The total impedance of the series R-L-C circuit of Fig. 20.2 at any frequency is determined by

$$\mathbf{Z}_T = R + jX_L - jX_C \quad \text{or} \quad \mathbf{Z}_T = R + j(X_L - X_C)$$

The magnitude of the impedance $\mathbf{Z}_T$ is

$$Z_T = \sqrt{R^2 + (X_L - X_C)^2}$$

The total-impedance-versus-frequency curve for the series resonant circuit of Fig. 20.2 can be found by applying the impedance-versus-frequency curve for each element of the equation just derived, written in the following form:

$$\boxed{Z_T(f) = \sqrt{[R(f)]^2 + [X_L(f) - X_C(f)]^2}} \qquad (20.13)$$

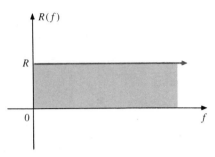

FIG. 20.8

Resistance vs. frequency.

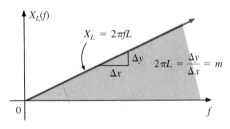

FIG. 20.9

Inductive reactance vs. frequency.

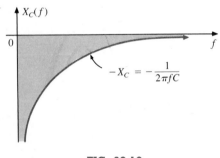

FIG. 20.10

Capacitive reactance vs. frequency.

where $Z_T(f)$ "means" the total impedance as a *function* of frequency. Ideally, the resistance R does not change with frequency, so its curve is a straight horizontal line with a magnitude R above the frequency axis (Fig. 20.8). The curve for the inductance, as determined by the reactance equation, is a straight line intersecting the origin with a slope equal to the inductance of the coil. The mathematical expression for any straight line in a two-dimensional plane is given by

$$y = mx + b$$

Thus, for the coil,

$$X_L = 2\pi fL + 0 = \underbrace{(2\pi L)}(f) + 0$$
$$\downarrow \qquad\qquad \downarrow \quad \downarrow \quad \downarrow$$
$$y = \qquad\qquad m \cdot x + b$$

(where $2\pi L$ is the slope), producing the results shown in Fig. 20.9. For the capacitor,

$$X_C = \frac{1}{2\pi fC} \quad \text{or} \quad X_C f = \frac{1}{2\pi C}$$

which becomes $yx = k$, the equation for a hyperbola, where

$$y \text{ (variable)} = X_C$$
$$x \text{ (variable)} = f$$
$$k \text{ (constant)} = \frac{1}{2\pi C}$$

Since the equation for the total impedance Z_T includes a term $-X_C(f)$, the curve plotted in Fig. 20.10 is that of $-1/2\pi fC$. Plotting all three quantities on the same set of axes gives the results shown in Fig. 20.11. Note first that resonance occurs at f where $X_L = X_C$. Since the capacitive reactance of the circuit is greater to the left of f than the

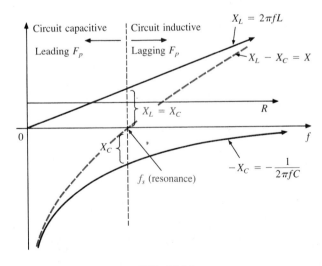

FIG. 20.11

inductive reactance, the circuit is predominantly *capacitive* to the left of f_s. To the right, the opposite is true, and the circuit is predominantly *inductive*. This also means that a leading power factor exists to the left of f_s, and a lagging power factor to the right. (Note the curve of Fig. 20.13.)

Applying

$$Z_T(f) = \sqrt{[R(f)]^2 + [X_L(f) - X_C(f)]^2}$$
$$= \sqrt{[R(f)]^2 + [X(f)]^2}$$

to the curves of Fig. 20.11 where $X(f)$ is as shown, we obtain the curve for $Z_T(f)$ as shown in Fig. 20.12. The minimum impedance occurs at the resonant frequency and is equal to the resistance R. Note that the curve is not symmetrical about the resonant frequency (especially at higher values of Z_T).

At very low frequencies, the circuit is almost purely capacitive, and the current leads the applied voltage by 90°. At very high frequencies, the circuit is almost purely inductive, and the current lags the voltage by 90°. The applied voltage and resulting current are in phase only at resonance, as indicated by the phase plot of Fig. 20.13.

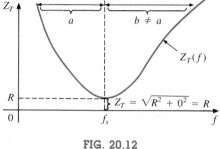

FIG. 20.12
Z_T vs. frequency for the series resonant circuit.

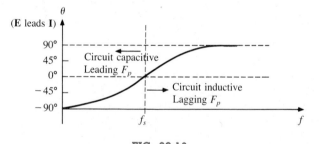

FIG. 20.13
Phase plot for the series resonant circuit.

20.5 SELECTIVITY

If we now plot the magnitude of the current $I = E/Z_T$ versus frequency for a *fixed* applied voltage E, we obtain the curve shown in Fig. 20.14, which rises from zero to a maximum value of E/R (where Z_T is a minimum) and then drops toward zero (as Z_T increases) at a slower rate than it rose to its peak value. The curve is actually the inverse of the impedance-versus-frequency curve. Since the Z_T curve is not absolutely symmetrical about the resonant frequency, the curve of the current versus frequency has the same property.

There is a definite range of frequencies at which the current is near its maximum value and the impedance is at a minimum. Those frequencies corresponding to 0.707 of the maximum current are called the *band frequencies, cutoff frequencies,* or *half-power frequencies.* They are indicated by f_1 and f_2 in Fig. 20.14. The range of frequencies between the two is referred to as the *bandwidth* (abbreviated BW) of the resonant circuit.

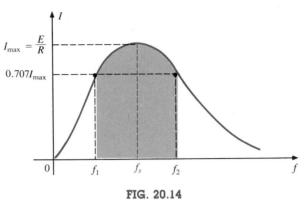

FIG. 20.14

I vs. frequency for the series resonant circuit.

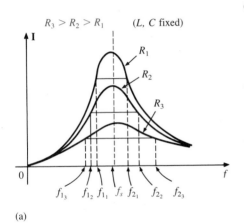

(a)

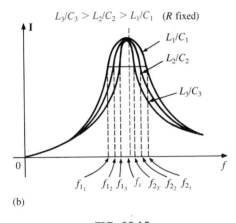

(b)

FIG. 20.15

Effect of R, L, and C on the selectivity curve for the series resonant circuit.

Half-power frequencies are those frequencies at which the power delivered is one-half that delivered at the resonant frequency; that is,

$$P_{\text{HPF}} = \frac{1}{2} P_{\text{max}}$$ **(20.14)**

The above condition is derived using the fact that

$$P_{\text{max}} = I_{\text{max}}^2 R$$

and

$$P_{\text{HPF}} = I^2 R = (0.707 I_{\text{max}})^2 R = 0.5 I_{\text{max}}^2 R = \frac{1}{2} P_{\text{max}}$$

Since the resonant circuit is adjusted to select a band of frequencies, the curve of Fig. 20.14 is called the *selectivity curve*. The term is derived from the fact that one must be selective in choosing the frequency to insure that it is in the bandwidth. The smaller the bandwidth, the higher the selectivity. The shape of the curve, as shown in Fig. 20.15, depends on each element of the series *R-L-C* circuit. If the resistance is made smaller with a fixed inductance and capacitance, the bandwidth decreases and the selectivity increases. Similarly, if the ratio *L/C* increases with fixed resistance, the bandwidth again decreases with an increase in selectivity.

In terms of Q_s, if R is larger for the same X_L, then Q_s is less, as determined by the equation $Q_s = \omega_s L/R$. *A small Q_s, therefore, is associated with a resonant curve with a large bandwidth and a small selectivity, while a large Q_s indicates the opposite.*

For circuits where $Q_s \geq 10$, a widely accepted approximation is that the resonant frequency bisects the bandwidth and that the resonant curve is symmetrical about the resonant frequency. These conditions are shown in Fig. 20.16, indicating that the cutoff frequencies are then equidistant from the resonant frequency.

For any Q_s, the preceding is not true. The cutoff frequencies f_1 and f_2 can be found for the general case (any Q_s) by first employing the fact that a drop in current to 0.707 of its resonant value corresponds to an increase in impedance equal to $1/0.707 = \sqrt{2}$ times the resonant value, which is R.

Substituting $\sqrt{2}R$ into the equation for the magnitude of $\mathbf{Z}_T$, we find that

$$\mathbf{Z}_T = \sqrt{R^2 + (X_L - X_C)^2}$$

becomes

$$\sqrt{2}R = \sqrt{R^2 + (X_L - X_C)^2}$$

or, squaring both sides, that

$$2R^2 = R^2 + (X_L - X_C)^2$$

and

$$R^2 = (X_L - X_C)^2$$

Taking the square root of both sides gives us

$$R = X_L - X_C$$

Let us first consider the case where $X_L > X_C$, which relates to f_2 or ω_2. Substituting $\omega_2 L$ for X_L and $1/\omega_2 C$ for X_C and bringing both quantities to the left of the equal sign, we have

$$R - \omega_2 L + \frac{1}{\omega_2 C} = 0 \quad \text{or} \quad R\omega_2 - \omega_2^2 L + \frac{1}{C} = 0$$

which can be written

$$\omega_2^2 - \frac{R}{L}\omega_2 - \frac{1}{LC} = 0$$

Solving the quadratic, we have

$$\omega_2 = \frac{-(-R/L) \pm \sqrt{[-(R/L)]^2 - [-(4/LC)]}}{2}$$

and

$$\omega_2 = +\frac{R}{2L} \pm \frac{1}{2}\sqrt{\frac{R^2}{L^2} + \frac{4}{LC}}$$

so, finally,

$$\boxed{\omega_2 = \frac{R}{2L} + \frac{1}{2}\sqrt{\left(\frac{R}{L}\right)^2 + \frac{4}{LC}}} \qquad \text{(rad/s)} \qquad \textbf{(20.15)}$$

The negative sign in front of the second factor was dropped because $(1/2)\sqrt{R^2/L^2 + 4/LC}$ is always greater than $R/(2L)$. If it were not

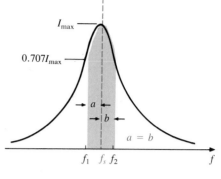

FIG. 20.16

Approximate series resonance curve for $Q_s \geq 10$.

dropped, there would be a negative solution for the radian frequency ω. Since $\omega_2 = 2\pi f_2$,

$$f_2 = \frac{1}{2\pi}\left[\frac{R}{2L} + \frac{1}{2}\sqrt{\left(\frac{R}{L}\right)^2 + \frac{4}{LC}}\right] \quad \text{(Hz)} \quad \textbf{(20.16)}$$

If we repeat the same procedure for $X_C > X_L$, which relates to ω_1 or f_1 such that $Z_T = \sqrt{R^2 + (X_C - X_L)^2}$, the solution for $\omega = \omega_1$ becomes

$$\omega_1 = -\frac{R}{2L} + \frac{1}{2}\sqrt{\left(\frac{R}{L}\right)^2 + \frac{4}{LC}} \quad \text{(rad/s)} \quad \textbf{(20.17)}$$

or, since $\omega_1 = 2\pi f_1$, it can be written

$$f_1 = \frac{1}{2\pi}\left[-\frac{R}{2L} + \frac{1}{2}\sqrt{\left(\frac{R}{L}\right)^2 + \frac{4}{LC}}\right] \quad \text{(Hz)} \quad \textbf{(20.18)}$$

The bandwidth (BW) is

$$\text{BW} = f_2 - f_1$$
$$= \left[\frac{R}{4\pi L} + \frac{1}{4\pi}\sqrt{\left(\frac{R}{L}\right)^2 + \frac{4}{LC}}\right]$$
$$- \left[-\frac{R}{4\pi L} + \frac{1}{4\pi}\sqrt{\left(\frac{R}{L}\right)^2 + \frac{4}{LC}}\right]$$

and

$$\text{BW} = f_2 - f_1 = \frac{R}{2\pi L} \quad \textbf{(20.19)}$$

Substituting $R/L = \omega_s/Q_s$ from $Q_s = \omega_s L/R$ and $1/2\pi = f_s/\omega_s$ from $\omega_s = 2\pi f_s$ gives us

$$\text{BW} = \frac{R}{2\pi L} = \left(\frac{f_s}{\omega_s}\right)\left(\frac{\omega_s}{Q_s}\right)$$

or

$$\text{BW} = \frac{f_s}{Q_s} \quad \textbf{(20.20)}$$

which is a very convenient form, since it relates the bandwidth to the Q_s of the circuit. As mentioned earlier, Eq. (20.20) verifies that the larger the Q_s, the smaller the bandwidth, and vice versa.

Written in a slightly different form, Eq. (20.20) becomes

$$\boxed{\frac{f_2 - f_1}{f_s} = \frac{1}{Q_s}}$$ **(20.21)**

The ratio $(f_2 - f_1)/f_s$ is sometimes called the *fractional bandwidth*.

20.6 V_R, V_L, AND V_C

Plotting the magnitude (effective value) of the voltages $\mathbf{V}_R$, $\mathbf{V}_L$, and $\mathbf{V}_C$ and the current $\mathbf{I}$ versus frequency for the series resonant circuit on the same set of axes, we obtain the curves shown in Fig. 20.17. Note that

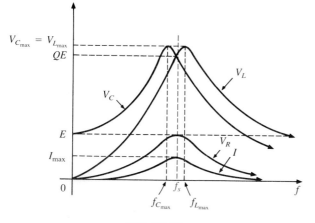

FIG. 20.17

V_R, V_L, V_C, and I vs. frequency for a series resonant circuit.

the V_R curve has the same shape as the I curve and a peak value equal to the magnitude of the input voltage E. The V_C curve builds up slowly at first from a value equal to the input voltage, since the reactance of the capacitor is infinite (open circuit) at zero frequency and the reactance of the inductor is zero (short circuit) at this frequency. As the frequency increases, $1/\omega C$ of the equation

$$V_C = IX_C = I\frac{1}{\omega C}$$

becomes smaller, but I increases at a rate faster than that at which $1/\omega C$ drops. Therefore, V_C rises and will continue to rise, due to the quickly rising current, until the frequency nears resonance. As it approaches the resonant condition, the rate of change of I decreases. When this occurs, the factor $1/\omega C$, which decreased as the frequency rose, will overcome the rate of change of I, and V_C will start to drop. The peak value will occur at a frequency just before resonance. After resonance, both V_C and I drop in magnitude, and V_C approaches zero.

The higher the Q_s of the circuit, the closer $f_{C_{max}}$ will be to f_s, and the closer $V_{C_{max}}$ will be to $Q_s E$. For circuits with $Q_s \geq 10$, $f_{C_{max}} \cong f_s$, and $V_{C_{max}} \cong Q_s E$.

The curve for V_L increases steadily from zero to the resonant frequency, since both quantities ωL and I of the equation $V_L = IX_L = \omega L I$ increase over this frequency range. At resonance, I has reached its maximum value, but ωL is still rising. Therefore, V_L will reach its maximum value after resonance. After reaching its peak value, the voltage V_L will drop toward E, since the drop in I will overcome the rise in ωL. It approaches E because X_L will eventually be infinite, and X_C will be zero.

As Q_s of the circuit increases, the frequency $f_{L_{max}}$ drops toward f_s, and $V_{L_{max}}$ approaches $Q_s E$. For circuits with $Q_s \geq 10$, $f_{L_{max}} \cong f_s$, and $V_{L_{max}} \cong Q_s E$.

The V_L curve has a greater magnitude than the V_C curve for any frequency above resonance, and the V_C curve has a greater magnitude than the V_L curve for any frequency below resonance. This again verifies that the series R-L-C circuit is predominantly capacitive from zero to the resonant frequency and predominantly inductive for any frequency above resonance.

For the condition $Q_s \geq 10$, the curves of Fig. 20.17 will appear as shown in Fig. 20.18. Note that they each peak (on an approximate basis) at the resonant frequency and have a similar shape.

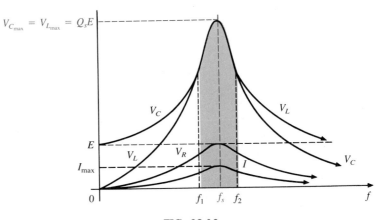

FIG. 20.18

Approximate curves for V_R, V_L, V_C, and I for a series resonant circuit where $Q_s \geq 10$.

20.7 EXAMPLES (SERIES RESONANCE)

EXAMPLE 20.1. For the resonant circuit shown in Fig. 20.19, find i, v_R, v_L, and v_C in phasor form. What is the Q_s of the circuit? If the resonant frequency is 5000 Hz, find the bandwidth. What is the power dissipated in the circuit at the half-power frequencies?

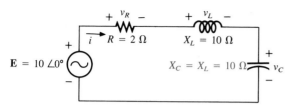

FIG. 20.19

Solution:

$$\mathbf{Z}_{T_s} = R = 2\,\Omega$$

$$\mathbf{I} = \frac{\mathbf{E}}{\mathbf{Z}_{T_s}} = \frac{10\,\angle 0^\circ}{2\,\angle 0^\circ} = \mathbf{5\,\angle 0^\circ}$$

$$\mathbf{V}_R = \mathbf{E} = 10\,\angle 0^\circ$$

$$\mathbf{V}_L = \mathbf{I}\mathbf{X}_L = (5\,\angle 0^\circ)(10\,\angle 90^\circ) = \mathbf{50\,\angle 90^\circ}$$

$$\mathbf{V}_C = \mathbf{I}\mathbf{X}_C = (5\,\angle 0^\circ)(10\,\angle -90^\circ) = \mathbf{50\,\angle -90^\circ}$$

$$Q_s = \frac{X_L}{R} = \frac{10}{2} = \mathbf{5}$$

$$\text{BW} = f_2 - f_1 = \frac{f_s}{Q_s} = \frac{5000}{5} = \mathbf{1000\,Hz}$$

$$P_{\text{HPF}} = \frac{1}{2}P_{\text{max}} = \frac{1}{2}I_{\text{max}}^2 R = \left(\frac{1}{2}\right)(5)^2(2) = \mathbf{25\,W}$$

EXAMPLE 20.2. The bandwidth of a series resonant circuit is 400 Hz. If the resonant frequency is 4000 Hz, what is the value of Q_s? If $R = 10\,\Omega$, what is the value of X_L at resonance? Find the inductance L and capacitance C of the circuit.

Solution:

$$\text{BW} = \frac{f_s}{Q_s} \quad \text{or} \quad Q_s = \frac{f_s}{\text{BW}} = \frac{4000}{400} = \mathbf{10}$$

$$Q_s = \frac{X_L}{R} \quad \text{or} \quad X_L = Q_s R = (10)(10) = \mathbf{100\,\Omega}$$

$$X_L = 2\pi f_s L \quad \text{or} \quad L = \frac{X_L}{2\pi f_s} = \frac{100}{(6.28)(4000)} = \mathbf{3.98\,mH}$$

$$X_C = \frac{1}{2\pi f_s C} \quad \text{or} \quad C = \frac{1}{2\pi f_s X_C} = \frac{1}{(6.28)(4000)(100)} = \mathbf{0.398\,\mu F}$$

EXAMPLE 20.3. A series R-L-C circuit has a series resonant frequency of 12,000 Hz. If $R = 5\,\Omega$ and X_L at resonance is 300 Ω, find the bandwidth. Find the cutoff frequencies.

Solution:

$$Q_s = \frac{X_L}{R} = \frac{300}{5} = 60$$

$$\text{BW} = \frac{f_s}{Q_s} = \frac{12,000}{60} = \mathbf{200\ Hz}$$

Since $Q_s \geq 10$, the bandwidth is bisected by f_s. Therefore,

$$f_2 = f_s + \frac{\text{BW}}{2} = 12,000 + 100 = \mathbf{12,100\ Hz}$$

$$f_1 = 12,000 - 100 = \mathbf{11,900\ Hz}$$

EXAMPLE 20.4.

a. Determine the Q_s and bandwidth for the response curve of Fig. 20.20.
b. For $C = 0.1\ \mu\text{F}$, determine L and R for the series resonant circuit.

Solutions:

a. The resonant frequency is 2800 Hz. At 0.707 times the peak value, BW = **200 Hz,** and

$$Q_s = \frac{f_s}{\text{BW}} = \frac{2800}{200} = \mathbf{14}$$

b. $f_s = \dfrac{1}{2\pi\sqrt{LC}}$ or $L = \dfrac{1}{4\pi^2 f_s^2 C}$

$$= \frac{1}{4\pi^2(2.8 \times 10^3)^2(0.1 \times 10^{-6})}$$

$$= \frac{1}{30.951} = \mathbf{32.31\ mH}$$

$Q_s = \dfrac{X_L}{R}$ or $R = \dfrac{X_L}{Q_s} = \dfrac{(17.58 \times 10^3)(32.31 \times 10^{-3})}{14}$

$$= \mathbf{40.572\ \Omega}$$

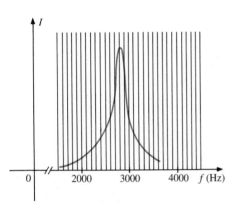

FIG. 20.20

PARALLEL RESONANCE

20.8 PARALLEL RESONANT CIRCUIT

The parallel resonant circuit has the basic configuration of Fig. 20.21. This circuit is often called the *tank circuit* due to the storage of energy by the inductor and capacitor. A transfer of energy similar to that discussed for the series circuit also occurs in the parallel resonant circuit. In the ideal case (no radiation losses, and so on), the capacitor absorbs

FIG. 20.21
Parallel resonant circuit.

energy during one half-cycle of the power curves at the same rate at which it is released by the inductor. During the next half-cycle of the power curves, the inductor absorbs energy at the same rate at which the capacitor releases it. The total reactive power at resonance is therefore zero, and the total power factor is 1.

Since tank circuits are frequently used with devices such as the transistor, which is essentially a constant-current source device, a current source will be used to supply the input to the parallel resonant circuits in the following analysis, as shown in Fig. 20.22. In addition, our analysis of parallel resonant circuits will be well served if we first replace the series R-L branch by an equivalent parallel combination using the technique of Section 15.9. That is,

$$\mathbf{Z}_{R\text{-}L} = R_l + jX_L$$

and

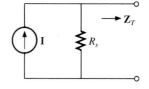

FIG. 20.22

$$\begin{aligned}
\mathbf{Y}_{R\text{-}L} &= \frac{1}{\mathbf{Z}_{R\text{-}L}} = \frac{1}{R_l + jX_L} = \frac{R_l}{R_l^2 + X_L^2} - j\frac{X_L}{R_l^2 + X_L^2} \\
&= \frac{1}{\dfrac{R_l^2 + X_L^2}{R_l}} + \frac{1}{j\left(\dfrac{R_l^2 + X_L^2}{X_L}\right)}
\end{aligned}$$

which is of the form

$$\mathbf{Y}_{R\text{-}L} = \frac{1}{R_p} + \frac{1}{jX_{L_p}} = G_p - jB_{L_p}$$

as shown in Fig. 20.23.

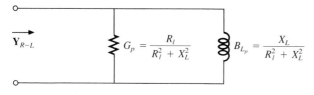

FIG. 20.23

Redrawing the network of Fig. 20.21 with the equivalent of Fig. 20.23 will result in the network of Fig. 20.24.

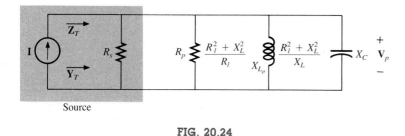

Source

FIG. 20.24

Recall that for series resonance, the condition of resonance was defined by that frequency which caused the reactive component of the total impedance to be zero. For the network of Fig. 20.24,

$$\mathbf{Y}_T = \frac{1}{\mathbf{R}_s} + \frac{1}{\mathbf{R}_p} + \frac{1}{\mathbf{X}_{L_p}} + \frac{1}{\mathbf{X}_C}$$

$$= \frac{1}{R_s} + \frac{1}{R_p} - j\left(\frac{1}{X_{L_p}}\right) + j\left(\frac{1}{X_C}\right)$$

and

$$\mathbf{Y}_T = \frac{1}{R_s} + \frac{1}{R_p} + j\left(\frac{1}{X_C} - \frac{1}{X_{L_p}}\right) \qquad \textbf{(20.22)}$$

For resonance, the reactive component must be zero as defined by

$$\frac{1}{X_C} - \frac{1}{X_{L_p}} = 0$$

Therefore,

$$\frac{1}{X_C} = \frac{1}{X_{L_p}}$$

and

$$X_{L_p} = X_C \qquad \textbf{(20.23)}$$

Substituting for X_{L_p} yields

$$\frac{R_l^2 + X_L^2}{X_L} = X_C \qquad \textbf{(20.24)}$$

and

$$R_l^2 + X_L^2 = X_C X_L = \left(\frac{1}{\omega C}\right)\omega L = \frac{L}{C}$$

or

$$X_L^2 = \frac{L}{C} - R_l^2$$

with

$$2\pi f_p L = \sqrt{\frac{L}{C} - R_l^2}$$

and

$$f_p = \frac{1}{2\pi L}\sqrt{\frac{L}{C} - R_l^2}$$

Multiplying the top and bottom of the factor within the square-root sign by C/L produces

$$f_p = \frac{1}{2\pi L}\sqrt{\frac{1 - R_l^2(C/L)}{C/L}} = \frac{1}{2\pi L\sqrt{C/L}}\sqrt{1 - \frac{R_l^2 C}{L}}$$

and

$$f_p = \frac{1}{2\pi\sqrt{LC}}\sqrt{1 - \frac{R_l^2 C}{L}} \qquad (20.25)$$

or

$$f_p = f_s\sqrt{1 - R_l^2\frac{C}{L}} \qquad (20.26)$$

where f_p is the resonant frequency of a parallel resonant circuit and f_s is the resonant frequency of a series resonant circuit of the same reactive elements. Note that unlike in a series resonant circuit, the resonant frequency of a parallel resonant circuit is dependent on the resistance R_l. Note also, however, the absence of the source resistance R_s in Eqs. (20.25) and (20.26).

20.9 SELECTIVITY CURVE FOR PARALLEL RESONANT CIRCUITS

At resonance, the reactive term of Eq. (20.22) dropped out, resulting in

$$Y_T = \frac{1}{R_s} + \frac{1}{R_p}$$

which is purely resistive, establishing a power factor of unity, as occurred for series resonance as well.

The total impedance is

$$Z_T = \frac{1}{Y_T} = \frac{1}{\dfrac{1}{R_s} + \dfrac{1}{R_p}} = \frac{R_s R_p}{R_s + R_p}$$

and

$$Z_{T_p} = R_s \parallel R_p \qquad (20.27)$$

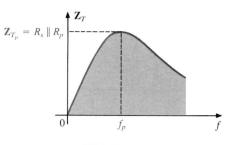

FIG. 20.25

Z_T vs. frequency for the parallel resonant circuit.

The Z_T-versus-frequency curve of Fig. 20.25 clearly reveals that a parallel resonant circuit exhibits maximum impedance at resonance, unlike the series resonant circuit which experiences minimum resistance levels at resonance.

The total impedance can be determined in terms of the reactive elements if we take the equation

$$R_p = \frac{R_l^2 + X_L^2}{R_l}$$

and substitute the fact that

$$R_l^2 + X_L^2 = X_L X_C \qquad \text{[Eq. (20.24)]}$$

which gives us

$$R_p = \frac{X_L X_C}{R_l} = \frac{(\omega L)(1/\omega C)}{R_l}$$

and

$$\boxed{R_p = \frac{L}{R_l C}} \qquad (20.28)$$

with

$$\boxed{Z_{T_p} = R_s \parallel R_p = R_s \parallel \frac{L}{R_l C}} \qquad (20.29)$$

Since the current I is constant (current source) for any value of Z_T, the voltage across the parallel circuit will have the same shape as the total impedance Z_T, as shown in Fig. 20.26.

For the parallel circuit, the resonance curve of interest is that of the voltage V_C across the capacitor. The reason for this interest in V_C derives from electronic considerations that often place the capacitor at the input to another stage of a network.

Since the voltage across parallel elements is the same,

$$\boxed{\mathbf{V}_C = \mathbf{V}_p = \mathbf{I}Z_T} \qquad (20.30)$$

The resonant value of $\mathbf{V}_C$ is therefore determined by the value of Z_{T_p} and the magnitude of the current source $\mathbf{I}$.

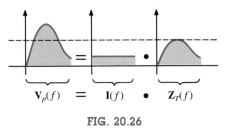

$$\mathbf{V}_p(f) \quad = \quad \mathbf{I}(f) \quad \bullet \quad \mathbf{Z}_T(f)$$

FIG. 20.26

20.10 THE QUALITY FACTOR Q_p

The quality factor of the parallel resonant circuit is determined by the ratio of the reactive power to the real power. That is,

$$Q_p = \frac{V_p^2/X_{L_p}}{V_p^2/R}$$

where $R = R_s \parallel R_p$, and V_p is the voltage across the parallel branches. The result is

$$Q_p = \frac{R}{X_{L_p}} \qquad \textbf{(20.31)}$$

or, since $X_{L_p} = X_C$ at resonance,

$$Q_p = \frac{R}{X_C} \qquad \textbf{(20.32)}$$

For a situation where R_s is considered to be sufficiently large and can be ignored,

$$R = R_s \parallel R_p \cong R_p$$

Then

$$Q_p = \frac{R_p}{X_{L_p}} = \frac{(R_l^2 + X_L^2)/R_l}{(R_l^2 + X_L^2)/X_L}$$

and

$$Q_p = \frac{X_L}{R_l} = Q \qquad \textbf{(20.33)}$$
$$R_s = \infty\ \Omega$$

which is simply the Q of the coil as defined for series resonance.

The bandwidth of the parallel resonant circuit is releated to the resonant frequency and Q_p in the same manner as for series resonant circuits. That is,

$$BW = f_2 - f_1 = \frac{f_p}{Q_p} \qquad \textbf{(20.34)}$$

The effect of R_l, L, and C on the shape of the parallel resonance curve, as shown in Fig. 20.27 for the input impedance, is quite similar to their effect on the series resonance curve. Whether or not R_l is zero,

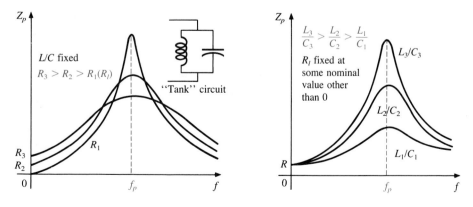

FIG. 20.27

Effect of R_l, L, and C on the parallel resonance curve.

the parallel resonant circuit will frequently appear in a network schematic as shown in Fig. 20.27.

At resonance, since $R_p = L/R_lC$, an increase in R_l or a decrease in the ratio L/C will result in a decrease in the resonant impedance, with a corresponding increase in the current. The bandwidth of the resonance curves is given by Eq. (20.34). For increasing R_l or decreasing L (or L/C for constant C), the bandwidth will increase as shown in Fig. 20.27.

At low frequencies, the capacitive reactance is quite high, and the inductive reactance is low. Since the elements are in parallel, the total impedance at low frequencies will therefore be inductive. At high frequencies, the reverse is true, and the network is capacitive. At resonance, the network appears resistive. These facts lead to the phase plot of Fig. 20.28. Note that it is the inverse of that appearing for the series resonant circuit in that at low frequencies the series resonant circuit was capacitive and at high frequencies inductive.

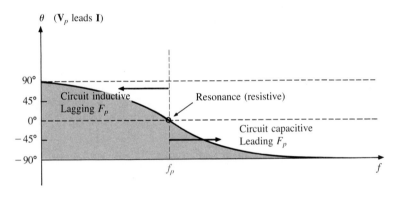

FIG. 20.28
Phase plot for the parallel resonant circuit.

20.11 EFFECT OF Q ≥ 10

The condition $Q = X_L/R_l \geq 10$ is a common one, and one that has some interesting effects on the equations just derived. The resulting equations will provide a good approximation for the desired quantities with a minimum of mathematical complexity.

Z_{T_p}:

$$R_p = \frac{R_l^2 + X_L^2}{R_l} = R_l + \frac{X_L^2}{R_l}\left(\frac{R_l}{R_l}\right) = R_l + \frac{X_L^2}{R_l^2}R_l$$
$$= R_l + Q^2R_l = (1 + Q^2)R_l$$

For $Q \geq 10$, $1 + Q^2 \cong Q^2$, and

$$\boxed{R_p \cong Q^2R_l} \tag{20.35}$$

with

$$Z_{T_p} = R_s \parallel R_p \cong R_s \parallel Q^2 R_l \qquad (20.36)$$

Of course, if $R_s \gg R_p$ or R_s is not included,

$$Z_{T_p} \cong Q^2 R_l \qquad (20.37)$$

an equation that frequently can be applied with excellent results.

X_{L_p}:

$$X_{L_p} = \frac{R_l^2 + X_L^2}{X_L} = \frac{R_l^2(X_L)}{X_L(X_L)} + X_L = \frac{X_L}{Q^2} + X_L$$

For $Q \geq 10$,

$$X_{L_p} \cong X_L \qquad (20.38)$$

Resonance condition, $X_{L_p} = X_C$:

Since $X_{L_p} \cong X_L$,

$$X_L = X_C \qquad (20.39)$$

and

$$f_p = \frac{1}{2\pi\sqrt{LC}} \qquad (20.40)$$

as derived for series resonance.

    Applying the approximations just derived to the network of Fig. 20.24 will result in the approximate equivalent circuit of Fig. 20.29.

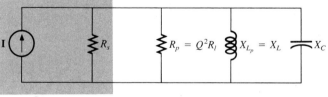

FIG. 20.29
Approximate equivalent circuit for $Q \geq 10$.

Q_p:

For $R_s \parallel R_p \cong R_p$ and $Q \geq 10$,

$$Q_p = \frac{R}{X_{L_p}} = \frac{R_p}{X_L} = \frac{Q^2 R_l}{X_L} = \frac{Q^2}{X_L/R_l} = \frac{Q^2}{Q}$$

and

$$\boxed{Q_p = Q} \qquad \qquad (20.41)$$

I_L, I_C:

You will recall that for the series resonant circuit, $V_L = V_C = QE$ at the resonant condition. A similar result can be obtained for the parallel resonant circuit if we carefully examine the network of Fig. 20.30. The current $\mathbf{I}_T$ is not the source current (due to R_s) but the current entering the tank circuit. Of course, for the condition $R_s \cong \infty \, \Omega$ (open circuit), which often occurs, $\mathbf{I}_T$ is equal to the source current $\mathbf{I}$.

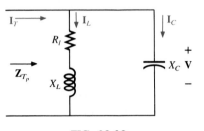

FIG. 20.30

At resonance, $\mathbf{Z}_{T_p} = Q^2 R_l$, as determined earlier, and through Ohm's law, $\mathbf{V} = \mathbf{I}_T \cdot Q^2 R_l$.

The current $\mathbf{I}_L$ is

$$\mathbf{I}_L = \frac{\mathbf{V}}{\mathbf{Z}_L} = \frac{\mathbf{I}_T Q^2 R_l}{R_l + jX_L}$$

Dividing by R_l in the numerator and denominator gives us

$$\mathbf{I}_L = \frac{\mathbf{I}_T Q^2}{1 + j\dfrac{X_L}{R_l}} = \frac{\mathbf{I}_T Q^2}{1 + jQ}$$

The magnitude of $\mathbf{I}_L$ is given by

$$I_L = \frac{I_T Q^2}{\sqrt{1 + Q^2}}$$

which for $Q \geq 10$ becomes

$$I_L = \frac{I_T Q^2}{Q}$$

and

$$\boxed{I_L \cong QI_T} \qquad \qquad (20.42)$$

In a parallel resonant circuit, therefore, the magnitude of the current through the inductive branch is Q times the current entering the tank circuit (at resonance only). Furthermore, since

$$\mathbf{I}_C = \frac{\mathbf{V}}{-jX_C} = \frac{\mathbf{I}_T Q^2 R_l}{-jX_C}$$

if we divide by R_l, we get

$$\mathbf{I}_C = \frac{\mathbf{I}_T Q^2}{-j\dfrac{X_C}{R_l}}$$

However, at resonance, for $Q \geq 10$,

$$X_L = X_C \quad \text{and} \quad \frac{X_C}{R_l} = \frac{X_L}{R_l} = Q$$

and therefore

$$\mathbf{I}_C = \frac{\mathbf{I}_T Q^2}{-jQ}$$

The magnitude is

$$I_C = \frac{I_T Q^2}{Q}$$

and

$$\boxed{I_C \cong QI_T} \qquad \text{(at resonance)} \qquad \textbf{(20.43)}$$

Table 20.1 is included as a review of the effect of $Q \geq 10$.

TABLE 20.1
Parallel resonant circuit.

	Any Q	$Q \geq 10$	$R_s = \infty\ \Omega$ ($Q \geq 10$)
Resonance	$\dfrac{R_l^2 + X_L^2}{X_L} = X_C$	$X_L = X_C$	$X_L = X_C$
f_p	$\dfrac{1}{2\pi\sqrt{LC}}\sqrt{1 - \dfrac{R_l^2 C}{L}}$	$\dfrac{1}{2\pi\sqrt{LC}}$	$\dfrac{1}{2\pi\sqrt{LC}}$
Z_{T_p}	$R_s \parallel \dfrac{R_l^2 + X_L^2}{R_l}, \; R_s \parallel \dfrac{L}{R_l C}$	$R_s \parallel Q^2 R_l$	$Q^2 R_l$
Q_p	$\dfrac{R}{X_{L_p}}, \; \dfrac{R}{X_C} \quad (R = R_s \parallel R_p)$	$\dfrac{R}{\omega_p L}$	Q
BW		$I_L = I_C = Q_p I_T$	$I_L = I_C = QI_T$
	$\dfrac{f_p}{Q_p}$	$\dfrac{f_p}{Q_p}$	$\dfrac{f_p}{Q}$

20.12 EXAMPLES (PARALLEL RESONANCE)

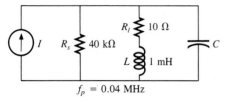

$f_p = 0.04$ MHz

FIG. 20.31

EXAMPLE 20.5. For the network of Fig. 20.31:

a. Determine Q.
b. Determine R_p.
c. Calculate Z_{T_p}.
d. Find C at resonance.
e. Find Q_p.
f. Calculate BW.

Solutions:

a. $Q = \dfrac{X_L}{R_l} = \dfrac{2\pi f_p L}{R_l} = \dfrac{(6.28)(0.04 \times 10^6)(10^{-3})}{10} = \mathbf{25.12}$

b. $Q \geq 10$. Therefore,

$$R_p = Q^2 R_l = (25.12)^2(10) = \mathbf{6.31\,k\Omega}$$

c. $Z_{T_p} = R_s \parallel R_p = 40\,k\Omega \parallel 6.31\,k\Omega = \mathbf{5.45\,k\Omega}$

d. $Q \geq 10$. Therefore,

$$f_p = \frac{1}{2\pi\sqrt{LC}}$$

and

$$C = \frac{1}{L(f2\pi)^2} = \frac{1}{(10^{-3})[(0.04 \times 10^6)2\pi]^2} = \mathbf{0.0159\,\mu F}$$

e. $Q \geq 10$. Therefore,

$$Q_p = \frac{R}{\omega_p L} = \frac{5.45\,k\Omega}{(6.28)(0.04 \times 10^6)(10^{-3})} = \mathbf{21.71}$$

f. BW $= \dfrac{f_p}{Q_p} = \dfrac{0.04 \times 10^6}{21.71} = \mathbf{1.84\,kHz}$

EXAMPLE 20.6. The equivalent network for the transistor configuration of Fig. 20.32 is shown in Fig. 20.33.

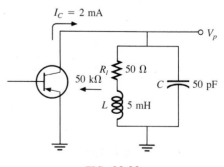

$I_C = 2$ mA

V_p

R_l 50 Ω

50 kΩ

C 50 pF

L 5 mH

FIG. 20.32

a. Find f_p.
b. Determine Q.
c. Determine Q_p.
d. Calculate BW.
e. Derive V_p at resonance.
f. Sketch the curve of V_C versus frequency.

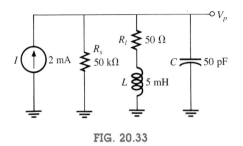

FIG. 20.33

Solutions:

a. $f_p = \dfrac{1}{2\pi\sqrt{LC}}\sqrt{1 - \dfrac{R_l^2 C}{L}} \cong \mathbf{318.5\,kHz}$

b. $Q = \dfrac{X_L}{R_l} = \dfrac{2\pi f_p L}{R_l} = \dfrac{(6.28)(318.5 \times 10^3)(5 \times 10^{-3})}{50}$

$= \dfrac{10,000.9}{50} \cong \mathbf{200}$

c. $Q_p = \dfrac{R}{X_L}$

$R = R_s \| R_p = R_s \| Q^2 R_l = 50\,k\Omega \| (200)^2(50)$
$= 50\,k\Omega \| 2\,M\Omega = 48.78\,k\Omega$
$X_L \cong 10\,k\Omega$

and

$Q_p = \dfrac{R}{X_L} = \dfrac{48.78\,k\Omega}{10\,k\Omega} = \mathbf{4.88}$

d. $BW = \dfrac{f_p}{Q_p} = \dfrac{318.5\,kHz}{4.88} = \mathbf{65.27\,kHz}$

e. At resonance,

$$Z_{T_p} = R_s \| R_p = R = 48.78\,k\Omega$$

and

$$V_p = IZ_{T_p} = (2 \times 10^{-3})(48.78 \times 10^3) = \mathbf{97.56\,V}$$

f. See Fig. 20.34.

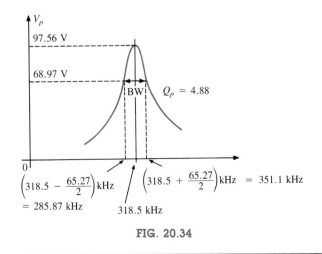

FIG. 20.34

EXAMPLE 20.7. Design a parallel resonant circuit to have the response curve of Fig. 20.35 using a 1-mH, 10-Ω inductor, with $R_s = 40\,\text{k}\Omega$.

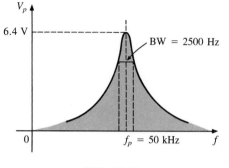

FIG. 20.35

Solution:

$$\text{BW} = \frac{f_p}{Q_p}$$

Therefore,

$$Q_p = \frac{f_p}{\text{BW}} = \frac{50{,}000}{2500} = 20$$

$$X_L = 2\pi f_p L = (6.28)(50 \times 10^3)(10^{-3}) = 314\,\Omega$$

and

$$Q = \frac{X_L}{R_l} = \frac{314}{10} = 31.4$$

$$R_p = Q^2 R_l = (31.4)^2 (10) = 9859.6\,\Omega$$

$$Q_p = \frac{R}{X_L} = \frac{R_s \parallel 9859.6}{314} = 20$$

$$\frac{(R_s)(9859.6)}{R_s + 9859.6} = 6280$$

$$9859.6 R_s = 6280 R_s + 61.92 \times 10^6$$

and

$$3579.6 R_s = 61.92 \times 10^6$$
$$R_s = 17.298\,\text{k}\Omega$$

However, the source resistance was given as 40 kΩ. We must therefore add a parallel resistor (R') that will reduce the 40 kΩ to approximately 17.298 kΩ. That is,

$$\frac{(40\,\text{k}\Omega)(R')}{40\,\text{k}\Omega + R'} = 17.298\,\text{k}\Omega$$

and

$$(40\,\text{k}\Omega)(R') = (17.298\,\text{k}\Omega)(R') + 691.92 \times 10^6$$

or

$$22.70R' = 691.92 \times 10^3$$

and

$$R' = \mathbf{30.481\,k\Omega}$$

At resonance, $X_L = X_C$, and

$$X_C = \frac{1}{\omega_p C}$$

or

$$314 = \frac{1}{(6.28)(50 \times 10^3)(C)} = \frac{1}{(314 \times 10^3)(C)}$$

or

$$C = \frac{1}{(314 \times 10^3)(314)} = \frac{10^{-6}}{98.596}$$

or, finally,

$$C \cong \mathbf{0.01\,\mu F}$$

The network appears in Fig. 20.36.

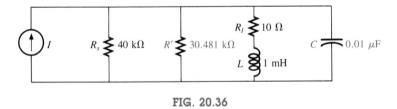

FIG. 20.36

20.13 LOG SCALES

The use of log scales permits a review of the response of a system for an extended range of frequencies. The low-, mid-, and high-frequency responses can all appear on the same graph without a significant loss in response for any frequency region.

Graph paper is available in the *semi-log* or *log-log* variety. Semi-log paper has only one log scale, with the other a linear scale. Both scales of log-log paper are log scales. A section of semi-log paper appears in Fig. 20.37. Note the linear (even-spaced-interval) vertical scaling and the repeating intervals of the log scale.

The spacing of the log scale is determined by taking the common log (base 10) of the number. The scaling starts with 1 since $\log_{10} 1 = 0$. The distance between 1 and 2 is determined by $\log_{10} 2 = 0.3010$ or approximately 30% of the full distance of a log interval, as shown on the graph. The distance between 1 and 3 is determined by $\log_{10} 3 = 0.4771$ or about 48% of the full width. For future reference, keep in

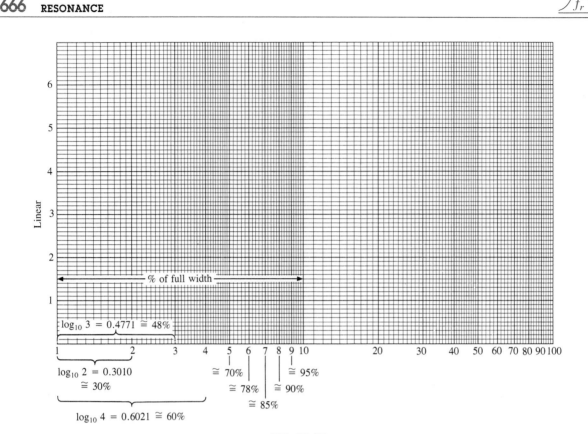

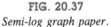

FIG. 20.37
Semi-log graph paper.

mind that almost 50% of the width of one log interval is represented by a 3 rather than by the 5 of a linear scale. The above is particularly useful when the various lines of the graph are left unnumbered.

Note how the log scale becomes compressed at the high end of each interval. With increasing frequency levels assigned to each interval, a single graph can provide a frequency plot extending from 1 Hz to 1 MHz, as shown in Fig. 20.38 with particular reference to the 30%, 50%, and 70% levels of each interval.

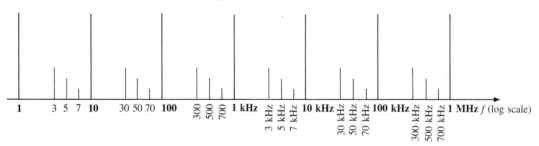

FIG. 20.38

The sections to follow that cover filters will make full use of the log scale to provide a clear picture of the full response of the network to be described.

20.14 R-C FILTERS

The *R-C* filter, incredibly simple in design, can be used as a *low-pass* or *high-pass* filter. If the output is taken off the resistor, as shown in Fig. 20.39, the circuit is referred to as a high-pass filter—a circuit that will have the general characteristics of Fig. 20.40. The *pass band* is that

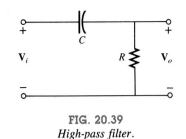

FIG. 20.39
High-pass filter.

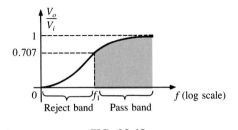

FIG. 20.40
High-pass filter characteristics.

range of frequencies that will result in a level of V_o equal to or greater than $0.707V_i$. In other words, the output will be at least 70.7% of the input. The *reject band* is that range of frequencies for which the output is less than 70.7% of the input. The cutoff frequency, that is, the frequency that defines the two regions, is denoted by f_1 in Fig. 20.40.

If the output is taken off the capacitor, as shown in Fig. 20.41, the *R-C* combination is a low-pass filter, and only frequencies less than the cutoff frequency f_2 will establish V_o greater than 70.7% of V_i, as shown in Fig. 20.42.

The *R-C* network of Fig. 15.44 was a low-pass filter as noted by the high level of V_o at low frequencies and the decreasing value of V_o with increasing frequencies. The plot provided in Chapter 15 employed a linear scale. The next section will demonstrate how the filter characteristics will appear on a log plot.

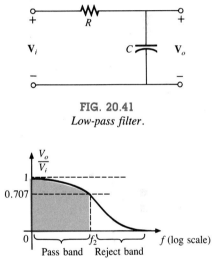

FIG. 20.41
Low-pass filter.

FIG. 20.42
Low-pass filter characteristics.

High-Pass R-C Filter

Let us now examine the high-pass filter of Fig. 20.39 in some detail and develop a procedure for plotting the frequency characteristics of Fig. 20.40.

At very high frequencies, the reactance is

$$X_C = \frac{1}{2\pi fC} \cong 0\,\Omega$$

and the short-circuit equivalent can be substituted for the capacitor as shown in Fig. 20.43, resulting in $V_o = V_i$ or $V_o/V_i = 1$, as shown in Fig. 20.40.

At $f = 0$ Hz,

$$X_C = \frac{1}{2\pi fC} = \infty\,\Omega$$

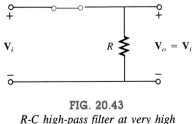

FIG. 20.43
R-C high-pass filter at very high frequencies.

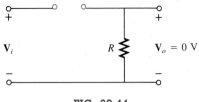

FIG. 20.44

R-C high-pass filter at f = 0 Hz.

and the open-circuit equivalent can be substituted for the capacitor as shown in Fig. 20.44, resulting in $V_o = 0$ V, as noted in Fig. 20.40.

At any intermediate frequency, the output voltage can be determined using the voltage divider rule:

$$\mathbf{V}_o = \frac{\mathbf{R}\mathbf{V}_i}{\mathbf{R} + \mathbf{X}_C}$$

or

$$\frac{\mathbf{V}_o}{\mathbf{V}_i} = \frac{\mathbf{R}}{\mathbf{R} + \mathbf{X}_C} = \frac{R \angle 0^\circ}{R - jX_C} = \frac{R \angle 0^\circ}{\sqrt{R^2 + X_C^2} \angle -\tan^{-1}(X_C/R)}$$

and

$$\frac{\mathbf{V}_o}{\mathbf{V}_i} = \frac{R}{\sqrt{R^2 + X_C^2}} \angle \tan^{-1}(X_C/R)$$

The magnitude of the ratio $\mathbf{V}_o/\mathbf{V}_i$ is therefore determined by

$$\boxed{\frac{V_o}{V_i} = \frac{R}{\sqrt{R^2 + X_C^2}} = \frac{1}{\sqrt{1 + \left(\dfrac{X_C}{R}\right)^2}}} \qquad \textbf{(20.44)}$$

and the phase angle θ by

$$\boxed{\theta = \tan^{-1}\frac{X_C}{R}} \qquad \textbf{(20.45)}$$

For the frequency at which $X_C = R$, the magnitude becomes

$$\frac{V_o}{V_i} = \frac{1}{\sqrt{1 + \left(\dfrac{X_C}{R}\right)^2}} = \frac{1}{\sqrt{1 + 1}}$$

and

$$\boxed{\frac{V_o}{V_i} = \frac{1}{\sqrt{2}} = 0.707} \qquad \textbf{(20.46)}$$
$$X_C = R, f = f_1$$

as appears in Fig. 20.40.

The frequency at which $X_C = R$ is determined by

$$X_C = \frac{1}{2\pi f_1 C} = R$$

and

$$\boxed{f_1 = \frac{1}{2\pi RC}} \qquad \textbf{(20.47)}$$

The impact of Eq. (20.47) extends beyond its relative simplicity. For any high-pass R-C filter, the application of any frequency greater than f_1 will result in an output voltage V_o that is at least 70.7% of the magnitude of the input signal. For any frequency below f_1, the output is less than 70.7% of the applied signal.

For the phase angle, high frequencies result in small values of X_C, and the ratio X_C/R will approach zero with $\tan^{-1}(X_C/R)$ approaching $0°$, as shown in Fig. 20.45. At low frequencies, the ratio X_C/R becomes quite large and $\tan^{-1}(X_C/R)$ approaches $90°$. For the case $X_C = R$, $\tan^{-1}(X_C/R) = \tan^{-1} 1 = 45°$. Assigning a phase angle of $0°$ to V_i such that $V_i = V_i \angle 0°$, the phase angle associated with V_o is θ, resulting in $V_o = V_o \angle \theta$ and revealing that θ is the angle by which V_o leads V_i. Since the angle θ is the angle by which V_o leads V_i throughout the frequency range of Fig. 20.45, the high-pass R-C filter is referred to as a *leading network*.

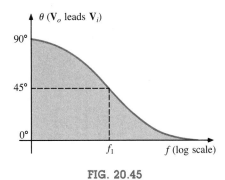

FIG. 20.45
Phase-angle response for the high-pass R-C filter.

Low-Pass R-C Filter

The analysis of the low-pass filter of Fig. 20.41 is similar in many respects to that of the high-pass filter of the same elements.

At high frequencies, $X_C \cong 0\,\Omega$ and the output approaches $0\,V$, as shown in Fig. 20.46.

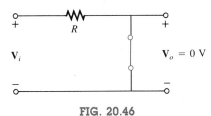

FIG. 20.46
R-C low-pass filter at high frequencies.

At low frequencies, X_C approaches $\infty\,\Omega$ and $V_o = V_i$, as shown in Fig. 20.47.

The cutoff frequency is again defined by

$$X_C = R$$

and

$$\boxed{f_2 = \frac{1}{2\pi RC}} \qquad (20.48)$$

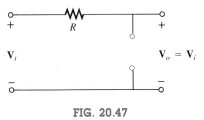

FIG. 20.47
R-C low-pass filter at low frequencies.

The equation for V_o changes somewhat, as shown below:

$$\mathbf{V}_o = \frac{\mathbf{X}_C \mathbf{V}_i}{\mathbf{R} + \mathbf{X}_C}$$

or

$$\frac{\mathbf{V}_o}{\mathbf{V}_i} = \frac{\mathbf{X}_C}{\mathbf{R} + \mathbf{X}_C} = \frac{X_C \angle -90°}{R - jX_C} = \frac{X_C \angle -90°}{\sqrt{R^2 + X_C^2}\,\angle -\tan^{-1}(X_C/R)}$$

and

$$\frac{\mathbf{V}_o}{\mathbf{V}_i} = \frac{X_C}{\sqrt{R^2 + X_C^2}} \ \underline{/-90° + \tan^{-1}(X_C/R)}$$

The magnitude of the ratio $\mathbf{V}_o/\mathbf{V}_i$ is now defined by

$$\frac{V_o}{V_i} = \frac{X_C}{\sqrt{R^2 + X_C^2}} = \frac{1}{\sqrt{\left(\dfrac{R}{X_C}\right)^2 + 1}} \qquad (20.49)$$

and the phase angle by

$$\theta = -90° + \tan^{-1}\frac{X_C}{R} \qquad (20.50)$$

At $X_C = R$,

$$\frac{V_o}{V_i} = \frac{1}{\sqrt{\left(\dfrac{R}{X_C}\right)^2 + 1}} = \frac{1}{\sqrt{1 + 1}}$$

and

$$\frac{V_o}{V_i} = \frac{1}{\sqrt{2}} = 0.707 \qquad (20.51)$$
$$X_C = R, f = f_2$$

as occurred for the high-pass filter.

At high frequencies, $\tan^{-1}(X_C/R)$ approaches 0°, and

$$\theta = -90° + \tan^{-1}\frac{X_C}{R} = -90° + 0°$$
$$= -90°$$

At low frequencies, $\tan^{-1}(X_C/R)$ approaches 90°, and

$$\theta = -90° + \tan^{-1}\frac{X_C}{R} = -90° + 90°$$
$$= 0°$$

At $X_C = R$, $\tan^{-1}(X_C/R) = \tan^{-1} 1 = 45°$, and

$$\theta = -90° + \tan^{-1}\frac{X_C}{R} = -90° + 45°$$
$$= -45°$$

A plot of θ versus frequency results in the phase plot of Fig. 20.48. The plot is of $\mathbf{V}_o$ leading $\mathbf{V}_i$, but since the phase angle is always negative, the phase plot of Fig. 20.49 is more appropriate. Note that a

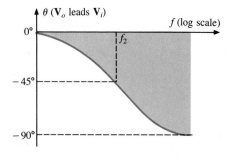

FIG. 20.48

Phase response for the low-pass R-C filter.

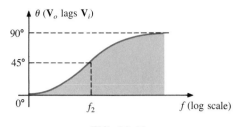

FIG. 20.49

Phase response for the low-pass R-C filter.

change in sign requires that the vertical axis be changed to the angle by which V_o lags V_i. The low-pass R-C filter is therefore a *lagging network*.

EXAMPLE 20.8. Given $R = 20\,k\Omega$ and $C = 1200\,pF$.

a. Sketch the magnitude plot if the filter is used as both a high- and a low-pass filter.
b. Sketch the phase plot for both filters of part (a).
c. Determine the magnitude and phase of V_o/V_i at $f = \frac{1}{2}f_1$ for the high-pass filter.

Solutions:

a. $f_1 = f_2 = \dfrac{1}{2\pi RC} = \dfrac{1}{(6.28)(20 \times 10^3)(1200 \times 10^{-12})}$

   $= \mathbf{6634.82\,Hz}$

The magnitude plots appear in Fig. 20.50.

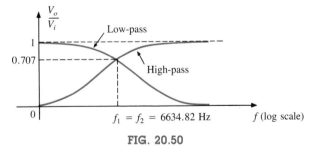

FIG. 20.50

b. The phase plots appear in Fig. 20.51.

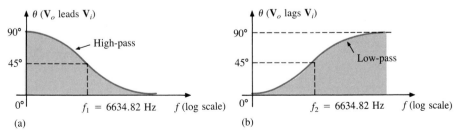

FIG. 20.51

c. $f = \dfrac{1}{2}f_1 = \dfrac{1}{2}(6634.82\,\text{Hz}) = 3317.41\,\text{Hz}$

$$X_C = \frac{1}{2\pi f C} = \frac{1}{(6.28)(3317.41)(1200 \times 10^{-12})}$$

$$\cong 40\,\text{k}\Omega$$

$$\frac{V_o}{V_i} = \frac{1}{\sqrt{1 + \left(\dfrac{X_C}{R}\right)^2}} = \frac{1}{\sqrt{1 + \left(\dfrac{40\,\text{k}\Omega}{20\,\text{k}\Omega}\right)^2}} = \frac{1}{\sqrt{1 + (2)^2}}$$

$$= \frac{1}{\sqrt{5}} = \mathbf{0.4472}$$

$$\theta = \tan^{-1}\frac{X_C}{R} = \tan^{-1}\frac{40\,\text{k}\Omega}{20\,\text{k}\Omega} = \tan^{-1} 2 = \mathbf{63.43°}$$

20.15 BODE PLOTS

It is standard practice in industry to plot the frequency response of filters, amplifiers, and systems in total against a *decibel* scale defined by the following equation:

$$\boxed{\text{Gain}_{\text{dB}} = 10\,\log_{10}\frac{P_2}{P_1}} \qquad (\text{decibels, dB}) \qquad \textbf{(20.52)}$$

Equation (20.52) permits the plotting of an extended range of interest on the same graph in much the same manner as demonstrated for frequency in Section 20.13. In addition, it permits dealing with gains of reasonable levels rather than very large magnitudes, as demonstrated in Table 20.2 of this section.

For the special case of $P_2 = 2P_1$, the gain in decibels is

$$G = 10\,\log_{10}\frac{P_2}{P_1} = 10\,\log_{10} 2 = 3\,\text{dB}$$

Therefore, for a speaker system, a 3-dB increase in output would require that the power level be doubled. In the audio industry, it is a generally accepted rule that an increase in sound level is accomplished with 3-dB increments in the output level. In other words, a 1-dB increase is barely detectable, and a 2-dB increase just discernible. A 3-dB increase normally results in a readily detectable increase in sound level. A further increase in the sound level is normally accomplished by simply increasing the output level another 3 dB. If an 8-W system were in use, a 3-dB increase would require a 16-W output, while a further increase of 3 dB (a total of 6 dB) would require a 32-W system, as demonstrated by the calculations below:

$$G = 10\,\log_{10}\frac{P_2}{P_1} = 10\,\log_{10}\frac{16}{8} = 10\,\log_{10} 2 = 3\,\text{dB}$$

$$G = 10 \log_{10} \frac{P_2}{P_1} = 10 \log_{10} \frac{32}{8} = 10 \log_{10} 4 = 6\,dB$$

For *similar load levels,* the power levels would be defined by

$$P_2 = \frac{V_2^2}{R} \quad \text{and} \quad P_1 = \frac{V_1^2}{R}$$

Substituting into Eq. (20.52),

$$G = 10 \log_{10} \frac{P_2}{P_1} = 10 \log_{10} \frac{V_2^2/R}{V_1^2/R}$$

$$= 10 \log_{10} \frac{V_2^2}{V_1^2} = 10 \log_{10} \left(\frac{V_2}{V_1}\right)^2$$

and

$$\boxed{G = 20 \log_{10} \frac{V_2}{V_1}} \qquad \text{(dB)} \qquad \textbf{(20.53)}$$

Our current interest lies in comparing an output voltage V_o to an input level V_i, resulting in

$$\boxed{G = 20 \log_{10} \frac{V_o}{V_i}} \qquad \textbf{(20.54)}$$

Table 20.2 compares the magnitude of specific gains to the resulting decibel level. In particular, note that when voltage levels are compared, a doubling of the level results in a change of 6 dB rather than 3 dB as obtained for power levels.

In addition, note that an increase in gain from 1 to 100,000 results in a change in decibels that can easily be plotted on a single graph. Also note that doubling the gain (from 1 to 2 and 10 to 20) results in a 6-dB increase in the decibel level, while a change of 10 to 1 (from 1 to 10, 10 to 100, and so on) always results in a 20-dB increase in the decibel level.

For any gain less than 1, the decibel level is negative, as demonstrated for $V_o/V_i = 0.707$ below:

$$G_{dB} = 20 \log_{10} \frac{V_o}{V_i} = 20 \log_{10} 0.707 = -3\,dB$$

For $V_o/V_i = (1/2)(0.707) = 0.3535$,

$$G_{dB} = 20 \log_{10} 0.3535 = -9\,dB$$

revealing, as above, that any change in level by 2 : 1 (whether increasing or decreasing) will result in a 6-dB change in the decibel level.

High-Pass R-C Filter

If we write the gain equation for the high-pass filter in the following manner:

TABLE 20.2

V_o/V_i	$G_{dB} = 20\,\log_{10}(V_o/V_i)$
1	0 dB
2	6 dB
10	20 dB
20	26 dB
100	40 dB
1,000	60 dB
100,000	100 dB

$$\mathbf{A}_v = \frac{\mathbf{V}_o}{\mathbf{V}_i} = \frac{R}{R - jX_C} = \frac{1}{1 - j\dfrac{X_C}{R}} = \frac{1}{1 - j\dfrac{1}{2\pi fCR}}$$

$$= \frac{1}{1 - j\left(\dfrac{1}{2\pi RC}\right)\dfrac{1}{f}}$$

and substitute

$$f_1 = \frac{1}{2\pi RC}$$

we find

$$\mathbf{A}_v = \frac{1}{1 - j(f_1/f)}$$

and

$$\boxed{\mathbf{A}_v = \frac{\mathbf{V}_o}{\mathbf{V}_i} = \frac{1}{\sqrt{1 + (f_1/f)^2}} \angle \tan^{-1}(f_1/f)} \qquad \textbf{(20.55)}$$

providing an equation for the magnitude and phase of the high-pass filter in terms of the frequency levels.

Using Eq. (20.54),

$$A_{v_{dB}} = 20 \log_{10} A_v$$

and, substituting the magnitude component of Eq. (20.55),

$$\boxed{A_{v_{dB}} = 20 \log_{10} \frac{1}{\sqrt{1 + (f_1/f)^2}}} \qquad \textbf{(20.56)}$$

In logs,

$$\log_{10} \frac{1}{x} = -\log_{10} x$$

and

$$A_{v_{dB}} = -20 \log_{10} \sqrt{1 + \left(\frac{f_1}{f}\right)^2}$$

Further,

$$\log_{10} \sqrt{x} = \frac{1}{2} \log_{10} x$$

and

$$A_{v_{dB}} = -\frac{1}{2}(20) \log_{10}\left[1 + \left(\frac{f_1}{f}\right)^2\right]$$

$$= -10 \log_{10}\left[1 + \left(\frac{f_1}{f}\right)^2\right]$$

For frequencies where $f \ll f_1$ or $(f_1/f)^2 \gg 1$,

$$1 + \left(\frac{f_1}{f}\right)^2 \cong \left(\frac{f_1}{f}\right)^2$$

and

$$A_{v_{dB}} = -10 \log_{10}\left(\frac{f_1}{f}\right)^2$$

but

$$\log_{10} x^2 = 2 \log_{10} x$$

with

$$\boxed{A_{v_{dB}} = -20 \log_{10}\frac{f_1}{f}} \qquad (20.57)$$
$$f \ll f_1$$

First note the similarities between Eq. (20.57) and the basic equation for gain in decibels: $G_{dB} = 20 \log_{10} V_o/V_i$. The comments regarding changes in decibel levels due to changes in V_o/V_i can therefore be applied here also, except now a change in frequency by a 2 : 1 ratio will result in a -6-dB change in gain due to the negative sign in Eq. (20.57). A change in frequency by a 10 : 1 ratio will result in a -20-dB change in gain. Any two frequencies separated by a 2 : 1 ratio are said to be an *octave* apart. Frequencies separated by a 10 : 1 ratio are said to be a *decade* apart. With this terminology, Eq. (20.57) will result in a 6-dB change in gain per octave and a 10-dB change per decade.

One may wonder about all the mathematical development to obtain an equation that initially appears confusing and of limited value. As specified, Eq. (20.57) is accurate only for frequency levels much less than f_1.

First, realize that the mathematical development of Eq. (20.57) will not have to be repeated for each configuration encountered. Second, the equation itself is seldom applied but simply used to define a straight line on a log plot that permits a sketch of the frequency response of a system with a minimum of effort and a high degree of accuracy. The resulting straight-line asymptotes are called a *Bode plot* of the frequency response.

To plot Eq. (20.57), consider the following levels:

For $f = f_1$, $f_1/f = 1$ and $-20 \log_{10} 1 = 0$ dB
For $f = f_1/2$, $f_1/f = 2$ and $-20 \log_{10} 2 = -6$ dB
For $f = f_1/4$, $f_1/f = 4$ and $-20 \log_{10} 4 = -12$ dB
For $f = f_1/10$, $f_1/f = 10$ and $-20 \log_{10} 10 = -20$ dB

A plot of these points on a log scale will result in the decibel plot of Fig. 20.52.

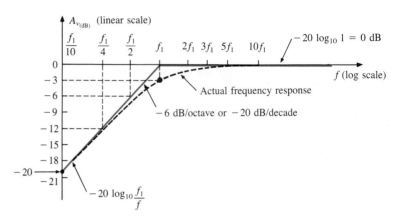

FIG. 20.52

Bode plot for the low-frequency region.

For the future, note that the resulting plot is a straight line intersecting the 0-dB line at f_1. It drops off to the left at a rate of -6 dB per octave or -10 dB per decade. In other words, once f_1 is determined, find $f_1/2$ and a plot point exists at -6 dB (or find $f_1/10$ and a plot point exists at -20 dB). The *actual* response curve will then pass through the -3-dB point at f_1 and approach the asymptote established by the straight line defined by Eq. (20.55). The curve will also approach an asymptote defined by $A_{v_{\mathrm{dB}}} = 0$ dB derived from the midband gain of $V_o = V_i$ ($V_o/V_i = 1$).

The phase response can be determined from Eq. (20.55) where

$$\theta = \tan^{-1}\frac{f_1}{f}$$

For frequencies where $f \ll f_1$, $\theta = \tan^{-1}(f_1/f)$ approaches $90°$, and for frequencies where $f \gg f_1$, $\theta = \tan^{-1}(f_1/f)$ approaches $0°$. At $f = f_1$, $\theta = \tan^{-1}(f_1/f) = \tan^{-1} 1 = 45°$. A plot of θ versus frequency appears in Fig. 20.53.

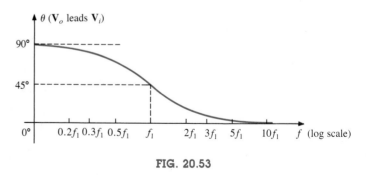

FIG. 20.53

EXAMPLE 20.9.

a. Sketch the frequency response for the high-pass R-C filter of Fig. 20.54.
b. Determine the decibel level at $f = 1$ kHz.

Solutions:

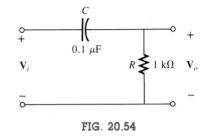

FIG. 20.54

a. $f_1 = \dfrac{1}{2\pi RC} = \dfrac{1}{(6.28)(1 \times 10^3)(0.1 \times 10^{-6})} = \mathbf{1592.36\ Hz}$

The frequency f_1 is identified on the log scale as shown in Fig. 20.55. A straight line is then drawn from f_1 with a slope that will intersect $-20\ dB$ at $f_1/10 = 159.24\ Hz$ or $-6\ dB$ at $f_1/2 = 796.18\ Hz$. The actual response curve can then be drawn through the -3-dB level at f_1 approaching the two asymptotes of Fig. 20.55.

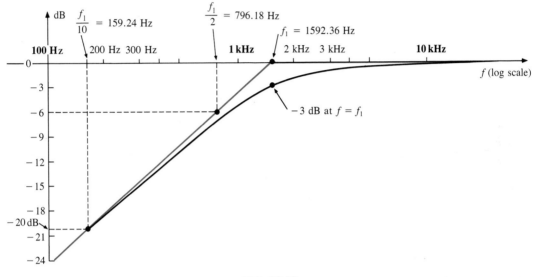

FIG. 20.55

Note in the above solution that there was no need to employ Eq. (20.57) or perform any extensive mathematical manipulations.

b. Eq. (20.56):

$$|A_{v_{dB}}| = 20\log_{10}\dfrac{1}{\sqrt{1 + \left(\dfrac{f_1}{f}\right)^2}} = 20\log_{10}\dfrac{1}{\sqrt{1 + \left(\dfrac{1592.36}{1000}\right)^2}}$$

$$= 20\log_{10}\dfrac{1}{\sqrt{1 + (1.592)^2}} = 20\log_{10} 0.5318$$

$$= \mathbf{-5.485\ dB}$$

Low-Pass R-C Filter

For the low-pass filter of Fig. 20.56,

$$\mathbf{A}_v = \dfrac{\mathbf{V}_o}{\mathbf{V}_i} = \dfrac{\mathbf{X}_C}{\mathbf{R} + \mathbf{X}_C} = \dfrac{-jX_C}{R - jX_C} = \dfrac{1}{\dfrac{R}{-jX_C} + 1}$$

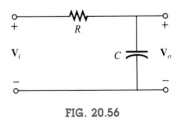

FIG. 20.56

$$= \frac{1}{1 + j\dfrac{R}{X_C}} = \frac{1}{1 + j\dfrac{R}{\dfrac{1}{2\pi f C}}} = \frac{1}{1 + j\dfrac{f}{\dfrac{1}{2\pi R C}}}$$

and

$$A_v = \frac{1}{1 + j\dfrac{f}{f_2}} \qquad \textbf{(20.58)}$$

with

$$f_2 = \frac{1}{2\pi R C}$$

as defined earlier.

Note that now the sign of the imaginary component in the denominator is positive and f_2 appears in the denominator of the frequency ratio rather than in the numerator as in the case of f_1 for the high-pass filter.

In terms of magnitude and phase,

$$\mathbf{A}_v = \frac{\mathbf{V}_o}{\mathbf{V}_i} = \frac{1}{\sqrt{1 + (f/f_2)^2}} \; \angle -\tan^{-1}(f/f_2) \qquad \textbf{(20.59)}$$

An analysis similar to that performed for the high-pass filter will result in

$$A_{v_{\text{dB}}} = -20 \log_{10} \frac{f}{f_2} \qquad \textbf{(20.60)}$$
$$f \gg f_2$$

Note in particular that the equation is exact only for frequencies much greater than f_2, but a plot of Eq. (20.60) does provide an asymp-

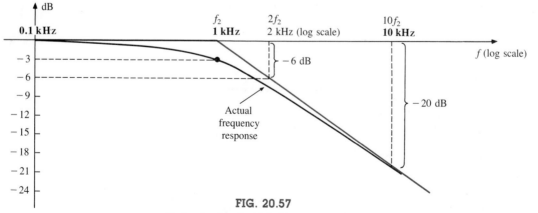

FIG. 20.57
Bode plot for the high-frequency region.

tote that performs the same function as the asymptote derived for the high-pass filter.

A plot of Eq. (20.60) appears in Fig. 20.57 for $f_2 = 1$ kHz. Note the 6-dB drop at $f = 2f_2$ and the 20-dB drop at $f = 10f_2$.

At $f \gg f_2$, the phase angle $\theta = -\tan^{-1}(f/f_2)$ approaches $-90°$, while for frequencies $f \ll f_2$, $\theta = -\tan^{-1}(f/f_2)$ approaches $0°$. At $f = f_2$, $\theta = -\tan^{-1} 1 = -45°$, confirming the plot of Fig. 20.48.

20.16 TUNED FILTERS

Series and parallel resonance can be utilized in the design of band-pass, band-stop, and double-tuned filters. Each of these filter types will be briefly described in this section.

Band-Pass Filter

The general characteristics of a band-pass filter are provided in Fig. 20.58.

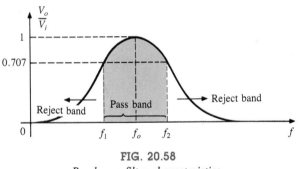

FIG. 20.58
Band-pass filter characteristics.

The response of Fig. 20.52 can be obtained using either network of Fig. 20.59.

For the series resonance configuration of Fig. 20.59(a), the quality factor of the network is

$$Q_s = \frac{X_L}{R} \qquad (20.61)$$

Applying the voltage divider rule yields

$$\mathbf{V}_o = \mathbf{V}_R = \frac{R\mathbf{V}_i}{R + j(X_L - X_C)}$$

At resonance, $X_L = X_C$, and

$$\mathbf{V}_o = \frac{R\mathbf{V}_i}{R} = \mathbf{V}_i$$

For frequencies to the right and left of resonance, the impedance of the L-C combination increases over that of R, resulting in a decreasing voltage across R. The resulting response curve is shown in Fig. 20.59(a).

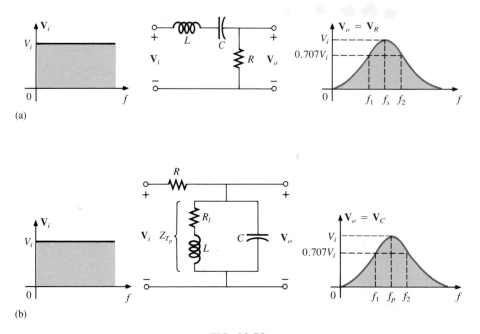

FIG. 20.59
Band-pass filters.

The parallel resonant circuit is employed in the network of Fig. 20.59(b). For this configuration, the impedance of the tank circuit is resistive at resonance, and

$$\mathbf{V}_o = \frac{Z_{T_p}\mathbf{V}_i}{Z_{T_p} + R} \qquad (20.62)$$

If $Q \geq 10$, $Z_{T_p} = Q^2 R_l$, and

$$\mathbf{V}_o = \frac{Q^2 R_l \mathbf{V}_i}{Q^2 R_l + R} \qquad \qquad (20.63)$$
$$Q \geq 10$$

For $Q^2 R_l \gg R$, $Q^2 R_l + R \cong Q^2 R_l$, and

$$\mathbf{V}_o \cong \mathbf{V}_i \qquad f = f_p, \ Q \geq 10, \ Q^2 R_l \gg R \qquad (20.64)$$

The cutoff frequencies are not simply defined by the resonant network. For the parallel resonant circuit, f_1 and f_2 are defined by $Z_T = 0.707 Z_{T_p}$, but V_o may not be $0.707 V_i$ for this impedance level due to the resistance R.

The cutoff frequency must be determined from the following condition:

$$\left| \frac{V_o}{V_i} \right| = 0.707 \quad \text{where} \quad \frac{\mathbf{V}_o}{\mathbf{V}_i} = \frac{\mathbf{Z}_T}{\mathbf{R} + \mathbf{Z}_T} = \frac{Z_T \angle \theta}{R + Z_T \angle \theta} \qquad \textbf{(20.65)}$$

In other words, the cutoff frequencies are those frequencies that satisfy the condition defined by Eq. (20.65).

As you review the network of Fig. 20.59(b), keep in mind that if $R = 0\,\Omega$, $\mathbf{V}_o = \mathbf{V}_i$ for all frequencies and the output is flat. The chosen value of R will therefore have an important impact on the shape and bandwidth of the tuned network. If R is too large, or $R \gg Q^2 R_l$, the output will stay almost flat at a low percentage of V_i.

Band-Stop Filter

The general characteristics of a band-stop filter are provided in Fig. 20.60.

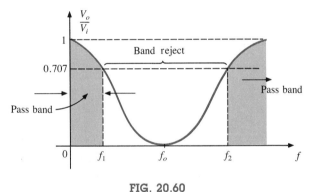

FIG. 20.60
Band-stop filter characteristics.

Since the characteristics of Fig. 20.60 are the inverse of the pattern obtained for the band-pass filters, we can employ the fact that at any frequency the sum of the magnitude of the two waveforms to the right of the equal sign in Fig. 20.61 will equal the applied voltage V_i.

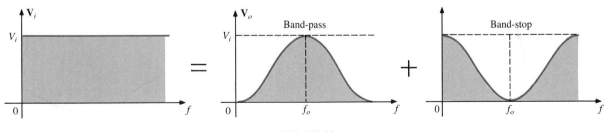

FIG. 20.61

For the band-pass filters of Fig. 20.59, therefore, if we take the output off the other series elements as shown in Fig. 20.62, a band-stop characteristic will be obtained as required by Kirchhoff's voltage law.

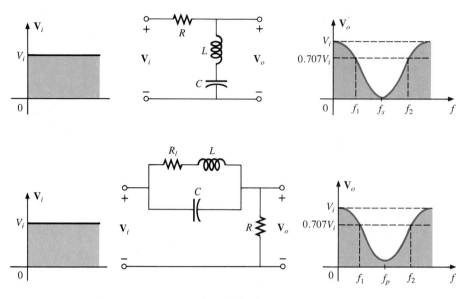

FIG. 20.62

The band-reject and band-pass characteristics of miniature filters manufactured by TRW/UTC inductive products appear in Fig. 20.63 along with a photograph of a typical unit. Note that the band-pass characteristics are a universal set since the resonant frequency is undefined and the horizontal axis is the ratio f/f_r. When examining the curves, remember that the vertical scale is a measure of the attenuation of the input signal. That is, at 0 dB the output is equal to the input, while at -30 dB the output is significantly less than the input. Since each of these units can be used for the pass-band or stop-band function, data about each application are provided. Note that the pass band is inserted by the center frequency for each unit and the type number of the unit reflects the center frequency. The stop band is implying essentially zero response with the -35-dB criterion. Note that the center frequency does not bisect the stop band since the resonance curve is not symmetrical about f_o.

Double-Tuned Filter

There are some network configurations that display both a pass-band and a band-stop characteristic, such as shown in Fig. 20.64. For the network of Fig. 20.64(a), the parallel resonant circuit will establish the band stop by resonating at the frequency not permitted to establish a $\mathbf{V}_L$. The greater part of the applied voltage $\mathbf{E}$ will appear across this parallel resonant circuit at this frequency due to its very high impedance compared with R_L.

For the pass band, the parallel resonant circuit is designed to be capacitive (inductive if L_s is replaced by C_s). The inductance L_s is chosen to cancel the effects of the resulting capacitive reactance at the resonant pass-band frequency of the tank circuit, thereby acting as a

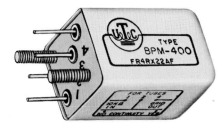

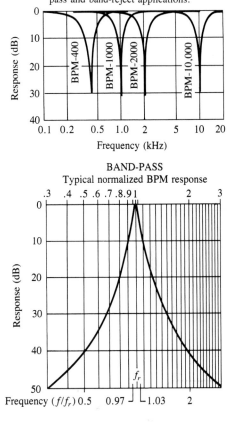

BAND-PASS
Typical normalized BPM response

Type No.	Center Frequency (Hz)	Pass Band (Less than 2 dB) (Hz)	Stop Band (More than 35 dB)	
			Below (Hz)	Above (Hz)
BPM 400	400	388–412	200	800
BPM 440	440	427–453	220	880
BPM 500	500	485–515	250	1,000
BPM 600	600	582–618	300	1,200
BPM 800	800	776–824	400	1,600
BPM 1000	1,000	970–1,030	500	2,000
BPM 1200	1,200	1,164–1,236	600	2,400
BPM 1500	1,500	1,455–1,545	750	3,000
BPM 1600	1,600	1,552–1,648	800	3,200
BPM 2000	2,000	1,940–2,060	1,000	4,000
BPM 2500	2,500	2,425–2,575	1,250	5,000
BPM 3000	3,000	2,910–3,090	1,500	6,000
BPM 3200	3,200	3,104–3,296	1,600	6,400
BPM 4000	4,000	3,880–4,120	2,000	8,000
BPM 4800	4,800	4,656–4,944	2,400	9,600
BPM 5000	5,000	4,850–5,150	2,500	10,000
BPM 6000	6,000	5,820–6,180	3,000	12,000
BPM 8000	8,000	7,760–8,240	4,000	16,000
BPM 10000	10,000	9,700–10,300	5,000	20,000
BPM 20000	20,000	19,400–20,600	10,000	40,000

FIG. 20.63
Band-pass and band-reject characteristics for BPM TRW/UTC filters.
(Courtesy of United Transformer Corp.)

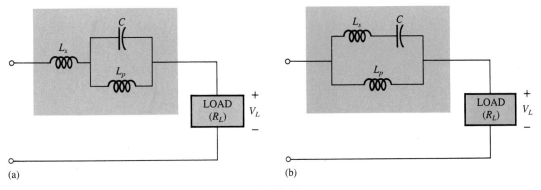

FIG. 20.64
Double-tuned networks.

series resonant circuit. The applied voltage **E** will then appear across R_L at this frequency.

For the network of Fig. 20.64(b), the series resonant circuit will still determine the pass band, acting as a very low impedance across the parallel inductor at resonance, establishing $V_L = E$. At the desired band-stop resonant frequency, the series resonant circuit is capacitive. The inductance L_p is chosen to establish parallel resonance at the resonant band-stop frequency. The high impedance of the parallel resonant circuit will result in a very low load voltage V_L.

For rejected frequencies below the pass band, the networks should appear as shown in Fig. 20.64. For the reverse situation, L_s in Fig. 20.64(a) and L_p in Fig. 20.64(b) are replaced by capacitors.

EXAMPLE 20.10. For the network of Fig. 20.64(b), determine L_s and L_p for a capacitance C of 500 pF if a frequency of 200 kHz is to be rejected and a frequency of 600 kHz accepted.

Solution: For series resonance, we have

$$f_s = \frac{1}{2\pi\sqrt{LC}}$$

and

$$L_s = \frac{1}{4\pi^2 f_s^2 C} = \frac{1}{4(3.14)^2(600 \times 10^3)^2(500 \times 10^{-12})}$$

$$= \mathbf{141\,\mu H}$$

At 200 kHz,

$$X_L = \omega L = 2\pi f_s L = (6.28)(200 \times 10^3)(141 \times 10^{-6})$$
$$= 177.1\,\Omega$$

and

$$X_C = \frac{1}{\omega C} = \frac{1}{(6.28)(200 \times 10^3)(500 \times 10^{-12})}$$

$$= 1.59\,\text{k}\Omega$$

For the series elements,

$$j(X_L - X_C) = j(177.1 - 1592) = -j1414.8 \text{ (cap.)}$$

At parallel resonance ($Q \geq 10$ assumed),

$$X_L = X_C$$

and

$$L_p = \frac{X_C}{\omega} = \frac{1412.9}{(6.28)(200 \times 10^3)} = \mathbf{1.125\,mH}$$

PROBLEMS

Sections 20.1 through 20.7

1. Find the resonant ω_s and f_s for the series circuit with the following parameters:
   a. $R = 10\,\Omega$, $L = 1\,\text{H}$, $C = 16\,\mu\text{F}$
   b. $R = 300\,\Omega$, $L = 0.5\,\text{H}$, $C = 0.16\,\mu\text{F}$
   c. $R = 20\,\Omega$, $L = 0.28\,\text{mH}$, $C = 7.46\,\mu\text{F}$

2. For the series circuit of Fig. 20.65:
   a. Find the value of X_C for resonance.
   b. Find the current **I** and the voltages $\mathbf{V}_R$, $\mathbf{V}_L$, and $\mathbf{V}_C$ in phasor form at resonance.
   c. Draw the phasor diagram of the voltages and current.
   d. Sketch the power triangle for the circuit at resonance.
   e. Find the Q_s of the circuit.

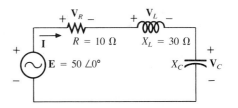

FIG. 20.65

3. Repeat Problem 2 for the circuit of Fig. 20.66.

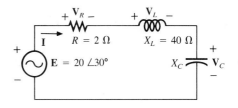

FIG. 20.66

4. For the circuit of Fig. 20.67:
   a. Find the value of L in millihenries if the resonant frequency is 1800 Hz.
   b. Repeat Problem 2, parts (b) through (e).
   c. Calculate the cutoff frequencies.
   d. Find the bandwidth of the series resonant circuit.

5. a. Find the bandwidth of a series resonant circuit having a resonant frequency of 6000 Hz and a Q_s of 15.
   b. Find the cutoff frequencies.
   c. If the resistance of the circuit at resonance is $3\,\Omega$, what are the values of X_L and X_C in ohms?
   d. What is the power dissipated at the half-power frequencies if the maximum current flowing through the circuit is 0.5 A?

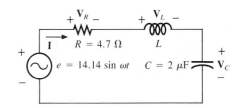

FIG. 20.67

6. A series circuit has a resonant frequency of 10 kHz. The resistance of the circuit is $5\,\Omega$, and X_C at resonance is $200\,\Omega$.
   a. Find the bandwidth.
   b. Find the cutoff frequencies.
   c. Find Q_s.
   d. If the input voltage is $30\,\angle 0°$, find the voltage across the coil and capacitor.
   e. Find the power dissipated at resonance.

7. a. The bandwidth of a series resonant circuit is 200 Hz. If the resonant frequency is 2000 Hz, what is the value of Q_s for the circuit?
  b. If $R = 2\,\Omega$, what is the value of X_L at resonance?
  c. Find the value of L and C at resonance.
  d. Find the cutoff frequencies.

8. The cutoff frequencies of a series resonant circuit are 5400 Hz and 6000 Hz.
  a. Find the bandwidth of the circuit.
  b. If Q_s is 9.5, find the resonant frequency of the circuit.
  c. If the resistance of the circuit is $2\,\Omega$, find the value of X_L and X_C at resonance.
  d. Find the value of L and C at resonance.

*9. Design a series resonant circuit with an input voltage of $5\,\angle 0°$ to have the following specifications:
  a. a peak current of 500 mA at resonance
  b. a bandwidth of 120 Hz
  c. a resonant frequency of 8400 Hz
  Find the value of L and C and the cutoff frequencies.

*10. Design a series resonant circuit to have a bandwidth of 400 Hz using a coil with a Q of 20 and a resistance of $2\,\Omega$. Find the value of L and C and the cutoff frequencies.

Sections 20.8 through 20.12

11. For the circuit of Fig. 20.68:
  a. Find the value of X_C at resonance.
  b. Find the total impedance Z_T at resonance.
  c. Find the currents I_L and I_C at resonance.
  d. If the resonant frequency is 20,000 Hz, find the value of L and C at resonance.
  e. Find Q_p and BW.

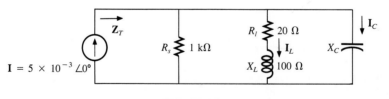

FIG. 20.68

12. Repeat Problem 11 for the circuit of Fig. 20.69.

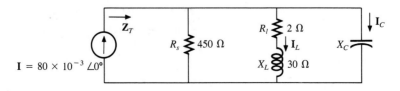

FIG. 20.69

13. For the circuit of Fig. 20.70:
  a. Find the resonant frequency.
  b. Find the value of X_L and X_C at resonance.
  c. Is the coil a high-Q or low-Q coil at resonance?
  d. Find the impedance Z_{T_p} at resonance.
  e. Find the currents I_L and I_C at resonance.
  f. Calculate Q_p and BW.

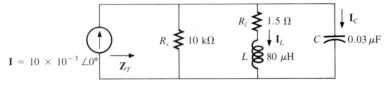

FIG. 20.70

14. Repeat Problem 13 for the circuit of Fig. 20.71.

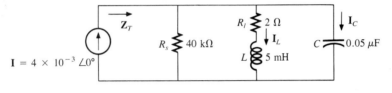

FIG. 20.71

15. It is desired that the impedance Z_T of the circuit of Fig. 20.72 be a resistor of 50 kΩ at resonance.
  a. Find the value of X_L.
  b. Compute X_C.
  c. Find the resonant frequency if $L = 16$ mH.
  d. Find the value of C.

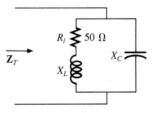

FIG. 20.72

16. For the network of Fig. 20.73:
  a. Find f_p.
  b. Calculate V_C at resonance.
  c. Determine the power absorbed at resonance.
  d. Find BW.

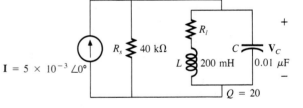

FIG. 20.73

*17. For the network of Fig. 20.74, the following are specified:

$$f_p = 100 \text{ kHz}$$
$$BW = 2500 \text{ Hz}$$
$$L = 2 \text{ mH}$$
$$Q = 80$$

Find R_s and C.

FIG. 20.74

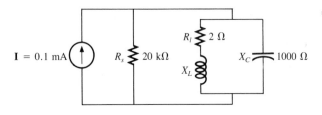

FIG. 20.75

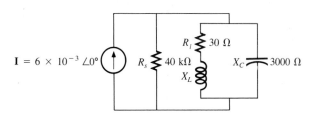

FIG. 20.76

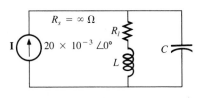

FIG. 20.77

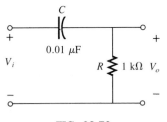

FIG. 20.78

*18. For the network of Fig. 20.75:
  a. Find the value of X_L for resonance.
  b. Find Q.
  c. Find the resonant frequency if the bandwidth is 1000 Hz.
  d. Find the maximum value of the voltage V_C.
  e. Sketch the curve of V_C versus frequency. Indicate its peak value, resonant frequency, and band frequencies.

*19. Repeat Problem 18 for the network of Fig. 20.76.

*20. Design the network of Fig. 20.77 to have the following characteristics:
  a. a bandwidth of 500 Hz
  b. $Q_p = 30$
  c. $V_{C_{max}} = 1.8$ V

Section 20.13

21. Using semi-log paper:
  a. Plot X_L versus frequency for a 10-mH coil for a frequency range of 1 Hz to 1 MHz. Set the vertical scale by the maximum value of X_L.
  b. Plot X_C versus frequency for a 1-μF capacitor for a frequency range of 10 Hz to 100 kHz. Choose an appropriate vertical scale for the frequency range of interest.

Section 20.14

22. For the high-pass R-C filter of Fig. 20.78:
  a. Determine the cutoff frequency.
  b. Plot V_o/V_i on semi-log paper for a frequency range of 1 Hz to 1 MHz.
  c. Plot the phase response for the frequency range in part (b).

23. For a low-pass R-C filter having the same elements as in Problem 22:
  a. Sketch the network identifying V_o and V_i.
  b. Determine the cutoff frequency.
  c. Plot V_o/V_i on semi-log paper for a frequency range of 1 Hz to 1 MHz.
  d. Plot the phase response for the frequency range of part (c).

24. Design a high-pass R-C filter having a cutoff frequency of 2 kHz given a capacitor of 0.1 μF. Sketch the response V_o/V_i on semi-log paper for a frequency range of 10 Hz to 100 kHz. Use commercial resistor values.

Section 20.15

*25. a. Sketch the Bode plot for the frequency response of a high-pass R-C filter if $R = 0.47 \, k\Omega$ and $C = 0.05 \, \mu F$.
   b. Using the results of part (a), sketch the actual frequency response for the same frequency range.
   c. Determine the decibel level at frequencies equal to one-half and twice the cutoff frequency.
   d. Determine the gain V_o/V_i at both frequencies of part (c).
   e. Sketch the phase response for the same frequency range.

*26. Repeat Problem 25 for an R-C low-pass filter if $R = 12 \, k\Omega$ and $C = 0.001 \, \mu F$.

27. Sketch the Bode plot of the following function:

$$\mathbf{A}_v = \frac{0.05}{0.05 - j100/f}$$

28. Sketch the Bode plot of the following function:

$$\mathbf{A}_v = \frac{200}{200 + j0.1f}$$

Section 20.16

29. For the pass-band filter of Fig. 20.79:
   a. Determine Q_s.
   b. Find the cutoff frequencies.
   c. Sketch the frequency characteristics.
   d. Find $Q_{s(loaded)}$ if a load of $200 \, \Omega$ is applied.
   e. Indicate on the curve of part (c) the change in the frequency characteristics with the load applied.

FIG. 20.79

*30. For the pass-band filter of Fig. 20.80:
   a. Determine Q_p ($R_L = \infty \, \Omega$, open circuit).
   b. Sketch the frequency characteristics.
   c. Find $Q_{p(loaded)}$ for $R_L = 100 \, k\Omega$, and indicate the effect of R_L on the characteristics of part (b).
   d. Repeat part (c) for $R_L = 20 \, k\Omega$.

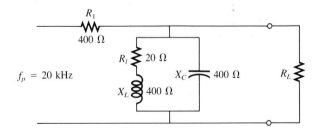

FIG. 20.80

31. For the band-stop filter of Fig. 20.81:
   a. Determine Q_s.
   b. Find the bandwidth and half-power frequencies.
   c. Sketch the frequency characteristics.
   d. What is the effect on the curve of part (c) if a load of $2 \, k\Omega$ is applied?

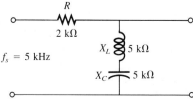

FIG. 20.81

32. The network of Problem 30 is used as a band-stop filter.
   a. Sketch the network when used as a band-stop filter.
   b. Sketch the band-stop characteristics.
   c. What is the effect on the band-stop characteristics if the 100-kΩ and 20-kΩ loads are applied?

33. a. For the network of Fig. 20.64(a), if $L_p = 400\,\mu$H $(Q > 10)$, $L_s = 60\,\mu$H, and $C = 120\,$pF, determine the rejected and accepted frequencies.
   b. Sketch the response curve for part (a).

34. a. For the network of Fig. 20.64(b), if the rejected frequency is 30 kHz, and the accepted, 100 kHz, determine the values of L_s and L_p $(Q \geq 10)$ for a capacitance of 200 pF.
   b. Sketch the response curve for part (a).

Computer Problems

35. Write a program to tabulate the impedance and current of the network of Fig. 20.2 versus frequency for a frequency range extending from $0.1f_s$ to $2f_s$ in increments of $0.1f_s$. For the first run, use the parameters defined by Example 20.1.

36. Write a program to provide a general solution for the network of Fig. 20.31. That is, given the parameters appearing in Fig. 20.31, determine the quantities appearing in parts (a) through (f) of Example 20.5. For the first run, use the parameters appearing in Example 20.5, and compare results.

37. Write a program that will tabulate the gain of Eq. (20.44) versus frequency for a frequency range extending from $0.1f_1$ to $2f_1$ in increments of $0.1f_1$. Note whether $f = f_1$ when $V_o/V_i = 0.707$.

38. Repeat Problem 37 for the phase angle of Eq. (20.45). Note whether $\theta = 45°$ when $f = f_1$.

39. Write a program to tabulate $A_{v_{dB}}$ as determined by Eq. (20.56) and $A_{v_{dB}}$ as calculated by Eq. (20.57). For a frequency range extending from $0.01f_1$ to f_1 in increments of $0.01f_1$, compare the magnitudes, and note whether the values are closer when $f \ll f_1$ and whether $A_{v_{dB}} = -3\,$dB at $f = f_1$ for Eq. (20.56) and zero for Eq. (20.57).

GLOSSARY

Band (cutoff, half-power, corner, −3-dB) frequencies Frequencies that define the points on the resonance curve that are 0.707 of the peak current or voltage value. In addition, they define the frequencies at which the power transfer to the resonant circuit will be half the maximum power level (−3 dB).

Band-stop filter A network designed to reject (not pass) signals within a particular range of frequencies.

Bandwidth The range of frequencies between the band, cutoff, or half-power frequencies.

Bode plot A plot (envelope) of the frequency response of a system using straight-line segments called asymptotes.

Double-tuned filter   A network having both a pass-band and a band-stop region.

Filter   Networks designed to either pass or reject the transfer of signals at certain frequencies to a load.

High-pass filter   A filter designed to pass high frequencies and reject low frequencies.

Low-pass filter   A filter designed to pass low frequencies and reject high frequencies.

Pass-band (band-pass) filter   A network designed to pass signals within a particular range of frequencies.

Quality factor (Q)   A ratio that provides an immediate indication of the sharpness of the peak of a resonance curve. The higher the Q, the sharper the peak and the more quickly it drops off to the right and left of the resonant frequency.

Reactance chart   A graph that is very useful in the analysis and design of series or parallel resonant circuits.

Resonance   A condition established by the application of a particular frequency (the resonant frequency) to a series or parallel R-L-C network. The transfer of power to the system is a maximum and, for frequencies above and below, the power transfer drops off to significantly lower levels.

Selectivity   A characteristic of resonant networks directly related to the bandwidth of the resonant system. High selectivity is associated with small bandwidth (high Q's), and low selectivity with larger bandwidths (low Q's).

Semi-log paper   Graph paper with one log scale and one linear scale.

21

Pulse Waveforms and the *R-C* Response

21.1 INTRODUCTION

Our analysis thus far has been limited to alternating waveforms that vary in a sinusoidal manner. This chapter will introduce the basic terminology associated with the pulse waveform and examine the response of an *R-C* circuit to a square-wave input. The importance of the pulse waveform to the electrical/electronics industry cannot be overstated. A vast array of instrumentation, communication systems, computers, radar systems, and so on, all employ pulse signals to control operation, transmit data, and display information in a variety of formats.

The response of the networks described thus far to a pulse signal is quite different from that obtained for sinusoidal signals. In fact, we will be returning to the dc chapter on capacitors to retrieve a few fundamental concepts and equations that will help us in the analysis to follow. The content of this chapter is quite introductory in nature, designed simply to provide the fundamentals that will be helpful when the pulse waveform is encountered in specific areas of application.

21.2 IDEAL VERSUS ACTUAL

The *ideal* pulse of Fig. 21.1 has vertical sides, sharp corners, a flat peak characteristic, and starts instantaneously at t_1 and ends just as abruptly at t_2.

The waveform of Fig. 21.1 will be applied in the analysis to follow in this chapter and will probably appear in the initial investigation of areas of application beyond the scope of this text. Once the fundamental

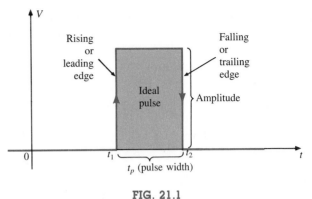

FIG. 21.1
Ideal pulse waveform.

operation of a device, package, or system is clearly understood using ideal characteristics, the effect of a *true, actual,* or *practical* pulse must be considered. If an attempt were made to introduce all the differences between an ideal and actual pulse in a single figure, the result would probably be complex and confusing. A number of waveforms will therefore be used to define the critical parameters.

The reactive elements of a network, in their effort to prevent instantaneous changes in voltage (capacitor) and current (inductor), establish a slope to both edges of the pulse waveform as shown in Fig. 21.2. The

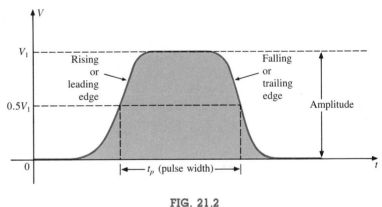

FIG. 21.2
Actual pulse waveform.

rising edge of the waveform of Fig. 21.2 is defined as the edge that increases from a lower to a higher level. The *falling* edge is defined by the region or edge where the waveform decreases from a higher to a lower level. Since the rising edge is the first to be encountered, it is also called the *leading* edge. The falling edge always follows the leading edge and is therefore often called the *trailing* edge. Both regions are defined in Figs. 21.1 and 21.2.

Amplitude

For most applications, the amplitude of a pulse waveform is defined as the peak-to-peak value. Of course, if the waveforms all start and return to the zero-volt level, then the peak and peak-to-peak values are synonymous. For the purposes of this text, the amplitude of a pulse waveform is the peak-to-peak value, as illustrated in Figs. 21.1 and 21.2.

Pulse Width

The *pulse width* (t_p), or *pulse duration*, is defined by a pulse level equal to 50% of the peak value. For the ideal pulse of Fig. 21.1, the pulse width is the same at any level, whereas t_p for the waveform of Fig. 21.2 is a very specific value.

Base-Line Voltage

The *base-line voltage* (V_b) is the voltage level from which the pulse is initiated. The waveforms of Figs. 21.1 and 21.2 both have a 0-V baseline voltage. In Fig. 21.3(a) the base-line voltage is 1 V, while in Fig. 21.3(b) the base-line voltage is −4 V.

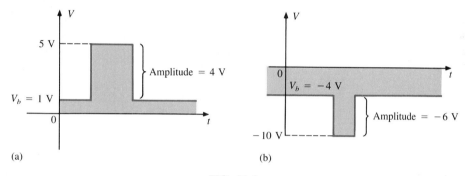

FIG. 21.3
Defining the base-line voltage.

Positive-Going and Negative-Going Pulses

A *positive-going* pulse increases positively from the base-line voltage. Quite obviously, therefore, a *negative-going* pulse increases in the negative direction from the base-line voltage. The waveform of Fig. 21.3(a) is a positive-going pulse, whereas the waveform of Fig. 21.3(b) is a negative-going pulse.

Even though the base-line voltage of Fig. 21.4 is negative, the waveform is positive-going (with an amplitude of 10 V) since the voltage increased in the positive direction from the base-line voltage.

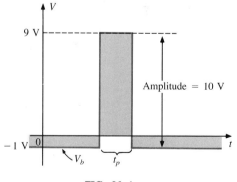

FIG. 21.4
Positive-going pulse.

Rise Time (t_r) and Fall Time (t_f)

Of particular importance is the time required for the pulse to shift from one level to another. The *rounding* (defined in Fig. 21.5) that occurs at the beginning and end of each transition makes it difficult to define the exact point at which the rise time should be initiated and terminated. For this reason, the *rise* and *fall* times are defined by the 10% and 90% levels as indicated in Fig. 21.5. Note that there is no requirement that t_r equal t_f.

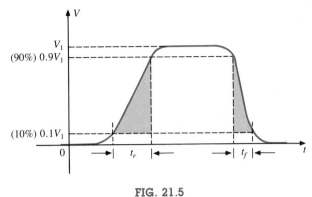

FIG. 21.5

Defining t_r and t_f.

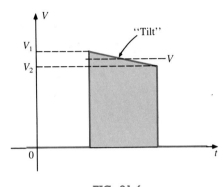

FIG. 21.6

Defining tilt.

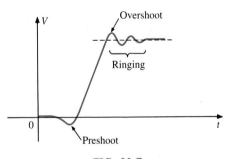

FIG. 21.7

Defining preshoot, overshoot, and ringing.

Tilt

An undesirable but common distortion normally occurring due to a poor low-frequency response characteristic of the system through which a pulse has passed appears in Fig. 21.6. The drop in peak value is called *tilt*, *droop*, or *sag*. The percent tilt is defined by

$$\% \text{ tilt} = \frac{V_1 - V_2}{V} \times 100\% \qquad \textbf{(21.1)}$$

where V is the average value of the peak amplitude as determined by

$$V = \frac{V_1 + V_2}{2} \qquad \textbf{(21.2)}$$

Naturally, the less the percent tilt or sag, the more ideal the pulse. Due to rounding, it may be difficult to define the value of V_1 and V_2. It is then necessary only to approximate the sloping region by a straight-line approximation and use the resulting values of V_1 and V_2.

Other distortions include the *preshoot* and *overshoot* appearing in Fig. 21.7, normally due to pronounced high-frequency effects of a system, and *ringing*, due to the interaction between the capacitive and inductive elements of a network at their natural or resonant frequency.

EXAMPLE 21.1. Determine the following for the pulse waveform of Fig. 21.8:
a. positive- or negative-going?
b. base-line voltage
c. pulse width
d. maximum amplitude
e. tilt

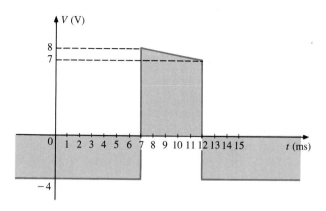

FIG. 21.8

Solutions:

a. **positive-going**

b. $V_b = \mathbf{-4\,V}$

c. $t_p = (12 - 7)\,\text{ms} = \mathbf{5\,ms}$

d. $V_{\max} = 8 + 4 = \mathbf{12\,V}$

e. $V = \dfrac{V_1 + V_2}{2} = \dfrac{12 + 11}{2} = \dfrac{23}{2} = 11.5$

$\%\text{ tilt} = \dfrac{V_1 - V_2}{V} \times 100\% = \dfrac{12 - 11}{11.5} \times 100\% = \mathbf{8.696\%}$

(Remember, V is defined by the average value of the peak ampli-tude.)

EXAMPLE 21.2. Determine the following for the pulse waveform of Fig. 21.9:

a. positive- or negative-going?

b. base-line voltage

c. tilt

d. amplitude

e. t_p

f. t_r and t_f

Solutions:

a. **positive-going**

b. $V_b = \mathbf{0\,V}$

c. $\%\text{ tilt} = \mathbf{0\%}$

d. amplitude $= (4\,\text{div.})(10\,\text{mV/div.}) = \mathbf{40\,mV}$

e. $t_p = (3.2\,\text{div.})(5\,\mu\text{s/div.}) = \mathbf{16\,\mu s}$

f. $t_r = (0.4\,\text{div.})(5\,\mu\text{s/div.}) = \mathbf{2\,\mu s}$

$t_f = (0.8\,\text{div.})(5\,\mu\text{s/div.}) = \mathbf{4\,\mu s}$

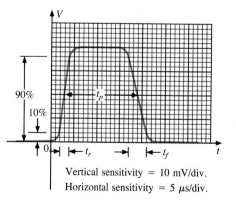

Vertical sensitivity $= 10\,\text{mV/div.}$
Horizontal sensitivity $= 5\,\mu\text{s/div.}$

FIG. 21.9

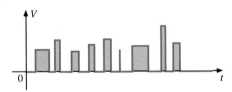

FIG. 21.10
Pulse train.

21.3 PULSE REPETITION RATE AND DUTY CYCLE

A series of pulses such as those appearing in Fig. 21.10 is called a *pulse train*. The varying widths and heights may contain information that can be decoded at the receiving end.

If the pattern repeats itself in a periodic manner as shown in Figs. 21.11(a) and (b), the result is called a *periodic pulse train*.

The *period (T)* of the pulse train is defined as the time differential between any two similar points on the pulse train as shown in Figs. 21.11(a) and (b).

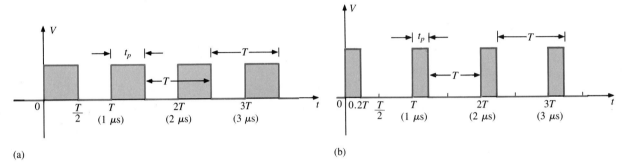

(a)

(b)

FIG. 21.11
Periodic pulse trains.

The *pulse repetition frequency* (prf), or *pulse repetition rate* (prr), is defined by

$$\text{prf (or prr)} = \frac{1}{T} \qquad \text{(Hz or pulses/s)} \qquad \textbf{(21.3)}$$

Applying Eq. (21.3) to each waveform of Fig. 21.11 will result in the same pulse repetition frequency since the periods are the same. The result clearly reveals that the shape of the periodic pulse has nothing to do with the pulse repetition frequency. The pulse repetition frequency is determined solely by the period of the repeating pulse. The factor that will reveal how much of the period is encompassed by the pulse is called the *duty cycle*, defined as follows:

$$\text{Duty cycle} = \frac{\text{pulse width}}{\text{period}} \times 100\%$$

or

$$\text{Duty cycle} = \frac{t_p}{T} \times 100\% \qquad \textbf{(21.4)}$$

For Fig. 21.11(a) (a square-wave pattern),

$$\text{Duty cycle} = \frac{0.5T}{T} \times 100\% = \textbf{50\%}$$

and for Fig. 21.11(b),

$$\text{Duty cycle} = \frac{0.2T}{T} \times 100\% = \mathbf{20\%}$$

The above results clearly reveal that the duty cycle provides a percent indication of the portion of the total period encompassed by the pulse waveform.

EXAMPLE 21.3.   Determine the pulse repetition frequency and the duty cycle for the periodic pulse waveform of Fig. 21.12.

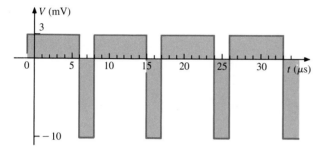

FIG. 21.12

Solution:

$$T = (15 - 6)\,\mu s = 9\,\mu s$$

$$\text{prf} = \frac{1}{T} = \frac{1}{9\,\mu s} \cong \mathbf{0.111\ MHz}$$

$$\text{Duty cycle} = \frac{t_p}{T} \times 100\% = \frac{(8 - 6)\,\mu s}{9\,\mu s} \times 100\%$$

$$= \frac{2}{9} \times 100\% \cong \mathbf{22.22\%}$$

EXAMPLE 21.4.   Determine the pulse repetition frequency and the duty cycle for the periodic waveform of Fig. 21.13.

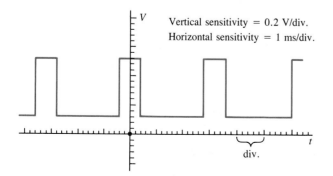

Vertical sensitivity = 0.2 V/div.
Horizontal sensitivity = 1 ms/div.

div.

FIG. 21.13

Solution:

$$T = (3.2\,\text{div.})(1\,\text{ms/div.}) = 3.2\,\text{ms}$$

$$t_p = (0.8\,\text{div.})(1\,\text{ms/div.}) = 0.8\,\text{ms}$$

$$\text{prf} = \frac{1}{T} = \frac{1}{3.2\,\text{ms}} = \mathbf{312.5\,Hz}$$

$$\text{Duty cycle} = \frac{t_p}{T} \times 100\% = \frac{0.8\,\text{ms}}{3.2\,\text{ms}} \times 100\% = \mathbf{25\%}$$

EXAMPLE 21.5. Determine the pulse repetition rate and duty cycle for the trigger waveform of Fig. 21.14.

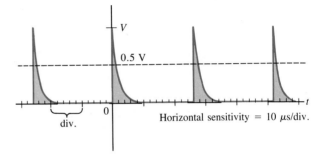

FIG. 21.14

Solution:

$$T = (2.6\,\text{div.})(10\,\mu\text{s/div.}) = 26\,\mu\text{s}$$

$$\text{prf} = \frac{1}{T} = \frac{1}{26\,\mu\text{s}} = \mathbf{38,462\,kHz}$$

$$t_p \cong (0.2\,\text{div.})(10\,\mu\text{s/div.}) = 2\,\mu\text{s}$$

$$\text{Duty cycle} = \frac{t_p}{T} \times 100\% = \frac{2\,\mu\text{s}}{26\,\mu\text{s}} \times 100\% = \mathbf{7.69\%}$$

21.4 AVERAGE VALUE

The average value of a pulse waveform can be determined using one of two methods. The first is the procedure outlined in Section 13.8 which can be applied to any alternating waveform. The second can be applied only to pulse waveforms since it utilizes terms specifically related to pulse waveforms. That is,

$$V_{\text{av}} = (\text{duty cycle})(\text{peak value}) + (1 - \text{duty cycle})(V_b) \qquad \textbf{(21.5)}$$

In Eq. (21.5), the peak value is the maximum deviation from the reference or zero-volt level and the duty cycle is in decimal form. Eq. 21.5 does not include the effect of any tilt pulse waveforms with sloping sides.

EXAMPLE 21.6. Determine the average value for the periodic pulse waveform of Fig. 21.15.

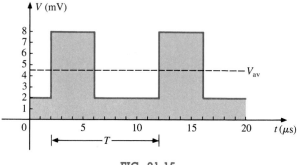

FIG. 21.15

Solution: By the method of Section 13.8,

$$G = \frac{\text{area under curve}}{T}$$

$$T = (12 - 2)\,\mu s = 10\,\mu s$$

$$G = \frac{(8\,\text{mV})(4\,\mu s) + (2\,\text{mV})(6\,\mu s)}{10\,\mu s} = \frac{32 \times 10^{-9} + 16 \times 10^{-9}}{10 \times 10^{-6}}$$

$$= \frac{44 \times 10^{-9}}{10 \times 10^{-6}} = \textbf{4.4\,mV}$$

By Eq. (21.5),

$$V_b = +2\,\text{mV}$$

$$\text{Duty cycle} = \frac{t_p}{T} = \frac{(6 - 2)\,\mu s}{10\,\mu s} = \frac{4}{10} = 0.4 \text{ (decimal form)}$$

$$\text{Peak value (from 0-V reference)} = 8\,\text{mV}$$

$$\begin{aligned} V_{\text{av}} &= (\text{duty cycle})(\text{peak value}) + (1 - \text{duty cycle})(V_b) \\ &= (0.4)(8\,\text{mV}) + (1 - 0.4)(2\,\text{mV}) \\ &= 3.2\,\text{mV} + 1.2\,\text{mV} = \textbf{4.4\,mV} \end{aligned}$$

as obtained above.

EXAMPLE 21.7. Given a periodic pulse waveform with a duty cycle of 28%, a peak value of 7 V, and a base-line voltage of -3 V:
a. Determine the average value.
b. Sketch the waveform.
c. Verify the result of part (a) using the method of Section 13.8.

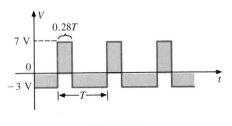

FIG. 21.16

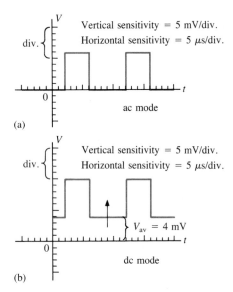

FIG. 21.17

*Determining the average value of a pulse
waveform using an oscilloscope.*

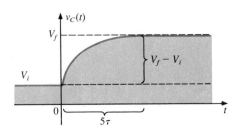

FIG. 21.18

Solutions:

a. By Eq. (21.5),

$$V_{av} = \text{(duty cycle)(peak value)} + (1 - \text{duty cycle})(V_b)$$
$$= (0.28)(7) + (1 - 0.28)(-3) = 1.96 + (-2.16)$$
$$= \mathbf{-0.2\,V}$$

b. See Fig. 21.16.

c. $G = \dfrac{(7\,\text{V})(0.28T) - (3\,\text{V})(0.72T)}{T} = 1.96\,\text{V} - 2.16\,\text{V}$

$$= \mathbf{-0.2\,V}$$

as obtained above.

The average value (dc value) of any waveform can be easily determined using the oscilloscope. If the mode switch of the scope is set in the ac position, the average or dc component of the applied waveform will be blocked by an internal capacitor from reaching the screen. The pattern can be adjusted to establish the display of Fig. 21.17(a). If the mode switch is then placed in the dc position, the vertical shift (positive or negative) will reveal the average or dc level of the input signal, as shown in Fig. 21.17(b).

21.5 *R-C* CIRCUITS WITH INITIAL VALUES

In the description of transients in an *R-C* circuit (Chapter 10), the capacitors were assumed to be initially uncharged and the final voltage was the supply voltage. The initial voltage V_i was defined for those cases where the transient voltage did not reach the final value and the decay phase began with an initial voltage less than the supply voltage. In this section, a more general solution for the transient behavior of the *R-C* circuit will be provided that includes the effect of an initial voltage on the capacitor and a final voltage that may not be the supply voltage.

In Fig. 21.18, a capacitor has an initial voltage V_i and will charge to a final voltage V_f. Using the conclusions of Chapter 10, we know that the equation for the transient waveform beginning at $t = 0$ s is $(V_f - V_i)(1 - e^{-t/RC})$. However, since V_i is the same throughout the transient phase, it can simply be added to the above mathematical expression to generate an equation for the voltage $v_C(t)$ of Fig. 21.18. That is,

$$\boxed{v_C(t) = V_i + (V_f - V_i)(1 - e^{-t/RC})} \tag{21.6}$$

For the case where V_i equals zero volts and V_f equals E volts, as described in detail in Chapter 10,

$$v_C(t) = 0 + (E - 0)(1 - e^{-t/RC}) = E(1 - e^{-t/RC})$$

For the case of Fig. 21.19, $V_i = -2$ V, $V_f = +5$ V, and

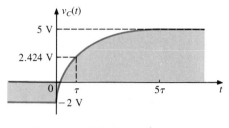

FIG. 21.19

$$v_C(t) = V_i + (V_f - V_i)(1 - e^{-t/RC})$$
$$= -2 + [5 - (-2)](1 - e^{-t/RC})$$
$$v_C(t) = -2 + 7(1 - e^{-t/RC})$$

For the case where $t = \tau = RC$,

$$v_C(t) = -2 + 7(1 - e^{-\tau/\tau}) = -2 + 7(1 - e^{-1})$$
$$= -2 + 7(1 - 0.368) = -2 + 7(0.632)$$
$$v_C(t) = 2.424 \text{ V}$$

as verified by Fig. 21.19.

EXAMPLE 21.8. The capacitor of Fig. 21.20 is initially charged to 2 V before the switch is closed. The switch is then closed.
a. Determine the mathematical expression for $v_C(t)$.
b. Determine the mathematical expression for $i_C(t)$.
c. Sketch the waveforms of $v_C(t)$ and $i_C(t)$.
Solutions:
a. $V_i = 2$ V
  V_f (after 5τ) $= E = 8$ V
  $\tau = RC = (100 \text{ k}\Omega)(1 \text{ }\mu\text{F}) = 100 \text{ ms}$

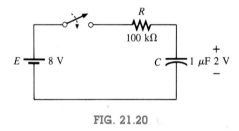

FIG. 21.20

By Eq. (21.6),

$$v_C(t) = V_i + (V_f - V_i)(1 - e^{-t/\tau})$$
$$= 2 + (8 - 2)(1 - e^{-t/\tau})$$

and

$$v_C(t) = \mathbf{2 + 6(1 - e^{-t/\tau})}$$

b. When the switch is first closed, the voltage across the capacitor cannot change instantaneously, and $V_R = E - V_i = 8 - 2 = 6$ V. The current therefore jumps to a level determined by Ohm's law:

$$I_{R_{max}} = \frac{V_R}{R} = \frac{6}{100 \text{ k}\Omega} = 0.06 \text{ mA}$$

The current will then decay to zero amperes with the same time constant calculated in part (a), and

$$i_C(t) = \mathbf{0.06 \times 10^{-3} e^{-t/\tau}}$$

c. See Fig. 21.21.

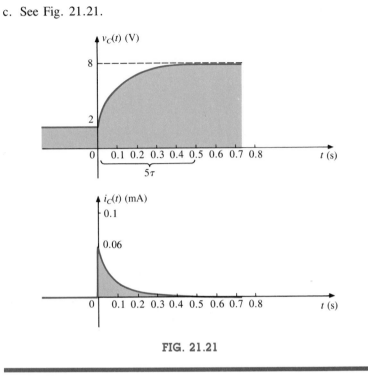

FIG. 21.21

EXAMPLE 21.9. Sketch $v_C(t)$ for the step input shown in Fig. 21.22. Assume the $-4\,\text{mV}$ has been present for a period of time in excess of five time constants of the network. Then determine when $v_C(t) = 0\,\text{V}$ if the step changes levels at $t = 0\,\text{s}$.

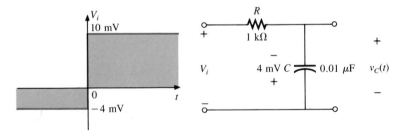

FIG. 21.22

Solution:

$$V_i = -4\,\text{mV} \qquad V_f = 10\,\text{mV}$$
$$\tau = RC = (1\,\text{k}\Omega)(0.01\,\mu\text{F}) = 10\,\mu\text{s}$$

By Eq. (21.6),

$$v_C(t) = V_i + (V_f - V_i)(1 - e^{-t/\tau})$$
$$= -4\,\text{mV} + [10\,\text{mV} - (-4\,\text{mV})](1 - e^{-t/\tau})$$

and

$$v_C(t) = -4 \times 10^{-3} + 14 \times 10^{-3}(1 - e^{-t/(10\times10^{-6})})$$

The waveform appears in Fig. 21.23.

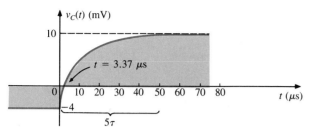

FIG. 21.23

Substituting $v_C(t) = 0\,\text{V}$ into the above equation yields

$$v_C(t) = 0 = -4 \times 10^{-3} + 14 \times 10^{-3}(1 - e^{-t/\tau})$$

and

$$\frac{4 \times 10^{-3}}{14 \times 10^{-3}} = 1 - e^{-t/\tau}$$

or

$$0.286 - 1 = -e^{-t/\tau}$$

and

$$0.714 = e^{-t/\tau}$$

but

$$\log_e 0.714 = \log_e(e^{-t/\tau}) = \frac{-t}{\tau}$$

and

$$t = -\tau \log_e 0.714 = -(10\,\mu s)(-0.337) = \mathbf{3.37\,\mu s}$$

as indicated in Fig. 21.23.

21.6 *R-C* RESPONSE TO SQUARE-WAVE INPUTS

The square wave of Fig. 21.24 is a particular form of pulse waveform. It has a duty cycle of 50% and an average value of zero volts, as calculated below:

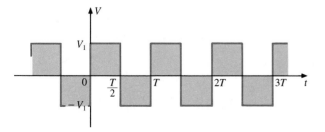

FIG. 21.24

$$\text{Duty cycle} = \frac{t_p}{T} \times 100\% = \frac{T/2}{T} \times 100\% = \mathbf{50\%}$$

$$V_{\text{av}} = \frac{(V_1)(T/2) + (-V_1)(T/2)}{T} = \frac{0}{T} = \mathbf{0\ V}$$

The application of a dc voltage V_1 in series with the square wave of Fig. 21.24 can raise the base-line voltage from $-V_1$ to zero volts, and the average value to V_1 volts.

If a square wave such as developed in Fig. 21.25 is applied to an R-C

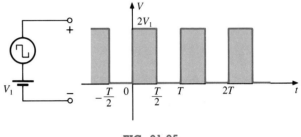

FIG. 21.25

circuit as shown in Fig. 21.26, the period of the square wave can have a pronounced effect on the resulting waveform for $v_C(t)$.

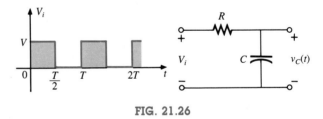

FIG. 21.26

For the analysis to follow, we will assume that steady-state conditions will be established after a period of five time constants has passed. The types of waveform developed across the capacitor can then be separated into three fundamental types.

$T/2 > 5\tau$

The condition $T/2 > 5\tau$ or $T > 10\tau$ establishes a situation where the capacitor can charge to its steady-state value in advance of $t = T/2$. The resulting waveforms for $v_C(t)$ and $i_C(t)$ will appear as shown in Fig. 21.27. Note how closely the voltage $v_C(t)$ shadows the applied waveform and how $i_C(t)$ is nothing more than a series of very sharp spikes. Note also that the change of V_i from V to zero volts during the trailing edge simply results in a rapid discharge of $v_C(t)$ to zero volts. In essence, when $V_i = 0$ the capacitor and resistor are in parallel and the capacitor simply discharges through R with a time constant equal to that encountered during the charging phase but with a direction of charge flow (current) opposite to that established during the charging phase.

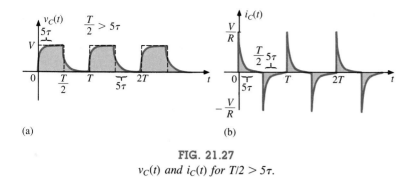

(a)   (b)

FIG. 21.27
$v_C(t)$ and $i_C(t)$ for $T/2 > 5\tau$.

$T/2 = 5\tau$

If the frequency of the square wave is chosen such that $T/2 = 5\tau$ or $T = 10\tau$, the voltage $v_C(t)$ will reach its final value just before beginning its discharge phase, as shown in Fig. 21.28. The voltage $v_C(t)$ no longer resembles the square-wave input and, in fact, has some of the characteristics of a triangular waveform. The increased time constant has resulted in a more rounded $v_C(t)$, and $i_C(t)$ has increased substantially in width to reveal the longer charging period.

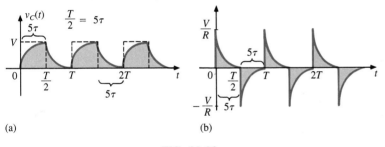

(a)   (b)

FIG. 21.28
$v_C(t)$ and $i_C(t)$ for $T/2 = 5\tau$.

$T/2 < 5\tau$

If $T/2 < 5\tau$ or $T < 10\tau$, the voltage $v_C(t)$ will not reach its final value during the first pulse (Fig. 21.29), and the discharge cycle will not

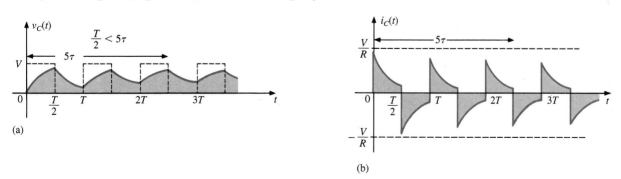

(a)   (b)

FIG. 21.29
$v_C(t)$ and $i_C(t)$ for $T/2 < 5\tau$.

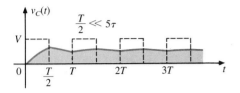

FIG. 21.30

$v_C(t)$ for $T/2 \ll 5\tau$ or $T \ll 10\tau$.

return to zero volts. In fact, the initial value for each succeeding pulse will change until steady-state conditions are reached. In most instances, it is a good approximation to assume that steady-state conditions have been established in 5 cycles of the applied waveform.

As the frequency increases and the period decreases, there will be a flattening of the response for $v_C(t)$ until a pattern like that in Fig. 21.30 results. Figure 21.30 begins to reveal an important conclusion regarding the response curve for $v_C(t)$: Under steady-state conditions, the average value of $v_C(t)$ will equal the average value of the applied square wave. Note in Figs. 21.29 and 21.30 how the waveform for $v_C(t)$ approaches an average value of $V/2$.

EXAMPLE 21.10. The 1000-Hz square wave of Fig. 21.31 is applied to the *R-C* circuit of the same figure.

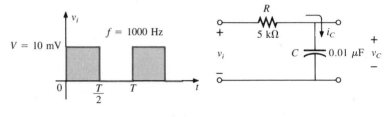

FIG. 21.31

a. Compare the pulse width of the square wave to the time constant of the circuit.
b. Sketch $v_C(t)$.
c. Sketch $i_C(t)$.

Solutions:

a. $T = \dfrac{1}{f} = \dfrac{1}{1000} = 1\text{ ms}$

$t_p = \dfrac{T}{2} = 0.5\text{ ms}$

$\tau = RC = (5 \times 10^3)(0.01 \times 10^{-6}) = 0.05\text{ ms}$

$\dfrac{t_p}{\tau} = \dfrac{0.5\text{ ms}}{0.05\text{ ms}} = 10$

and

$$t_p = 10\tau = \dfrac{T}{2}$$

The result reveals that $v_C(t)$ will charge to its final value in half the pulse width.

b. For the charging phase, $V_i = 0\text{ V}$ and $V_f = 10\text{ mV}$, and

$$v_C(t) = V_i + (V_f - V_i)(1 - e^{-t/\tau})$$
$$= 0 + (10\text{ mV} - 0)(1 - e^{-t/\tau})$$

and

$$v_C(t) = (10 \times 10^{-3})(1 - e^{-t/\tau})$$

For the discharge phase, $V_i = 10\,\text{mV}$ and $V_f = 0\,\text{V}$, and

$$v_C(t) = V_i + (V_f - V_i)(1 - e^{-t/\tau})$$
$$= 10\,\text{mV} + (0 - 10\,\text{mV})(1 - e^{-t/\tau})$$
$$v_C(t) = 10\,\text{mV} - 10\,\text{mV} + (10\,\text{mV})(e^{-t/\tau})$$

and

$$v_C(t) = (\mathbf{10 \times 10^{-3}})(e^{-t/\tau})$$

The waveform for $v_C(t)$ appears in Fig. 21.32.

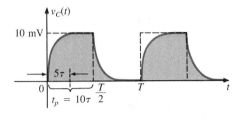

FIG. 21.32

c. For the charging phase at $t = 0\,\text{s}$, $V_R = V$ and $I_{R_{max}} = V/R = 10\,\text{mV}/5\,\text{k}\Omega = 2\,\mu\text{A}$, and

$$i_C(t) = I_{max}e^{-t/\tau} = (\mathbf{2 \times 10^{-6}})(e^{-t/\tau})$$

For the discharge phase, the current will have the same mathematical formulation but the opposite direction, as shown in Fig. 21.33.

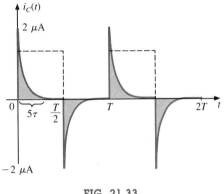

FIG. 21.33

EXAMPLE 21.11. Repeat Example 21.10 for $f = 10\,\text{kHz}$.
Solution:

$$T = \frac{1}{f} = \frac{1}{10\,\text{kHz}} = 0.1\,\text{ms}$$

and

$$\frac{T}{2} = 0.05\,\text{ms}$$

with

$$\tau = t_p = \frac{T}{2} = 0.05\,\text{ms}$$

In other words, the pulse width is exactly equal to the time constant of the network. The voltage $v_C(t)$ will not reach the final value before the first pulse of the square-wave input returns to zero volts.

For t in the range $t = 0$ to $t = T/2$, $V_i = 0\,\text{V}$ and $V_f = 10\,\text{mV}$, and

$$v_C(t) = (10 \times 10^{-3})(1 - e^{-t/\tau})$$

At $t = \tau$, we recall from Chapter 10 that $v_C(t) = 63.2\%$ of the final value. Substituting $t = \tau$ into the equation above yields

$$v_C(t) = (10 \times 10^{-3})(1 - e^{-1}) = (10 \times 10^{-3})(1 - 0.368)$$
$$= (10 \times 10^{-3})(0.632) = 6.32\,\text{mV}$$

as shown in Fig. 21.34.

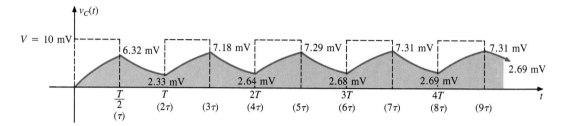

FIG. 21.34
R-C response for $t_p = \tau = T/2$.

For the discharge phase between $t = T/2$ and T, $V_i = 6.32\,\text{mV}$ and $V_f = 0\,\text{V}$, and

$$v_C(t) = V_i + (V_f - V_i)(1 - e^{-t/\tau})$$
$$= 6.32\,\text{mV} + (0 - 6.32\,\text{mV})(1 - e^{-t/\tau})$$
$$v_C(t) = (6.32\,\text{mV})(e^{-t/\tau})$$

with t now being measured from $t = T/2$ in Fig. 21.34. In other words, for each interval of Fig. 21.34, the beginning of the transient waveform is defined as $t = 0\,\text{s}$. The value of $v_C(t)$ at $t = T$ is therefore determined by substituting $t = \tau$ into the above equation, and not 2τ as defined by Fig. 21.34.

Substituting $t = \tau$,

$$v_C(t) = (6.32\,\text{mV})(e^{-1}) = (6.32\,\text{mV})(0.368)$$
$$= 2.33\,\text{mV}$$

as shown in Fig. 21.34.

For the next interval, $V_i = 2.33\,\text{mV}$ and $V_f = 10\,\text{mV}$, and

$$v_C(t) = V_i + (V_f - V_i)(1 - e^{-t/\tau})$$
$$= 2.33\,\text{mV} + (10\,\text{mV} - 2.33\,\text{mV})(1 - e^{-t/\tau})$$
$$v_C(t) = 2.33\,\text{mV} + (7.67\,\text{mV})(1 - e^{-t/\tau})$$

At $t = \tau$ (since $t = T = 2\tau$ is now $t = 0\,\text{s}$ for this interval),

$$v_C(t) = 2.33\,\text{mV} + (7.67\,\text{mV})(1 - e^{-1})$$
$$= 2.33\,\text{mV} + 4.85\,\text{mV}$$
$$v_C(t) = 7.18\,\text{mV}$$

as shown in Fig. 21.34.

For the discharge interval, $V_i = 7.18\,\text{mV}$ and $V_f = 0\,\text{V}$, and

$$v_C(t) = V_i + (V_f - V_i)(1 - e^{-t/\tau})$$
$$= 7.18\,\text{mV} + (0 - 7.18\,\text{mV})(1 - e^{-t/\tau})$$
$$v_C(t) = (7.18\,\text{mV})(e^{-t/\tau})$$

At $t = \tau$ (measured from 3τ of Fig. 21.34),

$$v_C(t) = (7.18\,\text{mV})(e^{-1}) = (7.18\,\text{mV})(0.368)$$
$$= 2.64\,\text{mV}$$

as shown in Fig. 21.34.

Continuing in the same manner, the remaining waveform for $v_C(t)$ will be generated as depicted in Fig. 21.34. Note that repetition occurs after $t = 8\tau$ and the waveform has essentially reached steady-state conditions in a period of time less than 10τ, or 5 cycles of the applied square wave.

A closer look will reveal that both the peak and lower levels continued to increase until steady-state conditions were established. Since the exponential waveforms between $t = 4T$ and $t = 5T$ have the same time constant, the average value of $v_C(t)$ can be determined from the steady-state 7.31-mV and 2.69-mV levels as follows:

$$V_{\text{av}} = \frac{7.31\,\text{mV} + 2.69\,\text{mV}}{2} = \frac{10\,\text{mV}}{2} = 5\,\text{mV}$$

which equals the average value of the applied signal as stated earlier in this section.

We can use the results of Fig. 21.34 to plot $i_C(t)$.

At any instant of time,

$$v_i = v_R + v_C$$

or

$$v_R = v_i - v_C$$

and

$$i_R = i_C = \frac{v_i - v_C}{R}$$

At $t = 0^+$, $v_C = 0\,\text{V}$, and

$$i_R = \frac{v_i - v_C}{R} = \frac{10\,\text{mV} - 0}{5\,\text{k}\Omega} = 2\,\mu\text{A}$$

as shown in Fig. 21.35.

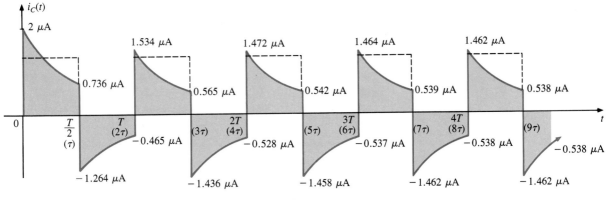

FIG. 21.35

As the charging process proceeds, the current i_C will decay at a rate determined by

$$i_C(t) = (2\,\mu\text{A})(e^{-t/\tau})$$

At $t = \tau$,

$$i_C(t) = (2\,\mu\text{A})(e^{-\tau/\tau}) = (2\,\mu\text{A})(e^{-1}) = (2\,\mu\text{A})(0.368)$$
$$= 0.736\,\mu\text{A}$$

as shown in Fig. 21.35.

For the trailing edge of the first pulse, the voltage across the capacitor cannot change instantaneously, resulting in the following when v_i drops to zero volts:

$$i_C = i_R = \frac{v_i - v_C}{R} = \frac{0 - 6.32\,\text{mV}}{5\,\text{k}\Omega} = -1.264\,\mu\text{A}$$

as illustrated in Fig. 21.35. The current will then decay as determined by

$$i_C(t) = (-1.264\,\mu\text{A})(e^{-t/\tau})$$

and at $t = \tau$ (actually $t = 2\tau$ in Fig. 21.35),

$$i_C(t) = (-1.264\,\mu\text{A})(e^{-\tau/\tau}) = (-1.264\,\mu\text{A})(e^{-1})$$
$$= (-1.264\,\mu\text{A})(0.368) = -0.465\,\mu\text{A}$$

as shown in Fig. 21.35.

At $t = T$ ($t = 2\tau$), $v_C = 2.33\,\text{mV}$ and v_i returns to $10\,\text{mV}$, resulting in

$$i_C = i_R = \frac{v_i - v_C}{R} = \frac{10\,\text{mV} - 2.33\,\text{mV}}{5\,\text{k}\Omega} = 1.534\,\mu\text{A}$$

The equation for the decaying current is now

$$i_C(t) = (1.534\,\mu\text{A})(e^{-t/\tau})$$

and at $t = \tau$ (actually $t = 3\tau$ in Fig. 21.35),

$$i_C(t) = (1.534\,\mu\text{A})(0.368) = 0.565\,\mu\text{A}$$

The process will continue until steady-state conditions are reached at the same time they were attained for $v_C(t)$. Note in Fig. 21.35 that the positive peak current decreased toward steady-state conditions while the negative peak became more negative. It is also interesting and important to realize that the current waveform becomes symmetrical about the axis when steady-state conditions are established. The result is that the net average current over one cycle is zero, as it should be in a series R-C circuit. Recall from Chapter 10 that the capacitor under dc steady-state conditions can be replaced by an open-circuit equivalent, resulting in $I_C = 0\,\text{A}$.

Although both examples provided above started with an uncharged capacitor, there is no reason that the same approach cannot be used effectively for initial conditions. Simply substitute the initial voltage on the capacitor as V_i in Eq. (21.6) and proceed as above.

21.7 OSCILLOSCOPE ATTENUATOR AND COMPENSATING PROBE

The ×10 attenuator probe employed with oscilloscopes is designed to reduce the magnitude of the input voltage by a factor of 10. If the input impedance to a scope is 1 MΩ, the ×10 attenuator probe would have an internal resistance of 9 MΩ, as shown in Fig. 21.36.

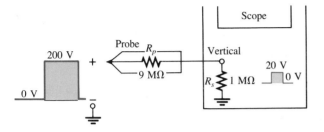

FIG. 21.36

×10 attenuator probe.

Applying the voltage divider rule,

$$V_{\text{scope}} = \frac{(1\,\text{M}\Omega)(V_i)}{1\,\text{M}\Omega + 9\,\text{M}\Omega} = \frac{1}{10}V_i$$

In addition to the internal resistance, oscilloscopes also have some internal input capacitance, represented by C_s in Fig. 21.37.

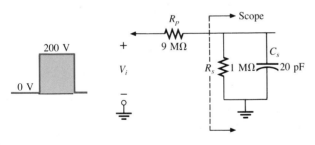

FIG. 21.37

For the analysis to follow, let us determine the Thevenin equivalent circuit for the capacitor C_s:

$$E_{Th} = \frac{(1\,\text{M}\Omega)(V_i)}{1\,\text{M}\Omega + 9\,\text{M}\Omega} = \frac{1}{10}V_i$$

and

$$R_{Th} = 9\,\text{M}\Omega \parallel 1\,\text{M}\Omega = 0.9\,\text{M}\Omega$$

The Thevenin network is shown in Fig. 21.38.

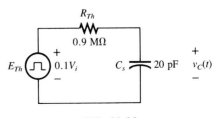

FIG. 21.38

For $V_i = 200\,\text{V}$ (peak),

$$E_{Th} = 0.1V_i = 20\,\text{V (peak)}$$

and for $v_C(t)$, $V_f = 20\,\text{V}$ and $V_i = 0\,\text{V}$, with

$$\tau = RC = (0.9 \times 10^6)(20 \times 10^{-12}) = 18\,\mu\text{s}$$

For an applied frequency of 5 kHz,

$$T = \frac{1}{f} = 0.2\,\text{ms}$$

and

$$\frac{T}{2} = 0.1\,\text{ms} = 100\,\mu\text{s}$$

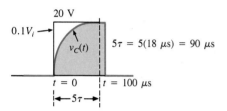

20 V

$0.1V_i \rightarrow$

$v_C(t)$

$5\tau = 5(18\,\mu\text{s}) = 90\,\mu\text{s}$

$t = 0 \quad\quad t = 100\,\mu\text{s}$

$\leftarrow 5\tau \rightarrow$

FIG. 21.39

The resulting waveform for $v_C(t)$ appears in Fig. 21.39, clearly revealing a severe rounding distortion of the square wave and a poor representation of the applied signal.

To improve matters, a variable capacitor is often added in parallel with the resistance of the attenuator to establish a probe referred to as a *compensated attenuator*. In Chapter 24, it will be demonstrated that a square wave can be generated by a summation of sinusoidal signals of particular frequency and amplitude. If we therefore design a network such as shown in Fig. 21.40 that will insure that V_{scope} is $\frac{1}{10}V_i$ for any frequency, then the rounding distortion will be removed and V_{scope} will have the same appearance as V_i.

Applying the voltage divider rule to the network of Fig. 21.40,

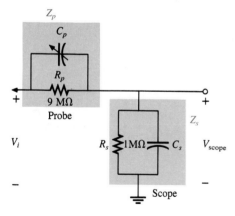

Z_p

C_p

R_p

$9\,\text{M}\Omega$

Probe

Z_s

V_i

$R_s \gtrless 1\text{M}\Omega \quad C_s$

V_{scope}

Scope

FIG. 21.40

$$\boxed{\mathbf{V}_{\text{scope}} = \frac{\mathbf{Z}_s \mathbf{V}_i}{\mathbf{Z}_s + \mathbf{Z}_p}} \tag{21.7}$$

If the parameters are chosen or adjusted such that

$$\boxed{R_p C_p = R_s C_s} \tag{21.8}$$

the phase angle of $\mathbf{Z}_s$ and $\mathbf{Z}_p$ will be the same and Eq. (21.7) will reduce to

$$\boxed{\mathbf{V}_{\text{scope}} = \frac{R_s \mathbf{V}_i}{R_s + R_p}} \tag{21.9}$$

which is insensitive to frequency.

In the laboratory, simply adjust the probe capacitance using a standard or known square-wave signal until the desired sharp corners of the square wave are obtained. If you avoid the calibration step, you may make a rounded signal look square since you assumed a square wave at the point of measurement.

Too much capacitance will result in an overshoot effect, while too little will continue to show the rounding effect.

PROBLEMS

Section 21.2

1. Determine the following for the pulse waveform of Fig. 21.41:
   a. positive- or negative-going?
   b. base-line voltage
   c. pulse width
   d. amplitude
   e. % tilt

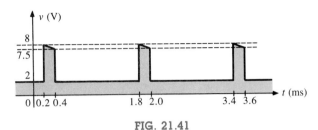

FIG. 21.41

2. Repeat Problem 1 for the pulse waveform of Fig. 21.42.

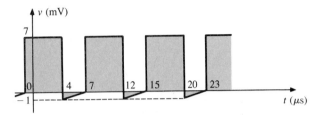

FIG. 21.42

3. Repeat Problem 1 for the pulse waveform of Fig. 21.43.

4. Determine the rise and fall time for the waveform of Fig. 21.43.

5. Sketch a pulse waveform that has a base-line voltage of $-5\,mV$, a pulse width of $2\,\mu s$, an amplitude of $15\,mV$, a 10% tilt, a period of $10\,\mu s$, vertical sides, and is positive-going.

Vertical sensitivity = 10 mV/div.
Horizontal sensitivity = 2 ms/div.

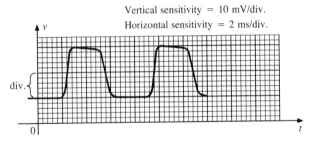

FIG. 21.43

6. For the waveform of Fig. 21.44, established by straight-line approximations of the original waveform:
   a. Determine the rise time.
   b. Find the fall time.
   c. Find the pulse width.
   d. Calculate the frequency.

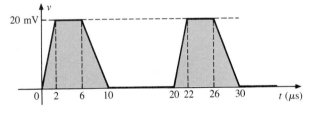

FIG. 21.44

7. For the waveform of Fig. 21.45:
   a. Determine the period.
   b. Find the frequency.
   c. Find the maximum and minimum amplitude.

Section 21.3

8. Determine the pulse repetition frequency and duty cycle for the waveform of Fig. 21.41.

9. Repeat Problem 8 for the waveform of Fig. 21.42.

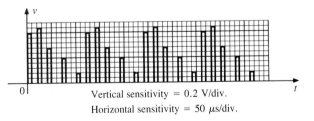

Vertical sensitivity = 0.2 V/div.
Horizontal sensitivity = 50 μs/div.

FIG. 21.45

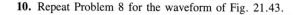

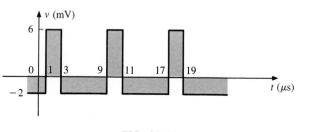

FIG. 21.46

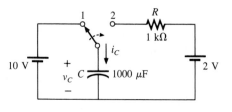

FIG. 21.47

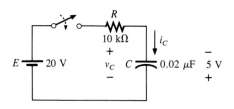

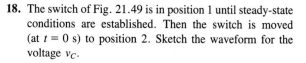

FIG. 21.48

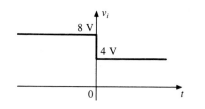

FIG. 21.49

10. Repeat Problem 8 for the waveform of Fig. 21.43.

Section 21.4

11. Determine the average value of the periodic pulse waveform of Fig. 21.46.

12. Determine the average value of the periodic pulse waveform of Fig. 21.41.

13. To the best accuracy possible, determine the average value of the waveform of Fig. 21.43.

14. Determine the average value of the waveform of Fig. 21.44.

15. Determine the average value of the periodic pulse train of Fig. 21.45.

Section 21.5

16. The capacitor of Fig. 21.47 is initially charged to 5 V, with the polarity indicated in the figure. The switch is then closed at $t = 0$ s.
   a. What is the mathematical expression for the voltage $v_C(t)$?
   b. Sketch v_C versus t.
   c. What is the mathematical expression for the current $i_C(t)$?
   d. Sketch i_C versus t.

17. For the input voltage v_i appearing in Fig. 21.48 sketch the waveform for v_o. Assume that steady-state conditions were established with $v_i = 8$ V.

18. The switch of Fig. 21.49 is in position 1 until steady-state conditions are established. Then the switch is moved (at $t = 0$ s) to position 2. Sketch the waveform for the voltage v_C.

19. Sketch the waveform for i_C for Problem 18.

Section 21.6

20. Sketch the voltage v_C for the network of Fig. 21.50 due to the square-wave input of the same figure with a frequency of
  a. 500 Hz   **b.** 100 Hz
  c. 5000 Hz

21. Sketch the current i_C for each frequency of Problem 20.

22. Sketch the response (v_C) of the network of Fig. 21.50 to the square-wave input of Fig. 21.51.

23. If the capacitor of Fig. 21.50 is initially charged to 20 V, sketch the response (v_C) to the same input signal at a frequency of 500 Hz.

24. Repeat Problem 23 if the capacitor is initially charged to -10 V.

Section 21.7

25. Given the network of Fig. 21.40 with $R_p = 9\,M\Omega$ and $R_s = 1\,M\Omega$, find $\mathbf{V}_{scope}$ in polar form if $C_p = 2\,pF$, $C_s = 18\,pF$, and $v_i = \sqrt{2}(100)\sin 10{,}000t$. That is, determine $\mathbf{Z}_s$ and $\mathbf{Z}_p$ and substitute into Eq. (21.7) and compare the results obtained with Eq. (21.9). Is it verified that the phase angle of $\mathbf{Z}_s$ and $\mathbf{Z}_p$ is the same under the condition $R_pC_p = R_sC_s$?

26. Repeat Problem 25 at $\omega = 10^5$ rad/s.

Computer Problems

27. Given a periodic pulse train such as that in Fig. 21.11, write a program to determine the average value, given the base-line voltage, peak value, and duty cycle.

28. Given the initial and final values and the network parameters (R and C), write a program to tabulate the values of v_C at each time constant (of the first five) of the transient phase.

29. For the case of $T/2 < 5\tau$ as defined by Fig. 21.29, write a program to determine the values of v_C at each half-period of the applied square wave. Test the solution by entering the conditions of Example 21.10.

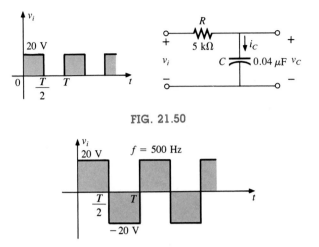

FIG. 21.50

FIG. 21.51

GLOSSARY

Actual (true, practical) pulse   A pulse waveform having a leading and trailing edge that are not vertical, along with other distortion effects such as tilt, ringing, or overshoot.

Attenuator probe   A scope probe that will reduce the strength of the signal applied to the vertical channel of a scope.

Base-line voltage   The voltage level from which a pulse is initiated.

Compensated attenuator probe   A scope probe that can reduce the applied signal and balance the effects of the input capacitance of a scope on the signal to be displayed.

Duty cycle   Reveals how much of a period is encompassed by the pulse waveform.

Fall time (t_f)   The time required for the trailing edge of a pulse waveform to drop from the 90% to the 10% level.

Ideal pulse   A pulse waveform characterized as having vertical sides, sharp corners, and a flat peak response.

Negative-going pulse   A pulse that increases in the negative direction from the base-line voltage.

Periodic pulse train   A sequence of pulses that repeats itself after a specific period of time.

Positive-going pulse   A pulse that increases in the positive direction from the base-line voltage.

Pulse amplitude   The peak-to-peak value of a pulse waveform.

Pulse repetition frequency   The frequency of a periodic pulse train.

Pulse train   A series of pulses that may have varying heights and widths.

Pulse width (t_p)   The pulse width defined by the 50% voltage level.

Rise time (t_r)   The time required for the leading edge of a pulse waveform to travel from the 10% to the 90% level.

Square wave   A periodic pulse waveform with a 50% duty cycle.

Step function   A waveform that abruptly changes from one level to another.

Tilt   The drop in peak value across the pulse width of a pulse waveform.

22
Polyphase Systems

22.1 INTRODUCTION

If an ac generator is designed to develop a single sinusoidal voltage for each rotation of the shaft (rotor), it is referred to as a *single-phase generator*. If the number of coils on the rotor is increased in a specified manner, the result is a *polyphase generator*, which develops more than one ac phase voltage per rotation of the rotor. In this chapter, the three-phase system will be discussed in detail since it is the most frequently used for power transmission.

In general, the three-phase system is more economical for transmitting power at a fixed power loss than the single-phase system. This economy is due primarily to the reduction of the I^2R losses of the transmission lines. A reduction in these losses permits the use of smaller conductors, which in turn reduces the weight of copper required.

The three-phase system is used in almost all commercial electric generators. This does not mean that single-phase and two-phase generating systems are obsolete. Most small emergency generators, such as the gasoline type, are one-phase generating systems.

One of the more common applications of the two-phase system is in servomechanisms, which are self-correcting control systems capable of detecting and adjusting their own operation. Servomechanisms are used in ships and aircraft to keep them on course automatically, or, in simpler devices such as a thermostatic circuit, to regulate heat output. In most cases, however, where single-phase and two-phase inputs are required, they are supplied by one and two phases of a three-phase generating system rather than generated independently.

The number of phase voltages that can be produced by a polyphase generator is not limited to three. Any number of phases can be obtained by spacing the windings for each phase at the proper angular position around the rotor. Some electrical systems operate more efficiently if more than three phases are used. One such system involves the process of rectification, which is used to convert alternating current to direct current. The greater the number of phases, the smoother the dc output of the system.

22.2 THE THREE-PHASE GENERATOR

The three-phase generator of Fig. 22.1(a) has three induction coils placed 120° apart on the rotor (armature), as shown symbolically by

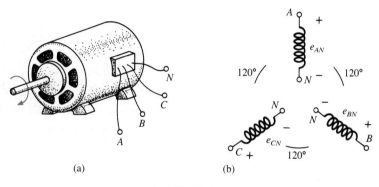

(a)   (b)

FIG. 22.1

(a) Three-phase generator; (b) induced voltages of a three-phase generator.

Fig. 22.1(b). Since the three coils have an equal number of turns, and each coil rotates with the same angular velocity, the voltage induced across each coil will have the same peak value, shape, and frequency. As the shaft of the generator is turned by some external means, the induced voltages e_{AN}, e_{BN}, and e_{CN} will be generated simultaneously, as shown in Fig. 22.2. Note the 120° phase shift between waveforms and the similarities in appearance of the three sinusoidal functions.

In particular, note that *at any instant of time, the algebraic sum of the three phase voltages is zero.* This is shown at $\omega t = 0$ in Fig. 22.2, where it is also evident that *when one induced voltage is zero, the other two are 86.6% of their positive or negative maximums. In addition, when any two are equal in magnitude and sign (at $0.5E_m$), the remaining induced voltage has the opposite polarity and a peak value.*

The sinusoidal expression for each of the induced voltages of Fig. 22.2 is

$$
\begin{aligned}
e_{AN} &= E_{m(AN)} \sin \omega t \\
e_{BN} &= E_{m(BN)} \sin(\omega t - 120°) \\
e_{CN} &= E_{m(CN)} \sin(\omega t - 240°) = E_{m(CN)} \sin(\omega t + 120°)
\end{aligned}
\tag{22.1}
$$

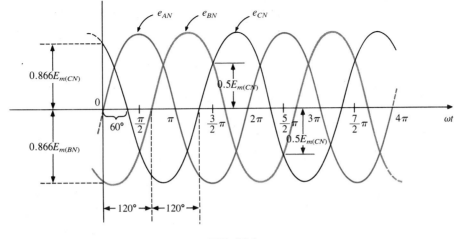

FIG. 22.2

The phasor diagram of the induced voltages is shown in Fig. 22.3, where

$$E_{AN} = 0.707E_{m(AN)}$$
$$E_{BN} = 0.707E_{m(BN)}$$
$$E_{CN} = 0.707E_{m(CN)}$$

and

$$\mathbf{E}_{AN} = E_{AN} \angle 0°$$
$$\mathbf{E}_{BN} = E_{BN} \angle -120°$$
$$\mathbf{E}_{CN} = E_{CN} \angle +120°$$

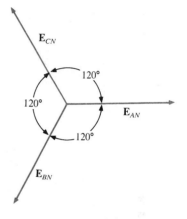

FIG. 22.3

By rearranging the phasors as shown in Fig. 22.4 and applying a law of vectors which states that *the vector sum of any number of vectors*

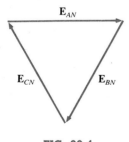

FIG. 22.4

drawn such that the "head" of each is connected to the "tail" of the next, and that the "tail" of the first vector is connected to the "head" of the last is zero, we can conclude that the phasor sum of the phase voltages in a three-phase system is zero. That is,

$$\Sigma \ (\mathbf{E}_{AN} + \mathbf{E}_{BN} + \mathbf{E}_{CN}) = 0 \qquad (22.2)$$

22.3 THE Y-CONNECTED GENERATOR

If the three terminals denoted N of Fig. 22.1(b) are connected together, the generator is referred to as a *Y-connected three-phase generator* (Fig. 22.5). As indicated in Fig. 22.5, the Y is inverted for ease of notation

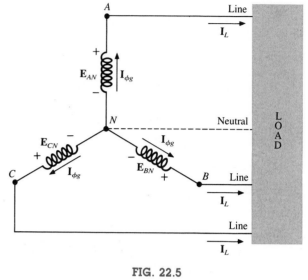

FIG. 22.5
Y-connected generator.

and for clarity. The point at which all the terminals are connected is called the *neutral point*. If a conductor is not attached from this point to the load, the system is called a *Y-connected, three-phase, three-wire generator*. If the neutral is connected, the system is a *Y-connected, three-phase, four-wire generator*. The function of the neutral will be discussed in detail when we consider the load circuit.

The three conductors connected from A, B, and C to the load are called *lines*. For the Y-connected system, it should be obvious from Fig. 22.5 that the line current equals the phase current for each phase; that is,

$$\boxed{I_L = I_{\phi g}} \qquad (22.3)$$

The voltage from one line to another is called a *line voltage*. On the phasor diagram (Fig. 22.6) it is the phasor drawn from the end of one phase to another in the counterclockwise direction.

Applying Kirchhoff's voltage law around the indicated loop of Fig. 22.6, we obtain

$$\mathbf{E}_{AB} - \mathbf{E}_{AN} + \mathbf{E}_{BN} = 0$$

or

$$\mathbf{E}_{AB} = \mathbf{E}_{AN} - \mathbf{E}_{BN} = \mathbf{E}_{AN} + \mathbf{E}_{NB}$$

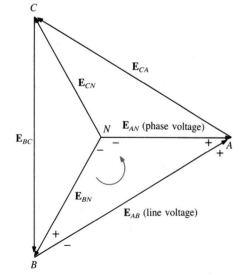

FIG. 22.6
Line and phase voltages of the Y-connected three-phase generator.

The phasor diagram is redrawn to find $\mathbf{E}_{AB}$ as shown in Fig. 22.7. Since each phase voltage when reversed ($\mathbf{E}_{NB}$) will bisect the other two, $\alpha =$

60°. The angle β is 30°, since a line drawn from opposite ends of a rhombus will divide in half both the angle of origin and the opposite angle. Lines drawn between opposite corners of a rhombus will also bisect each other at right angles.

The length x is

$$x = E_{AN} \cos 30° = \frac{\sqrt{3}}{2} E_{AN}$$

and

$$E_{AB} = 2x = (2)\frac{\sqrt{3}}{2} E_{AN} = \sqrt{3} E_{AN}$$

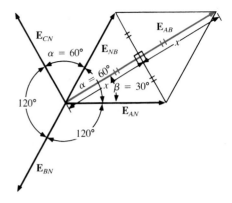

FIG. 22.7

Noting from the phasor diagram that θ of $\mathbf{E}_{AB} = \beta = 30°$, the result is

$$\mathbf{E}_{AB} = E_{AB} \angle 30° = \sqrt{3} E_{AN} \angle 30°$$

In words, the magnitude of the line voltage of a Y-connected generator is $\sqrt{3}$ times the phase voltage:

$$\boxed{E_L = \sqrt{3} E_\phi} \qquad \textbf{(22.4)}$$

In addition, the phase angle between any line voltage and the nearest phase voltage is 30°, as shown for $\mathbf{E}_{AB}$ and $\mathbf{E}_{AN}$ in Fig. 22.7 ($\beta = 30°$).

In sinusoidal notation,

$$e_{AB} = \sqrt{2} E_{AB} \sin(\omega t + 30°)$$

Repeating the same procedure for the other line voltages results in

$$e_{CA} = \sqrt{2} E_{CA} \sin(\omega t + 150°)$$

and

$$e_{BC} = \sqrt{2} E_{BC} \sin(\omega t + 270°)$$

The phasor diagram of the line and phase voltages is shown in Fig. 22.8. If the phasors representing the line voltages in Fig. 22.8(a) are

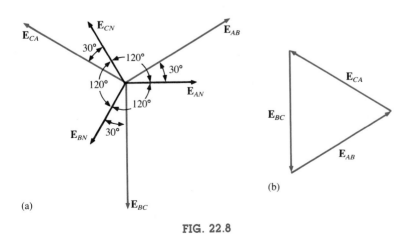

(a)

(b)

FIG. 22.8

rearranged slightly, they will form a closed loop [Fig. 22.8(b)]. There-fore, we can conclude that the sum of the line voltages is also zero; that is,

$$\Sigma \ (\mathbf{E}_{AB} + \mathbf{E}_{CA} + \mathbf{E}_{BC}) = 0 \qquad \qquad \textbf{(22.5)}$$

22.4 PHASE SEQUENCE (Y-CONNECTED GENERATOR)

The phase sequence is the order in which the phasors representing the phase voltages pass through a fixed point on the phasor diagram if the phasors are rotated in a counterclockwise direction. For example, in Fig. 22.9 the phase sequence is *ABC*. However, since the fixed point can be chosen anywhere on the phasor diagram, the sequence can also be written as *BCA* or *CAB*. The phase sequence is quite important in the three-phase distribution of power. In a three-phase motor, for example, if two phase voltages are interchanged, the sequence will change and the direction of rotation of the motor will be reversed. Other effects will be described when we consider the loaded three-phase system.

The phase sequence can also be described in terms of the line volt-ages. Drawing the line voltages on a phasor diagram in Fig. 22.10, we

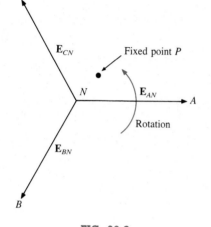

FIG. 22.9

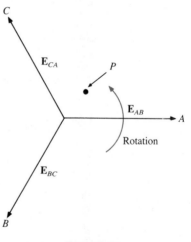

FIG. 22.10

are able to determine the phase sequence by again rotating the phasors in the counterclockwise direction. In this case, however, the sequence can be determined by noting the order of the passing first or second subscripts. In the system of Fig. 22.10, for example, the phase se-quence of the first subscripts passing point *P* is *ABC*, and of the second subscripts *BCA*. But we know that *BCA* is equivalent to *ABC*, so the sequence is the same for each. Note that the phase sequence is the same as that of the phase voltages described in Fig. 22.9.

If the sequence is given, the phasor diagram can be drawn by simply picking a reference voltage, placing it on the reference axis, and then drawing the other voltages at the proper angular position. For a sequence of *ACB*, for example, we might choose E_{AB} to be the reference [Fig. 22.11(a)] if we wanted the phasor diagram of the line voltages, or E_{NA} for the phase voltages [Fig. 22.11(b)]. For the sequence indicated, the phasor diagrams would be as in Fig. 22.11. In phasor notation,

$$\text{Line voltages} \begin{cases} \mathbf{E}_{AB} = E_{AB} \angle 0° \quad \text{(reference)} \\ \mathbf{E}_{CA} = E_{CA} \angle -120° \\ \mathbf{E}_{BC} = E_{BC} \angle +120° \end{cases}$$

$$\text{Phase voltages} \begin{cases} \mathbf{E}_{AN} = E_{AN} \angle 0° \quad \text{(reference)} \\ \mathbf{E}_{CN} = E_{CN} \angle -120° \\ \mathbf{E}_{BN} = E_{BN} \angle +120° \end{cases}$$

22.5 THE Y-CONNECTED GENERATOR WITH A Y-CONNECTED LOAD

Loads connected to three-phase supplies are of two types: the Y and the Δ.

If a Y-connected load is connected to a Y-connected generator, the system is symbolically represented by Y-Y. The physical setup of such a system is shown in Fig. 22.12.

If the load is balanced, the neutral can be removed without affecting the circuit in any manner. That is, if

$$\mathbf{Z}_1 = \mathbf{Z}_2 = \mathbf{Z}_3$$

then I_N will be zero. (This will be demonstrated in Example 22.1.) Note that in order to have a balanced load, the phase angle must also be the same for each impedance—a condition that was not necessary in dc circuits when we considered balanced systems.

In practice, if a factory, for example, had only balanced three-phase loads, the absence of the neutral would have no effect, since ideally the

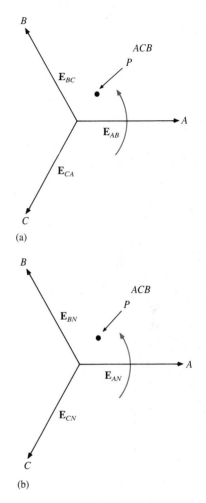

(a)

(b)

FIG. 22.11

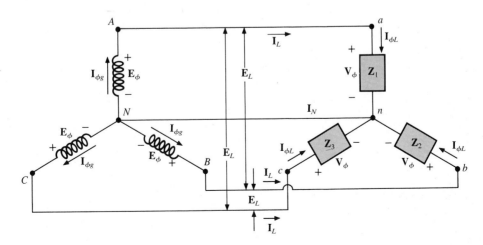

FIG. 22.12

Y-connected generator with a Y-connected load.

system would always be balanced. The cost would therefore be less since the number of required conductors would be reduced. However, lighting and most other electrical equipment will use only one of the phase voltages, and even if the loading is designed to be balanced (as it should be), there will never be perfect continuous balancing, since lights and other electrical equipment will be turned on and off, upsetting the balanced condition. The neutral is therefore necessary to carry the resulting current away from the load and back to the Y-connected generator. This will be demonstrated when we consider unbalanced Y-connected systems.

We shall now examine the *four-wire Y-Y-connected system*. The current passing through each phase of the generator is the same as its corresponding line current, which in turn for a Y-connected load is equal to the current in the phase of the load to which it is attached:

$$\boxed{\mathbf{I}_{\phi g} = \mathbf{I}_L = \mathbf{I}_{\phi L}} \qquad (22.6)$$

For a balanced or unbalanced load, since the generator and load have a common neutral point, then

$$\boxed{\mathbf{V}_\phi = \mathbf{E}_\phi} \qquad (22.7)$$

In addition, since $\mathbf{I}_{\phi L} = \mathbf{V}_\phi / \mathbf{Z}_\phi$, the magnitude of the current in each phase will be equal for a balanced load and unequal for an unbalanced load. You will recall that for the Y-connected generator, the magnitude of the line voltage is equal to $\sqrt{3}$ times the phase voltage. This same relationship can be applied to a balanced or unbalanced four-wire Y-connected load:

$$\boxed{E_L = \sqrt{3}V_\phi} \qquad (22.8)$$

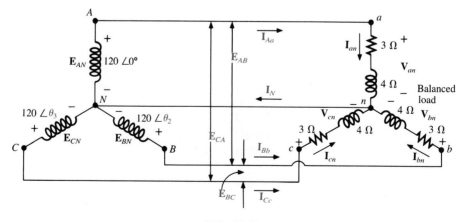

FIG. 22.13

For a voltage drop across a load element, the first subscript refers to that terminal through which the current enters the load element, and the second subscript to the terminal from which the current leaves. In other words, the first subscript is, by definition, positive with respect to the second for a voltage drop. Note Fig. 22.13, in which the standard double subscripts for a source of voltage and a voltage drop are indicated.

EXAMPLE 22.1. The phase sequence of the Y-connected generator in Fig. 22.13 is ABC.
a. Find the phase angles θ_2 and θ_3.
b. Find the magnitude of the line voltages.
c. Find the line currents.
d. Verify that since the load is balanced, $\mathbf{I}_N = 0$.

Solutions:

a. For an ABC phase sequence,

$$\theta_2 = \mathbf{-120°} \quad \text{and} \quad \theta_3 = \mathbf{+120°}$$

b. $E_L = \sqrt{3}E_\phi = (1.73)(120) = 208$ V. Therefore,

$$E_{AB} = E_{BC} = E_{CA} = \mathbf{208 \ V}$$

c. $\mathbf{V}_\phi = \mathbf{E}_\phi$. Therefore,

$$\mathbf{V}_{an} = \mathbf{E}_{AN} \qquad \mathbf{V}_{bn} = \mathbf{E}_{BN} \qquad \mathbf{V}_{cn} = \mathbf{E}_{CN}$$

$$\mathbf{I}_{\phi L} = \mathbf{I}_{an} = \frac{\mathbf{V}_{an}}{\mathbf{Z}_{an}} = \frac{120 \ \angle 0°}{3 + j4} = \frac{120 \ \angle 0°}{5 \ \angle 53.13°} = 24 \ \angle -53.13°$$

$$\mathbf{I}_{bn} = \frac{\mathbf{V}_{bn}}{\mathbf{Z}_{bn}} = \frac{120 \ \angle -120°}{5 \ \angle 53.13°} = 24 \ \angle -173.13°$$

$$\mathbf{I}_{cn} = \frac{\mathbf{V}_{cn}}{\mathbf{Z}_{cn}} = \frac{120 \ \angle +120°}{5 \ \angle 53.13°} = 24 \ \angle 66.87°$$

and, since $\mathbf{I}_L = \mathbf{I}_{\phi L}$,

$$\mathbf{I}_{Aa} = \mathbf{I}_{an} = \mathbf{24 \ \angle -53.13°}$$
$$\mathbf{I}_{Bb} = \mathbf{I}_{bn} = \mathbf{24 \ \angle -173.13°}$$
$$\mathbf{I}_{Cc} = \mathbf{I}_{cn} = \mathbf{24 \ \angle 66.87°}$$

d. Applying Kirchhoff's current law, we have

$$\mathbf{I}_N = \mathbf{I}_{Aa} + \mathbf{I}_{Bb} + \mathbf{I}_{Cc}$$

In rectangular form,

$$
\begin{aligned}
\mathbf{I}_{Aa} = 24 \ \angle -53.13° &= 14.40 - j19.20 \\
\mathbf{I}_{Bb} = 24 \ \angle -173.13° &= -23.83 - j2.87 \\
\mathbf{I}_{Cc} = 24 \ \angle 66.87° &= 9.43 + j22.07 \\
\hline
\Sigma \ (\mathbf{I}_{Aa} + \mathbf{I}_{Bb} + \mathbf{I}_{Cc}) &= 0 + j0
\end{aligned}
$$

and $\mathbf{I}_N$ is in fact equal to **zero,** as required for a balanced load.

22.6 THE Y-Δ SYSTEM

There is no neutral connection for the Y-Δ system of Fig. 22.14. Any variation in the impedance of a phase which produces an unbalanced system will simply vary the line and phase currents of the system.

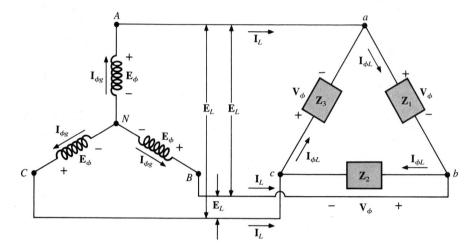

FIG. 22.14

Y-connected generator with a Δ-connected load.

For a balanced load,

$$\boxed{\mathbf{Z}_1 = \mathbf{Z}_2 = \mathbf{Z}_3} \tag{22.9}$$

The voltage across each phase of the load is equal to the line voltage of the generator for a balanced or unbalanced load:

$$\boxed{\mathbf{V}_\phi = \mathbf{E}_L} \tag{22.10}$$

The relationship between the line currents and phase currents of a balanced Δ load can be found using an approach very similar to that used in Section 22.3 to find the relationship between the line voltages and phase voltages of a Y-connected generator. For this case, however, Kirchhoff's current law is employed instead of Kirchhoff's voltage law.

The results obtained are

$$\boxed{I_L = \sqrt{3}I_\phi} \tag{22.11}$$

and the phase angle between a line current and the nearest phase current is 30°. A more detailed discussion of this relationship between the line and phase currents of a Δ-connected system can be found in Section 22.7.

For a balanced load, the line currents will be equal in magnitude, as will the phase currents.

EXAMPLE 22.2. For the three-phase system of Fig. 22.15:

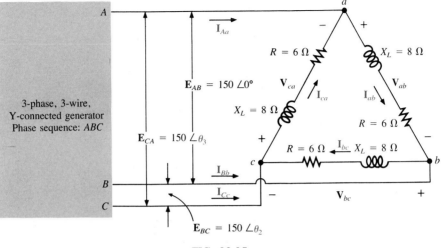

FIG. 22.15

a. Find the phase angles θ_2 and θ_3.
b. Find the current in each phase of the load.
c. Find the magnitude of the line currents.

Solutions:

a. For an *ABC* sequence,

$$\theta_2 = -120° \quad \text{and} \quad \theta_3 = +120°$$

b. $\mathbf{V}_\phi = \mathbf{E}_L$. Therefore,

$$\mathbf{V}_{ab} = \mathbf{E}_{AB} \qquad \mathbf{V}_{ca} = \mathbf{E}_{CA} \qquad \mathbf{V}_{bc} = \mathbf{E}_{BC}$$

The phase currents are

$$\mathbf{I}_{ab} = \frac{\mathbf{V}_{ab}}{\mathbf{Z}_{ab}} = \frac{150 \angle 0°}{6 + j8} = \frac{150 \angle 0°}{10 \angle 53.13°} = \mathbf{15 \angle -53.13°}$$

$$\mathbf{I}_{bc} = \frac{\mathbf{V}_{bc}}{\mathbf{Z}_{bc}} = \frac{150 \angle -120°}{10 \angle 53.13°} = \mathbf{15 \angle -173.13°}$$

$$\mathbf{I}_{ca} = \frac{\mathbf{V}_{ca}}{\mathbf{Z}_{ca}} = \frac{150 \angle +120°}{10 \angle 53.13°} = \mathbf{15 \angle 66.87°}$$

c. $I_L = \sqrt{3}I_\phi = (1.73)(15) = 25.95$ A. Therefore,

$$I_{Aa} = I_{Bb} = I_{Cc} = \mathbf{25.95\,A}$$

22.7 THE Δ-CONNECTED GENERATOR

If we rearrange the coils of the generator in Fig. 22.16(a) as shown in Fig. 22.16(b), the system is referred to as a *three-phase, three-wire, Δ-connected ac generator*. In this system, the phase and line voltages

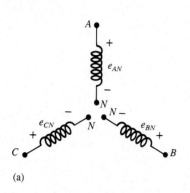

(a)

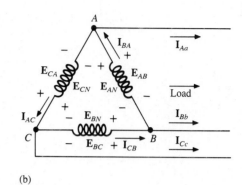

(b)

FIG. 22.16
Δ-connected generator.

are equivalent and equal to the voltage induced across each coil of the generator. That is,

$$
\left.
\begin{aligned}
\mathbf{E}_{AB} &= \mathbf{E}_{AN} \quad \text{and} \quad e_{AN} = \sqrt{2}E_{AN} \sin \omega t \\
\mathbf{E}_{BC} &= \mathbf{E}_{BN} \quad \text{and} \quad e_{BN} = \sqrt{2}E_{BN} \sin(\omega t - 120°) \\
\mathbf{E}_{CA} &= \mathbf{E}_{CN} \quad \text{and} \quad e_{CN} = \sqrt{2}E_{CN} \sin(\omega t + 120°)
\end{aligned}
\right\}
\begin{aligned}
&\text{Phase} \\
&\text{sequence} \\
&ABC
\end{aligned}
$$

or

$$\boxed{E_L = E_{\phi g}} \qquad (22.12)$$

Note that only one voltage (magnitude) is available instead of the two available in the Y-connected system.

Unlike the line current for the Y-connected generator, the line current for the Δ-connected system is not equal to the phase current. The relationship between the two can be found by applying Kirchhoff's current law at one of the nodes and solving for the line current in terms of the phase currents. That is, at node *A*,

$$\mathbf{I}_{BA} = \mathbf{I}_{Aa} + \mathbf{I}_{AC}$$

or

$$\mathbf{I}_{Aa} = \mathbf{I}_{BA} - \mathbf{I}_{AC} = \mathbf{I}_{BA} + \mathbf{I}_{CA}$$

The phasor diagram is shown in Fig. 22.17 for a balanced load.

Using the same procedure to find the line current as was used to find the line voltage of a Y-connected generator produces the following general results:

$$\boxed{I_L = \sqrt{3}I_{\phi g}} \qquad (22.13)$$

The phase angle between a line current and the nearest phase current is 30°. The phasor diagram of the currents is shown in Fig. 22.18.

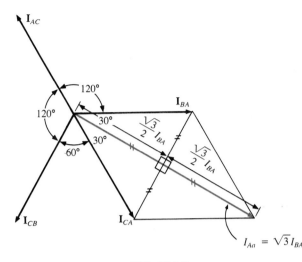

FIG. 22.17

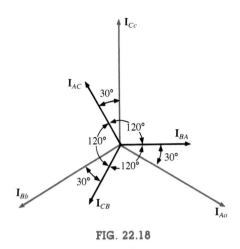

FIG. 22.18

It can be shown in the same manner employed for the voltages of a Y-connected generator that the phasor sum of the line currents or phase currents for Δ-connected systems with balanced loads is zero.

22.8 PHASE SEQUENCE (Δ-CONNECTED GENERATOR)

Even though the line and phase voltages of a Δ-connected system are the same, it is standard practice to describe the phase sequence in terms of the line voltages. The method used is the same as that described for the line voltages of the Y-connected generator. For example, the phasor diagram of the line voltages for a phase sequence ABC is shown in Fig. 22.19. In drawing such a diagram, one must take care to have the sequence of the first and second subscripts the same.

In phasor notation,

$$\mathbf{E}_{AB} = E_{AB} \angle 0°$$
$$\mathbf{E}_{BC} = E_{BC} \angle -120°$$
$$\mathbf{E}_{CA} = E_{CA} \angle 120°$$

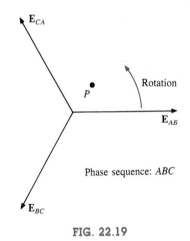

Phase sequence: ABC

FIG. 22.19

22.9 THE Δ-Δ, Δ-Y THREE-PHASE SYSTEMS

The basic equations necessary to analyze either of the two systems (Δ-Δ, Δ-Y) have been presented at least once in this chapter. We will therefore proceed directly to two descriptive examples, one with a Δ-connected load and one with a Y-connected load.

EXAMPLE 22.3. For the Δ-Δ system shown in Fig. 22.20:

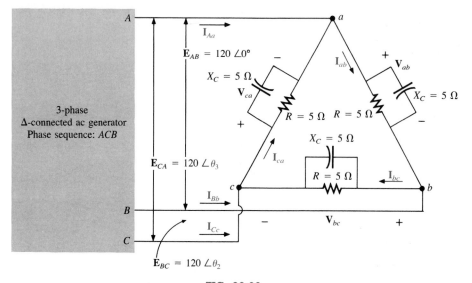

FIG. 22.20
Δ-Δ *system.*

a. Find the phase angles θ_2 and θ_3 for the specified phase sequence.
b. Find the current in each phase of the load.
c. Find the magnitude of the line currents.

Solutions:

a. For an *ACB* phase sequence,

$$\theta_2 = 120° \quad \text{and} \quad \theta_3 = -120°$$

b. $\mathbf{V}_\phi = \mathbf{E}_L$. Therefore,

$$\mathbf{V}_{ab} = \mathbf{E}_{AB} \qquad \mathbf{V}_{ca} = \mathbf{E}_{CA} \qquad \mathbf{V}_{bc} = \mathbf{E}_{BC}$$

The phase currents are

$$\mathbf{I}_{ab} = \frac{\mathbf{V}_{ab}}{\mathbf{Z}_{ab}} = \frac{120\ \angle 0°}{\dfrac{(5\ \angle 0°)(5\ \angle -90°)}{5 - j5}} = \frac{120\ \angle 0°}{\dfrac{25\ \angle -90°}{7.07\ \angle -45°}}$$

$$= \frac{120\ \angle 0°}{3.54\ \angle -45°} = \mathbf{33.9\ \angle 45°}$$

$$\mathbf{I}_{bc} = \frac{\mathbf{V}_{bc}}{\mathbf{Z}_{bc}} = \frac{120\ \angle 120°}{3.54\ \angle -45°} = \mathbf{33.9\ \angle 165°}$$

$$\mathbf{I}_{ca} = \frac{\mathbf{V}_{ca}}{\mathbf{Z}_{ca}} = \frac{120\ \angle -120°}{3.54\ \angle -45°} = \mathbf{33.9\ \angle -75°}$$

c. $I_L = \sqrt{3}I_\phi = (1.73)(34) = 58.82$ A. Therefore,

$$I_{Aa} = I_{Bb} = I_{Cc} = \mathbf{58.82\ A}$$

EXAMPLE 22.4. For the Δ-Y system shown in Fig. 22.21:

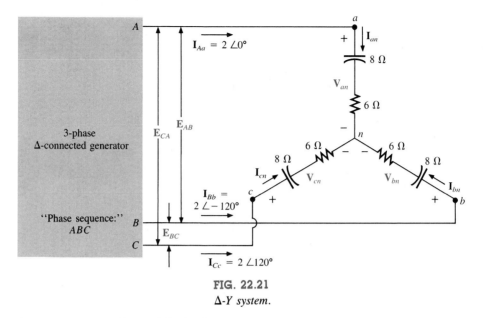

FIG. 22.21
Δ-Y system.

a. Find the voltage across each phase of the load.
b. Find the magnitude of the line voltages.

Solutions:

a. $\mathbf{I}_{\phi L} = \mathbf{I}_L$. Therefore,

$$\mathbf{I}_{an} = \mathbf{I}_{Aa} = 2 \angle 0°$$
$$\mathbf{I}_{bn} = \mathbf{I}_{Bb} = 2 \angle -120°$$
$$\mathbf{I}_{cn} = \mathbf{I}_{Cc} = 2 \angle 120°$$

The phase voltages are

$$\mathbf{V}_{an} = \mathbf{I}_{an}\mathbf{Z}_{an} = (2 \angle 0°)(10 \angle -53.13°) = \mathbf{20} \angle -\mathbf{53.13°}$$
$$\mathbf{V}_{bn} = \mathbf{I}_{bn}\mathbf{Z}_{bn} = (2 \angle -120°)(10 \angle -53.13°) = \mathbf{20} \angle -\mathbf{173.13°}$$
$$\mathbf{V}_{cn} = \mathbf{I}_{cn}\mathbf{Z}_{cn} = (2 \angle 120°)(10 \angle -53.13°) = \mathbf{20} \angle \mathbf{66.87°}$$

b. $E_L = \sqrt{3}V_\phi = (1.73)(20) = 34.6 \text{ V}$. Therefore,

$$E_{BA} = E_{CB} = E_{AC} = \mathbf{34.6 \text{ V}}$$

22.10 POWER

Y-Connected Balanced Load (Fig. 22.22)

Average power The average power delivered to each phase can be determined by any one of Eqs. (22.14) through (22.16).

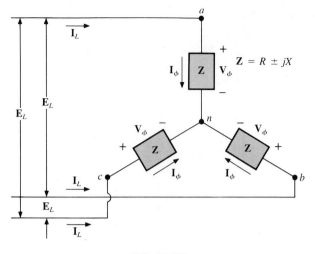

FIG. 22.22

$$P_\phi = V_\phi I_\phi \cos \theta_{I_\phi}^{V_\phi} = I_\phi^2 R_\phi = \frac{V_R^2}{R_\phi} \qquad \text{(watts, W)} \quad \textbf{(22.14)}$$

where $\theta_{I_\phi}^{V_\phi}$ indicates that θ is the phase angle between V_ϕ and I_ϕ. The total power to the balanced load is

$$P_T = 3P_\phi \qquad \text{(W)} \qquad \textbf{(22.15)}$$

or, since

$$V_\phi = \frac{E_L}{\sqrt{3}} \quad \text{and} \quad I_\phi = I_L$$

then

$$P_T = 3\frac{E_L}{\sqrt{3}} I_L \cos \theta_{I_\phi}^{V_\phi}$$

But

$$\left(\frac{3}{\sqrt{3}}\right)(1) = \left(\frac{3}{\sqrt{3}}\right)\left(\frac{\sqrt{3}}{\sqrt{3}}\right) = \frac{3\sqrt{3}}{3} = \sqrt{3}$$

Therefore,

$$P_T = \sqrt{3}E_L I_L \cos \theta_{I_\phi}^{V_\phi} = 3I_L^2 R_\phi \qquad \text{(W)} \qquad \textbf{(22.16)}$$

Reactive power The reactive power of each phase (in volt-amperes reactive) is

$$Q_\phi = V_\phi I_\phi \sin \theta_{I_\phi}^{V_\phi} = I_\phi^2 X_\phi = \frac{V_X^2}{X_\phi} \qquad \text{(VAR)} \quad \textbf{(22.17)}$$

The total reactive power of the load is

$$\boxed{Q_T = 3Q_\phi} \qquad \text{(VAR)} \qquad\qquad \textbf{(22.18)}$$

or, proceeding in the same manner as above, we have

$$\boxed{Q_T = \sqrt{3}E_L I_L \sin \theta_{I_\phi}^{V_\phi} = 3I_L^2 X_\phi} \qquad \text{(VAR)} \qquad \textbf{(22.19)}$$

Apparent power The apparent power of each phase is

$$\boxed{S_\phi = V_\phi I_\phi} \qquad \text{(VA)} \qquad\qquad \textbf{(22.20)}$$

The total apparent power of the load is

$$\boxed{S_T = 3S_\phi} \qquad \text{(VA)} \qquad\qquad \textbf{(22.21)}$$

or, as before,

$$\boxed{S_T = \sqrt{3}E_L I_L} \qquad \text{(VA)} \qquad\qquad \textbf{(22.22)}$$

Power factor The power factor of the system is given by

$$\boxed{F_p = \frac{P_T}{S_T} = \cos \theta \text{ (leading or lagging)}} \qquad \textbf{(22.23)}$$

EXAMPLE 22.5. See Fig. 22.23. Here,

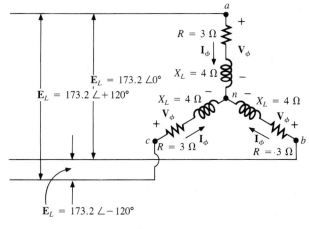

$$\mathbf{E}_L = 173.2 \angle 0°$$
$$\mathbf{E}_L = 173.2 \angle +120°$$
$$\mathbf{E}_L = 173.2 \angle -120°$$

FIG. 22.23

$$\mathbf{Z}_\phi = 3 + j4 = 5 \angle 53.13°$$
$$V_\phi = \frac{V_L}{\sqrt{3}} = \frac{173.2}{1.732} = 100 \text{ V}$$

$$I_\phi = \frac{V_\phi}{Z_\phi} = \frac{100}{5} = 20 \text{ A}$$

The *average power* is

$$P_\phi = V_\phi I_\phi \cos \theta_{I_\phi}^{V_\phi} = (100)(20) \cos 53.13° = (2000)(0.6)$$
$$= \mathbf{1200 \text{ W}}$$
$$P_\phi = I_\phi^2 R_\phi = (20)^2(3) = (400)(3) = \mathbf{1200 \text{ W}}$$
$$P_\phi = \frac{V_R^2}{R_\phi} = \frac{(60)^2}{3} = \frac{3600}{3} = \mathbf{1200 \text{ W}}$$
$$P_T = 3P_\phi = (3)(1200) = \mathbf{3600 \text{ W}}$$

or

$$P_T = \sqrt{3} E_L I_L \cos \theta_{I_\phi}^{V_\phi} = (1.732)(173.2)(20)(0.6) = \mathbf{3600 \text{ W}}$$

The *reactive power* is

$$Q_\phi = V_\phi I_\phi \sin \theta_{I_\phi}^{V_\phi} = (100)(20) \sin 53.13° = (2000)(0.8)$$
$$= \mathbf{1600 \text{ VAR}}$$

or

$$Q_\phi = I_\phi^2 X_\phi = (20)^2(4) = (400)(4) = \mathbf{1600 \text{ VAR}}$$
$$Q_T = 3Q_\phi = (3)(1600) = \mathbf{4800 \text{ VAR}}$$

or

$$Q_T = \sqrt{3} E_L I_L \sin \theta_{I_\phi}^{V_\phi} = (1.732)(173.2)(20)(0.8) = \mathbf{4800 \text{ VAR}}$$

The *apparent power* is

$$S_\phi = V_\phi I_\phi = (100)(20) = \mathbf{2000 \text{ VA}}$$
$$S_T = 3S_\phi = (3)(2000) = \mathbf{6000 \text{ VA}}$$

or

$$S_T = \sqrt{3} E_L I_L = (1.732)(173.2)(20) = \mathbf{6000 \text{ VA}}$$

The *power factor* is

$$F_p = \frac{P_T}{S_T} = \frac{3600}{6000} = \mathbf{0.6 \text{ lagging}}$$

Δ-Connected Balanced Load (Fig. 22.24)

Average power

$$\boxed{P_\phi = V_\phi I_\phi \cos \theta_{I_\phi}^{V_\phi} = I_\phi^2 R_\phi = \frac{V_R^2}{R_\phi}} \qquad \text{(W)} \qquad \textbf{(22.24)}$$

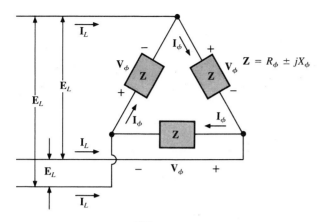

FIG. 22.24

$$\boxed{P_T = 3P_\phi} \quad \text{(W)} \qquad \textbf{(22.25)}$$

Reactive power

$$\boxed{Q_\phi = V_\phi I_\phi \sin \theta_{I_\phi}^{V_\phi} = I_\phi^2 X_\phi = \frac{V_X^2}{X_\phi}} \quad \text{(VAR)} \quad \textbf{(22.26)}$$

$$\boxed{Q_T = 3Q_\phi} \quad \text{(VAR)} \qquad \textbf{(22.27)}$$

Apparent power

$$\boxed{S_\phi = V_\phi I_\phi} \quad \text{(VA)} \qquad \textbf{(22.28)}$$

$$\boxed{S_T = 3S_\phi = \sqrt{3}E_L I_L} \quad \text{(VA)} \qquad \textbf{(22.29)}$$

Power factor

$$\boxed{F_p = \frac{P_T}{S_T}} \qquad \textbf{(22.30)}$$

EXAMPLE 22.6. Determine the total watts, volt-amperes reactive, and volt-amperes for the network of Fig. 22.25. In addition, calculate the total power factor of the load.

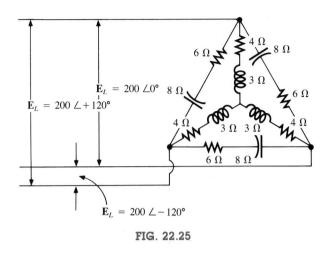

$E_L = 200 \angle +120°$

$E_L = 200 \angle 0°$

$E_L = 200 \angle -120°$

FIG. 22.25

Solution: Consider the Δ and Y separately.

For the Δ:

$$\mathbf{Z}_\Delta = 6 - j8 = 10 \angle -53.13°$$

$$I_\phi = \frac{E_L}{Z_\Delta} = \frac{200}{10} = 20 \text{ A}$$

$$P_{T_\Delta} = 3I_\phi^2 R_\phi = (3)(20)^2(6) = \textbf{7200 W}$$

$$Q_{T_\Delta} = 3I_\phi^2 X_\phi = (3)(20)^2(8) = \textbf{9600 VAR (cap.)}$$

$$S_{T_\Delta} = 3V_\phi I_\phi = (3)(200)(20) = \textbf{12,000 VA}$$

For the Y:

$$\mathbf{Z}_Y = 4 + j3 = 5 \angle 36.87°$$

$$I_\phi = \frac{E_L/\sqrt{3}}{Z_Y} = \frac{200/\sqrt{3}}{5} = \frac{116}{5} = 23.12 \text{ A}$$

$$P_{T_Y} = 3I_\phi^2 R_\phi = (3)(23.12)^2(4) = \textbf{6414.41 W}$$

$$Q_{T_Y} = 3I_\phi^2 X_\phi = (3)(23.12)^2(3) = \textbf{4810.81 VAR (ind.)}$$

$$S_{T_Y} = 3V_\phi I_\phi = (3)(116)(23.12) = \textbf{8045.76 VA}$$

For the total load:

$$P_T = P_{T_\Delta} + P_{T_Y} = 7200 + 6414.41 = \textbf{13,614.41 W}$$

$$Q_T = Q_{T_\Delta} - Q_{T_Y} = 9600 \text{ (cap.)} - 4810.81 \text{ (ind.)}$$
$$= \textbf{4789.19 VAR (cap.)}$$

$$S_T = \sqrt{P_T^2 + Q_T^2} = \sqrt{(13,614.41)^2 + (4789.19)^2}$$
$$= \textbf{14,432.2 VA}$$

$$F_p = \frac{P_T}{S_T} = \frac{13,614.41}{14,432.20} = \textbf{0.943 leading}$$

22.11 THE THREE-WATTMETER METHOD

The power delivered to a balanced or an unbalanced four-wire Y-connected load can be found using three wattmeters in the manner shown in Fig. 22.26. Each wattmeter measures the power delivered to

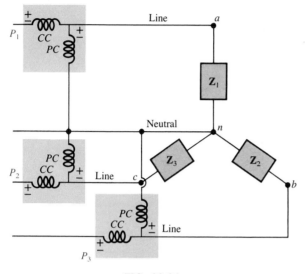

FIG. 22.26

each phase. The potential coil of each wattmeter is connected parallel with the load, while the current coil is in series with the load. The total average power of the system can be found by summing the three wattmeter readings; that is,

$$P_{T_Y} = P_1 + P_2 + P_3 \qquad (22.31)$$

For the load (balanced or unbalanced), the wattmeters are connected as shown in Fig. 22.27. The total power is again the sum of the three wattmeter readings:

$$P_{T_\Delta} = P_1 + P_2 + P_3 \qquad (22.32)$$

If in either of the cases just described the load is balanced, the power delivered to each phase will be the same. The total power is then just three times any one wattmeter reading.

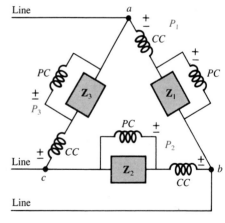

FIG. 22.27

22.12 THE TWO-WATTMETER METHOD

The power delivered to a three-phase, three-wire, Δ- or Y-connected balanced or unbalanced load can be found using only two wattmeters if the proper connection is employed and if the wattmeter readings are

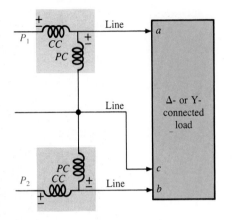

FIG. 22.28

properly interpreted. The basic connections are shown in Fig. 22.28. One end of each potential coil is connected to the same line. The current coils are then placed in the remaining lines.

The connection shown in Fig. 22.29 will also satisfy the requirements. A third hookup is also possible, but this is left to the reader as an exercise.

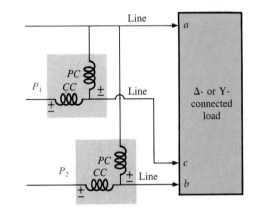

FIG. 22.29

The total power delivered to the load is the algebraic sum of the two wattmeter readings. For a *balanced* load, we will now consider two methods of determining whether the total power is the sum or the difference of the two wattmeter readings. The first method to be described requires that we know or be able to find the power factor (leading or lagging) of any one phase of the load. When this information has been obtained, it can be applied directly to the curve of Fig. 22.30.

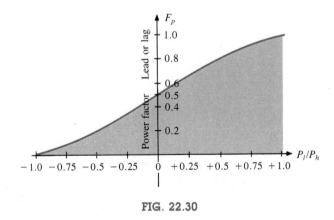

FIG. 22.30

The curve in Fig. 22.30 is a plot of the power factor of the load (phase) versus the ratio P_l/P_h, where P_l and P_h are the magnitudes of the lower- and higher-reading wattmeters, respectively. Note that for a power factor (leading or lagging) greater than 0.5, the ratio has a positive value. This indicates that both wattmeters are reading positive, and the total power is the sum of the two wattmeter readings; that is, $P_T =$

$P_l + P_h$. For a power factor less than 0.5 (leading or lagging), the ratio has a negative value. This indicates that the smaller-reading wattmeter is reading negative, and the total power is the difference of the two wattmeter readings; that is, $P_T = P_h - P_l$.

A closer examination will reveal that when the power factor is 1 (cos 0° = 1), corresponding to a purely resistive load, $P_l/P_h = 1$ or $P_l = P_h$, and both wattmeters will have the same wattage indication. At a power factor equal to 0 (cos 90° = 0), corresponding to a purely reactive load, $P_l/P_h = -1$ or $P_l = -P_h$, and both wattmeters will again have the same wattage indication but with opposite signs. The transition from a negative to a positive ratio occurs when the power factor of the load is 0.5 or $\theta = \cos^{-1} 0.5 = 60°$. At this power factor, $P_l/P_h = 0$, so that $P_l = 0$, while P_h will read the total power delivered to the load.

The second method for determining whether the total power is the sum or difference of the two wattmeter readings involves a simple laboratory test. For the test to be applied, both wattmeters must first have an up-scale deflection. If one of the wattmeters has a below-zero indication, an up-scale deflection can be obtained by simply reversing the leads of the current coil of the wattmeter. To perform the test, first remove the lead of the potential coil of the *low-reading* wattmeter from the line that has no current coil in it. Take this lead and touch it to the line that has the current coil of the *high-reading* wattmeter in it. If the pointer of the low-reading wattmeter deflects upward, the two wattmeter readings should be added. If the pointer deflects downward (below zero watts), the wattage reading of the low-reading wattmeter should be subtracted from that of the high-reading wattmeter.

For a *balanced system,* since

$$P_T = P_h \pm P_l = \sqrt{3}E_L I_L \cos \theta_{I_\phi}^{V_\phi}$$

the power factor of the load (phase) can be found from the wattmeter readings and the magnitude of the line voltage and current:

$$\boxed{F_p = \cos \theta_{I_\phi}^{V_\phi} = \frac{P_h \pm P_l}{\sqrt{3}E_L I_L}} \qquad \textbf{(22.33)}$$

22.13 UNBALANCED THREE-PHASE, FOUR-WIRE, Y-CONNECTED LOAD

For the three-phase, four-wire, Y-connected load of Fig. 22.31, conditions are such that *none* of the load impedances are equal. Since the neutral is a common point between the load and source, no matter what the impedance of each phase of the load, the voltage across each phase is the phase voltage of the generator:

$$\boxed{\mathbf{V}_\phi = \mathbf{E}_\phi} \qquad \textbf{(22.34)}$$

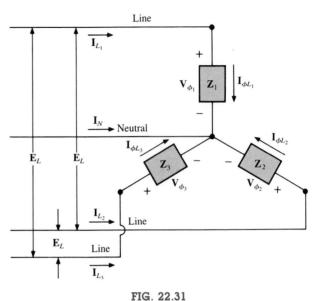

FIG. 22.31
Unbalanced Y-connected load.

The phase currents, therefore, can be determined by Ohm's law:

$$\boxed{\mathbf{I}_{\phi_1} = \frac{\mathbf{V}_{\phi_1}}{\mathbf{Z}_1} = \frac{\mathbf{E}_{\phi_1}}{\mathbf{Z}_1}, \text{ and so on}} \qquad \textbf{(22.35)}$$

The current in the neutral for any unbalanced system can then be found by applying Kirchhoff's current law at the common point n:

$$\boxed{\mathbf{I}_N = \mathbf{I}_{\phi_1} + \mathbf{I}_{\phi_2} + \mathbf{I}_{\phi_3} = \mathbf{I}_{L_1} + \mathbf{I}_{L_2} + \mathbf{I}_{L_3}} \qquad \textbf{(22.36)}$$

22.14 UNBALANCED THREE-PHASE, THREE-WIRE, Y-CONNECTED LOAD

For the system shown in Fig. 22.32, the required equations can be derived by first applying Kirchhoff's voltage law around each closed loop to produce

$$\mathbf{E}_{AB} - \mathbf{V}_{an} + \mathbf{V}_{bn} = 0$$
$$\mathbf{E}_{BC} - \mathbf{V}_{bn} + \mathbf{V}_{cn} = 0$$
$$\mathbf{E}_{CA} - \mathbf{V}_{cn} + \mathbf{V}_{an} = 0$$

Substituting, we have

$$\mathbf{V}_{an} = \mathbf{I}_{an}\mathbf{Z}_1 \qquad \mathbf{V}_{bn} = \mathbf{I}_{bn}\mathbf{Z}_2 \qquad \mathbf{V}_{cn} = \mathbf{I}_{cn}\mathbf{Z}_3$$

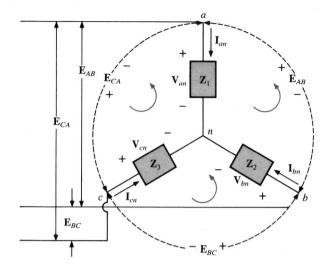

FIG. 22.32

$$\boxed{\begin{aligned}\mathbf{E}_{AB} &= \mathbf{I}_{an}\mathbf{Z}_1 - \mathbf{I}_{bn}\mathbf{Z}_2 \\ \mathbf{E}_{BC} &= \mathbf{I}_{bn}\mathbf{Z}_2 - \mathbf{I}_{cn}\mathbf{Z}_3 \\ \mathbf{E}_{CA} &= \mathbf{I}_{cn}\mathbf{Z}_3 - \mathbf{I}_{an}\mathbf{Z}_1\end{aligned}}$$

(22.37a)
(22.37b)
(22.37c)

Applying Kirchhoff's current law at node n results in

$$\mathbf{I}_{an} + \mathbf{I}_{bn} + \mathbf{I}_{cn} = 0 \quad \text{and} \quad \mathbf{I}_{bn} = -\mathbf{I}_{an} - \mathbf{I}_{cn}$$

Substituting for I_{bn} in Eqs. (22.37a) and (22.37b) yields

$$\mathbf{E}_{AB} = \mathbf{I}_{an}\mathbf{Z}_1 - [-(\mathbf{I}_{an} + \mathbf{I}_{cn})]\mathbf{Z}_2$$
$$\mathbf{E}_{BC} = -(\mathbf{I}_{an} + \mathbf{I}_{cn})\mathbf{Z}_2 - \mathbf{I}_{cn}\mathbf{Z}_3$$

which are rewritten as

$$\mathbf{E}_{AB} = \mathbf{I}_{an}(\mathbf{Z}_1 + \mathbf{Z}_2) + \mathbf{I}_{cn}\mathbf{Z}_2$$
$$\mathbf{E}_{BC} = \mathbf{I}_{an}(-\mathbf{Z}_2) + \mathbf{I}_{cn}[-(\mathbf{Z}_2 + \mathbf{Z}_3)]$$

Using determinants, we have

$$\mathbf{I}_{an} = \frac{\begin{vmatrix} \mathbf{E}_{AB} & \mathbf{Z}_2 \\ \mathbf{E}_{BC} & -(\mathbf{Z}_2 + \mathbf{Z}_3) \end{vmatrix}}{\begin{vmatrix} \mathbf{Z}_1 + \mathbf{Z}_2 & \mathbf{Z}_2 \\ -\mathbf{Z}_2 & -(\mathbf{Z}_2 + \mathbf{Z}_3) \end{vmatrix}}$$

$$= \frac{-(\mathbf{Z}_2 + \mathbf{Z}_3)\mathbf{E}_{AB} - \mathbf{E}_{BC}\mathbf{Z}_2}{-\mathbf{Z}_1\mathbf{Z}_2 - \mathbf{Z}_1\mathbf{Z}_3 - \mathbf{Z}_2\mathbf{Z}_3 - \mathbf{Z}_2^2 + \mathbf{Z}_2^2}$$

$$\mathbf{I}_{an} = \frac{-\mathbf{Z}_2(\mathbf{E}_{AB} + \mathbf{E}_{BC}) - \mathbf{Z}_3\mathbf{E}_{AB}}{-\mathbf{Z}_1\mathbf{Z}_2 - \mathbf{Z}_1\mathbf{Z}_3 - \mathbf{Z}_2\mathbf{Z}_3}$$

Apply Kirchhoff's voltage law to the line voltages:

$$\mathbf{E}_{AB} + \mathbf{E}_{CA} + \mathbf{E}_{BC} = 0 \quad \text{or} \quad \mathbf{E}_{AB} + \mathbf{E}_{BC} = -\mathbf{E}_{CA}$$

Substitute for $\mathbf{E}_{AB} + \mathbf{E}_{CB}$ in the above equation for $\mathbf{I}_{an}$:

$$\mathbf{I}_{an} = \frac{-\mathbf{Z}_2(-\mathbf{E}_{CA}) - \mathbf{Z}_3\mathbf{E}_{AB}}{-\mathbf{Z}_1\mathbf{Z}_2 - \mathbf{Z}_1\mathbf{Z}_3 - \mathbf{Z}_2\mathbf{Z}_3}$$

and

$$\mathbf{I}_{an} = \frac{\mathbf{E}_{AB}\mathbf{Z}_3 - \mathbf{E}_{CA}\mathbf{Z}_2}{\mathbf{Z}_1\mathbf{Z}_2 + \mathbf{Z}_1\mathbf{Z}_3 + \mathbf{Z}_2\mathbf{Z}_3} \qquad \textbf{(22.38)}$$

In the same manner, it can be shown that

$$\mathbf{I}_{cn} = \frac{\mathbf{E}_{CA}\mathbf{Z}_2 - \mathbf{E}_{BC}\mathbf{Z}_1}{\mathbf{Z}_1\mathbf{Z}_2 + \mathbf{Z}_1\mathbf{Z}_3 + \mathbf{Z}_2\mathbf{Z}_3} \qquad \textbf{(22.39)}$$

Substituting Eq. (22.39) for $\mathbf{I}_{cn}$ in the right-hand side of Eq. (22.37b), we obtain

$$\mathbf{I}_{bn} = \frac{\mathbf{E}_{BC}\mathbf{Z}_1 - \mathbf{E}_{AB}\mathbf{Z}_3}{\mathbf{Z}_1\mathbf{Z}_2 + \mathbf{Z}_1\mathbf{Z}_3 + \mathbf{Z}_2\mathbf{Z}_3} \qquad \textbf{(22.40)}$$

FIG. 22.33
Phase-sequence indicator. (Courtesy of General Electric Co.)

EXAMPLE 22.7. A *phase-sequence indicator* (Fig. 22.33) is an instrument that can determine the phase sequence of a polyphase circuit. The numbers 1-2-3 correspond to the terminals *A-B-C* described in this chapter.

A network that will perform the same function as the indicator of Fig. 22.33 appears in Fig. 22.34. As noted, the applied phase sequence is *ABC*. The bulb corresponding to this phase sequence will burn more

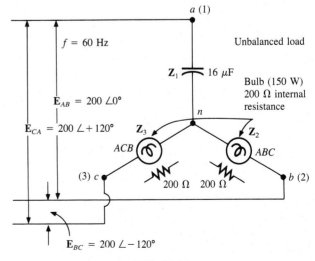

FIG. 22.34

brightly than the bulb indicating the *ACB* sequence because a greater current is passing through the *ABC* bulb. Calculating the phase currents will demonstrate that this situation does in fact exist:

$$Z_1 = X_C = \frac{1}{\omega C} = \frac{1}{(377)(16 \times 10^{-6})} = 166 \ \Omega$$

By Eq. (22.39),

$$\mathbf{I}_{cn} = \frac{\mathbf{E}_{CA}\mathbf{Z}_2 - \mathbf{E}_{BC}\mathbf{Z}_1}{\mathbf{Z}_1\mathbf{Z}_2 + \mathbf{Z}_1\mathbf{Z}_3 + \mathbf{Z}_2\mathbf{Z}_3}$$

$$= \frac{(200 \ \angle 120°)(200 \ \angle 0°) - (200 \ \angle -120°)(166 \ \angle -90°)}{(166 \ \angle -90°)(200 \ \angle 0°) + (166 \ \angle -90°)(200 \ \angle 0°) + (200 \ \angle 0°)(200 \ \angle 0°)}$$

$$\mathbf{I}_{cn} = \frac{40,000 \ \angle 120° + 33,200 \ \angle -30°}{33,200 \ \angle -90° + 33,200 \ \angle -90° + 40,000 \ \angle 0°}$$

Dividing the numerator and denominator by 1000 and converting both to the rectangular domain yields

$$\mathbf{I}_{cn} = \frac{(-20 + j34.64) + (28.75 - j16.60)}{40 - j66.4}$$

$$= \frac{8.75 + j18.04}{77.52 \ \angle -58.93°} = \frac{20.05 \ \angle 64.13°}{77.52 \ \angle -58.93°}$$

$$\mathbf{I}_{cn} = \mathbf{0.259} \ \angle \mathbf{123.06°}$$

By Eq. (22.40),

$$\mathbf{I}_{bn} = \frac{\mathbf{E}_{BC}\mathbf{Z}_1 - \mathbf{E}_{AB}\mathbf{Z}_3}{\mathbf{Z}_1\mathbf{Z}_2 + \mathbf{Z}_1\mathbf{Z}_3 + \mathbf{Z}_2\mathbf{Z}_3}$$

$$= \frac{(200 \ \angle -120°)(166 \ \angle -90°) - (200 \ \angle 0°)(200 \ \angle 0°)}{77.5 \times 10^3 \ \angle -59.3°}$$

$$\mathbf{I}_{bn} = \frac{33,200 \ \angle -210° - 40,000 \ \angle 0°}{77.5 \times 10^3 \ \angle -59.3°}$$

Dividing by 1000 and converting to the rectangular domain yields

$$\mathbf{I}_{bn} = \frac{-28.75 + j16.60 - 40.0}{77.52 \ \angle -58.93°}$$

$$= \frac{-68.75 + j16.60}{77.52 \ \angle -58.93°}$$

$$= \frac{70.73 \ \angle 166.43°}{77.52 \ \angle -58.93°}$$

$$\mathbf{I}_{bn} = \mathbf{0.91} \ \angle \mathbf{225.36°}$$

and $I_{bn} > I_{cn}$ by a factor of more than 3:1. Therefore, the bulb indicating an *ABC* sequence will burn more brightly due to the greater current. If the phase sequence were *ACB*, the reverse would be true.

22.15 COMPUTER TECHNIQUES

The program of Fig. 22.36 will determine all the quantities of interest

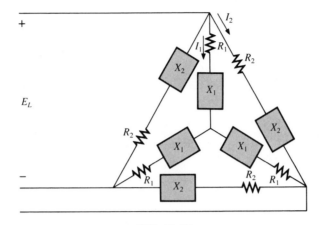

FIG. 22.35

for the Δ-Y-connected load of Fig. 22.35. As indicated by the INPUT bracket, lines 130 through 220 request the line voltage and the network

```
10 REM ***** PROGRAM  22-1 *****
20 REM ******************************
30 REM Program to analyze a 3-phase
40 REM delta-wye load.
50 REM ******************************
60 REM
100 PRINT "This program analyzes the 3-phase"
110 PRINT "delta-wye load."
120 PRINT
130 PRINT "Enter the following network information:"
140 PRINT
150 INPUT "Line voltage, E=";EL
160 PRINT "For the series-connected wye load:"
170 INPUT "R=";R1
180 INPUT "X=";X1 :REM Enter negative sign if capacitive
190 PRINT
200 PRINT "And for the series-connected delta load:"
210 INPUT "R=";R2
220 INPUT "X=";X2 :REM Enter negative sign if capacitive
230 REM Now do calculations and print results
240 PRINT:PRINT
250 Z1=SQR(R1^2+X1^2)
260 I1=EL/(SQR(3)*Z1)
270 P1=3*I1^2*R1
280 Q1=3*I1^2*X1
290 S1=3*EL*I1/SQR(3)
300 PRINT "For the wye connection:"
310 PRINT "The phase current I=";I1;"amps"
320 PRINT "The total power dissipated is";P1;"watts"
330 PRINT "With a net reactive power of Q=";ABS(Q1);"vars";
340 IF SGN(X1)=-1 THEN PRINT "(cap.)"
350 IF SGN(X1)=1 THEN PRINT "(ind.)"
360 PRINT "and apparent power of:";S1;"VA"
370 PRINT:PRINT
380 Z2=SQR(R2^2+X2^2)
390 I2=EL/Z2
400 P2=3*I2^2*R2
410 Q2=3*I2^2*X2
420 S2=3*EL*I2
```

Input — (bracket for lines 130–220)

Y — (bracket for lines 250–360)

```
Δ   430 PRINT "For the delta connected load:"
    440 PRINT "The phase current is, I=";I2;"amps"
    450 PRINT "The total power dissipated is";P2;"watts"
    460 PRINT "with a net reactive power of Q=";ABS(Q2);"vars";
    470 IF SGN(X2)=-1 THEN PRINT "(cap.)"
    480 IF SGN(X2)=1 THEN PRINT "(ind.)"
   └490 PRINT "and apparent power of:";S2;"VA"
    500 PRINT:PRINT
   ┌510 PT=P1+P2
    520 QT=Q1+Q2
    530 PRINT "For the combined system:"
P_T 540 PRINT "The total power dissipated is";PT;"watts"
Q_T 550 PRINT "and the net reactive power QT=";ABS(QT);"vars";
    560 IF SGN(QT)=-1 THEN PRINT "(cap.)"
   └570 IF SGN(QT)=1 THEN PRINT "(ind.)"
   ┌580 ST=SQR(PT^2+QT^2)
    590 FP=PT/ST
S_T 600 PRINT:PRINT "The total apparent power ST=";ST;"VA"
F_p 610 PRINT "with a network power factor Fp=";FP;
    620 IF SGN(QT)=-1 THEN PRINT "(leading)"
   └630 IF SGN(QT)=1 THEN PRINT "(lagging)"
    640 END


READY
RUN


This program analyzes the 3-phase
delta-wye load.

Enter the following network information:

Line voltage, E=? 200

For the series-connected wye load:
R=? 4

X=? 3


And for the series-connected delta load:
R=? 6

X=? -8




For the wye connection:
The phase current I= 23.094 amps
The total power dissipated is 6400 watts
With a net reactive power of Q= 4800 vars(ind.)
and apparent power of: 8000 VA


For the delta connected load:
The phase current is, I= 20 amps
The total power dissipated is 7200 watts
with a net reactive power of Q= 9600 vars(cap.)
and apparent power of: 1.2E+04 VA


For the combined system:
The total power dissipated is 1.36E+04 watts
and the net reactive power QT= 4800 vars(cap.)

The total apparent power ST= 1.4422E+04 VA
with a network power factor Fp= .943 (leading)


READY
```

FIG. 22.36
Program 22.1.

parameters of the series-connected phase impedances. Inductive and capacitive reactances are distinguished by the sign entered on lines 180 and 220.

The calculations for the Y-connected loads are made by lines 250 through 360 using equations introduced in Section 22.10. Lines 380 through 490 perform a detailed analysis of the Δ section. The total real and reactive power are calculated by lines 510 through 570, and the total apparent power and the power factor by lines 580 through 630.

For comparison purposes, the run employed the network parameters used in Example 22.6.

PROBLEMS

Section 22.5

1. A balanced Y load having a 10-Ω resistance in each leg is connected to a three-phase, four-wire, Y-connected generator having a line voltage of 208 V. Calculate the magnitude of
   a. the phase voltage of the generator.
   b. the phase voltage of the load.
   c. the phase current of the load.
   d. the line current.

2. Repeat Problem 1 if each phase impedance is changed to a 12-Ω resistor in series with a 16-Ω capacitive reactance.

3. Repeat Problem 1 if each phase impedance is changed to a 10-Ω resistor in parallel with a 10-Ω capacitive reactance.

4. The phase sequence for the Y-Y system of Fig. 22.37 is *ABC*.
   a. Find the angles θ_2 and θ_3 for the specified phase sequence.
   b. Find the voltage across each phase impedance in phasor form.
   c. Find the current through each phase impedance in phasor form.

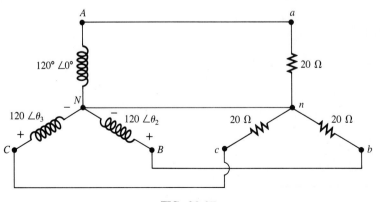

FIG. 22.37

d. Draw the phasor diagram of the currents found in part (c) and show that their phasor sum is zero.

e. Find the magnitude of the line currents.

f. Find the magnitude of the line voltages.

5. Repeat Problem 4 if the phase impedances are changed to a 9-Ω resistor in series with a 12-Ω inductive reactance.

6. Repeat Problem 4 if the phase impedances are changed to a 6-Ω resistance in parallel with an 8-Ω capacitive reactance.

7. For the system of Fig. 22.38, find the magnitude of the unknown voltages and currents.

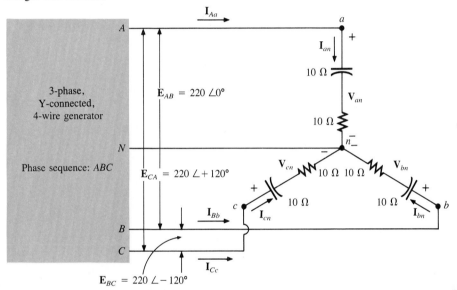

FIG. 22.38

*8. Compute the magnitude of the voltage E_{AB} for the balanced three-phase system of Fig. 22.39.

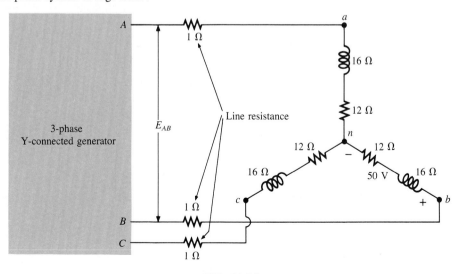

FIG. 22.39

Section 22.6

9. A balanced Δ load having a 20-Ω resistance in each leg is connected to a three-phase, three-wire, Y-connected generator having a line voltage of 208 V. Calculate the magnitude of
   a. the phase voltage of the generator.
   b. the phase voltage of the load.
   c. the phase current of the load.
   d. the line current.

10. Repeat Problem 9 if each phase impedance is changed to a 6.8-Ω resistor in series with a 14-Ω inductive reactance.

11. Repeat Problem 9 if each phase impedance is changed to an 18-Ω resistance in parallel with an 18-Ω capacitive reactance.

12. The phase sequence for the Y-Δ system of Fig. 22.40 is *ABC*.
   a. Find the angles θ_2 and θ_3 for the specified phase sequence.
   b. Find the voltage across each phase impedance in phasor form.
   c. Draw the phasor diagram of the voltages found in part (b) and show that their sum is zero around the closed loop of the Δ load.
   d. Find the current through each phase impedance in phasor form.
   e. Find the magnitude of the line currents.
   f. Find the magnitude of the generator phase voltages.

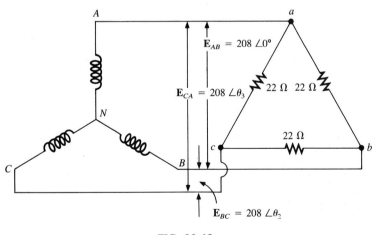

FIG. 22.40

13. Repeat Problem 12 if the phase impedances are changed to a 100-Ω resistor in series with a capacitive reactance of 100 Ω.

14. Repeat Problem 13 if the phase impedances are changed to a 3-Ω resistor in parallel with an inductive reactance of 4 Ω.

15. For the system of Fig. 22.41, find the magnitude of the unknown voltages and currents.

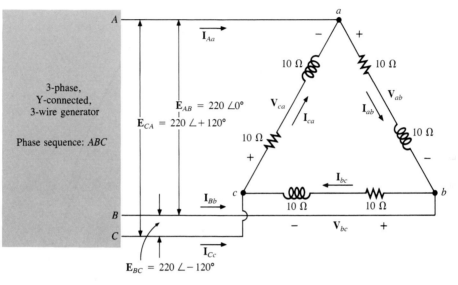

FIG. 22.41

Section 22.9

16. A balanced Y load having a 30-Ω resistance in each leg is connected to a three-phase Δ-connected generator having a line voltage of 208 V. Calculate the magnitude of
a. the phase voltage of the generator.
b. the phase voltage of the load.
c. the phase current of the load.
d. the line current.

17. Repeat Problem 16 if each phase impedance is changed to a 12-Ω resistor in series with a 12-Ω inductive reactance.

18. Repeat Problem 16 if each phase impedance is changed to a 15-Ω resistor in parallel with a 20-Ω capacitive reactance.

*19. For the system of Fig. 22.42, find the magnitude of the unknown voltages and currents.

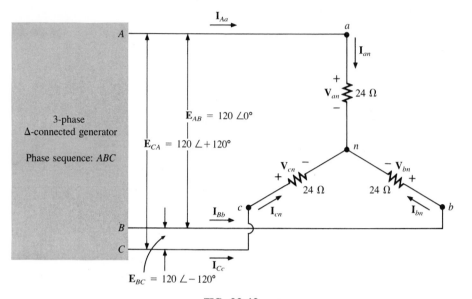

FIG. 22.42

20. Repeat Problem 19 if each phase impedance is changed to a 10-Ω resistor in series with a 20-Ω inductive reactance.

21. Repeat Problem 19 if each phase impedance is changed to a 20-Ω resistor in parallel with a 15-Ω capacitive reactance.

22. A balanced Δ load having a 220-Ω resistance in each leg is connected to a three-phase Δ-connected generator having a line voltage of 440 V. Calculate the magnitude of
  a. the phase voltage of the generator.
  b. the phase voltage of the load.
  c. the phase current of the load.
  d. the line current.

23. Repeat Problem 22 if each phase impedance is changed to a 12-Ω resistor in series with a 9-Ω capacitive reactance.

24. Repeat Problem 22 if each phase impedance is changed to a 22-Ω resistor in parallel with a 22-Ω inductive reactance.

25. The phase sequence for the Δ-Δ system of Fig. 22.43 is *ABC*.
  a. Find the angles θ_2 and θ_3 for the specified phase sequence.
  b. Find the voltage across each phase impedance in phasor form.
  c. Draw the phasor diagram of the voltages found in part (b) and show that their phasor sum is zero around the closed loop of the Δ load.

d. Find the current through each phase impedance in phasor form.

e. Find the magnitude of the line currents.

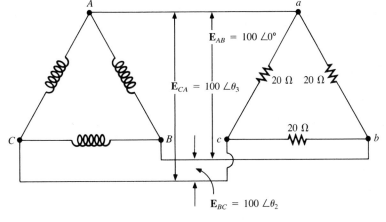

$\mathbf{E}_{AB} = 100 \angle 0°$

$\mathbf{E}_{CA} = 100 \angle \theta_3$

20 Ω 20 Ω

20 Ω

$\mathbf{E}_{BC} = 100 \angle \theta_2$

FIG. 22.43

26. Repeat Problem 25 if each phase impedance is changed to a 12-Ω resistor in series with a 16-Ω inductive reactance.

27. Repeat Problem 25 if each phase impedance is changed to a 20-Ω resistor in parallel with a 20-Ω capacitive reactance.

Section 22.10

28. Find the total watts, volt-amperes reactive, volt-amperes, and F_p of the three-phase system of Problem 2.

29. Find the total watts, volt-amperes reactive, volt-amperes, and F_p of the three-phase system of Problem 4.

30. Find the total watts, volt-amperes reactive, volt-amperes, and F_p of the three-phase system of Problem 7.

31. Find the total watts, volt-amperes reactive, volt-amperes, and F_p of the three-phase system of Problem 11.

32. Find the total watts, volt-amperes reactive, volt-amperes, and F_p of the three-phase system of Problem 13.

33. Find the total watts, volt-amperes reactive, volt-amperes, and F_p of the three-phase system of Problem 15.

34. Find the total watts, volt-amperes reactive, volt-amperes, and F_p of the three-phase system of Problem 18.

35. Find the total watts, volt-amperes reactive, volt-amperes, and F_p of the three-phase system of Problem 20.

36. Find the total watts, volt-amperes reactive, volt-amperes, and F_p of the three-phase system of Problem 24.

37. Find the total watts, volt-amperes reactive, volt-amperes, and F_p of the three-phase system of Problem 26.

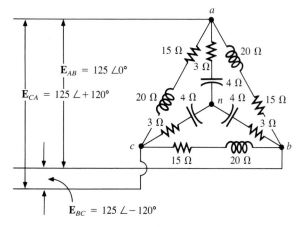

$E_{AB} = 125 \angle 0°$

$E_{CA} = 125 \angle +120°$

15 Ω   3 Ω   20 Ω

4 Ω

20 Ω   4 Ω n 4 Ω   15 Ω

3 Ω   3 Ω

c 15 Ω   20 Ω b

$E_{BC} = 125 \angle -120°$

FIG. 22.44

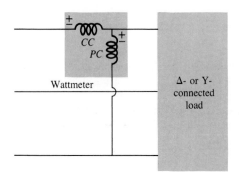

$+$

CC

PC $+$

Wattmeter

Δ- or Y-
connected
load

FIG. 22.45

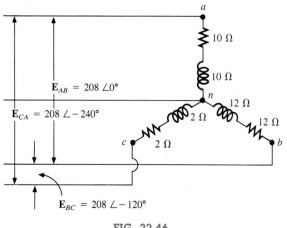

a

10 Ω

10 Ω

$E_{AB} = 208 \angle 0°$

n 12 Ω

$E_{CA} = 208 \angle -240°$ 2 Ω   12 Ω

c 2 Ω b

$E_{BC} = 208 \angle -120°$

FIG. 22.46

38. A balanced three-phase Δ-connected load has a line voltage of 200 and a total power consumption of 4800 W at a lagging power factor of 0.8. Find the impedance of each phase in rectangular coordinates.

39. A balanced three-phase Y-connected load has a line voltage of 208 and a total power consumption of 1200 W at a leading power factor of 0.6. Find the impedance of each phase in rectangular coordinates.

*40. Find the total watts, volt-amperes reactive, volt-amperes, and F_p of the system of Fig. 22.44.

Section 22.11

41. a. Sketch the connections required to measure the total watts delivered to the load of Fig. 22.38 using three wattmeters.
  b. Determine the total wattage dissipation and the reading of each wattmeter.

42. Repeat Problem 41 for the network of Fig. 22.40.

Section 22.12

43. a. For the three-wire system of Fig. 22.45, properly connect a second wattmeter so that the two will measure the total power delivered to the load.
  b. If one wattmeter has a reading of 200 W and the other a reading of 85 W, what is the total dissipation in watts if the total power factor is 0.8 leading?
  c. Repeat part (b) if the total power factor is 0.2 lagging and $P_l = 100$ W.

44. Sketch three different ways that two wattmeters can be connected to measure the total power delivered to the load of Problem 15.

Section 22.13

*45. For the system of Fig. 22.46:
  a. Calculate the magnitude of the voltage across each phase of the load.
  b. Find the magnitude of the current through each phase of the load.
  c. Find the total watts, volt-amperes reactive, volt-amperes, and F_p of the system.

Section 22.14

*46. For the three-phase, three-wire system of Fig. 22.47, find the magnitude of the current through each phase of the load and the total watts, volt-amperes reactive, volt-amperes, and F_p of the load.

Section 22.15

47. Given the magnitude of the line voltages and the impedance of each phase (in series or parallel), write a program to determine the magnitude of all the voltages and currents of a balanced Y-connected load.

48. Repeat Problem 47 for a Δ-connected load.

49. For a balanced Y-connected load, write a program to determine
 a. the magnitude of the load currents and voltages.
 b. the real, reactive, and apparent power to each phase.
 c. the total real, reactive, and apparent power to the load.
 d. the load power factor.

50. Repeat Problem 49 for a balanced Δ-connected load.

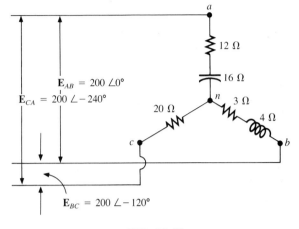

FIG. 22.47

GLOSSARY

Delta (Δ)-connected generator   A three-phase generator having the three phases connected in the shape of the capital Greek letter delta (Δ).

Line current   The current that flows from the generator to the load of a single-phase or polyphase system.

Line voltage   The potential difference that exists between the lines of a single-phase or polyphase system.

Neutral connection   The connection between the generator and the load that, under balanced conditions, will have zero current associated with it.

Phase current   The current that flows through each phase of a single-phase or polyphase generator load.

Phase sequence   The order in which the generated sinusoidal voltages of a polyphase generator will affect the load to which they are applied.

Phase voltage   The voltage that appears between the line and neutral of a Y-connected generator and from line to line in a Δ-connected generator.

Polyphase ac generator   An electromechanical source of ac power that generates more than one sinusoidal voltage per

rotation of the rotor. The frequency generated is determined by the speed of rotation and the number of poles of the rotor.

Single-phase ac generator   An electromechanical source of ac power that generates a single sinusoidal voltage having a frequency determined by the speed of rotation and the number of poles of the rotor.

Three-wattmeter method   A method for determining the total power delivered to a three-phase load using three wattmeters.

Two-wattmeter method   A method for determining the total power delivered to a Δ- or Y-connected three-phase load using only two wattmeters and considering the power factor of the load.

Unbalanced polyphase load   A load not having the same impedance in each phase.

WYE (Y)-connected generator   A three-phase source of ac power in which the three phases are connected in the shape of the letter Y.

23

Nonsinusoidal Circuits

NON

23.1 INTRODUCTION

Any waveform that differs from the basic description of the sinusoidal waveform is referred to as *nonsinusoidal*. The most obvious are the dc and square-wave voltage or current; others are voice patterns, such as those in Fig. 23.1, which were recorded as a person held continuous "A" and "O" sounds. Note that the waveforms are almost periodic, but that a variation in tone causes a slight change in the waveform. Voice patterns are unique to people in much the same manner as fingerprints.

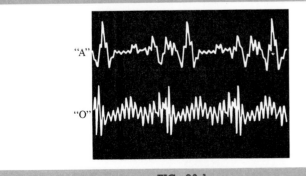

FIG. 23.1
Voice patterns.

The output of many electrical and electronic devices will be a non-sinusoidal quantity if a sinusoidal input is applied. For example, the nonlinear characteristics of a diode are shown in Fig. 23.2(a). If a sinusoidal voltage is applied across the diode, the current passing through the diode will have the nonsinusoidal wave shape shown in Fig. 23.2(b). Note, however, that the nonsinusoidal output is also periodic: It has the same period as the input voltage. A periodic nonsinusoidal output will result for any nonlinear system that has no residual effects associated with it if a sinusoidal quantity is applied at the input.

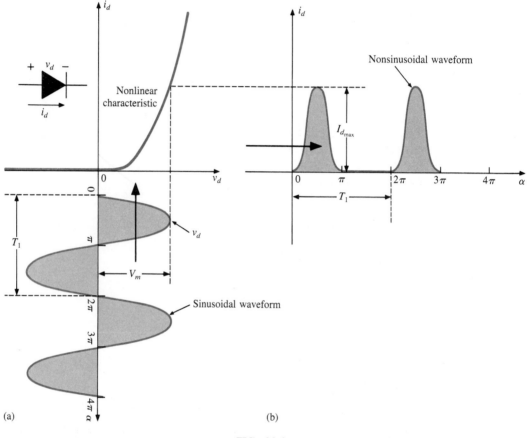

FIG. 23.2

In this chapter, we shall discuss one method of obtaining the response of a system to a periodic nonsinusoidal input.

23.2 FOURIER SERIES

Fourier series refers to a series of terms, developed in 1826 by Baron Jean Fourier, that can be used to represent a nonsinusoidal waveform. In the analysis of these waveforms, we solve for each term in the Fourier series:

$$f(\alpha) = \underbrace{A_0}_{\substack{\text{dc or} \\ \text{average value}}} + \underbrace{A_1 \cos \alpha + A_2 \cos 2\alpha + A_3 \cos 3\alpha + \cdots + A_n \sin n\alpha}_{\text{cosine terms}}$$

$$+ \underbrace{B_1 \sin \alpha + B_2 \sin 2\alpha + B_3 \sin 3\alpha + \cdots + B_n \sin n\alpha}_{\text{sine terms}} \qquad \textbf{(23.1)}$$

Depending on the waveform, a large number of these terms may be required to approximate the waveform closely for the purpose of circuit analysis.

The Fourier series has three basic parts. The first is the dc term A_0, which is the average value of the waveform over one full cycle. The second is a series of cosine terms. There are no restrictions on the values or relative values of the amplitudes of these cosine terms, but each will have a frequency that is an integer multiple of the frequency of the first cosine term of the series. The third part is a series of sine terms. There are again *no* restrictions on the values or relative values of the amplitudes of these sine terms, but each will have a frequency that is an integer multiple of the frequency of the first sine term of the series. For a particular waveform, it is quite possible that all of the sine *or* cosine terms are zero. Characteristics of this type can be determined by simply examining the nonsinusoidal waveform and its position on the horizontal axis.

If a waveform is symmetric about the vertical axis, it is called an *even* function or is said to have *axis symmetry* [Fig. 23.3(a)]. For all even functions, the $B_{1 \to n}$ constants will all be zero, and the function can be completely described by just the *dc and cosine terms*. Note that the cosine wave itself is also symmetrical about the vertical axis (ordinate) [Fig. 23.3(b)].

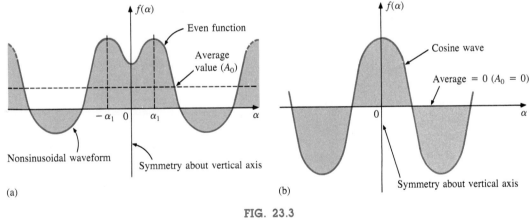

(a)

(b)

FIG. 23.3

Axis symmetry.

For both waveforms of Fig. 23.3, the following mathematical relationship is true:

$$\boxed{f(\alpha) = f(-\alpha)} \qquad \text{(even function)} \qquad \textbf{(23.2)}$$

In words, it states that the magnitude of the function is the same at $+\alpha$ as at $-\alpha$ [α_1 in Fig. 23.3(a)].

If the waveform is such that its value for $+\alpha$ is the negative of that for $-\alpha$ [Fig. 23.4(a)], it is called an *odd* function or is said to have *point symmetry* (about any point of intersection on the horizontal axis), and all of the constants $A_{1 \rightarrow n}$ will be zero. The function can then be represented by the *dc and sine terms*. Note that the sine wave itself is an odd function [Fig. 23.4(b)].

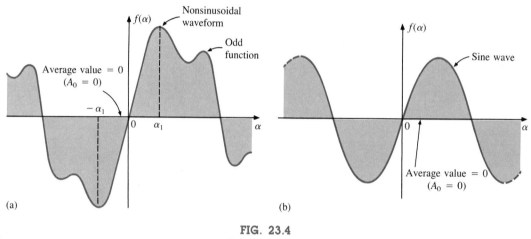

FIG. 23.4
Point symmetry.

For both waveforms of Fig. 23.4, the following mathematical relationship is true:

$$f(\alpha) = -f(-\alpha) \qquad \text{(odd function)} \qquad \textbf{(23.3)}$$

In words, it states that the magnitude of the function at $+\alpha$ is equal to the negative of the magnitude at $-\alpha$ [α_1 in Fig. 23.4(a)].

The first term of the sine and cosine series is called the *fundamental component*. It represents the minimum frequency term required to represent a particular waveform, and it also has the same frequency as the waveform being represented. A fundamental term, therefore, must be present in any Fourier series representation. The other terms with higher-order frequencies (integer multiples of the fundamental) are called the *harmonic terms*. A term that has a frequency equal to twice the fundamental is the second harmonic; three times, the third harmonic; and so on.

If the waveform is such that

$$f(t) = f\left(\frac{T}{2} + t\right) \qquad \textbf{(23.4)}$$

the odd harmonics of the series of cosine and sine terms are zero. Figure 23.5 is an example of this type of function.

Equation (23.4) states that the function repeats itself after each $T/2$ time interval (t_1 in Fig. 23.5). The waveform, however, will also repeat itself after each period T. In general, therefore, for a function of this type, if the period T of the waveform is chosen to be twice that of the minimum period ($T/2$), the odd harmonics will all be zero.

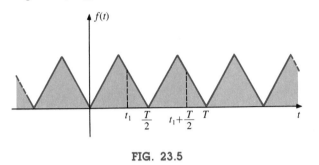

FIG. 23.5

If the waveform is such that

$$f(t) = -f\left(\frac{T}{2} + t\right) \qquad (23.5)$$

the waveform is said to have *half-wave* or *mirror symmetry* and *the even harmonics of the series of cosine and sine terms will be zero.* Figure 23.6 is an example of this type of function.

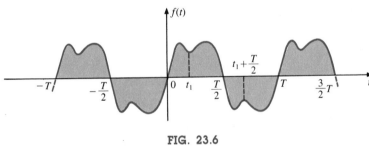

FIG. 23.6
Mirror symmetry.

Equation (23.5) states that the waveform encompassed in one time interval $T/2$ will repeat itself in the next $T/2$ time interval, but in the negative sense (t_1 in Fig. 23.6). For example, the waveform of Fig. 23.6 from zero to $T/2$ will repeat itself in the time interval $T/2$ to T, but below the horizontal axis.

An instrument for measuring harmonic frequencies and corresponding amplitudes is the wave analyzer shown in Fig. 23.7.

The constants A_0, $A_{1 \to n}$, $B_{1 \to n}$ can be determined by using the following integral formulas:

$$A_0 = \frac{1}{T} \int_0^T f(t)\, dt = \frac{1}{2\pi} \int_0^{2\pi} f(\alpha)\, d\alpha \qquad (23.6)$$

FIG. 23.7
Wave analyzer. (Courtesy of Hewlett Packard Co.)

$$A_n = \frac{2}{T} \int_0^T f(t) \cos n\omega t \, dt = \frac{1}{\pi} \int_0^{2\pi} f(\alpha) \cos n\alpha \, d\alpha \qquad \textbf{(23.7)}$$

$$B_n = \frac{2}{T} \int_0^T f(t) \sin n\omega t \, dt = \frac{1}{\pi} \int_0^{2\pi} f(\alpha) \sin n\alpha \, d\alpha \qquad \textbf{(23.8)}$$

These equations have been presented for recognition purposes only; they will not be used in the following analysis.

The following examples will demonstrate the use of the equations and concepts introduced thus far in this chapter.

EXAMPLE 23.1. Write the Fourier series expansion for the waveforms of Fig. 23.8.

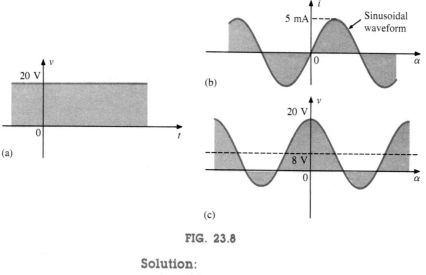

FIG. 23.8

Solution:
a. $A_0 = 20$, $A_{1 \to n} = 0$, $B_{1 \to n} = 0$
   $v = 20$
b. $A_0 = 0$, $A_{1 \to n} = 0$, $B_1 = 5 \times 10^{-3}$, $B_{2 \to n} = 0$
   $i = 5 \times 10^{-3} \sin \alpha$
c. $A_0 = 8$, $A_1 = 12$, $A_{2 \to n} = 0$, $B_{1 \to n} = 0$
   $v = 8 + 12 \cos \alpha$

EXAMPLE 23.2. Sketch the following Fourier series expansion:

$$v = 2 + 1 \cos \alpha + 2 \sin \alpha$$

Solution: Note Fig. 23.9.

The solution could be obtained graphically by first plotting all of the functions and considering a sufficient number of points on the horizontal axis; or phasor algebra could be employed as follows:

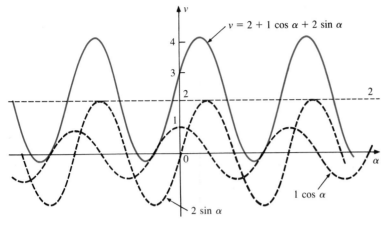

FIG. 23.9

$$1 \cos \alpha + 2 \sin \alpha = 1 \angle 90° + 2 \angle 0° = j1 + 2$$
$$= 2 + j1 = 2.236 \angle 26.57°$$
$$= 2.236 \sin(\alpha + 26.57°)$$

and

$$v = 2 + 2.236 \sin(\alpha + 26.57°)$$

which is simply the sine wave portion riding on a dc level of 2 V. That is, its positive maximum is $2 + 2.236 = 4.236$ V, and its minimum is $2 - 2.236 = -0.236$ V.

EXAMPLE 23.3. Sketch the following Fourier series expansion:

$$i = \sin \alpha + \sin 2\alpha$$

Solution: See Fig. 23.10. Note that in this case the sum of the two sinusoidal waveforms of different frequencies is *not* a sine wave. Recall that complex algebra can be applied only to waveforms having the *same* frequency. In this case the solution is obtained graphically.

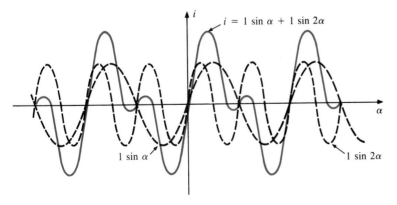

FIG. 23.10

As a further example in the use of the Fourier series approach, consider the square wave shown in Fig. 23.11. The average value is zero, so $A_0 = 0$. It is an odd function, so all the constants $A_{1 \to n}$ equal zero; only sine terms will be present in the series expansion. Since the waveform satisfies the criteria for $f(t) = -f(T/2 + t)$, the even harmonics will also be zero.

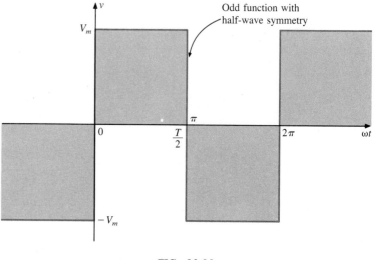

Odd function with half-wave symmetry

FIG. 23.11
Square wave.

The expression obtained after evaluating the various coefficients from Eq. (23.8) is

$$v = \frac{4}{\pi} V_m \left(\sin \omega t + \frac{1}{3} \sin 3\omega t + \frac{1}{5} \sin 5\omega t + \frac{1}{7} \sin 7\omega t + \cdots + \frac{1}{n} \sin n\omega t \right) \qquad \textbf{(23.9)}$$

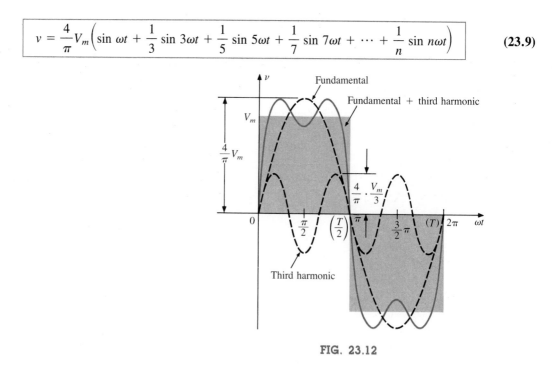

Fundamental

Fundamental + third harmonic

Third harmonic

FIG. 23.12

Note that the fundamental does indeed have the frequency of the square wave. If we add the fundamental and third harmonics, we obtain the results shown in Fig. 23.12.

Even with only the first two terms, the wave shape is beginning to look like a square wave. If we add the next two terms (Fig. 23.13), the width of the pulse increases, and the number of peaks increases.

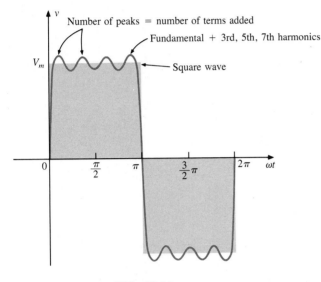

FIG. 23.13

As we continue to add terms, the series will better approximate the square wave. Note, however, that the amplitude of each succeeding term diminishes to the point at which it will be negligible compared with those of the first few terms. A good approximation would be to assume that the waveform is composed of the harmonics up to and including the ninth. Any higher harmonics would be less than one-tenth the fundamental. If the waveform just described were shifted above or below the horizontal axis, the Fourier series would be altered only by a change in the dc term. Figure 23.14(a), for example, is the sum of Figs.

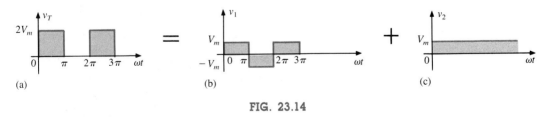

FIG. 23.14

23.14(b) and (c). The Fourier series for the complete waveform is therefore

$$v_T = V_m + \text{Eq. (23.9)}$$

$$= V_m + \frac{4}{\pi} V_m \left(\sin \omega t + \frac{1}{3} \sin 3\omega t + \frac{1}{5} \sin 5\omega t + \frac{1}{7} \sin 7\omega t + \cdots \right)$$

The equation for the pulsating waveform of Fig. 23.15(b) is

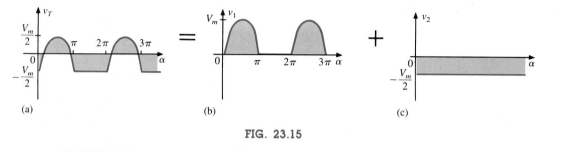

FIG. 23.15

$$v = 0.318V_m + 0.500V_m \sin \alpha - 0.212V_m \cos 2\alpha - 0.0424V_m \cos 4\alpha - \cdots \qquad (23.10)$$

The waveform in Fig. 23.15(a) is the sum of the two in Figs. 23.15(b) and (c). The Fourier series for the complete waveform is therefore

$$v_T = -\frac{V_m}{2} + \text{Eq. (23.10)}$$

$$= \underbrace{(-0.500 + 0.318)}_{-0.182}V_m + 0.500V_m \sin \alpha - 0.212V_m \cos 2\alpha - 0.0424V_m \cos 4\alpha + \cdots$$

If either waveform were shifted to the right or left, the phase shift would be subtracted or added, respectively, from the sine and cosine terms. The dc term would not change with a shift to the right or left.

If the half-wave rectified signal is shifted 90° to the left, as in Fig. 23.16, the Fourier series becomes

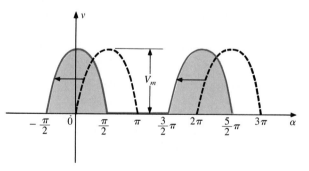

FIG. 23.16

$$v_o = 0.318V_m + 0.500V_m \underbrace{\sin(\alpha + 90°)}_{\cos \alpha} - 0.212V_m \cos 2(\alpha + 90°) - 0.0424V_m \cos 4(\alpha + 90°) + \cdots$$

$$= 0.318V_m + 0.500V_m \cos \alpha - 0.212V_m \cos(2\alpha + 180°) - 0.0424V_m \cos(4\alpha + 360°) + \cdots$$

$$v_o = 0.318V_m + 0.500V_m \cos \alpha + 0.212V_m \cos 2\alpha - 0.0424V_m \cos 4\alpha + \cdots$$

23.3 CIRCUIT RESPONSE TO A NONSINUSOIDAL INPUT

The Fourier series representation of a nonsinusoidal input can be applied to a linear network using the principle of superposition. Recall that this theorem allowed us to consider the effects of each source of a circuit independently. If we replace the nonsinusoidal input by the terms of the Fourier series deemed necessary for practical considerations, we can use superposition to find the response of the network to each term (Fig. 23.17).

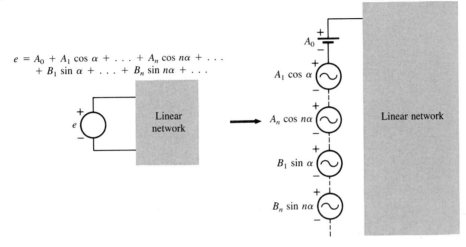

$$e = A_0 + A_1 \cos \alpha + \ldots + A_n \cos n\alpha + \ldots$$
$$+ B_1 \sin \alpha + \ldots + B_n \sin n\alpha + \ldots$$

FIG. 23.17

The total response of the system is then the algebraic sum of the values obtained for each term. The major change between using this theorem for nonsinusoidal circuits and using it for the circuits previously described is that the frequency will be different for each term in the nonsinusoidal application. Therefore, the reactances

$$X_L = 2\pi f L \quad \text{and} \quad X_C = \frac{1}{2\pi f C}$$

will change for each term of the input voltage or current.

In Chapter 13, we found that the effective value of any waveform was given by

$$\sqrt{\frac{1}{T} \int_0^T [f(t)]^2 \, dt}$$

If we apply this equation to the following Fourier series:

$$v(\alpha) = V_o + V_{m_1} \cos \alpha + \cdots + V_{m_n} \cos n\alpha + V'_{m_1} \sin \alpha + \cdots + V'_{m_n} \sin n\alpha$$

then

$$V_{\text{eff}} = \sqrt{V_o^2 + \frac{V_{m_1}^2 + \cdots + V_{m_n}^2 + V_{m_1}'^2 + \cdots + V_{m_n}'^2}{2}}$$

(23.11)

However, since

$$\frac{V_{m_1}^2}{2} = \left(\frac{V_{m_1}}{\sqrt{2}}\right)\left(\frac{V_{m_1}}{\sqrt{2}}\right) = (V_{1_{\text{eff}}})(V_{1_{\text{eff}}}) = V_{1_{\text{eff}}}^2$$

then

$$V_{\text{eff}} = \sqrt{V_o^2 + V_{1_{\text{eff}}}^2 + \cdots + V_{n_{\text{eff}}}^2 + V'^2_{1_{\text{eff}}} + \cdots + V'^2_{n_{\text{eff}}}}$$

(23.12)

Similarly, for

$$i(\alpha) = I_o + I_{m_1} \cos \alpha + \cdots + I_{m_n} \cos n\alpha + I'_{m_1} \sin \alpha + \cdots + I'_{m_n} \sin n\alpha$$

we have

$$I_{\text{eff}} = \sqrt{I_o^2 + \frac{I_{m_1}^2 + \cdots + I_{m_n}^2 + I'^2_{m_1} + \cdots + I'^2_{m_n}}{2}}$$

(23.13)

and

$$I_{\text{eff}} = \sqrt{I_o^2 + I_{1_{\text{eff}}}^2 + \cdots + I_{n_{\text{eff}}}^2 + I'^2_{1_{\text{eff}}} + \cdots + I'^2_{n_{\text{eff}}}}$$

(23.14)

The total power delivered is the sum of that delivered by the corresponding terms of the voltage and current. In the following equations, all voltages and currents are effective values:

$$P_T = V_0 I_0 + V_1 I_1 \cos \theta_1 + \cdots + V_n I_n \cos \theta_n + \cdots$$

(23.15)

$$P_T = I_0^2 R + I_1^2 R + \cdots + I_n^2 R + \cdots$$ **(23.16)**

or

$$P_T = I_{\text{eff}}^2 R$$ **(23.17)**

with I_{eff} as defined by Eq. (23.13), and, similarly,

$$P_T = \frac{V_{\text{eff}}^2}{R}$$ **(23.18)**

with V_{eff} as defined by Eq. (23.11).

EXAMPLE 23.4.   The input to the circuit of Fig. 23.18 is the following:

$$e = 12 + 10 \sin 2t$$

a. Find the current i and the voltages v_R and v_C.
b. Find the effective values of i, v_R, and v_C.
c. Find the power delivered to the circuit.

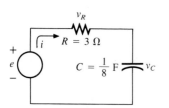

FIG. 23.18

Solutions:

a. Redraw the original circuit as shown in Fig. 23.19. Then apply superposition:

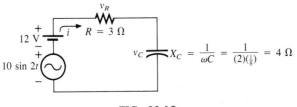

FIG. 23.19

1. *For the 12-V dc supply portion of the input, $I = 0$ since the capacitor is an open circuit to dc when v_C has reached its final (steady-state) value. Therefore,*

$$V_R = IR = 0 \text{ V}$$

and

$$V_C = 12 \text{ V}$$

2. *For the ac supply,*

$$\mathbf{Z} = 3 - j4 = 5 \angle -53.13°$$

and

$$\mathbf{I} = \frac{\mathbf{E}}{\mathbf{Z}} = \frac{\dfrac{10}{\sqrt{2}} \angle 0°}{5 \angle -53.13°} = \frac{2}{\sqrt{2}} \angle +53.13°$$

$$\mathbf{V}_R = \mathbf{IR} = \left(\frac{2}{\sqrt{2}} \angle +53.13°\right)(3 \angle 0°) = \frac{6}{\sqrt{2}} \angle +53.13°$$

and

$$\mathbf{V}_C = \mathbf{IX}_C = \left(\frac{2}{\sqrt{2}} \angle +53.13°\right)(4 \angle -90°)$$

$$= \frac{8}{\sqrt{2}} \angle -36.87°$$

In the time domain,

$$i = 0 + 2 \sin(2t + 53.13°)$$

Note that even though the dc term was present in the expression for the input voltage, the dc term for the current in this circuit is zero:

$$v_R = 0 + 6 \sin(2t + 53.13°)$$

and

$$v_C = 12 + 8 \sin(2t - 36.87°)$$

b. Eq. (23.14): $I_{\text{eff}} = \sqrt{(0)^2 + \dfrac{(2)^2}{2}} = \sqrt{2} = \textbf{1.414 A}$

Eq. (23.12): $V_{R_{\text{eff}}} = \sqrt{(0)^2 + \dfrac{(6)^2}{2}} = \sqrt{18} = \textbf{4.243 V}$

Eq. (23.12): $V_{C_{\text{eff}}} = \sqrt{(12)^2 + \dfrac{(8)^2}{2}} = \sqrt{176} = \textbf{13.267 V}$

c. $P = I_{\text{eff}}^2 R = (\sqrt{2})^2(3) = \textbf{6 W}$

EXAMPLE 23.5. Find the response of the circuit of Fig. 23.20 to the input shown.

$$e = 0.318E_m + 0.500E_m \sin \omega t - 0.212E_m \cos 2\omega t - 0.0424E_m \cos 4\omega t + \cdots$$

Solution: For discussion purposes, only the first three terms will be used to represent e. Converting the cosine terms to sine terms and substituting for E_m gives us

$$e = 63.60 + 100.0 \sin \omega t - 42.40 \sin(2\omega t + 90°)$$

Using phasor notation, the original circuit becomes as shown in Fig. 23.21. Apply superposition:

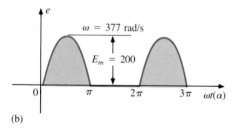

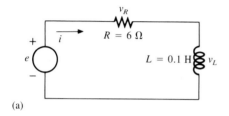

(a)

(b)

FIG. 23.20

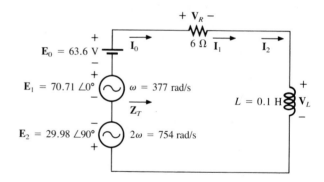

FIG. 23.21

For the dc term ($E_0 = 63.6$ V),

$$X_L = 0 \quad \text{(short for dc)}$$
$$\mathbf{Z}_T = 6 \angle 0° = \mathbf{R}$$
$$I_o = \frac{E_0}{R} = \frac{63.6}{6} = 10.60 \text{ A}$$
$$V_{R_o} = I_o R = E_0 = 63.60 \text{ V}$$
$$V_{L_o} = 0$$

The average power is

$$P_o = I_o^2 R = (10.60)^2(6) = 674.2 \text{ W}$$

For the fundamental term ($\mathbf{E}_1 = 70.71 \angle 0°$, $\omega = 377$),

$$X_{L_1} = \omega L = (377)(0.1) = 37.7 \text{ } \Omega$$

$$\mathbf{Z}_{T_1} = 6 + j37.7 = 38.17 \angle 80.96°$$

$$\mathbf{I}_1 = \frac{\mathbf{E}_1}{\mathbf{Z}_{T_1}} = \frac{70.71 \angle 0°}{38.17 \angle 80.96°} = 1.85 \angle -80.96°$$

$$\mathbf{V}_{R_1} = \mathbf{I}_1\mathbf{R} = (1.85 \angle -80.96°)(6 \angle 0°)$$
$$= 11.10 \angle -80.96°$$

$$\mathbf{V}_{L_1} = \mathbf{I}_1\mathbf{X}_L = (1.85 \angle -80.96°)(37.7 \angle 90°)$$
$$= 69.75 \angle 9.04°$$

The average power is

$$P_1 = I_1^2 R = (1.85)^2(6) = 20.54 \text{ W}$$

For the second harmonic ($\mathbf{E}_2 = 29.98 \angle -90°$, $\omega = 754$). The phase angle of $\mathbf{E}_2$ was changed to $-90°$ to give it the same polarity as the input voltages $\mathbf{E}_0$ and $\mathbf{E}_1$.

$$X_{L_2} = \omega L = (754)(0.1) = 75.4 \ \Omega$$

$$\mathbf{Z}_{T_2} = 6 + j75.4 = 75.64 \angle 85.45°$$

$$\mathbf{I}_2 = \frac{\mathbf{E}_2}{\mathbf{Z}_{T_2}} = \frac{29.98 \angle -90°}{75.64 \angle 85.45} = 0.396 \angle -174.45°$$

$$\mathbf{V}_{R_2} = \mathbf{I}_2\mathbf{R} = (0.396 \angle -174.45°)(6 \angle 0°)$$
$$= 2.38 \angle -174.45°$$

$$\mathbf{V}_{L_2} = \mathbf{I}_2\mathbf{X}_{L_2} = (0.396 \angle -174.45°)(75.4 \angle 90°)$$
$$= 29.9 \angle -84.45°$$

The average power is

$$P_2 = I_2^2 R = (0.396)^2(6) = 0.941 \text{ W}$$

The Fourier series expansion for i is

$$i = \mathbf{10.6} + \sqrt{2}(\mathbf{1.85}) \sin(\mathbf{377}t - \mathbf{80.96°}) + \sqrt{2}(\mathbf{0.396}) \sin(\mathbf{754}t - \mathbf{174.45°})$$

and

$$I_{\text{eff}} = \sqrt{(10.6)^2 + (1.85)^2 + (0.396)^2} = \mathbf{10.77 \text{ A}}$$

The Fourier series expansion for v_R is

$$v_R = \mathbf{63.6} + \sqrt{2}(\mathbf{11.10}) \sin(\mathbf{377}t - \mathbf{80.96°}) + \sqrt{2}(\mathbf{2.38}) \sin(\mathbf{754}t - \mathbf{174.45°})$$

and

$$V_{R_{\text{eff}}} = \sqrt{(63.6)^2 + (11.10)^2 + (2.38)^2} = \mathbf{64.61 \text{ V}}$$

The Fourier series expansion for v_L is

$$v_L = \sqrt{2}(\mathbf{69.75}) \sin(\mathbf{377}t + \mathbf{9.04°}) + \sqrt{2}(\mathbf{29.93}) \sin(\mathbf{754}t - \mathbf{84.45°})$$

and

$$V_{L_{\text{eff}}} = \sqrt{(69.75)^2 + (29.93)^2} = \mathbf{75.90 \text{ V}}$$

The total average power is

$$P_T = I_{\text{eff}}^2 R = (10.77)^2(6) = \mathbf{695.96 \text{ W}} = P_0 + P_1 + P_2$$

23.4 ADDITION AND SUBTRACTION OF NONSINUSOIDAL WAVEFORMS

The Fourier series expression for the waveform resulting from the addition or subtraction of two nonsinusoidal waveforms can be found using phasor algebra if the terms having the same frequency are considered separately.

For example, the sum of the following two nonsinusoidal waveforms is found using this method:

$$v_1 = 30 + 20 \sin 20t + \cdots + 5 \sin(60t + 30°)$$
$$v_2 = 60 + 30 \sin 20t + 20 \sin 40t + 10 \cos 60t$$

1. dc terms:

$$V_{T_o} = 30 + 60 = 90$$

2. $\omega = 20$:

$$V_{T_{1(max)}} = 30 + 20 = 50$$

and

$$V_{T_1} = 50 \sin 20t$$

3. $\omega = 40$:

$$V_{T_2} = 20 \sin 40t$$

4. $\omega = 60$:

$$5 \sin(60t + 30°) = (0.707)(5) \angle 30° = 3.54 \angle 30°$$
$$10 \cos 60t = 10 \sin(60t + 90°) \Rightarrow (0.707)(10) \angle 90°$$
$$= 7.07 \angle 90°$$
$$\mathbf{V}_{T_3} = 3.54 \angle 30° + 7.07 \angle 90°$$
$$= 3.07 + j1.77 + j7.07 = 3.07 + j8.84$$
$$\mathbf{V}_{T_3} = 9.36 \angle 70.85°$$

and

$$\mathbf{V}_{T_3} = 13.24 \sin(60t + 70.85°)$$

and

$$v_1 + v_2 = v_T = \mathbf{90 + 50 \sin 20t + 20 \sin 40t + 13.24 \sin(60t + 70.85°)}$$

23.5 COMPUTER ADDITION OF NONSINUSOIDAL WAVEFORMS

Figure 23.22 will add two nonsinusoidal signals of the form appearing on line 130. The dc terms for each are requested on lines 240 and 250, and the harmonics on lines 260 through 340. Note on line 190 that the number of terms N includes the dc term and the number of harmonics.

The algebraic addition of the dc terms occurs on line 370, while the sum of the sinusoidal terms is obtained by lines 380 through 440, using the indicated I loop. As an example, consider the case of finding the

```
   10 REM ***** PROGRAM 23-1 *****
   20 REM *********************************************
   30 REM Program to add two non-sinusoidal voltages.
   40 REM *********************************************
   50 REM
  100 DIM V1(15),A1(15),V2(15),A2(15),VS(15),VA(15)
  110 PRINT "This program add two non-sinusoidal voltages"
  120 PRINT "of the form:"
  130 PRINT "v(t)=V0 + V1sin(wt+TH1) + V2sin(2wt+TH2) + V3sin(3wt+TH3) + ..."
  140 PRINT
  150 PRINT "Enter the following signal information:"
  160 PRINT
  170 INPUT "Radian frequency, w=";W
  180 PRINT :PRINT "Enter the number of terms for the voltage"
  190 PRINT "where N=dc + harmonics, i.e. N=1 + 4 = 5 for up "
  200 PRINT "to and including the 4th harmonic."
  210 PRINT:INPUT "Number of terms, N=";N
  220 PRINT:PRINT "For the following, the number in the parentheses"
  230 PRINT "refers to the term in the Fourier series expansion."
  240 PRINT:INPUT "V1(0)=";V1(0)
  250 INPUT "V2(0)=";V2(0)
  260 FOR I=1 TO N-1
  270 PRINT
  280 PRINT "V1(";I;")=";
  290 INPUT V1(I)
  300 INPUT "at angle";A1(I)
  310 PRINT "V2(";I;")=";
  320 INPUT V2(I)
  330 INPUT "at angle";A2(I)
  340 NEXT I
  350 PRINT
  360 REM Calculate the sum terms
  370 VS(0)=V1(0)+V2(0)
  380 FOR I=1 TO N-1
  390 A1=A1(I)  :A2=A2(I)
  400 XS=V1(I)*COS(A1/57.296)+V2(I)*COS(A2/57.296)
  410 YS=V1(I)*SIN(A1/57.296)+V2(I)*SIN(A2/57.296)
  420 VS(I)=SQR(XS^2+YS^2)
  425 IF XS=0 THEN VA(I)=90 :GOTO 440
  430 VA(I)=57.296*ATN(YS/XS)
  440 NEXT I
  450 PRINT
  460 PRINT "Equation for the sum of the voltages is:"
  470 PRINT "v(t)=v1(t) + v2(t) =";VS(0);
  480 FOR I=1 TO N-1
  490 PRINT TAB(21);
  500 PRINT "+";VS(I);"sin(";W*I;"t";
  510 IF VA(I)>=0 THEN PRINT "+";
  520 PRINT VA(I);")"
  530 NEXT I
  540 END
```

READY

RUN

This program add two non-sinusoidal voltages
of the form:
v(t)=V0 + V1sin(wt+TH1) + V2sin(2wt+TH2) + V3sin(3wt+TH3) + ...

Enter the following signal information:

Radian frequency, w=? 60

Enter the number of terms for the voltage
where N=dc + harmonics, i.e. N=1 + 4 = 5 for up
to and including the 4th harmonic.

Number of terms, N=? 4

```
For the following, the number in the parentheses
refers to the term in the Fourier series expansion.

V1(0)=? 60

V2(0)=? 20


V1( 1 )=? 70

at angle? 0

V2( 1 )=? 30

at angle? 0


V1( 2 )=? 20

at angle? 90

V2( 2 )=? -20

at angle? 90


V1( 3 )=? 10

at angle? 60

V2( 3 )=? 5

at angle? 90



Equation for the sum of the voltages is:
v(t)=v1(t) + v2(t) = 80 + 100 sin( 60 t+ 0 )
                       + 0 sin( 120 t+ 90 )
                       + 14.5466 sin( 180 t+ 69.8961 )


READY
```

FIG. 23.22

Program 23.1.

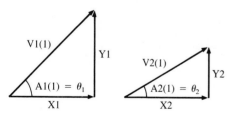

FIG. 23.23

sum of two fundamental frequency components for the provided print-out. Line 390 will define $A1 = A1(1) = \theta_1 = 30°$ and $A2 = A2(1) = \theta_2 = 40°$. Line 400 will then determine the sum of the real and imaginary components using the right-triangle relationships of Fig. 23.23. That is,

$$XS = X1 + X2 = V1(1) \cos \theta_1 + V2(1) \cos \theta_2$$

$$= V1(1) \cos \frac{A1}{57.296} + V2(1) \cos \frac{A2}{57.296}$$

$$= 2 \cos \frac{30}{57.296} + 5 \cos \frac{40}{57.296}$$

and, similarly,

$$YS = Y1 + Y2 = V1(1) \sin \theta_1 + V2(1) \sin \theta_2$$

$$= 2 \sin \frac{30}{57.296} + 5 \sin \frac{40}{57.296}$$

The magnitude and angle of the polar form of the sum are then determined by lines 420 and 430, respectively, and stored as VS(1) and VA(1). The same routine will then be repeated for each set of terms with the same frequency.

When the I loop is complete, the results will be printed out by lines 460 through 530, using an additional I loop from I = 1 to I = (N − 1).

Note in the program that the effective values are not employed in the calculations, because they were not to be printed out, and that the results are in the sinusoidal domain. The printout will also indicate the proper value of ω for each term and include each harmonic that appears in either or both of the nonsinusoidal signals.

The run shown here provides a solution of part (a) of Problem 23.16.

PROBLEMS

Section 23.2

1. For the waveforms of Fig. 23.24, determine whether the following will be present in the Fourier series representation:
   a. dc term
   b. cosine terms
   c. sine terms
   d. even-ordered harmonics
   e. odd-ordered harmonics

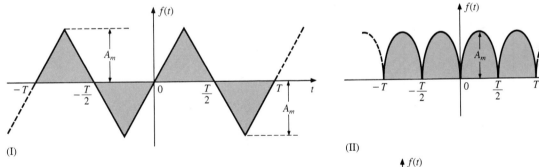

(I) (II)

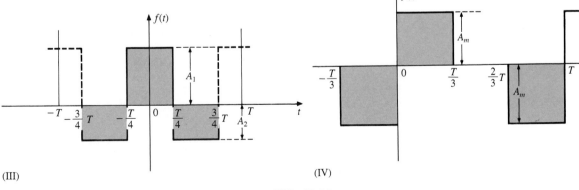

(III) (IV)

FIG. 23.24

2. If the Fourier series for the waveform of Fig. 23.25(a) is

$$i = \frac{2I_m}{\pi} \left(1 + \frac{2}{3} \cos 2\omega t - \frac{2}{15} \cos 4\omega t + \frac{2}{35} \cos 6\omega t + \cdots \right)$$

find the Fourier series representation for waveforms (b), (c), and (d).

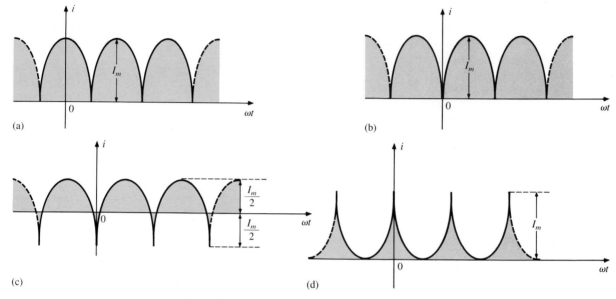

(a)

(b)

(c)

(d)

FIG. 23.25

3. Sketch the following nonsinusoidal waveforms with $\alpha = \omega t$ as the abscissa:
  a. $v = -4 + 2 \sin \alpha$ **b.** $v = (\sin \alpha)^2$
  c. $i = 2 - 2 \cos \alpha$

4. Sketch the following nonsinusoidal waveforms with α as the abscissa:
  a. $i = 3 \sin \alpha - 6 \sin 2\alpha$
  b. $v = 2 \cos 2\alpha + \sin \alpha$

5. Sketch the following nonsinusoidal waveforms with ωt as the abscissa:
  a. $i = 50 \sin \omega t + 25 \sin 3\omega t$
  b. $i = 50 \sin \alpha - 25 \sin 3\alpha$
  c. $i = 4 + 3 \sin \omega t + 2 \sin 2\omega t - 1 \sin 3\omega t$

Section 23.3

6. Find the average and effective values of the following nonsinusoidal waves:
  a. $v = 100 + 50 \sin \omega t + 25 \sin 2\omega t$
  b. $i = 3 + 2 \sin(\omega t - 53°) + 0.8 \sin(2\omega t - 70°)$

7. Find the average and effective values of the following nonsinusoidal waves:
  a. $v = 20 \sin \omega t + 15 \sin 2\omega t - 10 \sin 3\omega t$
  b. $i = 6 \sin(\omega t + 20°) + 2 \sin(2\omega t + 30°)$
        $- 1 \sin(3\omega t + 60°)$

8. Find the total average power to a circuit whose voltage and current are as indicated in Problem 6.

9. Find the total average power to a circuit whose voltage and current are as indicated in Problem 7.

10. The Fourier series representation for the input voltage to the circuit of Fig. 23.26 is

$$e = 18 + 30 \sin 400t$$

   a. Find the nonsinusoidal expression for the current i.
   b. Calculate the effective value of the current.
   c. Find the expression for the voltage across the resistor.
   d. Calculate the effective value of the voltage across the resistor.
   e. Find the expression for the voltage across the reactive element.
   f. Calculate the effective value of the voltage across the reactive element.
   g. Find the average power delivered to the resistor.

11. Repeat Problem 10 for

$$e = 24 + 30 \sin 400t + 10 \sin 800t$$

12. Repeat Problem 10 for the following input voltage:

$$e = -60 + 20 \sin 300t - 10 \sin 600t$$

13. Repeat Problem 10 for the circuit of Fig. 23.27.

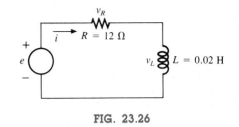

FIG. 23.26

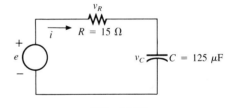

FIG. 23.27

*14. The input voltage [Fig. 23.28(a)] to the circuit of Fig. 23.28(b) is a full-wave rectified signal having the following Fourier series expansion:

$$e = \frac{(2)(100)}{\pi} \left(1 + \frac{2}{3} \cos 2\omega t - \frac{2}{15} \cos 4\omega t + \frac{2}{53} \cos 6\omega t + \cdots \right)$$

where $\omega = 377$.
   a. Find the Fourier series expression for the voltage v_o using only the first three terms of the expression.
   b. Find the effective value of v_o.
   c. Find the average power delivered to the 1-kΩ resistor.

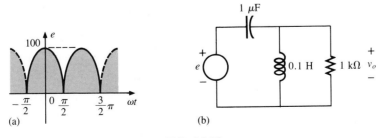

FIG. 23.28

*15. Find the Fourier series expression for the voltage v_o of Fig. 23.29.

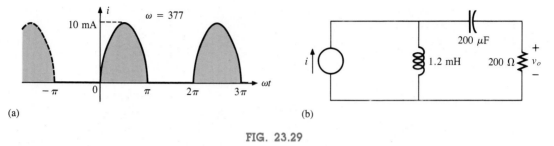

(a)

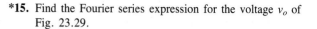

(b)

FIG. 23.29

Section 23.4

16. Perform the indicated operations on the following non-sinusoidal waveforms:

a.

$$[60 + 70 \sin \omega t + 20 \sin(2\omega t + 90°) + 10 \sin(3\omega t + 60°)] + [20 + 30 \sin \omega t - 20 \cos 2\omega t + 5 \cos 3\omega t]$$

b.

$$[20 + 60 \sin \alpha + 10 \sin(2\alpha - 180°) + 5 \cos(3\alpha + 90°)] - [5 - 10 \sin \alpha + 4 \sin(3\alpha - 30°)]$$

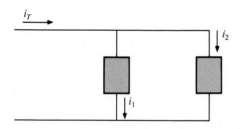

FIG. 23.30

17. Find the nonsinusoidal expression for the current i_T of the diagram of Fig. 23.30.

$$i_2 = 10 + 30 \sin 20t - 0.5 \sin(40t + 90°)$$
$$i_1 = 20 + 4 \sin(20t + 90°) + 0.5 \sin(40t + 30°)$$

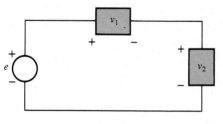

FIG. 23.31

18. Find the nonsinusoidal expression for the voltage e of the diagram of Fig. 23.31.

$$v_1 = 20 - 200 \sin 600t + 100 \cos 1200t + 75 \sin 1800t$$
$$v_2 = -10 + 150 \sin(600t + 30°) + 50 \sin(1800t + 60°)$$

Section 23.5

19. Write a program to determine the sum of the first 10 terms of Eq. (23.9) at $\omega t = \pi/2$, π, and $(3/2)\pi$, and compare to the values determined by Fig. 23.11. That is, enter Eq. (23.9) into memory and calculate the sum of the terms at the points listed above.

20. Given any nonsinusoidal function, write a program that will determine the average and effective values of the waveform. The program should request the data required from the nonsinusoidal function.

21. Write a program that will provide a general solution for the network of Fig. 23.18 for a single dc and ac term in the applied voltage. In other words, the parameter values are given along with the particulars regarding the applied signal, and the nonsinusoidal expression for the current and each voltage is generated by the program.

GLOSSARY

Axis symmetry A sinusoidal or nonsinusoidal function that has symmetry about the vertical axis.

Even harmonics The terms of the Fourier series expansion that have frequencies that are even multiples of the fundamental component.

Fourier series A series of terms, developed in 1826 by Baron Jean Fourier, that can be used to represent a nonsinusoidal function.

Fundamental component The minimum frequency term required to represent a particular waveform in the Fourier series expansion.

Half-wave (mirror) symmetry A sinusoidal or nonsinusoidal function that satisfies the relationship $f(t) = -f\left(\dfrac{T}{2} + t\right)$.

Harmonic The terms of the Fourier series expansion that have frequencies that are integer multiples of the fundamental component.

Nonsinusoidal waveform Any waveform that differs from the fundamental sinusoidal function.

Odd harmonics The terms of the Fourier series expansion that have frequencies that are odd multiples of the fundamental component.

Point symmetry A sinusoidal or nonsinusoidal function that satisfies the relationship $f(\alpha) = -f(-\alpha)$.

24

Transformers

24.1 INTRODUCTION

Chapter 12 discussed the *self-inductance* of a coil. We shall now examine the *mutual inductance* that exists between coils of the same or different dimensions. Mutual inductance is a phenomenon basic to the operation of the transformer, an electrical device used today in almost every field of electrical engineering. This device plays an integral part in power distribution systems and can be found in many electronic circuits and measuring instruments. In this chapter, we will discuss three of the basic applications of a transformer. These are its ability to build up or step down the voltage or current, to act as an impedance matching device, and to isolate (no physical connection) one portion of a circuit from another. In addition, we will introduce the dot convention and consider the transformer equivalent circuit. The chapter will conclude with a word about the effect of mutual inductance on the mesh equations of a network.

24.2 MUTUAL INDUCTANCE

The transformer is constructed of two coils placed so that the changing flux developed by one will link the other as shown in Fig. 24.1. This will result in an induced voltage across each coil. To distinguish between the coils, we will apply the transformer convention that the coil to which the source is applied is called the *primary,* and the coil to which the load is applied is called the *secondary*. For the primary of the

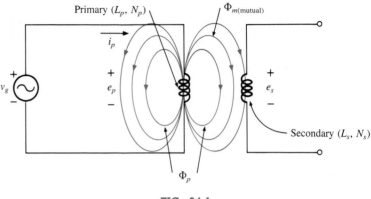

Primary (L_p, N_p)

$\Phi_{m(mutual)}$

i_p

v_g

$+$
e_p
$-$

$+$
e_s
$-$

Secondary (L_s, N_s)

Φ_p

FIG. 24.1

transformer of Fig. 24.1, an application of Faraday's law [Eq. (12.1)] will result in

$$e_p = N_p \frac{d\phi_p}{dt}$$

or, from Eq. (12.5),

$$e_p = L_p \frac{di_p}{dt}$$

The magnitude of e_s, the voltage induced across the secondary, is determined by

$$e_s = N_s \frac{d\phi_m}{dt} \tag{24.1a}$$

where ϕ_m is the portion of the primary flux ϕ_p that links the secondary. If all of the flux linking the primary links the secondary, then

$$\phi_p = \phi_m$$

and

$$e_s = N_s \frac{d\phi_p}{dt} \tag{24.1b}$$

The *coefficient of coupling* between two coils is determined by

$$k \text{ (coefficient of coupling)} = \frac{\phi_m}{\phi_p} \tag{24.2}$$

Since the maximum changing flux that can link the secondary is ϕ_p, the coefficient of coupling between two coils can never be greater than 1. The coefficient of coupling between various coils is indicated in Fig.

24.2. Note that for the iron core, $k \cong 1$, while for the air core, k is considerably less. Those coils with low coefficients of coupling are said to be *loosely coupled*.

For the secondary, we have

$$e_s = N_s \frac{d\phi_m}{dt} = N_s \frac{dk\phi_p}{dt}$$

and

$$\boxed{e_s = kN_s \frac{d\phi_p}{dt}} \qquad \textbf{(24.3)}$$

The mutual inductance between the two coils of Fig. 24.1 is determined by

$$\boxed{M = N_s \frac{d\phi_m}{di_p}} \qquad \text{(henries, H)} \qquad \textbf{(24.4a)}$$

or

$$\boxed{M = N_p \frac{d\phi_m}{di_s}} \qquad \text{(H)} \qquad \textbf{(24.4b)}$$

Note in the above equations that the symbol for mutual inductance is the capital letter M, and that its unit of measurement, like that of self-inductance, is the *henry*. In words, Eqs. (24.4a) and (24.4b) state that the mutual inductance between two coils is proportional to the instantaneous change in flux linking one coil due to an instantaneous change in current through the other coil.

In terms of the inductance of each coil and the coefficient of coupling, the mutual inductance is determined by

$$\boxed{M = k\sqrt{L_p L_s}} \qquad \text{(H)} \qquad \textbf{(24.5)}$$

The greater the coefficient of coupling (greater flux linkages), or the greater the inductance of either coil, the higher the mutual inductance between the coils. Relate this fact to the configurations of Fig. 24.2.

The secondary voltage e_s can also be found in terms of the mutual inductance if we rewrite Eq. (24.1) as

$$e_s = N_s \left(\frac{d\phi_m}{di_p} \right) \left(\frac{di_p}{dt} \right)$$

and, since $M = N_s(d\phi_m / di_p)$, it can also be written

$$\boxed{e_s = M \frac{di_p}{dt}} \qquad \textbf{(24.6a)}$$

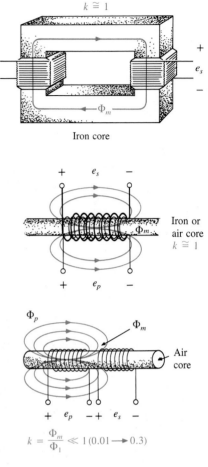

FIG. 24.2

Similarly,

$$e_p = M\frac{di_s}{dt} \qquad \textbf{(24.6b)}$$

$L_p = 200$ mH $\quad L_s = 400$ mH
$N_p = 50$ turns $\quad N_s = 100$ turns

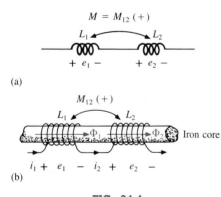

Φ_p $\qquad k = 0.6$

FIG. 24.3

EXAMPLE 24.1. For the transformer in Fig. 24.3:
a. Find the mutual inductance M.
b. Find the induced voltage e_p if the flux changes at the rate of 450 mWb/s.
c. Find the induced voltage e_s for the same rate of change indicated in part (b).
d. Find the induced voltages e_p and e_s if the current i_p changes at the rate of 2 A/s.

Solutions:

a. $M = k\sqrt{L_pL_s} = 0.6\sqrt{(200 \times 10^{-3})(400 \times 10^{-3})}$
$= 0.6\sqrt{8 \times 10^{-2}} = (0.6)(2.828 \times 10^{-1}) = \textbf{169.7 mH}$

b. $e_p = N_p\dfrac{d\phi_p}{dt} = (50)(450 \times 10^{-3}) = \textbf{22.5 V}$

c. $e_s = kN_s\dfrac{d\phi_p}{dt} = (0.6)(100)(450 \times 10^{-3}) = \textbf{27 V}$

d. $e_p = L_p\dfrac{di_p}{dt} = (200 \times 10^{-3})(2) = \textbf{400 mV}$

$e_s = M\dfrac{di_p}{dt} = (170 \times 10^{-3})(2) = \textbf{340 mV}$

24.3 SERIES CONNECTION OF MUTUALLY COUPLED COILS

In Chapter 12, we found that the total inductance of series isolated coils was determined simply by the sum of the inductances. For two coils that are connected in series but also share the same flux linkages, such as those in Fig. 24.4(a), a mutual term is introduced that will alter the total inductance of the series combination. The physical picture of how the coils are connected is indicated in Fig. 24.4(b). An iron core is included, although the equations to be developed are for any two mutually coupled coils with any value of coefficient of coupling k. When referring to the voltage induced across the inductance L_1 (or L_2) due to the change in flux linkages of the inductance L_2 (or L_1, respectively), the mutual inductance is represented by M_{12}. This type of subscript notation is particularly important when there are two or more mutual terms.

Due to the presence of the mutual term, the induced voltage e_1 is composed of that due to the self-inductance L_1 and that due to the mutual inductance M_{12}. That is,

$M = M_{12}\,(+)$

$L_1 \qquad L_2$

$+\ e_1\ - \qquad +\ e_2\ -$

(a)

$M_{12}\,(+)$

$L_1 \qquad L_2$

$\Phi_1 \qquad \Phi_2$ Iron core

$i_1 +\ e_1\ -\quad i_2 +\ e_2\ -$

(b)

FIG. 24.4
Mutually coupled coils connected in series.

$$e_1 = L_1\frac{di_1}{dt} + M_{12}\frac{di_2}{dt}$$

However, since $i_1 = i_2 = i$,

$$e_1 = L_1\frac{di}{dt} + M_{12}\frac{di}{dt}$$

or

$$e_1 = (L_1 + M_{12})\frac{di}{dt} \qquad\qquad \textbf{(24.7a)}$$

and, similarly,

$$e_2 = (L_2 + M_{12})\frac{di}{dt} \qquad\qquad \textbf{(24.7b)}$$

For the series connection, the total induced voltage across the series coils, represented by e_T, is

$$e_T = e_1 + e_2 = (L_1 + M_{12})\frac{di}{dt} + (L_2 + M_{12})\frac{di}{dt}$$

or

$$e_T = (L_1 + L_2 + M_{12} + M_{12})\frac{di}{dt}$$

and the total inductance is

$$L_{T(+)} = L_1 + L_2 + 2M_{12} \qquad \text{(H)} \qquad \textbf{(24.8a)}$$

The subscript (+) was included to indicate that the mutual terms have a positive sign. If the coils were wound such as shown in Fig. 24.5, where ϕ_1 and ϕ_2 are in opposition, the induced voltage due to the

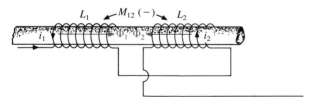

FIG. 24.5

Mutually coupled coils connected in series with negative mutual inductance.

mutual term would oppose that due to the self-inductance, and the total inductance would be determined by

$$L_{T(-)} = L_1 + L_2 - 2M_{12} \qquad \text{(H)} \qquad \textbf{(24.8b)}$$

Through Eqs. (24.8a) and (24.8b), the mutual inductance can be determined by

$$M = \frac{1}{4}(L_{T(+)} - L_{T(-)}) \qquad \text{(H)}$$

(24.9)

Equation (24.9) is very effective in determining the mutual inductance between two coils. It states that the mutual inductance is equal to one-quarter the difference between the total inductance with positive mutual inductance and with negative mutual inductance.

From the preceding, it should be clear that the mutual inductance will directly affect the magnitude of the voltage induced across a coil, since it will determine the net inductance of the coil. Further examination reveals that the sign of the mutual term for each coil of a coupled pair is the same. For $L_{T(+)}$ they were both positive, and for $L_{T(-)}$ they were both negative. On a network schematic where it is inconvenient to indicate the windings and the flux path, a system of dots is employed which will determine whether the mutual terms are to be positive or negative. The dot convention is shown in Fig. 24.6 for the series coils of Figs. 24.4 and 24.5.

If the current through *each* of the mutually coupled coils is going away from (or toward) the dot as it *passes through the coil,* the mutual term will be positive, as shown for the case in Fig. 24.6(a). If the arrow indicating current direction through the coil is leaving the dot for one coil and entering for the other, the mutual term is negative.

A few possibilities for mutually coupled transformer coils are indicated in Fig. 24.7(a). The sign of M is indicated for each. When determining the sign, be sure to examine the current direction within the coil itself. In Fig. 24.7(b), one direction was indicated outside for one coil and through for the other. It initially might appear that the sign should be positive since both currents enter the dot, but the current *through* coil 1 is leaving the dot; hence a negative sign is in order.

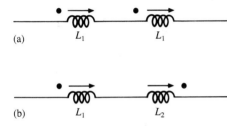

(a) L_1 L_1

(b) L_1 L_2

FIG. 24.6

Dot convention for the series coils of (a) Fig. 24.4 and (b) Fig. 24.5.

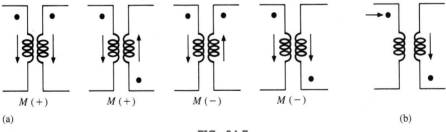

$M(+)$ $M(+)$ $M(-)$ $M(-)$

(a)   (b)

FIG. 24.7

The dot convention also reveals the polarity of the *induced* voltage across the mutually coupled coil. If the reference direction for the current *in* a coil leaves the dot, the polarity at the dot for the induced voltage of the mutually coupled coil is positive. In the first two figures of Fig. 24.7(a), the polarity at the dots of the induced voltages is positive. In the third figure of Fig. 24.7(a), the polarity at the dot of the right-hand coil is positive while the polarity at the dot of the left-hand coil is negative, since the current enters the dot (within the coil) of the right-hand coil. The comments for the third figure of Fig. 24.7(a) can also be applied to the last figure of Fig. 24.7(a).

EXAMPLE 24.2. Find the total inductance of the series coils of Fig. 24.8.

Solution:

Coil 1: $L_1 + M_{12} - M_{13}$

Current vectors leave dot.

One current vector enters dot while one leaves.

Coil 2: $L_2 + M_{12} - M_{23}$

Coil 3: $L_3 - M_{23} - M_{13}$

Note that each has the same sign above.

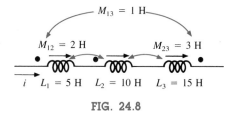

FIG. 24.8

and

$$L_T = (L_1 + M_{12} - M_{13}) + (L_2 + M_{12} - M_{23}) + (L_3 - M_{23} - M_{13})$$
$$= L_1 + L_2 + L_3 + 2M_{12} - 2M_{23} - 2M_{13}$$

Substituting values, we find

$$L_T = 5 + 10 + 15 + 2(2) - 2(3) - 2(1)$$
$$= 34 - 8 = \textbf{26 H}$$

EXAMPLE 24.3. Write the mesh equations for the transformer network in Fig. 24.9.

Solution: For each coil, the mutual term is positive, and the sign of M in $\mathbf{X}_m = \omega M \angle 90°$ is positive, as determined by the direction of $\mathbf{I}_1$ and $\mathbf{I}_2$. Thus,

$$\mathbf{E}_1 - \mathbf{I}_1\mathbf{R}_1 - \mathbf{I}_1\mathbf{X}_{L_1} - \mathbf{I}_2\mathbf{X}_m = 0$$

or

$$\mathbf{E}_1 - \mathbf{I}_1(\mathbf{R}_1 + \mathbf{X}_{L_1}) - \mathbf{I}_2\mathbf{X}_m = 0$$

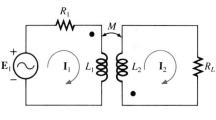

FIG. 24.9

For the other loop,

$$-\mathbf{I}_2\mathbf{X}_{L_2} - \mathbf{I}_1\mathbf{X}_m - \mathbf{I}_2\mathbf{R}_L = 0$$

or

$$\mathbf{I}_2(\mathbf{X}_{L_2} + \mathbf{R}_L) + \mathbf{I}_1\mathbf{X}_m = 0$$

24.4 THE IRON-CORE TRANSFORMER

An iron-core transformer under loaded conditions is shown in Fig. 24.10. The iron core will serve to increase the coefficient of coupling

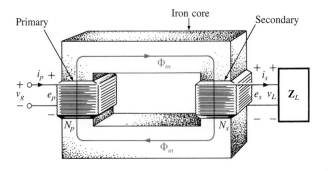

FIG. 24.10
Iron-core transformer.

between the coils by increasing the mutual flux ϕ_m. Recall from Chapter 11 that magnetic flux lines will always take the path of least reluctance, which in this case is the iron core.

We will assume in the analyses to follow in this chapter that all of the flux linking coil 1 will link coil 2. In other words, the coefficient of coupling is its maximum value, 1, and $\phi_m = \phi_1$. In addition, we will first analyze the transformer from an ideal viewpoint. That is, we will neglect such losses as the geometric or dc resistance of the coils, the leakage reactance due to the flux linking either coil that forms no part of ϕ_m, and the hysteresis and eddy current losses. This is not to convey the impression, however, that we will be far from the actual operation of a transformer. Most transformers manufactured today can be considered almost ideal. The equations we will develop under ideal conditions will be, in general, a first approximation to the actual response which will never be off by more than a few percent. The losses will be considered in greater detail in Section 24.6.

When the current i_1 through the primary circuit of the iron-core transformer is a maximum, the flux ϕ_m linking both coils is also a maximum. In fact, the magnitude of the flux is directly proportional to the current through the primary windings. Therefore, the two are in phase, and for sinusoidal inputs, the magnitude of the flux will vary as a sinusoid also. That is, if

$$i_p = \sqrt{2}I_p \sin \omega t$$

then

$$\phi_m = \Phi_m \sin \omega t$$

The induced voltage across the primary due to a sinusoidal input can be determined by Faraday's law:

$$e_p = N_p \frac{d\phi_p}{dt} = N_p \frac{d\phi_m}{dt}$$

Substituting for ϕ_m gives us

$$e_p = N_p \frac{d}{dt}(\Phi_m \sin \omega t)$$

and differentiating, we obtain

$$e_p = \omega N_p \Phi_m \cos \omega t$$

or

$$e_p = \omega N_p \Phi_m \sin(\omega t + 90°)$$

indicating that the induced voltage e_p leads the current through the primary coil by 90°.

The effective value of e_p is

$$E_p = \frac{\omega N_p \Phi_m}{\sqrt{2}} = \frac{2\pi f N_p \Phi_m}{\sqrt{2}}$$

and

$$\boxed{E_p = 4.44 f N_p \Phi_m} \qquad (24.10a)$$

which is an equation for the effective value of the voltage across the primary coil in terms of the frequency of the input current or voltage, the number of turns of the primary, and the maximum value of the magnetic flux linking the primary.

For the case under discussion, where the flux linking the secondary equals that of the primary, if we repeat the procedure just described for the induced voltage across the secondary, we get

$$\boxed{E_s = 4.44 f N_s \Phi_m} \qquad (24.10b)$$

Dividing Eq. (24.10a) by Eq. (24.10b), as follows:

$$\frac{E_p}{E_s} = \frac{4.44 f N_p \Phi_m}{4.44 f N_s \Phi_m}$$

we obtain

$$\boxed{\frac{E_p}{E_s} = \frac{N_p}{N_s}} \qquad (24.11)$$

Note that the ratio of the magnitudes of the induced voltages is the same as the ratio of the corresponding turns.

If we consider that

$$e_p = N_p \frac{d\phi_m}{dt}$$

and

$$e_s = N_s \frac{d\phi_m}{dt}$$

and divide one by the other; that is,

$$\frac{e_p}{e_s} = \frac{N_p(d\phi_m/dt)}{N_s(d\phi_m/dt)}$$

then

$$\frac{e_p}{e_s} = \frac{N_p}{N_s}$$

The *instantaneous* values of e_1 and e_2 are therefore related by a constant determined by the turns ratio. Since their instantaneous magnitudes are related by a constant, the induced voltages are in phase, and Eq. (24.11) can be changed to include phasor notation; that is,

$$\boxed{\frac{\mathbf{E}_p}{\mathbf{E}_s} = \frac{N_p}{N_s}} \qquad \textbf{(24.12)}$$

or, since $\mathbf{V}_g = \mathbf{E}_1$ and $\mathbf{V}_L = \mathbf{E}_2$ for the ideal situation,

$$\boxed{\frac{\mathbf{V}_g}{\mathbf{V}_L} = \frac{N_p}{N_s}} \qquad \textbf{(24.13)}$$

The ratio N_p/N_s, usually represented by the lowercase letter a, is referred to as the *transformation ratio:*

$$\boxed{a = \frac{N_p}{N_s}} \qquad \textbf{(24.14)}$$

If $a < 1$, the transformer is called a *step-up transformer,* since the voltage $E_s > E_p$:

$$\frac{E_p}{E_s} = \frac{N_p}{N_s} = a$$

so

$$E_s = \frac{E_p}{a}$$

and, if $a < 1$,

$$E_s > E_p$$

If $a > 1$, the transformer is called a *step-down transformer,* since $E_s < E_p$; that is,

$$E_p = aE_s$$

and, if $a > 1$, then

$$E_p > E_s$$

EXAMPLE 24.4. For the iron-core transformer of Fig. 24.11:
a. Find the maximum flux Φ_m.
b. Find the number of turns N_s.

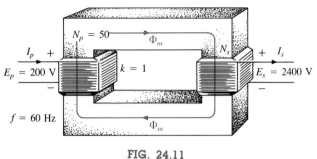

$N_p = 50$

Φ_m

N_s

I_p +

$k = 1$

$+$ I_s

$E_p = 200$ V

$E_s = 2400$ V

$-$

$-$

$f = 60$ Hz

Φ_m

FIG. 24.11

Solutions:

a. $E_p = 4.44 N_p f \, \Phi_m$

Therefore,

$$\Phi_m = \frac{E_p}{4.44 N_p f} = \frac{200}{(4.44)(50)(60)}$$

and

$$\Phi_m = \mathbf{15.02 \, mWb}$$

b. $\dfrac{E_p}{E_s} = \dfrac{N_p}{N_s}$

Therefore,

$$N_s = \frac{N_p E_s}{E_p} = \frac{(50)(2400)}{200}$$

$$= \mathbf{600 \ turns}$$

The induced voltage across the secondary of the transformer of Fig. 24.10 will establish a current i_s through the load Z_L and the secondary windings. This current and the turns N_s will develop an mmf $N_s i_s$ that would not be present under no-load conditions, since $i_s = 0$ and $N_s i_s = 0$. Under loaded or unloaded conditions, however, the net ampere-turns on the core produced by both the primary and the secondary must remain unchanged for the same flux ϕ_m to be established in the core. The flux ϕ_m must remain the same to have the same induced voltage across the primary to balance the voltage impressed across the primary. In order to counteract the mmf of the secondary, which is tending to change ϕ_m, an additional current must flow in the primary. This current is called the *load component of the primary current* and is represented by the notation i'_p.

For the balanced or equilibrium condition,

$$N_p i'_p = N_s i_s$$

The total current in the primary under loaded conditions is

$$i_p = i'_p + i_{\phi_m}$$

where i_{ϕ_m} is the current in the primary necessary to establish the flux ϕ_m. For most practical applications, $i'_p > i_{\phi_m}$. For our analysis, we will assume $i_p \cong i'_p$, so

$$N_p i_p = N_s i_s$$

Since the instantaneous values of i_p and i_s are related by the turns ratio, the phasor quantities $\mathbf{I}_p$ and $\mathbf{I}_s$ are also related by the same ratio:

$$N_p \mathbf{I}_p = N_s \mathbf{I}_s$$

or

$$\boxed{\frac{\mathbf{I}_p}{\mathbf{I}_s} = \frac{N_s}{N_p}} \tag{24.15}$$

Keep in mind that Eq. (24.15) holds true only if we neglect the effects of i_{ϕ_m}. Otherwise, the magnitudes of $\mathbf{I}_p$ and $\mathbf{I}_s$ are not related by the turns ratio, and $\mathbf{I}_p$ and $\mathbf{I}_s$ are not in phase.

For the step-up transformer, $a < 1$, and the current in the secondary, $I_s = aI_p$, is less in magnitude than that in the primary. For a step-down transformer, the reverse is true.

24.5 REFLECTED IMPEDANCE AND POWER

In the previous sections, we found that

$$\frac{\mathbf{V}_g}{\mathbf{V}_L} = \frac{N_p}{N_s} \quad \text{and} \quad \frac{\mathbf{I}_p}{\mathbf{I}_s} = \frac{N_s}{N_p}$$

Dividing one by the other, we have

$$\frac{\mathbf{V}_g/\mathbf{V}_L = N_p/N_s = a}{\mathbf{I}_p/\mathbf{I}_s \quad = N_s/N_p = 1/a}$$

or

$$\frac{\mathbf{V}_g/\mathbf{I}_p}{\mathbf{V}_L/\mathbf{I}_s} = a^2$$

and

$$\frac{\mathbf{V}_g}{\mathbf{I}_p} = a^2 \frac{\mathbf{V}_L}{\mathbf{I}_s}$$

However, since

$$\mathbf{Z}_p = \frac{\mathbf{V}_g}{\mathbf{I}_p} \quad \text{and} \quad \mathbf{Z}_L = \frac{\mathbf{V}_L}{\mathbf{I}_s}$$

then

$$\boxed{\mathbf{Z}_p = a^2 \mathbf{Z}_L} \tag{24.16}$$

which in words states that the impedance of the primary circuit of an ideal transformer is the transformation ratio squared times the impedance of the load. If a transformer is used, therefore, an impedance can be made to appear larger or smaller at the primary by placing it in the secondary of a step-down ($a > 1$) or step-up ($a < 1$) transformer, respectively. Note that if the load is capacitive or inductive, the reflected impedance will also be capacitive or inductive.

For the ideal iron-core transformer,

$$\frac{E_p}{E_s} = a = \frac{I_s}{I_p}$$

or

$$\boxed{E_p I_p = E_s I_s} \qquad (24.17)$$

and

$$\boxed{P_{\text{in}} = P_{\text{out}}} \qquad \text{(ideal case)} \qquad (24.18)$$

EXAMPLE 24.5. For the iron-core transformer of Fig. 24.12:

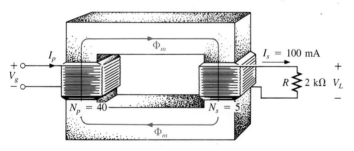

FIG. 24.12

a. Find the magnitude of the current in the primary and the impressed voltage across the primary.
b. Find the input resistance of the transformer.

Solutions:

a. $\dfrac{I_p}{I_s} = \dfrac{N_s}{N_p}$

or

$$I_p = \frac{N_s}{N_p} I_s = \left(\frac{5}{40}\right)(0.1)$$

and

$I_p = \mathbf{12.5\,mA}$

$V_L = I_s Z_L = (0.1)(2 \times 10^3)$

    $= 200\text{ V}$

and

$$\frac{V_g}{V_L} = \frac{N_p}{N_s}$$

or

$$V_g = \frac{N_p}{N_s} V_L = \left(\frac{40}{5}\right)(200)$$

and

$$V_g = \mathbf{1600\ V}$$

b. $Z_p = a^2 Z_L$

$$a = \frac{N_p}{N_s} = 8$$

$$Z_p = (8)^2 (2 \times 10^3)$$
$$= R_p = \mathbf{128\ k\Omega}$$

EXAMPLE 24.6. For the speaker in Fig. 24.13 to receive maximum power from the circuit, the internal resistance of the speaker should be 540 Ω. If a transformer is used, the speaker resistance of 15 Ω can be made to appear 540 Ω at the primary. Find the transformation ratio required and the number of turns in the primary if the secondary winding has 40 turns.

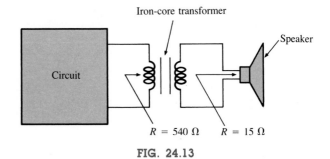

Iron-core transformer

Speaker

Circuit

$R = 540\ \Omega$ $R = 15\ \Omega$

FIG. 24.13

Solution:

$$Z_p = a^2 Z_L$$

or

$$a = \sqrt{\frac{Z_p}{Z_L}} = \sqrt{\frac{540}{15}} = \sqrt{36} = 6$$

Therefore,

$$a = 6 = \frac{N_p}{N_s}$$

or

$$N_p = 6 N_s = 6(40)$$

and

$$N_p = \textbf{240 turns}$$

EXAMPLE 24.7. For the residential supply appearing in Fig. 24.14, determine (assuming a totally resistive load) the following:

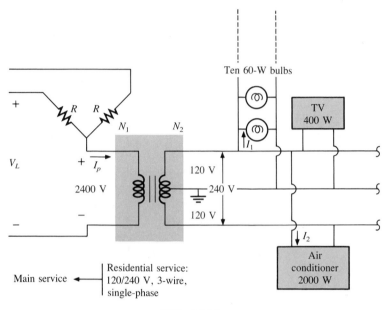

Ten 60-W bulbs

TV
400 W

N_1 N_2

I_1

V_L $+$ I_p

120 V

2400 V

240 V

120 V

I_2

Air conditioner
2000 W

Main service ◄── Residential service:
120/240 V, 3-wire,
single-phase

FIG. 24.14

a. the value of R to insure a balanced load
b. the magnitude of I_1 and I_2
c. the line voltage V_L
d. the total power delivered
e. the turns ratio N_1/N_2

Solutions:

a. $P_T = 600 + 400 + 2000 = 3000 \text{ W}$

   $P_{\text{in}} = P_{\text{out}}$

   $V_p I_p = V_s I_s = 3000 \text{ W}$ (purely resistive load)

   $2400 I_p = 3000$

   $I_p = 1.25 \text{ A}$

   $R = \dfrac{V}{I} = \dfrac{2400}{1.25} = \textbf{1920 } \boldsymbol{\Omega}$

b. $P = (10)(60) = 600 \text{ W} = VI_1 = 120 I_1$

   and

   $I_1 = \textbf{5 A}$

   $P = 2000 \text{ W} = VI_2 = 240 I_2$

   and

   $I_2 = \textbf{8}\tfrac{1}{3}\textbf{ A}$

$(120(26mA)$

$I_s = .183 \ A$

c. $V_L = \sqrt{3}V_\phi = 1.73(2400) = \mathbf{4152\ V}$

d. $P_T = 3P_\phi = 3(3000) = \mathbf{9\ kW}$

e. $\dfrac{N_1}{N_2} = \dfrac{V_p}{V_s} = \dfrac{2400}{240} = \mathbf{10}$

24.6 EQUIVALENT CIRCUIT (IRON-CORE TRANSFORMER)

For the nonideal or practical iron-core transformer, the equivalent circuit appears as in Fig. 24.15. As indicated, part of this equivalent cir-

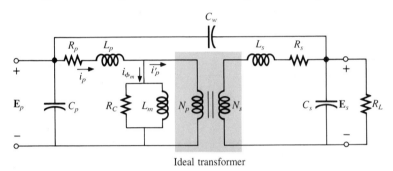

FIG. 24.15

Equivalent circuit for the practical iron-core transformer.

cuit is an ideal transformer. The remaining elements of Fig. 24.15 are those elements that contribute to the nonideal characteristics of the device. The resistances R_p and R_s are simply the dc or geometric resistance of the primary and secondary coils, respectively. For the primary and secondary coils of a transformer, there is a definite amount of flux that links each coil that does not pass through the core. This situation is shown in Fig. 24.16. This *leakage* flux, serving as a definite loss to the system since it employs an amount of input energy to be established but serves no useful purpose, is represented by an inductance L_p in the primary circuit and an inductance L_s in the secondary.

The resistance R_C represents the hysteresis and eddy current losses (core losses) within the core due to an ac flux through the core. The inductance L_m (magnetizing inductance) is the inductance associated with the magnetization of the core, that is, the establishing of the flux Φ_m in the core. The capacitances C_p and C_s are the lumped capacitances of the primary and secondary circuits, respectively, and C_w represents the equivalent lumped capacitances between the windings of the transformer.

Since i'_p is normally considerably larger than i_{ϕ_m}, we will ignore i_{ϕ_m} for the moment (set it equal to zero), resulting in the absence of R_C and L_m in the reduced equivalent circuit of Fig. 24.17. In addition, the capacitances C_p, C_w, and C_s do not appear in the equivalent circuit of Fig. 24.17, since their reactance in the present frequency range of inter-

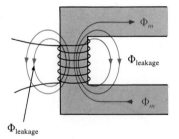

FIG. 24.16

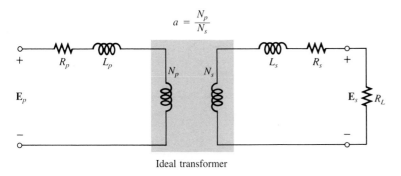

$$a = \frac{N_p}{N_s}$$

Ideal transformer

FIG. 24.17

Reduced equivalent circuit for the nonideal iron-core transformer.

est will not appreciably affect the transfer characteristics of the transformer.

If we now reflect the secondary circuit through the ideal transformer, as shown in Fig. 24.18(a), we will have the load and generator voltage

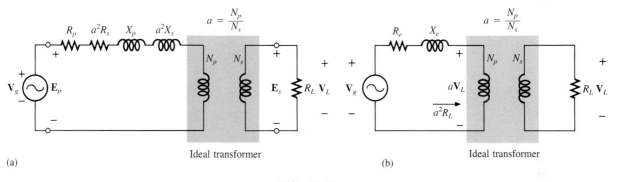

(a)

(b)

FIG. 24.18

in the same physical circuit. The total resistance and inductive reactance are determined by

$$\boxed{R_{\text{equivalent}} = R_e = R_p + a^2 R_s} \qquad \textbf{(24.19a)}$$

and

$$\boxed{X_{\text{equivalent}} = X_e = X_p + a^2 X_s} \qquad \textbf{(24.19b)}$$

which result in the useful equivalent circuit of Fig. 24.18(b). The load voltage can be obtained directly from the circuit of Fig. 24.18(b) through the voltage divider rule:

$$\boxed{a\mathbf{V}_L = \frac{a^2 R_L \mathbf{V}_g}{(R_e + a^2 R_L) + jX_e}} \qquad \textbf{(24.20)}$$

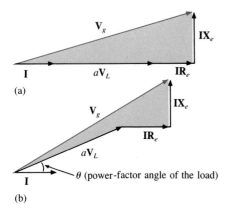

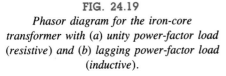

(a)

(b)

FIG. 24.19

Phasor diagram for the iron-core transformer with (a) unity power-factor load (resistive) and (b) lagging power-factor load (inductive).

In a different light, the generator voltage necessary to establish a particular load voltage can also be determined through Eq. (24.20). The voltages across the elements of Fig. 24.18(b) have the phasor relationship indicated in Fig. 24.19(a). Note that the current is the reference phasor for drawing the phasor diagram. That is, the voltages across the resistive elements are *in phase* with the current phasor, while the voltage across the equivalent inductance leads the current by 90°. The primary voltage, by Kirchhoff's voltage law, is then the phasor sum of these voltages, as indicated in Fig. 24.19(a). For an inductive load, the phasor diagram appears in Fig. 24.19(b). Note that aV_L leads I by the power-factor angle of the load. The remainder of the diagram is then similar to that for a resistive load. (The phasor diagram for a capacitive load will be left to the reader as an exercise.)

The effect of R_e and X_e on the magnitude of V_g for a particular V_L is obvious from Eq. (24.20) or Fig. 24.19. For increased values of R_e or X_e, an increase in V_g is required for the same load voltage. For R_e and $X_e = 0$, V_L and V_g are related by the turns ratio.

EXAMPLE 24.8. For a transformer having the equivalent circuit of Fig. 24.20:

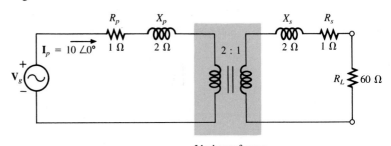

Ideal transformer

FIG. 24.20

a. Determine R_e and X_e.
b. Determine V_g.

Solutions:
a. $R_e = R_p + a^2R_s = 1 + (2)^2(1) = \textbf{5 Ω}$
   $X_e = X_p + a^2X_s = 2 + (2)^2(2) = \textbf{10 Ω}$

b. The transformed equivalent circuit appears in Fig. 24.21.
   $aV_L = (I_p)(a^2R_L) = 2400 \text{ V}$

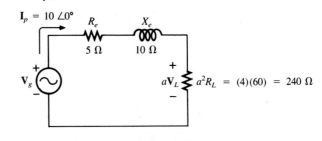

FIG. 24.21

Thus,

$$V_L = \frac{2400}{a} = \frac{2400}{2} = 1200 \text{ V}$$

and

$$\begin{aligned}\mathbf{V}_g &= \mathbf{I}_p(R_e + a^2 R_L + jX_e) \\ &= 10(5 + 240 + j10) = 10(245 + j10) \\ \mathbf{V}_g &= 2450 + j100\end{aligned}$$

and

$$V_g = \mathbf{2452.04 \text{ V}}$$

For R_e and $X_e = 0$, $V_g = aV_L = 2400$. Therefore, it is necessary to increase the generator voltage by 52.04 V (due to R_e and X_e) to obtain the same load voltage.

24.7 FREQUENCY CONSIDERATIONS

For certain frequency ranges, the effect of some parameters in the equivalent circuit of the iron-core transformer of Fig. 24.15 can be neglected. Since it is convenient to consider a low-, mid-, and high-frequency region, the equivalent circuits for each will now be introduced and briefly examined.

For the low-frequency region, the reactance ($2\pi f L$) of the primary and secondary leakage reactances can be ignored, and the reflected equivalent circuit will appear as shown in Fig. 24.22(a). The magnetizing inductance must be included, since it appears in parallel with the secondary reflected circuit. As the frequency approaches zero, the reactance of the magnetizing inductance will reduce in magnitude, causing a reduction in the voltage across the secondary circuit. For $f = 0$ Hz, L_m is ideally a short circuit, and $V_L = 0$. As the frequency increases, the reactance of L_m will eventually be sufficiently large compared with the reflected secondary impedance to be neglected. The mid-frequency reflected equivalent circuit will then appear as shown in Fig. 24.22(b). Note the absence of reactive elements, resulting in an *in-phase* relationship between load and generator voltages.

For higher frequencies, the capacitive elements and primary and secondary leakage reactances must be considered, as shown in Fig. 24.23.

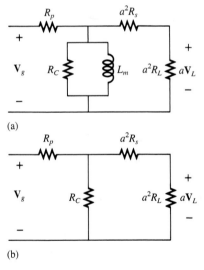

(a)

(b)

FIG. 24.22

(a) Low-frequency reflected equivalent circuit; (b) mid-frequency reflected circuit.

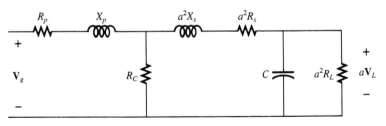

FIG. 24.23

High-frequency reflected equivalent circuit.

For discussion purposes, the effects of C_w and C_s appear as a lumped capacitor C in the reflected network of Fig. 24.23; C_p does not appear since the effect of C will predominate. As the frequency of interest increases, the capacitive reactance ($X_C = 1/2\pi fC$) will decrease to the point that it will have a shorting effect across the secondary circuit of the transformer, causing V_L to decrease in magnitude.

A typical transformer-frequency response curve appears in Fig. 24.24. For the low- and high-frequency regions, the primary element responsible for the drop-off is indicated. The peaking that occurs in the high-frequency region is due to the series resonant circuit of Fig. 24.23(b), composed of R_e, $X_e(L)$, and C. In the peaking region, the series resonant circuit is in, or near, its resonant or tuned state.

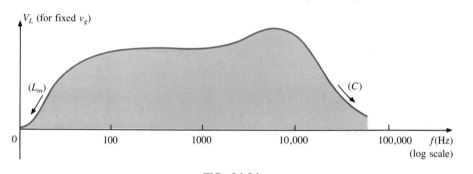

FIG. 24.24

Transformer-frequency response curve.

The network discussed in some detail earlier in this chapter was for the high mid-frequency region.

24.8 AIR-CORE TRANSFORMER

As the name implies, the air-core transformer does not have a ferromagnetic core to link the primary and secondary coils. Rather, the coils are placed sufficiently close to have a mutual inductance that will establish the desired transformer action. In Fig. 24.25, current direction and polarities have been defined for the air-core transformer. Note the pres-

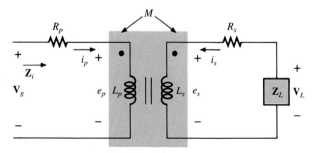

Ideal transformer

FIG. 24.25

Air-core transformer equivalent circuit.

ence of a mutual inductance term M, which will be positive in this case, as determined by the dot convention.

From past analysis in this chapter, we now know that

$$e_p = L_p \frac{di_p}{dt} + M \frac{di_s}{dt} \qquad \textbf{(24.21)}$$

for the primary circuit.

We found in Chapter 12 that for the pure inductor, with no mutual inductance present, the mathematical relationship

$$v_1 = L_1 \frac{di_1}{dt}$$

resulted in the following useful form of the voltage across an inductor:

$$\mathbf{V}_1 = \mathbf{I}_1 \mathbf{X}_{L_1} \quad \text{where } \mathbf{X}_{L_1} = \omega L_1 \angle 90° = j\omega L_1$$

Similarly, it can be shown, for a mutual inductance, that

$$v_1 = M \frac{di_2}{dt}$$

will result in

$$\mathbf{V}_1 = \mathbf{I}_2 \mathbf{X}_m \quad \text{where } \mathbf{X}_m = \omega M \angle 90° = j\omega M \qquad \textbf{(24.22)}$$

Equation (24.21) can then be written (using phasor notation)

$$\mathbf{E}_p = \mathbf{I}_p \mathbf{X}_{L_p} + \mathbf{I}_s \mathbf{X}_m \qquad \textbf{(24.23)}$$

and

$$\mathbf{V}_g = \mathbf{I}_p \mathbf{R}_p + \mathbf{I}_p \mathbf{X}_{L_p} + \mathbf{I}_s \mathbf{X}_m$$

or

$$\mathbf{V}_g = \mathbf{I}_p (\mathbf{R}_p + \mathbf{X}_{L_p}) + \mathbf{I}_s \mathbf{X}_m \qquad \textbf{(24.24)}$$

For the secondary circuit,

$$\mathbf{E}_s = \mathbf{I}_s \mathbf{X}_{L_s} + \mathbf{I}_p \mathbf{X}_m \qquad \textbf{(24.25)}$$

and

$$\mathbf{V}_L = \mathbf{I}_s \mathbf{R}_s + \mathbf{I}_s \mathbf{X}_{L_s} + \mathbf{I}_p \mathbf{X}_m$$

or

$$\mathbf{V}_L = \mathbf{I}_s (\mathbf{R}_s + \mathbf{X}_{L_s}) + \mathbf{I}_p \mathbf{X}_m \qquad \textbf{(24.26)}$$

Substituting

$$\mathbf{V}_L = -\mathbf{I}_s \mathbf{Z}_L$$

into Eq. (24.26) results in

$$0 = \mathbf{I}_s(\mathbf{R}_s + \mathbf{X}_{L_s} + \mathbf{Z}_L) + \mathbf{I}_p\mathbf{X}_m$$

Solving for $\mathbf{I}_s$, we have

$$\mathbf{I}_s = \frac{-\mathbf{I}_p\mathbf{X}_m}{\mathbf{R}_s + \mathbf{X}_{L_s} + \mathbf{Z}_L}$$

and, substituting into Eq. (24.24), we obtain

$$\mathbf{V}_g = \mathbf{I}_p(\mathbf{R}_p + \mathbf{X}_{L_p}) + \left(\frac{-\mathbf{I}_p\mathbf{X}_m}{\mathbf{R}_s + \mathbf{X}_{L_s} + \mathbf{Z}_L} \right)\mathbf{X}_m$$

Thus, the input impedance is

$$\mathbf{Z}_i = \frac{\mathbf{V}_g}{\mathbf{I}_p} = \mathbf{R}_p + \mathbf{X}_{L_p} - \frac{\mathbf{X}_m^2}{\mathbf{R}_s + \mathbf{X}_{L_s} + \mathbf{Z}_L}$$

or, defining

$$\mathbf{Z}_p = \mathbf{R}_p + \mathbf{X}_{L_p} \qquad \mathbf{Z}_s = \mathbf{R}_s + \mathbf{X}_{L_s} \qquad \mathbf{X}_m = j\omega M$$

we have

$$\mathbf{Z}_i = \mathbf{Z}_p - \frac{(+j\omega M)^2}{\mathbf{Z}_s + \mathbf{Z}_L}$$

and

$$\boxed{\mathbf{Z}_i = \mathbf{Z}_p + \frac{(\omega M)^2}{\mathbf{Z}_s + \mathbf{Z}_L}} \qquad \textbf{(24.27)}$$

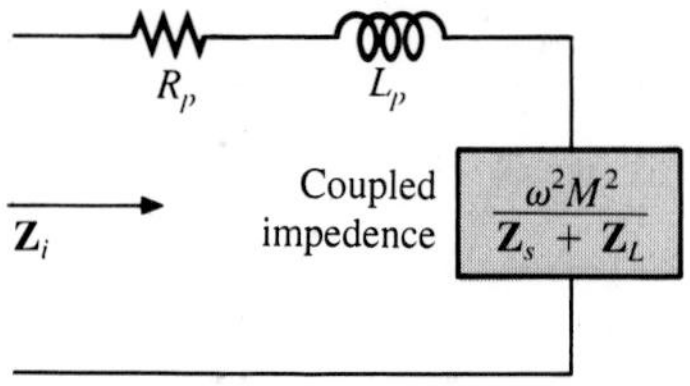

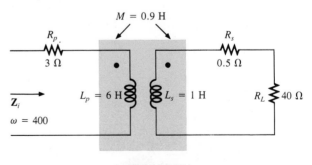

FIG. 24.26
Input characteristics for the air-core transformer.

The term $(\omega M)^2/(\mathbf{Z}_s + \mathbf{Z}_L)$ is called the *coupled impedance*. Note that it is independent of the sign of M. Consider also that since $(\omega M)^2$ is a constant with $0°$ phase angle, if $\mathbf{Z}_2$ is resistive, the resulting coupled impedance term will appear capacitive, due to division of $(\mathbf{R}_L + \mathbf{Z}_s)$ into $(\omega M)^2$. This resulting capacitive reactance will oppose the series primary inductance L_p, causing a reduction in $\mathbf{Z}_i$. Including the effect of the mutual term, the input impedance to the network will appear as shown in Fig. 24.26.

EXAMPLE 24.9. Determine the input impedance to the air-core transformer in Fig. 24.27.

FIG. 24.27

Solution:

$$\mathbf{Z}_i = \mathbf{Z}_p + \frac{(\omega M)^2}{\mathbf{Z}_s + \mathbf{Z}_L}$$

$$= R_p + jX_{L_p} + \frac{(\omega M)^2}{R_s + jX_{L_s} + R_L}$$

$$= 3 + j2.4 \times 10^3 + \frac{(360)^2}{0.5 + j400 + 40}$$

$$\cong j2.4 \times 10^3 + \frac{129.6 \times 10^3}{40.5 + j400}$$

$$= j2.4 \times 10^3 + \frac{129.6 \times 10^3}{402.05 \angle 84.22°}$$

$$= j2.4 \times 10^3 + 322.4 \angle -84.22° = j2.4 \times 10^3$$
$$+ \underbrace{(0.0325 \times 10^3 - j0.3208 \times 10^3)}_{\text{capacitive}}$$

$$= 0.0325 \times 10^3 + j(2.40 \times 10^3 - 0.3208 \times 10^3)$$
$$\mathbf{Z}_i = \mathbf{32.5} + \mathbf{j2079} = \mathbf{R}_i + \mathbf{jX}_{L_i} = \mathbf{2079.25} \ \angle \mathbf{89.10°}$$

24.9 THE TRANSFORMER AS AN ISOLATION DEVICE

The transformer is frequently used to isolate one portion of an electrical system from another. By *isolation,* we mean the absence of any direct physical connection. As a first example of its use as an isolation device, consider the measurement of line voltages on the order of 40,000 V (Fig. 24.28).

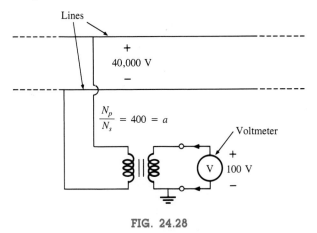

Lines

+
40,000 V
−

$\dfrac{N_p}{N_s} = 400 = a$

Voltmeter

V +
100 V
−

FIG. 24.28

To apply a voltmeter across 40,000 V would obviously be a danger-ous task due to the possibility of physical contact with the lines when making the necessary connections. By including a transformer in the transmission system as original equipment, one can bring the potential down to a safe level for measurement purposes and can determine the line voltage using the turns ratio. Therefore, the transformer will serve both to isolate and to step down the voltage.

As a second example, consider the application of the voltage v_x to the vertical input of the oscilloscope (a measuring instrument) in Fig. 24.29.

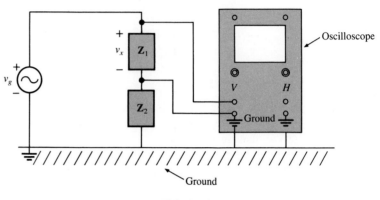

FIG. 24.29

If the connections are made as shown and the generator and oscilloscope have a common ground, the impedance $\mathbf{Z}_2$ has been effectively shorted out of the circuit by the ground connection of the oscilloscope. The input voltage to the oscilloscope will therefore be meaningless as far as the voltage v_x is concerned. In addition, if $\mathbf{Z}_2$ is the current-limiting impedance in the circuit, the current in the circuit may rise to a level that will cause severe damage to the circuit. If a transformer is used as shown in Fig. 24.30, this problem will be eliminated, and the input voltage to the oscilloscope will be v_x.

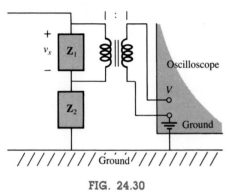

FIG. 24.30

24.10 NAMEPLATE DATA

A typical power transformer rating might be the following:

$$5\,\text{kVA, 2000/100 V, 60 Hz}$$

The 2000 V or the 100 V can be either the primary or the secondary voltage; that is, if 2000 V is the primary voltage, then 100 V is the secondary voltage, and vice versa. The 5 kVA is the apparent power ($S = VI$) rating of the transformer. If the secondary voltage is 100 V, then the maximum load current is

$$I_L = \frac{S}{V_L} = \frac{5000}{100} = 50 \text{ A}$$

and if the secondary voltage is 2000 V, then the maximum load current is

$$I_L = \frac{S}{V_L} = \frac{5000}{2000} = 2.5 \text{ A}$$

The transformer is rated in terms of the apparent power rather than the average power for the reason demonstrated by the circuit of Fig. 24.31.

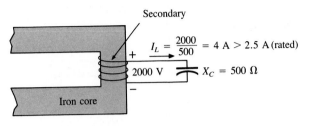

Secondary

$I_L = \dfrac{2000}{500} = 4\ A > 2.5\ A\ (\text{rated})$

2000 V

$X_C = 500\ \Omega$

Iron core

FIG. 24.31

Since the current through the load is greater than that determined by the apparent power rating, the transformer may be permanently damaged. Note, however, that since the load is purely capacitive, the average power to the load is zero. The wattage rating would therefore be meaningless regarding the ability of this load to damage the transformer.

The transformation ratio of the transformer under discussion can be either of two values. If the secondary voltage is 2000 V, the transformation ratio is $a = N_p/N_s = V_g/V_L = 100/2000 = 1/20$, and the transformer is a step-up transformer. If the secondary voltage is 100 V, the transformation ratio is $a = N_p/N_s = V_g/V_L = 2000/100 = 20$, and the transformer is a step-down transformer.

The rated primary current can be determined simply by applying Eq. (24.15):

$$I_1 = \frac{I_2}{a}$$

which is equal to $[2.5/(1/20)] = 50\ A$ if the secondary voltage is 2000 V, and $(50/20) = 2.5\ A$ if the secondary voltage is 100 V.

To explain the necessity for including the frequency in the nameplate data, consider Eq. (24.10a):

$$E_p = 4.44 f_p N_p \Phi_m$$

and the *B-H* curve for the iron core of the transformer (Fig. 24.32).

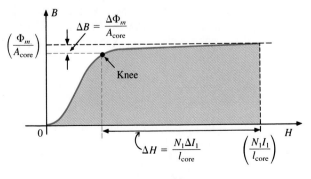

FIG. 24.32

The point of operation on the *B-H* curve for most transformers is at the knee of the curve. If the frequency of the applied signal should drop, and N_p and E_p remain the same, then Φ_m must increase in magnitude, as determined by Eq. (24.10a):

$$\Phi_m\uparrow = \frac{E_p}{4.44 f_p\downarrow N_p}$$

Note on the *B-H* curve that this increase in Φ_m will cause a very high current in the primary, resulting in possible damage of the transformer.

24.11 TYPES OF TRANSFORMERS

Transformers are available in many different shapes and sizes. Some of the more common types include the power transformer, audio transformer, I-F (intermediate-frequency) transformer, and R-F (radio-frequency) transformer. Each is designed to fulfill a particular requirement in a specific area of application. The symbols for some of the basic types of transformers are shown in Fig. 24.33.

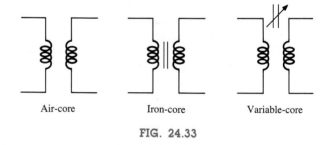

| Air-core | Iron-core | Variable-core |

FIG. 24.33

The method of construction varies from one transformer to another. Two of the many different ways in which the primary and secondary coils can be wound around an iron core are shown in Fig. 24.34. In either case, the core is made of laminated sheets of ferromagnetic material separated by an insulator to reduce the eddy current losses. The

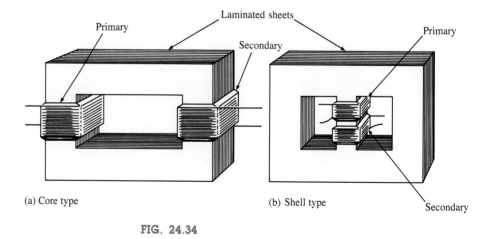

(a) Core type   (b) Shell type

FIG. 24.34

sheets themselves will also contain a small percentage of silicon to reduce the hysteresis losses.

A shell-type power transformer with its schematic representation is shown in Fig. 24.35.

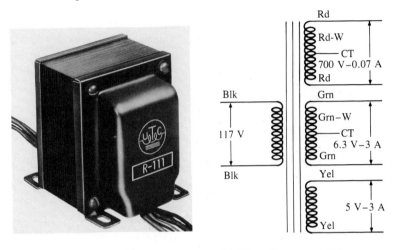

(a) Shell-type power transformer.
(b) Schematic representation.

FIG. 24.35
(Courtesy of United Transformer Co.)

The *autotransformer* [Fig. 24.36(b)] is a type of power transformer which, instead of employing the two-circuit principle (complete isolation between coils), has one winding common to both the input and output circuits. The induced voltages are related to the turns ratio in the same manner as that described for the two-circuit transformer. If the proper connection is used, a two-circuit power transformer can be employed as an autotransformer. The advantage of using it as an autotransformer is that a larger apparent power can be transformed. This can be demonstrated by the two-circuit transformer of Fig. 24.36(a). It is shown in Fig. 24.36(b) as an autotransformer.

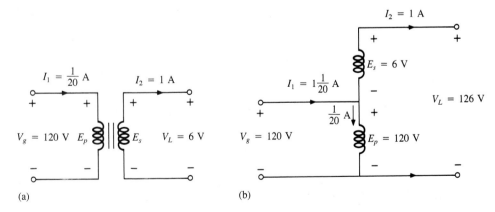

FIG. 24.36
(a) Two-circuit transformer; (b) autotransformer.

(a) I-F transformer

(b) R-F transformer

FIG. 24.37

For the two-circuit transformer, note that $S = (\frac{1}{20})(120) = 6\,\text{VA}$, while for the autotransformer, $S = (1\frac{1}{20})(120) = 126\,\text{VA}$, which is many times that of the two-circuit transformer. Note also that the current and voltage of each coil are the same as those for the two-circuit configuration. The disadvantage of the autotransformer is obvious: loss of the isolation between the primary and secondary circuits.

The R-F and I-F transformers used extensively in radio and television transmitters and receivers are shown in Fig. 24.37. Both are available with or without a shield. Permeability-tuned I-F transformers permit the changing of ϕ_m, and thereby k, by moving a ferromagnetic core within the primary and secondary coils of the transformer.

A dual-in-line pulse transformer package appears in Fig. 24.38 for use with integrated circuits and printed circuit board applications. Note the availability of four isolated transformers and the appearance of the dot convention. The data for one such unit include a $2:1$ primary-to-secondary turns ratio; a leakage inductance of $0.50\ \mu\text{H}$; a coupling capacitance of 7 pF; a primary dc resistance of $0.19\ \Omega$; and a secondary resistance of $0.13\ \Omega$.

FIG. 24.38
Dual-in-line pulse transformer package. (Courtesy of Bourns®, Inc.)

A miniature radio transformer designed for direct mounting on a printed circuit board appears in Fig. 24.39.

FIG. 24.39
(Courtesy of Microtran Company, Inc.)

24.12 TAPPED AND MULTIPLE-LOAD TRANSFORMERS

For the center-tapped (primary) transformer of Fig. 24.40, where the voltage from the center top to either outside lead is defined as $E_p/2$, the relationship between E_p and E_s is

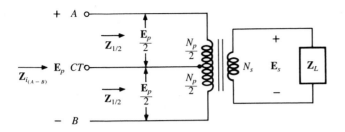

FIG. 24.40
Ideal transformer with a center-tapped primary.

$$\frac{E_p}{E_s} = \frac{N_p}{N_s} \qquad (24.28)$$

For each half-section of the primary,

$$\mathbf{Z}_{1/2} = \left(\frac{N_p/2}{N_s}\right)^2 \mathbf{Z}_L$$

$$\mathbf{Z}_{1/2} = \frac{1}{4}\left(\frac{N_p}{N_s}\right)^2 \mathbf{Z}_L$$

and

$$\mathbf{Z}_{1/2} = \frac{1}{4}\mathbf{Z}_i \qquad (24.29)$$

as indicated in Fig. 24.40.

For the multiple-load transformer of Fig. 24.41, the following equations apply:

$$\frac{E_1}{E_2} = \frac{N_1}{N_2} \qquad \frac{E_1}{E_3} = \frac{N_1}{N_3} \qquad \frac{E_2}{E_3} = \frac{N_2}{N_3} \qquad (24.30)$$

The total input impedance can be determined by first noting that for the ideal transformer, the power delivered to the primary is equal to the power dissipated by the load; that is,

$$P_1 = P_{L_2} + P_{L_3}$$

and, for resistive loads ($\mathbf{Z}_1 = R_1$, $\mathbf{Z}_2 = R_2$, and $\mathbf{Z}_3 = R_3$),

$$\frac{E_1^2}{R_1} = \frac{E_2^2}{R_2} + \frac{E_3^2}{R_3}$$

or, since

$$E_2 = \frac{N_2}{N_1}E_1 \quad \text{and} \quad E_3 = \frac{N_3}{N_1}E_1$$

then

$$\frac{E_1^2}{R_1} = \frac{[(N_2/N_1)E_1]^2}{R_2} + \frac{[(N_3/N_1)E_1]^2}{R_3}$$

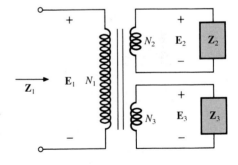

FIG. 24.41
Ideal transformer with multiple loads.

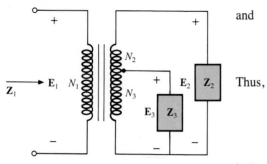

FIG. 24.42

Ideal transformer with a tapped secondary and multiple loads.

and

$$\frac{E_1^2}{R_1} = \frac{E_1^2}{(N_1/N_2)^2 R_2} + \frac{E_1^2}{(N_1/N_3)^2 R_3}$$

Thus,

$$\frac{1}{R_1} = \frac{1}{(N_1/N_2)^2 R_2} + \frac{1}{(N_1/N_3)^2 R_3} \qquad (24.31)$$

indicating that the load resistances are reflected in parallel.

For the configuration of Fig. 24.42 with E_2 and E_3 defined as shown, Eqs. (24.30) and (24.31) are applicable.

24.13 NETWORKS WITH MAGNETICALLY COUPLED COILS

For multiloop networks with magnetically coupled coils, the mesh-analysis approach is most frequently applied. A firm understanding of the dot convention discussed earlier should make the writing of the equations quite direct and free of errors. Before writing the equations for any particular loop, first determine whether the mutual term is positive or negative, keeping in mind that it will have the same sign as that for the other magnetically coupled coil. For the two-loop network of Fig. 24.43, for example, the mutual term has a positive sign since the current through each coil leaves the dot. For the primary loop,

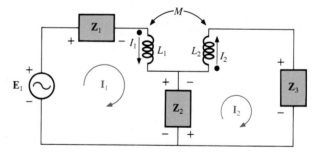

FIG. 24.43

$$\mathbf{E}_1 - \mathbf{I}_1\mathbf{Z}_1 - \mathbf{I}_1\mathbf{Z}_{L_1} - \mathbf{I}_2\mathbf{Z}_m - \mathbf{Z}_2(\mathbf{I}_1 - \mathbf{I}_2) = 0$$

where M of $\mathbf{Z}_m = \omega M \angle 90°$ is positive and

$$\mathbf{I}_1(\mathbf{Z}_1 + \mathbf{Z}_{L_1} + \mathbf{Z}_2) - \mathbf{I}_2(\mathbf{Z}_2 - \mathbf{Z}_m) = \mathbf{E}_1$$

Note in the above that the mutual impedance was treated as if it were an additional inductance in series with the inductance L_1 having a sign determined by the dot convention and the voltage across which is determined by the current in the magnetically coupled loop.

For the secondary loop,

$$-\mathbf{Z}_2(\mathbf{I}_2 - \mathbf{I}_1) - \mathbf{I}_2\mathbf{Z}_{L_2} - \mathbf{I}_1\mathbf{Z}_m - \mathbf{I}_2\mathbf{Z}_3 = 0$$

or

$$\mathbf{I}_2(\mathbf{Z}_2 + \mathbf{Z}_{L_2} + \mathbf{Z}_3) - \mathbf{I}_1(\mathbf{Z}_2 - \mathbf{Z}_m) = 0$$

For the network of Fig. 24.44, we find a mutual term between L_1 and L_2 and L_1 and L_3 labeled M_{12} and M_{13}, respectively.

For the coils with the dots (L_1 and L_3), since each current through the coils leaves the dot, M_{13} is positive for the chosen direction of I_1 and I_3. However, since the current I_1 leaves the dot through L_1, and I_2 enters the dot through coil L_2, M_{12} is negative. Consequently, for the input circuit,

$$\mathbf{E}_1 - \mathbf{I}_1\mathbf{Z}_1 - \mathbf{I}_1\mathbf{Z}_{L_1} - \mathbf{I}_2(-\mathbf{Z}_{m_{12}}) - \mathbf{I}_3\mathbf{Z}_{m_{13}} = 0$$

or

$$\mathbf{E}_1 - \mathbf{I}_1(\mathbf{Z}_1 + \mathbf{Z}_{L_1}) + \mathbf{I}_2\mathbf{Z}_{m_{12}} - \mathbf{I}_3\mathbf{Z}_{m_{13}} = 0$$

For loop 2,

$$-\mathbf{I}_2\mathbf{Z}_2 - \mathbf{I}_2\mathbf{Z}_{L_2} - \mathbf{I}_1(-\mathbf{Z}_{m_{12}}) = 0$$
$$-\mathbf{I}_1\mathbf{Z}_{m_{12}} + \mathbf{I}_2(\mathbf{Z}_2 + \mathbf{Z}_{L_2}) = 0$$

and for loop 3,

$$-\mathbf{I}_3\mathbf{Z}_3 - \mathbf{I}_3\mathbf{Z}_{L_3} - \mathbf{I}_1\mathbf{Z}_{m_{13}} = 0$$

or

$$\mathbf{I}_1\mathbf{Z}_{m_{13}} + \mathbf{I}_3(\mathbf{Z}_3 + \mathbf{Z}_{L_3}) = 0$$

In determinant form,

$$
\begin{matrix}
\mathbf{I}_1(\mathbf{Z}_1 + \mathbf{Z}_{L_1}) - \mathbf{I}_2\mathbf{Z}_{m_{12}} & + \mathbf{I}_3\mathbf{Z}_{m_{13}} & = \mathbf{E}_1 \\
-\mathbf{I}_1\mathbf{Z}_{m_{12}} & + \mathbf{I}_2(\mathbf{Z}_2 + \mathbf{Z}_{L_2}) + 0 & = 0 \\
\mathbf{I}_1\mathbf{Z}_{m_{13}} & + 0 & + \mathbf{I}_3(\mathbf{Z}_3 + \mathbf{Z}_{13}) = 0
\end{matrix}
$$

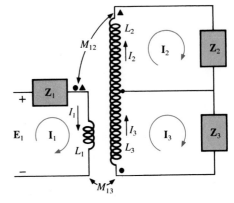

FIG. 24.44

PROBLEMS

Section 24.2

1. For the air-core transformer of Fig. 24.45:
   a. Find the value of L_s if the mutual inductance M is equal to 80 mH.
   b. Find the induced voltages e_p and e_s if the flux linking the primary coil changes at the rate of 0.08 Wb/s.
   c. Find the induced voltages e_p and e_s if the current i_p changes at the rate of 0.3 A/ms.

2. a. Repeat Problem 1 if k is changed to 1.
   b. Repeat Problem 1 if k is changed to 0.2.
   c. Compare the results of parts (a) and (b).

3. Repeat Problem 1 for $k = 0.9$, $N_p = 300$ turns, and $N_s = 25$ turns.

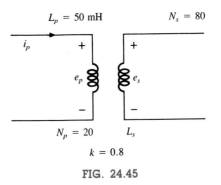

FIG. 24.45

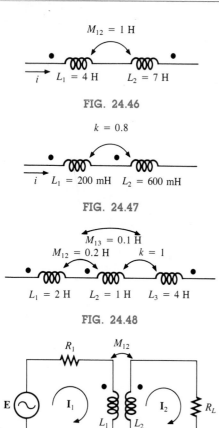

$M_{12} = 1$ H

i $L_1 = 4$ H $L_2 = 7$ H

FIG. 24.46

$k = 0.8$

i $L_1 = 200$ mH $L_2 = 600$ mH

FIG. 24.47

$M_{13} = 0.1$ H

$M_{12} = 0.2$ H $k = 1$

$L_1 = 2$ H $L_2 = 1$ H $L_3 = 4$ H

FIG. 24.48

R_1 M_{12}

E I_1 I_2 R_L

L_1 L_2

FIG. 24.49

Section 24.3

4. Determine the total inductance of the series coils of Fig. 24.46.

5. Determine the total inductance of the series coils of Fig. 24.47.

6. Determine the total inductance of the series coils of Fig. 24.48.

7. Write the mesh equations for the network of Fig. 24.49.

Section 24.4

8. For the iron-core transformer ($k = 1$) of Fig. 24.50:
   a. Find the magnitude of the induced voltage E_s.
   b. Find the maximum flux Φ_m.

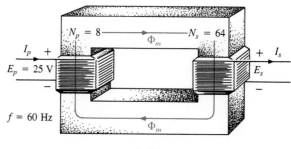

$N_p = 8$ Φ_m $N_s = 64$

I_p +   + I_s

$E_p = 25$ V E_s

−   −

$f = 60$ Hz

Φ_m

FIG. 24.50

9. Repeat Problem 8 for $N_p = 240$ and $N_s = 30$.

10. Find the applied voltage of an iron-core transformer with a secondary voltage of 240 and $N_p = 60$ with $N_s = 720$.

11. If the maximum flux passing through the core of Problem 8 is 12.5 mWb, find the frequency of the input voltage.

Section 24.5

12. For the iron-core transformer of Fig. 24.51:
    a. Find the magnitude of the current I_L and the voltage V_L if $a = 1/5$, $I_p = 2$ A, and $Z_L = 2$-Ω resistor.
    b. Find the input resistance for the data specified in part (a).

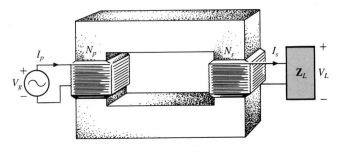

FIG. 24.51

13. Find the input impedance for the iron-core transformer of Fig. 24.51 if $a = 2$, $I_p = 4$ A, and $V_g = 1600$ V.

14. Find the voltage V_g and the current I_p if the input impedance of the iron-core transformer of Fig. 24.51 is 4 Ω and $V_L = 1200$ V with $a = 1/4$.

15. If $V_L = 240$ V, $Z_L = 20$-Ω resistor, $I_p = 0.05$ A, and $N_s = 50$, find the number of turns in the primary circuit of the iron-core transformer of Fig. 24.51.

16. **a.** If $N_p = 400$, $N_s = 1200$, and $V_g = 100$ V, find the magnitude of I_p for the iron-core transformer of Fig. 24.51 if $Z_L = 9 + j12$.
    b. Find the magnitude of the voltage V_L and the current I_L for the conditions of part (a).

17. **a.** For the circuit of Fig. 24.52, find the transformation ratio required to deliver maximum power to the speaker.
    b. Find the maximum power delivered to the speaker.

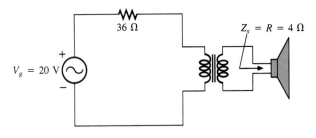

FIG. 24.52

Section 24.6

18. For the transformer of Fig. 24.53, determine
   a. the equivalent resistance R_e.
   b. the equivalent reactance X_e.
   c. the equivalent circuit reflected to the primary.
   d. the primary current for $V_g = 50 \angle 0°$.
   e. the load voltage V_L.
   f. the phasor diagram of the reflected primary circuit.
   g. the new load voltage if we assume the transformer to be ideal with a 4:1 turns ratio. Compare the result with that of part (e).

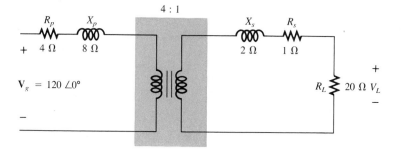

Ideal transformer

FIG. 24.53

19. For the transformer of Fig. 24.53, if the resistive load is replaced by an inductive reactance of $20\,\Omega$:
   a. Determine the total reflected primary impedance.
   b. Calculate the primary current.
   c. Determine the voltage across R_e, X_e, and the reflected load.
   d. Draw the phasor diagram.

20. Repeat Problem 19 for a capacitive load having a reactance of $20\,\Omega$.

Section 24.7

21. Discuss in your own words the frequency characteristics of the transformer. Employ the applicable equivalent circuit and frequency characteristics appearing in this chapter.

Section 24.8

22. Determine the input impedance to the air-core transformer of Fig. 24.54. Sketch the reflected primary network.

Section 24.10

23. An ideal transformer is rated 10 kVA, 2400/120 V, 60 Hz.
   a. Find the transformation ratio if the 120 V is the secondary voltage.
   b. Find the current rating of the secondary if the 120 V is the secondary voltage.

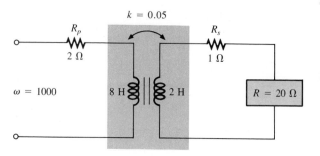

FIG. 24.54

c. Find the current rating of the primary if the 120 V is the secondary voltage.

d. Repeat parts (a) through (c) if the 2400 V is the secondary voltage.

Section 24.11

24. Determine the primary and secondary voltages and currents for the autotransformer of Fig. 24.55.

Section 24.12

25. For the center-tapped transformer of Fig. 24.40 where $N_p = 100$, $N_s = 25$, $\mathbf{Z}_L = \mathbf{R} = 5 \angle 0°$, and $\mathbf{E}_p = 100 \angle 0°$:
a. Determine the load voltage and current.
b. Find the impedance $\mathbf{Z}_i$.
c. Calculate the impedance $\mathbf{Z}_{1/2}$.

26. For the multiple-load transformer of Fig. 24.41 where $N_1 = 90$, $N_2 = 15$, $N_3 = 45$, $\mathbf{Z}_2 = \mathbf{R} = 8 \angle 0°$, $\mathbf{Z}_3 = \mathbf{R}_L = 5 \angle 0°$, and $\mathbf{E}_1 = 60 \angle 0°$:
a. Determine the load voltages and currents.
b. Calculate $\mathbf{Z}_1$.

27. For the multiple-load transformer of Fig. 24.42 where $N_1 = 120$, $N_2 = 40$, $N_3 = 30$, $\mathbf{Z}_2 = \mathbf{R} = 12 \angle 0°$, $\mathbf{Z}_3 = \mathbf{R}_C = 10 \angle 0°$, and $\mathbf{E}_1 = 120 \angle 60°$:
a. Determine the load voltages and currents.
b. Calculate $\mathbf{Z}_1$.

Section 24.13

28. Write the mesh equations for the network of Fig. 24.56.

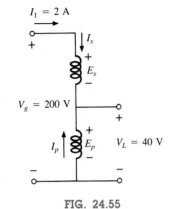

FIG. 24.55

29. Write the mesh equations for the network of Fig. 24.57.

Computer Problems

30. Write a program to provide a general solution to the problem of impedance matching as defined by Example 24.6. That is, given the speaker impedance and the internal resistance of the source, determine the required turns ratio and the power delivered to the speaker. In addition, calculate the load and source current and the primary and secondary voltages. The source voltage will have to be provided with the other parameters of the network.

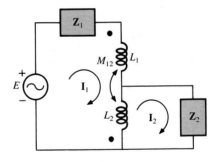

FIG. 24.56

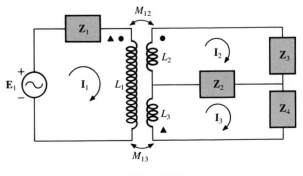

FIG. 24.57

31. Given the equivalent model of an iron-core transformer appearing in Fig. 24.20, write a program to calculate the magnitude of the voltage $\mathbf{V}_g$.

32. Given the parameters of Example 24.9, write a program to calculate the input impedance in polar form.

GLOSSARY

Autotransformer A transformer with one winding common to both the primary and the secondary circuits. A loss in isolation is balanced by the increase in its kilovolt-ampere rating.

Coefficient of coupling (k) A measure of the magnetic coupling of two coils which ranges from a minimum of zero to a maximum of 1.

Dot convention A technique for labeling the effect of the mutual inductance on a net inductance of a network or system.

Leakage flux The flux linking the coil that does not pass through the ferromagnetic path of the magnetic circuit.

Loosely coupled A term applied to two coils that have a low coefficient of coupling.

Multiple-load transformers Transformers having more than a single load connected to the secondary winding or windings.

Mutual inductance The inductance that exists between magnetically coupled coils of the same or different dimensions.

Nameplate data Information such as the kilovolt-ampere rating, voltage transformation ratio, and frequency of application that is of primary importance in choosing the proper transformer for a particular application.

Primary The coil or winding to which the source of electrical energy is normally applied.

Reflected impedance The impedance appearing at the primary of a transformer due to a load connected to the secondary. Its magnitude is controlled directly by the transformation ratio.

Secondary The coil or winding to which the load is normally applied.

Step-down transformer A transformer whose secondary voltage is less than its primary voltage. The transformation ratio a is greater than 1.

Step-up transformer A transformer whose secondary voltage is greater than its primary voltage. The magnitude of the transformation ratio a is less than 1.

Tapped transformer A transformer having an additional connection between the terminals of the primary or secondary windings.

Transformation ratio (a) The ratio of primary to secondary turns of a transformer.

25

Two-Port Parameters (z, y, h)

25.1 INTRODUCTION

In the broad spectrum of courses to follow, there will be an increasing need for the ability to *model* both devices and systems and employ these models in the analysis and synthesis of combined and enlarged systems.

This chapter will be limited to the configuration most frequently subject to the modeling technique: the two-port network shown in Fig. 25.1.

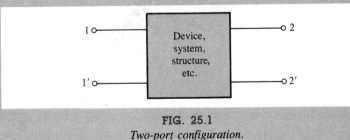

FIG. 25.1
Two-port configuration.

Note that in Fig. 25.1 there are two ports of entry or interest, each having a pair of terminals. For some devices, the two-port configuration of Fig. 25.1 may appear as shown in Fig. 25.2(a). The block diagram of Fig. 25.2(a) simply indicates that terminals 1′ and 2′ are in common—a particular case of the general two-port network. A single-port and a multiport network appear in Fig. 25.2(b). The former has been analyzed throughout the text, while the characteristics of the latter will be left for a more advanced course.

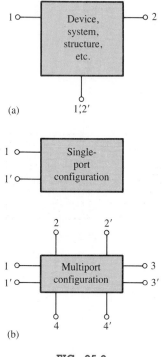

(a)

(b)

FIG. 25.2

The primary purpose of this chapter is to develop a set of equations (and, subsequently, networks) that will allow us to model the device or system appearing within the enclosed structure of Fig. 25.1. That is, we will be able to establish a network that will display the same terminal characteristics as those of the original system, device, and so on. In Fig. 25.3, for example, a transistor appears between the four external terminals. Through the analysis to follow, we will find a combination of network elements that will allow us to replace the transistor by a network that will behave very much like the original device for a specific set of operating conditions. Methods such as mesh and nodal analysis can then be applied to determine any unknown quantities. The models, when reduced to their simplest forms as determined by the operating conditions, can also provide very quick estimates of network behavior without a lengthy mathematical derivation. In other words, someone well-versed in the use of models can analyze the operation of large, complex systems in short order. The results may be only approximate in most cases, but this quick return for a minimum of effort is often worthwhile.

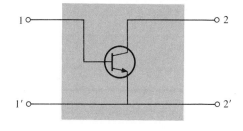

FIG. 25.3

The following analysis may initially appear very mathematical and devoid of any practical meaning. However, you will note that the parameters are determined by the ratio of electrical quantities under very specific network conditions. The resulting quantities also have a terminology that will be applied very frequently in the electronics course to follow. The analysis of this chapter will be limited to linear (fixed-value) systems with bilateral elements. Three sets of parameters will be developed for the two-port configuration, referred to as the *impedance* ($\mathbf{z}$), *admittance* ($\mathbf{y}$), and *hybrid* ($\mathbf{h}$) parameters. A table will be provided at the end of the chapter relating the three sets of parameters.

25.2 IMPEDANCE (z) PARAMETERS

For the two-port configuration of Fig. 25.4, four variables are specified. For most situations, if any two are specified, the remaining two variables can be determined. The four variables can be related by the following equations:

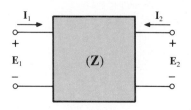

FIG. 25.4
*Two-port impedance parameter
configuration.*

$$\boxed{\mathbf{E}_1 = \mathbf{z}_{11}\mathbf{I}_1 + \mathbf{z}_{12}\mathbf{I}_2}$$

(25.1a)

$$\boxed{E_2 = z_{21}I_1 + z_{22}I_2} \qquad (25.1b)$$

The *impedance parameters* z_{11}, z_{12}, and z_{22} are measured in ohms.

To model the system, each impedance parameter must be determined. They are determined by setting a particular variable to zero.

z_{11}

For z_{11}, if I_2 is set to zero, as shown in Fig. 25.5, Eq. (25.1a) becomes

$$E_1 = z_{11}I_1 + z_{12}(0)$$

and

$$\boxed{z_{11} = \frac{E_1}{I_1}}\Big|_{I_2 = 0} \qquad (\text{ohms, } \Omega) \qquad (25.2)$$

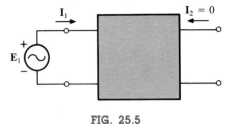

FIG. 25.5
z_{11} *determination.*

Equation (25.2) dictates that with I_2 set to zero, the impedance parameter is determined by the resulting ratio of E_1 to I_1. Since E_1 and I_1 are both input quantities, with I_2 set to zero, the parameter z_{11} is formally referred to in the following manner:

$z_{11} = $ *open-circuit, input-impedance parameter*

z_{12}

For z_{12}, I_1 is set to zero, and Eq. (25.1a) results in

$$\boxed{z_{12} = \frac{E_1}{I_2}}\Big|_{I_1 = 0} \qquad (\Omega) \qquad (25.3)$$

For most systems where input and output quantities are to be compared, the ratio of interest is usually that of the output quantity divided by the input quantity. In this case, the *reverse* is true, resulting in the following:

$z_{12} = $ *open-circuit, reverse-transfer impedance parameter*

The term *transfer* is included to indicate that z_{12} will relate an input and output quantity (for the condition $I_1 = 0$). The network configuration for determining z_{12} is shown in Fig. 25.6.

For an applied source E_2, the ratio E_1/I_2 will determine z_{12} with I_1 set to zero.

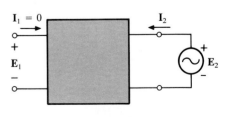

FIG. 25.6
z_{12} *determination.*

z_{21}

To determine z_{21}, set I_2 to zero and find the ratio E_2/I_1 as determined by Eq. (25.1b). That is,

$$\boxed{z_{21} = \frac{E_2}{I_1}}\Big|_{I_2 = 0} \qquad (\Omega) \qquad (25.4)$$

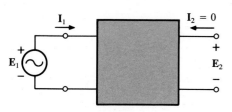

FIG. 25.7
z_{21} *determination.*

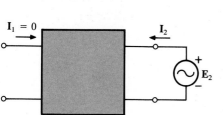

FIG. 25.8
z_{22} *determination.*

In this case, input and output quantities are again the determining variables, requiring the term *transfer* in the nomenclature. However, the ratio is that of an output to an input quantity, so the descriptive term *forward* is applied, and

$\quad z_{21} = $ *open-circuit, forward-transfer impedance parameter*

The determining network is shown in Fig. 25.7. For an applied voltage $\mathbf{E}_1$, it is determined by the ratio $\mathbf{E}_2/\mathbf{I}_1$ with $\mathbf{I}_2$ set to zero.

z_{22}

The remaining parameter, z_{22}, is determined by

$$\boxed{z_{22} = \frac{\mathbf{E}_2}{\mathbf{I}_2}\Bigg|_{I_1 = 0}} \qquad (\Omega) \qquad \textbf{(25.5)}$$

as derived from Eq. (25.1b) with $\mathbf{I}_1$ set to zero. Since it is the ratio of the output voltage to the output current with $\mathbf{I}_1$ set to zero, it has the terminology

$\quad z_{22} = $ *open-circuit, output-impedance parameter*

The required network is shown in Fig. 25.8. For an applied voltage $\mathbf{E}_2$, it is determined by the resulting ratio $\mathbf{E}_2/\mathbf{I}_2$ with $\mathbf{I}_1 = 0$.

EXAMPLE 25.1. Determine the impedance (**z**) parameters for the T network of Fig. 25.9.

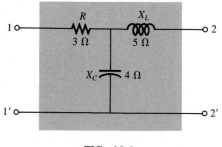

FIG. 25.9
T configuration.

Solution: For z_{11}, the network will appear as shown in Fig. 25.10, with $\mathbf{Z}_1 = 3 \angle 0°$, $\mathbf{Z}_2 = 5 \angle 90°$, and $\mathbf{Z}_3 = 4 \angle -90°$:

$$\mathbf{I}_1 = \frac{\mathbf{E}_1}{\mathbf{Z}_1 + \mathbf{Z}_3}$$

Thus,

$$z_{11} = \frac{\mathbf{E}_1}{\mathbf{I}_1}\Bigg|_{I_2 = 0}$$

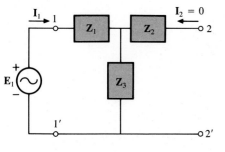

FIG. 25.10

and

$$\boxed{\mathbf{z}_{11} = \mathbf{Z}_1 + \mathbf{Z}_3} \qquad (25.6)$$

For $\mathbf{z}_{12}$, the network will appear as shown in Fig. 25.11:

$$\mathbf{E}_1 = \mathbf{I}_2\mathbf{Z}_3$$

Thus,

$$\mathbf{z}_{12} = \frac{\mathbf{E}_1}{\mathbf{I}_2}\bigg|_{\mathbf{I}_1 = 0} = \frac{\mathbf{I}_2\mathbf{Z}_3}{\mathbf{I}_2}$$

and

$$\boxed{\mathbf{z}_{12} = \mathbf{Z}_3} \qquad (25.7)$$

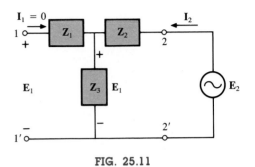

FIG. 25.11

For $\mathbf{z}_{21}$, the required network appears in Fig. 25.12:

$$\mathbf{E}_2 = \mathbf{I}_1\mathbf{Z}_3$$

Thus,

$$\mathbf{z}_{21} = \frac{\mathbf{E}_2}{\mathbf{I}_1}\bigg|_{\mathbf{I}_2 = 0} = \frac{\mathbf{I}_1\mathbf{Z}_3}{\mathbf{I}_1}$$

and

$$\boxed{\mathbf{z}_{21} = \mathbf{Z}_3} \qquad (25.8)$$

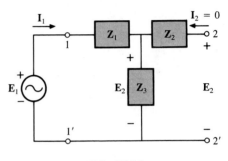

FIG. 25.12

For $\mathbf{z}_{22}$, the determining configuration is shown in Fig. 25.13:

$$\mathbf{I}_2 = \frac{\mathbf{E}_2}{\mathbf{Z}_2 + \mathbf{Z}_3}$$

Thus,

$$\mathbf{z}_{22} = \frac{\mathbf{E}_2}{\mathbf{I}_2}\bigg|_{\mathbf{I}_1 = 0} = \frac{\mathbf{I}_2(\mathbf{Z}_2 + \mathbf{Z}_3)}{\mathbf{I}_2}$$

and

$$\boxed{\mathbf{z}_{22} = \mathbf{Z}_2 + \mathbf{Z}_3} \qquad (25.9)$$

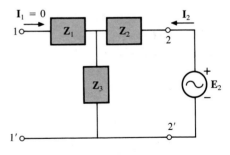

FIG. 25.13

Note that for the T configuration, $\mathbf{z}_{12} = \mathbf{z}_{21}$. For $\mathbf{Z}_1 = 3 \angle 0°$, $\mathbf{Z}_2 = 5 \angle 90°$, and $\mathbf{Z}_3 = 4 \angle -90°$, we have

$$\mathbf{z}_{11} = \mathbf{Z}_1 + \mathbf{Z}_3 = \mathbf{3} - j\mathbf{4}$$
$$\mathbf{z}_{12} = \mathbf{z}_{21} = \mathbf{Z}_3 = 4 \angle -90° = -j\mathbf{4}$$
$$\mathbf{z}_{22} = \mathbf{Z}_2 + \mathbf{Z}_3 = 5 \angle 90° + 4 \angle -90° = 1 \angle 90° = j\mathbf{1}$$

For a set of impedance parameters, the terminal (external) behavior of the device or network within the configuration of Fig. 25.1 is determined. An *equivalent circuit* for the system can be developed using the impedance parameters and Eqs. (25.1a) and (25.1b). Two possibilities for the impedance parameters appear in Fig. 25.14.

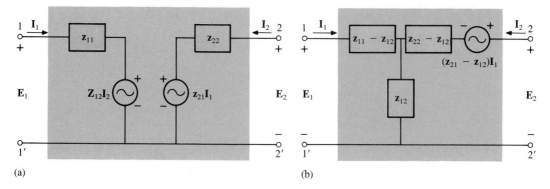

FIG. 25.14

Two-port, z-parameter equivalent networks.

Applying Kirchhoff's voltage law to the input and output loops of the network of Fig. 25.14(a) results in

$$\mathbf{E}_1 - \mathbf{z}_{11}\mathbf{I}_1 - \mathbf{z}_{12}\mathbf{I}_2 = 0$$

and

$$\mathbf{E}_2 - \mathbf{z}_{22}\mathbf{I}_2 - \mathbf{z}_{21}\mathbf{I}_1 = 0$$

which, when rearranged, become

$$\mathbf{E}_1 = \mathbf{z}_{11}\mathbf{I}_1 + \mathbf{z}_{12}\mathbf{I}_2$$
$$\mathbf{E}_2 = \mathbf{z}_{21}\mathbf{I}_1 + \mathbf{z}_{22}\mathbf{I}_2$$

corresponding exactly to Eqs. (25.1a) and (25.1b).

For the network of Fig. 25.14(b),

$$\mathbf{E}_1 - \mathbf{I}_1(\mathbf{z}_{11} - \mathbf{z}_{12}) - \mathbf{z}_{12}(\mathbf{I}_1 + \mathbf{I}_2) = 0$$

and

$$\mathbf{E}_2 - \mathbf{I}_1(\mathbf{z}_{21} - \mathbf{z}_{12}) - \mathbf{I}_2(\mathbf{z}_{22} - \mathbf{z}_{12}) - \mathbf{z}_{12}(\mathbf{I}_1 + \mathbf{I}_2) = 0$$

which, when rearranged, are

$$\mathbf{E}_1 = \mathbf{I}_1(\mathbf{z}_{11} - \mathbf{z}_{12} + \mathbf{z}_{12}) + \mathbf{I}_2\mathbf{z}_{12}$$
$$\mathbf{E}_2 = \mathbf{I}_1(\mathbf{z}_{21} - \mathbf{z}_{12} + \mathbf{z}_{12}) + \mathbf{I}_2(\mathbf{z}_{22} - \mathbf{z}_{12} + \mathbf{z}_{12})$$

and

$$\mathbf{E}_1 = \mathbf{z}_{11}\mathbf{I}_1 + \mathbf{z}_{12}\mathbf{I}_2$$
$$\mathbf{E}_2 = \mathbf{z}_{21}\mathbf{I}_1 + \mathbf{z}_{22}\mathbf{I}_2$$

Note in each network the necessity for a current-controlled voltage source, that is, a voltage source the magnitude of which is determined by a particular current of the network.

The usefulness of the impedance parameters and the resulting equivalent networks can best be described by considering the system of Fig. 25.15(a), which contains a device (or system) for which the impedance parameters have been determined. As shown in Fig. 25.15(b), the equivalent network for the device (or system) can then be substituted, and methods such as mesh analysis, nodal analysis, and so on, can be employed to determine required unknown quantities. The device itself can then be replaced by an equivalent circuit and the desired solutions obtained more directly and with less effort than is required using only the characteristics of the device.

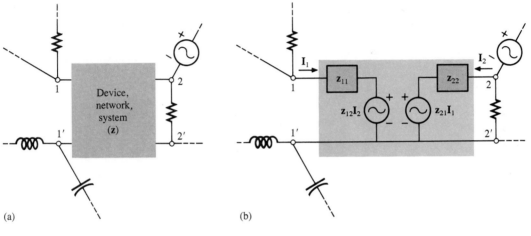

FIG. 25.15

EXAMPLE 25.2. Draw the equivalent circuit in the form shown in Fig. 25.14(a) using the impedance parameters determined in Example 25.1.

Solution: The circuit appears in Fig. 25.16.

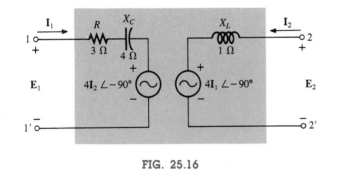

FIG. 25.16

25.3 ADMITTANCE (y) PARAMETERS

The equations relating the four terminal variables of Fig. 25.1 can also be written in the following form:

$$\boxed{I_1 = y_{11}E_1 + y_{12}E_2} \qquad \textbf{(25.10a)}$$

$$\boxed{I_2 = y_{21}E_1 + y_{22}E_2} \qquad \textbf{(25.10b)}$$

Note that in this case each term of each equation has the units of current, as compared to voltage for each term of Eqs. (25.1a) and (25.1b). In addition, the unit of each coefficient is siemens, compared with the ohm for the impedance parameters.

The impedance parameters were determined by setting a particular current to zero through an open-circuit condition. For the *admittance parameters* of Eqs. (25.10a) and (25.10b), a voltage is set to zero through a short-circuit condition.

The terminology applied to each of the admittance parameters follows directly from the descriptive terms applied to each of the impedance parameters. The equations for each are determined directly from Eqs. (25.10a) and (25.10b) by setting a particular voltage to zero.

$\mathbf{Y}_{11}$

$$\boxed{y_{11} = \dfrac{I_1}{E_1}}_{\;E_2 = 0} \qquad \text{(siemens, S)} \qquad \textbf{(25.11)}$$

y_{11} = *short-circuit, input-admittance parameter*

The determining network appears in Fig. 25.17.

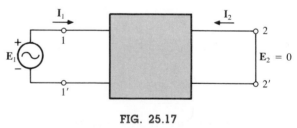

FIG. 25.17

y_{11} *determination.*

$\mathbf{Y}_{12}$

$$\boxed{y_{12} = \dfrac{I_1}{E_2}}_{\;E_1 = 0} \qquad \text{(S)} \qquad \textbf{(25.12)}$$

y_{12} = *short-circuit, reverse-transfer admittance parameter*

The network for determining y_{12} appears in Fig. 25.18.

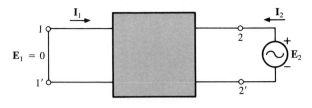

FIG. 25.18
y_{12} *determination*.

$\mathbf{Y}_{21}$

$$\boxed{\mathbf{y}_{21} = \frac{\mathbf{I}_2}{\mathbf{E}_1}\Bigg|_{\mathbf{E}_2 = 0}} \qquad \text{(S)} \qquad \qquad \textbf{(25.13)}$$

$\mathbf{y}_{21}$ = *short-circuit, forward-transfer admittance parameter*

The network for determining $\mathbf{y}_{21}$ appears in Fig. 25.19.

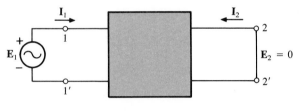

FIG. 25.19
y_{21} *determination*.

$\mathbf{Y}_{22}$

$$\boxed{\mathbf{y}_{22} = \frac{\mathbf{I}_2}{\mathbf{E}_2}\Bigg|_{\mathbf{E}_1 = 0}} \qquad \text{(S)} \qquad \qquad \textbf{(25.14)}$$

$\mathbf{y}_{22}$ = *short-circuit, output-admittance parameter*

The required network appears in Fig. 25.20.

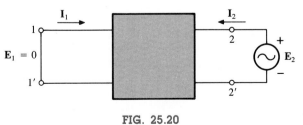

FIG. 25.20
y_{22} *determination*.

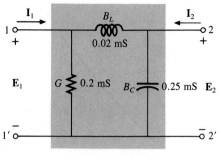

FIG. 25.21
π network.

EXAMPLE 25.3. Determine the admittance parameters for the π network of Fig. 25.21.

Solution: The network for $\mathbf{y}_{11}$ will appear as shown in Fig. 25.22,

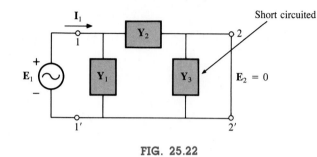

FIG. 25.22

with

$$\mathbf{Y}_1 = 0.2 \text{ mS } \angle 0° \qquad \mathbf{Y}_2 = 0.02 \text{ mS } \angle -90° \qquad \mathbf{Y}_3 = 0.25 \text{ mS } \angle 90°$$

We use

$$\mathbf{I}_1 = \mathbf{E}_1 \mathbf{Y}_T = \mathbf{E}_1 (\mathbf{Y}_1 + \mathbf{Y}_2)$$

Thus,

$$\mathbf{y}_{11} = \left. \frac{\mathbf{I}_1}{\mathbf{E}_1} \right|_{\mathbf{E}_2 = 0}$$

and

$$\boxed{\mathbf{y}_{11} = \mathbf{Y}_1 + \mathbf{Y}_2} \tag{25.15}$$

The determining network for $\mathbf{y}_{12}$ appears in Fig. 25.23. $\mathbf{Y}_1$ is short circuited; so $\mathbf{I}_{\mathbf{Y}_2} = \mathbf{I}_1$, and

$$\mathbf{I}_{\mathbf{Y}_2} = \mathbf{I}_1 = -\mathbf{E}_2 \mathbf{Y}_2$$

The minus sign results because the defined direction of $\mathbf{I}_1$ in Fig. 25.23 is opposite to the actual flow direction due to the applied source $\mathbf{E}_2$; and

$$\mathbf{y}_{12} = \left. \frac{\mathbf{I}_1}{\mathbf{E}_2} \right|_{\mathbf{E}_1 = 0}$$

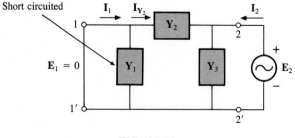

FIG. 25.23

and

$$\boxed{\mathbf{y}_{12} = -\mathbf{Y}_2} \qquad (25.16)$$

The network employed for $\mathbf{y}_{21}$ appears in Fig. 25.24. In this case, $\mathbf{Y}_3$ is short circuited, resulting in

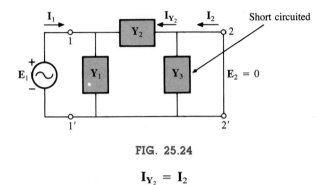

FIG. 25.24

$$\mathbf{I}_{\mathbf{Y}_2} = \mathbf{I}_2$$

and

$$\mathbf{I}_{\mathbf{Y}_2} = \mathbf{I}_2 = -\mathbf{E}_1\mathbf{Y}_2$$

Thus,

$$\mathbf{y}_{21} = \left.\frac{\mathbf{I}_2}{\mathbf{E}_1}\right|_{\mathbf{E}_2 = 0}$$

and

$$\boxed{\mathbf{y}_{21} = -\mathbf{Y}_2} \qquad (25.17)$$

Note that for the π configuration, $\mathbf{y}_{12} = \mathbf{y}_{21}$, which was expected, since the impedance parameters for the T network were such that $\mathbf{z}_{12} = \mathbf{z}_{21}$. A T network can be converted directly to a π network using the Y-Δ transformation.

The determining network for $\mathbf{y}_{22}$ appears in Fig. 25.25:

$$\mathbf{Y}_T = \mathbf{Y}_2 + \mathbf{Y}_3$$

and

$$\mathbf{I}_2 = \mathbf{E}_2(\mathbf{Y}_2 + \mathbf{Y}_3)$$

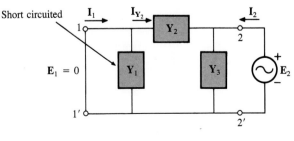

FIG. 25.25

Thus,

$$\mathbf{y}_{22} = \left.\frac{\mathbf{I}_2}{\mathbf{E}_2}\right|_{\mathbf{E}_1 = 0}$$

and

$$\boxed{\mathbf{y}_{22} = \mathbf{Y}_2 + \mathbf{Y}_3} \qquad \textbf{(25.18)}$$

Substituting values, we have

$$\mathbf{Y}_1 = 0.2 \times 10^{-3} \angle 0°$$
$$\mathbf{Y}_2 = 0.02 \times 10^{-3} \angle -90°$$
$$\mathbf{Y}_3 = 0.25 \times 10^{-3} \angle 90°$$

$$\mathbf{y}_{11} = \mathbf{Y}_1 + \mathbf{Y}_2$$
$$= \mathbf{0.2 \times 10^{-3} - j0.02 \times 10^{-3}} \textbf{ (ind.)}$$

$$\mathbf{y}_{12} = \mathbf{y}_{21} = -\mathbf{Y}_2 = -(-j0.02 \times 10^{-3})$$
$$= \mathbf{j0.02 \times 10^{-3}} \textbf{ (cap.)}$$

$$\mathbf{y}_{22} = \mathbf{Y}_2 + \mathbf{Y}_3 = -j0.02 \times 10^{-3} + j0.25 \times 10^{-3}$$
$$= \mathbf{j0.23 \times 10^{-3}} \textbf{ (cap.)}$$

Note the similarities between the results for $\mathbf{y}_{11}$ and $\mathbf{y}_{22}$ for the π network compared with $\mathbf{z}_{11}$ and $\mathbf{z}_{22}$ for the T network.

Two networks satisfying the terminal relationships of Eqs. (25.10a) and (25.10b) are shown in Fig. 25.26. Note the use of parallel branches, since each term of Eqs. (25.10a) and (25.10b) has the units of current, and the most direct route to the equivalent circuit is an application of Kirchhoff's current law in reverse. That is, find the network that satisfies Kirchhoff's current law relationship. For the impedance parameters, each term had the units of volts, so Kirchhoff's voltage law was applied in reverse to determine the series combination of elements in the equivalent circuit of Fig. 25.14(a).

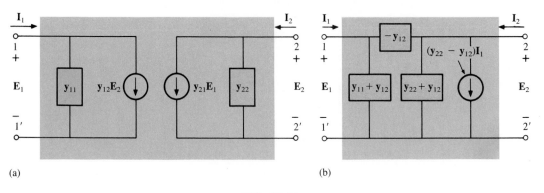

(a)   (b)

FIG. 25.26
*Two-port, **y**-parameter equivalent networks.*

Applying Kirchhoff's current law to the network of Fig. 25.26(b), we have

$$\text{Node } a: \overbrace{\mathbf{I}_1}^{\text{entering}} = \overbrace{\mathbf{y}_{11}\mathbf{E}_1 - \mathbf{y}_{12}\mathbf{E}_2}^{\text{leaving}}$$

$$\text{Node } b: \mathbf{I}_2 = \mathbf{y}_{22}\mathbf{E}_2 - \mathbf{y}_{21}\mathbf{E}_1$$

which, when rearranged, are Eqs. (25.10a) and (25.10b).

25.4 HYBRID (h) PARAMETERS

The *hybrid* (**h**) *parameters* are employed extensively in the analysis of transistor networks. The term *hybrid* is derived from the fact that the parameters have a mixture of units (a hybrid set) rather than a single unit of measurement such as ohms or siemens, used for the **z** and **y** parameters, respectively. The defining hybrid equations have a mixture of current *and* voltage variables on one side, as follows:

$$\boxed{\mathbf{E}_1 = \mathbf{h}_{11}\mathbf{I}_1 + \mathbf{h}_{12}\mathbf{E}_2} \qquad \textbf{(25.19a)}$$

$$\boxed{\mathbf{I}_2 = \mathbf{h}_{21}\mathbf{I}_1 + \mathbf{h}_{22}\mathbf{E}_2} \qquad \textbf{(25.19b)}$$

To determine the hybrid parameters, it will be necessary to establish both the short-circuit and the open-circuit condition, depending on the parameter desired.

$\mathbf{h}_{11}$

$$\boxed{\mathbf{h}_{11} = \frac{\mathbf{E}_1}{\mathbf{I}_1}}_{\mathbf{E}_2 = 0} \qquad (\Omega) \qquad \textbf{(25.20)}$$

$\mathbf{h}_{11}$ = *short-circuit, input-impedance parameter*

The determining network is shown in Fig. 25.27.

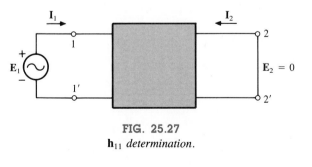

FIG. 25.27
$\mathbf{h}_{11}$ *determination.*

$\mathbf{h}_{12}$

$$\boxed{\mathbf{h}_{12} = \frac{\mathbf{E}_1}{\mathbf{E}_2}}_{\mathbf{I}_2 = 0} \qquad \text{(dimensionless)} \qquad \textbf{(25.21)}$$

$\mathbf{h}_{12}$ = *open-circuit, **reverse**-transfer voltage ratio parameter*

The network employed in determining $\mathbf{h}_{12}$ is shown in Fig. 25.28.

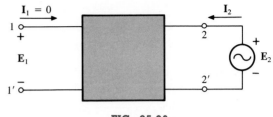

FIG. 25.28
$\mathbf{h}_{12}$ *determination.*

$\mathbf{h}_{21}$

$$\boxed{\mathbf{h}_{21} = \frac{I_2}{I_1}}_{E_2 = 0} \qquad \text{(dimensionless)} \qquad \textbf{(25.22)}$$

$\mathbf{h}_{21}$ = *short-circuit, **forward**-transfer current ratio parameter*

The determining network appears in Fig. 25.29.

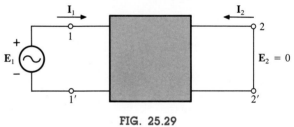

FIG. 25.29
$\mathbf{h}_{21}$ *determination.*

$\mathbf{h}_{22}$

$$\boxed{\mathbf{h}_{22} = \frac{I_2}{E_2}}_{I_1 = 0} \qquad \text{(S)} \qquad \textbf{(25.23)}$$

$\mathbf{h}_{22}$ = *open-circuit, **output** admittance parameter*

The network employed to determine $\mathbf{h}_{22}$ is shown in Fig. 25.30.

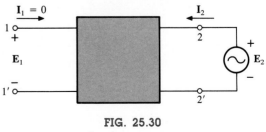

FIG. 25.30
$\mathbf{h}_{22}$ *determination.*

The subscript notation for the hybrid parameters is reduced to the following for most applications. The letter chosen is that letter appearing in boldface in the preceding description of each parameter:

$$\mathbf{h}_{11} = \mathbf{h}_i \qquad \mathbf{h}_{12} = \mathbf{h}_r \qquad \mathbf{h}_{21} = \mathbf{h}_f \qquad \mathbf{h}_{22} = \mathbf{h}_o$$

The hybrid equivalent circuit appears in Fig. 25.31. Since the unit of measurement for each term of Eq. (25.19a) is the volt, Kirchhoff's voltage law was applied in reverse to obtain the series input circuit indicated. The unit of measurement of each term of Eq. (25.19b) has the units of current, resulting in the parallel elements of the output circuit as obtained by applying Kirchhoff's current law in reverse.

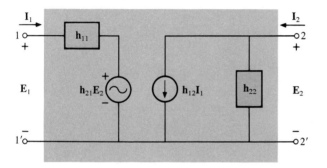

FIG. 25.31

Two-port, hybrid-parameter equivalent network.

Note that the input circuit has a voltage-controlled voltage source whose controlling voltage is the output terminal voltage, while the output circuit has a current-controlled current source whose controlling current is the current of the input circuit.

EXAMPLE 25.4.   For the hybrid equivalent circuit of Fig. 25.32:

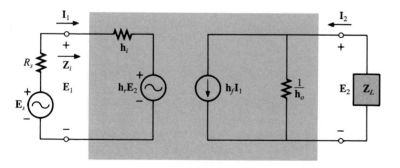

FIG. 25.32

a. Determine the current ratio (gain) $\mathbf{I}_2/\mathbf{I}_1$.
b. Determine the voltage ratio (gain) $\mathbf{E}_2/\mathbf{E}_1$.

Solutions:

a. Using the current divider rule, we have

$$\mathbf{I}_2 = \frac{(1/\mathbf{h}_o)\mathbf{h}_f\mathbf{I}_1}{(1/\mathbf{h}_o) + \mathbf{Z}_L} = \frac{\mathbf{h}_f\mathbf{I}_1}{1 + \mathbf{h}_o\mathbf{Z}_L}$$

and

$$\boxed{A_i = \frac{\mathbf{I}_2}{\mathbf{I}_1} = \frac{\mathbf{h}_f}{1 + \mathbf{h}_o\mathbf{Z}_L}} \qquad (25.24)$$

b. Applying Kirchhoff's voltage law to the input circuit gives us

$$\mathbf{E}_1 - \mathbf{h}_i\mathbf{I}_1 - \mathbf{h}_r\mathbf{E}_2 = 0$$

and

$$\mathbf{I}_1 = \frac{\mathbf{E}_1 - \mathbf{h}_r\mathbf{E}_2}{\mathbf{h}_i}$$

Apply Kirchhoff's current law to the output circuit:

$$\mathbf{I}_2 = \mathbf{h}_f\mathbf{I}_1 + \mathbf{h}_o\mathbf{E}_2$$

However,

$$\mathbf{I}_2 = -\frac{\mathbf{E}_2}{\mathbf{Z}_L}$$

so

$$-\frac{\mathbf{E}_2}{\mathbf{Z}_L} = \mathbf{h}_f\mathbf{I}_1 + \mathbf{h}_o\mathbf{E}_2$$

Substituting for $\mathbf{I}_1$ gives us

$$-\frac{\mathbf{E}_2}{\mathbf{Z}_L} = \mathbf{h}_f\left(\frac{\mathbf{E}_1 - \mathbf{h}_r\mathbf{E}_2}{\mathbf{h}_i}\right) + \mathbf{h}_o\mathbf{E}_2$$

or

$$\mathbf{h}_i\mathbf{E}_2 = -\mathbf{h}_f\mathbf{Z}_L\mathbf{E}_1 + \mathbf{h}_r\mathbf{h}_f\mathbf{Z}_L\mathbf{E}_2 - \mathbf{h}_i\mathbf{h}_o\mathbf{Z}_L\mathbf{E}_2$$

and

$$\mathbf{E}_2(\mathbf{h}_i - \mathbf{h}_r\mathbf{h}_f\mathbf{Z}_L + \mathbf{h}_i\mathbf{h}_o\mathbf{Z}_L) = -\mathbf{h}_f\mathbf{Z}_L\mathbf{E}_1$$

with the result that

$$\boxed{A_v = \frac{\mathbf{E}_2}{\mathbf{E}_1} = \frac{-\mathbf{h}_f\mathbf{Z}_L}{\mathbf{h}_i(1 + \mathbf{h}_o\mathbf{Z}_L) - \mathbf{h}_r\mathbf{h}_f\mathbf{Z}_L}} \qquad (25.25)$$

EXAMPLE 25.5. For a particular transistor, $\mathbf{h}_i = 1$ kΩ, $\mathbf{h}_r = 4 \times 10^{-4}$, $\mathbf{h}_f = 50$, and $\mathbf{h}_o = 25 \times 10^{-6}$ S. Determine the current and the voltage gain if $\mathbf{Z}_L$ is a 2-kΩ resistive load.

Solution:

$$A_i = \frac{\mathbf{h}_f}{1 + \mathbf{h}_o \mathbf{Z}_L} = \frac{50}{1 + (25 \times 10^{-6})(2 \times 10^3)}$$

$$A_i = \frac{50}{1 + (50 \times 10^{-3})} = \frac{50}{1.050} = \mathbf{47.62}$$

$$A_v = \frac{-\mathbf{h}_f \mathbf{Z}_L}{\mathbf{h}_i(1 + \mathbf{h}_o \mathbf{Z}_L) - \mathbf{h}_r \mathbf{h}_f \mathbf{Z}_L}$$

$$= \frac{-(50)(2 \times 10^3)}{(1 \times 10^3)(1.050) - (4 \times 10^{-4})(50)(2 \times 10^3)}$$

$$A_v = \frac{-100 \times 10^3}{(1.050 \times 10^3) - (0.04 \times 10^3)} = -\frac{100}{1.01} = \mathbf{-99}$$

The minus sign simply indicates a phase shift of 180° between $\mathbf{E}_2$ and $\mathbf{E}_1$ for the defined polarities in Fig. 25.32.

25.5 INPUT AND OUTPUT IMPEDANCES

The input and output impedances will now be determined for the hybrid equivalent circuit and a **z** parameter equivalent circuit. The input impedance can always be determined by the ratio of the input voltage to the input current with or without a load applied. The output impedance is always determined with the source voltage or current set to zero. We found in the previous section that for the hybrid equivalent circuit of Fig. 25.32,

$$\mathbf{E}_1 = \mathbf{h}_i \mathbf{I}_1 + \mathbf{h}_r \mathbf{E}_2$$
$$\mathbf{E}_2 = -\mathbf{I}_2 \mathbf{Z}_L$$

and

$$\frac{\mathbf{I}_2}{\mathbf{I}_1} = \frac{\mathbf{h}_f}{1 + \mathbf{h}_o \mathbf{Z}_L}$$

By substituting for $\mathbf{I}_2$ in the second equation (using the relationship of the last equation), we have

$$\mathbf{E}_2 = -\left(\frac{\mathbf{h}_f \mathbf{I}_1}{1 + \mathbf{h}_o \mathbf{Z}_L}\right)\mathbf{Z}_L$$

so the first equation becomes

$$\mathbf{E}_1 = \mathbf{h}_i \mathbf{I}_1 + \mathbf{h}_r\left(-\frac{\mathbf{h}_f \mathbf{I}_1 \mathbf{Z}_L}{1 + \mathbf{h}_o \mathbf{Z}_L}\right)$$

and

$$\mathbf{E}_1 = \mathbf{I}_1\left(\mathbf{h}_i - \frac{\mathbf{h}_r \mathbf{h}_f \mathbf{Z}_L}{1 + \mathbf{h}_o \mathbf{Z}_L}\right)$$

Thus,

$$\boxed{\mathbf{Z}_i = \frac{\mathbf{E}_1}{\mathbf{I}_1} = \mathbf{h}_i - \frac{\mathbf{h}_r \mathbf{h}_f \mathbf{Z}_L}{1 + \mathbf{h}_o \mathbf{Z}_L}} \qquad (25.26)$$

For the output impedance, we will set the source voltage to zero but preserve its internal resistance R_s as shown in Fig. 25.33.

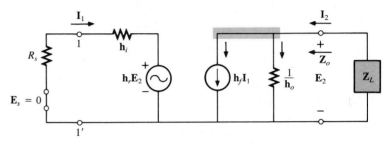

FIG. 25.33

Since

$$\mathbf{E}_s = 0$$

then

$$\mathbf{I}_1 = -\frac{\mathbf{h}_r \mathbf{E}_2}{\mathbf{h}_i + R_s}$$

From the output circuit,

$$\mathbf{I}_2 = \mathbf{h}_f \mathbf{I}_1 + \mathbf{h}_o \mathbf{E}_2$$

or

$$\mathbf{I}_2 = \mathbf{h}_f \left(-\frac{\mathbf{h}_r \mathbf{E}_2}{\mathbf{h}_i + R_s} \right) + \mathbf{h}_o \mathbf{E}_2$$

and

$$\mathbf{I}_2 = \left(-\frac{\mathbf{h}_r \mathbf{h}_f}{\mathbf{h}_i + R_s} + \mathbf{h}_o \right) \mathbf{E}_2$$

Thus,

$$\boxed{\mathbf{Z}_o = \frac{\mathbf{E}_2}{\mathbf{I}_2} = \frac{1}{\mathbf{h}_o - \dfrac{\mathbf{h}_r \mathbf{h}_f}{\mathbf{h}_i + R_s}}} \qquad (25.27)$$

EXAMPLE 25.6. Determine $\mathbf{Z}_i$ and $\mathbf{Z}_o$ for the transistor having the parameters of Example 25.5 if $R_s = 1 \text{ k}\Omega$.

Solution:

$$\mathbf{Z}_i = \mathbf{h}_i - \frac{\mathbf{h}_r \mathbf{h}_f \mathbf{Z}_L}{1 + \mathbf{h}_o \mathbf{Z}_L} = 1 \times 10^3 - \frac{0.04 \times 10^3}{1.050}$$

$$= 1 \times 10^3 - 0.0381 \times 10^3 = \mathbf{961.9 \, \Omega}$$

$$\mathbf{Z}_o = \frac{1}{\mathbf{h}_o - \dfrac{\mathbf{h}_r \mathbf{h}_f}{\mathbf{h}_i + R_s}} = \frac{1}{25 \times 10^{-6} - \dfrac{(4 \times 10^{-4})(50)}{1 \, \mathrm{k\Omega} + 1 \, \mathrm{k\Omega}}}$$

$$= \frac{1}{25 \times 10^{-6} - \dfrac{200 \times 10^{-4}}{2 \times 10^3}}$$

$$= \frac{1}{25 \times 10^{-6} - 10 \times 10^{-6}}$$

$$\mathbf{Z}_o = \frac{1}{15 \times 10^{-6}} = \mathbf{66.67 \, k\Omega}$$

For the **z** parameter equivalent circuit of Fig. 25.34,

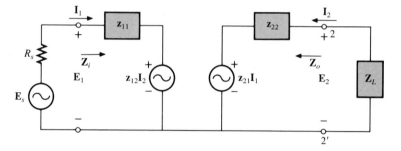

FIG. 25.34

$$\mathbf{I}_2 = \frac{-\mathbf{z}_{21} \mathbf{I}_1}{\mathbf{z}_{22} + \mathbf{Z}_L}$$

and

$$\mathbf{I}_1 = \frac{\mathbf{E}_1 - \mathbf{z}_{12} \mathbf{I}_2}{\mathbf{z}_{11}}$$

or

$$\mathbf{E}_1 = \mathbf{z}_{11} \mathbf{I}_1 + \mathbf{z}_{12} \mathbf{I}_2 = \mathbf{z}_{11} \mathbf{I}_1 + \mathbf{z}_{12} \left(\frac{-\mathbf{z}_{21} \mathbf{I}_1}{\mathbf{z}_{22} + \mathbf{Z}_L} \right)$$

and

$$\boxed{\mathbf{Z}_i = \frac{\mathbf{E}_1}{\mathbf{I}_1} = \mathbf{z}_{11} - \frac{\mathbf{z}_{12} \mathbf{z}_{21}}{\mathbf{z}_{22} + \mathbf{Z}_L}} \qquad \textbf{(25.28)}$$

For the output impedance, $E_s = 0$, and

$$I_1 = -\frac{z_{12}I_2}{R_s + z_{11}} \quad \text{and} \quad I_2 = \frac{E_2 - z_{21}I_1}{z_{22}}$$

or

$$E_2 = z_{22}I_2 + z_{21}I_1 = z_{22}I_2 + z_{21}\left(-\frac{z_{12}I_2}{R_s + z_{11}}\right)$$

and

$$E_2 = z_{22}I_2 - \frac{z_{12}z_{21}I_2}{R_s + z_{11}}$$

Thus,

$$\boxed{Z_o = \frac{E_2}{I_2} = z_{22} - \frac{z_{12}z_{21}}{R_s + z_{11}}} \qquad (25.29)$$

25.6 CONVERSION BETWEEN PARAMETERS

The equations relating the z and y parameters can be determined directly from Eqs. (25.1) and (25.10). For Eqs. (25.10a) and (25.10b),

$$I_1 = y_{11}E_1 + y_{12}E_2$$
$$I_2 = y_{21}E_1 + y_{22}E_2$$

The use of determinants will result in

$$E_1 = \frac{\begin{vmatrix} I_1 & y_{12} \\ I_2 & y_{22} \end{vmatrix}}{\begin{vmatrix} y_{11} & y_{12} \\ y_{21} & y_{22} \end{vmatrix}} = \frac{y_{22}I_1 - y_{12}I_2}{y_{11}y_{22} - y_{12}y_{21}}$$

Substituting the notation

$$\Delta_y = y_{11}y_{22} - y_{12}y_{21}$$

we have

$$E_1 = \frac{y_{22}}{\Delta_y}I_1 - \frac{y_{12}}{\Delta_y}I_2$$

which, when related to Eq. (25.1a),

$$E_1 = z_{11}I_1 + z_{12}I_2$$

indicates that

$$z_{11} = \frac{y_{22}}{\Delta_y} \quad \text{and} \quad z_{12} = \frac{y_{12}}{\Delta_y}$$

and, similarly,

$$\mathbf{z}_{21} = \frac{\mathbf{y}_{21}}{\Delta_{\mathbf{y}}} \quad \text{and} \quad \mathbf{z}_{22} = \frac{\mathbf{y}_{11}}{\Delta_{\mathbf{y}}}$$

For the conversion of $\mathbf{z}$ parameters to the admittance domain, determinants are applied to Eqs. (25.1a) and (25.1b). The impedance parameters can be found in terms of the hybrid parameters by first forming the determinant for $\mathbf{I}_1$ from the hybrid equations,

$$\mathbf{E}_1 = \mathbf{h}_{11}\mathbf{I}_1 + \mathbf{h}_{12}\mathbf{E}_2$$
$$\mathbf{I}_2 = \mathbf{h}_{21}\mathbf{I}_1 + \mathbf{h}_{22}\mathbf{E}_2$$

That is,

$$\mathbf{I}_1 = \frac{\begin{vmatrix} \mathbf{E}_1 & \mathbf{h}_{12} \\ \mathbf{I}_2 & \mathbf{h}_{22} \end{vmatrix}}{\begin{vmatrix} \mathbf{h}_{11} & \mathbf{h}_{12} \\ \mathbf{h}_{21} & \mathbf{h}_{22} \end{vmatrix}} = \frac{\mathbf{h}_{22}}{\Delta_{\mathbf{h}}}\mathbf{E}_1 - \frac{\mathbf{h}_{12}}{\Delta_{\mathbf{h}}}\mathbf{I}_2$$

and

$$\frac{\mathbf{h}_{22}}{\Delta_{\mathbf{h}}}\mathbf{E}_1 = \mathbf{I}_1 - \frac{\mathbf{h}_{12}}{\Delta_{\mathbf{h}}}\mathbf{I}_2$$

or

$$\mathbf{E}_1 = \frac{\Delta_{\mathbf{h}}\mathbf{I}_1}{\mathbf{h}_{22}} - \frac{\mathbf{h}_{12}}{\mathbf{h}_{22}}\mathbf{I}_2$$

TABLE 25.1

Conversions between $\mathbf{z}$, $\mathbf{y}$, *and* $\mathbf{h}$ *parameters.*

FROM / TO	$\mathbf{z}$		$\mathbf{y}$		$\mathbf{h}$	
$\mathbf{z}$	$\mathbf{z}_{11}$	$\mathbf{z}_{12}$	$\dfrac{\mathbf{y}_{22}}{\Delta_{\mathbf{y}}}$	$\dfrac{-\mathbf{y}_{12}}{\Delta_{\mathbf{y}}}$	$\dfrac{\Delta_{\mathbf{h}}}{\mathbf{h}_{22}}$	$\dfrac{\mathbf{h}_{12}}{\mathbf{h}_{22}}$
	$\mathbf{z}_{21}$	$\mathbf{z}_{22}$	$\dfrac{-\mathbf{y}_{21}}{\Delta_{\mathbf{y}}}$	$\dfrac{\mathbf{y}_{11}}{\Delta_{\mathbf{y}}}$	$\dfrac{-\mathbf{h}_{21}}{\mathbf{h}_{22}}$	$\dfrac{1}{\mathbf{h}_{22}}$
$\mathbf{y}$	$\dfrac{\mathbf{z}_{22}}{\Delta_{\mathbf{z}}}$	$\dfrac{-\mathbf{z}_{12}}{\Delta_{\mathbf{z}}}$	$\mathbf{y}_{11}$	$\mathbf{y}_{12}$	$\dfrac{1}{\mathbf{h}_{11}}$	$\dfrac{-\mathbf{h}_{12}}{\mathbf{h}_{11}}$
	$\dfrac{-\mathbf{z}_{21}}{\Delta_{\mathbf{z}}}$	$\dfrac{\mathbf{z}_{11}}{\Delta_{\mathbf{z}}}$	$\mathbf{y}_{21}$	$\mathbf{y}_{22}$	$\dfrac{\mathbf{h}_{21}}{\mathbf{h}_{11}}$	$\dfrac{\Delta_{\mathbf{h}}}{\mathbf{h}_{11}}$
$\mathbf{h}$	$\dfrac{\Delta_{\mathbf{z}}}{\mathbf{z}_{22}}$	$\dfrac{\mathbf{z}_{12}}{\mathbf{z}_{22}}$	$\dfrac{1}{\mathbf{y}_{11}}$	$\dfrac{-\mathbf{y}_{12}}{\mathbf{y}_{11}}$	$\mathbf{h}_{11}$	$\mathbf{h}_{12}$
	$\dfrac{-\mathbf{z}_{21}}{\mathbf{z}_{22}}$	$\dfrac{1}{\mathbf{z}_{22}}$	$\dfrac{\mathbf{y}_{21}}{\mathbf{y}_{11}}$	$\dfrac{\Delta_{\mathbf{y}}}{\mathbf{y}_{11}}$	$\mathbf{h}_{21}$	$\mathbf{h}_{22}$

which, when related to the impedance parameter equation,

$$E_1 = z_{11}I_1 + z_{12}I_2$$

indicates that

$$z_{11} = \frac{\Delta_h}{h_{22}} \quad \text{and} \quad z_{12} = -\frac{h_{12}}{h_{22}}$$

The remaining conversions are left as an exercise. A complete table of conversions appears in Table 25.1.

PROBLEMS

Section 25.2

1. a. Determine the impedance (z) parameters for the π network of Fig. 25.35.
  b. Sketch the z parameter equivalent circuit (using either form of Fig. 25.14).

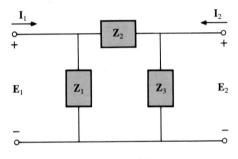

FIG. 25.35

2. a. Determine the impedance (z) parameters for the network of Fig. 25.36.
  b. Sketch the z parameter equivalent circuit (using either form of Fig. 25.14).

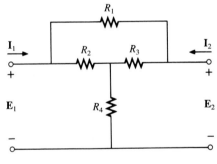

FIG. 25.36

Section 25.3

3. a. Determine the admittance (y) parameters for the T network of Fig. 25.37.
  b. Sketch the y parameter equivalent circuit (using either form of Fig. 25.26).

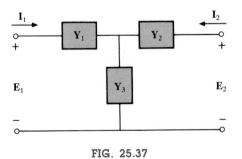

FIG. 25.37

4. a. Determine the admittance (**y**) parameters for the net-work of Fig. 25.38.
  b. Sketch the y parameter equivalent circuit (using either form of Fig. 25.26).

Section 25.4

5. a. Determine the **h** parameters for the network of Fig. 25.35.
  b. Sketch the hybrid equivalent circuit.

6. a. Determine the **h** parameters for the network of Fig. 25.36.
  b. Sketch the hybrid equivalent circuit.

7. a. Determine the **h** parameters for the network of Fig. 25.37.
  b. Sketch the hybrid equivalent circuit.

8. a. Determine the **h** parameters for the network of Fig. 25.38.
  b. Sketch the hybrid equivalent circuit.

9. For the hybrid equivalent circuit of Fig. 25.39:
  a. Determine the current gain $A_i = \mathbf{I}_2/\mathbf{I}_1$.
  b. Determine the voltage gain $A_v = \mathbf{E}_2/\mathbf{E}_1$.

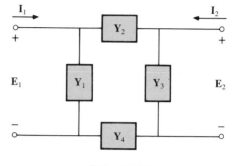

FIG. 25.38

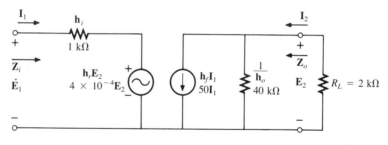

FIG. 25.39

Section 25.5

10. For the hybrid equivalent circuit of Fig. 25.39:
  a. Determine the input impedance.
  b. Determine the output impedance.

11. Determine the input and output impedances for the **z** parameter equivalent circuit of Fig. 25.40.

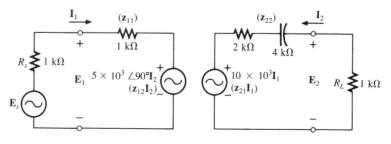

FIG. 25.40

12. Determine the expression for the input and output imped-
    ance of the **y** parameter equivalent circuit.

Section 25.6

13. Determine the **h** parameters for the following **z** para-
    meters:

$$z_{11} = 4\,k\Omega$$
$$z_{12} = 2\,k\Omega$$
$$z_{21} = 3\,k\Omega$$
$$z_{22} = 4\,k\Omega$$

14. **a.** Determine the **z** parameters for the following **h**
    parameters:

$$h_{11} = 1\,k\Omega$$
$$h_{12} = 2 \times 10^{-4}$$
$$h_{21} = 100$$
$$h_{22} = 20 \times 10^{-6}\,S$$

b. Determine the **y** parameters for the hybrid parameters
indicated in part (a).

Computer Problems

15. Given a T configuration such as that shown in Fig. 25.9,
    write a program to generate the **z** and **y** parameters. Each
    branch of the T configuration can be a resistor, capacitor,
    or inductor.

16. Given the hybrid parameters of a system, write an equa-
    tion to determine the current gain, voltage gain, input
    impedance, and output impedance.

17. Write a program to perform the conversions indicated in
    Table 25.1. That is, given the **z, y,** or **h** parameters, de-
    termine the **z, y,** or **h** parameters.

GLOSSARY

Admittance (y) parameters A set of parameters, having the
units of siemens, that can be used to establish a two-port
equivalent network for a system.

Hybrid (h) parameters A set of mixed parameters (ohms,
siemens, some unitless) that can be used to establish a two-
port equivalent network for a system.

Impedance (z) parameters A set of parameters, having the
units of ohms, that can be used to establish a two-port
equivalent network for a system.

Input impedance The impedance appearing at the input terminals of a system.

Output impedance The impedance appearing at the output
terminals of a system with the energizing source set to zero.

Single-port network A network having a single set of access
terminals.

Two-port network A network having two pairs of access terminals.

Appendixes

Appendix A

CONVERSION FACTORS

To Convert from	To	Multiply by
Btus	Calorie-grams	251.996
	Ergs	1.054×10^{10}
	Foot-pounds	777.649
	Hp-hours	0.000393
	Joules	1054.35
	Kilowatthours	0.000293
	Wattseconds	1054.35
Centimeters	Angstrom units	1×10^{8}
	Feet	0.0328
	Inches	0.3937
	Meters	0.01
	Miles (statute)	6.214×10^{-6}
	Millimeters	10
Circular mils	Square centimeters	5.067×10^{-6}
	Square inches	7.854×10^{-7}
Cubic inches	Cubic centimeters	16.387
	Gallons (U.S. liquid)	0.00433
Cubic meters	Cubic feet	35.315
Days	Hours	24
	Minutes	1440
	Seconds	86,400
Dynes	Gallons (U.S. liquid)	264.172
	Newtons	0.00001
	Pounds	2.248×10^{-6}
Electronvolts	Ergs	1.60209×10^{-12}
Ergs	Dyne-centimeters	1.0
	Electronvolts	6.242×10^{11}
	Foot-pounds	7.376×10^{-8}
	Joules	1×10^{-7}
	Kilowatthours	2.777×10^{-14}
Feet	Centimeters	30.48
	Meters	0.3048
Foot-candles	Lumens/square foot	1.0
	Lumens/square meter	10.764
Foot-pounds	Dyne-centimeters	1.3558×10^{7}
	Ergs	1.3558×10^{7}
	Horsepower-hours	5.050×10^{-7}
	Joules	1.3558
	Newton-meters	1.3558

To Convert from	To	Multiply by
Gallons (U.S. liquid)	Cubic inches	231
	Liters	3.785
	Ounces	128
	Pints	8
Gauss	Maxwells/square centimeter	1.0
	Lines/square centimeter	1.0
	Lines/square inch	6.4516
Gilberts	Ampere-turns	0.7958
Grams	Dynes	980.665
	Ounces	0.0353
	Pounds	0.0022
Horsepower	Btus/hour	2547.16
	Ergs/second	7.46×10^9
	Foot-pounds/second	550.221
	Joules/second	746
	Watts	746
Hours	Seconds	3600
Inches	Angstrom units	2.54×10^8
	Centimeters	2.54
	Feet	12
	Meters	0.0254
Joules	Btus	0.000948
	Ergs	1×10^7
	Foot-pounds	0.7376
	Horsepower-hours	3.725×10^{-7}
	Kilowatthours	2.777×10^{-7}
	Wattseconds	1.0
Kilograms	Dynes	980,665
	Ounces	35.2
	Pounds	2.2
Lines	Maxwells	1.0
Lines/square centimeter	Gauss	1.0
Lines/square inch	Gauss	0.1550
	Webers/square inch	1×10^{-8}
Liters	Cubic centimeters	1000.028
	Cubic inches	61.025
	Gallons (U.S. liquid)	0.2642
	Ounces (U.S. liquid)	33.815
	Quarts (U.S. liquid)	1.0567
Lumens	Candle power (spher.)	0.0796
Lumens/square centimeter	Lamberts	1.0

To Convert from	To	Multiply by
Lumens/square foot	Foot-candles	1.0
Maxwells	Lines	1.0
	Webers	1×10^{-8}
Meters	Angstrom units	1×10^{10}
	Centimeters	100
	Feet	3.2808
	Inches	39.370
	Miles (statute)	0.000621
Miles (statute)	Feet	5280
	Kilometers	1.609
	Meters	1609.344
Miles/hour	Kilometers/hour	1.609344
Newton-meters	Dyne-centimeters	1×10^{7}
	Kilogram-meters	0.10197
Oersteds	Ampere-turns/inch	2.0212
	Ampere-turns/meter	79.577
	Gilberts/centimeter	1.0
Quarts (U.S. liquid)	Cubic centimeters	946.353
	Cubic inches	57.75
	Gallons (U.S. liquid)	0.25
	Liters	0.9463
	Pints (U.S. liquid)	2
	Ounces (U.S. liquid)	32
Radians	Degrees	57.2958
Slugs	Kilograms	14.5939
	Pounds	32.1740
Watts	Btus/hour	3.4144
	Ergs/second	1×10^{7}
	Horsepower	0.00134
	Joules/second	1.0
Webers	Lines	1×10^{8}
	Maxwells	1×10^{8}
Years	Days	365
	Hours	8760
	Minutes	525,600
	Seconds	3.1536×10^{7}

Appendix B

THIRD-ORDER DETERMINANTS

Consider the three following simultaneous equations:

$$a_1x + b_1y + c_1z = d_1$$
$$a_2x + b_2y + c_2z = d_2$$
$$a_3x + b_3y + c_3z = d_3$$

The determinant configuration for x, y, and z can be found in a manner similar to that for two simultaneous equations. That is, to solve for x, obtain the determinant in the numerator by replacing the first column by the elements on the right of the equal sign, and the denominator is the determinant of the coefficients of the variables (the same approach applies to y and z). Again, the denominator is the same for each variable. Therefore,

$$x = \frac{\begin{vmatrix} d_1 & b_1 & c_1 \\ d_2 & b_2 & c_2 \\ d_3 & b_3 & c_3 \end{vmatrix}}{D = \begin{vmatrix} a_1 & b_1 & c_1 \\ a_2 & b_2 & c_2 \\ a_3 & b_3 & c_3 \end{vmatrix}} \qquad y = \frac{\begin{vmatrix} a_1 & d_1 & c_1 \\ a_2 & d_2 & c_2 \\ a_3 & d_3 & c_3 \end{vmatrix}}{D = \begin{vmatrix} a_1 & b_1 & c_1 \\ a_2 & b_2 & c_2 \\ a_3 & b_3 & c_3 \end{vmatrix}} \qquad z = \frac{\begin{vmatrix} a_1 & b_1 & d_1 \\ a_2 & b_2 & d_2 \\ a_3 & b_3 & d_3 \end{vmatrix}}{D = \begin{vmatrix} a_1 & b_1 & c_1 \\ a_2 & b_2 & c_2 \\ a_3 & b_3 & c_3 \end{vmatrix}}$$

There is more than one expanded format for the third-order determinant. Each, however, will give the same result. One expansion of the determinant (D) is the following:

$$D = \begin{vmatrix} a_1 & b_1 & c_1 \\ a_2 & b_2 & c_2 \\ a_3 & b_3 & c_3 \end{vmatrix} = a_1 \left(+ \begin{vmatrix} b_2 & c_2 \\ b_3 & c_3 \end{vmatrix} \right) + b_1 \left(- \begin{vmatrix} a_2 & c_2 \\ a_3 & c_3 \end{vmatrix} \right) + c_1 \left(+ \begin{vmatrix} a_2 & b_2 \\ a_3 & b_3 \end{vmatrix} \right)$$

$$\underbrace{\text{Minor}} \qquad \underbrace{\text{Minor}} \qquad \underbrace{\text{Minor}}$$

Cofactor   Cofactor   Cofactor

Multiplying factor   Multiplying factor   Multiplying factor

This expansion was obtained by multiplying the elements of the first row of D by their corresponding cofactors. It is not a requirement that the first row be used as the multiplying factors. In fact, any *row* or *column* (not diagonals) may be used to expand a third-order determinant.

The sign of each cofactor is dictated by the position of the multiplying factors (a_1, b_1, and c_1 in this case) as in the following standard format:

$$\begin{vmatrix} + \rightarrow & - & + \\ \downarrow & & \\ - & + & - \\ + & - & + \end{vmatrix}$$

Note that the proper sign for each element can be obtained by simply assigning the upper left element a positive sign and then changing sign as you move horizontally or vertically to the neighboring position.

For the determinant D, the elements would have the following signs:

$$\begin{vmatrix} a_1^{(+)} & b_1^{(-)} & c_1^{(+)} \\ a_2^{(-)} & b_2^{(+)} & c_2^{(-)} \\ a_3^{(+)} & b_3^{(-)} & c_3^{(+)} \end{vmatrix}$$

The minors associated with each multiplying factor are obtained by covering up the row and column in which the multiplying factor is located and writing a second-order determinant to include the remaining elements in the same relative positions that they have in the third-order determinant.

Consider the cofactors associated with a_1 and b_1 in the expansion of D. The sign is positive for a_1 and negative for b_1 as determined by the standard format. Following the procedure outlined above, we can find the minors of a_1 and b_1 as follows:

$$a_{1(minor)} = \begin{vmatrix} \not{a}_1 & \not{b}_1 & \not{c}_1 \\ \not{a}_2 & b_2 & c_2 \\ \not{a}_3 & b_3 & c_3 \end{vmatrix} = \begin{vmatrix} b_2 & c_2 \\ b_3 & c_3 \end{vmatrix}$$

$$b_{1(minor)} = \begin{vmatrix} \not{a}_1 & \not{b}_1 & \not{c}_1 \\ a_2 & \not{b}_2 & c_2 \\ a_3 & \not{b}_3 & c_3 \end{vmatrix} = \begin{vmatrix} a_2 & c_2 \\ a_3 & c_3 \end{vmatrix}$$

It was pointed out that any row or column may be used to expand the third-order determinant, and the same result will still be obtained. Using the first column of D, we obtain the expansion

$$D = \begin{vmatrix} a_1 & b_1 & c_1 \\ a_2 & b_2 & c_2 \\ a_3 & b_3 & c_3 \end{vmatrix} = a_1 \left(+ \begin{vmatrix} b_2 & c_2 \\ b_3 & c_3 \end{vmatrix} \right) + a_2 \left(- \begin{vmatrix} b_1 & c_1 \\ b_3 & c_3 \end{vmatrix} \right) + a_3 \left(+ \begin{vmatrix} b_1 & c_1 \\ b_2 & c_2 \end{vmatrix} \right)$$

The proper choice of row or column can often effectively reduce the amount of work required to expand the third-order determinant. For example, in the following determinants, the first column and third row, respectively, would reduce the number of cofactors in the expansion:

$$D = \begin{vmatrix} 2 & 3 & -2 \\ 0 & 4 & 5 \\ 0 & 6 & 7 \end{vmatrix} = 2 \left(+ \begin{vmatrix} 4 & 5 \\ 6 & 7 \end{vmatrix} \right) + 0 + 0 = 2(28 - 30)$$

$$= -4$$

$$D = \begin{vmatrix} 1 & 4 & 7 \\ 2 & 6 & 8 \\ 2 & 0 & 3 \end{vmatrix} = 2 \left(+ \begin{vmatrix} 4 & 7 \\ 6 & 8 \end{vmatrix} \right) + 0 + 3 \left(+ \begin{vmatrix} 1 & 4 \\ 2 & 6 \end{vmatrix} \right)$$

$$= 2(32 - 42) + 3(6 - 8) = 2(-10) + 3(-2)$$

$$= -26$$

EXAMPLES Expand the following third-order determinants:

a. $D = \begin{vmatrix} 1 & 2 & 3 \\ 3 & 2 & 1 \\ 2 & 1 & 3 \end{vmatrix} = 1 \left(+ \begin{vmatrix} 2 & 1 \\ 1 & 3 \end{vmatrix} \right) + 3 \left(- \begin{vmatrix} 2 & 3 \\ 1 & 3 \end{vmatrix} \right) + 2 \left(+ \begin{vmatrix} 2 & 3 \\ 2 & 1 \end{vmatrix} \right)$

$$= 1[6 - 1] + 3[-(6 - 3)] + 2[2 - 6]$$

$$= 5 + 3(-3) + 2(-4)$$

$$= 5 - 9 - 8$$

$$= -12$$

b. $D = \begin{vmatrix} 0 & 4 & 6 \\ 2 & 0 & 5 \\ 8 & 4 & 0 \end{vmatrix} = 0 + 2 \left(- \begin{vmatrix} 4 & 6 \\ 4 & 0 \end{vmatrix} \right) + 8 \left(+ \begin{vmatrix} 4 & 6 \\ 0 & 5 \end{vmatrix} \right)$

$$= 0 + 2[-(0 - 24)] + 8[(20 - 0)]$$

$$= 0 + 2(24) + 8(20)$$

$$= 48 + 160$$

$$= 208$$

Appendix C

COLOR CODING OF MOLDED MICA CAPACITORS (PICOFARADS)

RETMA *and standard* MIL *specifications*

Color	Sig-nificant Figure	Decimal Multiplier	Tolerance ± %	Class	Temp. Coeff. PPM/°C Not More than	Cap. Drift Not More than
Black	0	1	20	A	± 1000	± (5% + 1 pF)
Brown	1	10	—	B	± 500	± (3% + 1 pF)
Red	2	100	2	C	± 200	± (0.5% + 0.5 pF)
Orange	3	1000	3	D	± 100	± (0.3% + 0.1 pF)
Yellow	4	10,000	—	E	+100 − 20	± (0.1% + 0.1 pF)
Green	5	—	5	—	—	—
Blue	6	—	—	—	—	—
Violet	7	—	—	—	—	—
Gray	8	—	—	I	+150 − 50	± (0.03% + 0.2 pF)
White	9	—	—	J	+100 − 50	± (0.2% + 0.2 pF)
Gold	—	0.1	—			
Silver	—	0.01	10			

Courtesy of Sprague Electric Co.

NOTE: If both rows of dots are not on one face, rotate capacitor about axis of its leads to read second row on side or rear.

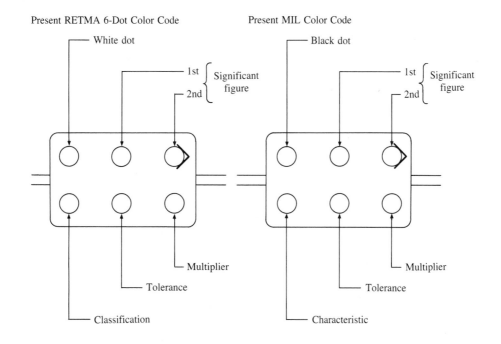

Present RETMA 6-Dot Color Code — White dot — 1st, 2nd Significant figure — Multiplier — Tolerance — Classification

Present MIL Color Code — Black dot — 1st, 2nd Significant figure — Multiplier — Tolerance — Characteristic

Appendix D

COLOR CODING OF MOLDED TUBULAR CAPACITORS (PICOFARADS)

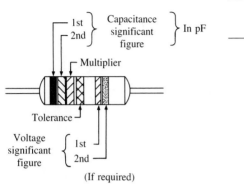

FIG. D.1

Color	Significant Figure	Decimal Multiplier	Tolerance ± %
Black	0	1	20
Brown	1	10	—
Red	2	100	—
Orange	3	1000	30
Yellow	4	10,000	40
Green	5	10^5	5
Blue	6	10^6	—
Violet	7	—	—
Gray	8	—	—
White	9	—	10

Courtesy of Sprague Electric Co.

NOTE: Voltage rating is identified by a single-digit number for ratings up to 900 V; a two-digit number above 900 V. Two zeros follow the voltage figure.

Appendix E

THE GREEK ALPHABET

Letter	Capital	Lowercase	Used to Designate
Alpha	A	α	Area, angles, coefficients
Beta	B	β	Angles, coefficients, flux density
Gamma	Γ	γ	Specific gravity, conductivity
Delta	Δ	δ	Density, variation
Epsilon	E	ϵ	Base of natural logarithms
Zeta	Z	ζ	Coefficients, coordinates, impedance
Eta	H	η	Efficiency, hysteresis coefficient
Theta	Θ	θ	Phase angle, temperature
Iota	I	ι	
Kappa	K	κ	Dielectric constant, susceptibility
Lambda	Λ	λ	Wavelength
Mu	M	μ	Amplification factor, micro, permeability
Nu	N	ν	Reluctivity
Xi	Ξ	ξ	
Omicron	O	o	
Pi	Π	π	3.1416
Rho	P	ρ	Resistivity
Sigma	Σ	σ	Summation
Tau	T	τ	Time constant
Upsilon	Υ	υ	
Phi	Φ	ϕ	Angles, magnetic flux
Chi	X	χ	
Psi	Ψ	ψ	Dielectric flux, phase difference
Omega	Ω	ω	Ohms, angular velocity

Appendix F

MAGNETIC PARAMETER CONVERSIONS

	SI (MKS)		CGS		English
Φ	webers (Wb)		maxwells		lines
	1 Wb	=	10^8 maxwells	=	10^8 lines
B	Wb/m^2		gauss (maxwells/cm^2)		lines/in.^{2}
	1 Wb/m^2	=	10^4 gauss	=	6.452×10^4 lines/in.^{2}
A	1 m^2	=	10^4 cm^2	=	1550 in.^{2}
μ_o	$4\pi \times 10^{-7}$ Wb/Am	=	1 gauss/oersted	=	3.20 lines/Am
$\mathscr{F}$	NI (ampere-turns, At)		$0.4\pi NI$ (gilberts)		NI (At)
	1 At	=	1.257 gilberts		1 gilbert = 0.7958 At
H	NI/l (At/m)		$0.4\pi NI/l$ (oersteds)		NI/l (At/in.)
	1 At/m	=	1.26×10^{-2} oersted	=	2.54×10^{-2} At/in.
H_g	$7.97 \times 10^5 B_g$ (At/m)		B_g (oersteds)		$0.313 B_g$ (At/in.)

Appendix G

ANSWERS TO SELECTED ODD-NUMBERED PROBLEMS

Chapter 1

1. 4400 ft
3. 737.8 ft-lb
5. 45.72 cm
7. a. 15×10^3 **b.** 30×10^{-3}
  c. 7.4×10^6 **d.** 6.8×10^{-6}
  e. 402×10^{-6} **f.** 200×10^{-12}
9. a. 10^{-1} **b.** 10^{-4} **c.** 10^9 **d.** 10^{-9}
  e. 10^{42} **f.** 10^3
11. a. 10^{-6} **b.** 10^{-3} **c.** 10^{-6} **d.** 10^9
  e. 10^{-16} **f.** 10^{-1}
13. a. 90 s **b.** 144 s **c.** 50×10^3 μs
  d. 160 mH **e.** 120 ns **f.** 41.9 days
  g. 1.02 m
15. a. 2.54 m **b.** 1.219 m **c.** 26.7 N
  d. 0.1348 lb **e.** 4921 ft **f.** 3.218 m
  g. 8530.17 yd
17. 2.045 s
19. 2.682 m/s
21. 40.391 min
23. a. 4.74×10^{-3} Btu
  b. 7.098×10^{-4} m^3
  c. 1.2096×10^5 s
  d. 2113.38 pints

Chapter 2

3. a. 111.197×10^{-6} N
  b. 288×10^4 N
  c. 114.195×10^6 N
5. a. 0.072 N
  b. 2×10^{-5} C, 4×10^{-5} C
7. 3.1 A
9. 90 C
11. 0.5 A
13. 195 mA
15. 252 J
17. 4 C
19. 3.533 V
21. 60.8 Ah
35. 600 C

Chapter 3

1. a. 500 mils **b.** 10 mils **c.** 4 mils
  d. 1000 mils **e.** 240 mils
  f. 3.937 mils

3. a. 0.04 in. **b.** 0.03 in. **c.** 0.2 in.
  d. 0.025 in. **e.** 0.00278 in.
  f. 0.009 in.
5. 73.3 Ω
7. a. 376.69 ft **b.** 394.57 ft
9. a. 1244.40 cm **b.** larger
  c. smaller
11. nickel
13. a. #11 = 0.567 Ω,
    #14 = 1.136 Ω
  b. #14 : #11 = 2 : 1
  c. #14 : #11 = 1 : 2
15. a. #2 **b.** #0
17. 58.7 m
19. 2.5×10^{-6} cm
21. 2.409 Ω
23. 3.334 Ω
25. a. 1.842×10^{-3} Ω
  b. 2.628×10^{-3} Ω
27. a. 40.29°C **b.** −195.61°C
29. 39.371 Ω
31. a. 2× **b.** 4×
33. 10,150 Ω
35. 6.25 kΩ, 18.75 kΩ
37. yes
39. yes
41. a. 0.1566 S **b.** 95.5 mS
  c. 21.95 mS
47. 10 foot-candles = 3 kΩ,
  100 foot-candles = 0.4 kΩ

Chapter 4

1. 15 V
3. 4 kΩ
5. 72 mV
7. 54.55 Ω
9. 1.68 kV
11. 12.63 Ω
13. 7.5 V
17. 800 V
19. 16 s
21. 250 W
23. 4.8 W
25. 27 μW
27. 22.36 mA
29. 40 W

31. 120 V, 32 Ω
33. a. 12 kW **b.** 10,130 W (yes)
35. 16.34 A
37. 14.03 A
39. 56.52 A
41. 38.4 J
43. 40%, 80%
45. a. 1350 J
  b. energy doubled, power the
    same
47. 6.67 h
49. $1.70

Chapter 5

1. a. 20 Ω, 3 A
  b. 1.63 MΩ, 6.135 μA
  c. 110 Ω, 318.2 mA
  d. 10 kΩ, 12 mA
3. a. 16 V **b.** 1.56 V
5. a. 4.18 mA **b.** 0.388 A
7. a. $V_2 = 10$ V, $V_1 = 4$ V
  b. $V_1 = 14$ V, $V_2 = 16$ V
9. a. $R_T = 6$ kΩ, $I = 20$ mA,
    $V_1 = 60$ V, $V_2 = 20$ V,
    $V_3 = 40$ V
  d. 3 kΩ = 2 W, 1 kΩ = $\frac{1}{2}$ W,
    2 kΩ = 1 W
11. a. $V = 60$ V, $I = 2$ A, $R = 30$ Ω
  b. $I = 3.636$ mA, $V_1 = 17.09$ V,
    $V_2 = 24.73$ V
  c. $R = 21$ Ω, $V_1 = 2$ V, $V_2 = 1$ V,
    $V_3 = 21$ V, $E = 24$ V
  d. $I = 2$ A, $R_1 = 2$ Ω, $R_2 = 13$ Ω,
    $E = 32$ V
13. a. 66.67 V **b.** −8 V **c.** 20 V
  d. 0.18 V
15. $R_1 = 3$ kΩ, $R_2 = 15$ kΩ
17. a. $V_a = 13.33$ V, $V_1 = 5.33$ V
  b. $V_a = 10$ V, $V_1 = 12$ V
19. $V_L = 11.821$ V, $P = 0.642$ W
21. 7.14%

Chapter 6

1. a. 2, 3, and 4 in parallel
  b. 2 and 3 in parallel

c. 2 and 3 in series, 1 and 4 in parallel

3. a. $R = 7.5\,\Omega$ **b.** $R = 8\,k\Omega$

5. a. $G_T = 0.167\,mS$, $R_T = 6\,k\Omega$
   b. $I_T = 8\,mA$, $I_1 = 6\,mA$, $I_2 = 2\,mA$
   e. both $\frac{1}{2}\,W$

7. a. $G_T = 0.8144\,mS$, $R_T = 1.2279\,k\Omega$
   b. $I_T = 9.7728\,mA$, $I_1 = 5.4545\,mA$, $I_2 = 2.5532\,mA$, $I_3 = 1.7647\,mA$
   e. all $\frac{1}{2}\,W$

9. a. $I_{bulbs} = 5\,A$, $I_{washer} = 3.33\,A$, $I_{TV} = 3\,A$
   b. 11.33 A, no
   c. $R_T = 10.59\,\Omega$
   d. $R_{del.} = 1360\,W$

11. a. $I_1 = 25\,A$, $I_2 = 23\,A$, $I_3 = 20\,A$
   b. $I_1 = 11\,A$, $I_2 = 6\,A$, $I_3 = 14\,A$, $I_4 = 10\,A$

13. a. $R_1 = 5\,\Omega$, $R_2 = 10\,\Omega$
   b. $E = 12\,V$, $I_2 = 1.33\,A$, $I_3 = 1\,A$, $R_3 = 12\,\Omega$, $I = 4.33\,A$
   c. $I_1 = 220\,mA$, $I_3 = 55\,mA$, $I_2 = 725\,mA$, $R = 303.45\,\Omega$
   d. $V_1 = 30\,V$, $E = 30\,V$, $I_1 = 1\,A$, $I_2 = 0.5\,A$, $R_2 = R_3 = 60\,\Omega$

15. a. $I = 4\,A$, $I_1 = 3\,A$
   b. $I_3 = I = 6\,\mu A$, $I_2 = 2\,\mu A$, $I_1 = 2\,\mu A$, $R = 9\,\Omega$

17. $I_1 = 1.714\,A$, $I_2 = 0.857\,A$

19. a. $V_L = 6.13\,V$
   b. $I_{sc} = 1.2776\,mA$
   c. $V_L = 9\,V$
   d. $I_{sc} = 1.6364\,mA$

Chapter 7

1. a. yes (KCL) **b.** $I_2 = 3\,A$ **c.** yes (KCL) **d.** $V_2 = 4\,V$ **e.** $R_T = 2\,\Omega$
   f. $I = 5\,A$
   g. $P_E = 50\,W$, $P_1 = 12\,W$, $P_2 = 18\,W$

3. a. $R_T = 4\,\Omega$
   b. $I_T = 9\,A$, $I_1 = 6\,A$, $I_2 = 3\,A$
   c. $V_a = 6\,V$

5. a. $I = 22\,A$, $I_1 = 6\,A$, $I_3 = 0.8\,A$

7. a. $I_1 = 4\,A$

b. $I_2 = 1.333\,A$, $I_3 = 0.6665\,A$
   c. $V_a = 8\,V$, $V_b = 4\,V$

9. a. $I = 16\,A$ **b.** $I_3 = I_9 = 4\,A$
   c. $I_8 = 1\,A$ **d.** $V_{ab} = 14\,V$

11. a. $V_G = 1.9\,V$, $V_S = 3.65\,V$
   b. $I_1 = I_2 = 7.05\,\mu A$, $I_D = I_S = 2.433\,mA$
   c. $V_{DS} = 6.268\,V$
   d. $V_{DG} = 8.02\,V$

13. a. $I = 2\,A$
   b. $I_1 = I_3 = \frac{2}{3}\,A$, $I_8 = \frac{2}{9}\,A$
   c. $P = 1.037\,W$

15. a. $I_2 = 1\frac{2}{3}\,A$, $I_6 = 1\frac{1}{9}\,A$, $I_8 = 0\,A$
   b. $V_4 = 10\,V$, $V_8 = 0\,V$

17. $V = 12\,V$, $I = 3\,A$

19. a. $R_T = 5.532\,\Omega$ **b.** $I = 361.5\,mA$
   c. $I_8 = 18.78\,mA$

21. a. $I_{CS} = 1\,mA$ **b.** $R_{shunt} = 5\,m\Omega$

23. a. $R_s = 300\,k\Omega$ **b.** $\Omega/V = 20,000$

25. $I_{CS} = 0.05\,\mu A$

Chapter 8

1. $V_{ab} = 28\,V$

3. a. $I_1 = 2\,A$, $I_T = 1\,A$
   b. $V_S = 12\,V$, $V_3 = 6\,V$

5. a. $I = 3\,A$, $R_p = 6\,\Omega$
   b. $I = 4.091\,mA$, $R_p = 2.2\,k\Omega$

7. a. $I = 8\,A$ **b.** $I = 8\,A$

9. $V_2 = 9.6\,V$, $I_1 = 2.4\,A$

11. a. $I = 5.4545\,mA$, $R_p = 2.2\,k\Omega$
   b. $V_1 = 17.375\,V$
   c. $V_2 = 5.375\,V$
   d. $I_2 = 2.443\,mA$

13. a. 29 **b.** 18

15. a. $x = 1.705$, $y = 2.591$, $z = 0.546$
   b. $x = 6$, $y = -7.33$, $z = -2.5$

17. a. $I_{R_1} = 1.445\,mA$, $I_{R_2} = 8.513\,mA$, $I_{R_3} = 9.958\,mA$
   b. $I_{R_1} = 0.6763\,mA$, $I_{R_2} = 0.7725\,mA$, $I_{R_3} = 1.4488\,mA$

19. a. $I_1\,(CW—R_1) = -\frac{1}{7}\,A$, $I_2\,(CW—R_2) = -\frac{5}{7}\,A$
   b. $I_1\,(CW—R_1) = -3.0625\,A$, $I_2\,(CW—R_3) = 0.1875\,A$

21. a. $I_2 = -8.548\,A$, $V_{ab} = -22.74\,V$
   b. $I_2 = 1.274\,A$, $V_{ab} = -0.904\,V$

23. a. $I_1 = 1.2059\,mA$, $I_2 = -0.4806\,mA$, $I_3 = -0.6206\,mA$

b. $I_1 = -0.2385\,mA$, $I_2 = -0.5169\,A$, $I_3 = -1.268\,A$

25. a. $I_1 = -\frac{1}{7}\,A$, $I_2 = -\frac{5}{7}\,A$
   b. $I_1 = -3.0625\,A$, $I_2 = 0.1875\,A$

27. a. $I_1 = 1.871\,A$, $I_2 = -8.548\,A$
   b. $I_2 = 1.274\,A$, $I_3 = 0.26\,A$

29. $I_1 = 1.2059\,mA$, $I_2 = -0.4806\,mA$, $I_3 = -0.6206\,mA$

31. a. $V_1 = 8.077\,V$, $V_2 = 9.385\,V$
   b. $V_1 = 4.8\,V$, $V_2 = 6.4\,V$

33. a. $V_1 = -2.653\,V$, $V_2 = 0.952\,V$
   b. $V_1 = 8.877\,V$, $V_2 = 9.831\,V$, $V_3 = -3.005\,V$

35. a. $V_1 = -5.311\,V$, $V_2 = 0.6219\,V$, $V_3 = 3.751\,V$
   b. $V_1 = -6.917\,V$, $V_2 = 12\,V$, $V_3 = 2.3\,V$

37. b. $V_{R_5} = 0.1967\,V$
   c., d. no

39. b. $V_{R_5} = 0\,V$
   c., d. yes

41. a. $I_{R_5} = 3.33\,mA$
   b. $I_{R_5} = 1.177\,A$

43. a. $I = 133.33\,mA$
   b. $I = 7\,A$

Chapter 9

1. a. $I_1 = \frac{5}{6}\,A$, $I_2 = 0\,A$, $I_3 = \frac{5}{6}\,A$
   b. E_1: 5.333 W, E_2: 0.333 W
   c. 8.333 W **d.** no

3. a. 4.4545 mA **b.** 3.6324 mA

5. a. $R_{Th} = 6\,\Omega$, $E_{Th} = 6\,V$
   b. $I_{2\Omega} = 0.75\,A$, $I_{30\Omega} = 0.1667\,A$, $I_{100\Omega} = 0.0566\,A$

7. (I) $R_{Th} = 2\,\Omega$, $E_{Th} = 84\,V$
   (II) $R_{Th} = 1.579\,k\Omega$, $E_{Th} = 1.149\,V$

9. (I) $R_{Th} = 45\,\Omega$, $E_{Th} = 5\,V$
   (II) $R_{Th} = 2.055\,k\Omega$, $E_{Th} = 16.772\,V$

11. (I) $R_N = 14\,\Omega$, $I_N = 2.571\,A$
   (II) $R_N = 7.5\,\Omega$, $I_N = 1.333\,A$

13. (I) $R_N = 9.756\,\Omega$, $I_N = 0.95\,A$
   (II) $R_N = 2\,\Omega$, $I_N = 30\,A$

15. (I) $R_N = 10\,\Omega$, $I_N = 0.2\,A$
   (II) $R_N = 4.033\,k\Omega$, $I_N = 2.9758\,mA$

17. a. (I) $R = 14\,\Omega$

(II) $R = 7.5\,\Omega$
b. (I) $23.14\,\text{W}$
(II) $3.332\,\text{W}$
19. a. (I) $R = 9.756\,\Omega$
(II) $R = 2\,\Omega$
b. (I) $2.2\,\text{W}$
(II) $450\,\text{W}$
21. $R_1 = 0\,\Omega$
27. $V_L = 0.22\,\text{V}$, $I_L = 39.3\,\mu\text{A}$
29. $I_L = 2.25\,\text{A}$, $V_L = 6.075\,\text{V}$
35. a. $I = 0.357\,\text{mA}$ **b.** same **c.** yes

Chapter 10

1. $9 \times 10^3\,\text{N/C}$
3. $70\,\mu\text{F}$
5. $50\,\text{V/m}$
7. $8 \times 10^3\,\text{V/m}$
9. $937.5\,\text{pF}$
11. mica
13. a. $10^6\,\text{V/m}$ **b.** $4.96\,\mu\text{C}$
c. $0.0248\,\mu\text{F}$
15. $29{,}035\,\text{V}$
17. a. $0.5\,\text{s}$
b. $v_C = 20(1 - e^{-t/0.5})$
c. $1\tau = 12.64\,\text{V}$, $3\tau = 19\,\text{V}$,
$5\tau = 19.87\,\text{V}$
d. $i_C = (0.2 \times 10^{-3})e^{-t/0.5}$,
$v_R = 20e^{-t/0.5}$
19. a. $5.5\,\text{ms}$
b. $v_C = 100(1 - e^{-t/(5.5 \times 10^{-3})})$
c. $1\tau = 63.21\,\text{V}$, $3\tau = 95.02\,\text{V}$,
$5\tau = 99.33\,\text{V}$
d. $i_C = (18.18 \times 10^{-3})$
$\times\, e^{-t/(5.5 \times 10^{-3})}$,
$v_{R_2} = 60e^{-t/(5.5 \times 10^{-3})}$
21. a. $10\,\text{ms}$
b. $v_C = 50(1 - e^{-t/(10 \times 10^{-3})})$
c. $i_C = (10 \times 10^{-3})e^{-t/(10 \times 10^{-3})}$
d. $v_C = 50\,\text{V}$, $i_C = 0\,\text{mA}$
e. $v_C = 50e^{-t/(4 \times 10^{-3})}$,
$i_C = (25 \times 10^{-3})e^{-t/(4 \times 10^{-3})}$
23. a. $v_C = 80(1 - e^{-t/(1 \times 10^{-6})})$
b. $i_C = (0.8 \times 10^{-3})e^{-t/(1 \times 10^{-6})}$
c. $v_C = 80e^{-t/(4.9 \times 10^{-6})}$,
$i_C = (0.163 \times 10^{-3})$
$\times\, e^{-t/(4.9 \times 10^{-6})}$
25. a. $10\,\mu\text{s}$ **b.** $3 \times 10^3\,\text{A}$ **c.** yes
27. $0.6\,\mu\text{s}$
29. $v_C = 3.2751(1 - e^{-t/(52.68 \times 10^{-3})})$,
$i_C = (1.216 \times 10^{-3})e^{-t/\pi}$
31. 0 to $2\,\mu\text{s}$: $90\,\text{mA}$,
$2\,\mu\text{s}$ to $4\,\mu\text{s}$: $-180\,\text{mA}$,

$4\,\mu\text{s}$ to $6\,\mu\text{s}$: $90\,\text{mA}$,
$6\,\mu\text{s}$ to $8\,\mu\text{s}$: $60\,\text{mA}$,
$8\,\mu\text{s}$ to $10\,\mu\text{s}$: $-30\,\text{mA}$,
$10\,\mu\text{s}$ to $15\,\mu\text{s}$: $0\,\text{mA}$,
$15\,\mu\text{s}$ to $16\,\mu\text{s}$: $-60\,\text{mA}$
33. a. C_1: $10\,\text{V}$, $60\,\mu\text{C}$,
C_2: $6.67\,\text{V}$, $40\,\mu\text{C}$,
C_3: $3.33\,\text{V}$, $40\,\mu\text{C}$
b. C_1: $8\,\text{V}$, $9600\,\text{pC}$,
C_2: $16\,\text{V}$, $3200\,\text{pC}$,
C_3: $16\,\text{V}$, $6400\,\text{pC}$,
C_4: $16\,\text{V}$, $9600\,\text{pC}$
35. $8640\,\text{pJ}$
37. a. $5\,\text{J}$ **b.** $0.1\,\text{C}$ **c.** $200\,\text{A}$
d. $10{,}000\,\text{W}$ **e.** $10\,\text{s}$

Chapter 11

1. CGS: 5×10^4 maxwells, 8 gauss
English: 5×10^4 lines, 51.616
lines/in.^{2}
3. $0.04\,\text{T}$
5. $952.4 \times 10^3\,\text{At/Wb}$
7. $2624.67\,\text{At/m}$
9. $2.133\,\text{A}$
11. a. 60 turns
b. $13.34 \times 10^{-4}\,\text{Wb/Am}$
13. $2.7432\,\text{A}$
15. $1.35\,\text{N}$
17. a. $2.028\,\text{A}$ **b.** $\cong 2\,\text{N}$
19. $59.4 \times 10^{-4}\,\text{Wb}$

Chapter 12

1. $4.25\,\text{V}$
3. 14 turns
5. $15.65\,\mu\text{H}$
7. a. $2.5\,\text{V}$ **b.** $0.3\,\text{V}$ **c.** $200\,\text{V}$
9. 0 to $3\,\text{ms}$: $0\,\text{V}$,
$3\,\text{ms}$ to $8\,\text{ms}$: $1.6\,\text{V}$,
$8\,\text{ms}$ to $13\,\text{ms}$: $-1.6\,\text{V}$,
$13\,\text{ms}$ to $14\,\text{ms}$: $0\,\text{V}$,
$14\,\text{ms}$ to $15\,\text{ms}$: $8\,\text{V}$,
$15\,\text{ms}$ to $16\,\text{ms}$: $-8\,\text{V}$,
$16\,\text{ms}$ to $17\,\text{ms}$: $0\,\text{V}$
11. a. $12.5\,\mu\text{s}$
b. $i_L = (2 \times 10^{-6})$
$\times\, (1 - e^{-t/(12.5 \times 10^{-6})})$
c. $v_L = (40 \times 10^{-3})e^{-t/(12.5 \times 10^{-6})}$,
$v_R = (40 \times 10^{-3})$
$\times\, (1 - e^{-t/(12.5 \times 10^{-6})})$
d. i_L: $1\tau = 1.264\,\mu\text{A}$,
$3\tau = 1.9\,\mu\text{A}$,

$5\tau = 1.987\,\mu\text{A}$,
v_L: $1\tau = 14.72\,\text{V}$,
$3\tau = 1.99\,\text{V}$,
$5\tau = 0.2695\,\text{V}$
13. a. $i_L = (2 \times 10^{-3})$
$\times\, (1 - e^{-t/(1 \times 10^{-6})})$,
$v_L = 20e^{-t/(1 \times 10^{-6})}$
b. $i_L = (2 \times 10^{-3})e^{-t/(0.5 \times 10^{-6})}$,
$v_L = -40e^{-t/(0.5 \times 10^{-6})}$
15. a. $i_L = (6 \times 10^{-3})$
$\times\, (1 - e^{-t/(0.5 \times 10^{-6})})$,
$v_L = 12e^{-t/(0.5 \times 10^{-6})}$
b. $i_L = (5.188 \times 10^{-3})$
$\times\, e^{-t/(0.0833 \times 10^{-6})}$,
$v_L = -62.256e^{-t/(0.0833 \times 10^{-6})}$
17. a. $i_L = (3.638 \times 10^{-3})$
$\times\, (1 - e^{-t/(6.676 \times 10^{-6})})$,
$v_L = 5.45e^{-t/(6.676 \times 10^{-6})}$
b. $i_L = 2.825\,\text{mA}$,
$v_L = 1.2186\,\text{V}$
c. $i_L = (2.825 \times 10^{-3})$
$\times\, e^{-t/(2.128 \times 10^{-6})}$,
$v_L = -13.278e^{-t(2.128 \times 10^{-6})}$
19. a. $L_T = 10\,\text{mH}$, $C_T = 18\,\mu\text{F}$
b. $L_T = 25\,\text{mH}$, $C_T = 18\,\mu\text{F}$
21. $I_1 = 2\,\text{A}$, $V_1 = 12\,\text{V}$
23. $2\,\text{H}$: $16\,\mu\text{J}$, $3\,\text{H}$: $24\,\mu\text{J}$

Chapter 13

1. a. $10\,\text{ms}$ **b.** 2 cycles **c.** $100\,\text{Hz}$
d. $5\,\text{V}$ (amplitude), $6.67\,\text{V}$
(peak-to-peak)
3. $T = 10\,\text{ms}$, $f = 100\,\text{Hz}$
5. a. $60\,\text{Hz}$ **b.** $100\,\text{Hz}$ **c.** $29.41\,\text{Hz}$
d. $40\,\text{kHz}$
7. $0.25\,\text{s}$
11. a. $45°$ **b.** $30°$ **c.** $18°$ **d.** $210°$
e. $540°$ **f.** $99°$
13. a. $314\,\text{rad/s}$ **b.** $3768\,\text{rad/s}$
c. $12.56 \times 10^3\,\text{rad/s}$
d. $25.12 \times 10^3\,\text{rad/s}$
15. $2.08\,\text{ms}$
17. a. 20, $60.03\,\text{Hz}$
b. 5, $120.06\,\text{Hz}$
c. 10^6, $1592.36\,\text{Hz}$
d. 0.001, $150\,\text{Hz}$
e. 7.6, $6.94\,\text{Hz}$
f. $\frac{1}{42}$, $1\,\text{Hz}$
21. $20\,\text{ms}$
23. $-11.76\,\text{V}$
25. $v = 80 \sin 523.12t$
29. a. in phase

b. i leads v by 90°
c. i leads v by 190°
31. a. $v = 0.01 \sin(157t - 110°)$
   b. $i = 2 \times 10^{-3}$
      $\sin(62.8 \times 10^3 t + 135°)$
33. a. 1.875 V **b.** −4.778 V
35. a. 14.14 V **b.** 5 V **c.** 4.24 mA
   d. 11.312 mA
37. 1.43 V
39. $V_{av} = 0$ V, $V_{eff} = 10$ V
41. 10 V

Chapter 14

3. a. $3770 \cos 377t$
   b. $-452.4 \sin 754t$
   c. $-7.85 \sin(157t - 10°)$
   d. $-500 \sin(20t - 150°)$
5. a. $v = 210 \sin 754t$
   b. $v = 84 \sin(400t - 120°)$
   c. $v = 42 \times 10^{-3} \sin(\omega t + 88°)$
   d. $v = 28 \sin(\omega t + 180°)$
7. a. 1.592 H **b.** 2.654 H **c.** 0.841 H
9. a. $v = 100 \sin(\omega t + 90°)$
   b. $v = 8 \sin(\omega t + 150°)$
   c. $v = 120 \sin(\omega t - 120°)$
   d. $v = 60 \sin(\omega t + 190°)$
11. a. $i = 1 \sin(\omega t - 90°)$
   b. $i = 0.6 \sin(\omega t - 70°)$
   c. $i = 0.8 \sin(\omega t + 10°)$
   d. $i = 1.6 \sin(377t + 130°)$
13. a. $\infty \Omega$ **b.** 530.79 Ω **c.** 265.39 Ω
   d. 17.693 Ω **e.** 1.327 Ω
15. a. 9.31 Hz **b.** 4.66 Hz
   c. 18.62 Hz **d.** 1.59 Hz
17. a. $i = 6 \times 10^{-3} \sin(200t + 90°)$
   b. $i = 33.96 \times 10^{-3}$
      $\sin(377t + 90°)$
   c. $i = 44.94 \times 10^{-3}$
      $\sin(374t + 300°)$
   d. $i = 56 \times 10^{-3} \sin(\omega t + 160°)$
19. a. $v = 1334 \sin(300t - 90°)$
   b. $v = 37.17 \sin(377t - 90°)$
   c. $v = 127.2 \sin 754t$
   d. $v = 100 \sin(1600t - 170°)$
21. a. $X_C = 400 \, \Omega$ **b.** $L = 254.78$ mH
   c. $R = 5 \, \Omega$
25. 318.47 mH
27. 5.067 nF
29. a. 0 W **b.** 0 W **c.** 122.5 W
31. 192 W

33. $i = 40 \sin(\omega t - 50°)$
35. a. $i = 2 \sin(157t - 60°)$
   b. 318.47 mH **c.** 0 W
37. a. $5 \angle 36.87°$ **b.** $2.83 \angle 45°$
   c. $16.38 \angle 77.66°$
   d. $806.23 \angle 82.87°$
   e. $1077.03 \angle 21.8°$
   h. $8.94 \angle 153.43°$
   i. $61.85 \angle -104.04°$
   j. $101.53 \angle -39.81°$
   k. $4326.66 \angle 123.69°$
   l. $25.495 \times 10^{-3} \angle -78.69°$
39. a. $15.033 \angle 86.19°$
   b. $60.208 \angle 4.76°$ **c.** $0.3 \angle 88.09°$
   d. $2002.5 \angle 87.14°$
   e. $86.182 \angle 93.73°$
   f. $38.694 \angle -94°$
41. a. $11.8 + j7$ **b.** $151.9 + j49.9$
   c. $4.72 \times 10^{-6} + j71$
   d. $5.2 + j1.6$ **e.** $209.30 + j311$
   f. $-21.2 + j12$ **g.** $7.03 + j9.93$
43. q. $6 \angle -50°$ **r.** $0.2 \times 10^{-3} \angle 140°$
   s. $109 \angle -230°$ **t.** $76.471 \angle -80°$
   u. $4 \angle 0°$ **v.** $0.71 \angle -16.49°$
   w. $4.21 \times 10^{-3} \angle 161.1°$
   x. $18.191 \angle -50.91°$
45. a. $100 \angle 30°$ **b.** $0.25 \angle -40°$
   c. $70.71 \angle -90°$ **d.** $29.69 \angle 0°$
   e. $4.242 \times 10^{-6} \angle 90°$
   f. $2.546 \times 10^{-6} \angle 70°$
47. $v_a = 41.769 \sin(377t + 29.43°)$
49. $v_c = 61.59 \sin(\omega t + 76.94°)$

Chapter 15

1. a. $Z = 6.8 \angle 0° = 6.8$
   b. $Z = 754 \angle 90° = j754$
   c. $Z = 15.7 \angle 90° = j15.7$
   d. $Z = 265.25 \angle -90°$
      $= -j265.25$
   e. $Z = 318.47 \angle -90°$
      $= -j318.47$
   f. $Z = 200 \angle 0° = 200$
3. a. $V = 62.216 \times 10^{-3} \angle 0°$
   b. $V = 6.397 \angle 150°$
   c. $V = 1.8 \times 10^3 \angle -50°$
5. a. $Z_T = 3 - j3 = 4.24 \angle -45°$
   b. $Z_T = 0.5$ kΩ $+ j3$ kΩ
      $= 3.04$ kΩ $\angle 80.54°$
   c. $Z_T = 47 + j1616.88$
      $= 1617.56 \angle 88.33°$

7. a. $Z_T = 10 \angle 36.87°$
   c. $I = 10 \angle -36.87°$,
      $V_R = 80 \angle -36.87°$,
      $V_L = 60 \angle 53.13°$
   f. $P = 800$ W
   g. $F_p = 0.8$ lagging
9. a. $Z_T = 1660.27 \angle -73.56°$
   b. $I = 8.517 \times 10^{-3} \angle 73.56°$
   c. $V_R = 4.003 \angle 73.56°$,
      $V_L = 13.562 \angle -16.44°$
   d. $P = 34.09$ mW
11. a. $Z_T = 3.16 \times 10^3 \angle 18.43°$
   c. $C = 3.18 \, \mu$F, $L = 6.37$ H
   d. $I = 1.3424 \times 10^{-3} \angle 41.57°$,
      $V_R = 4.027 \angle 41.57°$,
      $V_L = 2.6848 \angle 131.57°$,
      $V_C = 1.3424 \angle -48.43°$
   g. $P = 5.406$ mW
   h. $F_p = 0.9487$ lagging
   i. $i = 1.898 \times 10^{-3} \sin(\omega t + 41.57°)$,
      $v_R = 5.6942 \sin(\omega t + 41.57°)$,
      $v_L = 3.7963 \sin(\omega t + 131.57°)$,
      $v_C = 1.8982 \sin(\omega t - 48.43°)$
13. a. $V_1 = 112.92 \angle 12.432°$,
      $V_2 = 58.66 \angle -139.936°$
15. a. $I = 0.132 \angle -38.552°$,
      $V_R = 3.96 \angle -38.552°$,
      $V_C = 0.35 \angle -128.552°$
   b. $F_p = 0.198$ lagging
   c. $P = 0.523$ W
   f. $V_R = 3.969 \angle -38.552°$,
      $V_C = 0.351 \angle -128.552°$
   g. $Z_T = 30 + j148.147$
17. $Z = 31.34 + j22.10$
21. a. $Y_T = 0.0411 - j0.1095$,
      $Z_T = 3 + j8$
   b. $Y_T = 0.0071 + j0.0083$,
      $Z_T = 60 - j70$
   c. $Y_T = 4 \times 10^{-3} + j2 \times 10^{-3}$,
      $Z_T = 200 - j100$
23. a. $Y_T = 0.539 \angle -21.8°$
   c. $E = 3.71 \angle 21.8°$,
      $I_R = 1.855 \angle 21.8°$,
      $I_L = 0.742 \angle -68.2°$
   f. $P = 6.88$ W
   g. $F_p = 0.928$ lagging
   h. $e = 5.25 \sin(377t + 21.8°)$,
      $i_R = 2.62 \sin(377t + 21.8°)$,
      $i_L = 1.049 \sin(377t - 68.2°)$,
      $i_T = 2.828 \sin 377t$
25. a. $Y_T = 0.13 \angle -50.31°$

c. $I_T = 7.8 \angle -50.31°$,
  $I_R = 5 \angle 0°$,
  $I_L = 6 \angle -90°$
f. $P = 300$ W
g. $F_p = 0.638$ lagging
h. $e = 84.84 \sin 377t$,
  $i_R = 7.07 \sin 377t$,
  $i_L = 8.484 \sin(377t - 90°)$,
  $i_T = 11.03 \sin(377t - 50.31°)$
27. a. $Y_T = 0.4164 \times 10^{-3} \angle 36.897°$
  c. $L = 10.610$ H, $C = 1.326 \,\mu$F
  d. $E = 8.498 \angle -56.897°$,
  $I_R = 2.833 \times 10^{-3} \angle -56.897°$,
  $I_L = 2.125 \times 10^{-3} \angle -146.897°$,
  $I_C = 4.249 \times 10^{-3} \angle 33.103°$
  g. $P = 24.078$ mW
  h. $F_p = 0.8$ lagging
  i. $e = 12.016 \sin(377t - 56.897°)$,
  $i_R = 4.006 \sin(377t - 56.897°)$,
  $i_L = 3.005 \sin(377t - 146.897°)$,
  $i_C = 6.008 \sin(377t + 33.103°)$
29. a. $I_1 = 18.09 \angle 65.241°$,
  $I_2 = 8.528 \angle -24.759°$
  b. $I_1 = 11.161 \angle 0.255°$,
  $I_2 = 6.656 \angle 153.690°$
31. a. $Z_T = 4.454$ k$\Omega - j1.047$ kΩ
  b. $Z_T = 17.481 + j29.717$
33. a. $E = 75.6 \angle -70.11°$,
  $I_R = 0.3436 \angle -70.11°$,
  $I_L = 12.04 \times 10^{-3} \angle -160.11°$
  b. $F_p = 0.3401$ leading
  c. $P = 25.9734$ W
  f. $I_C = 0.4748 \angle 19.63°$
  g. $Z_T = 25.72 - j71.08$

Chapter 16

1. a. $Z_T = 1.2 \angle 90°$
  b. $I = 10 \angle -90°$
  c. $I_1 = 10 \angle -90°$
  d. $I_2 = 6 \angle -90°$,
  $I_3 = 4 \angle -90°$
  e. $V_L = 60 \angle 0°$
3. a. $Z_T = 3.87 \angle -11.817°$,
  $Y_T = 0.258 \angle 11.817°$
  b. $I_T = 15.504 \angle 41.817°$
  c. $I_2 = 3.985 \angle 82.826°$
  d. $V_C = 47.809 \angle -7.174°$
  e. $P = 910.705$ W
5. a. $I = 0.375 \angle 25.346°$
  b. $V_C = 70.711 \angle -45°$
  c. $P = 33.9$ W

7. a. $I_1 = 1.423 \angle 18.259°$
  b. $V_1 = 26.574 \angle 4.763°$
  c. $P = 54.074$ W
9. a. $Y_T = 0.099 \angle -9.709°$
  b. $V_1 = 20.4 \angle 30°$,
  $V_2 = 10.887 \angle 58.124°$
  c. $I_3 = 1.933 \angle 11.109°$
11. $I = 33.201 \angle 38.889°$
13. $P = 139.707$ mW

Chapter 17

3. a. $Z_s = 21.93 \angle -46.85°$,
  $E = 10.97 \angle 13.15°$
  b. $Z_s = 5.15 \angle 59.04°$,
  $E = 10.30 \angle 179.04°$
5. a. $5.15 \angle -24.5°$
  b. $0.442 \angle 143.48°$
7. a. $13.07 \angle -33.71°$
  b. $9.06 \angle 0.75°$
9. 3.165×10^{-3}V $\angle 137.29°$
11. a. $V_1 = 14.68 \angle 68.89°$,
  $V_2 = 12.97 \angle 155.88°$
  b. $V_1 = 5.12 \angle -79.36°$,
  $V_2 = 2.71 \angle 39.96°$
13. a. $V_1 = 5.74 \angle 122.76°$,
  $V_2 = 4.04 \angle 145.03°$,
  $V_3 = 25.94 \angle 78.07°$
  b. $V_1 = 15.13 \angle 1.29°$,
  $V_2 = 17.24 \angle 3.73°$,
  $V_3 = 10.59 \angle -0.11°$
15. 171.63I $\angle -120.70°$
17. a. yes b. 0 A c. 0 V
19. $L_x = 5$ mH, $R_x = 5 \,\Omega$
23. a. $12.02 \angle -38.61°$
  b. $7.02 \angle 20.56°$

Chapter 18

1. a. $6.095 \angle -32.115°$
  b. $3.77 \angle -93.8°$
3. $178.55 \times 10^{-3} \angle -26.57°$
5. $70.61 \times 10^{-3} \angle -11.31°$
7. $2.944 \times 10^{-3} \angle 0°$
9. a. $Z_{Th} = 18.035 + j11.355$,
  $E_{Th} = 2.131 \angle 32.196°$
  b. $Z_{Th} = 3.983 - j5.528$,
  $E_{Th} = 46.95 \angle 4.673°$
11. $Z_{Th} = 5.1 \times 10^3 \angle -11.31°$,
  $E_{Th} = 10$V
13. $Z_{Th} = 20 \times 10^3 \angle 0°$,
  $E_{Th} = -3990 \angle 0°$

15. $Z_{Th} = 25 \times 10^3 \angle 0°$,
  $E_{Th} = -1800 \angle 0°$
17. $Z_{Th} = 105 \times 10^3 \angle 0°$,
  $E_{Th} = 315 \angle 0°$
19. a. $Z_N = 18.035 + j11.355$,
  $I_N = 0.1 \angle 0°$
  b. $Z_N = 3.983 - j5.528$,
  $I_N = 6.891 \angle 58.901°$
21. a. $Z_N = 9.66 \angle 14.93°$,
  $I_N = 2.15 \angle -42.87°$
  b. $Z_N = 4.37 \angle 55.67°$,
  $I_N = 22.83 \angle -34.65°$
23. $Z_N = 4.44 \times 10^3 \angle -0.031°$,
  $I_N = 100$I $\angle 0.286°$
25. $Z_N = 25 \times 10^3 \angle 0°$,
  $I_N = 72 \times 10^{-3} \angle 0°$
27. $Z_N = 1.9625$ kΩ,
  $I_N = 2 \times 10^{-3} \angle 0°$
29. a. $Z_L = 11.035 \angle 77.03°$, 90 W
  b. $Z_L = 5.71 \angle 64.30°$, 618.33 W
31. $I_{ab} = 1.333 \times 10^{-3} \angle 0°$,
  $V_{ab} = 10.67 \angle 0°$
33. $25.77 \times 10^{-3} \angle 104.04°$

Chapter 19

1. a. R: 300 W, L: 0 W, C: 0 W
  b. R: 0 VAR, L: 900 VAR, C:
  500 VAR
  c. R: 300 VA, L: 900 VA, C:
  500 VA
  d. $P_T = 300$ W,
  $Q_T = 400$ VAR (L),
  $S_T = 500$ VA,
  $F_p = 0.6$ lagging
  f. $W_R = 5$ J
  g. $W_C = 1.327$ J,
  $W_L = 2.389$ J
3. a. $P_T = 1200$ W,
  $Q_T = 1200$ VAR (L),
  $S_T = 1697$ VA,
  $F_p = 0.7071$ lagging
  c. $I_T = 8.485 \angle -45°$
5. a. $P_R = 180$ W,
  $Q_R = 0$ VAR,
  $S = 180$ VA
  b. $P_L = 0$ W,
  $Q_L = 360$ VAR,
  $S = 360$ VA
  c. $P_T = 580$ W,
  $Q_T = 960$ VAR (L),
  $S_T = 1121.61$ VA,

$F_p = 0.52$ lagging
  d. $\mathbf{I}_T = 18.69 \angle -28.67°$
7. a. $P_R = 300\,\text{W}$,
  $P_L = 0\,\text{W}$,
  $P_C = 0\,\text{W}$
  b. $Q_R = 0\,\text{VAR}$,
  $Q_L = 400\,\text{VAR}$,
  $Q_C = 250\,\text{VAR}$
  c. $S_R = 300\,\text{VA}$,
  $S_L = 400\,\text{VA}$,
  $S_C = 250\,\text{VA}$
  d. $P_T = 300\,\text{W}$,
  $Q_T = 150\,\text{VAR}\,(L)$,
  $S_T = 335.41\,\text{VA}$,
  $F_p = 0.89$ lagging
  f. $W_R = 8.33\,\text{W}$
  g. $W_L = 1.06\,\text{J}$,
  $W_C = 0.66\,\text{J}$
9. a. $\mathbf{Z}_T = 2 - j3.464$
  b. $P_T = 5000\,\text{W}$
11. a. $P_T = 300\,\text{W}$,
  $Q_T = 400\,\text{VAR}\,(C)$,
  $S_T = 500\,\text{VA}$,
  $F_p = 0.6$ leading
  b. $\mathbf{I}_T = 16.67 \angle 53.13°$
  d. $C\colon X_C = 2.159\,\Omega$,
  $L\colon R = 1.0796\,\Omega$,
  $X_L = 0.7197\,\Omega$
13. a. $P_T = 1100\,\text{W}$,
  $Q_T = 2366.26\,\text{VAR}\,(C)$,
  $S_T = 2609.44\,\text{VA}$,
  $F_p = 0.4215$ leading
  b. $\mathbf{E} = 521.89 \angle -65.07°$
  c. $100\,\text{W}\colon R = 1743.38\,\Omega$,
  $X_C = 1307.53\,\Omega$
  $1000\,\text{W}\colon R = 43.59\,\Omega$,
  $X_C = 99.88\,\Omega$
15. a. $438\,\mu\text{F}$ **b.** $230\,\mu\text{F}$
17. a. $S_T = 37.441\,\text{kVA}$
  b. $F_p = 0.935$ lagging
  c. $Q_T = 13.299\,\text{kVAR}\,(L)$
  d. $I = 37.441\,\text{A}$
  e. $C = 35\,\mu\text{F}$
  f. $I = 35\,\text{A}$

Chapter 20

1. a. $\omega_s = 250\,\text{rad/s}$,
  $f_s = 39.81\,\text{Hz}$
  b. $\omega_s = 3535.53\,\text{rad/s}$,
  $f_s = 562.98\,\text{Hz}$
  c. $\omega_s = 21{,}880\,\text{rad/s}$,
  $f_s = 3484.08\,\text{Hz}$

3. a. $X_C = 40\,\Omega$
  b. $\mathbf{I} = 10 \angle 30°$,
  $\mathbf{V}_R = 20 \angle 30°$,
  $\mathbf{V}_L = 400 \angle 120°$,
  $\mathbf{V}_C = 400 \angle -60°$
  d. $P = 200\,\text{W}$,
  $Q_L = 4\,\text{kVAR}$,
  $Q_C = 4\,\text{kVAR}$
  e. $Q = 20$
5. a. $\text{BW} = 400\,\text{Hz}$
  b. $f_1 = 5800\,\text{Hz}$,
  $f_2 = 6200\,\text{Hz}$
  c. $X_L = X_C = 45\,\Omega$
  d. $P_{\text{HPF}} = 375\,\text{mW}$
7. a. $Q_s = 10$
  b. $X_L = 20\,\Omega$
  c. $L = 1.59\,\text{mH}$, $C = 3.98\,\mu\text{F}$
  d. $f_1 = 1900\,\text{Hz}$,
  $f_2 = 2100\,\text{Hz}$
9. $R = 10\,\Omega$, $L = 13.27\,\text{mH}$,
  $C = 27.08\,\text{nF}$
11. a. $X_C = 104\,\Omega$
  b. $Z_T = 342.11\,\Omega$
  c. $\mathbf{I}_L = 16.78 \times 10^{-3} \angle -78.69°$,
  $\mathbf{I}_C = 16.45 \times 10^{-3} \angle 90°$
  d. $L = 0.796\,\text{mH}$, $C = 76.56\,\text{nF}$
  e. $Q_p = 3.29$, $\text{BW} = 6079.03\,\text{Hz}$
13. a. $f_p = 102.7\,\text{kHz}$
  b. $X_L = 51.6\,\Omega$, $X_C = 51.68\,\Omega$
  c. $Q = 34.40$ (high)
  d. $Z_{T_p} = 1.51\,\text{k}\Omega$
  e. $\mathbf{I}_L = 292.52 \times 10^{-3} \angle -88.33°$,
  $\mathbf{I}_C = 292.18 \times 10^{-3} \angle 90°$
  f. $Q_p = 29.22$, $\text{BW} = 3514.72\,\text{Hz}$
15. a. $X_L = 1581\,\Omega$
  b. $X_C = 1581\,\Omega$
  c. $f_p = 15{,}735\,\text{Hz}$
  d. $C = 6.4\,\text{nF}$
17. $R_s = 100.44\,\text{k}\Omega$, $C = 1.268\,\text{nF}$
19. a. $X_L = 3000\,\Omega$
  b. $Q = 100$
  c. $f_p = 11.764 \times 10^3\,\text{Hz}$
  d. $V_{C_{\max}} = 211.764\,\text{V}$
  e. $f_1 = 11.264\,\text{kHz}$,
  $f_2 = 12{,}264\,\text{kHz}$
23. b. $f_2 = 15{,}923.57\,\text{Hz}$
  c. $f = 1\,\text{Hz}\colon V_o/V_i = 1$,
  $f = 1000\,\text{Hz}\colon V_o/V_i = 0.847$,
  $f = 100\,\text{kHz}\colon V_o/V_i = 0.157$,
  $f = 1\,\text{MHz}\colon V_o/V_i = 0$
  d. $f = 1\,\text{Hz}\colon \theta = 0°$,
  $f = 1000\,\text{Hz}\colon \theta = -3.593°$,
  $f = 100\,\text{kHz}\colon \theta = -80.966°$,

$f = 1\,\text{MHz}\colon \theta = -89.089°$
25. a. $f_1 = 6756.76\,\text{Hz}$
  c. $f = \frac{1}{2}f_1\colon -4.771\,\text{dB}$,
  $f = 2f_1\colon -1.761\,\text{dB}$
  d. $f = \frac{1}{2}f_1\colon 0.447$,
  $f = 2f_1\colon 0.894$
  e. $f = \frac{1}{2}f_1\colon 63.435°$,
  $f = 2f_1\colon 26.565°$
27. high-pass R-C filter, $f_1 = 2\,\text{kHz}$
29. a. $Q_s = 20$
  b. $f_1 = 9{,}750\,\text{Hz}$, $f_2 = 10{,}250\,\text{Hz}$
  d. $Q_s = 25$, $f_1 = 9800\,\text{Hz}$,
  $f_2 = 10{,}200\,\text{Hz}$
31. a. $Q_s = 2.5$
  b. $\text{BW} = 2\,\text{kHz}$, $f_1 = 4\,\text{kHz}$,
  $f_2 = 6\,\text{kHz}$
  d. negligible
33. a. $f_p = 0.727\,\text{mHz}$ (band-stop)
  $f = 2.013\,\text{mHz}$ (pass-band)

Chapter 21

1. a. positive-going
  b. $V_b = 2\,\text{V}$
  c. $t_P = 0.2\,\text{ms}$
  d. amplitude $= 6\,\text{V}$
  e. tilt $= 6.5\%$
3. a. positive-going
  b. $V_b = 10\,\text{mV}$
  c. $t_P = 3.2\,\text{ms}$
  d. amplitude $= 20\,\text{mV}$
  e. tilt $= 3.4\%$
7. a. $T = 120\,\mu\text{s}$
  b. $f = 8.333\,\text{kHz}$
  c. maximum $= 440\,\text{mV}$,
  minimum $= 80\,\text{mV}$
8. prf $= 125\,\text{kHz}$, duty cycle $=$
  62.5%
11. $V_{\text{av}} = 0\,\text{V}$
13. $V_{\text{av}} = 18.88\,\text{mV}$
15. $V_{\text{av}} = 117\,\text{mV}$
17. $v_o = 4(1 + e^{-t/(20 \times 10^{-3})})$
19. $i_C = (-8 \times 10^{-3})e^{-t}$
21. $i_C = (4 \times 10^{-3})e^{-t/(0.2 \times 10^{-3})}$
  a. $5\tau = T/2$
  b. $25\tau = T/2$
  c. $0.25\tau = T/2$
23. $\tau = 0.2\,\text{ms}$,
  0 to $T/2$: $v_C = 20\,\text{V}$,
  $T/2$ to T: $v_C = 20e^{-t/\tau}$,
  T to $\frac{3}{2}T$: $v_C = 20(1 - e^{-t/\tau})$,
  $\frac{3}{2}T$ to $2T$: $v_C = 20e^{-t/\tau}$
25. yes

Chapter 22

1. a. $E_\phi = 120.1$ V
  b. $V_\phi = 120.1$ V
  c. $I_\phi = 12.01$ A
  d. $I_L = 12.01$ A
3. a. $E_\phi = 120.1$ V
  b. $V_\phi = 120.1$ V
  c. $I_\phi = 16.98$ A
  d. $I_L = 16.98$ A
5. a. $\theta_2 = -120°$, $\theta_3 = 120°$
  b. $V_{an} = 120 \angle 0°$,
    $V_{bn} = 120 \angle -120°$,
    $V_{cn} = 120 \angle 120°$
  c. $I_{an} = 8 \angle -53.13°$,
    $I_{bn} = 8 \angle -173.13°$,
    $I_{cn} = 8 \angle 66.87°$
  e. $I_L = 8$ A
  f. $E_L = 207.85$ V
7. $V_{an} = V_{bn} = V_{cn} = 127$ V
  $I_{an} = I_{bn} = I_{cn} = 8.98$ A
  $I_{Aa} = I_{Bb} = I_{Cc} = 8.98$ A
9. a. $E_\phi = 120.1$ V
  b. $V_\phi = 208$ V
  c. $I_\phi = 10.4$ A
  d. $I_L = 18$ A
11. a. $E_\phi = 120.092$ V
  b. $V_\phi = 208$ V
  c. $I_\phi = 16.342$ A
  d. $I_L = 28.304$ A
13. a. $\theta_2 = -120°$, $\theta_3 = 120°$
  b. $V_{ab} = 208 \angle 0°$,
    $V_{bc} = 208 \angle -120°$,
    $V_{ca} = 208 \angle 120°$
  d. $I_{ab} = 1.471 \angle 45°$,
    $I_{bc} = 1.471 \angle -75°$,
    $I_{ca} = 1.471 \angle 165°$
  e. $I_L = 2.548$ A
  f. $E_\phi = 120.1$ V
15. $V_{ab} = V_{bc} = V_{ca} = 220$ V
  $I_{ab} = I_{bc} = I_{ca} = 15.56$ A
  $I_{Aa} = I_{Bb} = I_{Cc} = 26.95$ A
17. a. $E_\phi = 208$ V
  b. $V_\phi = 120.092$ V
  c. $I_\phi = 7.076$ A
  d. $I_L = 7.076$ A
19. $V_{an} = V_{bn} = V_{cn} = 69.28$ V
  $I_{an} = I_{bn} = I_{cn} = 2.89$ A
  $I_{Aa} = I_{Bb} = I_{Cc} = 2.89$ A
21. $V_{an} = V_{bn} = V_{cn} = 69.28$ V
  $I_{an} = I_{bn} = I_{cn} = 5.77$ A
  $I_{Aa} = I_{Bb} = I_{Cc} = 5.77$ A
23. a. $E_\phi = 440$ V

b. $V_\phi = 440$ V
  c. $I_\phi = 29.33$ A
  d. $I_L = 50.8$ A
25. a. $\theta_2 = -120°$, $\theta_3 = 120°$
  b. $V_{ab} = 100 \angle 0°$,
    $V_{bc} = 100 \angle -120°$,
    $V_{ca} = 100 \angle 120°$
  d. $I_{ab} = 5 \angle 0°$,
    $I_{bc} = 5 \angle -120°$,
    $I_{ca} = 5 \angle 120°$
  e. $I_{Aa} = I_{Bb} = I_{Cc} = 8.66$ A
27. a. $\theta_2 = -120°$, $\theta_3 = 120°$
  b. $V_{ab} = 100 \angle 0°$,
    $V_{bc} = 100 \angle -120°$,
    $V_{ca} = 100 \angle 120°$
  d. $I_{ab} = 7.072 \angle 45°$,
    $I_{bc} = 7.072 \angle -75°$,
    $I_{ca} = 7.072 \angle 165°$
  e. $I_{Aa} = I_{Bb} = I_{Cc} = 12.25$ A
29. $P_T = 2160$ W,
  $Q_T = 0$ VAR,
  $S_T = 2160$ VA,
  $F_p = 1$
31. $P_T = 7210.667$ W,
  $Q_T = 7210.667$ VAR (C),
  $S_T = 10{,}197.423$ VA,
  $F_p = 0.707$ leading
33. $P_T = 7.263$ kW,
  $Q_T = 7.263$ kVAR (L),
  $S_T = 10.272$ kVA,
  $F_p = 0.7071$ lagging
35. $P_T = 287.93$ W,
  $Q_T = 575.86$ VAR (L),
  $S_T = 643.83$ VA,
  $F_p = 0.4472$ lagging
37. $P_T = 900$ W,
  $Q_T = 1200$ VAR (L),
  $S_T = 1500$ VA,
  $F_p = 0.6$ lagging
39. $Z_\phi = 12.98 - j17.31$
45. a. $V_\phi = 120.092$ V
  b. $I_{an} = 8.492$ A,
    $I_{bn} = 7.076$ A,
    $I_{cn} = 42.465$ A
  c. $P_T = 4928.53$ W,
    $Q_T = 4928.53$ VAR (L),
    $S_T = 6969.994$ VA,
    $F_p = 0.707$ lagging

Chapter 23

1. (I) a. no **b.** no **c.** yes **d.** no
  e. yes

(II) a. yes **b.** yes **c.** no **d.** yes
  e. yes
(III) a. yes **b.** yes **c.** no **d.** yes
  e. yes
(IV) a. no **b.** no **c.** yes **d.** yes
  e. yes
7. a. $V_{av} = 0$ V, $V_{eff} = 19.04$ V
  b. $I_{av} = 0$ A, $I_{eff} = 4.53$ A
9. $P = 71.872$ W
11. a. $i = 2 + 2.08 \sin(400t - 33.69°) + 0.5 \sin(800t - 53.13°)$
  b. $I_{eff} = 2.508$ A
  c. $v_R = 24 + 24.96 \sin(400t - 33.69°) + 6 \sin(800t - 53.13°)$
  d. $V_{R_{eff}} = 30.092$ V
  e. $v_L = 16.64 \sin(400t + 56.31°) + 8 \sin(800t + 36.87°)$
  f. $V_{L_{eff}} = 13.055$ V
  g. $P_T = 75.481$ W
13. a. $i = 1.2 \sin(400t + 53.13°)$
  b. $I_{eff} = 0.849$ A
  c. $v_R = 18 \sin(400t + 53.13°)$
  d. $V_{R_{eff}} = 12.73$ V
  e. $v_C = 18 + 23.98 \sin(400t - 36.87°)$
  f. $V_{C_{eff}} = 24.73$ V
  g. $P_T = 10.79$ W
15. $v_o = 2.257 \times 10^{-3} \sin(377t + 93.66°) + 1.923 \times 10^{-3} \sin(754t + 1.64°)$
17. $i_T = 30 + 30.27 \sin(20t + 7.59°) + 0.5 \sin(40t - 30°)$

Chapter 24

1. a. $L_s = 0.2$ H
  b. $e_p = 1.6$ V, $e_s = 5.12$ V
  c. $e_p = 15$ V, $e_s = 24$ V
3. a. $L_s = 158.02$ mH
  b. $e_p = 24$ V, $e_s = 1.8$ V
  c. $e_p = 15$ V, $e_s = 24$ V
5. $L_{T(+)} = 1.354$ H
7. $I_1(R_1 + X_{L_1}) + I_2(X_m) = E_1$,
  $I_1(X_m) + I_2(X_{L_2} + R_L) = 0$,
  $X_m = -\omega M \angle 90°$
9. a. $E_s = 3.125$ V
  b. $\Phi_m = 391.02 \times 10^{-6}$ Wb
11. $f = 56.31$ Hz
13. $Z_p = 400 \,\Omega$
15. $N_p = 12{,}000$ turns
17. a. $a = 3$

b. $P = 2.78\,\text{W}$
19. a. $Z_p = 20 + j360$
  b. $I_p = 332.82 \times 10^{-3} \angle -86.82°$
  c. $V_{R_e} = 6.656 \angle -86.82°$,
    $V_{X_e} = 13.313 \angle 3.18°$,
    $V_{X_L} = 106.50 \angle 3.18°$
23. a. $a = 20$
  b. $I_s = 83.33\,\text{A}$
  c. $I_p = 4.167\,\text{A}$
  d. $a = \frac{1}{20}$, $I_s = 4.167\,\text{A}$, $I_p = 83.33\,\text{A}$
25. a. $V_L = 25 \angle 0°$, $I_L = 5 \angle 0°$
  b. $Z_i = 80 \angle 0°$
  c. $Z_{1/2} = 20 \angle 0°$
27. a. $E_2 = 40 \angle 60°$,
    $I_2 = 3.33 \angle 60°$,
    $E_3 = 30 \angle 60°$,

$I_3 = 3 \angle 60°$
  b. $Z_i = 64.478 \angle 0°$

Chapter 25

1. a. $z_{11} = (Z_1Z_2 + Z_1Z_3)/(Z_1 + Z_2 + Z_3)$
    $z_{12} = (Z_1Z_3)/(Z_1 + Z_2 + Z_3)$
    $z_{21} = (Z_1Z_3)/(Z_1 + Z_2 + Z_3)$
    $z_{22} = (Z_1Z_3 + Z_2Z_3)/(Z_1 + Z_2 + Z_3)$
3. a. $y_{11} = (Y_1Y_2 + Y_1Y_3)/(Y_1 + Y_2 + Y_3)$
    $y_{12} = -(Y_1Y_2)/(Y_1 + Y_2 + Y_3)$
    $y_{21} = -(Y_1Y_2)/(Y_1 + Y_2 + Y_3)$

$y_{22} = (Y_1Y_2 + Y_2Y_3)/(Y_1 + Y_2 + Y_3)$
5. a. $h_{11} = (Z_1Z_2)/(Z_1 + Z_2)$
    $h_{12} = Z_1/(Z_1 + Z_2)$
    $h_{21} = -Z_1/(Z_1 + Z_2)$
    $h_{22} = (Z_1 + Z_2 + Z_3)/(Z_1Z_3 + Z_2Z_3)$
7. a. $h_{11} = (Y_1 + Y_2 + Y_3)/(Y_1Y_2 + Y_1Y_3)$
    $h_{12} = Y_2/(Y_2 + Y_3)$
    $h_{21} = -Y_2/(Y_2 + Y_3)$
    $h_{22} = Y_2Y_3/(Y_2 + Y_3)$
9. a. $A_i = 47.62$ **b.** $A_v = -99$
11. $Z_i = -7 \times 10^3 - j6 \times 10^3$,
    $Z_o = 2 \times 10^3 - j29 \times 10^3$
13. $h_{11} = 2.5\,\text{k}\Omega$, $h_{12} = 0.5$,
    $h_{21} = -0.75$, $h_{22} = 0.25 \times 10^{-3}\,\text{S}$

Index

Index